AF457100

Tribology Issues and Opportunities in MEMS

Tribology Issues and Opportunities in MEMS

Proceedings of the NSF/AFOSR/ASME Workshop on
Tribology Issues and Opportunities in MEMS
held in Columbus, Ohio, U.S.A., 9-11 November 1997

Edited by

Bharat Bhushan
Computer Microtribology and Contamination Laboratory,
Department of Mechanical Engineering,
The Ohio State University,
Columbus, Ohio, U.S.A.

SPRINGER-SCIENCE+BUSINESS MEDIA, B.V.

A C.I.P. Catalogue record for this book is available from the Library of Congress.

DOI 10.1007/978-94-011-5050-7

Printed on acid-free paper

Originally published by Kluwer Academic Publishers in 1998
MyCopy version of the original edition 1998

TABLE OF CONTENTS

5. Tribology of MEMS Components and Materials

6. Mechanical Property Measurements

7. Modification and Characterization of Surfaces

8. Breakout Session Reports

9. Panel Discussion Report

PREFACE

Micro Electro Mechanical Systems (MEMS) is already about a billion dollars a year industry and is growing rapidly. So far major emphasis has been placed on the fabrication processes for various devices. There are serious issues related to tribology, mechanics, surface chemistry and materials science in the operation and manufacturing of many MEMS devices and these issues are preventing an even faster commercialization. Very little is understood about tribology and mechanical properties on micro- to nanoscales of the materials used in the construction of MEMS devices. The MEMS community needs to be exposed to the state-of-the-art of tribology and vice versa.

Fundamental understanding of friction/stiction, wear and the role of surface contamination and environmental debris in micro devices is required. There are significant adhesion, friction and wear issues in manufacturing and actual use, facing the MEMS industry. Very little is understood about the tribology of bulk silicon and polysilicon films used in the construction of these microdevices. These issues are based on surface phenomena and cannot be scaled down linearly and these become increasingly important with the small size of the devices. Continuum theory breaks down in the analyses, e.g. in fluid flow of micro-scale devices. Mechanical properties of polysilicon and other films are not well characterized. Roughness optimization can help in tribological improvements. Monolayers of lubricants and other materials need to be developed for ultra-low friction and near zero wear. Hard coatings and ion implantation techniques hold promise.

Better tribological understanding of MEMS will advance the state-of-the-art in micromachining and the IC industry, in general. For example, better understanding will contribute to better performance prediction for micromachined pressure sensors, accelerometers and gyro's as well as a better understanding of stiction behavior of micro-mirrors and micromotors and of the influence of roughness on micro-fluids.

The objective of this workshop was to bring together researchers, manufacturers and potential users of MEMS and experts in tribology (including mechanics, mechanical properties and surface modification) so that the MEMS community could better understand the tribology expertise and the tribology community could better understand the major problems on hand and identify critical research issues facing MEMS industry. There were three specific objectives of the workshop. The first was to provide tutorials on MEMS technology and state-of-the-art of tribology for education of tribology and MEMS community, respectively. The second objective was to share whatever tribological understanding of MEMS devices exists. The third objective was to identify tribology research issues and opportunities and general directions for tribology in MEMS research, which were accomplished via breakout sessions and a panel discussion. Since the objective of the first part of the workshop was to provide tutorials, we had a large number of lectures of short durations followed by lectures on tribology of MEMS, mechanical property measurements, modification and characterization of surfaces, and breakout sessions and a panel discussion. A side benefit of this workshop was to bring macro- and microtribologists together.

We assembled individuals who are active in research and manufacturing of MEMS devices or are potential users of this technology. We also invited experts in macro- and microtribology, surface modification and mechanical property measurement arenas from industry, government and academia. The micro/nanotribology community can play a pivotal role. Tribological issues in MEMS are closely related to those faced today by magnetic storage devices community, which were also invited to attend. Furthermore, a special effort was made to attract students and younger researchers from the U.S. and overseas. The response to the course workshop was overwhelming. We had 122 participants from eleven countries including Belarus, Belgium, Germany, India, Japan, Netherlands, Norway, Poland, Switzerland, U.K. and U.S.A. Participants ranged from undergraduate students in engineering to very senior researchers. A total of 46 oral and 27 poster presentations were made at this workshop in addition to four breakout sessions and a panel discussion.

We thank Dr. Jorn Larsen-Basse of NSF and Major Hugh C. De Long of AFOSR for the financial support. Some of the administrative costs were borne by the Computer Microtribology and Contamination Laboratory of The Ohio State University.

Finally, I would like to thank the following members of the Organizing Committee for their assistance in the organization of the workshop: Prof. Steven Danyluk of the Georgia Institute of Technology at Atlanta; Prof. Steven Granick of the University of Illinois at Urbana-Champaign; Prof. Kyriakos Komvopoulos of University of California at Berkeley; and Prof. Marc J. Madou of the Ohio State University. Membership for the ASME Research Committee on Tribology provided general guidance. We hope that participants had a productive and enjoyable workshop.

Invited lectures and selected contributed articles are being published in this hard volume by Kluwer Academic Publishers B.V. The workshop proceedings contains 46 articles, 4 breakout session reports, and a panel discussion report.

Professor Bharat Bhushan
Workshop Director
Columbus, Ohio, U.S.A.
Nov. 1997

MEMS R&D IN EUROPE

L. HERMANS
IMEC
Kapeldreef 75, B-3001 Leuven
Belgium

1. Introduction

The present article tries to give an overview of the major axis of the European R&D activities in the field of Micro System Technology (MST). The activity in Europe on MST has grown considerably during the last years and many new players have entered the field recently. Groups working on IC technology have shifted to MST and laboratories active on the development of specific applications have built up their own MST prototyping technology. Therefore it is impossible to give an exhaustive overview of all the work going on in this rapidly expanding field. Choices had to be made. These choices do not reflect a judgement on the quality of the work of those mentioned or not. The examples have been chosen because the information was available and are to be representative examples of some recent trends. Other examples might have done as well. The goal was to be representative, not exhaustive.

The European R&D community prefers to speak about MST rather than MEMS. Micro Electro Mechanical Systems (MEMS) is considered to be a too narrow definition to describe the full potential of this extension of the IC technology. A microsystem is defined as "an intelligent miniaturised system comprising sensing, signal processing and/or actuating functions". The microsystem may be integrated on a single semiconductor chip or be a multi-chip hybrid. The driving force for microsystems R&D is the conviction that microsystems can make existing products smaller, cheaper, better and more reliable, but also that they allow the creation of new products. A good example of these new type of products are the microtips in all kinds of scanning microscopes, the digital micromiror devices (DMD) and microbolometer uncooled IR detector arrays.

As far as general trends are concerned the European situation is not very different from the US, except for the fact that in the US there are much more start-ups in the MST field compared to Europe. But the number of MST related start-ups is growing. Most of these start-ups have close links with the European MST R&D centres, from which most of them originated.

B. Bhushan (ed.), Tribology Issues and Opportunities in MEMS, 1-16.

2. European MST

It is very difficult to make general statements about MST industry and R&D in Europe. However some clear observations can be made.

The automotive industry (1) is the driving force behind the industrial development of MST in Europe. The prime driving force for using MST in cars is price reduction of components and systems, followed by the improvement of reliability and size and weight reduction. Well known applications in a car are emission control, safety, comfort and entertainment. Most of the industrial players have built up there own MST manufacturing capability.

A lot of universities and research centres have moved in to the MST field during the last years. They can be grouped in four classes. First there is the group of veterans including organisations like MESA (Enschede, The Netherlands), DIMES (Delft, The Netherlands) and IMT (Neuchatel, Switzerland). These organisations were active on MST form the early beginnings and contributed to the early R&D work. There is the group of MST laboratories which entered the field from the IC technology side. Before getting involved in MST these centers either active in IC process development, X-ray lithography or miniaturisation by "classical" methods. The available know-how and equipment allowed them the fast development of in house MST technology. Typical examples for this group are the Forschungszentrum Karlsruhe (Germany) for the LIGA technology, LETI (Grenoble, France) for Silicon-On-Insulator (SOI) bulk and surface micromachining and, CETEHOR (Besançon, France) based on their experience in miniaturisation for the watch industry and IMEC (Leuven, Belgium) for sub micron CMOS based MST. On the other hand laboratories specialised in certain application areas started to work in the MST field because it offered the potential for new and interesting developments and innovations in their domain of expertise. A good example is the Fraunhofer Institute for Biomedical Engineering (St Ingbert, Germany) which has a long track record in the development of medical devices and recognised in an early stage the potential of MST for that domain. The fourth group comprises the new MST labs, set-up with government support under the direction of experienced staff attracted from existing MST groups. Representatives of this last group are mainly found in Germany. Examples are IMSAS (Bremen, Germany) and IMIT (Villingen-Schwenningen, Germany). These new centers received substantial funding mainly from the regional governments (Länder).

More and more MST component manufacturers are opening there production facilities for external users. The driving forces are overcapacity and cost of exploitation. Although tens of millions of MST based components are currently made, the corresponding number of wafers is not comparable to the volumes of wafers handled by the IC industry. The offering of these manufacturing services to external users is heavily supported by the European Commission as part of the Europractice project.

2.1 INDUSTRY

2.1.1. Automotive

The major industrial players in the European MST scene are found in the automotive sector. The classical applications are pressure sensors and accelerometers. New products under development include mass flow and yaw rate sensors as well as chemical sensors for exhaust and cabin atmosphere control. The most important players in the automotive sector have their own silicon processing facility. The mayor players, their MST technology and current products are listed below.

2.1.1.1.. SensoNor (Horten, Norway)

SensoNor is an independent OEM supplier. Their business philosophy is to develop, manufacture and market microsensors based on silicon micromechanics for high volume applications. The company is currently the European leader for air bag sensors and employs over 300 people. The state-of-the art crash sensor is the SA20 of which more than 25 million units have been sold. This product represents 85% of their business and a European market share of 75 %. SensoNor's strong point is that has invested in state-of-art plastic packaging lines characterised by a high level of automation. The other important business areas are medical and aerospace. The currently offered crash sensors are based on bulk Si micromaching. SensoNor also intends to open their production for external users. Within the framework of Europractice SensoNor intends to organise for one of their bulk Si micromachining processes a multi project wafer (MPW) service for pressure sensors and accelerometers. The target date for starting that service is mid 1999.

2.1.1.2. Bosch (Reutlingen, Germany)

Robert Bosch GmbH is one of the leading independent suppliers of electronics and electromechanical systems for the automotive industry. The Si microsensor fabrication is performed on 6" manufacturing line, also used to produce CMOS and bipolar IC's. Bosch performs as well bulk Si as Si surface micromachining. The surface manufacturing process is since about one year open for external users. In order to lower the level of entry and financial risk a MPW service has been organised. The standard die size is 4.6x4.9 mm2 and the available sensor/actuator area is 2.8x2.5 mm2. The basic financial contribution is +/- $ 5500 per design plus $11 per untested chip supplied. For 1998 three MPW runs have been scheduled. Bosch MST products include a bulk Si pressure sensor with integrated electronics and a surface micromachined accelerometer. The surface micromachining process is characterised by the availability of a 20 μm thick polysilicon layer.

2.1.1.3. Sagem

Sagem SA is a 5800 employee, French company focused on electronic equipment for defence, telecommunications and the automotive sector. Sagem has a 1000 m^2 bulk Si micromachining facility. The company concentrates on pressure and acceleration sensors ranging from low cost for the automotive market to high accuracy and high cost for the aerospace market. Sagem is also interested in the application of MST for ink-jet printing applications due to it's involvement in the fax business.

Besides the three companies mentioned there many other players in the field including Siemens (München, Germany), Daimler Benz (Stuttgart, Germany), TEMIC (Heilbronn, Germany), Vaisala (Finland). Some of them already offer products for the automotive market, others still are in the development or prototyping phase.

2.1.2 Aerospace/Defence

A mayor player in this market segment is the French company Sextant Avionique. Sextant has in total 6250 employees an a turnover of M $830. The two main activities are avionics for aerospace and defence including flight control, data processing, man-machine interfaces and displays, navigation and sensors and an industrial components branch (brand name Crouzet Automatismes) including products such as switching, motors, micromotors, control, pneumatic control, microsensors, actuators. The sensor unit has 2000 m^2 of clean rooms and a production capacity of 5000 4" wafers per year. The MST activity concentrates on high accuracy pressure sensors, accelerometers, rate gyros, magnetometers, hydrophones and geophones based on silicon and quartz bulk micromachining and the integration aspects such as ASIC development, hybridisation and packaging. Related to the aerospace and defence market is the instrumentation and process control market. A company with a strong involvement is Schlumberger (France), active in sensors for harsh environments mainly to be used in oil well drilling. These sensors have to be operated at high temperatures and also require high temperature electronics.

2.1.3. Medical

In the medical field existing products including MST are mainly miniaturised pressure sensors. New types of ultrasonic probes with large arrays of piezoelectric transducers also exploit the potential of MST. A major player in the field is the Belgian company Melexis. Melexis has staff of over 200 people and realised in 1996 a sales figure over M $ 15. The bulk Si pressure sensors are fabricated in the German X-Fab facility. This is a 5" line also used for the production of CMOS ASICs with feature size down to 0.8 μm. Melexis also fabricates Hall sensors and sensor front-end IC's for pressure sensors and accelerometers.

2.1.4. Ink jet printing/nozzles

Little public information is available on the development of MST based ink jet printing components and nozzles. The reason is that mostly system companies are involved in this field and that the ongoing developments are mainly intended for in house use. Some companies active in this field are member of a NEXUS User Supplier Club (USC). Current members of this USC are Image, OCE, Stork, Videojet, Heidelberger and Sagem Fax. In the field of nozzles for drug delivery devices Microparts has announced some products.

2.1.5. New initiatives/start-ups

During the last years the number of start-up companies active in MST has increased significantly. In most cases these companies are fables and use the cleanroom facilities of MST R&D centres from which they often are a spin-off. These companies are at least in the early phase of their life more service, than product oriented. Some recent examples are:

2.1.5.1. Applied Microengineering Ltd (AML, Abingdon, UK).
Created in 1992, AML focuses on development contracts for industry, low and medium volume production and licensing. Higher volumes can be accessed through technology transfer. AML uses partially its own infrastructure but also makes use of the facilities of organisations such as AEA Harwell.

2.1.5.2 Twente Micro Products (TMP, Enschede, The Netherlands).
TMP was started in 1995 as a spin-off of MESA, a MST laboratory associated to the University of Twente (The Netherlands). TMP offers technical project management, production and project co-ordination from the initial idea to the microsystem production. For development and small production volumes TMP uses the MESA clean room facilities. For larger volumes TMP co-operates with CSEM.

2.1.5.3. Tronic's Microsystems (Grenoble, France)
Tronic's is a spin-off company of LETI. Tronic's was created in May 1997 to commercialise micro components based on the LETI Silicon-On-Insulator (SOI) based MST technology. Bulk as well as surface SOI micromachining processing will be offered. In addition the company offers thin film Pt filaments on quartz and SnPb based flip-chip ball bonding. Tronic's will rent LETI cleanroom infrastructure for production. Tronic's technology is based on the new Unibond SOI wafers supplied by SOITEC, another LETI spin-off and now a mayor supplier of SIMOX SOI wafers.

2.2. ACADEMIA/RESEARCH CENTRES

It is impossible to give a complete overview of all European universities and R&D centres involved in MST. According to NEXUS office there are already more than 125 independent or university linked R&D centres member of NEXUS. Therefore we will limit the discussion to some general remarks

At this moment each European country has al least one mayor R&D laboratory focusing on MST. The leading labs are found in Germany, The Netherlands, Switzerland and France. Other important MST activity can be found in Norway, Sweden, Denmark, Spain, Italy, Belgium, UK and Ireland. In addition there is a growing interest and activity in MST in Central and Eastern Europe. R&D centres active on MST either have moved to MST from form IC related R&D, entered the MST field through the applications (medical, optical, chemical, ...) or decided to restructure in to MST because a lack of public interest and funding for their previous nuclear R&D activities.

It should be noticed that most research centres in Europe work in close collaboration with industry. The emphasis on the ongoing R&D work is now more on application than on technology development.

3. Government support

Support for MST is offered in Europe on the national as well on the international level (2). The main funding source for MST R&D on the international level is the Commission of the European Union. MST related R&D work is also funded to a much smaller extent by the European Space Agency (ESA) and focussed on the specific application of MST in space projects.

On the national level the situations differ strongly from country to country. Germany provided massive funding for MST through specials programs running since 1990, whereas in a country such a the UK no dedicated MST support has been offered up to now.

In the following sections a rather elaborate description of EU supported initiatives will be given completed by some case studies on Germany and Switzerland, both having a dedicated MST support program. These two countries are considered to be the European MST leaders.

3.1. NATIONAL PROGRAMS

3.1.1 Germany

Public support for MST is provided in Germany as well on the federal as regional level. At the federal level a first innovation support programme "Microsystem Technology" ran from 1990 to 1993 with a total funding of +/- M $ 250 over four years. The programme was divided in three parts. One part targeted the support the application of SMT and ASIC technologies by small and medium sized enterprises (SME) and represented 30 % of the total budget, the precompetitive co-operation projects accounted for about 55% of the total available budget of which half was for industry and the other half for universities or research centres. The remaining 15 % were used to support technology transfer from research to industry and promotional activities. In 1994 a new MST program was launched. The project will run until the end of the year 1999 with an average yearly funding level of M$ 60. Emphasis is put on joint research projects with clear industrial perspectives, development of standard components and technologies taking into account manufacturing requirements and the development of MST demonstrators in areas where the government has public responsibilities such as health, environment and public transportation infrastructure. Both projects were managed by VDE/VDI the German society of engineers.

In addition to the federal Government the Länder (States) run their own MST initiatives. These initiatives are very diverse and include support to existing R&D centres (for example the restructuring of the KfK from nuclear to MST R&D), the co-ordination of MST activities on the regional level (Mikro-Struktur Iniative NRW), the creation of new MST R&D centres (IMM, Mainz) to specific R&D programs tailored to the industrial structure of the particular region . The state of Thüringen started a special R&D initiative in micro-optics and MST for bioprocesses.

3.1.2 Switzerland

MST is considered to be very important for the Swiss industry. There a strong tradition in co-operation between industrial companies and research centres. This co-operation has assured a rapid transfer of this rather new technology to the Swiss industry in particular to the instrumentation and medical sector. The Swiss Government strongly support MST related activities. In 1996 a priority programme named MINAST was started. Half of the funding of the projects is coming from the government, the other half should be contributed by industry. In addition the Swiss government strongly supports the participation of Swiss companies and R&D centres in EU funded MST projects on the basis of an association agreement with the Commission of EU.

3.2 COMMISSION OF EUROPEAN UNION

During the past years the Commission of the European Union has continuously increased its support for establishing in Europe a microsystem technology industry. The Commission expects that the application of a microsystems will give a decisive added value to many products, difficult to achieve by other means. The stimulation of the development of microsystems technologies is supported by many different programs. Here under we present only the most important ones

3.2.1 Information Technology (IT) programme

The IT programme supports the development of technologies required needed for the development of the global information society of today and tomorrow. Microsystems technology is considered to be one of these.

EUROPRACTICE.

A new M$ 35 ECU ESPRIT initiative in the field of microelectronics was launched by the European Commission (DGIII). EUROPRACTICE (Promoting Access to Components, subsystems and microsystems Technologies for Industrial Competitiveness in Europe) aims to stimulate the wider application of state-of-the-art microelectronics technologies by European Industry in order to enhance European competitiveness in the global market-place.

It is generally accepted that by incorporating modern microelectronics technologies in their products and processes companies can enhance production efficiency, reduce costs, improve product performance, achieve reduced size, weight and power consumption and differentiate products through new functionality's. Although the use of ASICs is already common, the technology is yet to be fully exploited by European industry. MCMs and Microsystems technologies on the other hand are not in widespread use and their application need to be stimulated.

Through EUROPRACTICE, the EC has created a range of Basic Services that offer users a cost-effective and flexible means of accessing three main microelectronics-based technologies:

- Application Specific Integrated Circuits (ASICs)
- Multi-Chip Modules (MCMs)
- Microsystems

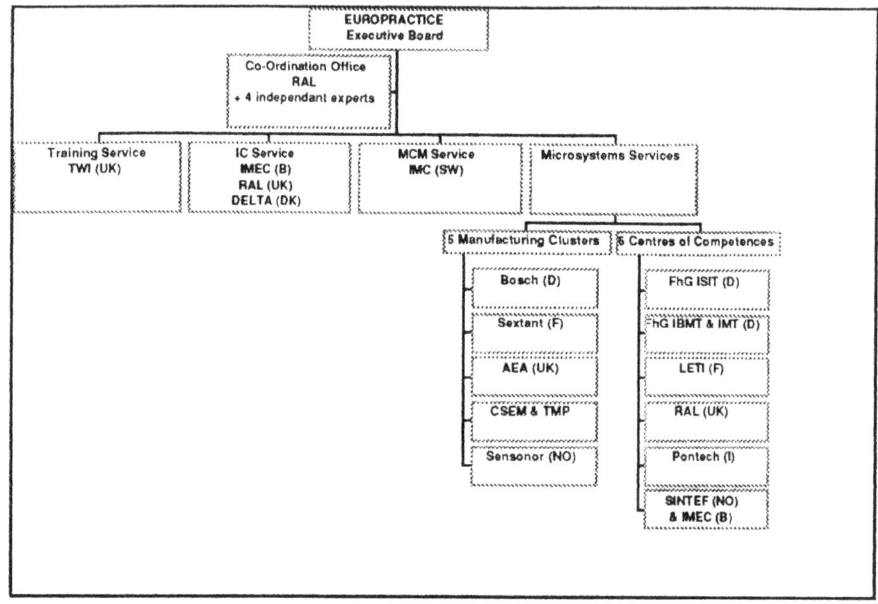

Figure 1. Organisation of Europractice II (Oct. 97 to Oct. 99).

EUROPRACTICE also provides users with the advice, training and CAD software needed to design and develop the ASICs, MCMs or Microsystems. EUROPRACTICE Basic Services will reduce the cost and risk for companies wishing to begin using these technologies.

EUROPRACTICE services are available to both industrial and academic users although the main emphasis is on industry and in particular on Small-Medium sized Enterprises (SMEs). Currently, many European SMEs could benefit from the exploitation of ASICs, MCMs and Microsystems in their products but many are unaware of the benefits offered by the use of these technologies. EUROPRACTICE aims to provide the impetus and technical support to encourage these SMEs to develop ASICs, MCMs and Microsystems for use in their products.

EUROPRACTICE Microsystems

The EUROPRACTICE Microsystems Service helps by providing information and advice on whether the products of a particular enterprise can benefit from the application of MST. If a decision to proceed is made, then the EUROPRACTICE Microsystems Service provides a simplified design interface to a choice of prototyping and manufacturing foundries. Costs of entry are lowered by the sharing of development and fabrication costs using multi-project wafers in the same way as with the EUROPRACTICE ASIC Service.

In the first two year phase of Europractice support was provided on a regional basis through a network of partners giving detailed advice on the services available, conducted feasibility studies and routed application projects into the design, prototyping and manufacturing stages. This network was called the Demonstration Activity and consisted of 11 Microsystems consultants, located all over Europe.

In the second phase of Europractice this approach has been replaced by the concept of Competence Centres. Six competence centres have been selected, each having expertise in six different and promising MST technical areas: micro fluidics, physical measurements, micro opto mechanical systems (MOEMS), micro actuators and machines, bio-medical and bio-chemical, radiation and imaging sensors. IMEC is a competence centre for smart imaging systems. It is the task of a competence centre to promote the application of microsystems technology in their field of expertise and to support interested user. In contrast to the demonstration activity which was organised on a regional basis the competence centres will be active of all of the EU. but limited to their specific field of expertise.

Technical skills	Micro Fluidic	Physical measure-ments	MOEMS	Micro-Actuators & Micro-Machines	Bio-Medical & Bio-Chemical	Radiation & Imaging sensors
Organisation	PONTECH (I)	FhG ISIT (D)	LETI (F)	RAL (UK)	FhG IBMT + IMT (CH)	SINTEF (NO) + IMEC (B)
Applications skills	Micro Fluidic	Automotive	Peripheral & Telecom	Process control & Machine tools manufact.	Medical & Environ-mental	Aerospace & Scientific Instruments

Figure 2. Overview of Europractice Competence Centres.

The EUROPRACTICE Microsystems Manufacturing Service is organised as a set of five manufacturing clusters (MC) situated across Europe. Each of these MCs is offering a more or less unique type of microsystems fabrication technology. The services offered by the MCs are open to all interested parties. Most of them offer or will offer in the near future a MPW fabrication service in order to reduce the non recurring engineering (NRE) costs. In general the cluster consists of one or more industrial companies supported by one or more research centres. The task of the research centres is to support the industrial companies in the development of special process modules or to support potential customers during the feasibility and design phase of the project. All MC's have received considerable European funding to open their infrastructure and production processes for outside use.

MC1 is the Germany based manufacturing cluster lead by Bosch. Bosch offers a multi project wafer surface micromaching service. The process is characterised by the availability of a 20 μm thick polysilicon layer. Other partners in MC1 are Microparts, GMA, HL Planar and the Fraunhofer Institutes IMS and ISiT, offering a very broad range of manufacturing services such as electroforming and plastic moulding, CMOS based bulk and surface micromachining, thin film microstructures and suitable packaging techniques.

The recently restructured French MC2 is now lead by Sextant Avionique. Other partners are the already mentioned start-up company Tronic's Microsystems and the R&D centres LETI and LAAS (Toulouse, France). LETI and LAAS mainly focus on modelling, design and prototyping activities, whereas the industrial partners focus on volume production. MC2 intends to offer a MPW service for SOI bulk and surface micromachining.

MC 3 is a Swiss/Dutch manufacturing cluster led by CSEM. Other partners involved are Twente Micro Products (TMP) and Hollandse Signalen. The technologies offered are silicon surface and bulk micromachining, electro-plating, double sided wafer processing, hybrid packaging and processing of quartz and fused silica. With respect to the last technology a MPW service for the fabrication of Diffractive Optical Elements (DOE) is offered on a regular basis. Standard die size is 6.9x6.9 mm^2 with a clear aperture of 6.5x6.5 mm^2. Processing costs range from 5,500 CHF to 7,500 CHF depending on the number of levels. Starting from November 1997 CSEM will also offer a MPW service for the fabrication of custom microelectrodes in amorphous or diamond like carbon. In relation to the topic of the present workshop it is interesting to note CSEM experience in the field of tribology and surface measurements. The Tribology Service Centre of CSEM offers a competitive range of tribological characterisations of materials or mechanical components related to the friction and wear behaviour in precision mechanisms and microsystems. These services aim to increase the lifetime and performance of mechanisms and/or tribosystems operating under normal or severe environmental. PVD and CVD coating services include customised developments for hard, wear resistant or self-lubricating films. CSEM has also developed a complete family of instruments to control the mechanical properties of material surfaces. They are designed for use by laboratories and the manufacturing industry to control hardness, friction, adhesion and scratch resistance. The following product families are available: tribometers, scratch-testers, indenters and cracking testers and atomic-force tribometers.

MC4 is the new British manufacturing cluster lead by AEA Harwell. AEA Harwell will offer high volume fabrication capacity. Other industrial partners are the already mentioned company AML, the company Graseby responsible for packaging related issues and the University of Herefordshire. At the present moment no concrete information on the specific technologies offered by this MC is available.

MC5 is a Scandinavian based cluster led by SensoNor. The other industrial partner is DELTA (Denmark) responsible for testing. The industrial partners are supported by a number of Scandinavian R&D centres SINTEF (Oslo, Norway), MIC (Denmark),

IMC (Sweden) and VTT (Finland). MC5 also has a Mediterranean antenna namely CNM, a Spanish university laboratory involved in MST (Barcelona, Spain). SensoNor also plans to offer a MPW service for bulk micromaching process for the fabrication of accelerometers and pressure sensors by mid 1999.

First User Action (FUSE)

The goal of FUSE is funding scheme to stimulate more enterprises to incorporate electronic technologies into their production, thereby increasing their competitiveness. This must be done in such a way that the enterprise acquires the necessary know-how and experience to access and use new technologies and is able to repeat the process in its future product development. The measure of success for a participating First User will be both the degree to which the experiment provides a more competitive prototype product and the degree to which the enterprise acquires the know-how to access and use the new technologies and to repeat the exercise in the future.

The main instrument for achieving the goals of FUSE is the Application Experiment. In the Application Experiment, the enterprise will use technologies which are new to that enterprise to carry out the design, manufacture and test of a component which is relevant for the improvement of it's products. The enterprise should sub-contract to third parties and demonstrate that the necessary know-how to repeat the exercise can be transferred from the sub-contractor to the enterprise.

NEXUS

NEXUS started in 1992 as a network driven mainly by academic institutions, with an industrial working group providing an advisory role. Since the founding of NEXUS, the commitment and interest from industry has grown steadily. This development is also reflected by the changing nature of NEXUS membership: The initial NEXUS proposal was submitted by a consortium comprising 11 institutes and only one company . The network now includes 93 industrial companies and 125 institutes in 17 (West) European countries NEXUS includes the following types of 'MST players': producers of microsystems , users of microsystems, R&D laboratories working on MST, manufacturers of equipment for MST production and technology transfer organisations The new leading role of industries is reflected by a new organisational structure which was established upon the transition from Phase 1 to Phase 2 of NEXUS. It includes various boards and working groups whose role, composition, and achievements are described in the following sections:

Executive Board

A small industrial Executive Board (EB) has been established by the NEXUS Board to define the policy of NEXUS to achieve the network's general objectives. The EB is in charge of launching, delegating and monitoring specific tasks, it controls the utilisation of the financial resources, and it prepares the major decisions of the NEXUS Board. The EB reports to the NEXUS Board twice a year.

NEXUS Board

The NEXUS Board is a broader representation of the industrial MST community. It provides and nominates the candidates for the Executive Board and the various task forces and User-Supplier Clubs. The NEXUS Board provides consultancy to the Executive Board regarding strategic orientation of the network.

NEXUS Office

The NEXUS Office is responsible for the technical, administrative, and financial management of the network, including contractual and reporting issues, preparation and reporting on meetings, information dissemination (mailings, newsletter, press releases), almanacs, and various accompanying measures. The NEXUS Office closely collaborates with the Executive Board and the other working groups in NEXUS. It is the common contact point for the Commission, the NEXUS community, and anybody interested in the network's activities and services.

Academic Working Group (AWG)

Recently re-launched is the Academic Working Group where Universities and Institutes meet to propagate the very latest MST ideas, addressing tasks of primarily scientific character, to provide a forum to discuss problems raised by industry and to provide consultancy to the NEXUS Board on scientific and educational matters. The AWG includes representatives of the major European MST laboratories in universities and research institutes. Major tasks of AWG are:

- Development of long-term perspectives for MST and their integration into future European programmes.
- Establishment of 'virtual centres' by networking in the field of advanced technologies in order to increase the critical mass to accelerate innovative but practical MST solutions for an industrial problem.
- To contribute to training in the field of MST in collaboration with EUROPRACTICE.
- To organise exchange programmes for scientists.

To meet these objectives, the AWG closely collaborates with the various other groups in NEXUS; in particular with the User Supplier Clubs on technical issues, and it also participates in task forces launched by the Executive Board.

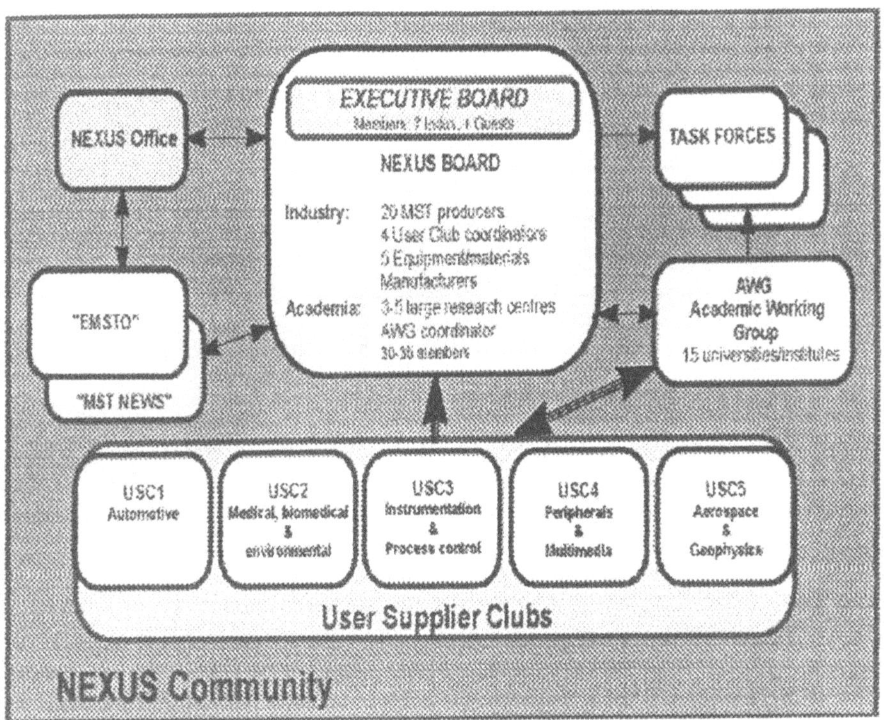

Figure 3. NEXUS organisation.

User Supplier Clubs (USC)

For many MST related issues (for example market perspectives, quality and performance standards, technological interfaces etc.), an application driven approach is considered to be most helpful. The idea of USCs is, for each application area, to bring together industrial technology providers, system houses focused on applications, and R&D institutes. USCs are currently operational in the automotive, medical, instrumentation, computer peripherals and aerospace sectors.

	Application Area	Co-ordinator
USC 1	Automotive applications	R. Turner (KBIC, GB)
USC 2	Medical, biomedical, and environmental monitoring	C. Legge (Glaxo-Wellcome, GB)
USC 3	Instrumentation and process control	A. Zumsteg (CSEM. CH)
USC 4	Peripherals and multimedia	A. Bellone (Balteadisk, I)
USC 5	Aerospace and geophysics	J. Suski (Schlumberger, F)

Figure 4. Overview of existing USCs/.

The co-ordinator is in charge of the organisation of their respective user clubs and requests anyone interested in joining to contact him directly. The user clubs define their objectives and tasks in close co-operation with the Executive Board. Short-form activity reports of the User Supplier Clubs are forwarded to all members of NEXUS, and industry specific meetings and workshops are organised as part of the USC programme, usually in conjunction with major exhibitions or conferences. Detailed reporting of USC activities is only available for the active USC members.

Task Forces

NEXUS has set up expert task forces focused on specific actions. Operational task Forces are:

The Market Analysis Task Force focuses on the gathering of detailed MST market information to make more accurate prediction of the future. Initial findings for the year 2002 indicate a world-wide market (of the smallest saleable part utilising MST) of $34 billion. Expert input from the User Supplier Clubs is used to update and verify this information. Recently the MST market analysis task force published an interim report.

The International Relations task Force establishes and maintains relations to MST related institutions and authorities outside Europe, like DARPA (USA) and MMC (Japan) and thus promoting) NEXUS as the European reference body in the world-wide dialogue on microsystems. One of the issues of this task force is benchmarking of MST activities world-wide.

The Eastern Countries Task Force maintains relations with the East European MST community.

The MST Training Task Force promotes MST specific training and education activities in co-operation with other EU programmes like EUROPRACTICE, Socrates, or TMR.

The Long-Term Perspectives Task Force, which is primarily based on the Academic Working Group, is now preparing a study which outlines what will be achievable with MST in 10 years from now.

The EMSTO task force is in charge of defining the requirements on the EMSTO (3) website and of monitoring its implementation.

Future task forces are likely to tackle MST design software and Standards.

3.2.2. Industrial and Materials Technology program (Brite/Euram III)

The Brite-Euram programme aims at increasing the competitiveness of European industry through concentrated and targeted R&D actions on industrial objectives in three technical areas : production technologies, materials and technologies for product innovation and technologies for transport such as aeronautics, automotive, train and ships. Within these three areas there is an important emphasis on the development of microtechnologies. About 7 % of the total of 1,722 MECU Brite/Euram budget (1994-1998) was allocated to MST related projects. In total at least 50 MST projects have been or are still running within Brite/Euram. More information can be found at the CORDIS WEB site (4).

3.3. EUROPEAN SPACE AGENCY (ESA).

One also notes a growing interest from ESA for microtechnologies. The interest was triggered by the enormous potential of MST for the reduction of volume, weight and power consumption on payload, platform and launcher level. Lower volume, weight and power consumption mean lower cost and/or more missions. The launch of about 1kg of payload costs approximately K$ 28.

In addition MST opens new technical grounds, makes cluster missions or satellite fleets possible and allows for more frequent access to space. Miniaturisation also allows for far-reaching redundancy resulting in more long term reliability and autonomy.

There are however a number of specific issues to be addressed with respect to the utilisation of MST in space such as a culture chance with respect to system designs, the limited proven lifetime of current MEMS components, production and test of small batches, lack of understanding of many MST characteristics (e.a. friction), packaging and interconnections and the use embedded software which will require a new product assurance philosophy.

ESA has defined the following strategy for MST space developments. Considering the large initial investments required for MST product development a synergy with the non space MST industry is required, taking in to account a strong dependency. For the non space market mass production and a high innovation speed are the key. This is in contrast with the space industry with it's typical low volume requirements and long product development cycles. In order to cope with this difference new system design approaches will be needed as well as a new product assurance philosophy.

In order to support the incorporation of MST in spacecrafts ESA has foreseen in it's new Technological Research Program (TRP) (5) a budget of about M$ 3.5. Topics to be supported are system studies, micro robotics, micro imagers, gyros, micro mechanical switches, vibration damping, micromachined RF front end, lubrification for micromotors, etc..

Additional MST related ESA activities are funded through the optional R&D programs. The most important with respect to MST is the General Support Technology Program (GSTP). One should also note that the national space agencies have their own space programmes.

4. Conclusions

Europe sees MST as a very promising and strategic industrial field in which Europa wants to be an important player. Europe also gives a broader definition to MEMS. Integration of different functions is important, of which the mechanical function is only one. It is clear that MST related activities in Europe will continue to grow strongly during the coming years. This growth will be fuelled by a continuing government support and new breakthroughs in the commercial exploitation. The manufacturing technologies and facilities will be opened for external users and the barriers of entrance will be lowered through the offering of MPW services for different types of MST fabrication technologies. A growth in the number of start-ups and spin-offs will be observed. The industrial base is still narrow and mainly in the automotive sector. Through easy access to manufacturing services this base will certainly be broadened.

5. References

(1) Proceedings International Conference on Advanced Microsystems for Automotive Applications, December 1996, Berlin
(2) Micro System Technologies in Europe today, MUST (Esprit 8520) study report 2nd ed. April 1995, CME, Veenendaal, The Netherlands
(3) http://www.emsto.com
(4) http://www.cordis.lu
(5) The ESA Technology R&D Programme 1997-1999

INTEGRATED MEMS IN CONVENTIONAL CMOS

GARY K. FEDDER
Department of Electrical and Computer Engineering
and The Robotics Institute
Carnegie Mellon University
Pittsburgh, PA, 15213-3890
E-mail: fedder@ece.cmu.edu

1. Abstract

This paper provides an overview of fabrication and design of CMOS-based microelectromechanical systems with emphasis on inertial sensor and data storage applications. High-aspect-ratio (4.4:1) microstructures can be fabricated using conventional CMOS processing followed by a sequence of maskless dry-etching steps. The CMOS dielectric and metallization layers, normally used for electrical interconnect, serve a dual function as a composite metal/dielectric structural material. Reactive-ion etching produces near vertical sidewalls, enabling micromechanical beam widths and gap spacings down to 1.2 μm. The process is tailored for design of lateral electrostatic actuators as well as capacitive position and motion sensors. Tight integration of the microstructures with CMOS provides an opportunity to make low-noise sensor interface circuitry, and to include the signal processing needed to manage arrayed sensor-and-actuator systems-on-a-chip. Novel actuator and sensor topologies can be designed by embedding multiple isolated conductors into the microstructures. An additional post-CMOS processing sequence produces platinum tips on the movable microstructures. These tips are being explored for use in probe-based data storage and tunneling sensor applications.

2. Introduction

Future applications of MEMS are requiring microsystems that are increasingly complex and that are embedded with greater computational power. Applications include inertial measurement systems, digital light processors, infrared imagers, ultrasonic imagers, probe-based data storage, RF communications, and RF switching arrays. One promising approach to low-cost manufacturable integration of

B. Bhushan (ed.), Tribology Issues and Opportunities in MEMS, 17-29.

MEMS and electronics is to start with a conventional CMOS process and build MEMS capabilities into it.

Direct integration of MEMS with electronics is desirable for systems with arrayed microsensors and microactuators on a single chip. On-chip electronics removes the interconnect bottleneck. For example, an array of 10,000 microsensors without integrated electronics requires at least 10,000 external connections. However, the signal processing needed to manage arrayed sensor-and-actuator systems can be placed on the chip in a CMOS-compatible MEMS process. Output signals can be electronically multiplexed or combined to reduce the number of off-chip connections.

Several key benefits to leveraging conventional CMOS processing for MEMS are that fabrication is fast, reliable, repeatable, and economical. There exist material limitations; most notable is the large vertical residual stress gradient in the microstructures, which must be overcome through design techniques. The CMOS-MEMS processes provide two specific performance advantages over similar technologies, such as integrated polysilicon MEMS. First, active devices can be placed directly next to microstructures, thereby virtually eliminating interconnect parasitic capacitance on high-impedance capacitive position sensor nodes. Second, multiple isolated conductors can be placed inside of suspended microstructures to create new capacitive sensors and electrostatic actuators for application in inertial sensors, resonator structures, and micropositioners.

3. Prior Work in CMOS-MEMS

In 1989, researchers at Simon-Frasier University were the first to report microstructures made from metal and dielectric thin films and directly integrated in standard CMOS processes [1]. In the basic process, illustrated in Figure 1 for two-metal CMOS, microstructures are made from combinations of aluminum, silicon oxide, and silicon nitride thin films. The metallization and dielectric layers, normally used for electrical interconnect, now serve a dual function as structural layers. Microstructural sidewalls are defined by stacking the drain/source contact cut and metal via cuts, and removing the metallization layers above the cuts. At the end of the CMOS processing, the substrate is exposed in the cut regions. Composite metal/dielectric (usually aluminum/oxide) microstructures are released by undercutting the silicon substrate, which acts as the sacrificial material. Wet-etch release may be accomplished using KOH, ethylene diamine pyrocatechol (EDP), or tetramethyl ammonium hydroxide (TMAH). Alternatively, researchers have used a dry etch release in a XeF_2 plasma [2] or in a SF_6:O_2 plasma [3] to prevent problems with sticking during release.

The contact-cut and metal-etch steps in the foundry CMOS process are tailored for planarized interconnect, not for creation of micromechanical structures. Since the contact cuts used to define the microstructures are not covered with

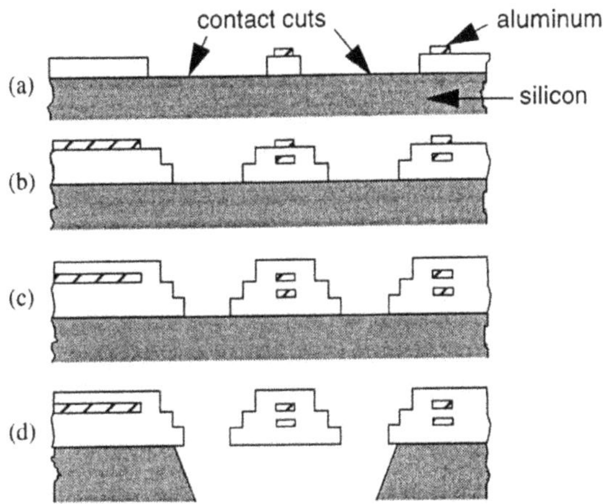

Figure 1. Cross sections of the stacked-contact-cut process. (a) After first metal, showing exposed oxide cuts. (b) After second metal, illustrating sequential stacking of cuts. (c) After pad overglass cut; microstructural sidewalls are fully defined. (d) After anisotropic wet silicon etch to release the microstructures.

metal, the layout violates the CMOS design rules. Incomplete etching of subsequent layers deposited on top of the cuts can result in formation of undesired metal and dielectric sidewalls. These sidewall films adversely affect the effective beam width by 1 μm or more. Very narrow lateral gaps cannot be fabricated because of the large topography of the oxide cuts, with the structure spacing being limited to about 10 μm for acceptable yield. Therefore, this process is primarily used for the manufacture of thermally isolated structures and vertically actuated structures. Current active research includes work on inertial sensors [4][5], biological cell force sensors [6], thermally operated gas flow sensors [7], and infrared detectors [8].

Various research groups have altered the basic post-CMOS process steps to form other kinds of MEMS. Micromechanical structures have been made using the gate polysilicon as the structural layer and field oxide as the sacrificial material [5]. Electrochemical wet etching has been used to fabricate thermally isolated transistors for true-rms power sensors [9]. The electrochemistry allows p-doped silicon to be preferentially etched, while not attacking n-doped silicon, thus resulting in suspended active devices in well regions undercut by the release etch. Microfluidic structures have been formed by wet etching the aluminum interconnect to form silicon dioxide capillaries on top of the substrate [10].

Currently, two foundry process services provide fabrication of CMOS-MEMS devices. In the U.S., MOSIS (the MOS Implementation Service) provides limited support to fabricate MEMS devices in certain CMOS processes (primarily AMI 1.2 μm) [11]. Future plans include full support of the post-CMOS silicon etch for release of microstructures. *Circuits Multi-Projets* (CMP) of France is the first

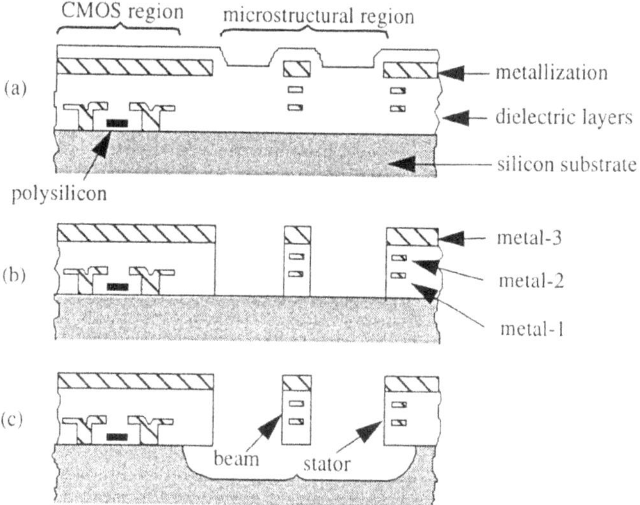

Figure 2. Flow for the high-aspect-ratio CMOS-MEMS process. (a) Conventional foundry CMOS. (b) After reactive-ion etch of the dielectric layers. (c) After dry plasma etch of the silicon substrate.

fabrication service outside of the U.S. to offer MEMS manufacturing in standard CMOS [12]. Their activity started in 1995, and has ramped to 16 MEMS multi-project runs in the ATMEL-ES2 1.0 μm CMOS process.

4. High-Aspect-Ratio CMOS-MEMS

One of the limitations of most CMOS-MEMS processes is the inability to create microstructures with narrow beam widths and narrow gaps for lateral electrostatic actuation and capacitive sensing. This drawback motivated our research group to develop a new set of post-CMOS process steps designed to create relatively high-aspect-ratio beams and gaps [3][13].

A simplified process flow is shown in Figure 2. First, standard CMOS is fabricated using the Hewlett-Packard 3-metal 0.5 μm n-well CMOS process available through MOSIS (Figure 2(a)). Separate areas must be partitioned for the CMOS electronics and for the microstructures. The first post-CMOS step is a CHF_3:O_2 reactive-ion etch (RIE), shown in Figure 2(b). The topmost metal layer acts as a highly selective mask which defines the microstructures. The RIE etches any dielectric (*i.e.*, overglass, intermetal oxide/nitride, and field oxide) that is not covered with metal. The Hewlett-Packard 0.5 μm process employs aluminum as the conductor material with tungsten-plug vias between metal layers. The top aluminum layer is partially eroded by ion milling during the RIE. The last process step,

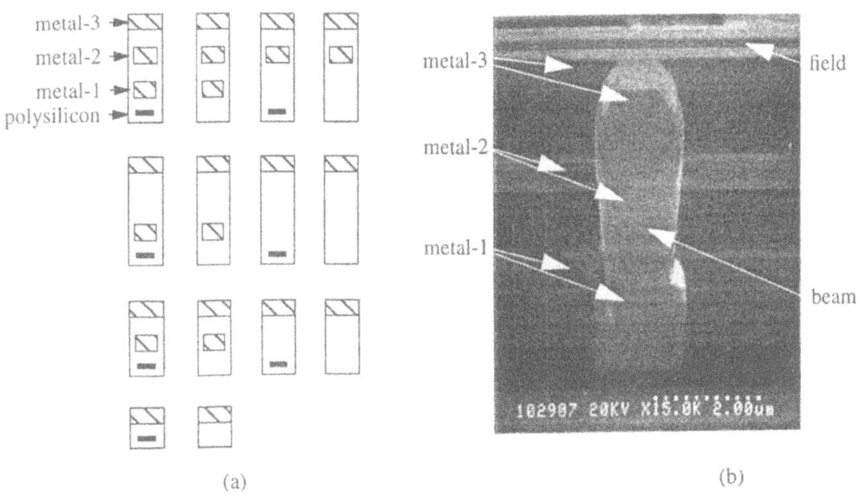

Figure 3. (a) The fourteen possible beam cross sections in the CMOS-MEMS process using 3-metal CMOS. An overlap between the top metal and underlying conductors is shown. (b) SEM of the end of a minimum width, metal-3/metal-2/metal-1 beam with no overlap (as drawn) between metal layers. Note the misalignment in the metal-1 layer causes a step in the sidewall. The 3-metal field can be seen in the background.

shown in Figure 2(c), is a dry SF_6:O_2 plasma etch of the silicon substrate. The plasma chemistry etches silicon without attacking the microstructural sidewalls. The etch is timed to undercut structures up to 16 μm wide. Larger structures must have etch holes for proper release.

The resulting microstructures are made of composite beams with near vertical sidewalls and beam widths and gaps down to 1.2 μm. As shown in Figure 3(a), it is possible to design fourteen different beam cross sections using combinations of the three metal layers and the gate polysilicon. The second-thickest beam section, shown in Figure 3(b), includes all three metal layers: metal-3, metal-2, metal-1; these beams are around 5.3 μm tall with an maximum aspect ratio of around 4.4:1. In principle, conventional electrical design rules dictate the minimum metal width and spacing, and therefore dictate the minimum beam widths and air gaps. However, the air gap is currently limited to 1.2 μm instead of the metal-to-metal spacing of 0.9 μm, because of polymerization issues stemming from the dielectric RIE. Narrow beams enable design of very compliant lateral suspensions, and the small air gaps enable design of lateral electrostatic actuators with relatively high force. These capabilities are useful in accelerometers, vibratory-rate gyroscopes, micropositioners, and micromechanical resonators.

One corner of a released electrostatically actuated lateral resonator, is shown in Figure 4. This device has very narrow beam widths and gap spacings, which are features normally associated with polysilicon microresonators [14]. The interdigi-

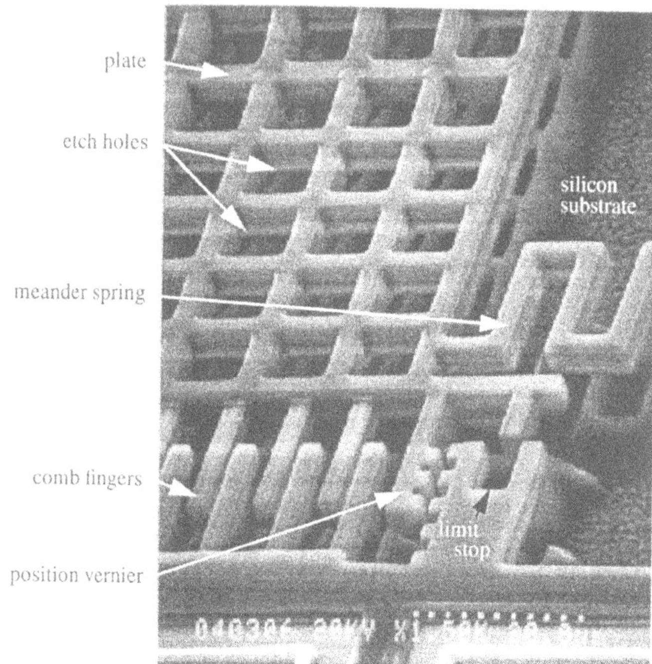

Figure 4. SEM of a section of a lateral comb-drive microresonator after the silicon etch has released the structure. The noted features demonstrate the successful fabrication of very small beams and gaps in conventional CMOS.

tated comb fingers and compliant meander springs are 2.4 μm wide and 4.8 μm thick with gaps of 1.6 μm between fingers. Etch holes allow the large plate to be released. Silicon ridges are present about 10 μm below the suspended structure and result from the slightly anisotropic release etch. Embedded aluminum traces wind through the meander springs to provide interconnect between the comb fingers and bond pads. The resonator plate mass experiences an electrostatic attractive force upon application of a potential across the gap between comb fingers.

Mechanical material properties are not a primary concern for CMOS foundries, which focus on reliably producing electronic circuits, not MEMS. As a result, CMOS-based microstructures can suffer from large residual stresses and stress gradients, which may vary from run to run. Residual stress has been measured using bent-beam strain test structures [15]. The 3-metal composite beam has an in-plane residual stress of 69 MPa, corresponding to a strain of 1.1×10^{-3} and a buckling length of 65 μm for a 1.2 μm-wide fixed-fixed beam.

Vertical stress gradients due to the composite nature of the structures cause curling out of the plane of the substrate. Curl of eight different kinds of beams are

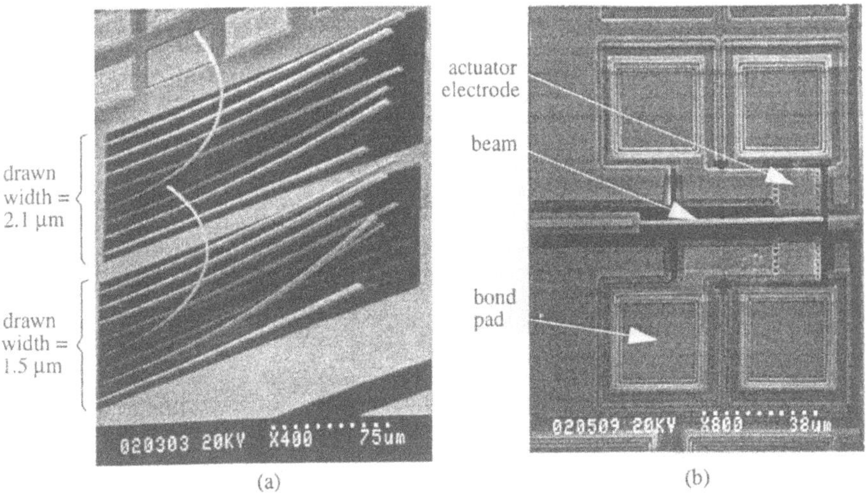

Figure 5. Initial CMOS-MEMS characterization structures. (a) Cantilever structures showing various amounts of curling depending on beam type. The beam type is from top to bottom: m1-m2-m3, m2-m3, m1-m3, m1-m2, m1, m2, m3, poly-m1-m2-m3. The metal-1 beam exhibits the maximum curl. Curling increases as the beams width decreases. (b) Cantilever beam resonator for stiffness measurement.

shown in Figure 5(a). The radius of curvature of these beams varies between 1 to 14 mm and depends on both the beam cross section and the beam width. The measured tip displacement of several 1.4 mm-long, 3.6 μm-wide, 4.8 μm-thick 3-metal beams over several process runs was 200 ± 30 μm, which corresponds to a nominal radius of curvature of 4.9 mm and a strain gradient of $2x10^{-4}$/μm. The variation is due in part to modification in the post-processing over time, which can alter the beam cross section. However, curling of beams on a given die is matched. A more systematic investigation of strain gradient variation from run to run is an important next step in our research.

The 3-metal composite beam in the HP 0.5 μm CMOS-MEMS process has an effective Young's modulus value of 52 GPa. This value was determined by fitting the resonant frequency of three matched cantilever beams of differing lengths. A typical beam resonator test structure is shown in Figure 5(b). The density of the aluminum, oxide, and nitride films are assumed to be 2700 kg/m^3, 2500 kg/m^3, and 3100 kg/m^3, respectively.

We have qualitatively observed deflected cantilever beams sticking to laterally adjacent surfaces. However, the beams can be pulled away from the surface using a micromechanical electrostatic actuator generating on the order of micronewtons of force. If this action can be shown to be reliable, then devices can be designed to

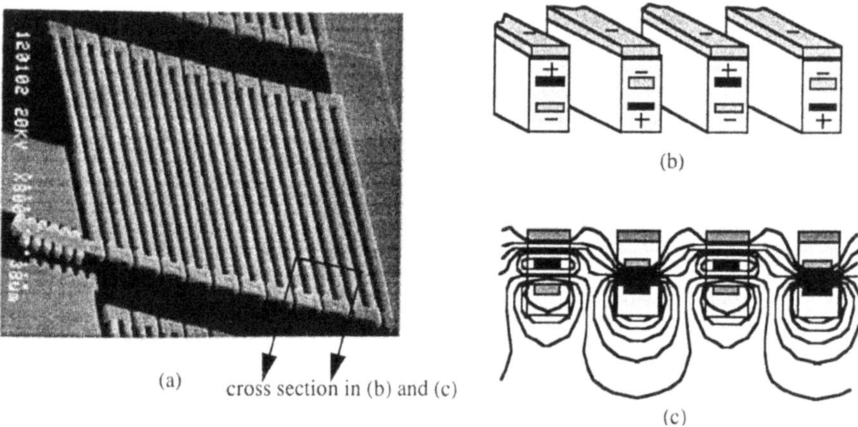

Figure 6. The self-actuating spring. (a) SEM of a self-actuating spring with 22 meanders. A vernier, shown at the left of the SEM, is used to measure position. (b) Schematic cross section showing voltage polarity of the embedded conductors. (c) Cross section showing electric equipotentials.

recover from contact. Future characterization and understanding of surface charging and lateral contact forces will be essential in developing robust devices.

5. CMOS-MEMS Design

As discussed previously, the CMOS-MEMS fabrication technology is a relatively low cost method for fabricating integrated MEMS. The primary drawback of CMOS-MEMS is the inability to optimize the mechanical properties of the microstructures, which can limit the performance of certain devices. The disadvantage of variability of the mechanical parameters can be mitigated through design in many important cases. In particular, microstructures can be designed with matched curling to ensure a high value of lateral air-gap capacitance between sensors and actuators. Although the mechanical structures pose challenging design problems, the CMOS-MEMS process does have two important performance advantages: multi-conductor structures, and ultra-low parasitic capacitance. For some complex MEMS applications, these advantages may outweigh the disadvantages.

One performance advantage of CMOS-MEMS is that independent, isolated conductors may be embedded into a single suspended microstructure. This ability opens up new design possibilities like the self-actuating spring, shown in Figure 6. Two differential voltages (+ and -), applied to conductors embedded in the meanders, alternate connection between the metal-2 and metal-1 layers by swapping the wiring at the ends of each meander. The resulting electrostatic force laterally com-

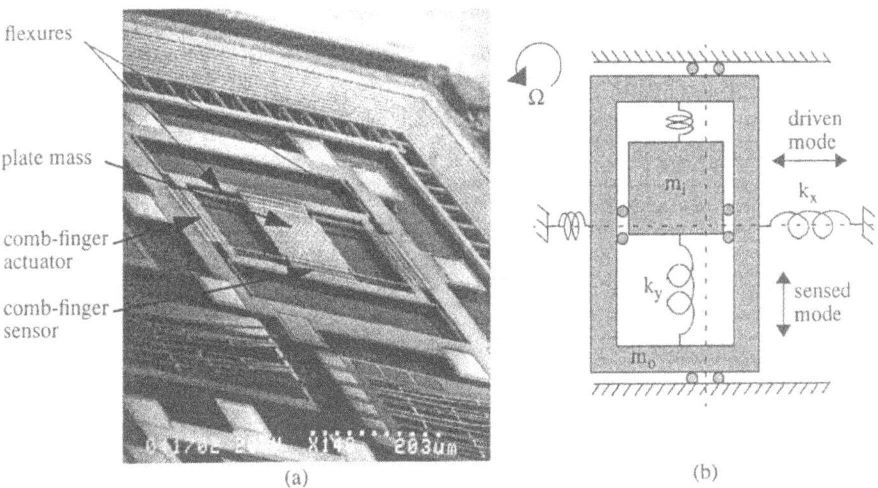

Figure 7. The elastically gimbaled gyroscope. (a) SEM. (b) Schematic of basic operation.

presses the spring with increasing applied voltage. Such actuators and sensors are impossible to design using homogeneous polysilicon microstructures.

In another example, completely independent electrostatic actuators and sensors are designed into the elastically gimbaled vibratory-rate gyroscope shown in Figure 7 [16]. The inner proof mass, m_i, and springs k_y are nested within a surrounding outer frame, m_o, and driven in resonance in the x direction by an electrostatic comb drive. A Coriolis force proportional to external rotational rate, Ω, is produced in the y direction of the rotating reference frame of the gyroscope. The resulting motion of the inner proof mass in the y direction is sensed with an independent comb-finger sensor. The actuator and sensor are mechanically decoupled, thereby eliminating mechanical crosstalk and reducing errors coupled into the sensor from the driven mode. The decoupled design also allows for independent optimization of the driving and sensing elements.

The second performance advantage of CMOS-MEMS is the very low parasitic capacitance of interconnect from the micromechanical device to the electronics. Current integrated polysilicon MEMS technologies suffer from junction capacitance of heavily doped silicon interconnect, or from anchor pad capacitance to substrate. Parasitic capacitance from external bond-wire or solder-bump flip-chip connections is between 100 fF-1 pF. For comparison, a moderate-sized micromechanical lateral capacitive sensor in a surface micromachining technology (including CMOS-MEMS) has a total capacitance value of around 10 fF. Therefore, even a few femtofarads of parasitic capacitance will degrade the sensitivity substantially.

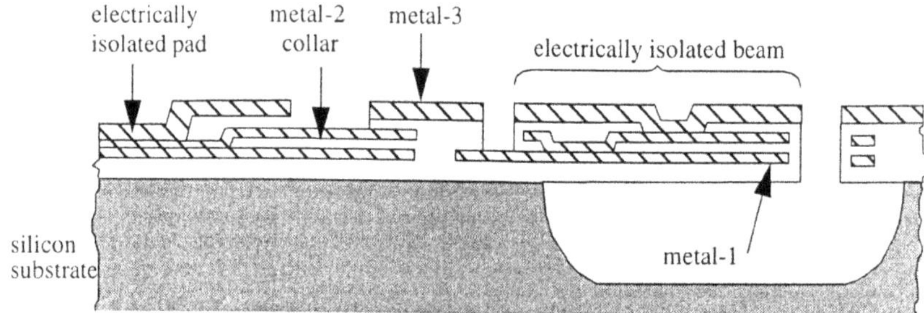

Figure 8. Cross-section of interconnect for electrical isolation.

In CMOS-MEMS processes, the minimum spacing between active devices (*e.g.*, transistors and diodes) and micromechanical components is dictated by the silicon release etch, since the active devices are made from the silicon. An anisotropic wet silicon release etch can preserve the silicon almost up to the edge of the etch pit, due to the high selectivity to <111> crystal planes. The isotropic dry plasma release in the high-aspect-ratio CMOS-MEMS process undercuts the silicon at the anchored edges. Typically, active devices can be placed as close as 12 μm from the microstructures, corresponding to a parasitic capacitance of 1.7 fF. This very low capacitance can be exploited to make ultra-sensitive displacement sensors.

Because a protective metal layer (usually metal-3) is required to prevent the CMOS electronics from etching, at most two unconstrained metal layers are left available for electrical interconnect. Measurements of transistor threshold voltage verify that there is no degradation from either the dielectric or silicon etch.

A gap in metal-3 is necessary for electrical isolation between external pads and the rest of the metal-3 plane. The interconnect arrangement is shown in Figure 8. A metal-1 or metal-2 collar is inserted underneath the break in metal-3 to prevent the silicon etch from reaching the substrate surface. Such gaps in metal are also used to electrically isolate different conducting areas on suspended structures.

6. Probe-Based Data Storage

The post-CMOS processing sequence may be modified to fabricate platinum tips on the movable microstructures as shown in Figure 9(a). These tips are being explored for use in probe-based data storage and tunneling sensor applications. In the probe-based storage application, the tips are used to write and detect pits on the surface of an amorphous carbon media, as illustrated in Figure 9(b). Initial tests using a commercial scanning tunneling microscope have been successful in writing pits as small as 3 nm in size by pulsing voltage in the range of 4-6 V. The pit can be subsequently read by detecting the change in tunneling current as the tip is scanned

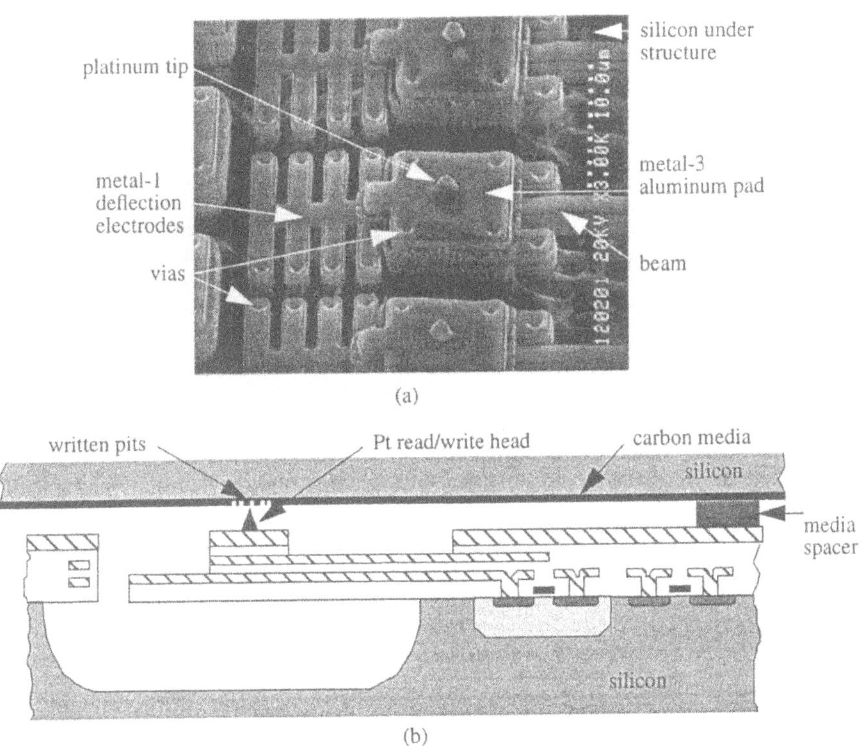

Figure 9. Probe-based data storage application. (a) Platinum tips deposited on CMOS-MEMS structures. (b) Schematic cross-section of the write-once-read-many (WORM) system.

over the media. By controlling up to 10,000 tips on a single chip with each tip being scanned in a 10 μm by 10 μm area, data storage densities of 10 GB/cm^2 may be possible. Presently, we are studying narrow-gap lateral electrostatic actuators for microstages having a high percentage of swept area to device area. Cantilevered stages have moved up to ±5 μm linear static displacement with 15 V applied to the actuators. Reliable operation of future MEMS probe-based storage systems will require an understanding of the physical mechanisms underlying the reading and writing of data, as well as knowledge of the tribology of the contact surfaces between the tip and media.

7. Conclusions

Many of the manufacturing difficulties with fabricating MEMS in a standard CMOS process stem from the inability to control the microstructural sidewall geometry when using the stacked-contact-cut technique. These problems are solved by moving the microstructural etch from the main CMOS flow to an optimized post-processing step. The resulting process can produce microelectromechanical devices with narrow beam width and gap spacing as required for lateral electrostatic actuation.

We have developed several CMOS-MEMS that take advantage of the multi-conductor structures and low parasitic capacitance, and demonstrate the potential for inertial sensor and probe-based data storage applications. Further work must be done on characterizing the composite material of the CMOS-MEMS structures, however design matching techniques can compensate for problems with out-of-plane curling and residual stress. Future designs will integrate sense electronics with efficient actuation mechanisms for higher performance.

The probe-based data storage is an important future MEMS application. WORM data storage can be accomplished using the platinum-tip and carbon media system, but other combination of tip and media must be developed for read-write storage. Research in read-write physical mechanisms and in tribology of such tip and media systems is critical for implementation of reliable MEMS data storage systems.

8. Acknowledgment

The author gratefully acknowledges the work of L. R. Carley, D. Guillou, M. Kranz, M. S.-C. Lu, S. Santhanam, and M. L. Reed. The fabrication work was supported in part by the National Science Foundation under grant ECD-8907068. The research effort was sponsored by the Defense Advanced Research Projects Agency and the Air Force Materiel Command, USAF, under the Air Force Office of Scientific Research, grant F49620-94-1-0192; and under Rome Laboratory, cooperative agreement F30602-97-2-0323. The U.S. Government is authorized to reproduce and distribute reprints for Governmental purposes notwithstanding any copyright notation thereon. The views and conclusions contained herein are those of the authors and should not be interpreted as necessarily representing the official policies or endorsements, either expressed or implied, of the Air Force Office of Scientific Research, Rome Laboratory, or the U.S. Government.

9. Bibliography

[1] Parameswaran, M., Baltes, H. P., Ristic, L., Dhaded, A. C., and Robinson, A. M. (1989) A new approach for the fabrication of micromechanical structures, *Sensors and Actuators* **19**(3), 289-307.

[2] Hoffman, E., Warneka, B., Kruglick, E., Weigold, J., and Pister, K. S. J. (1995) 3D structures with piezoresistive sensors in standard CMOS, in *Proc. IEEE Micro Electro Mech. Sys. Workshop*, Amsterdam, The Netherlands, 288-293.

[3] Fedder, G. K., Santhanam, S., Reed, M. L., Eagle, S. C., Guillou, D. F., Lu, M. S.-C., and Carley, L. R. (1997) Laminated high-aspect-ratio structures in a conventional CMOS process, *Sensors & Actuators A* **57**(2), 103-110.

[4] Seidel, H., Fritsch, U., Gottinger, R., Schalk, J., Walter, J., and Ambaum, K. (1995) A piezoresistive silicon accelerometer with monolithically integrated CMOS-circuitry, in *Tech. Digest of the 8th Int. Conf. on Solid-State Sensors and Actuators (Transducers '95)*, Stockholm, Sweden, **1**, 597-600.

[5] Biebl, M., Scheiter, T., C. Hierold, Philipsborn, H. v., and Klose, H. (1995) Micromechanics compatible with an 0.8 μm CMOS process, *Sensors and Actuators A* **46-47**, 593-597.

[6] Lin, G., Pister, K. S. J., and Roos, K. P. (1996) Standard CMOS piezoresistive sensor to quantify heart cell contractile forces, in *Proc. IEEE Micro Electro Mech. Sys. Workshop (MEMS '96)*, San Diego, CA, 150-155.

[7] Mayer, F., Paul, O., and Baltes, H. (1995) Influence of design geometry and packaging on the response of thermal CMOS flow sensors, in *Tech. Digest 8th Int. Conf. on Solid-State Sensors and Actuators (Transducers '95), Stockholm, Sweden,* **1**, 528-531.

[8] Syrzycki, M. J., Carr, L., Chapman, G. H., and Parameswaran, M. (1993) A wafer scale visual-to-thermal converter, *IEEE Trans. Compon. Hybrids Manuf. Technol.*, **16**(7) 665-673.

[9] Reay, R. J., Klaasen, H., and Kovacs, G. T. A. (1994) Thermally and electrically isolated single crystal silicon structures in CMOS technology, *IEEE Electron Device Letters*, **15**(10) 399-401.

[10] Westburg, D., Paul, O., Andersson, G. I., and Baltes, H. (1997) A CMOS-compatible device for fluid density measurements, in *Proc. IEEE Micro Electro Mech. Sys. Workshop (MEMS '97)*, Nagoya, Japan, 278-283.

[11] MOSIS - VLSI Fabrication Service, USC/Information Sciences Institute, 4676 Admiralty Way, Marina del Rey, CA 90292-6695, http://www.isi.edu/mosis/.

[12] *Circuits Multi-Projets*, CMP (1996) *French MPC Activity Report*, 46 Avenue Felix Viallet, 38031 Grenoble, France (http://tima-cmp.imag.fr).

[13] Carley, L. R., Reed, M. L., Fedder, G. K., and Santhanam, S. (1996) Microelectromechanical structure and process of making same, U.S. patent pending.

[14] Tang, W. C., Nguyen, T.-C. H., Judy, M. W., and Howe, R. T. (1990) Electrostatic-comb drive of lateral polysilicon resonators, *Sensors and Actuators A* **21**(1-3), 328-331.

[15] Gianchandani, Y. B., and Najafi, K. (1996) Bent-beam strain sensors, *J. of MicroElectroMech. Sys.*, **5**(1), 52-58.

[16] Kranz, M. S., and Fedder, G. K. (1997: in press) Micromechanical vibratory rate gyroscopes fabricated in conventional CMOS, in *Proc. Symposium Gyro Technology*, Stuttgart, Germany.

Facilitating Choices of Machining Tools and Materials for 'Miniaturization Science': A Review

Dr. Marc Madou,
Center for Materials Scholar, The Ohio State University
Columbus, OH 43210-1178

1. Abstract

After an introduction to terminology and classification methods in micromachining the different manufacturing techniques suitable to making very small devices are considered. They are grouped according to the energy source involved in removal or addition of materials at the workpiece. Subsequently a method is proposed on how to go about deciding upon the ideal tool and substrate material for the micromachining job at hand.
In this review traditional precision engineering and more modern micromachining techniques, based on lithography methods, are presented as complementary. Making a manufacturing decision is presented not so much as a question of which manufacturing technique is technologically superior (invariably, since the sixties, Silicon has been the answer, often independent of the question) but which one is appropriate to use for the micromachining problem and the intended market. A development strategy for micromachines starting from the package and the interface with the real world as well as good understanding of the specifications is outlined as a guide to future micromachine design.

2. Terminology

2.1. DEFINITIONS

In manufacturing forming and removing are the two primary processes. Primary forming processes create an original shape from a molten mass, gaseous state, or from solid particles. During such processes, cohesion is created among particles [Chryssolouris, 1991], examples include plastic molding, metal deposition by evaporation or sputtering and electroforming. Removing processes, remove material, destroying cohesion among particles, examples are wet chemical etching, electrodischarge machining (EDM) and traditional mechanical turning and drilling. In this review we are interested in those forming and removing processes enabling the manufacture of precision micromachines.

Precision machining has been defined as machining in which the relative accuracy is 10^{-4} or less of a feature/part size [Slocum, 1992]. For comparison, a relative accuracy of 10^{-3} in the construction of a house is considered excellent. It is important to realize that,

B. Bhushan (ed.), Tribology Issues and Opportunities in MEMS, 31-51.

while silicon micromachining can achieve excellent absolute tolerances, relative tolerances are rather poor compared to those achieved by more traditional techniques. The decrease in manufacturing accuracy with decreasing size is rarely mentioned in discussions of Si micromachines, this probably is related to the fact that most Si micromachinists are electrical engineers rather than mechanical engineers. The contrast is striking; in traditional machining relative tolerances of 10^{-6} (ppm) are becoming standard whereas in the IC industry a 10^{-2} relative tolerance is considered good. The definition of precision machining, with relative tolerances of 10^{-4}, does actually exclude micromachining. To capture micromachining with the same terminology it must be expanded to cover all machining methods where the relative accuracies are at least 10^{-4} or where absolute size in one or more dimensions is in the micrometer range.
The term precision machining, when used by mechanical engineers, has typically been reserved for removal processes only, whereas 'micromachining', as used to describe IC based fabrication technology, covers both removal and forming processes. Precision machining and Si micromachining are complementary; both are striving to improve relative tolerances, with Si based technology better at obtaining smaller features and traditional methods better at obtaining tighter relative tolerances. In this review precision machining encompasses removal and forming processes, with micromachining as one of its newest embodiments.
The terminology 'non-traditional' or 'non-conventional machining' has been used by mechanical engineers for at least the last 30 years [van Osenbruggen, 1969)] to describe non mechanical machining methods such as thermal, electrochemical and chemical machining of high strength and corrosion-and-wear resistant materials. It often involves, small, precise and complex structures and may incorporate computer integrated machining methods [Boothroyd, 1989]. The name non-traditional in this sense is becoming obsolete though as the so-called non-conventional machining methods have found a wide range of commercial applications even in more readily machined materials [Snoeys, 1986]. In what follows we will refer to all non-IC manufacturing techniques as traditional machining techniques. In this nomenclature, IC based micromachining techniques are the non-traditional machining methods of today.

2.2. CLASSIFICATION METHODS

Jargon from different disciplines clings to sensors, rendering discussions difficult. A useful scheme for classifying sensors would make communication easier between sensor researchers from different disciplines, be they manufacturers, immunologists, mechanical engineers or market researchers. Several attempts to classify sensors occurred. We are distilling the most important aspects here.
White [1987] describes a sensor classification scheme derived from a Hitachi Research Laboratory communication. He distinguishes ten different measurand domains:
Acoustic, Biological, Chemical, Electric, Magnetic, Mechanical, Optical , Radiation, Thermal and Other. Middlehoek [1989] follows Lion [1969] and contracts this list into six signal domains ; Radiant, Mechanical, Thermal, Electrical, Magnetic and Chemical.
For simplicity we adopted the latter list here. Based on Göpel et al. [1989] and Habekotte [1993] Table 1 exemplifies sensing principles in these six signal domains. Since the

listed energy domains can be energy input as well as energy output a six by six matrix results. We may thus have a mechanical sensor converting mechanical energy into thermal energy i.e. a thermo-mechanical sensor (e.g. a friction calorimeter) as well as a thermal sensor converting thermal energy in mechanical energy i.e. a mechano-thermal sensor (e.g. a bimetallic strip). In the literature no logic seems to be applied concerning the use of 'thermo-optical' or 'opto-thermal'. We have taken on the convention that the 'o' goes to the output energy form.

Middlehoek et al. further distinguish between self-generating and modulating sensors. A sensor based on a modulating principle requires an auxiliary energy source e.g. a fiberoptic magnetic-field sensor in which a magnetostrictive jacket is used to convert a magnetic field into an induced strain in the optical fiber; one based on a self-generating principle does not e.g. a thermocouple in which a change in temperature directly results into an electrical signal that can be measured. The difference between the two is illustrated in Figure 1. Since the modulating sensor needs another energy input one also refers to such sensors as "passive" whereas the self-generating sensors are "active".

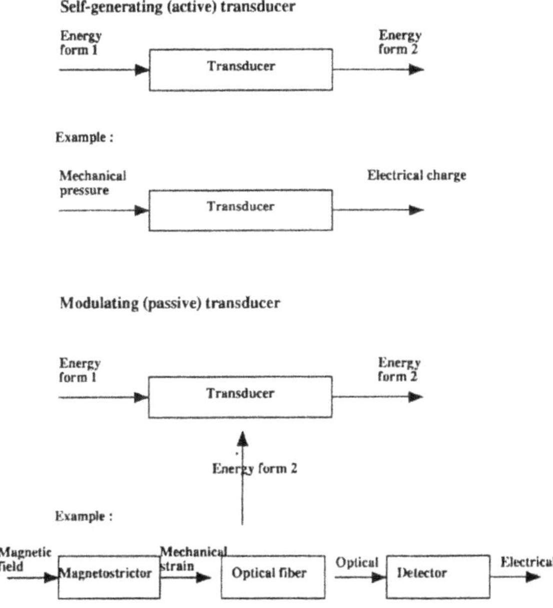

Figure 1 . Comparison of self-generating (active) transducer and modulating transducer (passive); the piezoelectric sensor represents a self-generating transducer and the fiber-optic magnetic field sensor represents a modulating transducer. After National Research Council [1995].

3. Absolute and Relative Tolerances in non-Traditional (lithography Based) and Traditional Machining Technology

In Figure 2 the application field of precision machining in terms of absolute size and absolute and relative tolerances are illustrated. In traditional mechanical precision machining, typically used to machine the biggest objects, and therefore setting the upperbound one is mostly concerned with objects of an absolute size of 10 cm or less, although a parabolic astronomic mirror with a diameter of several meters must obviously be considered a precision machined object as well [Zuurveen, 1994]. Absolute tolerances of the dimensions dealt with in precision engineering are 10 μm or below, depending on the technique. Traditional machining methods are excellent at obtaining very small relative tolerances (sub ppm is common now) but non-traditional machining methods, based on lithography, are better at obtaining small absolute tolerances while being poor at obtaining good relative tolerances. When the relative tolerances are very important for the operation of a given micromachine and there is no opportunity for compensation schemes after the manufacture is finished, Figure 2 teaches us that we might have to make the overall machine larger and resort to traditional machining methodology. To illustrate this point, consider the microfabrication of the magnetic poles for a miniaturized mass spectrometer with a required 0.15 % relative tolerance on the poles radius and length. Using ultraprecision diamond milling of poles 1 cm long and 0.2 cm diameter constitutes a relatively straightforward task but attempting to make poles 100 μm tall and 20 μm diameter with the same relative tolerance is quite a challenge in the case of lithography techniques. The application domains of normal machining, precision machining and ultra precision machining are depicted in Figure 3. This illustration combines the Taniguchi curves [Taniguchi, 1983], dramatically illustrating the improvement in traditional manufacturing accuracies achieved in the second half of this century, and Moore's law, demonstrating the logarithmic progress in density of transistors on a chip over the last few decades of this century. Traditional machining and IC-technology based machining methods enabling those accuracies are listed on the left hand side of this figure (see also Evans [1989]). From this figure we can expect ultra-precision diamond machining to reach a manufacturing accuracy of better than 1 nm by the year 2000. To obtain those accuracies one cannot, of course, work with materials such as wood or brick but one must employ form stable and workable materials such as aluminum, stainless steel, ceramics or glasses in other words material and machining accuracy are intertwined. From Figure 3 we also observe that the slope of both the Taniguchi and Moore plot decreases as the year 2000 approaches. Progress in mechanical machining is to a large part based on continued improvement in positioning stages, computer control and metrology techniques. Photolithography equipment is based on the same ruling engine design philosophy from mechanical technology, reasons for the slowdown in the Moore curve are a bit more complicated though. In lithography, subsequent layers need to be aligned , an operation not needed when diamond milling a high precision lens. The alignment operation may be performed by laser-interferometer controlled stages to an absolute positioning accuracy of <0.1 μm, thereby allowing the registration of successive layers of a semiconductor device to a similar precision. Alignment errors (A) depict only one part of the total device ground-rule tolerances. Tolerance (T) or the ability to overlay (align) spatially one masking layer with the previously etched or deposited pattern is given by the standard deviations on the mask (M), etch (E), and alignment (A) errors [Moreau, 1988]. Typical mask errors range from 0.1 to 0.3 μm, etch errors from 0.2 to 0.4 μm, and alignment errors from 0.2 to 0.5 μm. For the next generation of submicrometer ground rules, a fourfold reduction in alignment and other errors will be necessary. Since interferometric schemes enable sensitivities to below 10 nm, it appears feasible that aligners compatible with linewidths of 0.1 μm or even 50 nm can be developed [Smith, 1992].

Precision Machining Application Domain

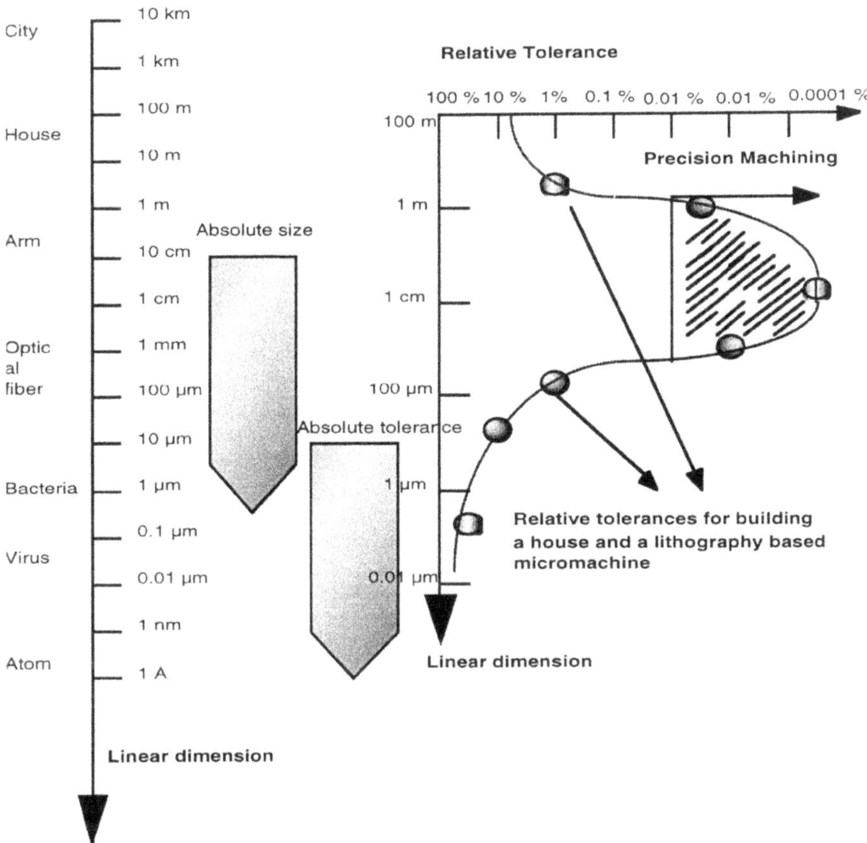

Figure 2. Application field for precision-machining in terms of absolute sizes and absolute and relative tolerances.

Unfortunately, there is a second reason for the slowdown in electronic device development; below 0.1μm quantum effects start playing a role. To further improve manufacturing accuracy in electronics one has to look beyond current manufacturing methodologies and consider methods such as molecular engineering and nanofabrication. Micromachining plays a pivotal role in those new machining approaches. For example, the scanning tunneling microscope (STM), invented in 1982, creates machining opportunities with accuracies on the atomic level, well beyond the predictions of the Taniguchi and Moore curves in Figure 3. The crucial component in an STM is a micromachined cantilever with an integrated sharp tip. For mechanical structures operating in the microdomain imposes new design philosophies; one must adopt more error insensitive designs with simpler mechanism, smaller numbers of parts and, wherever possible, flexible members rather than rigid ones. Nature, an excellent teacher in nanomachining, builds living organisms this way [Hayashi, 1994]

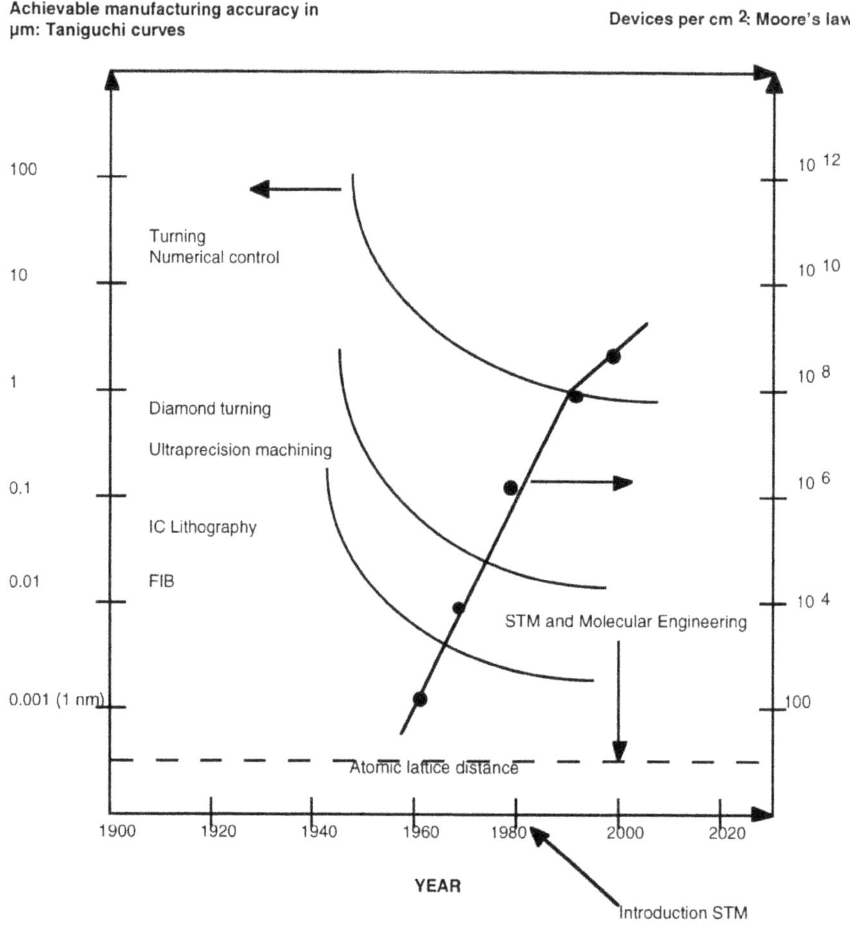

Figure 3. Definition for normal, precision and ultra-precision machining. Left side ordinate: Increase of manufacturing accuracy over time according to Taniguchi [1983]. Right side ordinate : increase in transistor density over time according to Moore's law.

4. Comparison of Micromachining Techniques

4.1. DESIGN FROM THE PACKAGE INWARDS

The choice of processes for manufacturing a 3D micro device can best be made after studying the detailed requirements of the application. Since the package serves as the interface between the microstructure and the 'macro' world and since it is the major contributor to cost and size, one should start the design with the package and work toward the best machining process for the micromachine inside. The package is often made with a traditional machining method such as turning and grinding or possibly with ultrasonic or wire electrical-discharge machining.

For the micromachine inside the package there may be many machining options as both batch IC and traditional serial production techniques must be considered. Several new, batch Si micromachining techniques derived from the IC industry, are gaining acceptance, but serial, precision machining techniques, with specialized tools (e.g. ultrasonic machining or laser beam machining) are still often the only commercially available method to make intricate small 3D micromachines .

4.2. BATCH VS. SERIAL

Most non-traditional micromachining carried out today, like IC fabrication, relies on batch fabrication i.e. repetitive features are simultaneously photographically defined on a silicon wafer, and many wafers are then processed further to fabricate desired structures. Other micromachining techniques, such as laser drilling or chemical jet-etching are serial techniques. Processing requires N times as long for N features on a wafer as it does for one feature, and M times as long for M wafers as it does for one wafer. Batch fabrication techniques lend themselves to economies of scale which are unavailable with serial techniques. Although micromachining is often compared to microelectronics it lacks the high degree of generality characteristic of microelectronics. ICs can be grouped in a limited number of classes, within which design and production follows well defined and common steps. As a consequence the price-performance ratio allows industry to make profits on complex ICs and it makes ICs suited for mass production. Sensors and actuators on the contrary are very specific. They have to be in contact with their surroundings, and each environment imposes its own constraints. This makes the packaging more important further illustrating our earlier claim that the design should start from the packaging constraints and work its way to the miniature sensor inside rather than the other way around. For micromachines the economics dictating profitable IC manufacture, luckily, do not necessarily apply; a microinstrument, say a micro gas chromatograph, may cost considerably more than a typical IC. Moreover sensors and actuators often use quite esoteric non-IC type materials, are produced in relatively low volumes and are application specific.

4.3. COMPARISON TABLE

Micromachining is emerging as a set of new manufacturing tools to solve specific industrial problems rather than as a monolithic new industry with generic solutions for every manufacturing problem. In Table 2 some of the more popular machining techniques to craft micromachines are compared. From this Table, Si micromachining emerges as only one of the many options for precision machining. The renewed interest in some of the non-Si based machining methods often stems from two major deficiencies of IC based machining techniques i.e. the difficulty to create truly 3-D microstructures and the problem of interfacing microstructures with the macroworld. A machining method with a range that covers macroscopic to microscopic scale devices is needed as

Table 1. Example of Transduction Principles

Input (Primary Signal)	Output (Secondary Signal) Mechanical (Mechano-)	Thermal (Thermo-)	Electrical (Electro-)	Magnetic (Magneto-)	Radiant (Photo- or Radio-)	Chemical (Chemo-)
Mechanical	Acoustics Fluidics	Friction calorimeter Cooling effect	Piezo electricity Piezo resistivity	Piezo magnetic effect	Photo-elasticity	
Thermal	Thermal expansion Bimetallic strip		Pyroelectricity Seebeck effect		Radiant emission	Reaction
Electrical	Piezoelectricity Electrometer	Joule heating, Peltier effect	Langmuir probe		Electro-luminescence	Electrolysis
Magnetic	Magneto-striction Magnetometer	Thermo-magnetic	Magneto resistance			
Radiant	Radiation pressure	Thermopile Bolometer	Photo-electric Dember			Photo-reactions
Chemical	Hygrometer	Calorimeter Thermal conductivity	Amperometry Flame ionization Volta effect	Nuclear magnetic resonance	Chemi-luminescence	

Table 2. Precision Engineering Machining Tools

Group : Traditional-Techniques(not involving photo lithography defined masks)

Machining Method S=subtractive A=additive Batch=Ba, Serial=Se Continuous = C	Material/Appl.	Typical Min/Max Size	IC Comp.	Tolerance	Aspect Ratio (depth/ width)	Initial investment cost/Access
Chemical Milling (S),(Ba)	Almost all metals	From sub mm to a few meters (x,y)) , max thickness (z) ± 1cm	Yes	Lateral tol. 0.25 to 0.5 mm	±1	Low/ Good
Electrochemical Machining (S/A), (Se)	Hard and soft metals, turbin blades,pistons, fuel-injection nozzles	(S)Minimum size devices larger than in chemical milling because of the contacting need	Fair	Lateral tol. < 10 μm	(S) 100	± $ 400,000

Electrodischarge Machining (EDM) (S),(Se)	Hard brittle conductive materials used for tools and dies	Min holes of 0.3 mm in 20 mm thick plate	No	Lateral tol. 5-20 μm	100	High/Good Extensive line of equipment available, numerical control is common
Electrodischarge Wire Cutting (EDWC) (S), (Se)	Hard brittle materials many punch and die appl.	Min rods of 20 μm in diameter and 3 mm long	No	Lateral tol. 1 μm	>100	High/Good Doesn't require a variety of electrodes as in the case of EDM
Electron-Beam Machining (EBM) (S/A),(Se)	Hard to machine materials	(S) most suited for large numbers of simple holes (< 0.1 mm)	Fair	(S)~ 10 percent of feature size (5 μm on a 50 μm hole)	(S) 10 is typical but 100 is possible	Very high/Fair But only ± $ 100,000 with modified SEM
Continous deposition (A), (C) e.g. doctor's blade technology	Materials available in inks, e.g. glucose sensors	Most suited for inexpensive disposables, from 100 μm to a few mm	No	15 μm	-	Low/Good
Focused Ion Beam on a lathe (S)/(A),(Se)	Very pure IC materials	From sub microns to mm	Yes	(S) 50-100 nm	-	High/ Poor
Hybrid thick film (A),(Ba)	Wide variety of materials available in inks	Minimum feature size 90 μm	Fair	12 μm	-	±$ 30,000/ Good
Laser Beam Machining (LBM) (S/A),(Se)	Complex profiles in hard materials	(S) Holes from 10 μm to 1.5 mm at all angles	Fair	1 μm	(S) 50	± $ 50,000 five-axis systems up to $ 400,000 Good
Plasma Beam Machining (PBM) (S/A),(Se/Ba)	Very high temp. materials	(A) only used for thick films > 25 μm, (S) for very thick films > 2.5 mm	No	(A) 20 μm for a 25 μm thick film (S) typical ±3 mm but 0.8 mm is possible	-	±$ 600,000/ Fair
Stereo Lithography (A),(Se)	Polymeric photo-sensitive materials	Max 10 by 10 by 10 mm(X,Y,Z)	Yes	min solidification unit 5,5,3[μm](X,Y,Z)	-	Low/ Good
Ultra Precision Mechanical Machining (S),(Se)	Form stable materials	From sub mm (e.g 0.2 mm hole) to meters	No	1 nm by the year 2000	-	$400 k/ Good
Ultrasonic Machining (S), (Se)	Hard and brittle materials	Holes from 50 μm to 75 mm	No	Lateral tolerance 10 μm	2.5 μm for a 250 μm hole	$ 20 k /Good

Group : Non-Traditional (involving photo lithography defined masks)

Machining Method, all Batch S=subtractive A=additive	Material Appl.	Typical Min/Max Size Feature	IC Comp.	Tol.	Aspect Ratio (depth/ width)	X,Y,Z shape and Z height/depth	Inital investment cost/ Acess
Photofabrication (S)	Plastic, glass (ceramic) e.g. fluidic elements	Max.x,y = 40 by 40 cm and max z=0.6 cm	Yes	Lat. tolerance 20 µm	-3 for photo plastics -20 for photoglass	x,y is free z= up to 6 mm	Medium/ Poor to medium
Photochemical milling (S)	Printed circuit boards, lead frames, shadow masks	Max 60 by 60 cm, max thickness < 0.5 mm.	Yes	13 µm (printed circuits)	± 1	x,y is free z= up to 0.5 mm	Medium/ Good
Wet Etching of Anisotropic Materials (S)	Crystal Si,GaAS, Quartz, SiC, InP,	Max wafer size, minimum feature a few microns	Fair	1 µm	100	x,y,z shape locked in by crystallography, z height of the wafer	Low/ Good
Dry Etching (S)	Most solids	Max wafer size, minimum feature sub micron	Good	0.1 µm	10	x,y shape free, z= up to 200 µm	High/ Good
Poly-Si Surface Micromachining (A)	Poly-Si, Al, Ti, etc.	Max wafer size, minimum feature sub micron	Good	0.5 µm	-	x,y free, z height 0.1 to 10 µm but preferably 1- 2 µm	High/ Fair
SOI (S)	Crystalline Si	Max wafer size, min feature sub micron	Good	0.1 µm	-	x,y free, z height depending on the epi layer e.g. 100 µm	High/Poor
LIGA (S)/(A)	Ni, PMMA, Au , ceramic, etc	10 by 10 cm or more, 0.2 µm	Fair	0.3 µm	>100	x,y free,z up to several cm's	Very high /Poor
UV Transparent Resists (S)/(A)	Polyimide, SU-8, AZ-4000	Max wafer size, min	Good	0.5 µm	10	x,y free,z up to 100 µm	High/ Good
Molded Poly-Si Hexsil (Keller) (S)/(A)	Poly-Si, Ni, etc.		Good	0.5 µm	10	x,y free,z up to 100 µm	High/ Poor
Erect poly-Si (Pister) (S)/(A)	Poly-Si		Good	0.5 µm	10	x,y free, z up to mm	High/ Poor

often micromachines cannot easily be accessed. Laser beam lithography, electro-discharge machining, ultra precision machining, LIGA (from the German acronym for Lithography (X-ray lithography to be more specific), electrodeposition and molding) and pseudo-LIGA techniques are steps in that direction, forming so-called 'hand-shake 'technologies; bridging micro- to macro-world. These hand-shake technologies should be of importance beyond micromachines, for example, in packaging of ICs where one is faced with the same dilemma of packaging, mounting and wiring LSIs and VLSIs. Several of these new technologies are maskless and enable very fast prototyping.

Silicon-based micromachining techniques, often resulting in stunningly beautiful SEM micrographs, are the most visual and 'sexy' micromachining tools used today (the M(usic)T(ele)V(ision) of micromachining), but in the trend toward miniaturization, almost a law of nature ever since the oil-crises and the IC revolution, many other precision machining techniques on many different materials have found much larger commercial applications. Wet bulk Si micromachining, the most traditional form of Si micromachining, dominated by electrical engineers, is gaining acceptance in traditional precision machining circles, dominated by mechanical engineers, and vice versa; Si micromachinists are starting to look at precision engineering techniques as a way to extend their tool kit. The most important characteristic for both disciplines is the ever increasing quest for improved accuracies, higher aspect ratios and reduced cost, and practitioners are starting to recognize the complementary nature of their respective fields Whereas mechanical sensor manufacture is moving towards more integration of sensing function with electronics, embodied in increased reliance on surface micromachining, chemical and biosensors are moving away from integration toward hybrid technology. This trend in chemical and biological sensors is caused by the need for modularity and the tremendous compatibility problems involved when attempting to integrate chemical sensor materials with ICs. In the case of instrumentation there is a trend toward using laser machining and some early exploration of LIGA use and very recently a rediscovery of photoformed glass.

By applying the right tool to the machining job at hand we hope that micromachining will lead to much more successful commercial applications than we have seen sofar. A better insight of the characteristics of the various machining options as compared in Table 2 should be most useful to that end.

4.4. FROM PERCEPTION TO REALIZATION

Often Si micromachining seem to be applied to prove to the world that group 'X' also has a cleanroom or can make a yet smaller micro-motor. In order for micromachining to loose the stigma of a technology looking for an application and used in experimental back-rooms and clean-rooms only, it will be important to describe most new results in academia on micromachinery as fine-tuning micromachining skills rather than to proclaim each new result as the latest breakthrough in sensor technology or analytical chemistry.

Micromachining must be seen as a new set of tools to solve practical problems in sub-miniaturized 3D structures. Once a thorough understanding of the micromachining application is reached, it will be easier to decide upon the correct manufacturing technique. The micromachining tool used in the early prototyping phase does not need to

be the same as the eventual manufacturing technique decided upon. In many cases a method enabling faster prototyping is preferred.
Micromachining is thus very engineering oriented and application specific, and therefore it can best be carried out in industry or in very applied R&D groups. Indeed all recent examples of triumphs of precision engineering were realized in industry and the Si micromachining community often failed to appreciate those accomplishments of the more traditional machining methods. The HP Kittyhawk Personal Storage Module (PSM), a 1.3 inch disk drive and the' Schwarzschild Objective' CD pick-up, with a distance from bottom of pick-up to disc of 6.5 mm, shown in Figure 4, are triumphs of precision engineering, but neither involve Si. Romankiw's thin film-head, shown in Figure 5 is another example and constitutes probably the micromachine with the highest sales today (3 billion is predicted for 1996). The manufacture of the thin-film head is principally based on electroplating and does not involve any Si either.
Because there are no good engineering practices established on deciding upon an ideal manufacturing technique for a given micro-application, manufacturing techniques often seem to be rediscovered. The latter is illustrated by the evolution of manufacturing methods which have been attempted for the construction of microfluidic elements as illustrated in Figure 6. Microfluidic elements in the sixties were principally made from photo glass. After attempting to craft the same devices by dry etching in Si in the eighties, LIGA was employed in the early nineties but buy the mid-nineties photo glass was rediscovered as the best machining option for this application [Ehrfeld, 1994] .

5. Micromachine Development

5.1. 'MINIATURIZATION SCIENCE'

When approaching the subject of downscaling a device we suggest using the terminology "miniaturization science" rather than "micromachining", a term too loaded with the notion that silicon technology represents the only solution. Applied miniaturization starts with a thorough understanding of the sensor or microsystem specification list, interviews with potential users, understanding of the application environment and most importantly; a firm market appreciation. Different markets actually dictate different machining approaches. For example, chemical sensors, serving a fragmented market, are better constructed in a modular hybrid thick film approach, but automotive accelerometers, serving a huge and uniform market, are better built in an integrated polysilicon design. The choice of materials for miniaturization of sensors, actuators and power sources will depend also on the particular sensing principle employed. Available transduction choices are summarized in a six by six matrix in Table 1.

5.2. APPLYING THE SPECIFICATIONS LIST TO SENSOR PRINCIPAL MATRIX

The transduction principle matrix as listed in Table 1 needs to be challenged with a detailed sensor specification list. While discussing the specifications required for measuring a certain measurand jargon confusion comes about. A specification list should include static and dynamic performance characteristics such as:

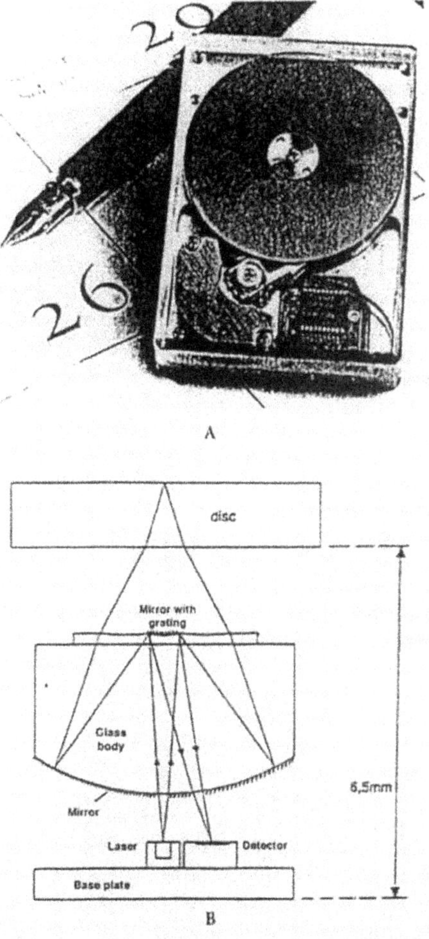

Figure 4. Triumphs of mechatronics
A . HP's Kittyhawk
B. Schematic of CD Pick-up.

Figure 4. Triumphs of mechatronics
A . HP's Kittyhawk
B. Schematic of CD Pick-up.

Static

- Sensitivity
- Dynamic range of measurand
- Resolution
- Selectivity/Specificity
- Stability (long term and short term)
- Response time
- Ambient conditions (e.g. temperature, under or above water, pressure, etc.)
- Operating life
- Shelf live

- Cost, size, weight
- Number of devices needed
- Operating voltage, current, power
- Quality, reliability, MTBF (mean time between failures)
- Accuracy
- Output format
- Hysteresis
- Threshold
- Nonlinearity
- Minimum detectable signal, etc.

Dynamic

- Dynamic error response
- Hysteresis
- Instability and drift
- Noise
- Dynamic range (operating range), etc.

A good specification lists also application-related intricacies well beyond the sensor itself. An example is a pH sensor for sensing pH in the secondary flow loop of a nuclear reactor. The presence of radiation and water at high temperature and pressure will significantly impact the sensor design in terms of safety measures mandated by the NRC (Nuclear Regulatory Committee). In the case of a biomedical sensor an important application related intricacy is the requirement for FDA (Food and Drug Administration) approval.

Figure 7 depicts the use of the specification list in sensor realization. Illustrated is the search for a sensor to measure glucose in a drop of blood (chemical domain). The specifications: very low cost (< 15 cents), disposable, millions of devices per month, easy to read, etc., lead us to an electrochemical approach (electro-chemical conversion option in Table 1). After this brain storming phase the preliminary design phase begins. For implementation of the preferred sensor materials and manufacturing technology the electrochemical conversion option is challenged with the list of manufacturing options and a check-off list in Table 2, introduced below, is used as a control gate for the design process. Several iterations later we end up deciding on a final design with thick film sensor electrodes and chemistry layers deposited on a disposable plastic strip in a continuous printing process, the option industry is pursuing.

5.3. MATERIALS AND MACHINING OPTIONS CHOICE: CHECK-OFF LIST

Some ten years ago, Senturia [1987] posed a rhetorical question about building micromachines: “Can we design microrobotic devices without knowing the mechanical properties of the materials involved?.” In subsequent years characterizing MEMS materials became popular. However, an equally important question: "Would you build a microstructure without knowing all the available construction methods?" remained largely unaddressed. Choices of materials and machining processes are usually intertwined and both must be confronted early on in the design phase. Table 2 gives a good idea about the many available machining options. Quite a few criteria need consideration before deciding on an ideal substrate and machining method for a given sensor

application. Following we introduce a more complete check-off list to help in the decision process of substrate and machining tool.
In a miniaturization project one should start with the design of the interface of the micromachine with the real world, in other words, with its package. Once the package design is established one can look for the best manufacturing technique for the

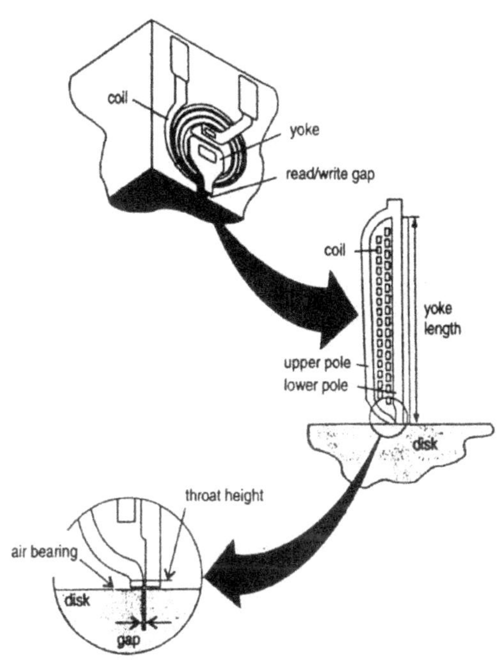

Read and write thin film head

A

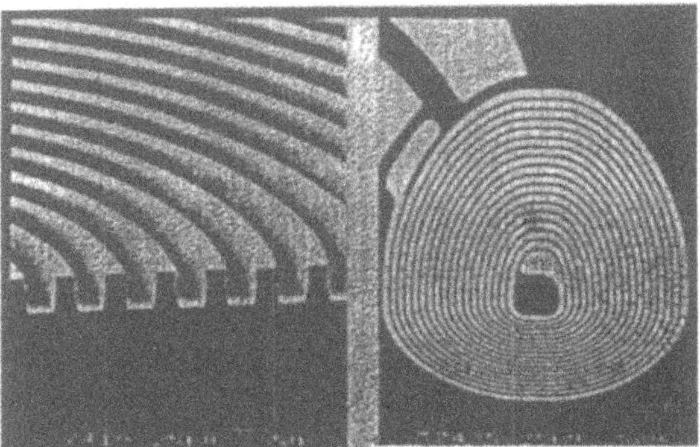

B

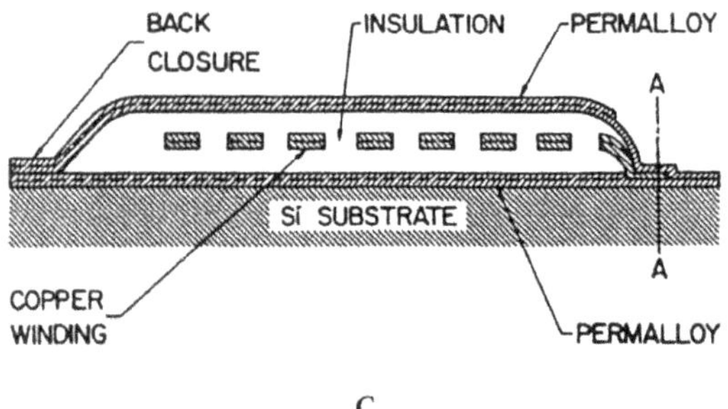

C

Figure 5. Thin film head
A Schematic of a multiturn thin film head with inset of pole tip structure on the air-bearing surface
B Resist mold and electroplated copper coil after resist removal
C Schematic cross section of the head [18].

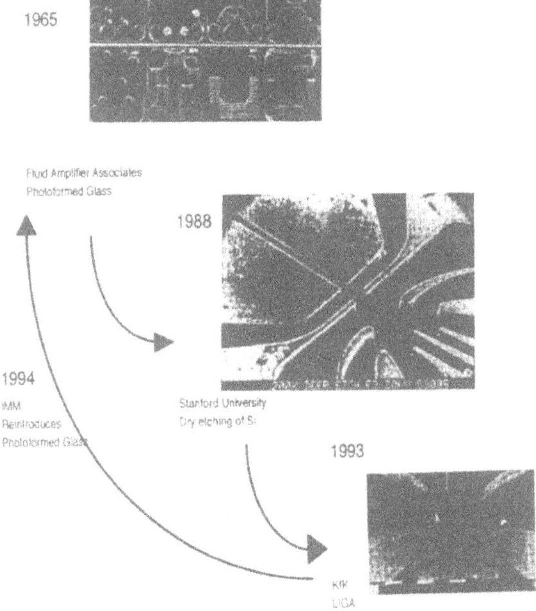

Figure 6. The rediscovery of photoformed glass for the manufacture of microfluidics.

micromachined part inside. This strategy dictates itself because the packaging cost contributes largely to the overall cost of any micromachined product and often overwhelms size specifications as well. The reason for expensive packaging can be ascribed to the many serial labor processes involved. Also, each device may need individual attention. The size of the package is either dictated by the need to interface to a bigger structure, by limitations of traditional machining methods or by the need for manual handling of the structure. Micromachining techniques such as fusion bonding, anodic bonding and other integrated encapsulation techniques, commonly used in bulk and surface micromachining, might help transform more and more of the serial packaging steps into parallel batch steps even for Ics which will prove advantageous in keeping costs of the total structure down. In the author's opinion, micromachined packaging solutions represent one of the most unmined potentials for new micromachining applications.

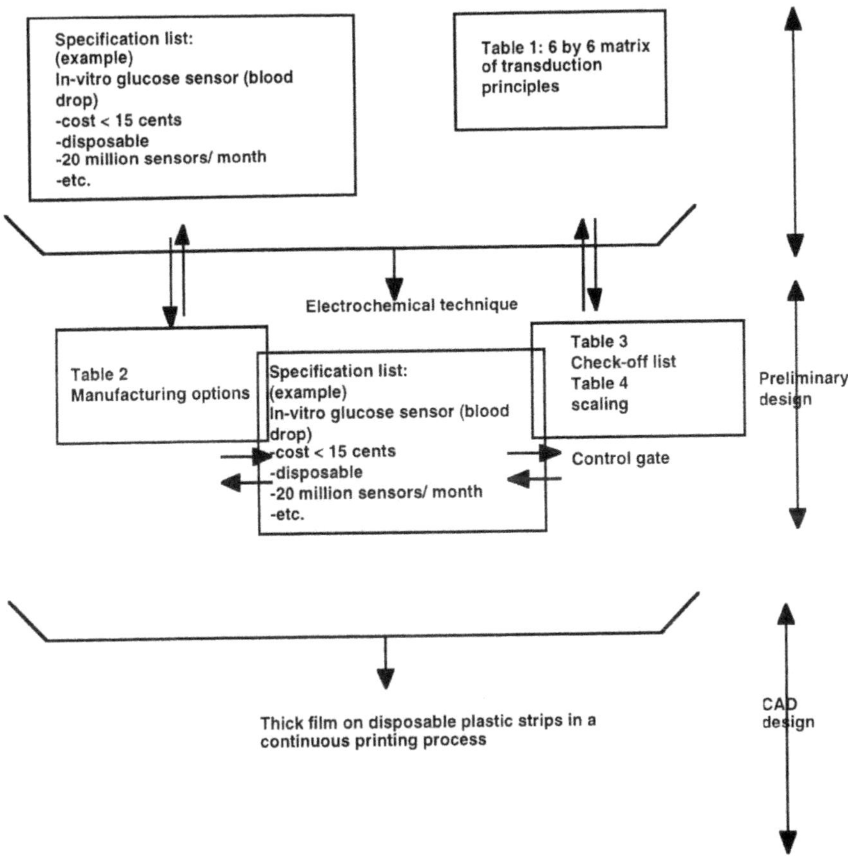

Figure 7. For implementation of the preferred sensor materials and manufacturing technology we challenge the electrochemical conversion option with the list of manufacturing options presented in Table 2 and follow the check-off list in Table 3.

Next one needs to establish whether to integrate active electronics on the substrate and confirm the expected temperature range of operation (if too high, the latter might void the integration option). A further consideration concerns number and cost of the devices and which substrate will make packaging easier. When faced with a new micromachining problem one should consider all available precision manufacturing options i.e. zero-base the approach. Zero-basing the technological approach to a micromachining problem should include a good check-off list based on the previous questions as well as the questions summarized in Table 3.

Table 3. Check-Off List To Determine Susbstrate for a New Micromachining Application

(1) How will the package of the sensor or system most likely look and how does it interface with the real world? The package and the interface with the environment determine size and cost of the total product, the nature of the micro device inside as well as the answers to most of the following questions.

(2) Is there a need for integration of electronic functions in the micromachine?
Such need (e.g. because the sensor has a very high impedance) necessitates a semiconductor substrate.

(3) How many of the micro devices will be made (production volume (# of units)) and unit complexity (# of devices in a unit)?
The number might suggest a serial (small number), hybrid (large number) or batch (very large number) approach. High complexity might point to a batch approach.

(4) Cost per device?
The cost may suggest a serial (high cost > $40.00), or Si batch (low cost < $2.00) approach. For very low cost (glucose sensor strips < 20 cents) a continuous fabrication process becomes mandatory.

(5) Does the substrate merely function as a support?
If so, glass, ceramic or even plastic and cardboard all become options. If the substrate has a mechanical function Si is an excellent candidate. If the substrate must have good optical properties, GaAs or PMMA are candidates.

(6) Is there a need for modularity?
Modularity is important with chemical sensor arrays, where integration often is counter productive because of the incompatibility issues faced when depositing different chemical sensor coatings on an integrated chip.

(7) What are the expected relative tolerances on the lateral dimensions and what is the required depth to width aspect ratios of features built into the substrate?
Very small relative tolerances on lateral dimensions cannot be achieved yet; 1 % lateral tolerance on a 100 µm line, in optical lithography, is considered good, but 10^{-3} on 1 cm in diamond milling is very poor (see Figure 3). Large aspect ratios (say > 20) might necessitate wet anisotropic etching of a single crystalline material, room temperature and cryogenic dry etching or LIGA and pseudo LIGA processes.

(8) What environment (air, water or other) will the device be exposed to?
Sensors exposed to aqueous environments such as blood pose more packaging issues and make integration of electronics difficult.

(9) Which substrate makes the packaging requirements less stringent?
For a sensor in aqueous solutions, for example, a ceramic substrate requires no protection of the sides of the ceramic strip; for a Si sensor on the other hand, it is difficult to insulate and package, since a conductive medium might short out the chemical sensor signal via the conductive Si side-walls.

(10) Is the desired microstructure a truly three dimensional part (3D) (e.g. a contact lens) or is it a projected form?
A truly 3D part suggest traditional precision engineering e.g. diamond turning or a molding process; the projected form allows for a lithography process.

(11) Thermal requirements?
-Maximum temperature of exposure (temperatures > 150 °C can make integrated Si functions impossible). If integration and high temperature are needed, specially processed Si, SOI or GaAs become candidates.
-Is thermal conductivity needed?
-Is a thermal match with other materials important ?

(12) Flatness requirements (often in connection with the optical properties of the substrate?
-Average roughness, R_a?
-One or both sides polished?

(13) Optical requirements?
-Transparency in certain wavelength regions?
-Index of refraction?
-Reflectivity?

(14) Electrical and magnetic requirements?
-Conductor vs. insulator?
-Dielectric constant?
-Magnetic properties?

(15) Process compatibility?
- Is the substrate part of the process?
- Chemical compatibility?
- Ease of metallization?
- Machinability?

(16) Strain-dependent properties
-Piezoresistivity?
-Piezoelectricity?
-Fracture behavior?
-Young's modulus?

5.4. SCALING CHARACTERISTICS OF MICROMECHANISMS

Superimposed on answers to the listed questions in Table 3 is a required understanding of the scaling characteristics of micromechanisms. Some intuitive guidance, summarized in Table 4, may suffice [Goemans, 1994]:

Table 4. Some Important Effects of Scaling on Micromachine Design

(1) Size effect on forces. Weak forces are preferred in the micro domain.Electrostatic forces (with a quadratic dependency on dimension (l^2)), under certain conditions , outperform magnetic torque (with a cubic dependency on dimension (l^3)) since they become relatively stronger in small devices,
(2) Increasing strength of materials in the micro-domain. Single crystal whiskers may be more than 1000 times stronger than the bulk material,
(3) The dominance of surface effects in the micro world. For example, due to an increase in S/V (surface to volume ratio), better heat dissipation result (thermal isolation is difficult and cooling is excellent) and frictional forces increase,
(4) Decrease of manufacturing accuracy of micromachines (see Figure2),
(5) The need of error insensitive design (fewer parts, flexible materials) and the use of nature as a guide in design philosophy for very small devices,
(6) Mechanical, thermal, electromagnetic and many other time constants shrink with advanced miniaturization, generally resulting in shorter response times,
(7) Microsystems with their small inertial mass are faster, leading to higher speed or frequency but higher frequency of a cyclic process generally leads to larger losses and consequently to lower efficiency,
(8) Power consumption is often dramatically reduced especially if good thermal isolation is possible for elements with small heat capacity,
(9) Energy sources scale very poorly (l^3) making on-board energy sources (e.g. batteries and fuel cells) less attractive for powering micromachines than radiating in energy (e.g. solar or laser light with micromachined photovoltaic converters or microwaves if extremely small receivers and converters were available),
(10) Most actuators, which rely on power for inducing a movement, also scale poorly and it is often better to consider a 'macro-actuator' which can produce a micro-motion than a micromachined actuator with too small a stroke.

For the use of the check-off Table 3 and the scaling laws in Table 4 we again refer to Figure 7 which exemplifies the process for a glucose sensor design. The confrontation of the electrochemical approach with the manufacturing options of Table 2 leads, after several iterations, to the suggestion of a thick film manufacturing process with a continuous printing process on a plastic substrate. In this iterative process the check-off list in Table 3 functions as a control gate.

6. Design Software

After brainstorming about sensor specifications and sensor transduction principles and making preliminary designs on paper and the white board, it is a good idea to initiate a computer-aided design (CAD) of the overall microsystem in order to better grasp how all the components, including the package, fit together.

Especially for microparts involving 'regular' IC manufacture some very evolved CAD software packages are now available (e.g. the well known SUPREM 3 [Ho et al., 1983] and SAMPLE [Oldham et al.,1979, 1980]). A good introduction to the available CAD programs for IC processes comes from Fichtner [1988]. In general, these CAD software packages operate as sketched in Figure 8 with the design aids used to create the design, simulation to develop the technology and verification to check the design. The final verification of course always happens in the lab. The goal though is to avoid wasteful and slow experiments by carrying out less costly computer work in order to get the fabrication 'right' the first time. These IC development software packages are being expanded with mechanical modeling programs and materials data bases. There is indeed a growing need for the ability to perform mechanical analysis of microelectronic devices, both in assuring structural reliability against failure of thin film layers, and in evaluating the effects of various external loads including temperature and humidity effects. In addition, with the development of increasingly high aspect ratio micromachines, including pumps, valves, and micromotors, and with the increasing performance demands being placed on these devices, notably in precision and accuracy, there is a critical need for computer-aided-design (CAD) tools which will permit rational design of these devices. In order to construct the required data base for these new CAD tools one must understand all the parameters influencing the microstructure. At this point in time a good understanding of the processes and materials properties required to ensure reproducibility for commercial MEMS products slowly is emerging. Several CAD systems, which might facilitate the wider acceptance of MEMS, are under development. Most of them include a materials data base which can be updated by the user.

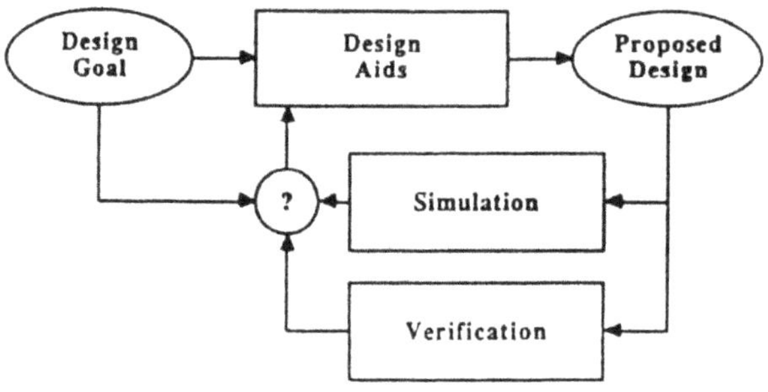

Figure 8.Design, simulation and verification.

REFERENCES

Boothroyd, G. and Knight, W.A. (1989) *Fundamentals of Machining and Machine Tools*, Marcel Dekker, Inc., New York and Basel.

Chryssolouris, G. (1991), *Laser Machining*, Mechanical Engineering Series, Ling, F. (Ed.), Spinger Verlag, New York.

Ehrfeld, W. (1994), LIGA at IMM, hardcopy of viewgraphs, Banff, Canada.

Evans, C. (1989), *Precision Engineering: An Evolutionary View*, Cranfield Press, Cranfield, Bedford, England.

Fichtner, W. (1988) Process Simulation, in S.M. Sze (ed.), *VLSI Technology*, McGraw-Hill Book Company: New York, pp. 422-463.

Goemans, P.A.F.M. (1994) Microsystems and Energy, in G.K. Lebbink (ed), *Microsystem Technology*, Samsom BedrijfsInformatie bv, Alphen aan den Riijn/Zaventem, The Netherlands, pp.50-64.

Gopel, W., Hesse,J., and Zemel,J.N. (1989), *Sensors: A Comprehensive Survey, Vol. 1* VCH: New York.

Habekotte, E. (1993) De Technologie van Sensoren en Actuatoren in de Werktuigbouw, in A.W. van Schadewijk and R.E.M. Kerver (eds.), *Sensoren and Actuatoren in de Werktuigbouw/Machinebouw*, Centrum voor Micro-Electronica, Den Haag, pp. 15-96.

Harris, T.W. (1976) *Chemical Milling*, Clarendon Press, Oxford.

Hayashi, T. (1994) Micromechanism and Their Characteristics, *Micro Electro Mechanical Systems* **94**, 39-44, Oiso, Japan.

Ho, C. P., Plummer, J.D., Hansen,S.E., and Dutton,R.W. (1983), VLSI process modeling SUPREM 3, *IEEE Trans.Electron.Dev.* **ED-30**, 1438-1452.

Lion, K..S. (1969), Transducers-Problems and Prospects, *IEEE Transactions on Industrial Electronics* **16**, 2-5.

Middlehoek S. and Audet, S. (1989), *Silicon Sensor,* Academic Press, London.

Moreau, W.M (1988), *Semiconductor Lithography*, Plenum Press, New York and London.

Oldham, W.G., Neureuther, A.R., Reynolds, J.L., Nandgaonkar, S.N. and Sung, C. (1980), *A General Simulator for VLSI Lithography and Etching Processes: Part 2-Application to Deposition and Etching.* IEEE Trans. Electron. Dev. **ED-27**: 1455-1559.

Oldham, W.G., Nandgaonkar, S.N., Neureuther, A.R., and O'Toole, M. (1979), *A General Simulator for VLSI Lithography and Etching Processes: Part 1-Application to Projection Lithography.* IEEE Trans.Electron.Dev. **ED-26**: 717-722.

National Research Council (1995), *Expanding the Vision of Sensor Materials,* Washington D.C.

Romankiw, L.T.,Croll,I.M. and Hatzakis,M. (1970), Batch-Fabricated Thin-Film Magnetic Recording Heads, *IEEE Transcations on Magnetics, **MAG-6**(3): 597-601.*

Senturia, S. (1987) Can we Design Microrobotic Devices Without Knowing the Mechanical Properties of Materials? , *Micro Robots and Teleoperators Workshop*, Hyannis, Massachusetts, IEEE-87 (third article).

Slocum, A., H. (1992) Precision Machine Design: Macromachine Design Philosophy and its Applicability to the Design of Micromachines, *Micro Electro Mechanical Systems*, Travemunde, Germany.

Smith, H. I. and Schattenburg,M.L. (1192)., Why Bother with X-Ray Lithography ? *SPIE, **1671**, 282-298.*

Snoeys, R. (1986) Non-Conventional Machining Techniques, *The State of the Art in Advances in Non-Traditional Machining*, The American Society of Mechanical Engineers, Anaheim, California.

Taniguchi, N. (1983), Current Status in, and Future Trends of, Ultraprecision Machining and Ultrafine Materials Processing, *Annals of the CIRP **32**(2): S.573-582.*

van Osenbruggen, C., (1969) High-Precision Spark Machining, *Philips Technical Review* **30**(6/7): 195-208.

White, R., M. *(1987)* A Sensor Classification Scheme, *IEEE Transactions on Ultrasonics, Ferroelectrics, and Frequency Control. **UFFC-34**(2): 124-126.*

Zuurveen, F. (1994), Precisie-Technologie: Kennis en Kunde, Inzicht en Uitzicht, in Precisie-Technologie-Jaarboek., NVFT: Eindhoven, pp.99-118.

MICROFABRICATION TECHNOLOGIES FOR HIGH PERFORMANCE MICROACTUATORS

F. MICHEL, W. EHRFELD
Institut für Mikrotechnik Mainz GmbH
Carl-Zeiss-Strasse 18-20,
D-55129 Mainz, Germany

Abstract

Silicon processes, which are the basis for micro electronics and also for a large number of MEMS, show several drawbacks when applied for high performance actuators. In particular, there is still a lack in suitable materials for bearings or magnetic elements. Alternatively, other micro fabrication technologies have recently enabled a status of industrial appliance hardly without any restriction in design and material. With LIGA metals such as NiFe or WCo are electroplated; Foturan®, a photo-sensitive glass, can be processed; steel, gem stones and Cu alloys can be micro cut; hard alloys are very common for micro spark erosion; and with micro moulding e.g. plastics and ceramics are micro structured in mass production. This calls, of course, for the assembly of so-called hybrid micro systems. The present contribution gives an introduction to the mentioned technologies and shows their application for high performance microactuators developed at the Institute of Microtechnology in Mainz (IMM), e.g. a fiber optical switch, a self-filling micropump generating a pressure of 2.000 hPa and a micro motor with integrated gear box appropriate for mechanical interface. In discussing the micromotor bearings, tribology needs are indicated. On the other side the potentials of microfabrication technologies, e.g. for generating porous bearings by LIGA or dynamic air bearings by micro cutting or spark erosion, are demonstrated. The contribution will show the importance and potentials of tribology research for MEMS.

B. Bhushan (ed.), Tribology Issues and Opportunities in MEMS, 53-72.

1. Introduction

Microfabrication technologies gain a rapidly increasing importance for the fabrication of components and systems in nearly every field of modern technology. By now, a number of MEMS, e.g. read/write heads of hard disks, inkjet printer nozzles, recording heads for CD players, microchips with some hundred thousands of mirrors for high resolution projection TV and, of course, the large number of miniaturized sensors, have a multi-billion $ presence on the market and huge yearly growth rates. The success of these devices has initiated a number of developments for the application of miniaturized actuators, including the related microtribological problems.

The key prerequisite for the fabrication of miniaturized actuators is the access to a broad palette of microfabrication materials. The huge importance of material selection for micro tribology systems is demonstrated in figure 1. A steel axle has run against a planar body made of silicon, Foturan® (see section 2.2) and ruby (see section 2.3). Lubrication is not applied. The axle is forced in axial direction. The experiment simulates extremely operating conditions of the micromotor described in section 3.3.

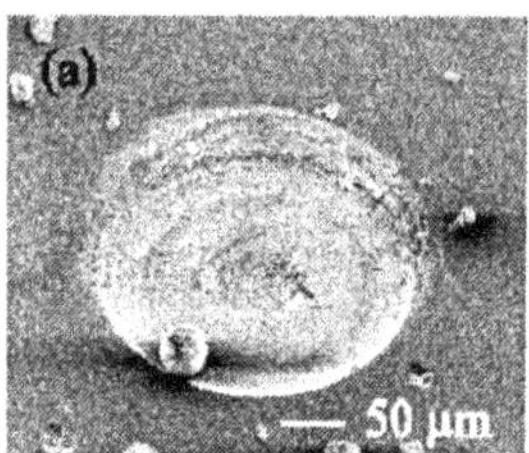

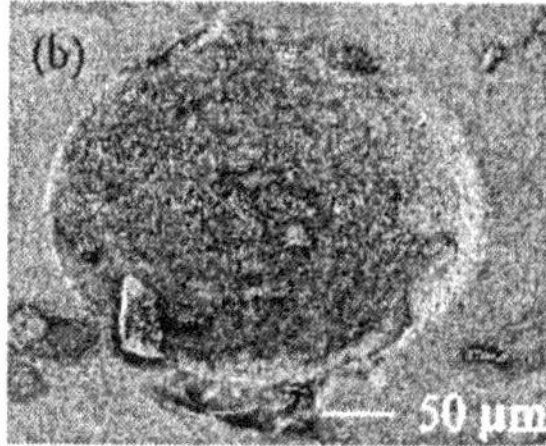

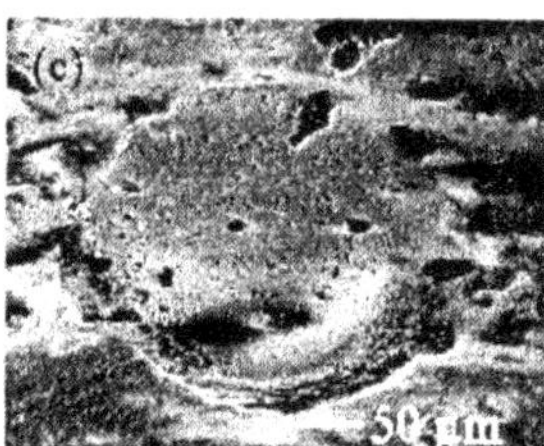

Figure 1: Wear on a loaded axial bearing consisting of a steel shaft (Ø 240 µm) and a planar plate made of (a) silicon, (b) Foturan® and (c) ruby, run-time 1 hour, speed 10.000 rpm, axial force 0.5 N, no lubrication.

Therefore, this contribution will not deal with silicon surface or bulk micro machining which have an outstanding importance on one and a couple of disadvantages for MEMS on the other side. Instead, the presentation focuses on those micro fabrication technologies with which a much more broader material palette appropriate particularly for microactuators and their tribology systems can be processed. Especially for high performance actuators delivering high forces and speeds, a combination of lithography based micro techniques and improved processes of precision engineering proved to be very successful. Some of them are introduced in the next chapter.

2. Microfabrication technologies

Figure 2 gives a flow chart overview of selected microfabrication technologies relevant for MEMS production and described in this chapter. As mentioned before, silicon technologies are not explicitely included, but will naturally be touched since lithography techniques include standard wafer processes like sputtering or vapor deposition.

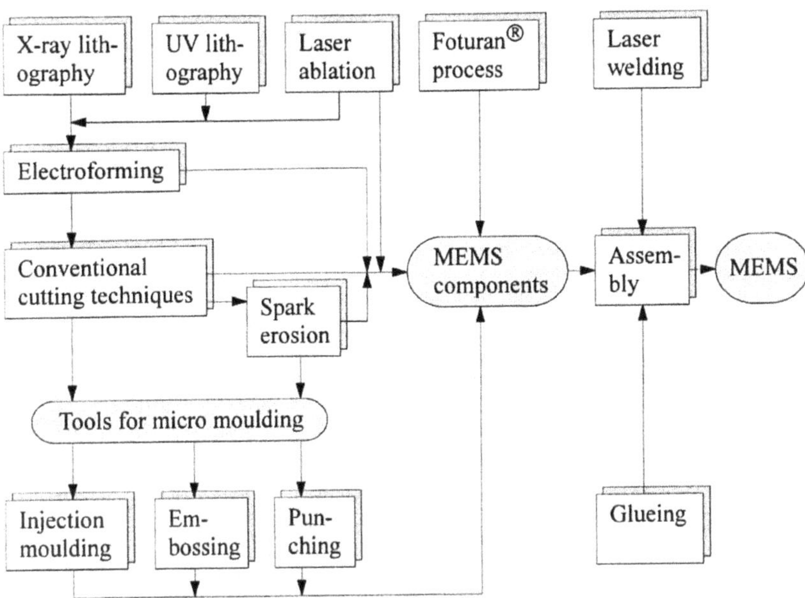

Figure 2: Overview of „non silicon" microfabrication technologies for MEMS

The technologies at the top of figure 2 have their roots in micro electronics, those at the bottom have been adapted mainly from precision engineering. However, there are various interferences.

2.1. LIGA PROCESS

The abbreviation LIGA originates from the German expressions for the major process steps: „LIthography", „Galvanoformung" (electroforming) and „Abformung" (moulding) [4]. Figure 3 shows the considerable number of involved technologies. By now, particularly moulding technologies are also applied for replicating cut or spark eroded micro structures. For that reason, moulding is discussed in section 2.5. This section only deals with lithography and electroforming delivering metallic MEMS components or mould tools.

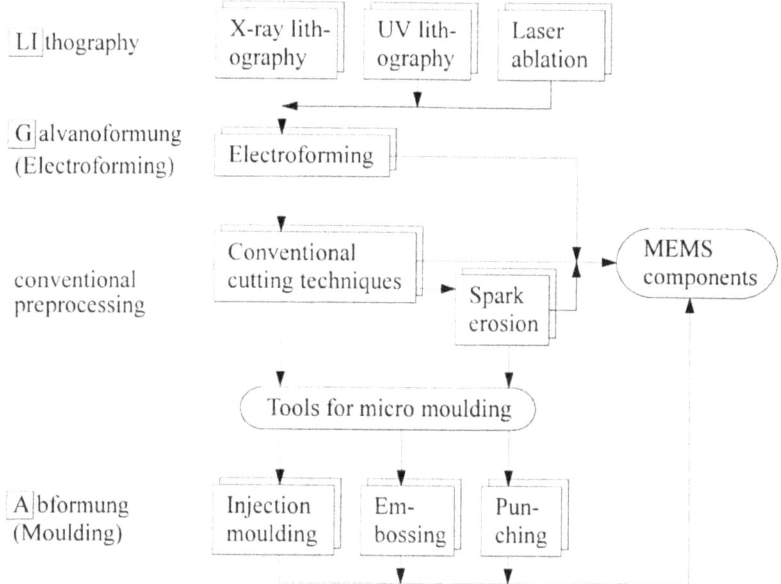

Figure 3: Microfabrication technologies involved in the LIGA process

Prerequisites for lithography are a suitable mask and a substrate coated with a resist layer. As substrate, which has to be electrically conductive, IMM uses titanium plates, but also silicon wafers or thin polymer foils with a metallic coating have been applied. The structurization of the radiation-sensitive polymer can be achieved by utilizing a laser, an electron or ion beam, as well as by optical (UV) or deep X-ray (DXRL) lithography.

The DXRL exposure process shows the most accurate results due to the enormous progress in understanding the physical and technical basics [7]. The shadow printing process uses the high energy part of the radiation spectrum from synchrotron radiation sources. Typically, mask and resist/substrate are scanned vertically through the radiation beam. The extremely precise motion with tilt errors less than 0.05 mrad has to be carried out with high velocities up to 50 mm/s in order to avoid thermoelastic deformations. Finally, microstructures with a typical height of 500 μm, a lateral accuracy of about 0.2 μm and a wall roughness R_a of 40 nm are obtained. But IMM has also generated microstructures of a height of 10 mm by DXRL.

In order to enlarge design variety a multi-layer process generating step-like LIGA structures has been developed. Figure 4 shows the process flow chart and a demonstrator structure. A diamond mask membrane has been developed which is transparent both for optical and X-ray radiation.

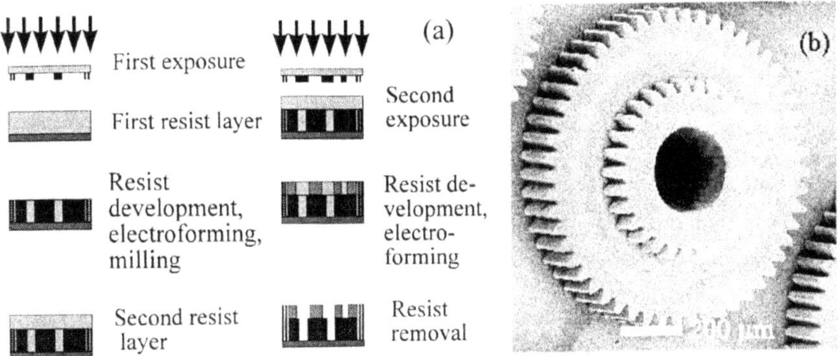

Figure 4: LIGA multi layer process. (a) Principle, (b) staged gear wheel.

In another process variation mask and resist/substrate are tilted with respect to the radiation beam (see figure 5). This technique could be very attractive for porous bearings (see section 4.2).

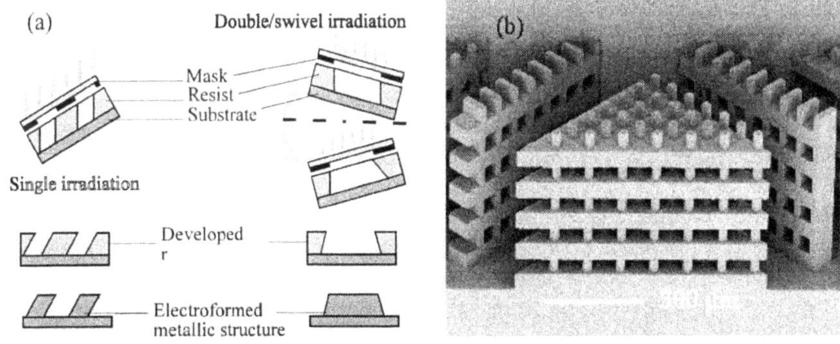

Figure 5: LIGA process with tilted exposure. (a) Principle, (b) demonstrator.

For DXRL usually PMMA (polymethylmethacrylate) is cast. When developing, the exposed material is removed. At IMM also a negative resist is under development that proves to be much more radiation-sensitive.

Over the last few years, world-wide efforts have caused an enormous progress in UV lithography. The resist height rose from a few micrometers typical in semiconductor industry to some hundreds of micrometers in the laboratory stage by now. The lateral accuracy does not reach the DXRL level, but is still suitable, e.g. for coils or groove structures (see section 4.2).

Laser LIGA uses laser light for resist ablation [1]. Lateral accuracy and aspect ratio also lie under the quality level of DXRL. On the other side, laser ablation offers more design possibilities particular for 3D shapes. In contrast to DXRL and UV lithography a metallic or metal coated substrate is not necessary. Resist development is replaced by a metal coating of the ablated resist. At IMM, Laser LIGA is mostly applied for the production of prototypes, particularly for fluid systems.

During the next step - the electroforming - the exact negative shape of the resist micro structures is generated. Suitable process parameters ensure that the metallic bulk arises with nanometer accuracy and without inner stresses and hydrogen bubbles. Whether components or mould inserts are required, the process is stopped until the resist height is reached or continued until a uniform metal plate has grown.

Standard processes for Au, Cu, Ni and NiCo have been established for a long time. For soft magnetic functional elements, IMM developed an electroforming process for NiFe reaching a saturation remanence of 1.4 V/m^2 [11]. In order to get harder moulding tools, micro structures made of WCo and WNi have been electroformed successfully provided with a micro hardness of about 600 HV. Another development aims at additionally friction reduced materials. Therefore, compound materials containing micro- and nanoparticles made of diamond, CBN, Al_2O_3 or PTFE are applied [5].

Unlike the accurate side walls, the electroformed top surface is rather vague. Consequently, mechanical preprocessing is necessary (see also section 2.3). At IMM a combination of grinding, lapping and sometimes polishing has been established. Based on the intensive use in semiconductor industry, CMP may play an important role in further developments.

Separation of substrate and resist/metal is achieved by thermo-mechanical methods or - more carefully - by selective etching of the thinned substrate or a sacrificial layer (see also section 3.1). For the removal of the undeveloped resist also a selective etching technique has been applied.

A suitable technique for the separation of the electroformed plate into mould inserts is wire electro discharge machining (Wire EDM, see section 2.4).

2.2. FOTURAN® GLASS

Wet etching methods of semiconductor industry can also be applied for structuring a completely other material - a photo-sensitive glass called Foturan® [2]. Wafers with a usual size of 5 inches in diameter and a thickness of about 1 mm can be delivered by Schott Glaswerke, Mainz, Germany. The main process steps are as follows:

- Exposure: The Foturan® wafer is exposed to UV light through a quartz mask with chromium absorber pattern. Both for illumination (310 nm) and mask (conventional 6 inch mask) equipment and deliverer of semiconductor industry can be used.

- Crystallization and polishing: The exposed structures are crystallized by a heat treatment of about 500 and 600°C. Due to different properties of crystals and glass a subsequent grinding and polishing process is recommended. At this point, some products such as optical reticules are finished.

- Etching with a 10% hydrofluoric acid (HF): Due to an etching ratio of ceramic and glass parts of 20:1, holes or deep pocket with almost perpendicular walls (1-4°) can be achieved. Side walls show an roughness R_a of about 1 µm. The anisotropic etching process can be followed by an isotropic etching in order to smooth the surfaces or to generate additional flat pockets. For that purpose, a second lithography step has to be included.

- Postprocessing: Foturan® is compatible to a considerable number of other microfabrication technologies. Metal coatings can be obtained, for exeample, by sputtering or vapor deposition, while throughplatings can be produced by electroforming and two Foturan® wafers can be bonded (400°C) or soldered.

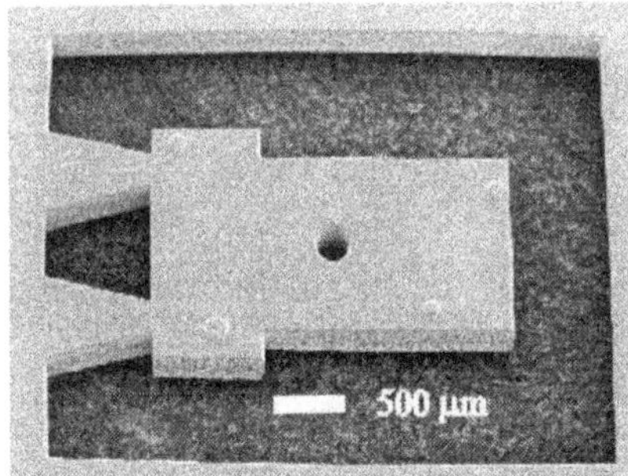

Figure 6: Foturan® front end of a suction gripper still within the glass wafer.

Depending on the structure size, the lateral accuracy lies in the range of some micrometers. However, Foturan® glass and ceramics show outstanding mechanical, chemical and thermal properties. Hardness and breaking strength are considerably higher compared to silicon and there are absolutely no pores. Temperature limits are as high as 750°C.

A part of an assembly tool realized by the Foturan® process is shown in figure 6.

2.3. MICRO CUTTING TECHNIQUES

Unlike the LIGA and Foturan® techniques, the roots both of micro cutting and spark erosion processes lie in precision engineering. There is such a large number of cutting techniques, probably better known as so-called mechanical manufacturing, that a complete description would exceed the possibilities of this contribution. For that reason three technology fields of eminent importance for the fabrication of high performance microactuators are chosen.

Grinding, lapping and polishing are obviously a substantial step, for example, of the LIGA technique (see section 2.1). They are applied for the planishing of the mask membrane, the substrate, the cast resist and the electroformed metal microstructures. In order to remove an additional bulk layer a couple of moulding techniques require mechanical manufacturing, too (see sections 2.5 and 2.6). In most cases planar surfaces with a flatness in the sub-micrometer range is required. Resist or structure height also have to be determined with micrometer precision. When processing a polymer/metal, polymer/ceramic or polymer/polymer compound, burr has to be avoided and edge sharpness has to be ensured. The required surface roughness ranges from a R_a value of a few to some hundreds of nanometers. Smoother surfaces as applied in optics are also fabricated by the so-called fly-cutting process using milling tools made of mono-crystal diamond. Rougher surfaces improve the adhesion of resist and substrate. The wide-spread CMP equipment of semiconductor industry, which is used for the planishing of processed silicon wafers, has not been applied to LIGA, yet, due to the different material compositions.

In watch industry mechanical processing has reached such a high level that e.g. miniaturized axles and bearings can be fabricated with an accuracy of a few micrometers. Figure 7 shows a ruby slider bearing with a bore diameter of 250 µm, figure 8 shows a centerless ground axle (diameter: 240 µm) and the components of a ball bearing with an outer diameter as small as 1.6 mm.

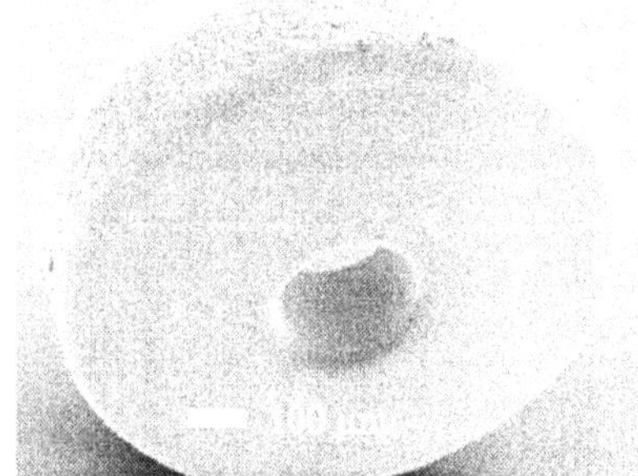

Figure 7: Ruby bearing. Source: Audemars S.A. (CH)

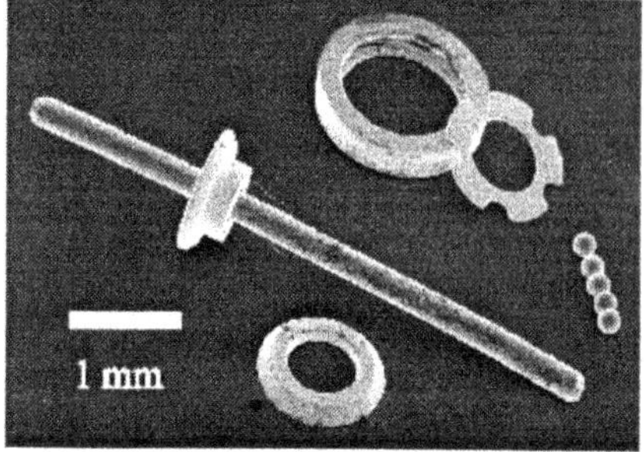

Figure 8: Component of a ball bearing, outer diameter 1.6 mm. Source: RMB (CH)

Machine tools in watch industry are as efficient that the shown bearings only cost the fraction of a dollar when ordered in millions of pieces.

Ultra-precision manufacturing traditionally aims for the fabrication of optical components [18]. With tools made of mono-crystal diamond particularly work pieces of copper alloys, aluminum and polymers are turned or milled. Actual developments are also directed towards micro structuring tasks. Sub-micron accuracy, roughness R_a in the nanometer range and cutting groove widths of some ten micrometers represent the state of the art. Explorations of steel manufacturing, e.g. with CBN tools, is still under way. Another technology, the so-called ductile grinding, deals with the manufacturing of glass. Figure 9 shows a micro triple mirror array. For manufacturing the grooves in figure 10 a modified turning technology was used. A so-called fast tool server axially moves the diamond very quickly and synchronously to the angle position of the spindle.

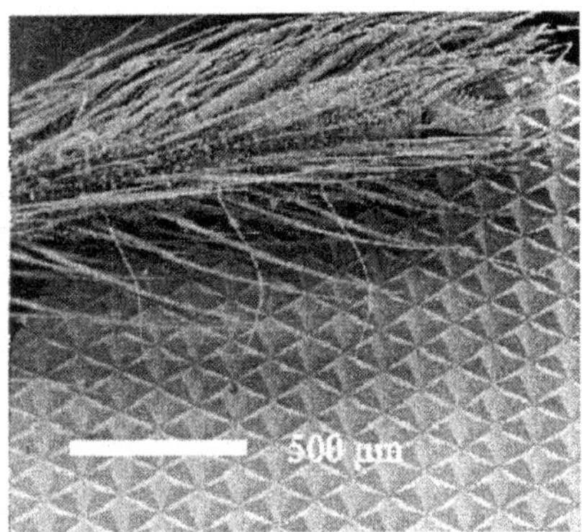

Figure 9: Micro triple mirror array, compared with the leg of a spider.

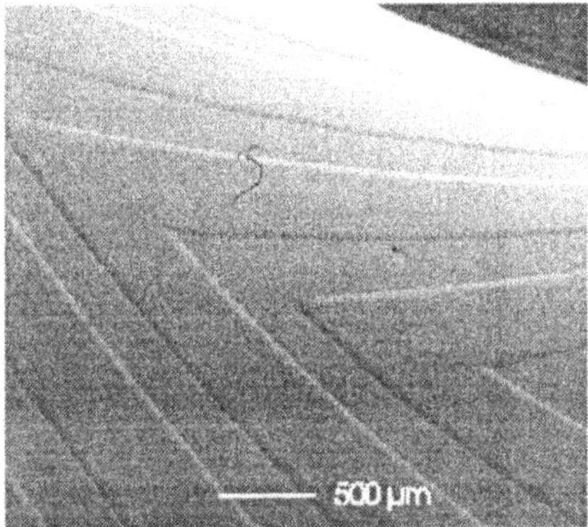

Figure 10: Grooves of a dynamic air bearing. Source: FhG IPT, Germany

2.4. MICRO SPARK EROSION

Spark erosion, also called Electro Discharge Machining (EDM), is a well-known and established technology in precision industry, particularly for the fabrication of moulding or punching tools. The technology is based on an erosion process, whereby conductive material is removed due to a high energy discharge between electrode and a work piece which both are surrounded by a dielectric fluid. In recent years, the machine facilities have been considerably improved. All conductive materials, e.g. stainless steel, titanium, hard alloys, ceramics like TiB_2 and even silicon, can be processed almost without any manufacturing forces. This, an accuracy of a few micrometers and a surface roughness R_a of up to 100 nm open the door for micro system applications. In close cooperation, IMM and the Swiss EDM machine manufacturer AGIE are developing adapted strategies and equipment [6], [8].

There are two main EDM technologies which require different machine equipment.

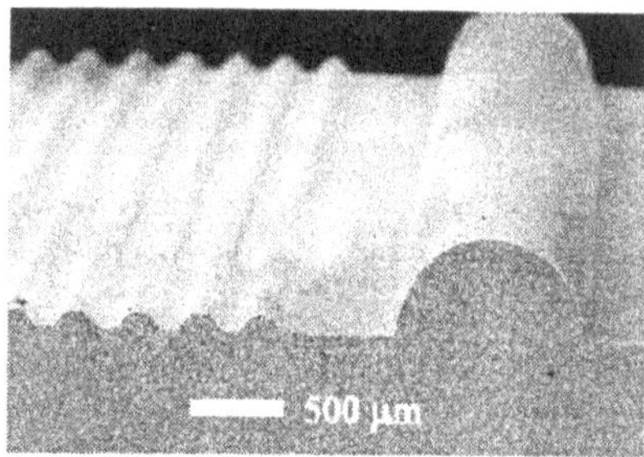

Figure 11: Wire-cut mould insert for a multiple optical fiber connector.

Wire-EDM uses thin wires as electrode made, for instance, of brass or tungsten. The thinnest wire commercially applied has a diameter of 30 µm and allows the generation of gaps as small as 50µm and inner radii up to 20 µm. Figure 11 shows a wire cut mould insert for an optical fiber connector.

Die sinking transfers the electrode shape into the work piece by vertical electrode movements. Usually many electrodes are needed due to the electrode wear. Especially for micro structuring additional electrode movements are applied.

Conventionally, electrodes are manufactured subsequently by milling or Wire-EDM. Figure 12 shows a toothed rack milled with a diamond tool (see section 2.3). By a combined horizontal and rotating electrode movement the cylindrical workpiece was provided with screw-like micro grooves (see figure 13).

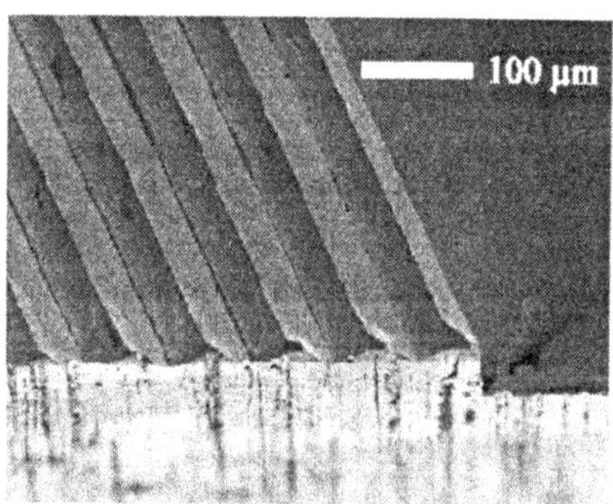

Figure 12: Ultra precision milled toothed rack. Material: brass.

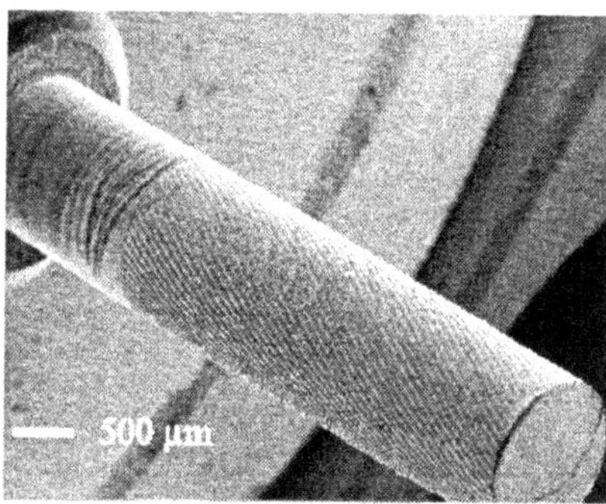

Figure 13: Steel axle with 1 mm diameter, provided with screw-like micro grooves

For micro die sinking cylincrical electrodes are often manufactured directly on the EDM machine. The quickly rotating electrode shaft is eroded with a hard alloy bar or wire. That way electrode pins with a diameter as small as 20 µm can be achieved. Subsequently, the electrode can be moved vertically to the work piece generating holes (EDM drilling), for example, or horizontally forming grooves (EDM milling) [14]. Figure 14 demonstrates the process principle and figure 15 shows a groove as small as 40µm that was eroded into TiB_2.

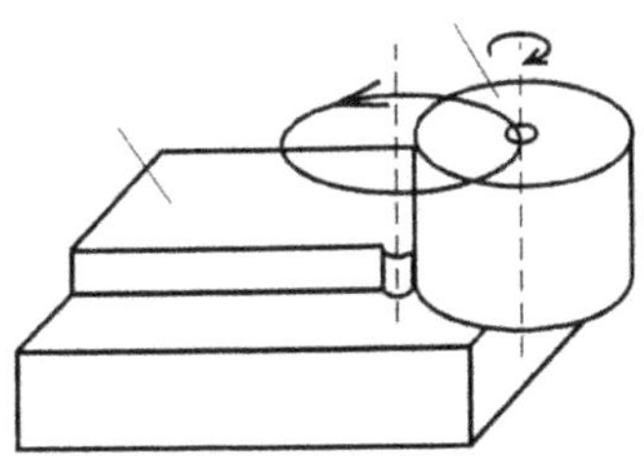

Figure 14: Principle of EDM milling

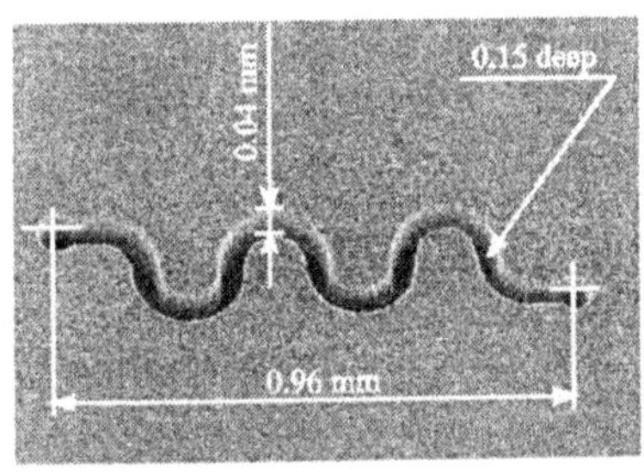

Figure 15: Workpiece made of TiB_2 with an EDM-milled twisted groove

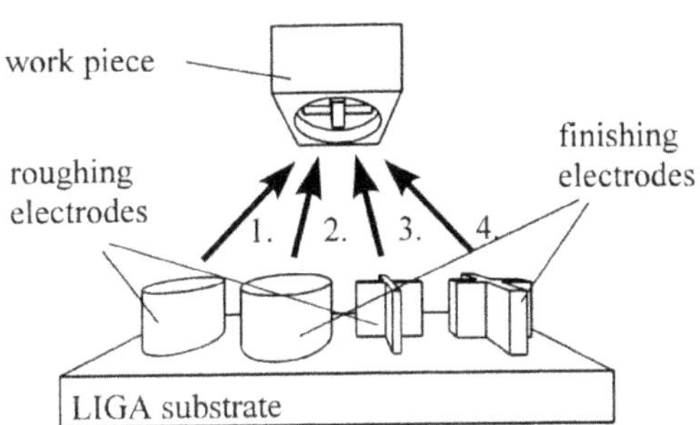

Figure 16: Principle of LIGA-EDM

Micro structures generated with the technique mentioned above are rather simple. For that reason, a new EDM technology combining the advantages of LIGA and die sinking has been developed [19]. The principle is explained in figure 16. By placing all the needed roughing and finishing electrodes onto a substrate, LIGA is utilized to generate electrodes of complex shape and very exact relative position. Via micro die sinking, materials like hard alloys can be performed which cannot be processed by LIGA electroforming. As an example a LIGA heat exchanger structure was used as electrode to erode an embossing tool (see figure 17).

Figure 17: Micro heat exchangers realized by LIGA-EDM. (a) LIGA electrode made of Ni, (b) die sunk embossing steel tool, (c) embossed Al heat exchanger.

2.5. MICRO MOULDING

Mass production in precision engineering is unthinkable without moulding techniques. Especially plastic pieces with sizes from some cubic millimeters to cubic meters are formed by injection moulding. Traditionally, aluminum chassis are die-cast. Recent developments have enlarged the material palette to ceramics and even to stainless steel via Metal Injection Moulding (MIM) while embossing of polymer foils plays an important role for the production of holographic structures. Textile fibers or hollow fibers for medical applications are extruded using a polymer or glass melting mass. Punching and bending of metal foils are wide-spread for small electro mechanical devices such as contacts and springs in relays, leadframes in integrated circuits, housings and even gear wheels for watches. Two powerful micro moulding techniques shall be described in more detail:

With the use of the LIGA process for tool fabrication a new quality for injection moulding of tiny micro structures with a huge aspect ratio has been obtained [17]. In contrast to precision engineering, the smooth LIGA surfaces allow the separation of the moulded part by perpendicular walls. The material palette contains standard materials like PA, PBT, PC, PE, PMMA, POM, PP, PPE, PS and PSU, but also engineering plastics like LCP, PEEK, PEI and biodegradable materials can be processed.

There are two main strategies to improve process efficiency. With the isotherm process cycle times of a few seconds are achieved. Mould inserts fabricated by various micro techniques can be combined. Figure 18 shows assembled mould inserts fabricated by LIGA and Wire-EDM for the injection moulding of a multiple optical fiber connector (figure 19). On the other hand only few parts with deep micro structures can be moulded simultaneously due to the fact that a relatively cold mould insert has to be filled. No problems occur when the structures are relatively flat like on compact disks.

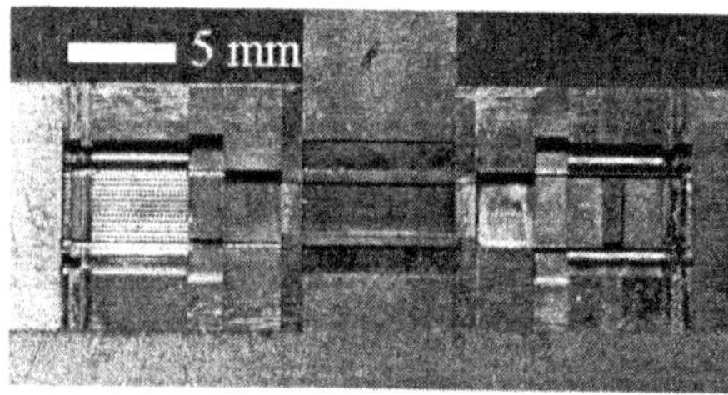

Figure 18: Assembly of mould inserts fabricated by LIGA and Wire-EDM

Figure 19: Micro injection moulded multiple fiber connector

In using a variotherm injection process cycle times are rather in the minute range, but thousands of deep micro structures can be generated simultaneously. The basis is given by LIGA technique with its ability to generate thousands of mould inserts in one batch process.

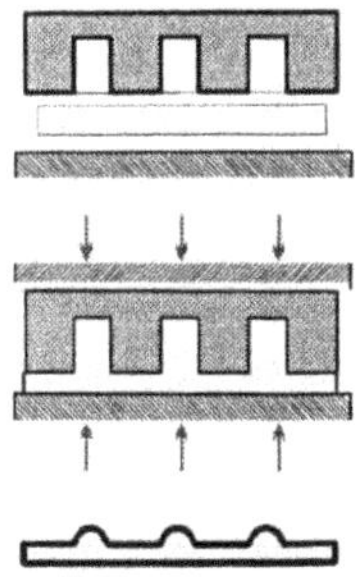

Figure 20: Principle of contactless embossing

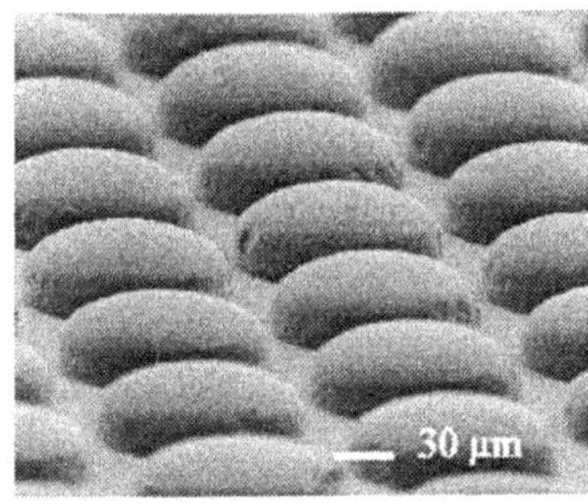

Figure 21: Contactless embossed micro lenses. Material: PC

Hot embossing processes have mostly been applied for the replication of micro optics, usually made of PC or PMMA. An interesting contactless embossing process of optical lenses is introduced in figures 20 and 21.

Besides powder injection moulding or slurry casting using lost forms, embossing is a suitable technology for the micro structuring of ceramics, too. Al_2O_3, ZrO, Si_3N_4 or PZT-ceramics, like the fluid sensor in figure 22, can be structured by green-tape casting, embossing, debindering and final sintering [15]. But there are still some drawbacks, in particular the high shrinkage during sintering and - due to the size of the ceramic powder - the high surface roughness R_a of a few micrometers. For that reason, an alternative technology was developed reaching considerablybetter surface qualities. It starts with the moulding or embossing of ceramic precursor material, followed by a partial pyrolysis and sintering process generating siliconcarbonitride ceramic micro structures as shown in figure 23.

Figure 22: Fluid sensor, assembled of two embossed Al_2O_3 ceramic parts

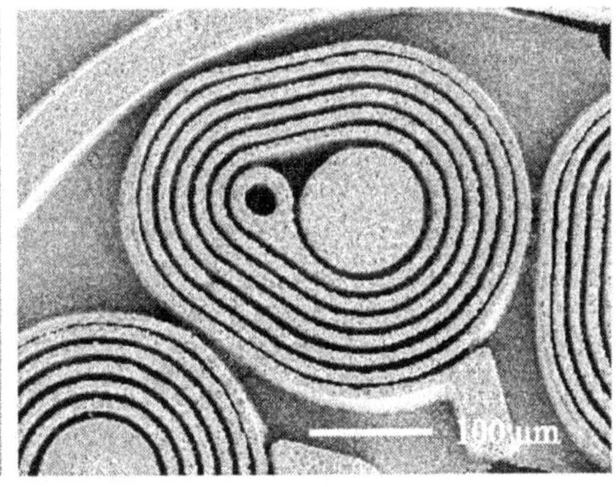

Figure 23: Embossed micro coil structures. Material: Si-C-N ceramic

2.6. MICRO ASSEMBLY

For prototypes manual micro assembly has proved to be sufficient. On the one hand micro components made of metal, polymer or ceramics are much more stable than those of brittle semiconductor materials. On the other hand human tactile abilities, supported by suitable suction or tong grippers, are enormous. However, for mass production automated processes with reliable and fast assembly operations have to be installed. Since this is commonly a time-consuming serial process, assembly is more and more becoming the bottleneck. New developments, therefore, use the arrangement of the micro parts on wafers or substrates and aim for simultaneous mounting [13].

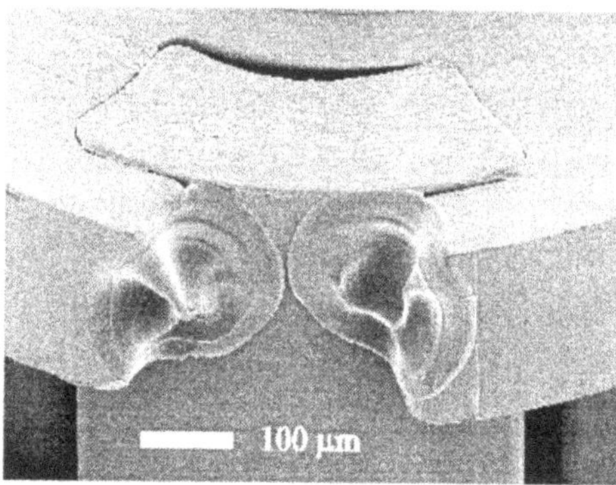

Figure 24: Mounted and welded components of a micro gear system. Material: POM

The mounted micro parts often have to be connected via welding, soldering or glueing. Figure 24 shows two welded polymer parts of the gear system introduced in section 2.3. For micro systems press fits are hardly suitable so that snap-in fits are preferred.

Assembly equipment of semiconductor industry, especially for die-bonding or the SMD assembly, has to be improved, for example in positioning accuracy, and to be adapted more to the three-dimensional mounting tasks [3].

3. High Performance Microactuators

In this chapter three microactuators - a fiber optical switch, micro pumps and a micro motor with size adapted gear system - are introduced. All of them deliver considerable power data. For the actuator fabrication a mixture of the shown micro technologies is applied. Inner friction in micro structured materials, possibilities for an active lubrication transport and - of course - the friction zones inside the motor and the gear box are demonstrated.

3.1. FIBER OPTICAL SWITCH

Fiber systems become more and more important, in particular for tele-communication purposes. In order to switch an interfered to a reserve connection, so-called fiber switches are demanded. Especially for monomode fibers an enormous mechanical switching accuracy is required. Two demonstrators, that switch the light of one incoming to one of the two outgoing fibers, have been successfully realized [11]. They are the basis for further developments of n x m switches, which will be able to connect hundreds of input fibers to any desired output fiber.

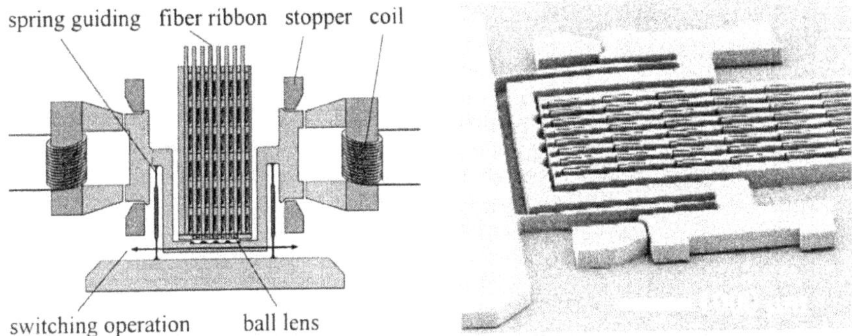

Figure 25: Principle of fiber switch.

Figure 26: LIGA part of the fiber switch. Sacrificial layer technique, material NiFe.

The principle and the design of one of the demonstrators are shown in figure 25. The fiber holder, the spring guiding and the mirror slider are integrated on a LIGA carrier. The glass fiber ribbon and the ball lenses can easily be mounted into the grooves and pockets. The yokes are completed by wire coils winded on a special machine and fixed onto the carrier. The LIGA part, which is made of a NiFe alloy,

- ensures the exact position of the mirrors, fiber grooves, lens pockets and stoppers,
- leads the magnetic flux (which is calling for low coercitivity, high permeability and high saturation induction)
- and guides the mirror slider by a parallel spring system.

According to the last requirement, the electroformed material additionally has to resist considerable mechanical stresses in the range of 100 MPa. In life time tests of more than 10^8 oscillations no fatigue was observed.

Another requirement touching tribology aspects is the contact of slider and stopper. On the one hand the contact has to be precise and long time stable without any wear. On the other hand the impacts should be damped and rebounds have to be avoided.

3.2. MICRO PUMP

Various applications, for example in medicine, demand for miniaturized pumps for the transport of gas and liquids. The main requirements of most applications are similar:

- A pressure of at least 1,000 hPa is needed.
- When liquids have to be transported, the pump has to be self-filling.
- A size in the region of 1 cm^3 is acceptable in most cases.
- The required flow rate is rather marginal, often much smaller than 1 ml/min.

According to these demands, two types of micro pumps were developed [3].

The membrane pump (see figure 27) delivers a maximum pressure of 2,000 hPa and a maximum flow rate of 0.5 ml/min when water is transported. The pump is self-filling, i.e. also gas can be transported. In that case the pressure value of maximum pressure degrees and of the maximum flow rate is considerably increased. In mass production costs of a few $ are estimated. The membrane pump consists of

- two micro injection moulded parts presently made of PC in order to ensure medical applications. The parts integrate the pump chamber, two valves and the fluid in- and outlet. One part is made of transparent, the other of colored polymer in order to allow laser welding. The mould inserts are fabricated by a combination of LIGA, spark erosion and precision manufacturing techniques;
- a membrane as thin as 2 μm inserted between the moulded parts. The membrane, also made of PC, is perforated directly above the valve seats by laser ablation;
- and the linear actuator driving the membrane up and down above the pump chamber, for example a commercially available bimorph piezo plate.

Figure 27: Micro membrane pump.

Figure 28: Micro gear pump.

When more viscous fluids like oil have to be transported, the micro gear pump in figure 28 can be applied delivering a pressure of 8,000 hPa.

3.3. MICRO MOTOR WITH INTEGRATED GEAR BOX

In close cooperation with the motor manufacturer Dr. Fritz Faulhaber GmbH & Co. KG in Schönaich, Germany, a micro motor with a size adapted gear box has been developed, which is - in small series - commercially available by now. Emphasis was given to the optimization of mechanical power, torque ripple and mechanical interface [12].

For that reason, a motor suitable principle was chosen with a diametrically attracted rare-earth magnet in the rotor while the coil, a yoke and stone bearings (see chapter 4) in the stator (figure 29, [9]). All parts are fabricated by mechanical manufacturing with the exception that for the mass production of the coils UV LIGA is likely to be more efficient than the now exercised winding of an isolated wire. With a size of 1.9 mm in diameter and 5.5 mm in length the motor can rotate permanently with a speed of 100,000 rpm and a torque of 7.5 μNm.

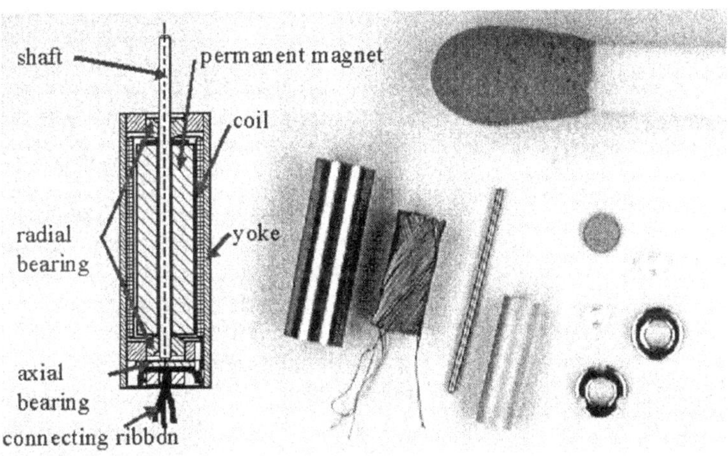

Figure 29: Design and components of the 1.9 mm micro motor

The gear box with the same diameter and a length of 3.7 mm contains 17 micro injection moulded gear components made of POM, a turned brass bearing and a centerless ground steel shaft (figure 30, [16]). The mould inserts carrying numerous wheel structures are fabricated by LIGA process. The three-staged epicyclic gear system reduces the motor speed with a ratio of 47. For short times torques of 300 μNm can be delivered. Life times of 1,000 h have been achieved with the whole gear box lubricated either with oil or grease (chapter 4.). Figure 31 shows that dealing with these small actuators can also be fun.

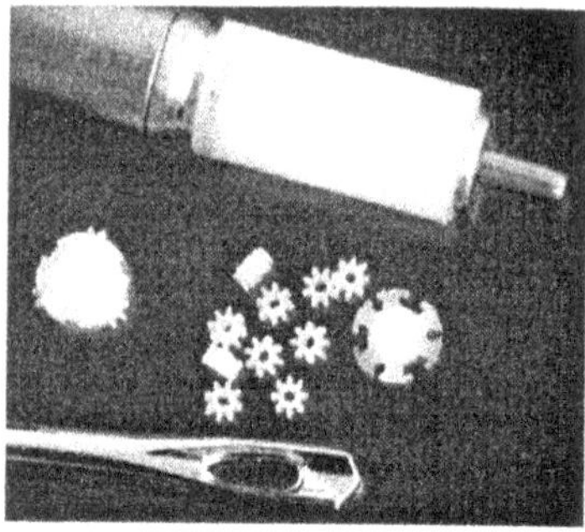

Figure 30: Micro gear box, mounted on the micro motor, and moulded POM components in comparison to a needle.

Figure 31: Tiny racing car with 1 inch of length on a German „1 DM“ coin. The rear axle is driven by the gear motor.

4. Tribology aspects

As already mentioned above, exploration of inner friction and adhesive forces is of major interest for MEMS. However, considerations in this chapter will focus on bearing problems. Based on the systematy in [10], four bearing types are observed.

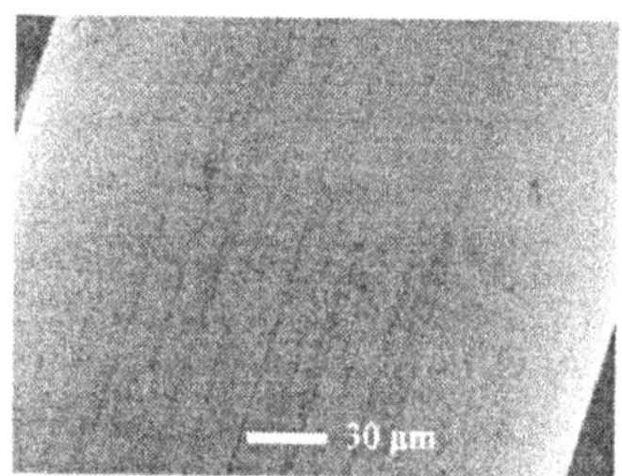

Figure 32: Hardened steel axle after 10^9 revolutions.

Due to the microscopic size leading to small absolute velocities and relatively high gaps, the hydrodynamic state cannot be reached in MEMS. Therefore, micro friction bearings are applied calling for hard bearing materials and compatible lubricants. Gemstone bearing from watch industry have been utilized sucessfully (see figures 1c, 7, 29), the wear of the steel wheel is negligible (figure 32). Some micro techniques are able to use extremely hard materials (see figures 1b, 22). Coatings, for example generated by plasma treatment, can also be applied. Additionally they may reduce the friction coefficient or keep the lubricant inside the bearing (epilams). Micro structures like polygon bushesform pockets for lubricant and wear particles [10], [16].

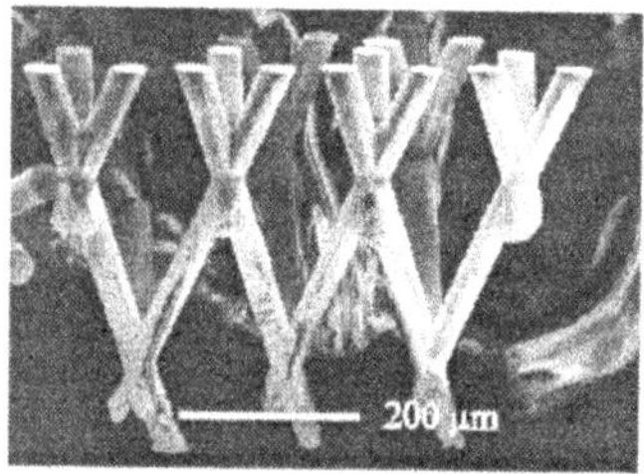

Figure 33: Bandgap filter. 3D-LIGA, Material: Cu.

Porous bearings are often applied in precision engineering. The use of 3D-LIGA (see section 2.1) may deliver metallic or ceramic materials with a pore portion of 90% and more.

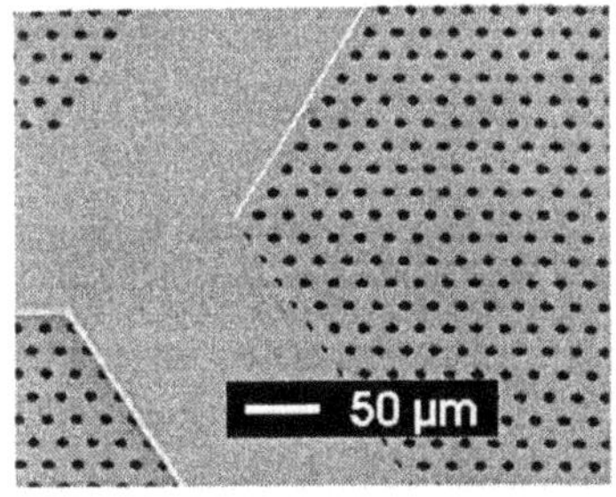

Figure 34: Metallic micro pore filter with supporting structures. Two-layer UV lithography.

By now, rolling-contact bearings are exclusively fabricated by mechanical manufacturing. The minimum diameter seems to be limited to a value of about 1 mm due to the fact that suitable balls are delivered with a diameter not smaller than 0.2 mm (see figure 8). For smaller roller bearings, the LIGA process with its ability to generate precise prismatic micro structures with any likely cross-section seems may offer further potentials. In precision engineering micro pumps, integrated into friction or roller bearings, would guarantee an outstanding lubrication (see section 3.3.)

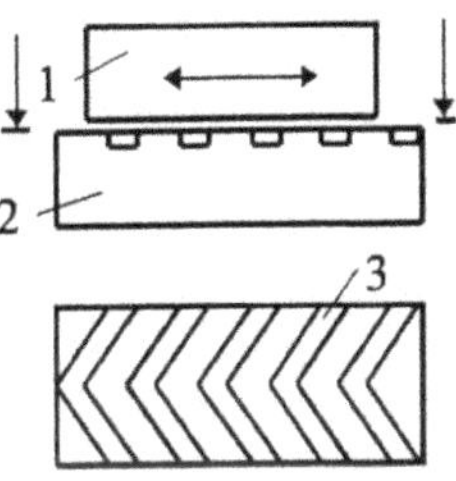

Figure 35: Principle of a dynamic fluid bearing. The slider 1 is moved to body 2 with arrow-like grooves 3.

Fluid bearings are usually of a considerable size. Especially static air bearings need a large bearing area due to the limited air pressure commonly available. On the other hand, a regular pattern of deep microholes could improve the bearing properties. Figure 34 shows a micro filter fabricated by two layer UV lithography. Regularly, arrow-like micro grooves serve as pump structures generating the fluid film in dynamic bearings. Figure 35 shows the principle. Microfabricaton technologies have sucessfully been used for generating these grooves. Ultra-precision turning was used for the toroid bearing shown in figure 10. LIGA, in turn, proves to be an ideal technology for the planar groove structures needed for axial bearings while LIGA-EDM, which was used for the screw-like grooves in figure 13, can be applied for radial bearings.

5. Conclusion

This consideration illuminates that tribology tasks occuring in MEMS have to be inquired systematically. Results of tribology research have already given important impulses for MEMS development. Microfabrication technologies presently have reached a level that they can be sucessfully applied for bearings both on the micro and macro scale.

6. References

[1] Arnold, J.; Dasbach, U.; Ehrfeld, W.; Hesch, K.; Löwe, H.: *„Combination of Excimer Laser Micromachining and Replication Processes Suited for Large Scale Production (Laser-LIGA)"*. Applied Surface Science, 86. 1995, 1/4; S.251-ff

[2] Dietrich, T. R.; Ehrfeld, W.; Lacher, M.; Speit, B.: *„Fabrication technologies for microsystems utilizing photoetchable glass"*. Microelectronic Engineering 30 (1996) 497-504

[3] Döpper, J.: *„ Untersuchungen zur Auslegung und Fertigung von Mikropumpen"*. Fortschrittberichte VDI, VDI-Verlag Reihe 1, Nr.287, Groß-Gerau, 1997

[4] Ehrfeld, W.; Lehr, H.: *„Deep X-ray Lithography for the Production of Three-dimensional Microstructures from Metals, Polymers and Ceramics"*. Radiat., Phys. and Chemistry, 45, 3 (1995), 349-365

[5] Ehrfeld, W.; Hessel, V.; Löwe, H.; Schulz, C.; Weber, L.: *„Materials of LIGA Technology"*. to be published in Microsystem Technologies

[6] Ehrfeld, W.; Lehr, H.; Michel, F.; Wolf, A.; Bertholds, A.; Gruber, H. P.: „Micro Electro discharge Machining as a Technology in Micromachining", Proc. of SPIE's 1996 Symposium on Micromachining and Microfabrication, 14./15.10.1996, Austin, Texas, Vol. 2879, pp. 332-337

[7] Feiertag, G.; Ehrfeld, W.; Lehr, H.; Schmidt, M.; Schmidt, A.: *„Accuracy of structure transfer in deep x-ray lithography"*. International Conference on Micro- and Nanofabrication (MNE), 1996

[8] Gruber, H.P.; Wolf, A.: „Komponenten für µSysteme durch die µEDM ? " , Proc. of Microengineering 1997, Stuttgart, 24.-26.09.1997

[9] Kämper, K.-P.; Ehrfeld, W.; Hagemann, B.; Lehr, H.; Michel, F.; Schirling, A.: *„Electromagnetic permanent micromotor with integrated micro gear box"*. Proc. Actuator 96, Bremen, 1996

[10] Krause, W.: *„Konstruktionselemente der Feinmechanik"*.Hanser-Verlag, 2.Auflage, 1993

[11] Lehr, H.; Abel, S.; Ehrfeld, W.; Gerner, M.; Hagemann, B.; Kämper, K.-P.; Michel, F.; Nopper, M.; Löwe, H.; Thürigen, C.: *„ Towards the Commercialisation of Microactuators"*. Proceedings of the Europe-Asia-Workshop on Mechatronics (MICRO-ENGINEERING 95), 21-26

[12] Lehr, H.; Abel, S.; Döpper, J.; Ehrfeld, W.; Hagemann, B.; Kämper, K.-P.; Michel, F.; Schulz, C.; Thürigen, C.: *„Microactuators as driving units for microrobotic systems"*. Proceeedings of: Microrobotics: Components and Applications, Boston, Massachusetts, 21-22 November 1996

[13] Michel, F.; Ehrfeld, W.; Kämper, K.-P.; Lehr, H.; Nienhaus, M.; Schappert, T.; Soultan, H.; Thürigen, C.; Weber, L.; Wieduwilt, M.: *„Development of Hybrid Micro System Assembly for Mechanical, Optical and Electronical Systems"*. Proceedings of the Symposium on Handling and Assembly of Microparts, Vienna, 28 - 29 November 1997

[14] Richter, T.; Ehrfeld, W.; Wolf, A.; Gruber, H. P.; Wörz, O.: „Fabrication of microreactor components by electro discharge machining"; Proc. of 1st Int. Conference on Microreaction Technology; Feb. 23-25 1997, Frankfurt/M./Germany; to be published in Springer Verlag 1997

[15] Stadel, M.; Freimuth, H.; Hessel, V.; Lacher, M.: *„Abformung keramischer Mikrostrukturen durch die LIGA-Technik"*. Keramische Zeitschrift, Vol.48 (1996)

[16] Thürigen, C.; Ehrfeld, W.; Hagemann, B.; Lehr, H.; Michel, F.: *„Development, Fabrication and Testing of a multi-stage Micro Gear System"*. Proc. Tribology Issues Oppotunities in MEMS, Columbus, Ohio, 1997

[17] Weber, L.; Ehrfeld, W.; Freimuth, H.; Lacher, M.; Lehr, H.; Pech, B.: *„Micro molding - a powerful tool for the large scale production of precise microstructures"*. Proceedings of Micromachining and Microfabrication Process Technology 2, Austin, Texas, 14-15 October 1996,156-167

[18] Weck, M.; Luderich, J.; Vos, M.; Schroeder, H. B.: *„Manufacturing of Micromechanical Components by Diamond Cutting -Limits and Chances-"*. Proceedings of the 8th International Percision Engineering Seminar, Compiègne, France, May 1995, 431-434

[19] Wolf, A.; Ehrfeld, W.; Lehr, H.; Michel, F.; Nienhaus, M.; Gruber, H.P.: *„Combining LIGA and Electro Discharge machining for the generation of complex microstructures in hard materials"*; Proc. of 9-IPES/UME4 1997 (26.-30.05.97), Braunschweig, Germany, Vol.2; pp. 657-660

SURFACE CHARACTERIZATION OF NON-LITHOGRAPHIC MICROMACHINING

C. R. FRIEDRICH and R.O. WARRINGTON
Michigan Technological University
Houghton, MI 49931 USA

1. Introduction

Non-lithographic micromachining, or micromechanical machining, is a set of processes for the fabrication of microcomponents using direct material cutting or removal by energy beams. The micromechanical processes are not intended for mass fabrication but rather to aid in the evolution of design concepts and to produce small quantities of components. One potential area for applicability to mass fabrication of these processes is fabricating lithography masks and masters for molding processes which can result in mass replication.

The micromechanical processes include milling, drilling, and diamond machining (force processes) and electrical discharge machining, focused ion beam, and laser machining and polymerization (forceless processes), all at the microscale. Because each of these processes is able to create functional mechanical microcomponents, part of the process characterization has been to quantify surface roughness and to compare these figures with the generally accepted lithographic process for high aspect ratio microstructures, namely deep x-ray lithography.

2. X-ray Lithography

As a basis for comparison of generated surfaces, x-ray lithography will be used. This technology can generate high aspect ratio microstructures and a height in the centimeter range has been demonstrated [1]. The technology is well developed after work by many researchers and groups the past 20 years, or so [2-4]. X-ray lithography is used for deep penetration of the radiation into the photoresist which allows tall structures, vertical walls, and good surface smoothness of those walls.

B. Bhushan (ed.), Tribology Issues and Opportunities in MEMS, 73-84.

The basic process is shown in Figure 1 which includes electroforming to produce final metal structures.

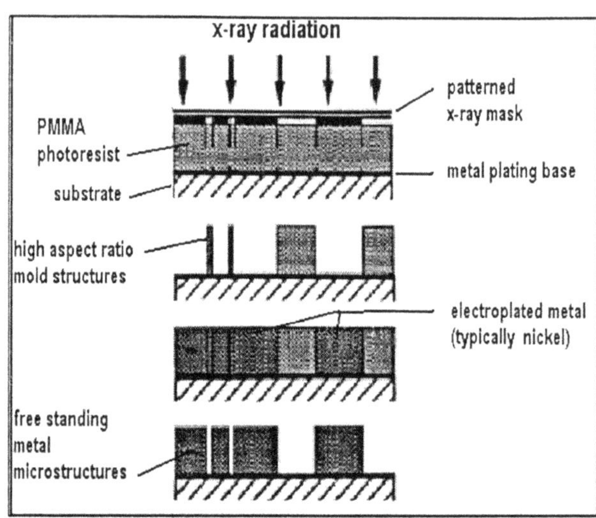

Figure 1. Basic steps in deep x-ray lithography process including electroforming to produce high aspect ratio metal microstructures.

Polymethyl methacrylate (PMMA) is the photoresist of choice for deep x-ray work because of its ability to faithfully replicate intricate detail from the photomask. In thin layers it can be spun on the substrate like most photoresists while for very thick layers, in the sub-millimeter to millimeter range, it can be adhered to the substrate as a pre-cast sheet.

A few studies have been performed to measure the surface roughness of walls produced by deep x-ray lithography. Generally, this parameter has apparently been less important than others because of few actual applications involving parts with relative motion produced by x-ray lithography. The wall roughness will be a function of several variables including the quality of the edges on the x-ray mask and the aggressiveness of the resist developer used.

One study compared the walls of structures made with a "conventional" gold mask and a micromilled composite graphite-gold-titanium mask. The conventional mask was on loan from *IMM-Mainz* and the micromilled mask was produced at the *IfM-Louisiana Tech University*. In this study, both masks were used to expose 300 um thick PMMA sheet at the LSU-CAMD synchrotron operating at 1.3 GeV for 8 hours. The samples were developed in GG solution (60vol% 2-[2-butoxyethoxyl] ethanol, 20vol% tetrahydri-1:4-oxazin {morpholine}, 5vol% 2-aminoethanol, and 15vol% water) for 90 minutes at 36C. The samples then underwent a double rinse procedure. The samples were sectioned and the wall roughness measured with a WYKO RST-Plus, white light interferometric microscope and the average roughness was measured to be 0.075 micrometers[5]. This value is slightly higher than the generally accepted range of 0.030 to 0.050 micrometers. A comparison of surface roughness of the exposed PMMA walls by the micromilled mask and the conventional mask is shown in Table 1, along with a summary of other roughness data still to be discussed.

Table 1. Comparison of wall roughness for various microfabrication processes

Process	Average Roughness (um)
x-ray lithography with conventional mask	0.03 - 0.05
x-ray lithography with micromilled mask	0.15
direct micromilling	0.1 or less
microdrilling	0.014 best
focused ion beam (surface tangent to beam)	0.005 in hard metal
laser ablation (surface tangent to beam)	0.1
laser polymerization (perpendicular to beam)	<0.5

In each case shown in Table 1, the roughness is not meant to imply it was the best unless noted. In the case of microdrilling, a procedure was developed to give the smoothest wall surface possible and that will be discussed subsequently.

3. Micromechanical Milling and Milled X-ray Masks

Although the thrust of this paper is on non-lithographic processing, a short overview will be given which describes a non-lithographic process (micromechanical milling) for fabricating x-ray lithography masks.

3.1 MICROMECHANICAL MILLING

Micromechanical milling (micromilling) is a direct machining process whereby the tools and machine have been specially designed for the very small cutting environment. A milling tool acts as a cantilever beam subject to both torsional and lateral loading. The strength of the cutting tool is a function of the stresses induced by the

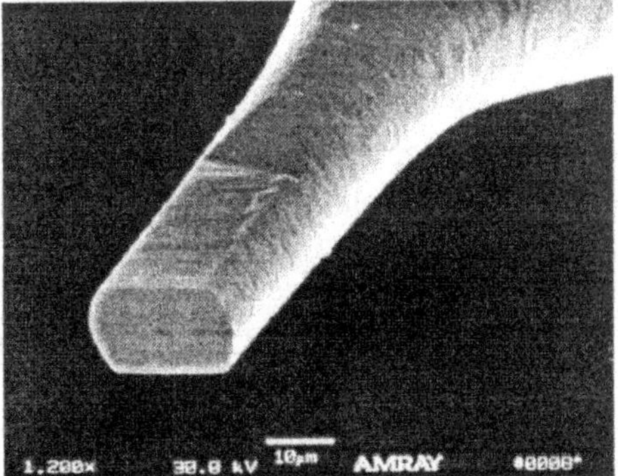

Fig. 2. 22-micrometer diameter micromilling tool made of high speed steel using focused ion beam process. (Scale Bar = 10 um) [6].

forces at the cutting edge, and the precision of the cut will be influenced by the deflection of the tool. For a given magnitude of a cutting force acting at the periphery of the milling tool, the shear and bending stresses are proportional to the cube of the diameter. Because the tool geometry is stress limited, any imprecision due to tool deflection can be ignored. Considering that milling tools as small as 22 micrometers in diameter have been successfully made and used, it is readily apparent that the machining conditions (feed and speed) must be carefully controlled. Such a tool is shown in Figure 2 and examples of microstructures machined in polymethyl methacrylate (PMMA) with this tool are shown in Figures 3 and 4.

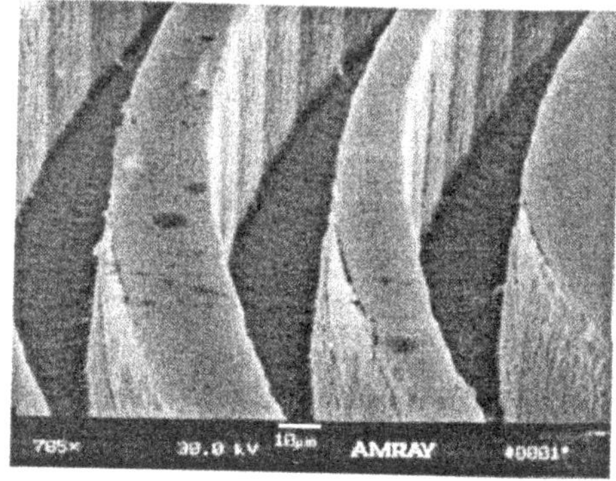

Fig.3. Vertical walls milled in PMMA. Trench width is 22 um and thinnest wall is 8 um wide, about the same width as a human red blood cell. (Scale Bar = 10 um) [6].

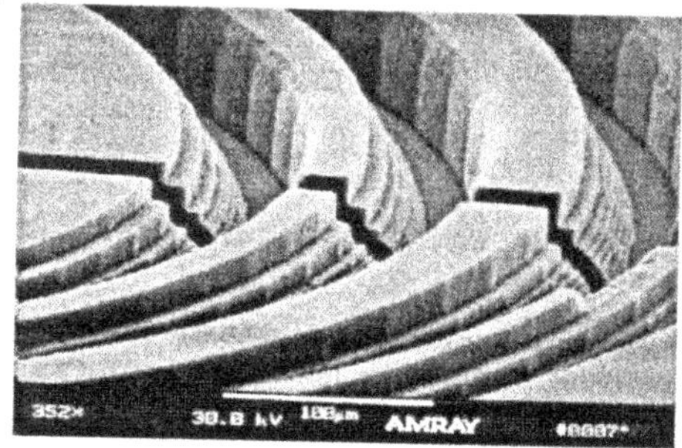

Fig. 4. Portion of step structures in PMMA. Total depth is 62 um and is composed of 15 successive passes. This is a mold which can be electroformed to create a metallic micro-structure. (Scale Bar = 100 um) [6].

Micromilling must be performed on a high precision machine tool with very smooth motions. In such a tool, all motions should be on air bearings with closed-loop positional control. This results in very smooth movements, particularly at slow speed where slip-stiction could result in tool breakage due to a higher level of impact loading on the cutting edges. High resolution feedback also allows for continuous work table velocity at slow speeds.

3.2 MICROMILLED X-RAY LITHOGRAPHY MASKS

Micromilling has been used to rapidly fabricate composite x-ray masks. The mask architecture has a graphite substrate 250 micrometers thick upon which sputtered x-ray absorber films of gold and titanium have been placed with a thickness of 8 micrometers and 3 micrometers, respectively. Graphite is used because it has a low x-ray absorption coefficient and is capable of being machined. Gold acts as the primary x-ray absorber but is too ductile to machine without excessive burrs. The titanium top layer nearly eliminates all of the machining-induced burrs.

A portion of a test mask is shown in Figure 5. The trenches were milled with a 100 micrometer nominal diameter tungsten-carbide endmill. After compensating for tool radial runout, the trenches were machined with the intended absorber wall widths being 50, 20, 15, and 10 micrometers. From measurements, the machined absorber wall widths were 50.4, 20.7, 15.4, and 11.1 micrometers, respectively with an average one-sigma standard deviation of 0.55 micrometers. The mask was used to expose PMMA to x-rays and the patterned walls in the PMMA were 48.2, 16.0, 13.6, and 9.6 micrometers, respectively. A closeup of the 10 micrometer nominal wide wall is shown in Figure 6. The wall microroughness can also be seen as the vertical striations caused by a combination of machining burrs and the structure of the sputtered gold film[7,8].

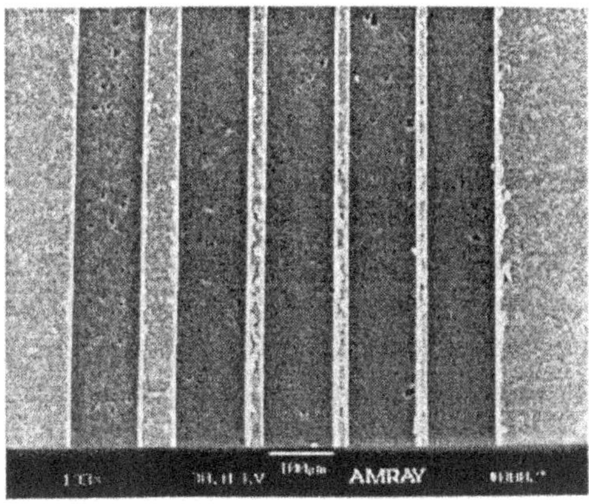

Fig. 5. Micromilled mask absorber walls of gold-titanium (Scale bar = 100 um) [7].

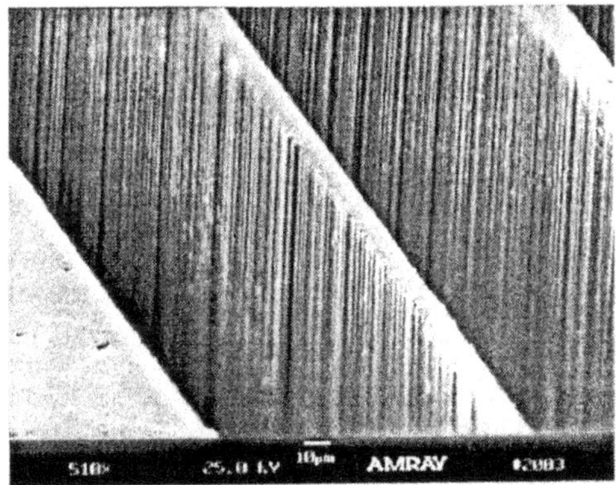

Fig. 6. Closeup of PMMA wall created by exposure with mask of Fig. 5 (Scale bar = 10 um) [7].

4. Microdrilling

Microdrilling differs from conventional drilling in that the tools do not have helical shapes to help with down-hole chip removal. The smallest twist drills available are 50 micrometers in diameter. Microdrills smaller than this size are made with the simpler geometry of the spade drill.

Microdrilling is always performed with a vee-block arrangement. This is because the drill can not tolerate eccentricity of rotation which would place very large bending loads on it, in addition to axial columnar loads. The length-to-diameter ratio depends on the drill diameter with an L/D of 4 to 7 being normal in the 25 to 100 um range, and an L/D of more than 11 in larger drills made of tungsten carbide.

Because a microdrill can not tolerate side loads, and since it is structurally weak, it is very important to start a hole with minimal drill "walking" and deflection which could result in drill breakage after the drill has reached a particular depth. It is therefore necessary to have a small initial feed per cutting edge, typically in the range of less than one micrometer. However, it is important not to let the drill dwell while in contact with the work material as hardening can take place resulting in higher drilling forces and possible drill breakage. It is therefore important to have a machine tool which can provide high vertical acceleration to the drill head during retraction. This is consistent with the practice of peck drilling which is necessary because spade drills do not have helical flutes.

When microdrilling, a thin fluid is always recommended to help remove chips from the hole, in addition to the pecking action of the drill. This fluid can be air or an air-oil mist. A moving fluid is preferred over a stagnant fluid drop because chips can flow back into the hole unless effectively removed from the drilling site. The air-oil mist provides an additional level of friction reduction and is therefore preferred over air alone. When not using a fluid, drill thrust forces are several times higher than when air alone is used.

Microdrilled holes can be made to high precision if the drills are inspected and sized prior to use. Typically, the tolerance on precision microdrills is 2.5 micrometers which may be excessive in critical applications. Testing was performed to determine the quality of microdrilled holes in terms of diameter, straightness, and wall smoothness in a semi-production situation. Working parts were being made but the process cycle was slowed to obtain satisfactory results. In these tests, holes were drilled in polymer as straight and as close to size as possible. Drills were inspected prior to use, except for a rapid break-in process which consisted of drilling a few holes to small depth. This was done to remove any residual burrs left on the drill due to the grinding process. The straightness of the holes was measured by inserting a glass optical fiber and allowing the fibers to extend some distance beyond the part surface. By measuring the misalignment of the two fibers, an indication of hole straightness, coupled with hole oversize, was found. The results were that the two holes were repeatedly drilled in a number of parts with a non-parallelism of 0.08 degrees (1.5 milliradians) and the oversize of the hole was estimated to be 0.5 micrometers. The holes had an L/D of 8.

Microdrills suffer from the same small size and high machining forces as do micromills. Commercial microdrills have a cutting edge sharpness typically 1 to 1.5 micrometers while the chisel edge radius is typically several micrometers. This, coupled with a small feed per cutting edge, results in an effective rake angle of -80 to -85 degrees at the cutting edges. The thrust forces in microdrilling are very high relative to the material removal rate. In fact, the change in thrust force can be used as an indicator of drill sharpness.

Because chips must be cleared form the hole, microdrilling uses the peck-drilling cycle where the drill is periodically retracted out of the hole as shown in Figure 7. This motion, coupled with the cutting fluid viscosity, helps extract chips from the hole. In addition, the turbulence created when the drill is inserted for the next peck cycle helps clear remaining chips as the fluid is forced back out of the hole.

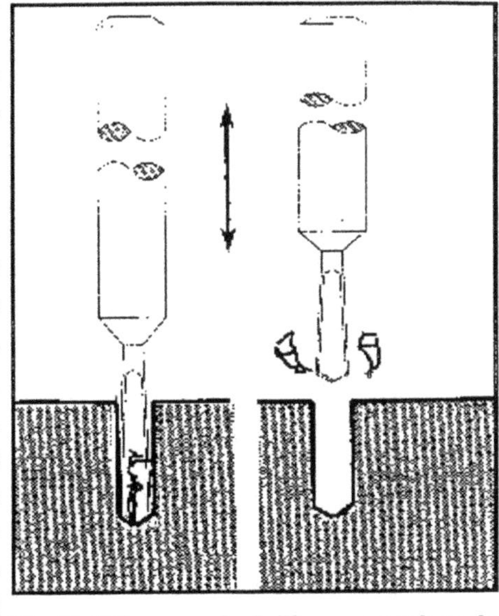

Fig. 7. Micro-spade drills use a peck cycle to help clear chips from the hole. The speed of the final retract greatly affects wall roughness.

During drilling, the drill is normally retracted at a speed much larger than the feed rate during drilling. The repetitive burnishing of the hole wall by the drill edges provides the very good surface smoothness of this operation, as previously shown in Table 1. However, as the drill is retracted it will leave a spiral indentation on the wall surface each time. At a high retraction speed, this helix is measurable and detracts from the smoothness. Therefore, to drill a hole with very smooth walls, it is necessary to retract the drill from the hole at the drilling feed rate for the last several peck cycles. This adds to process time, but results in a much smoother surface [9].

5. Focused Ion Beam

Focused ion beam (FIB) is a process widely used as a process diagnostic tool in the semiconductor industry. FIB can also be used as a machining process for the fabrication of microstructures. The process uses an accelerated beam of metal ions which impact the surface of the work material. The kinetic energy of the ions results in atoms of the work piece being dislodged from the surface thus providing atomic-scale machining. The beam of ions can be focused to a spot with a diameter as small several nanometers thus greatly concentrating the ionic energy to levels of 50 Amperes per square centimeter, or more.

FIB is particularly useful for the fabrication of micromechanical cutting tools. The tool shown previously in Figure 2 was made by FIB [10]. The tool started as a cylinder and then the beam was rastered within a rectangular area where material was removed. Because the incident ion beam has a Gaussian energy distribution, the upstream edge of the tool becomes well rounded and is not used for cutting. However, as the large rake face is developed, it acts as a physical filter which eliminates one side of the Gaussian beam. The downstream edge is thus very sharp and is used as a cutting edge. After one edge is completed, the tool is rotated by 180 degrees and the process is repeated. Although the milling tool may look symmetrical and able to be used with either sense of rotation, only two of the edges can be used for cutting.

When the incident beam strikes a perpendicular surface, atoms of the sample are ejected or driven into the subsurface. This will tend to leave a very small cratered effect where the crater dimensions may be on the order of atomic spacings, depending on the sputter yield. The average roughness of the surface reflects these dimensions and is well less than 10 nanometers. As the beam becomes more tangent to the surface, as in the tool making process, the surface is less effected by the direct incidence of the beam as the sputter yield goes back down. In this mode, tools with cutting edge radii less than a few nanometers is possible. To produce such extremely sharp intersections, the surface must have a very smooth texture. Such a tool is shown in Figure 8.

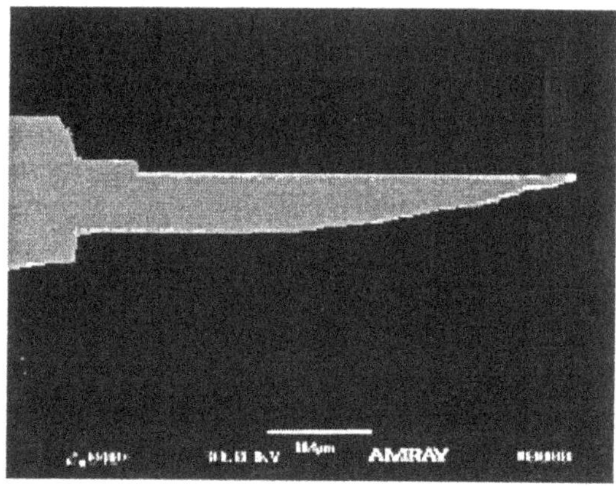

Fig. 8. Microscalpel fabricated by FIB with a cutting edge radius of a few nanometers. (Scale bar = 10 um).

6. Diamond Micromachining

Fabricating microparts by direct diamond turning is a seldom-used process and therefore no data is included in Table 1. However, enough work has been done to characterize the process. Diamond machining is commonly used to produce very smooth surfaces with highly precise geometry for optical, and many other, applications. As a cutting tool material, diamond is superb because of its high hardness, stiffness, thermal conductivity, low friction (in air) and relative inertness. This inertness ceases when the work material is readily able to absorb carbon and these type materials have been identified [11]. With these materials the diamond will be readily absorbed by the work piece unless the diffusion rate can be significantly lowered, such as by cryogenic cooling.

A variety of microstructures can be made by diamond machining because the tool geometry can have small features. One such example is micro thermal surfaces and applications. By machining parallel channels into high conductivity copper, micro compact heat exchangers can be realized with very high volumetric heat transfer coefficients [12]. Other channel geometries are possible depending on the shape of the diamond cutting tool [13].

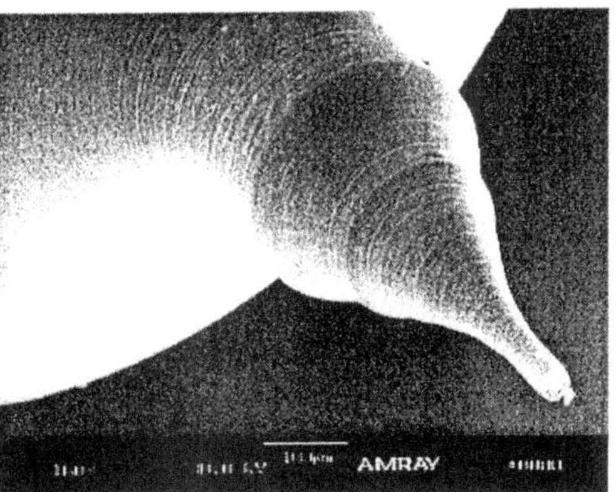

Fig. 9. Diamond-turned microshaft of aluminum. Minimum diameter is 25 micro-meters and has L/D of 4. (Scale Bar = 100 um).

A variety of stand alone microparts can be manufactured by diamond machining. One such part, a microshaft, is shown in Figure 9. The part was turned with the aid of a compensation algorithm to adjust the height of the turning tool as the shaft was made smaller. In this application, the part is a cantilever beam only 25 micrometers (0.001 inches) in diameter and its stiffness decreases toward the free end. Therefore, the algorithm had to predict the cutting forces based on the shaft deflection which is a function of the tool location along the axial length of the shaft. Nevertheless, the shaft shown in the figure has a straight section with an L/D of four and an average surface finish of less than 0.1 micrometers (4 microinches) [14]. Work at NIST in diamond turning electroless nickel has resulted in a surface finish as smooth as 0.005 micrometers.

7. Laser Ablation Micromachining

Laser ablation uses a focused and pulsed excimer laser with a short wavelength for high power and good absorption by the material to be machined. Certainly, not all materials can be processed by direct cutting methods. Yet, materials such as ceramics and diamond have use in high temperature or corrosive applications and these materials must be given complex geometries. Laser micromachining typically uses a pulsed excimer laser with a wavelength of 248nm (KrF). The pulse width lasts tens-of-nanoseconds and the pulse rate is normally 2 kHz or less. This gives a relatively low duty cycle which, coupled with the short pulse width, results in little thermal damage to the material away from the ablated region. The average laser power is typically tens-of-milliwatts and the peak power during the pulse is typically in the 1kW range. These number do not seem exceedingly high but the power is focused into a spot only a few micrometers in diameter so a very high power flux results.

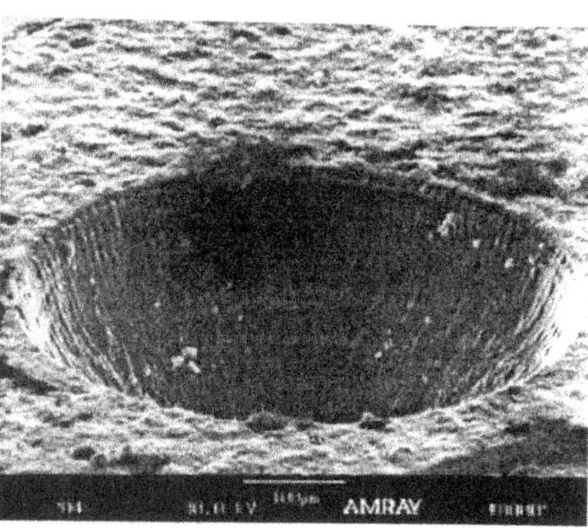

Fig. 10. Converging nozzle made with laser ablation in polycrystalline diamond (Scale Bar = 100 um) [15,17].

The lasers used in this type of micro-machining have a non-symmetrical energy distribution. For this reason, it is important to shape, or chop, the beam so only the central portion is allowed to pass to the final focusing optics. This helps produce better sidewalls and a more predictable ablated profile as shown in Figure 10. The focusing optics also have a short depth of focus so the distance control of the work piece surface relative to the beam focal point is important, otherwise there may be insufficient energy density to ablate the material, or the converging or diverging beam may give poor results.

The depth to which the material will be removed per pass of the laser spot depends on the pulse rate, the spot size, the absorption of the light by the material, and the temperature and energy of vaporization of the work material. All things regarded, the typical depth of cut per pass is a few tenths of a micrometer. Therefore, many passes are required to remove the volume of material layer by layer. This requires a relatively long machining time, but it also allows for reasonably good feature control in the vertical direction. Normal overlap of the beam in the traverse, as well as the lateral directions, is about 50%.

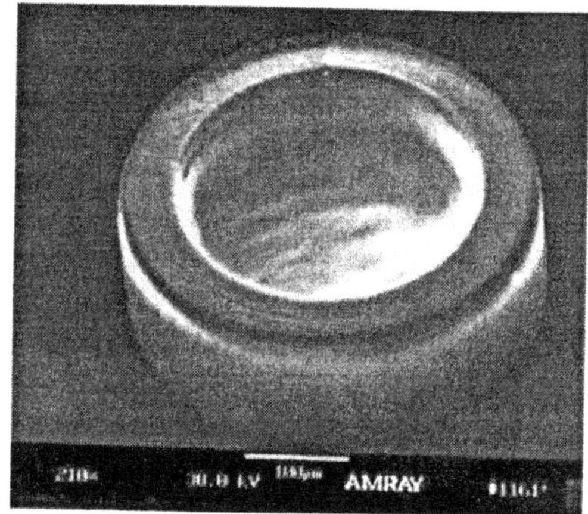

Fig. 11. Concave lens-type structure made by laser polymerization (Scale Bar = 100 um) [16,17].

8. Laser Polymerization

Photopolymers can be polymerized by ultraviolet radiation. This process in similar to conventional stereo-lithography except the spot is typically much smaller, a few micrometers, and this difference makes the microscale "writing" process slower but is required because of the size of the features involved. In the microscale process, the laser wavelength is typically 351 nm from a XeF excimer. This process requires additional capabilities and hardware so that after a layer of the resist is "written" by the laser with the desired pattern, the partially solidified material is immersed in the liquid polymer and raised so the surface of the previously written layer is just micrometers below the surface of the liquid. This is not easily accomplished considering the viscosity and surface tension effects of the liquid. The addition of a thin, transparent cover plate may aid in control of the liquid layer but the adhesion of the polymerized layer to the plate can be problematic. Examples of structures made by laser photopolymerization are shown in Figures 11 and 12.

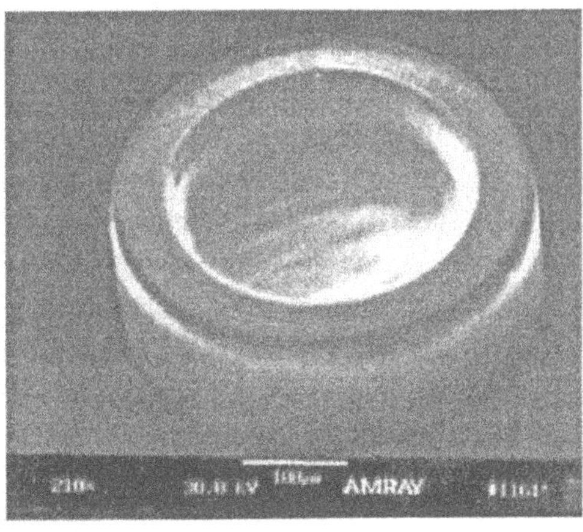

Fig. 12. Variable cross section ring structure made by laser polymerization (Scale Bar = 100 um) [16,17].

9. Summary

Non-lithographic processing does not compete with lithography in terms of fine features or mass production. However, if structures 5 to 10 micrometers and larger are required, the non-lithographic processes provide a very fast and direct method for small batch fabrication. For tribology studies, wherein surface wear and friction are the parameters, the non-lithographic processes may be a suitable alternative.

10. References

1. Guckel, H. (1996) LIGA and LIGA-like processing with high energy photons, *Microsystem Technologies* **(2)3**, 153-156.

2. Becker, EW, Ehrfeld, W, Hagmann, P, Maner, A, and Munchmeyer, D (1986) Fabrication of microstructures with high aspect ratios and great structural heights by synchrotron radiation lithography, galvanoforming, and plastic molding, *Microelectronic Engineering* **(4)**, 35-56.
3. Bley, P and Mohr, J (1994) The LIGA process - a microfabrication technology, *FED Journal* **(5)**, Suppl.1, 34-48.
4. Ballandras, S (1995) Deep etch x-ray lithography using silicon gold masks fabricated by deep etch uv lithography and electroforming, *J. Micromech. Microengr.* **(5)**, 203-208.
5. Gopinathin, N (1997) Direct fabrication of x-ray masks for LIGA by micromilling, MS Thesis, Louisiana Tech University.
6. Friedrich, C.and Vasile, M (1996), Development of the micromilling process for high aspect ratio microstructures, *J. MEMS* **(5)1**, .33-38.
7. Coane,P and Friedrich, C (1996) Fabrication of composite x-ray masks by micromilling, in *Microlithography and Metrology in Micromachining II,* SPIE. 2880, 130-141.
8. Friedrich, C, Coane, P, and Vasile, M (1997) Micromilling development and applications for microfabrication, *Microelectronic Engineering* **(35)**, 367-372.
9. Friedrich,C, She,Y, Muthu,R, and Dave,C., (1995) Thrust force and wall roughness in microdrilling, *Proceedings of 1995 ASPE Annual Conference,* Austin, TX, October 1995, 280-283.
10. Vasile, M, Friedrich, C, Kikkeri, B, and McElhannon, R (1997) Micron-scale machining: tool fabrication and initial results, *Precision Engineering* **(119)2/3**,180-186.
11. Paul, E, Evans,C., Mangamelli, A., and Polvani, R. (1996) Chemical aspects of tool wear in single point diamond turning, *Precision Engineering, (18)1*, 4-19.
12. Bier, W, Keller, W, Linder, G, Schubert, K, and Seidel, D (1990) Manufacturing and testing of compact heat exchangers with high volumetric heat transfer coefficients, in *Microstructures, Sensors, and Actuators, ASME DSC-Vol.19.*
13. Friedrich,C, and Kang, S.(1994), Micro heat exchangers fabricated by diamond machining, *Precision Engineering (16)1*, 56-59.
14. Friedrich,C, and Kithiganahalli,B, (1994) Deflection compensation model for the machining of microshafts, *Proceedings of 1994 ASPE Annual Conference*, Cincinnati, 461-464.
15. Gu,R. (1996), Fabrication of microstructures by UV laser ablation," MS Thesis, Louisiana Tech University.
16. Huang,X. (1996), Fabrication of three-dimensional microstructures by UV laser-induced polymerization," MS Thesis, Louisiana Tech University.
17. Friedrich, C, and Warrington, R (1997) Techniques for micromechanical machining, *Proceedings Tooling '97*, Southampton, England.

BIOSENSORS AND MICROFLUIDIC SYSTEMS

P. J. HESKETH, S. ZIVANOVIC, S. PAK, B. ILIC, L. ST.CLAIR,
B. SHIH, K.Y. CHUNG, J. C. CUNNEEN, S. CARIFFINI* and
J. G. BOYD*,

Microfabrication Applications Laboratory,
Department of Electrical Engineering and Computer Science,
**Department of Mechanical Engineering,*
University of Illinois at Chicago, IL 60607.

J. R. STETTER,

Biological, Chemical and Physical Sciences Department
Illinois Institute of Technology
Chicago, IL 60616

S. M. LUNTE and G. S. WILSON

Department of Chemistry and Pharmaceutical Chemistry
University of Kansas,
Lawrence, KS 66045

1. Introduction

Tribological issues have received little attention in microfabricated system. Although work on miniature mechanical devices was initiated in the early 1970's with the suspended beam and rotating electrostatic motor, issues of surface interactions in these structures has prevented their reliable operation. The application of the technology for the miniaturization of integrated circuits has also found utility in the miniaturization of chemical analysis systems (Manz et al., 1995). Semi-automated chemical analysis systems have been commercialized (Ruzicka and Hansen, 1988) and the key advantages of further reduction in size is the higher through-put of analyses can be achieved resulting in lower cost per analysis. Therefore application of microfabrication technologies facilitates the fabrication of miniature analysis systems. Furthermore, the economic benefits from batch fabrication processes will make miniature, light weight, portable, low cost analysis systems possible. In fluidic devices, the range of scaling does not offer the benefits as the volume of analyte becomes too small. Two issues arise, that of increased surface tension and pressure drop in miniature channels, and the fact that statistically there needs to be enough of the analyte molecule for the sensor to function. Many assays in biomedical analysis are at low concentration and hence volumes of at least 10 nL are

B. Bhushan (ed.), Tribology Issues and Opportunities in MEMS, 85-94.

required. The issues of control of sample volume without loss due to evaporation and sample reproducibility are important to solve in designing these systems. Miniature fluid handling systems can be applied to a variety of applications. Those in which the fluid is the primary component in microdialysis and drug delivery, for miniature hydraulics for the transmission of power, and in precision manufacturing for dispensing fluids in a controlled manner. The range of chemical analysis systems that can be miniaturized is likewise very broad.

The purpose of this paper is not to provide a comprehensive survey of the field, but to highlight some of the work in our laboratory in the application of microfabrication technologies to biosensors and fluidic systems. The pioneering work of Terry et al. (1979) in gas chromatography led the way to work on other analytical techniques including electrophoresis Manz (1995), Harrison (1993), Ramsey (1995) both with optical (Harrison and Glavina, 1993) and electrochemical detection (Hogan and Lunte 1994). More recently HPLC has been addressed and a great deal of effort has also gone into the analysis of DNA molecules specifically oglionucleotides for sequencing by hybridization (Woudenberg et al., 1996) with fluorescence detection (Lipshutz et al, 1995) in addition to other techniques.

2. Microfabrication Techniques

Microfabrication is a process of etching, patterning and layering materials into microstructures for miniature mechanical devices, optical devices, bio- or chemical sensors and electromechanical systems. Micromachining is realized by either a bulk etching process, surface micromachining process, or the LIGA process which involves electroplating into a thick polyimide or PMMA mold. An excellent review of these techniques is given in the recent book by Madou (1997).

There are two important processes that distinguish microfabrication from IC fabrication: (1) the broad range of materials that are used including materials that have improved structural properties and catalytic metals for chemical sensors and (2) the use of the z-direction in designing structures. In particular, bonding of these layered substrates together is a key process for the fabrication of three dimensional structures. Multilayers of material are deposited for surface micromachined polysilicon devices (Ristic 1994) and for out of plane actuation and assembly (Yeh et al. 1996). The use of polishing for planarization has become a key process in the deposition of polysilicon layers and is indispensable due to its ability to produce extremely smooth surfaces (Sniegowski and Garcia 1996). The range of materials and processes continues to expand the horizon for useful applications of this technology. For example in chemical analysis systems, the use of polyimide (Beebe et al. 1995) and molding process (Olsson et al. 1996) have demonstrated improved chemical compatibility and much lower cost.

A review of microfabricated sensors is given in Sze (1994) and Gardener (1994). There are several examples of clear successes in the application of microfabrication in particular accelerometers, pressure sensors, and gas flow sensors. In addition to chemical sensors for pH and hydrogen. There are increased challenges in building reliable miniature chemical sensors related to keeping the electronics dry while exposing the sensing interface to the analyte often in an aqueous environment. Clearly miniature chemical analysis systems have to deal with many similar issues and there is a need for new materials with improved compatibility such as plastics, glasses and ceramics. For new materials, the issues of dimensional control and applicability of batch processing are important. In some cases compatibility with silicon processing is required for integrated electronics. This however is not always necessary, and careful consideration should be given to design of the sensing system, i.e. how much of the device is fabricated in silicon.

2.1 MICROCHAMBER FOR DNA

The objective of this study is to develop a miniature reaction chamber for DNA sequencing by hybridization. The gel-immobilized oligonucleotides are defined in arrays on the sequencing Microchips for SHOM (sequencing by hybridization with oligonucleotides matrix) (Yershov et al., 1996). The patterned gel micromatrix is loaded with microvolumes of oligonucleotide solution and the oligonucleotides are immobilized on the matrix. The micromachined chamber is a microdevice which is designed to enclose the acrylamide gel array coated substrates and to control the distribution of small liquid samples over the array. Figure 1 shows a diagram of the proposed 5 µL chamber. The fabrication sequence for the microchambers includes anisotropic wet-chemical etching of (100) silicon followed by anodic bonding to a pyrex glass cover plate.

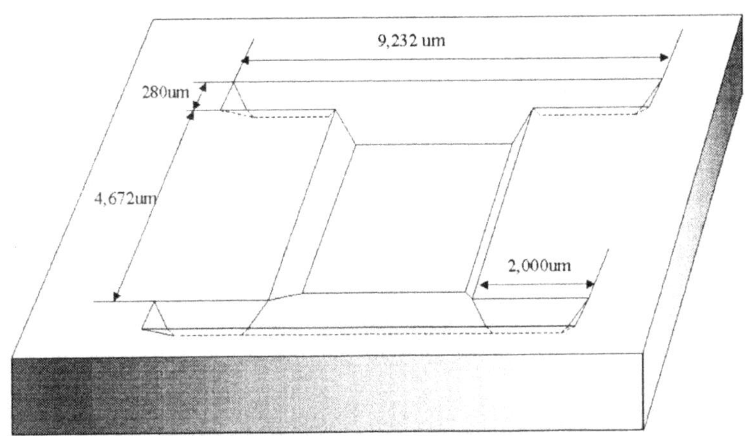

Figure 1: Diagram of microchamber for reaction with DNA.

During the anisotropic etching process, high index planes are exposed at the junction of the channel and chamber which etch faster than the (100) plane. Therefore, during anisotropic etching convex corner compensation is required in the form of a beam added at the convex corners which are removed during etching (Shih et al, 1997).

3. Biosensors

There has been a great deal of effort in the development of biosensors for a variety of analytes over the past twenty years (Turner et al, 1987). The heart of the biosensor comprises two major components: a biological element that will be selective to the analyte of interest, and a transducer that will detect the binding or reaction products formed (Figure 2). Much of the work on microfabricated biosensors has been directed to enzyme based systems and blood gas (O_2, CO_2, pH) measurements (Hesketh et al., 1989). We have developed a novel glucose sensor based on an impedance method (Kassapbasioglu et al. 1993). Our recent work has been directed at utilizing impedance methods for the measurement of antibody/antigen(Ab/Ag) binding (DeSilva et al, 1995). A key advantage of this method over traditional enzyme-, flurochrome-, or radiolabeling methods is that it alleviates the need for an Ab or Ag label and the multiple wash steps necessary in most immunoassays.

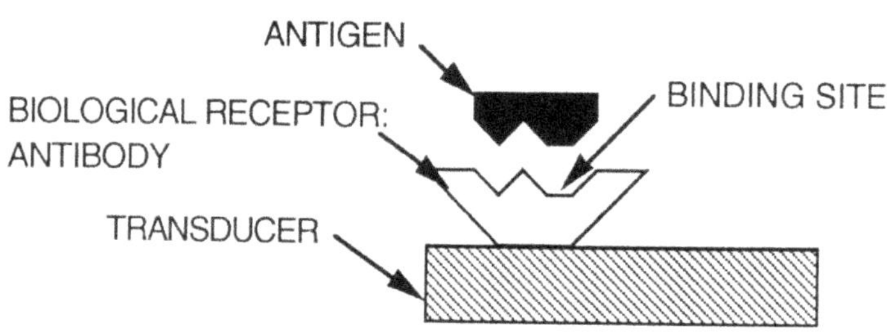

Figure 2: Schematic of a biosensor interface.

3.1 SENSOR FABRICATION

The sensor comprises an ultrathin 25 Å platinum film with an immobilized layer of staphylococcal enterotoxin B (SEB) antibody. The ultrathin metal film has a high resistivity and is characterized by an interconnected metal island structure, as evidenced by atomic force microscope (AFM) see Figure 3.

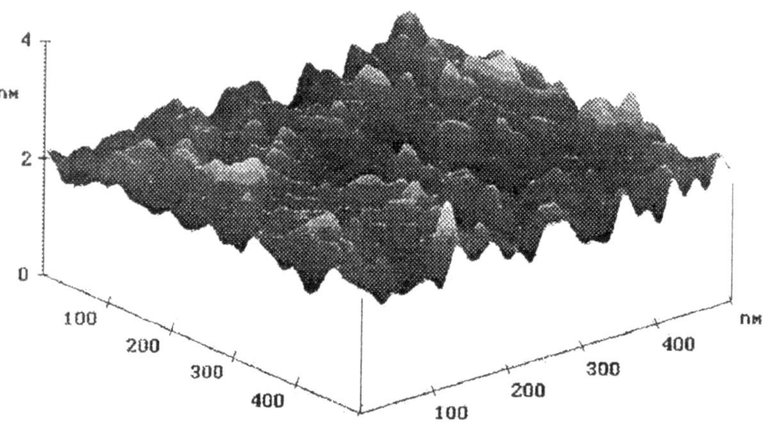

Figure 3: AFM image of the sputtered Pt thin film taken with a Nanoscope-III in tapping mode.

The impedance responds specifically to SEB antigen in the concentration range 0.4 to 10 ng/mL (DeSilva et al, 1995). We have focused on determining the mechanism of the sensor response. A decrease in film impedance was observed after anti-SEB immobilization and during titrations with 50 ng/mL addition of SEB in phosphate buffered saline (PBS) (Zivanovic et al, 1996). Experiments with radiolabelled anti-SEB indicate that there was 2 pmoles of anti-SEB on the electrode surface, approximately one monolayer (Hesketh et al, 1997).

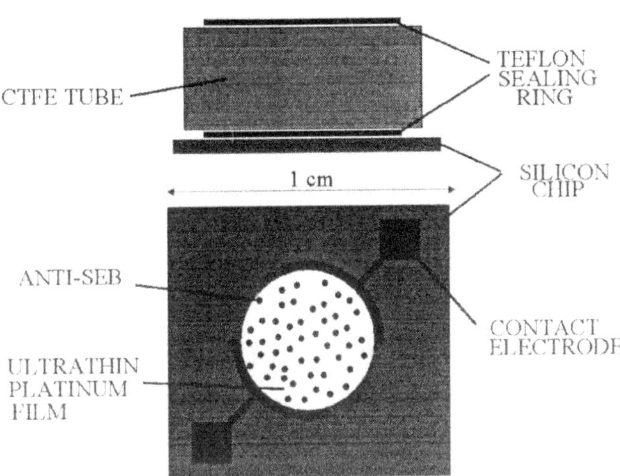

Figure 4: Schematic diagram of the platinum ultrathin film sensor in the measurement cell.

Details of the sensor fabrication process was described previously (Zivanovic et al, 1996). Two contact pads of 500 Å gold with 50 Å titanium were deposited by electron beam evaporation and patterned by a lift-off process. The wafer was extensively cleaned and annealed in the vacuum chamber at 350 °C before the ultrathin Pt/Ti film was RF sputtered to a thickness of 25 Å at 28 W power at a pressure of 5.5mTorr Argon at room temperature. Electrical connections were made to the Au contacts with Au wires and silver epoxy cured at 70°C for 3 hours. The surface of sensors was treated with first sulfuric acid and deionized water at 70°C. These sensors are vapor phase treated in 3-GOP (3-glycidoxypropyldimethylethoxy silane) by putting into an oven at 50 °C and baked for 30-60 min. The sensors, after vapor treatment, are rinsed in 10mM mercaptoethanol and stored in mercaptoethanol before anti-SEB (Sigma Corp., St. Louis) immobilization.

3.2 SENSOR CHARACTERISTICS

The sensor was placed into a Kel-F measurement cell as shown in Figure 4. An area of 0.5 cm^2 of the ultrathin film is exposed to the PBS solution. Immobilization is at room temperature for 1 to 2 hours at a concentration of 0.73mg/ml anti-SEB. Complex impedance was measured with an HP4284 LCR meter over the frequency range of 20 Hz to 1 MHz using an AC drive voltage of 10 mV rms. The sensor impedance is stabilized in PBS for a few hours. Measurements at 10kHz indicate impedance increases as the anti-SEB is immobilized onto the electrode surface (Figure 5). The next step is to proceed with the anti-SEB bonding characterization for which some preliminary results have been published (DeSilva et al, 1995).

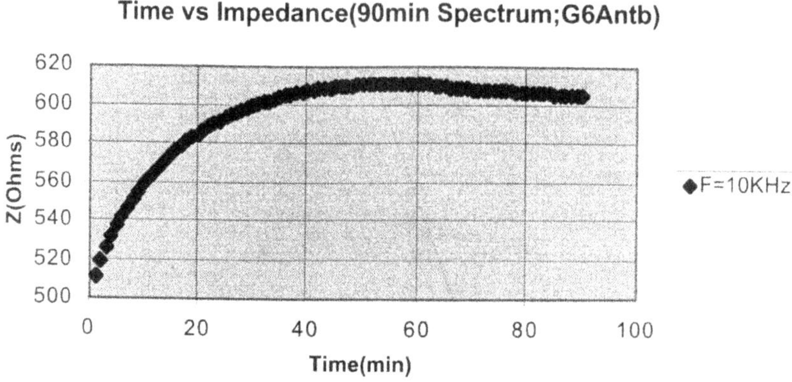

Figure 5: Impedance versus time for the immobilization of the anti-SEB in PBS.

4. Micropump

Several micropumps have been built with MEMS technology which are small, light-weight, and low-power. There are numerous applications that can benefit from this technology and our focus at present is in microdialysis sampling. Microdialysis as an *in vivo* sampling technique that permits continuous sampling of drugs and endogenous compounds in awake, freely moving animals, and is accomplished by implantation of a small probe into the tissue of interest. This probe, which is composed of a short length of dialysis tubing, is perfused with an isotonic saline solution. The use of a precisely controlled flow rate allows chemicals to be predictably removed from or introduced into the extracellular space by establishment of a steady-state flux across the dialysis membrane. Pumps have been developed for micro-chemical analysis systems (Shoji and Esashi 1994; Rapp et al, 1997). Various pump volumes, flow rates and drive mechanisms have been investigated including electrostatic (Zengerle and Geiger, et al. 1995), magnetic (Zhang and Ahn, 1996), piezoelectric (Zengerle and Ulrich, et al. 1995) and pneumatic (Rapp et al, 1994).

4.1 FABRICATION PROCESS

Details of the fabrication process of the positive displacement micropump have been published previously (Lin et al., 1994). The positive displacement pump consists of a silicon component into which the fluid channels have been etched anodically bonded to a pyrex glass wafer. A photograph of the micropump without fluid pipes connected is shown in Figure 6. The flexible membrane in the pump is polyimide. The pump also includes two passive check valves made of polyimide, 10 μm thick. The pump diaphragm is deflected by externally applied compressed air or with a piezoelectric bimorph as shown in Figure 7. The pump chamber volume is 1 μL.

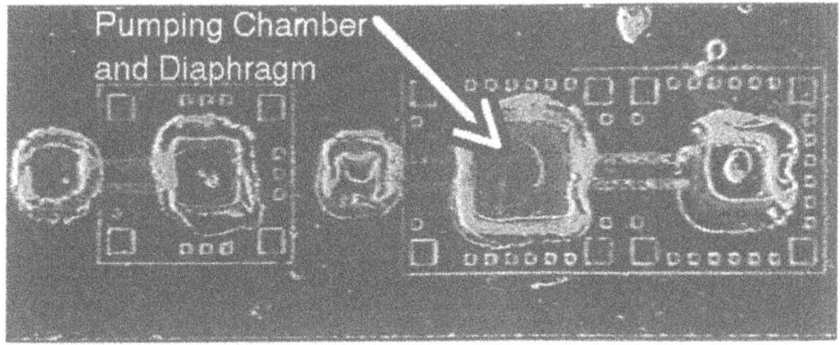

Figure 6: Photograph of a microfabricated pump, dimensions 15 mm x 4 mm.

4.2 MICROPUMP CHARACTERISTICS

The flow rate, maximum pressure and stop flow characteristics of the micropump were determined with DI water as the working solution. Measurements to characterize the passive check valves have indicated that they function very effectively as fluidic diodes (Lin et al, 1996). The diaphragm was driven by an external source of compressed air and vacuum controlled by a two way solenoid valve. Several loads have been connected to the pump including a dialysis probe type CMA/11 (CMA Microdialysis, Acton MA) of external diameter 0.38 mm and the flow rate determined. The weight of water was measurements at fifteen minute intervals.

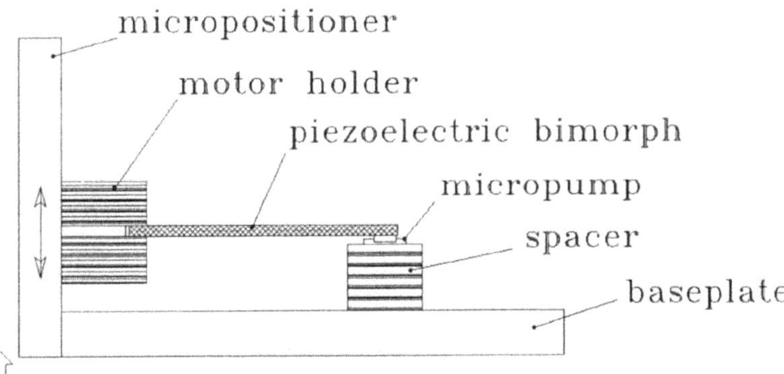

Figure 7: Schematic diagram of piezobimorph attached to micropump

Figure 7 shows a schematic illustration how the piezobimorph is attached to the microfabricated pump. The piezoelectric motor is a bimorph made of two 500 micron thick strips of piezoceramic PZT-5H poled for parallel connection. The depoling electric field for this material is larger than 300 kV/m, the maximum operating voltage is therefore 150 V. The piezoelectric constant d_{31} is equal to -2.74×10^{-10} m/V, giving a maximum longitudinal non-elastic strain of 82.2 μstrain. The motor has been modeled as a bending bimorph according to the Euler-Bernoulli beam theory. The predictions of the model agree within an error of 15% with the measured values. Figure 8 shows the volume flow rate as a function of drive frequency from 2 to 25 Hz. The mass flow rate range was 0.1-11 μL/min. The chamber is therefore not totally emptied each pump cycle. Water may also be lost through the dead volume of the valve and the time interval when reverse flow occurs during the closing of the valve. All of the pumps showed a saturation effect, in which the volume flow rate per pump cycle decreased at higher flow rates. The motor presents very low hysteresis, a virtually unlimited lifetime and a power consumption smaller than 10 mW at an operating voltage of 150 V.

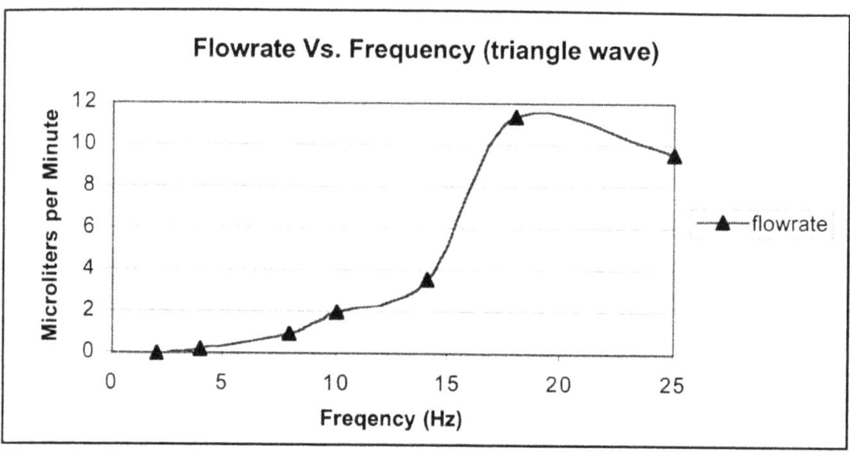

Figure 8: Volume flow rate versus drive frequency with the piezoelectric drive using a triangular waveform of 100 V peak.

5. Conclusions

Examples have been given illustrating the variety of systems to which micromachined elements have a key impact in miniature systems. A wider range of materials and processes are brought to bear for the realization of these systems compatible with IC processing. The integration of these miniature components into microfluidic systems is an area of current research in our laboratory and at other institutions around the world. One of the key issues that has not received enough attention is that of obtaining a fluidic seal between the components. Minimization of deal volume between the various components in the microfluidic system is important to avoid sample dispersion in miniature chemical analysis systems. The electronic revolution brought about by the miniaturization and integration of circuit elements is self evident. The future of miniature chemical and biological analysis systems is rapidly advancing and the future looks very bright for further levels of miniaturization and integration of such devices. Microassembly is an area receiving renewed interest to bring together disparate component sizes and technologies into an integrated system. The future looks very promising for the continued development of products based upon these technologies and devices provided that the issues of tribology, packaging, assembly and manufacturability can be addresses in our research laboratories.

Acknowledgments

The work on micropumps and biosensors is in collaboration with William Penrose at Sensors Solutions, Naperville, IL. The support of the Whitaker Foundation is acknowledged for the immunobiosensor studies. The micropump and miniature analysis systems work was supported by the Center for Bioanalytical Research at the University of Kansas. Technical assistance of Tony Cocco and Professor Jordan Maclay is gratefully acknowledged.

References

Beebe, D.J. Hsieh, A.S., Denton, D.D., Radwin, R.G., (1995) A Silicon Force Sensor For Robotics And Medicine, *Sensors andActuators A* **50**, 55-65.

Feng, Chang-Dong, Nelson, T. E., Hardman, S., Hesketh, P. J., Maclay, G. J., Gendel, S. M. and J. R. Stetter, (1995) Impedance analysis of ultrathin platinum film immunosensors with different thickness and macrogeometry, in *Proceedings of Transducers '95*, Stockholm, Sweden, June.

DeSilva, M., Zhang, Yu, Hesketh, P. J., Maclay, G. J., Gendel, S. and J. R. Stetter, (1995) A Novel Biosensor for Staphylococcal Enterotoxin," *Biosensors & Bioelectronics*. **10**, 675-682.

Gardner, J.W. (1994), *Microsensors: Principles and Applications*, (Wiley, Chichester, England).

Haririson, D.J., Fluri, K., Seiler, K., Fan, Z., Effenhauser, C.S. and A. Manz, (1993) Micromachining a miniaturized capillary electrophoresis based chemical analysis system on a chip, *Science* **261**, 895.

Harrison, D. J. and P. G. Glavina, (1993) Towards Miniaturized Electrophoresis and Chemical Analysis Systems on Silicon: an Alternative to Chemical Sensors, *Sensors and Actuators B* **10**, 107-116.

Hesketh, P., Madou, M. , Otagawa, T. , Joseph, J. and A. Saaman, (1989) "A Microelectrochemical Blood Gas Sensing Probe," Electrochemical Society 175th Meeting, Los Angeles, CA, May, pp. .

Hesketh, P. J., Zivanovic, S., Ming, Y., Pak, S., Svojanovsky, S., Cunneen, J. C., Caraffini, S., Boyd, J.G., Stetter, J.R., Lunte, S.M. and G.S. Wilson, (1997) Microfabricated Biosensors and Microsystems, in *Proceedings of the 21st International Conference on Microelectronics, (MIEL'97)* Niš, Yugoslavia, (14-17 September) pp.63-69.

Hogan B. L. and S. M. Lunte (1994) On-line Coupling of In Vivo Microdialysis Sampling with Capillary Electrophoresis, *Analytical-Chemistry* **66**, 596-602.

Kasapbasioglu, B., Hesketh, P. J., Hanly, W. C., Maclay, G. J. and R. N. Esfahani, (1993) A Novel Ultra-Thin Film Glucose Sensor, *Sensors and Actuators* **14**, 749-751.

Lipshutz, R.J., Morris, D., Chee, M., Hubbell, E., Kozal, M.J., Shen, N., Yang, R. and S.P.A. Fodor, (1995) Using Oglionucleotide Probe Arrays to Access Genetic Diversity, *BioTechniques* **19**, 442-447.

Lin, Y.-Cheng., Hesketh, P.J. , Lunte, S.M. and G.S. Wilson, (1995) A micromachined micropump, in *Microstructures and Microfabricated Systems - II* Editors: P.J. Hesketh, D.D. Denton and H.Hughes, Hardbound Proceedings of the ECS, Vol. **95-27** (The Electrochemical Society, Pennington, New Jersey) pp. 67-72.

Lin, Y.-C., Hesketh, P.J., Boyd, J., Lunte, S.M. and G.S. Wilson, (1996) Characteristics of a polyimide microvalve, Sensors and Actuator Workshop, Hilton Head, South Carolina, June.

Manz, A., Verpoorte, E., Raymond, D.E., Effenhauser, C.S., Burggraf, N., Widmer, H.M. (1995) µ-TAS - Miniaturized total chemical analysis systems, in *Micro Total Analysis Systems,* A. van den Berg, P. Bergveld, eds., MESA Monograph, Vol., 1, Kluwer Academic Publishers, Dordrecht, pp. 2-5.

Madou, M. (1997), *Microfabrication*, (CRC Press, Boca Raton)

Olsson, A., Larsson, O., Holm, J., Lundbladh, L., Öhman, O. and G. Stemme, (1996) Valve-less diffuser micropumps fabricated using thermoplastic replication, in *Proceedings of the IEEETenth Annual International Workshop on Micro Electro Mechanical Systems,* Nagoya, Japan, January. pp. 305-310.

Ramsey, J.M., Jacobson, S.C. and K.R. Knapp, (1995), Microfabricated chemical measurement systems, *Nature Medicine,* **1**, 1093-1096.

Rapp, R., Bley, P., Menz, W. and W.K. Schomburg, (1994) Micropump fabricated with LIGA process, *Proceedings of IEEE International Workshop on MEMS*, Fort Lauderdale, FL, February, p. 123.

Rapp, R. Hoffmann, W., Süss, W., Ache, H.J., and H. Gölz (1997) Performance of an electrochemical microanalysis system, *Electrochimica Acta* **42**, 3391-3398.

Ristic, L. J. (1994) *Sensor Technology and Devices*, Artec House, Boston.

Ruzicka, J. and E. H. Hansen, (1988) *Flow Injection Analysis*, (John Wiley, New York)

Shih, B., St.Clair, L., Hesketh, P.J., Naylor, D.L. and G. M. Yershov, (1997) A Micromachined Chamber to Enclose Acrylamide Gel Array Coated Substrates and to Control and Distribute Small Liquid Samples Over the Array, in *Microstructures and Microfabricated Systems - III*, Hardbound Proceedings of the ECS, PV. **97-5**, (The Electrochemical Society, New Jersey), pp. 94-101.

Shoji, S. and M. Esashi, (1995) Micro Flow Devices and Systems, *J. Micromech. Microeng.* **4**, 157-171.

Sniegowski J.J., Garcia E.J., (1996) Surface-Micromachined Gear Trains Driven by an On-Chip Electrostatic Microengine, *IEEE Electron Device Letters* **17**, 366-368.

Sze, S. M. (1994) Editor, *Semiconductor Sensors*, John Wiley, New York.

Terry, S.C., Jerman, J. H., and J.B. Angell, (1979), A gas chromatographic air analyzer fabricated on a silicon wafer, *IEEE Trans. Elect. Dev.* **26**, 1880-1886.

Turner, A. P. F., Karube, I. and G. S. Wilson, (1987) *Biosensors - Fundamentals and Applications*, (Oxford Scientific Press, Oxford, UK),

Woudenberg, T.M., Winn-Deen, E.S. and M. Albin (1996) High -Density PCR and Beyond, in *Proceedings of the 2nd International Symposium on Miniatureized Total Analysis Systems, μTAS96,* Basel, Switzerland, November, pp. 55-59.

Yeh, R. Kruglick, E.J.J., and K.S.J. Pister (1996) Surface Micromachined Components For Articulated Microrobots, *Journal of Microelectromechanical Systems* **5**, 10-17.

Yershov, G.M., Belgovsky, A.I., Drobyshev, L.D., Sushkov, V.N., Mologina, N.V., Guschih, D., Steele, J., Femmel, A., Zaslavsky, A., Naylor, D. and A.D.Mirzabekov, (1996) Robot for manufacturing SHOM olignucleotide microchips, in *Technical Abstracts of Workshop for Microfabrication,* San Francisco, CA, March.

Zhang, W., and C.H. Ahn, (1996) A Bi-directional Micropump on a Silicon Wafer, in *Digest of Technical Papers from the Solid-State Sensors and Actuators Workshop,* Hilton Head, South Carolina, June, pp. 94-97.

Zengerle, R., Geiger, W., Richter, M., Ulrich, J., Kluger, S. and A. Richter, (1995) Transient measurements on miniaturized diaphragm pumps in microflow systems, *Sensors and Actuators A* **50**, 557-561.

Zengerle, R., Ulrich, J., Kluger, S., Richter, M. and A. Richter, (1995) A bi-directional silicon micropump, *Sensors and Actuators A* **50**, 81-86.

Zivanovic, S., Feng, C.-D., Ming, Y.-D., Hanly, C.W., Hesketh, P.J., Penrose, W.R. and J.R. Stetter, (1996) Efficient method for screening ultrathin metal sensor with largest biological response, in *Proceedings of 6th International Meeting on Chemical Sensors*, Washington, DC, July, p. 258.

POWER MEMS MATERIALS AND STRUCTURES

S. M. Spearing and K. S. Chen
Massachusetts Institute of Technology
Cambridge, MA 02139

Abstract

The favorable scaling of the strength of brittle materials at small scales has the potential to enable a new class of devices: *power MEMS*. In this paper the effect of scale on the structural design of such devices is discussed with particular reference to the MIT microengine project. The initial goal of this project is to produce a turbine generator by deep reactive ion etching single crystal silicon. The design and fabrication of such a device offers significant challenges and opportunities. The major structural challenges arise from the very high stress levels (~ 1 GPa) required to achieve acceptable turbomachinery performance and the associated need to measure accurately the material properties. The task is further complicated by the need to achieve a good structural design within the constraints imposed by microfabrication processes. The major opportunities arise from the use of silicon at very small lengthscales. In particular, the use of microfabrication techniques offers the potential to control the processing-induced flaw size such that very high strengths can be obtained.

Keywords: Turbomachinery, ceramic, silicon, lengthscales, MEMS

1. Introduction

The favorable scaling of the mechanical properties of materials at the microscale is potentially enabling for a new class of *power MEMS* (microelectromechanical systems). These devices would be characterized by thermal, electrical and mechanical power densities equivalent to the best large-sized machines produced today. Specific applications could include portable electrical power production, localized cooling and propulsion units. One example of such a device is the "Microengine" currently under development at MIT [1-3]. This is designed to produce between ten and one hundred watts of electrical power in a package occupying less than one cubic centimeter, while consuming seven to eight grams of hydrocarbon fuel per hour. Such a device would represent a major advance in portable electrical power sources, with the potential to achieve more than ten times the power and energy density of current batteries at competitive costs.

The concept on which this effort is based is the ability to micromachine refractory, structural ceramics such as silicon nitride and silicon carbide. Much work has been done on optimizing processes for conventional semiconductor materials, such as silicon. Although they are not as mature, analogous processes

B. Bhushan (ed.), Tribology Issues and Opportunities in MEMS, 95-107.

are being developed for silicon carbide [4] and silicon nitride, as well as methods for producing single crystal wafers of these materials. SiC and Si_3N_4 potentially have excellent mechanical, thermal and chemical properties for gas turbine applications, in principle they would permit uncooled operation in the range of 1500-1700 K. For many years efforts have been underway to introduce these materials into macroscale engines however, they have generally failed to realize their potential, largely due to issues stemming from their inherently low toughness [5]. In the context of performing the preliminary structural design of the microengine it is interesting to understand the differences between attempts to realize large scale ceramic turbomachinery and microengines, and why in some cases the scaling of material behavior leads to favorable performance on the microscale.

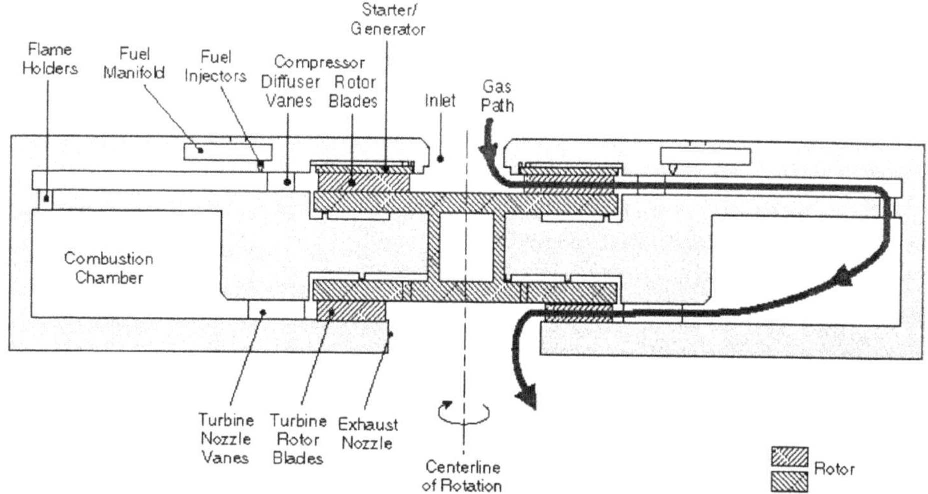

Figure 1 Schematic of a micro gas turbine engine cross-section [1].

In order to appreciate the effects of material scaling on structural design it is important to have a basic understanding of the functional requirements of a microengine. The most challenging component of the device from a structural viewpoint is the turbine rotor. A schematic of the turbine is shown in Fig. 1, and a micrograph showing a stator/rotor pair etched from silicon is shown in Fig. 2. The geometry of the device is quite different from conventional turbomachinery, as it is largely determined by the capabilities of the microfabrication processes. The use of etching and deposition processes in conjunction with lithographically produced masks limits the components to planar, or cylindrical geometries.

The demands of efficient turbomachinery operation drive the design toward operating at the maximum possible temperature and the maximum possible tip speed of the blades. The stress levels scale with the square of the speed and linearly with the density of the material [4]. Thus the strength at temperature of the material defines the performance of gas turbine engines. This consideration is common to all turbomachinery irrespective of lengthscale.

Figure 2. Scanning Electron Micrograph of a turbine rotor and stator etched in silicon. (Courtesy of Mr. Chuang-Chia Lin and Prof. M. Schmidt)

The initial focus of the microengine program is to design and fabricate a turbine-generator from silicon, for which fabrication processes have already been demonstrated. Concurrently development is underway of processes suitable for more refractory materials. This paper describes the progress to date on the structural design and associated material testing for the silicon turbogenerator.

2. Microfabrication

The heart of the fabrication process used to create the microengine structures, such as the rotor shown in Fig 2., is the deep reactive ion etching process. This can produce very high-aspect ratio, parallel-sided features with dimensions, in the etch direction, of several hundred micrometers. This capability is essential for achieving the necessary geometries for the aerodynamic requirements of the blades and the close tolerances required to obtain effective bearings.

The material scale which governs the fabrication process is a timescale rather than a lengthscale. The rates of etching and deposition processes are controlled by chemical reaction rates and mass diffusion rates. These processes act on the surface of the part being fabricated. This introduces a cube-square scaling law, i.e. the volume of material to be removed to achieve a particular shape scales with the linear dimensions raised to the third power, but the volume of material that can be removed in a given time scales with the linear dimensions to the second power (i.e. the surface area). The characteristic time to etch a particular shape out of a blank is, therefore, proportional to the linear dimensions of the part, which imposes an economic limit on the use of microfabrication to devices below a certain lengthscale. An important consequence of employing microfabrication is that it can produce a rotor consisting of a single monolithic piece of ceramic.

This contrasts with the macroscale where joining components, such as blades, leads to a significant reduction in reliability, as well as being a source of increased cost [7]

The fabrication processes are important in ensuring the feasibility of microturbomachinery. The characteristic timescales of these processes and the resulting lengthscales are responsible for the difference between fabrication at the micro- and macroscales. As will become apparent in the next sections, the use of single crystals and microfabrication processes play a key role in achieving high strengths at the microscale.

3. The Scaling of Material Properties

The role of the structural dimensions in determining the strength of brittle materials was recognized in the first half of the twentieth century [8,9]. Many experimental studies have shown that small specimens, on average, exhibit higher strengths than larger ones. It is this observation that has led to the development of very high strength fibers for use in structural composites at the macroscale. In the particular case of MEMS there are two factors which contribute to a significant strength enhancement at small lengthscales; firstly, the processing route and secondly the statistics of the strength controlling flaw population.

The use of microfabrication processes on single crystal materials results in very high quality surfaces on the finished structure. The reduction in the mean size of surface flaws can result in very high material strengths [10], significantly higher than those usually found in polycrystalline ceramics at the macroscale. The second source of strength enhancement is the statistics of the flaw population. Processing-induced flaw sizes and locations are random variables which combine with a deterministic value of the fracture toughness to make strength a stochastic quantity. Use of the empirically-determined Weibull probability density function [9, 11] allows a comparison of the ratio of the characteristic stresses (s_1/s_2) to give equal probability of failure for two volumes (V_1,V_2) with geometrically similar stress distributions:

$$(\sigma_1 / \sigma_2) = (V_2 / V_1)^{\frac{1}{m}} \tag{1}$$

where m is the Weibull modulus. Thus, when comparing micro- and macroscale devices which have characteristic dimensions of order 1mm and 0.1m respectively, the volume ratio is 1×10^6. A value of m of twelve is typical for sintered ceramics [12] and this gives an expected strength ratio of about three between the micro and macroscale. A caveat needs to be placed on this argument since extrapolation to this degree cannot usually be justified. Using a Weibull distribution with the same fitting parameters (reference strength s_0 and modulus m) assumes that the flaw population is the same between the two cases, which is unlikely to be the case since the processing routes are different at the micro and macro scales. Nevertheless, the comparison gives some indication of the grounds for expecting an improved mechanical performance at the microscale.

The significance of the increased strength achievable at the microscale is that it permits higher operating speeds, and hence improved specific performance of the engine. The stress levels in rotating components scale with the density and the square of the peripheral speed. For identical geometries operating at the same speed, the stress levels will be identical, regardless of scale. It follows, therefore, that microengines should be capable of significantly higher rotor speeds than their macroscale equivalents.

An important additional consideration is the scaling of impact loading. The low toughness of ceramics make this a key concern at the macro scale. The stress generated by an impact event is independent of the lengthscale. However, for a brittle material failure is due to the force of the impact event generating cracks, which reduce the strength of the material below the operating stress levels. The force generated by geometrically similar objects colliding at a given speed scales with the square of the linear dimensions. For damage introduced by a sharp indenter the residual strength can be shown to depend on the reciprocal of the cube root of the indentation load [13]. This scaling, therefore, significantly reduces the concern regarding impact events at the microscale compared to the macroscale.

Not all of the consequences of the small lengthscale are to increase the blade speed. As previously noted, the constraints of the microfabrication processes make it relatively difficult to achieve three dimensional structures. As a consequence, stress relieving strategies such as tapering the turbine disk thickness and blade cross sections are not as readily achieved at the microscale.

For the design of high temperature turbomachinery the scaling of material behavior also has important implications. The high surface area to volume ratio causes surface heat transfer to dominate, and results in nearly isothermal structures. Thermal shock, a key issue for ceramics at the macroscale, becomes much less of a concern at the microscale. In contrast, the cube-square scaling increases the potential sensitivity to oxidation at the microscale, even for ceramics such as SiC that are relatively insensitive at the macroscale. Detailed analysis and preliminary material testing in the engine combustion enviroment suggests that the overall effect of the scaling of materials and structural performance is to permit the design of high performance micro turbomachinery. A summary of the effects of scaling on the material selection and structural design is shown in Table 1.

Table 1. The effect of lengthscale on structural design

Property	Effect when Lengthscale Decreases	Comments
Mechanical Strength	Increases	Permits increased operating speed
Structural Stress	Unchanged	Scales with density and $(\text{velocity})^2$
Heat Transfer Rate	Increases	Structures become nearly isothermal
Thermal Shock Resistance	Increases	Allows consideration of materials which are not feasible at the macroscale
Resonant frequencies	Increase	But operating frequencies also increase
Oxidation resistance	Decreases	Due to surface area to volume ration
Structural impact tolerance	Increases	Scaling of fracture behavior
Creep Resistance	Unchanged	So long as structural scale significantly greater than material microstructure

4. Material Strength Testing

In order to proceed with the detailed design of the highly stressed rotating components an accurate characterization of material strength must be obtained. Since the surface flaw population depends on the micromachining process, it is important to obtain strength data from specimens produced by the same fabrication route. Furthermore the volumetric scaling of material strength necessitates the use of specimens with a similar size to the rotor in order to avoid excessive extrapolation of the test data.

Two specimen types were employed. Biaxial flexure specimens were used to obtain the strength of uniformly etched flat surfaces. Radiused hub flexure specimens were used to probe the local strength of material at stress concentrations such as at the roots of blades or the central hub.

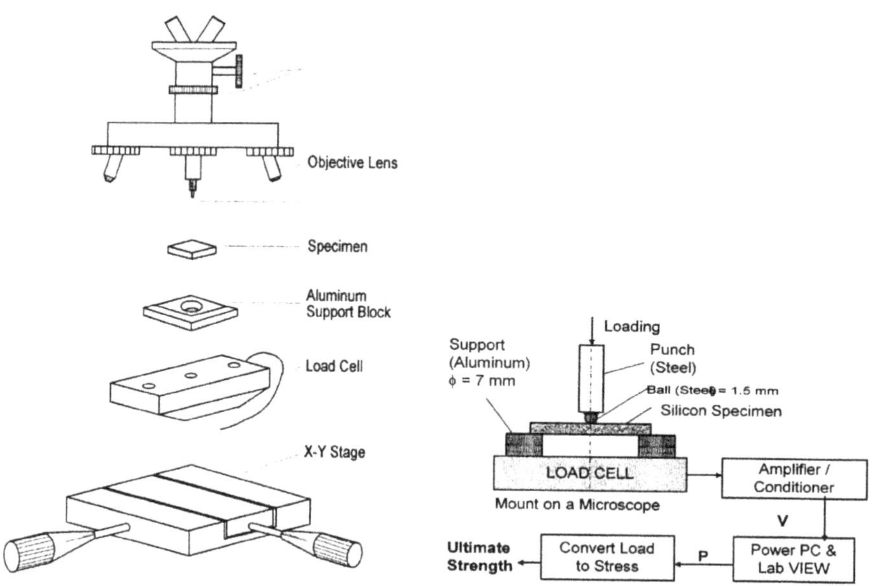

Figure 3. (a) Schematic of strength testing apparatus. (b) Flow chart of testing procedure.

BIAXIAL FLEXURE TESTING

Figure 3 shows a schematic of the apparatus used to determine the material strength. 10 mm by 10 mm squares of silicon wafer (250 - 300 mm thick) were cut using a dicing saw from uniformly etched silicon wafers. The specimens were placed over a circular hole (7mm in diameter) and loaded to fracture using a spherical steel indentor head. A 150 N capacity load cell was used to measure the applied force. The test rig was mounted under a microscope fitted with an X-Y stage. The advantage of this test method is that incidental edge damage to the specimen would not contribute to the failure stress. The microscope and the micrometers on the X-Y stage permitted accurate positioning of the indentor on the specimen. A finite element analysis has been used to generate a calibration curve for the relationship between fracture load and material strength. In addition the sensitivity to positioning errors, contact friction and indentor deformation have been examined [14]. The controlled surface of the specimen is placed facing away from the indentor in order to load this surface in tension.

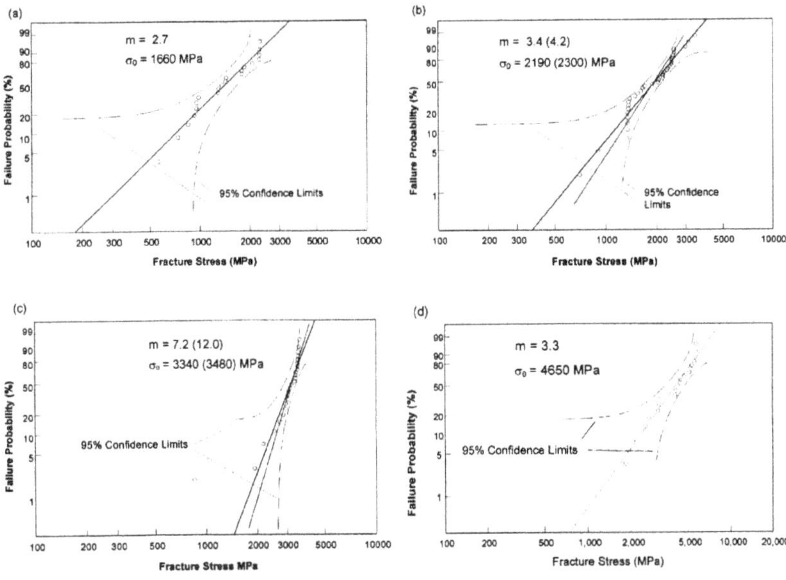

Figure 4. Weibull strength plots of silicon specimens with four surface roughness values (see Table 2)

Five sets of silicon specimens with different surface qualities have been tested. Surface roughness was measured using a profilometer. Weibull statistics has been used to characterize the strength distribution [9]. Table 2 is summarizes the Weibull parameters. Figure 4 shows Weibull strength plots of four of the sets of specimens. The specimen dimensions were chosen such that the probability of failure at a given load level is comparable to that of the rotor under centrifugal loading. Figure 5 shows the stress ratio between the test specimen and the rotating disk. If a volume dependent Weibull model is assumed, with m=10, to obtain the same probability of failure, the maximum stress in the test specimen should be twice the allowable maximum stress in the rotating disk.

Table 2. Strength characteristics of silicon with five surface conditions.

	Mechanically polished (a)	Mechanically polished (b)	KOH Etched Silcon (c)	Deep Reactive Ion Etched (d)	Chemically Polished
Sample size	19	30	25	20	4
P-P Surface roughness (μm)	~ 3	~ 1	~ 0.3	0.3	~0.1 μm
Reference Strength σ_0 (GPa	1.2	2.2	3.5	4.6	> 4
Weibull modulus m	2.7	3.4-4.2	7.2-12	3.5	?

Clearly the processing route and resulting surface finish have a strong influence on strength. Chemically polished silicon has a better surface roughness (~0.1 μm) and an even higher strength than the four surface conditions listed in Table 2 and shown in Figure 4. The strength exceeds the current load capabilities of the test apparatus (the reference strength, σ_0, is greater than 4 GPa).

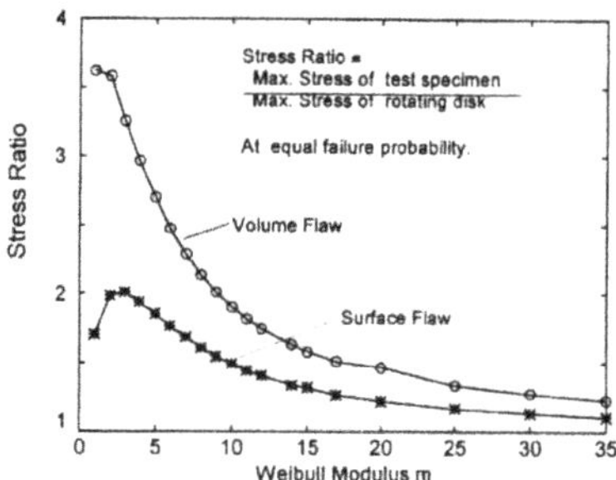

Figure 5. The relationship between the Weibull modulus m and the stress ratio between the mechanical test specimen and a rotating disk.

RADIUSED HUB FLEXURE SPECIMENS

Figure 6 shows a magnified view of the root of the central hub on a turbine rotor such as that shown in figure 2. The horizontal surface of the rotor disk is relatively smooth, the vertical surface of the hub is covered by vertical striations and the transition region at the root is rougher than either. In addition the hub root is a stress concentration, therefore, the reduced local surface quality, and the associated reduction in strength, have potentially serious consequences for the design and operation of the device. To characterize the local strength at such features the radiused hub flexure specimen (RHFS) has been developed [14].

A schematic of the RHF specimen is shown in figure 7. The specimen consists of a square plate of silicon with a central raised hub. The specimen is produced by a timed deep reactive ion etch using a single mask. The specimen dimensions can be varied to replicate the fabrication conditions likely to be encountered in a particular application. For the data presented here the central hub had a height of 200 µm, and the root radius was 15 µm. The load is applied centrally to the flat surface of the specimen, in the same apparatus as used for the biaxial flexure specimen. The resulting stress is obtained via a calibration curve generated by a finite element analysis of the particular specimen geometry [14].

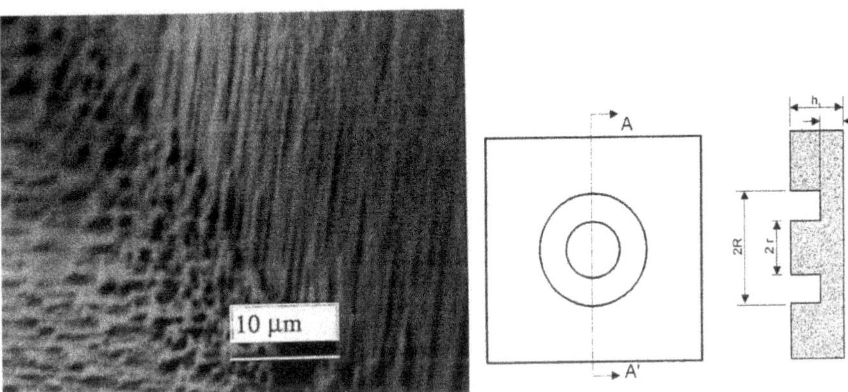

Figure 6. Micrograph of hub-disk transition

Figure 7. Diagram of radiused hub flexure specimen (not to scale)

Table 3 presents test data from RHFS specimens with three surface conditions. The reference strength for specimens fabricated by deep reactive ion etching was 1.45 GPa, which is only 32% of the reference strength of DRIE biaxial flexure specimens. This result confirms the influence of the inferior surface quality at horizontal-vertical transitions. In order to explore approaches to restoring the strength to a value closer to that obtained on flat surfaces RHFS specimens were treated with short secondary etches to attempt to reduce the strength-limiting surface roughness. An aqueous isotropic etch (5 % (by volume) HF, 55% HNO_3 and

40% deionized water) and an isotropic plasma etch (SF_6) were used for this purpose. In each case 2-3 μm of material were removed uniformly from the specimen surfaces. As shown in table 3, the secondary etch resulted in a significant recovery of strength, approaching the value obtained for flat biaxial flexure specimens in the case of the plasma etch. These results indicate the importance of adequately specifying the fabrication route in order to achieve the requisite material properties for power MEMS devices.

Table 3. Strength characteristics of silicon obtained from radiused hub flexure specimens, showing the effects of secondary etching.

	STS DRIE	DRIE + HNO_3/HF Etch	DRIE + SF_6 Plasma Etch
Sample Size	20	16	18
Secondary Etch Depth (μm)	Not Applicable	1.8	2.7
Reference Strength, σ_0 (GPa)	1.45	3.0	4.0
Weibull Modulus	7.5	3.9 - 5.7	3.0 - 6.0

Notwithstanding the promising mechanical properties of microfabricated silicon other factors can contribute to the strength achieved in situ. In particular to the low toughness of implies a low tolerance to any surface damage. Experiments show that the strength of silicon is reduced to the order of 100 MPa after indenting by a Vickers micro indentor at an applied load of only 5 N. It is anticipated that damage introduced by wear processes will be less severe than this, however further experimentation and analysis will be required.

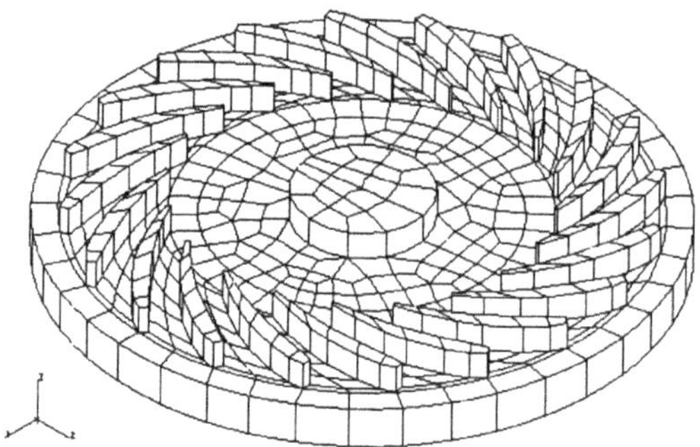

Figure 8. FE Mesh of 3-D turbine rotor model

5. Microengine Structural Design

The structural design tasks associated with the turbine rotor include calculation of the deflections due to the centrifugal loading, the structural dynamic response and stress analysis. Each of these tasks is described in more detail elsewhere [15]. A short outline of the stress analysis and the connection to the mechanical test results described in section 4 are included here.

Analysis of the turbine rotor under centrifugal loading indicates that there are three critical points [16] where the stress levels may be of concern; at the center of the disk, at the transition from the hub to the disk and at the root of the blade trailing edges. The emphasis of the stress analysis task has been to quantify the stress levels at these points and to perform parametric analyses to quantify the potential for design trades in the rotor geometry.

Elasticity solutions indicate that the maximum stress in a flat single crystal silicon disk is 240 MPa for the design speed of 500 m/s [6]. Due to the additional inertia of the blades, the maximum stress at the center of the turbine disk is higher than this value. Finite element calculations, using meshes such as that shown in Figure 8, show the maximum stress at the disk center is approximately 400 MPa.

The second stress-critical location is the fillet radius at the interface between the central hub and the disk. The height of the hub and the fillet radius are critical in defining the magnitude of the stresses at this location. A parametric study has been performed using an axisymmetric finite element analysis. Figure 9 shows these results. The dependence of the maximum tensile stress at the fillet on the fillet radius and hub height is clearly illustrated. The stress level is strongly dependent on the fillet radius. With a 10 mm fillet radius, a maximum tensile stress of 1 GPa will be reached. The stress concentration also depends on the hub height as shown in Figure 9b.

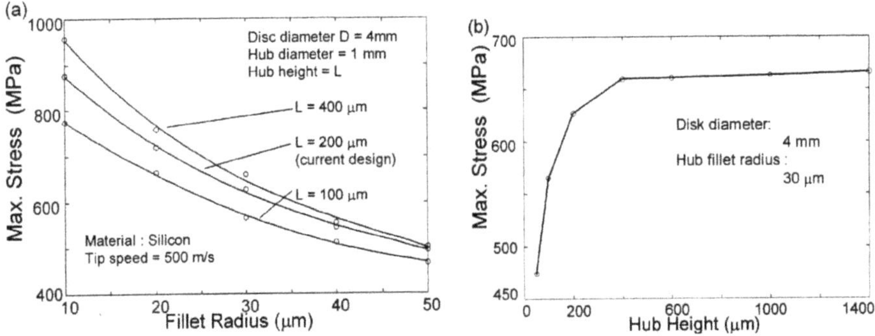

Figure 9. Parametric study of hub root stress concentration vs. (a) hub height and (b) fillet radius

Longer span blades are preferred for turbomachinery design as they can extract more power from the fluid for a given rotational speed. However, the bending stress at the blade root increases with the blade height. The effects of varying the blade height and the blade root fillet radius are shown in Figure 10. The stress level increases from approximately 600 MPa to 1.7 GPa as the height is doubled from 200 μm to 400 μm.

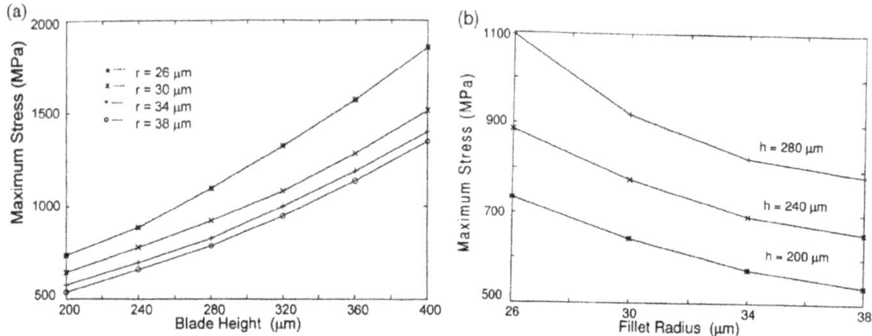

Figure 10. Parametric Study of blade root stress concentration vs. (a) blade height and (b) fillet radius

The stress at all three critical points is relatively high compared with conventional macro-scale turbomachinery. However, the material strength is also very high. Based on the results of section 4, an allowable stress of 1 GPa is a conservative value for structural design purposes which nevertheless permits adequate performance of the turbomachinery. Detailed analysis is underway employing probabilistic methods to properly account for the stochastic nature of material. Achieving adequate fillet radii to limit the stresses to this level is a key fabrication task.

6. Concluding Remarks

Preliminary results have been presented for material testing and structural design of a silicon turbogenerator. The use of microfabrication processes and single crystal silicon results in a favorable scaling of strength at the microscale. The high material strength permits the design of a viable device, despite the restrictions on the geometries that can be achieved by microfabrication, for aerodynamic performance and stress relief. Material and structural scaling has important consequences for the realization of high temperature microengines. Testing and analysis has commenced to understand these issues.

References

1. A. H. Epstein, S. D. Senturia, O. Al-Midani, G. Anathasuresh, A. Ayon, K. Breuer, K-S. Chen, F. E. Ehrich, E. Esteve, L. Frechette, G. Gauba, R. Ghodssi, C. Groshenry, S. Jacobson, J. L. Kerrebrock, J. H. Lang, C-C. Lin, A. London, J. Lopata, A. Mehra, J. O. Mur Miranda, S. Nagle, D. J. Orr, E. Piekos, M. A. Schmidt, G. Shirley, S. M. Spearing, C. S. Tan, Y-S. Tzeng, I. A. Waitz, (1997), Micro-Heat Engines, Gas Turbines, and Rocket Engines - the MIT Microengine Project. AIAA Paper AIAA-97-1773, Presented at 28th AIAA Fluid Dynamics Conference/4th AIAA Shear Flow Control Conference, Snowmass Village, CO, June 29 -July 2 1997.
2. A. H. Epstein, S. D. Senturia, G. Anathasuresh, A. Ayon, K. Breuer, K-S. Chen, F. E. Ehrich, G. Gauba, R. Ghodssi, C. Groshenry, S. Jacobson, J. H. Lang, C-C. Lin, A. Mehra, J. M. Miranda, S. Nagle, D. J. Orr, E. Piekos, M. A. Schmidt, G. Shirley, S. M. Spearing, C. S. Tan, Y-S. Tzeng, I. A. Waitz, (1997) Power MEMS and Microengines, Presented at IEEE Transducers '97 Conference, Chicago, IL.
3. Epstein, A. H., "Micro Gas Turbine Generators", Second Interim Technical Progress Report, ARO 33888CH-MUR, Army Research Office, N.C., 1997
4. Wolf, R. and Helibig, R., "Reactive Ion Etching of 6H-SiC in SF_6/O_2 and CF_4/O_2 with N_2 Additive for Device Fabrication," *J. Electrochem. Soc.*, Vol. 143, pp. 1037-1042
5. Neuberger, M., "Physical and Electronic Properties of SiC", *Mat. Res. Bull.*, Vol 4, pp s365-370, 1969
6. Timoshenko, S. and Goodier, J.: *"Theory of Elasticity*, 3rd edition", McGraw-Hill , New York, 1970.
7. ATTAP : *Advanced Turbine Technology Application Project 1994 Annual Report*
8. Griffith, A. A., "The Phenomena of Rupture and Flow in Solids," *Phil. Trans. Roy. Soc.* A221, 163-198, 1920
9. Weibull, W., "A Statistical Theory of the Strength of Materials," Ing. Vetenskaps Akad. Handl., No. 151, pp. 45, 1939
10. Hu, S. M.: "Critical stress in silicon brittle fracture, and effect of ion implantation and other surface treatments", *J. of Applied Physics*, 53(5), pp. 3576-3580, 1982
11. Ashby M. F. and Jones, D., *Engineering Materials II, An Introduction to Microstructures, Processing and Design*, pp. 169-174, Pergamon Press, Oxford, U.K., 1987.
12. Nakakado, K., Machida, T., Miyata, H., Hisamatsu, T., Mori, N., and Yuri, I.: "Strength Design and Reliability Evaluation of a Hybrid Ceramic Stator Vane for Industrial Gas Turbines", *J. Eng., for Gas Turbine and Power*, Vol. 117, pp 245 -250, 1995
13. Lawn, B. R. Facture of Brittle Solids, Second Edition, Cambridge University Press, 1993, pp. 249-300
14. K-S. Chen, A. Ayon and S. M. Spearing, Controlling and Testing the Fracture Strength of Silicon at the Mesoscale, (1997), Submitted to *J. Am. Ceram. Soc.*
15. K-S. Chen, S. M. Spearing and N. N. Nemeth, Structural Design of a Silicon MicroTurboGenerator, in preparation.
16. S. M. Spearing and K. S. Chen, Micro Gas Turbine Engine Materials and Structures, *Ceramic Engineering and Science Proceedings*, **18** (4), 11-18, 1997

ACKNOWLEDGMENTS

The authors wish to thank the Army Research Office for financial support under ARO Grant DAAH04-95-1-0093, technical manager Dr. Richard Paur. Discussions with colleagues at MIT, in particular, Prof. Alan Epstein are gratefully acknowledged.

The authors would like to thank Dr. Arturo Ayon for his assistance in fabricating the mechanical test specimens.

MEMS OPPORTUNITIES IN ACCELEROMETERS AND GYROS AND THE MICROTRIBOLOGY PROBLEMS LIMITING COMMERCIALIZATION

R.E. SULOUFF
Analog Devices, Inc.
Micromachined Products Division
21 Osborn Street
Cambridge, MA 02139

Abstract

This paper discusses the application of surface micromachined technology to opportunities in accelerometers and gyros. Consideration is from the design and development of two micron polysilicon mechanical structures (LPCVD) that are integrated with a BIMOS process on a single IC chip. The limitations of commercialization due to high static friction (stiction) is discussed especially with consideration of the conflicting manufacturing processes required to build an integrated MEMS device and minimize high static friction. The product trends are to structures that have larger compliance and therefore require "nonsticky surfaces". This is further complicated by device robustness expectations and very low ppm failure rates in the applications. The gyro places additional challenges on the technology with the need for high vibrational Q's achieved by vacuum environments. This paper concludes with identification of critical research issues and how the MEMS industry can be enhanced with a better understanding of Microtribology.

1. Introduction

There has been considerable interest in Micro-Electro Mechanical Systems (MEMS) in the last 5 years due in large part to the rapid growth of accelerometer products and the potential for vibratory rate sensors or gyroscopes. (Walsh, '96) The infrastructure developed by the integrated circuit industry over the last 20 years has provided a low cost means to manufacture electromechanical devices. Using batch fabrication, thou-

B. Bhushan (ed.), Tribology Issues and Opportunities in MEMS, 109-120.

sands of devices can be built on one silicon wafer where typical manufacturing lots exceed 20 wafers. Since motion is a very valuable input to a system doing work, the applications are quite wide and growing as the price per accelerometer drops below the $5.00 level and performance continues to improve. Automotive air bags have helped to drive the demand for such devices where safety and reliability are very important attributes. In addition automotive applications for vehicle dynamic control and navigation are also large potential markets. Gyros for image stabilization of camcorders as well as heading information for virtual reality and numerous mission critical controls for missiles and fuzing can use MEMS devices. It is obvious that system level performance can be degraded or perhaps nonfunctional if these MEMS devices stick or operate intermittently due to surface effects of the mechanical structures.

The question being asked in many laboratories and research groups is if the problems are solvable in a commercial environment and therefore will remain trade secrets and differentiating inventions. Analog Devices has invested in a clear understanding of the factors contributing to released and "nonsticky" devices and has achieved a significant level of success in the production of reliable accelerometers. As with many aspects of surface science and commercial practice, the measurement and control of variables that directly relate and interact together to produce a reliable part with single digit ppm's is a continuing quest. Larger devices with lower stiffness open up new markets and drive for even better nonsticky surfaces. Since an atom layer or two can change the properties of microgram devices that require fractional micronewton forces to come in contact, coatings and tribology are expected to be limiters to commercialization for some time to come.

2. Opportunities In Accelerometers and Gyros

The air bag market has over 20 million axis of acceleration per year of demand and can not tolerate sticking or intermittent failures. The total for navigation as well as other low g applications like computer inputs or seismic detection could account for over 60 million axis. With industrial as well as military missile and fuzing applications, a $450 million accelerometer market can be identified with growth rates greater than 40% possible over the next 5 to 7 years. If the gyro market (Song, '97) is also enabled with MEMS devices an additional billion dollar market is envisioned resulting in a $1.5 billion inertial market by the end of 1999.

(Walsh '96) These projections are consistent with studies performed by the Systems Planning Corporation, and SEMI. (Walsh'96, Seeley '96) If the total of all MEMS devices are considered, the revenue exceeds \$6 billion per year of devices which further enables \$30 billion to \$40 billion of system level business.

The application of accelerometers defines the full scale range of the device. Sensors for automotive air bags measure acceleration from a few tenths of a g ($9.8 m/sec^2 = 1g$) up to 300g's. Navigation and vehicle dynamics have full scale ranges of a few g's and rate sensor applications from 200 degrees per second down to a few tenths of a degree per second. Likewise computer inputs like a 3D mouse, virtual reality and earthquake detection have a similar range of a few g's. Missile controls and fuzing can span a much larger range from 40 g's up to 100,000 g's. The low g accelerometers can be over-ranged with handling shock and result in mechanical elements being thrust into contact. This creates the potential for adhesion and failure. The higher g devices many times operate in higher shock environments and therefore can have adhesion failures in operation.

3. Integrated Surface Micromachined Technology

The implementation of micromachined structures integrated with electronics on the same silicon chip (Chau, '96; Core, '93) is the technology used by Analog Devices (figure 1). In this approach, electronics are processed as BIMOS (bipolar and MOS) with the mechanical structure inserted in the middle of the fabrication process prior to the metalization and passivation. When the layers are completed, the sacrificial layer is etched resulting in mechanically free structures.

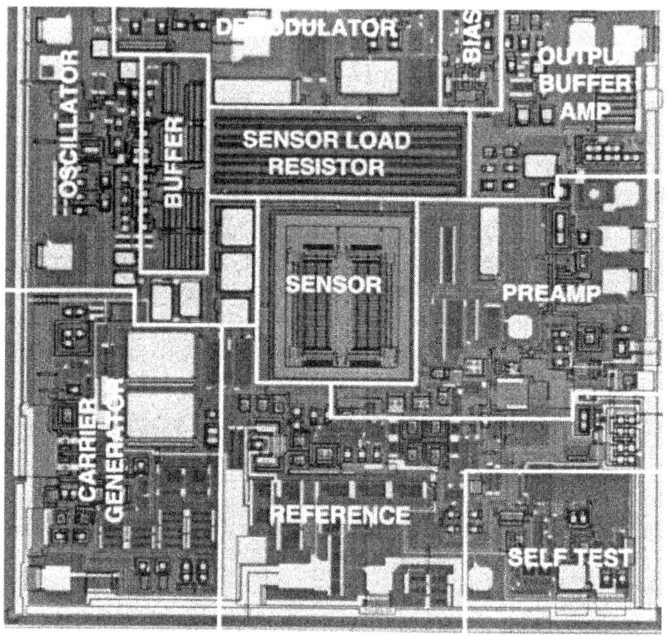

Figure 1. Integrated surface micromchined accelerometer die

Figure 2 shows the buildup and etching away of the structures. The mechanical structures are 2 micron thick polysilicon and are spaced 1.6 microns above the substrate surface. To eliminate the surface tension effects that occur when the wet etch of the sacrificial oxide releases the mechanical polysilicon, supports of organic material are built under the structure.(Core,'94) Therefore, after the wet etching, the structure is held up off the surface by these pillars. The removal of the photoresist standoffs then occurs with a plasma oxide dry etch. This method places significant limitations on the surface treatment options for adhesion control. Coatings that are applied with liquid solutions must either be able to survive the oxygen plasma or provide no surface tension that will adhere the beams. The integration of IC processes are thus an important consideration when addressing tribology solutions. (Maboudian, '97) In IC processing, efforts are made many times to increase adhesion of films and layers. This can have undesirable effects on the micromachined structures.

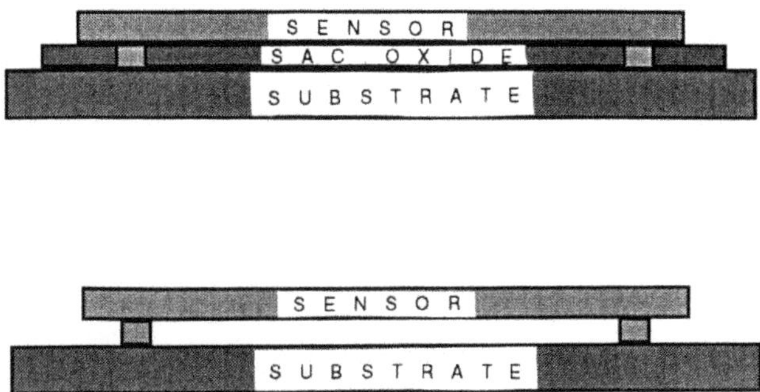

Figure 2. Selective etching of a polysilicon structure to produce a surface micromachined device

There are additional means of minimizing the contact area by fashioning bumps on the surface or building stops into the moving mechanical structures. These structures are visible in figure 3 which show the spring and mass system and the corresponding capacitive structure used to measure displacement.

The packaging of these devices can provide many opportunities for changing of the surface properties and forcing of the moving structure into contact. The devices are manufactured as 150mm wafers that contain over 4400 sensors per wafer. The protection of the mechanical structures in wafer form is accomplished with layers of plastic film that create a cavity around the structure. The wafer is then san from the backside and pushed off the film and placed in the package. (Roberts, '94)

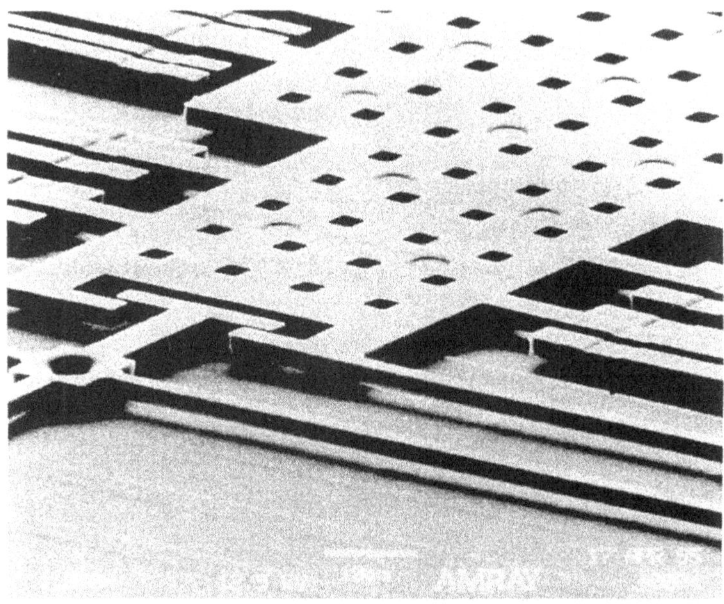

Figure 3. Accelerometer spring and mass system with stops and bumps to minimize area

There are two types of packages used by Analog Devices to hermetically seal the MEMS devices. The metal can with glass to metal seals as shown in figure 4 is sealed at room ambient temperatures after a dry nitrogen bake.

The second type of package is a ceramic package with glass seal of the ceramic and leads. This second type is often called a cerdip package. The high temperature sealing (430°C) and oxygen atmosphere of the cerdip package can place additional difficult requirements on the coatings that go through the process. In the case of the gyro, (Juneau, '97) a package that has a vacuum rather than an atmospheric internal pressure is used. The coatings and surface finishes that are used for these types of devices

Figure 4. Hermetic package, metal can with glass to metal seals.

require stability in a vacuum and low out-gassing to ensure that the vacuum is stable. The coatings must also perform reliably over time that could exceed 10 years in some applications. The materials used for die attach and other typical packaging materials such as lead finishes can not be a source of atomic level coatings. Manufacturing processes in the IC industry have been optimized to not show a sensitivity to coatings by using passivation layers and ultrasonic bonding for bonding through surfaces.

When the coatings are present, it is possible that electrical properties of exposed high electrical impedance mechanical elements can contribute to instabilities due to charge in the coatings or time dependent changes in conductivity. The hermetic package is also sealed in low moisture ambients (<5000 ppm) with dry packages considered less than 300 ppm. The presence of adsorbed layers of moisture can lubricate surfaces (Bhushan, '95) but at the same time hydrates of silicon dioxide are known to produce gummy surfaces that contribute to adhesion. (Ikler, '79)

4. Microtribology Problems Limiting Commercialization

The structures commonly used for surface micromachined accelerometers and gyros are designed to avoid contact in normal use.(Chau,'96) The mechanical structures typically use a factor of two for safety and have contact occur at forty times the full scale range. Thus an 80X acceleration of the full scale range is required to produce contact. As stated in section 3 above, adhesion can occur at several stages in the product life cycle. The failure of the moving structures can occur by one of two general modes. The lateral adhesion mode is a result of excessive acceleration in the direction of measurement. This is shown in Figure 5. Here the forces produce a movement of the structure that drives the side-walls of the sensor in contact.

The second failure mode that can occur with the mechanical structures is referred to as vertical adhesion. Here the moving structure is in contact with the substrate. Since the polysilicon is two microns thick and the feature sizes for spring dimensions are approximately two microns wide the cross-section of the spring for maximum compliance is as stiff in the vertical direction as in the horizontal direction. The impact of squeeze film damping due to gases between the moving mass and the substrate help to cushion the mating of the surfaces. Nonetheless, forces can act to drive the surfaces in contact. Principal among these is surface tension during the fabrication of the wafer or moisture during the assembly operations.

Figure 5. The lateral adhesion of the mechanical structure.

With failure modes as described in figures 5 and 6 as the result of adhesion. The problems limiting commercialization in surface micromachined accelerometers and gyros occur in four different categories. They are the etching and release, the packaging, coatings and measurement and reliability. Each of these categories have limiting factors as follows:

ETCHING AND RELEASE

- Wet etches and liquid cleaning produce surface tension
- Residues promote adhesion
- Dry release methods are aggressive and remove coatings
- Unwanted coatings from chemical processing prevent desired coatings from bonding
- Etches also oxidize or alter the surface
- Semiconductor chemicals are optimized for wafer processing not low adhesion (safety, OSHA, waste treatment)

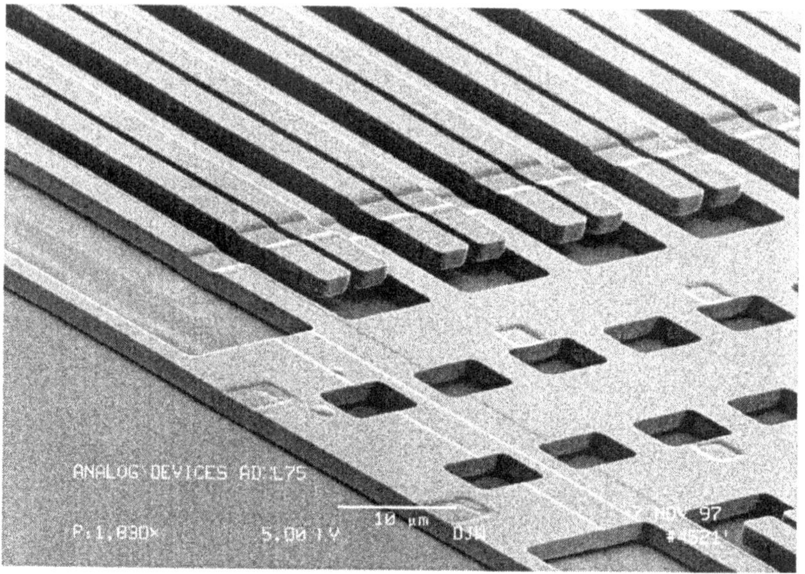

Figure 6. The vertical adhesion of the mechanical structure.

PACKAGING AND PACKAGE AMBIENTS

- Sealing processes alter surface properties and coatings
- Package ambients can coat or interact with surfaces and impact adhesion and durability in high shock
- Hard packages (ceramic, metal) can amplify or transmit acceleration and drive structures into contact
- Nonconductive packages can charge and produce electrostatic fields resulting in structural movement
- Nonhermetic packages are low cost but not effective for surface micromachined structures

COATINGS

- Selection of a compatible coating to limit adhesion is largely trial and error
- Coatings can introduce electrical instabilities, especially with capacitive, high impedance circuits

- Coatings must be stable in vacuum (gyro) and thermal exposures
- Coatings of micromachined structures must occur out of the line of sight (under and behind structures)

MEASUREMENT AND RELIABILITY

- In-line methods to measure adhesion in microstructures is not production orientated
- Acceleration methods for reliability testing are not correlated
- Surface characteristics and properties for low adhesion in semiconductor processing are not correlated in a production environment
- Intermittent adhesion is difficult to measure
- Service life of 10 to 15 years have not been demonstrated
- Coating quality and reliability are difficult to quantify
- Failure rates in the single digit ppm's are sufficient to impact commercial acceptance in some applications (high reliability)

5. Conclusions

This paper has shown that significant opportunities exist in surface micromachined sensors of accelerometers and gyros. The challenge is to apply methods to control the adhesion and at the same time process the devices using IC fabrication methods that have not considered mechanical surface properties. Structures that have higher over-range performance and greater sensitivity can be built in this technology if the probability of adhesion can be reduced when surfaces come in contact. To meet the commercial expectations, improvements in adhesion must show a consistent quality level of less than 10 ppm. Low adhesion in a manufacturing environment will, therefore, benefit from a better understanding of the factors that cause and control adhesion.

6. References

Bhushan, B., Israelachvili, J., Landman, U., "Nanotechnology: Friction, Wear, and Lubrication at the Atomic Scale", Nature, Vol. 374, April 1995, pp.607-616.

Chau, K.H.-L, Lewis, S., Zhao,Y., Howe, R., Bart, S., Marcheselli, R., "An Integrated Force-Balanced Capacitive Accelerometer for Low-G Applications," Sensors and Actuators, vol. A. 54, 1996, pp. 472-476.

Core,T., Tsang, W., Sherman, S., "Fabrication Technology for an Integrated Surface-Micromachined Sensor", Solid State Technology, Oct. 1993, pp.39-47.

Core, T., Howe, R., 1994, Method for Fabricating Microstructures, U.S. Patent 5,314,572.

Iler, R., 1979, The Chemistry of Silica, John Wiley & Sons, New York, p.3.

Juneau,T., Pisano, A.P., Smith, J.H., "Dual Axis Operation of a Micromachined Rate Gyroscope", Proc. Transducers '97, Chicago, pp. 883-886.

Maboudian, R., Howe, R., 1997, "Stiction Reduction Processes for Surface Micromachines", Tribology Letters, vol. 3, 1997, pp.215-221.

Roberts, C., Long, L., Ruggerio, P., 1994, Method for Separating Circuit Dies from a Wafer, U.S. Patent 5,362,681.

Seeley, R., 1996, "Micromachines Rev Up for Fast Growth", Electronic Business Today, April, pp.33-38.

Song, C., 1997, "Commercial Vision of Silicon Based Inertial Sensors", Proc. Transducers '97, Chicago, pp.839-842.

Walsh, S., Thukral, I., 1996, "Creating Competitive Advantage with MEMS Technology", Commercialization of Microsystems '96, Kona, Hawaii.

NEW TECHNOLOGY AND APPLICATIONS AT LUCAS NOVASENSOR

KIRT R. WILLIAMS
Lucas NovaSensor
Fremont, California, USA

Abstract

We give a tutorial on two commercial micromachining processes used at Lucas NovaSensor, orientation- and doping- dependent etching and silicon fusion bonding. Combining these with a relatively new technology, deep reactive ion etching (DRIE), new or improved microelectromechanical devices can be made, including accelerometers, thermal actuators, and micro flow channels. While DRIE is capable of relatively rapidly etching vertical trenches, reductions in etch-rate variation with trench width and increases of etch rate can still be made.

1. Introduction

In this paper we provide a brief introduction to micromachining and a tutorial on two commercial micromachining processes, then discuss a new process, deep reactive ion etching, being used at Lucas NovaSensor.

Many different techniques are available when designing a fabrication process for a micromachined device. These including various deposition, lithographic patterning, etching, bonding, and packaging steps, as well as a selection of starting materials for a substrate. Examples are given in texts by Madou (1997) and Muller et al. (1990).

Certain of these steps are classically grouped into those used for "bulk" micromachining, in which patterns are etched *into* a substrate, or for "surface" micromachining, in which patterned layers reside *on* the surface of a substrate. (Many other steps can be used for either and thus defy such classification.) Bulk micromachining generally relies on the doping- or crystal-orientation- dependent differential etch rates of some combinations of etchant and substrate (e.g., single-crystal silicon etched in potassium hydroxide) to form three-dimensional structures. Because selective doping can only be done a limited depth from the surface, and because the surfaces of orientation-dependent-etched structures follow certain crystal planes, bulk micromachining is limited in the patterns that can be produced (Fig. 1). Surface

B. Bhushan (ed.), Tribology Issues and Opportunities in MEMS, 121-134.

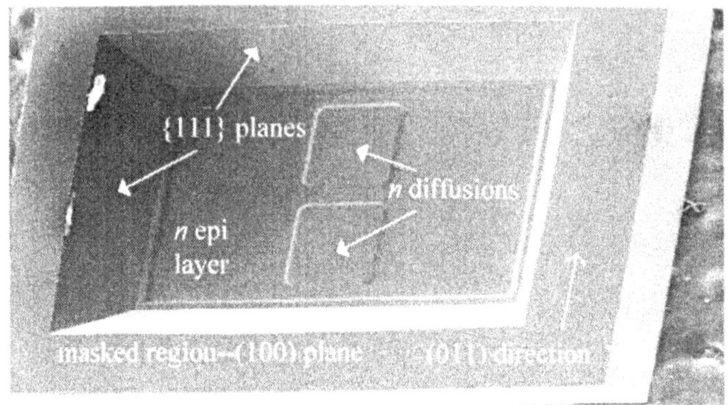

Figure 1 *Bulk micromachining example (a pressure sensor)*
The KOH etch has stopped on silicon {111} planes to form a frame, on a doped epitaxial layer form a membrane, and on diffusions to form thick bosses

micromachining, on the other hand, can be used to make arbitrary patterns (when viewed from the top) in many different materials, but typically is limited in the thickness dimension to a few micrometers (Fig. 2). Surface micromachining most commonly uses polycrystalline silicon as a structural layer and silicon dioxide as a sacrificial layer.

Bulk micromachining is the older of these two and is still used for the most commercial products (e.g., NovaSensor's pressure sensors, IC Sensors' accelerometers, and ink-jet nozzles). Surface micromachining has perhaps been the subject of the most research recently, and is used in a few commercial products (e.g., Analog Devices' accelerometers).

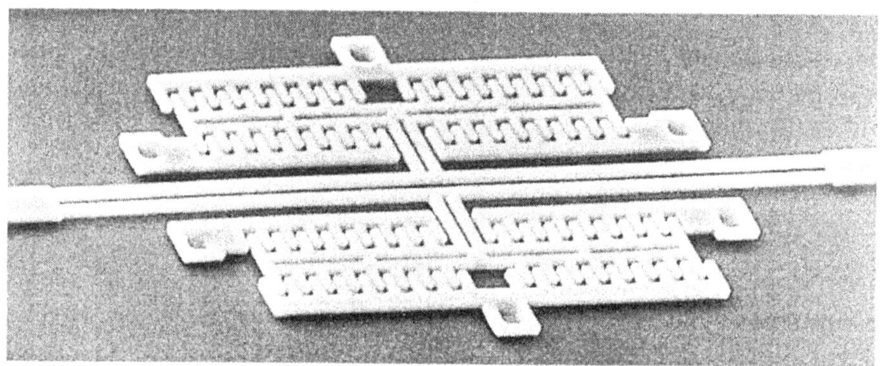

Figure 2 *Surface micromachining example*
The structure is made of 2-µm-thick polycrystalline silicon;
the sacrificial layer, etched from under the poly, was silicon dioxide

In the following sections we review two of the steps available for bulk micromachining, orientation-dependent etching and silicon fusion bonding (SFB), then present a relatively new micromachining technique, deep reactive ion etching (DRIE), which is used to etch deep vertical holes and channels in substrates (typically silicon). We then present some applications of the combination of cavity etching, SFB, and DRIE.

2. Orientation- and Doping- Dependent Etching

Several alkaline solutions, particularly potassium hydroxide (KOH), ethylenediamine-pyrocatechol (EDP), and tetramethyl ammonium hydroxide (TMAH), are used to selectively etch single-crystal silicon, as discussed by many authors (Seidel et al., 1990; Raley et al., 1984; Schnakenberg et al., 1991; Madou, 1997). These solutions etch heavily doped *p*-type silicon much more slowly than lower-doped or *n*-type silicon. Using an appropriate bias, either type silicon can be etched, stopping at the junction with the other type. These solutions also etch {111}-type and some higher-order crystal planes more slowly than {100} planes. Silicon nitride or thick silicon dioxide can be used as a masking layer.

When rectangular windows, aligned to <110> directions, are made in a masking layer on (100)-oriented silicon wafers, pyramidal pits are etched, with the etch terminating on {111} and higher-order planes (as in Fig. 1). Many shapes other than pyramidal pits can be made in (100) silicon, but these require special attention to the rates at which the mask is undercut, particularly at convex corners. A major limitation of orientation-dependent etching is that the resulting structures have walls that are sloped relative to the surface, consuming more die area than would a vertical etch.

Thin structures of arbitrary shape in the plane of the wafer, such as membranes and cantilevers, can be made by selectively diffusing a dopant into a low-doped (100) silicon wafer, then etching. Membranes can also be made by stopping an etch on an appropriately doped epitaxial layer. This is used at Lucas NovaSensor in the production of commercial piezoresistive pressure sensors, etching through *p*-type silicon to stop at the junction with *n*-type silicon by using an electrochemical etch stop (also seen in Fig. 1).

3. Silicon Fusion Bonding

Silicon fusion bonding is used to permanently attach two or more silicon wafers together. The bond is strong enough that when attempts are made to separate bonded wafers, the wafers themselves usually break.

SFB requires that the surfaces to be bonded be exceptionally clean, flat, and smooth. They can be bare silicon or may have a grown oxide on one or both (Petersen et al., 1988). Surface preparation for SFB consists of cleaning, then forming a hydrated layer with OH groups at the surface. Upon contact, the surfaces spontaneously stick due to hydrogen bonding (Madou, 1997). Heating above 1000°C for a few hours turns Si-OH-HO-Si bonds into stronger Si-O-Si bonds with the release of water molecules.

Applications of silicon fusion bonding include the formation of silicon-on-insulator (SOI) wafers, fabricating pressure sensors with a smaller die area, and making micro flow channels, as described below.

One method of making SOI wafers is to grow a layer of thermal oxide on one wafer, use SFB to bond another wafer to it, grind one of the wafers down to the near the desired thickness (on the order of a micrometer), and etch and polish to the final thickness (Madou, 1997).

SFB can be used to make a piezoresistive pressure sensor with the same membrane area as the sensor shown earlier (Figs. 1 and 3a), but with a smaller die area. This is done by etching a pyramidal cavity from the *top* rather than the back side of the wafer, bonding an epi wafer on top so as to cover the cavity, then grinding, polishing, and etching back this wafer to the epi layer to form a thin membrane (Fig. 3b). If the pyramidal cavity extends all the way through the wafer, a gauge pressure sensor is formed. If, instead, the cavity is sealed, an absolute pressure sensor is created (Fig. 3b). Even with the extra processing steps, the reduction in die area results in a cheaper device (Bryzek et al., 1991).

Micro flow channels can be simply formed by using an orientation-dependent or isotropic etchant to form grooves in one or two wafers, then mating the wafers with SFB (Fig. 4). If patterns are made in both wafers, aligned bonding must be done. Equipment for aligned bonding is available commercially from several manufacturers, including Electronic Visions and Karl Suss.

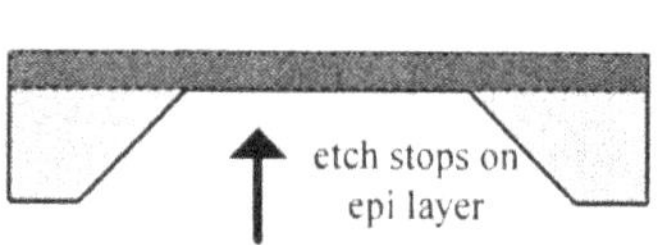

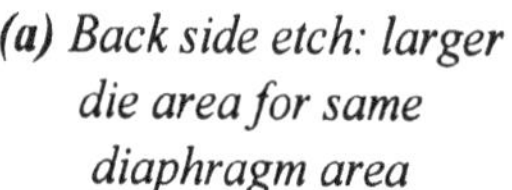

(a) Back side etch: larger die area for same diaphragm area

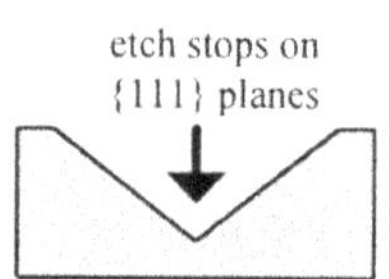

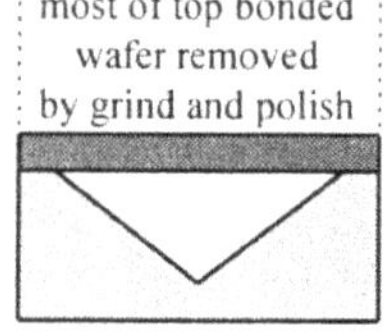

(b) Top side etch, followed by SFB and grind and polish: smaller die area

Figure 3 *Using silicon fusion bonding to make a smaller-die-area pressure sensor*

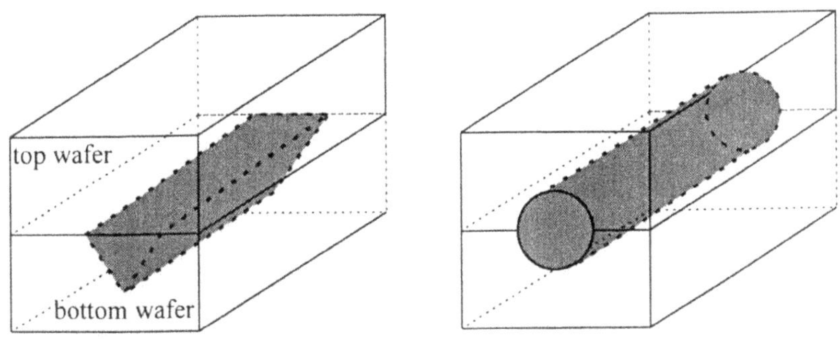

(a) Orientation-dependent etch; SFB *(b) Isotropic etch two wafers; aligned SFB*

Figure 4 *Using silicon fusion bonding to make flow channels*

4. Deep Reactive Ion Etching (DRIE)

4.1 INTRODUCTION TO DRIE

So-called reactive ion etching (RIE) has been used for years to perform vertical (as opposed to isotropic) plasma etching. Most applications have been in the IC industry, which called for etching through thin films with thicknesses of at most a few micrometers. Surface-micromachined devices are typically limited to similar thicknesses. Recently, researchers in the micromachining field have blurred the lines between surface and bulk micromachining by pushing the limits of their standard RIE machines, etching holes tens of microns deep into silicon wafers (see, for example, Keller, 1994).

In conventional silicon RIE, halogen (fluorine, chlorine, or bromine) radicals generated in the plasma combine chemically with the silicon to form a volatile etch product such as SiF_4 (Williams and Muller, 1996). At the same time, however, halocarbon polymers formed from the plasma products coat all surfaces, both vertical and horizontal, on the wafer (Bhardwaj et al., 1997). This polymer coating inhibits etching. Ions generated in the bulk plasma (e.g., SF_x^+) are accelerated through the plasma sheath to the wafer surface. The main velocity component is vertical, so few ions strike the sidewalls of trenches that have been etched (Fig. 5). The ions do, however, bombard the horizontal surfaces at the bottoms of trenches, supplying energy to remove the polymer and allow etching by radicals such as $F^{\cdot}$ (Lieberman and Lichtenberg, 1994). In this way, vertical etching is done with little undercutting of the mask.

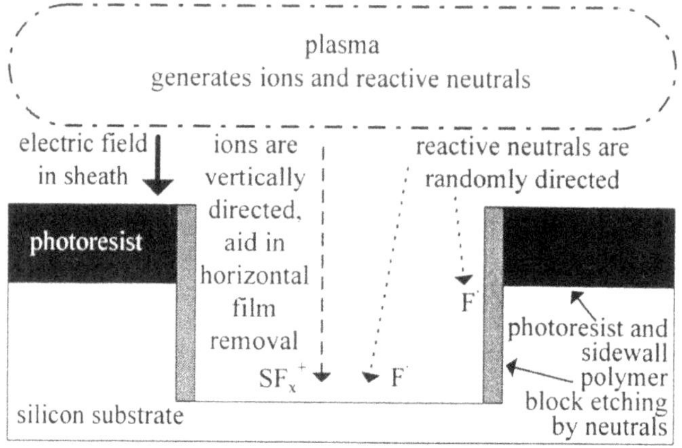

Figure 5 *Vertical plasma etching (RIE)*
Ions accelerated through sheath are vertically directed and result in polymer removal mainly from horizontal surface. Neutrals actually do silicon etching.

In the past few years, several manufacturers have developed new equipment and processes geared specifically toward the micromachining community (examples can be found in Walsh, 1996). Deep RIE (DRIE) is the result of efforts at several companies. Alcatel has concentrated on a high-density inductively coupled plasma (ICP) using a SF_6 + O_2 chemistry at cryogenic temperatures to achieve high-aspect-ratio etching (Vintro, 1995). The low temperature reduces the volatility of the sidewall passivation layer (Bhardwaj et al., 1997). Surface Technology Systems (STS) and Plasma-Therm also use a high-density ICP source, but etch at and above room temperature using a two-step process licensed from Bosch (Laermer and Schilp, 1996).

In the Bosch process, separate etching and polymer-deposition steps are performed (Bhardwaj and Ashraf, 1995; Bhardwaj et al., 1997): a fluoropolymer passivation layer (nCF_2) is deposited for a short period, then the polymer and silicon at the bottoms of trenches are etched for short time with SF_6. This sequence is then repeated.

The polymer-deposition gas is carefully selected to supply sufficient C and F to form the polymer film, but not generate excessive F, as this results in spontaneous film removal (Bhardwaj et al., 1997). As of this writing, the deposition gases used in STS deep silicon etchers are proprietary.

The use of a low-pressure inductively coupled plasma source (Fig. 6), rather than a traditional capacitive source, is one key to the success of these etchers. A higher plasma density (defined as the density of electrons or ions per unit volume in the bulk plasma) can be achieved with an ICP. This results in a higher etch rate and a thinner sheath (Bhardwaj et al., 1997). At low pressures, the ion mean-free path is also larger.

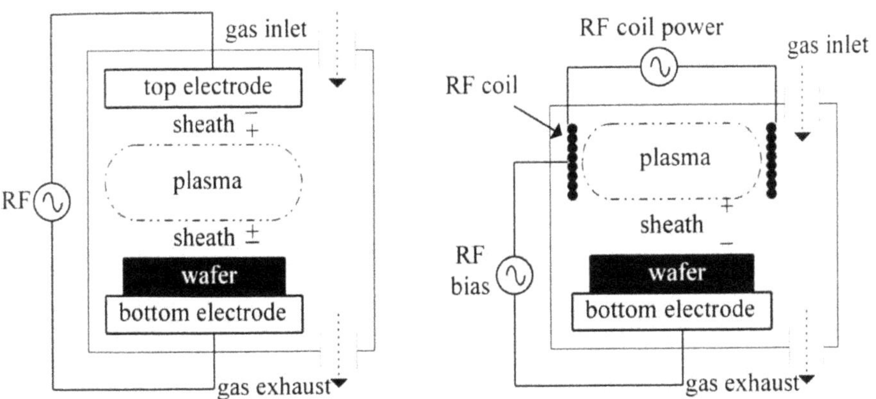

(a) *Capacitively coupled plasma etcher Single RF supply determines plasma input power and sheath voltages*

(b) *Inductively coupled plasma etcher Separate power supplies allow independent control of sheath voltage*

Figure 6 *Comparison of capacitively coupled and inductively coupled plasma etchers*

These two factors result in a reduced sheath ion-collision probability, a very vertical ion directionality, and therefore a vertical etch. Further, with an ICP, the voltage across the sheath can be controlled independently of input power generating the plasma, greatly increasing the control of the process.

4.2 DRIE RESULTS

With NovaSensor's STS DRIE etcher, typical silicon etch rates are about 3 μm/min. While this is rapid by IC industry standards, it takes over two hours to etch through a wafer at this rate. The selectivity over photoresist is about 75:1 and twice that for silicon dioxide. Etch verticality is within 2° of normal.

Unfortunately, DRIE exhibits an "RIE lag," in which narrower trenches etch more slowly than wide ones, as seen in Figure 7 (note that this is the opposite effect of the typical etch loading, in which small areas etch more rapidly than large ones). This can be attributed to the greater difficulty etch reactants and products have diffusing into and out of low-aspect-ratio trenches. Sidewall angle is also dependent on the width of the trench being etched, as shown in the graphs in Figure 8.

Process parameters, particularly plasma power, substrate bias, and the etch and deposition times, affect the etch. Faster etch rates tend to reduce the anisotropy.

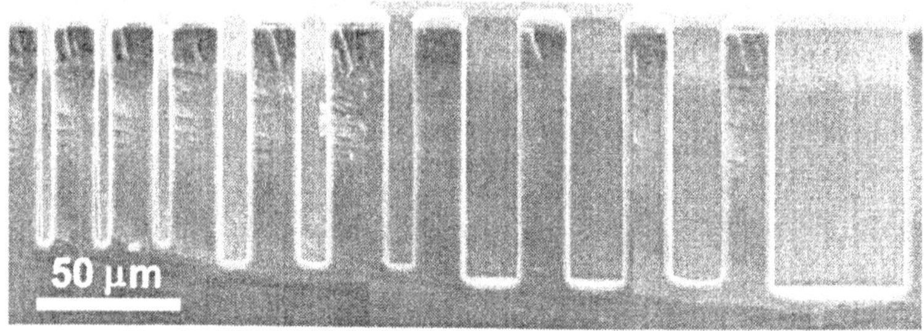

Figure 7 *SEM of trenches of various widths that have been DRIE etched. Due to RIE lag, narrower trenches etch more slowly.*

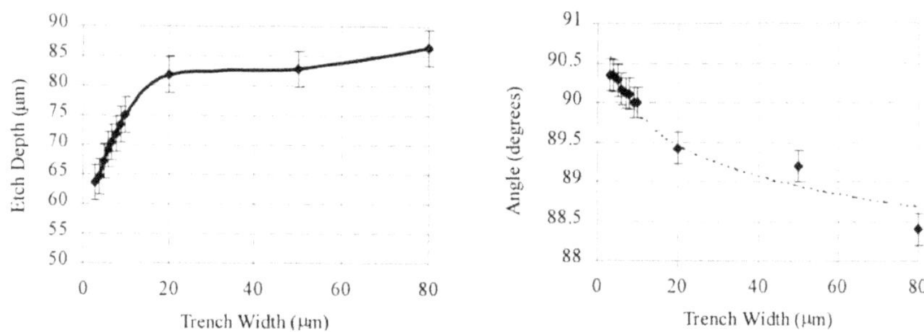

Figure 8 *Plots of etch depth and sidewall angle vs. trench width*

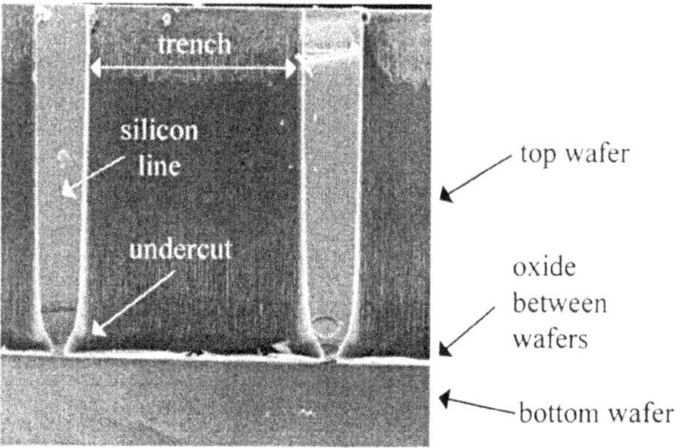

Figure 9 *SEM showing notching of silicon lines at oxide interface*

The aforementioned etch rate and profile variation requires that all trenches be made the same width if their etching is to be completed at the same time. If trenches must be different widths, one might consider using an oxide etch stop, as the selectivity of silicon over oxide is high, but with much overetching this tends to result in notching at the sides of the bottom of the trench as seen in Figure 9. It has been suggested that this is due to charging of the nonconductive oxide by incident positive ions, which then repel subsequently arriving ions into the sides of the trench.

STS reports that newer etchers are able to stop on oxide without the notching effect (Bhardwaj et al., 1997). This is accomplished by detecting when oxide has been reached, then altering the etch parameters to avoid lateral etching. This is done at the expense of the vertical etch rate.

5. Combining Cavity Etching, Silicon Fusion Bonding, and Deep RIE

By etching a cavity in a bottom wafer (using wet or dry etching), silicon fusion bonding on a top wafer, optionally grinding and polishing the top wafer back to a desired thickness, then DRIE etching a pattern though the top wafer and terminating at the cavity, high-aspect-ratio microelectromechanical structures can be formed (Fig. 10). Lucas NovaSensor has applied for a patent (still pending) on this process sequence.

These structures can have layouts very much like polysilicon surface-micromachined structures, but offer many advantages. They can be of arbitrary thickness (up to the thickness of a wafer), leading to much larger capacitance and therefore higher sensitivity with greater out-plane stiffness and therefore lower cross-axis sensitivity in devices such as micromachined interlaced-comb accelerometers and gyros. A large thickness also results in a larger proof mass, which gives an accelerometer a lower mechanical noise floor. Because they are made of single-crystal silicon, residual stresses are almost nonexistent (as compared to polysilicon), there are no grain boundaries and therefore negligible fatigue and structural damping mechanisms, and no grain-size variation leading to spring-constant variation from device to device. Because the wafer away from the structures is also device-quality single-crystal silicon, circuitry can be fabricated beside the structure before the DRIE etch, as demonstrated in Figure 11. Because both mechanical structure and substrate are the same material, the thermal expansion rates are matched, nearly eliminating the effect of ambient-temperature fluctuation on sensors and actuators.

In the next sections we present some structures made at Lucas NovaSensor using SFB and DRIE.

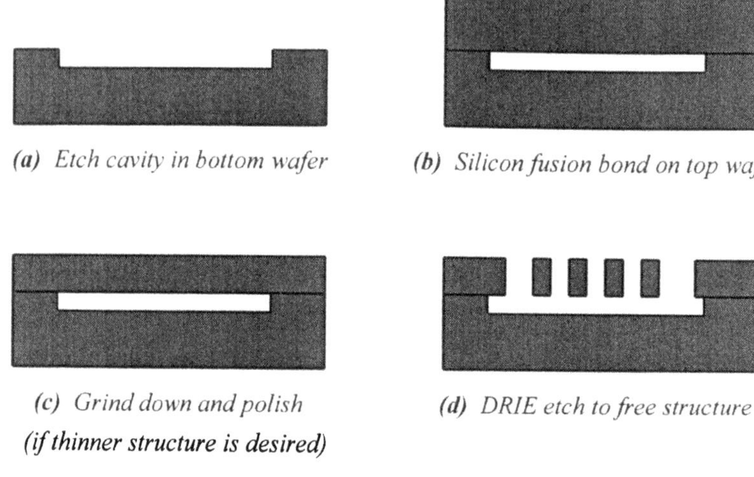

(a) Etch cavity in bottom wafer

(b) Silicon fusion bond on top wafer

(c) Grind down and polish (if thinner structure is desired)

(d) DRIE etch to free structure

Figure 10 *Cavity etch + SFB + grind and polish + DRIE process flow*

Figure 11 *DRIE structure with integrated circuitry*

5.1 FORCE-BALANCED ACCELEROMETER USING INTERLACED COMBS

Force-balanced accelerometers using interlaced combs (Fig. 12) were made using the process shown in Figure 10 (van Drieënhuizen et al., 1997). The comb fingers are 60 μm thick and 6 μm wide with a 4-μm spacing. Using off-chip circuitry, the accelerometers had a 60-dB dynamic range and milli-G resolution (over 100-500 Hz), limited by $1/f$ noise in the electronics. The noise-limited resolution of the mechanical device itself is estimated to be 200 μG (rms over 1 kHz bandwidth).

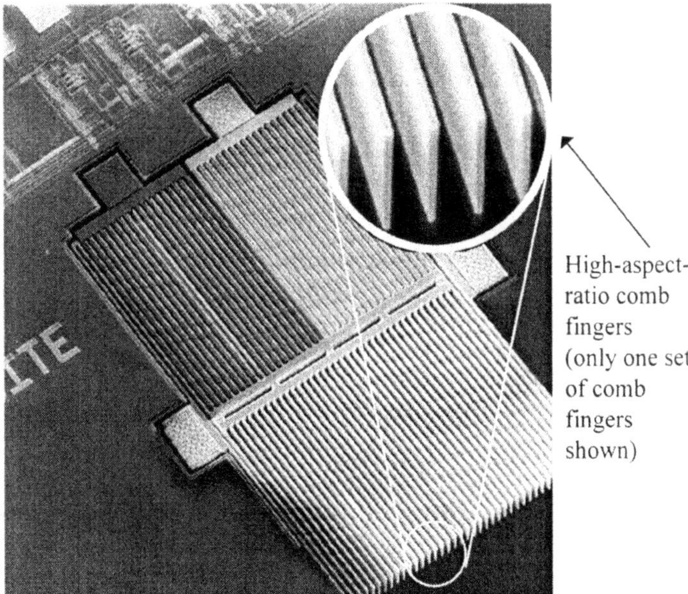

Figure 12 *DRIE accelerometer*

5.2 THERMAL ACTUATOR

Thermal actuators employing symmetrical pairs of angled "ribs" can be used to generate large forces (Klaassen et al., 1995): over a newton for a temperature rise of 100°C. Figure 13 shows an SEM of such an actuator, which is a full 385-μm wafer in height. In operation, current is forced from one side of the ribs to the other, resulting in ohmic heating. As the ribs try to expand in length (mainly in the y direction), the center bar constrains movement to the x direction.

5.3 MICROFLUIDIC CONNECTORS AND FLOW CHANNELS

By alternately DRIE etching channels and silicon fusion bonding wafers together, multi-level microfluidic flow channels can be formed, as shown in Figure 14. If high-aspect-ratio connectors suitable for sliding in capillaries and sealing with epoxy (Fig. 15) are included at the surface, a "microfluidic printed-circuit board" is formed, with the channels corresponding to copper traces and connectors and epoxy corresponding to wiring holes and solder. Such fluidic PC boards might be used for interconnecting various components in a microfluidic system, such as pumps, valves, and pressure

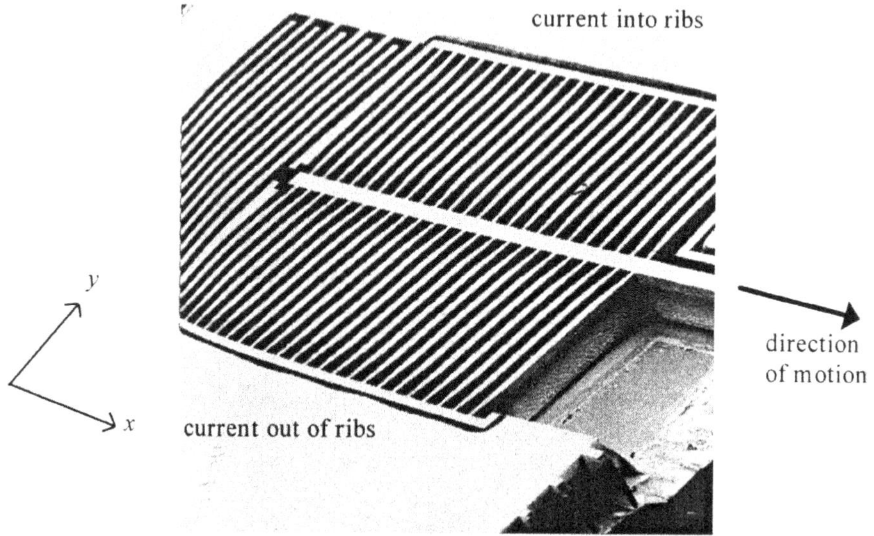

Figure 13 *DRIE thermal actuator*

sensors. In addition to making connections, mixers and other fluidic devices can be incorporated on the same wafer stack.

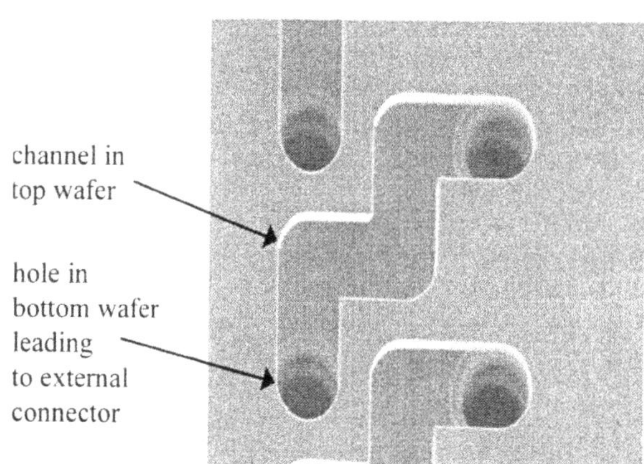

Figure 14 *DRIE multilevel fluidic interconnect (a third wafer seals the channels)*

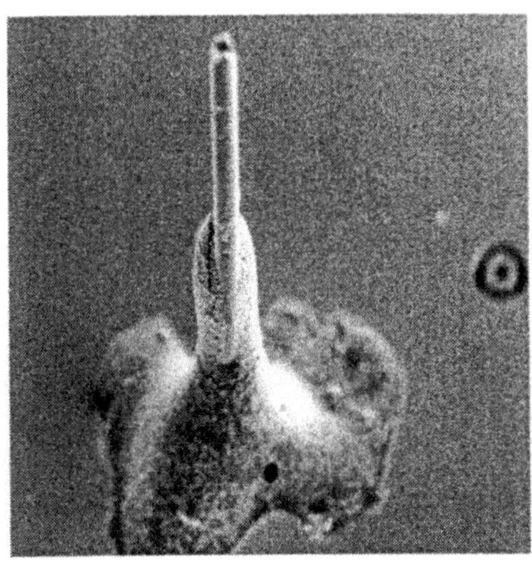

Figure 15 *SEM of capillary held into DRIE connector with epoxy*

6. Summary

We have reviewed two commercial bulk micromachining processes, orientation- and doping- dependent etching and silicon fusion bonding, then discussed the function and advantages of deep silicon RIE over polysilicon surface micromachining and conventional bulk micromachining. Improved devices made at Lucas NovaSensor include higher-sensitivity accelerometers and high-force thermal actuators. A new microfluidic device is the "fluidic PC board" with a straightforward interface to the macroscopic world via tubular connectors.

While DRIE provides a fairly vertical etch, improvements can still be made in providing etch uniformity and speed.

Acknowledgments

Many people at Lucas NovaSensor took part in the research and fabrication described in this paper, including Nadim Maluf, Bert van Drieënhuizen, Bonnie Gray, Nick Mourlas, and Rose Scimeca. This work was partially funded by DARPA.

References

Bhardwaj, J.K. and Ashraf, H. (1995) Advanced silicon etching using high density plasmas, *Proc. SPIE Micromachining and Microfabrication Process Technology*, **2639**, 224-232.

Bhardwaj, J., Ashraf, H., and McQuarrie, A. (1997) Dry silicon etching for MEMS, *Tech. Dig. Symposium on Microstructures and Microfabricated Systems, Annual Meeting of the Electrochemical Society.*

Bryzek, J., Petersen, K., Mallon, J.R., Christel, L., and Pourahmadi, F. (1988) *Silicon Sensors and Microstructures*, Lucas NovaSensor, Fremont, California.

Keller, C., and Ferrari, M. (1994) Milli-scale polysilicon structures, *Tech. Dig. 1994 Solid State Sensor and Actuator Workshop (Hilton Head '94)*, 136-139.

Klaassen, E.H., Petersen, K., Noworolski, J.M., Logan, J., Maluf, N.I., Brown, J., Storment, C., McCulley, W., and Kovacs, G.T.A. (1995) Silicon fusion bonding and deep reactive ion etching: a new technology for microstructures, *Tech. Dig. 8th Int. Conf. on Solid-State Sensors and Actuators (Transducers '95)*, **1**, 556-559.

Laermer, F. and Schilp, A. (1996) Method of anisotropically etching silicon, *U.S. Patent 5,501,893.*

Lieberman, M. and Lichtenberg, A. (1994) *Principles of Plasma Discharges and Materials Processing*, Wiley, New York.

Madou, M. (1997) *Fundamentals of Microfabrication*, CRC Press, Boca Raton, Florida.

Muller, R.S., Howe, R.T., Senturia, S.D., Smith, R.L., and White, R.M. (1990) *Microsensors*, IEEE Press, New York.

Petersen, K., Barth, P., Poydock, J., Brown, J., Mallon, J., and Bryzek, J. (1988) Silicon fusion bonding for pressure sensors, *Tech. Dig. 1998 Solid State Sensor and Actuator Workshop (Hilton Head '88)*, 144.

Raley, N.F., Sugiyama, Y., and Van Duzer, T. (1984), (100) silicon etch-rate dependence on boron concentration in ethylenediamine-pyrocatechol-water solutions, *J. Electrochem. Soc.*, **131**, 161-171.

Schnakenberg, U. Benecke, W. and Lange, P. (1991), TMAHW etchants for silicon micromachining, *Tech. Dig. 1991 Int. Conf. on Solid-State Sensors and Actuators (Transducers '91)*, 815-818.

Seidel, H., Csepregi, L., Heuberger, A., and Baumgartel, H. (1990a) Anisotropic etching of crystalline silicon in alkaline solutions, I., Orientation dependence and behavior of passivation layers, *J. Electrochem. Soc.*, **137**, 3612-3626.

Seidel, H., Csepregi, L., Heuberger, A., and Baumgartel, H. (1990b) Anisotropic etching of crystalline silicon in alkaline solutions, II., Influence of dopants, *J. Electrochem. Soc.*, **137**, 3626-3632.

Van Drieënhuizen, B.P., Maluf, N.I., Opris, I.E., and Kovacs, G.T.A. (1997) Force-balanced accelerometer with mG resolution, fabricated using silicon fusion bonding and deep reactive ion etching, *Tech. Dig. 1997 Int. Conf. on Solid-State Sensors and Actuators (Transducers '97)*, 1229-1230.

Vintro, L. (1995) Talk on high-aspect-ratio etching of silicon, *Bay Area MEMS Journal Club Meeting*, 1 March 1995.

Walsh, S., Editor (1996) *Tech. Dig. Commercialization of Microsystems '96*, Kona, Hawaii.

Williams, K.R., and Muller, R.S. (1996) Etch rates for micromachining processing, *J. Microelectromechanical Systems*, **5**, 256-269.

ROUGH SURFACE CHARACTERIZATION

J.H. TRIPP
SKF Engineering & Research Centre
Nieuwegein, The Netherlands

Abstract

The present work covers three aspects of surface characterization necessary for the correct use and interpretation of descriptive topographical roughness parameters.. The first is the development of the traditional parametric description, evolving from early 1-D surface profiles to the more recent 2-D surface maps, often (though incorrectly) referred to as 3-D. This accounts for the current position with regard to surface parameterization. Next, consideration is given to the problems of using such descriptions in modelling the physical behavior of surfaces. Thirdly, a solution of these problems is offered involving the fractal nature of the surface and the functions it is expected to perform.

1. Introduction

The roughness h of a manufactured surface is usually taken as the distance of the actual surface, $\varsigma(\mathbf{r}) = 0$, from a specified nominal surface $\varsigma_n(\mathbf{r}) = 0$, this distance being measured along the local normal to the nominal surface. In the case of a nominal plane, taken as $z = 0$, h will be equal to $\eta(x,y)$, where $z = \eta(x,y)$ is the equation $\varsigma(\mathbf{r}) = 0$ written in the (x,y,z) Cartesian system. The problem with such a definition of roughness is that, while $\varsigma_n(\mathbf{r})$ will usually be a continuous differentiable function with a well-defined normal, the physical surface $\varsigma(\mathbf{r})$ and hence also the roughness h are by nature continuous but nowhere differentiable. Thus, although the two surface ς and ς_n may approach arbitrarily closely (i.e. $h \rightarrow 0$), when viewed on a sufficiently small scale, their slopes and curvatures will never coincide. Consequently, features of the real surface described by the roughness h, such as summits and pits, persist down to the smallest scales. Physical surfaces thus have a geometry different from classical 2-D Euclidean geometry, known as ***fractal geometry*** (Mandelbrot, 1982).

Surfaces of materials are commonly prepared either by the repetition of many small random material-moving operations, as in grinding and polishing, or by single-point operations, as in diamond turning and laser texturing. From the point of view of characterizing the surface geometry, all such processes produce a random topography, the difference between them being only the length scales over which randomness

B. Bhushan (ed.), Tribology Issues and Opportunities in MEMS, 135-148.

prevails. For features much larger or smaller than the typical material element moved by the operation, the surface appears random, while on scales comparable to the element, it shows deterministic behavior. Hence, the characterization of surface topography is largely the problem of finding appropriate parameters to describe the random aspects of $h(x,y)$. For the deterministic features, either a specification may be given in terms of their typical size and spacing or, alternatively, they may be added to the nominal surface $\varsigma_n(r)$ and thus subtracted from h. The terms ***patterned surfaces*** or ***transitional topographies*** (Williamson *et al*, 1970) both refer to surfaces having distinguishable features at more than one distinct size scale, for example, laser pits on a honed surface. In all cases, specification of the range of length scale for which the roughness parameters apply is a vital part of surface characterization. This of course is similar to the traditional partition of h into the three scale ranges: form, waviness and roughness. The novelty lies in regarding this as an intrinsic property of surfaces, rather than as merely a practical convenience.

The scale problem presented by physical surfaces is mirrored in the problem of their measurement. Within a sample of a surface of linear dimension L, a measuring instrument aims to give the height $\eta(x,y)$ at each point (x,y), although physically of course it can only yield this value averaged over the area of its sensing element of linear dimension l. Associated with the observation of surface topography, then, two length scale limits L and l are imposed from the outset. With the almost universal adoption of digital, rather than analog, data recording, the sampling interval δ enters as a third length with a determining role in surface characterization.

2. Standard Approach

2.1 EARLY WORK

Amongst the earliest recordings of surface topography were the traces made by drawing a fine diamond stylus along a straight line across the surface. The vertical motion of the tip was coupled to a coil in a magnetic field, producing a signal proportional to the vertical speed of the stylus, i.e. to the slope of the surface rather than to its height, so that some additional circuitry was needed to represent the actual roughness features. The output of this Abbott ***Profilometer*** (Abbott and Firestone, 1933), as it was aptly named, was proportional to the root-mean-square (rms) deviation, R_q, of the profile from its mean. The signal was considered to be more-or-less sinusoidal so that division by $(\pi^2/8)^{1/2}$ served to give the **C**enter **L**ine **A**verage (CLA) or arithmetic average deviation, R_a, in those days the preferred measure of roughness. Thus, the first parameter ever used to characterize random roughness was a simple measure of its mean amplitude measured from a mean surface. If the approximately Gaussian nature of the roughness height distribution of many surfaces had been known, the conversion factor of $(\pi/2)^{1/2}$ would have been used but, from the present viewpoint, this difference is unimportant: both R_a and R_q give only a measure of height, with no information whatsoever on the shapes of surface features. What will develop here is that, since the physical behavior of a surface is governed by the actual geometrical shape of its features, height alone is an inadequate characterization.

2.2 AMPLITUDE, SPATIAL AND HYBRID PARAMETERS

Just a few years after the Abbott and Firestone profilometer was launched, Taylor Hobson introduced their line of Talysurf instruments, a line continuing until this day. In the 1960s, the output of this profilometer, used by many effectively as a standard, consisted of a surface profile on a strip chart, together with its R_a value. In their classic paper, Greenwood and Williamson (1966) describe how this output was first sampled and digitized. Obtaining the trace as a sequence of numbers immediately allowed computer power to be applied and, for the first time, surface theory to be compared more quantitatively with experiment. For example, the conjecture that the statistical height distribution of many surfaces would be approximately Gaussian was easily confirmed, as shown in Fig.1, which at once favored the use of R_q over R_a. It also became straightforward to generate the Abbott-Firestone bearing area curve, proposed 30 years earlier, giving the ratio of material-to-air as a function of surface height. This curve still plays a key role in some proposed systems for roughness characterization.

The well-known contact model of Greenwood and Williamson (1966) required 3 statistical parameters in order to predict, for example, the stiffness of a rough contact or the real contact area v. load relationship. Since contact of a rough surface with an ideal smooth surface involves only the higher points, not surprisingly these 3 parameters are derived principally from the upper part of the height distribution. The stresses developed in any contact as two surfaces approach depend both on the height of the point of first contact of each feature or asperity and also on the shape of the surface in the vicinity of this point, given by the heights of neighboring points. In the Greenwood and Williamson model, this shape is described by the local curvature of the ***profile peaks***, so that the first problem is to define a peak, by no means so straightforward a task as it might seem. Simply choosing a peak as any point higher than its 2 nearest neighbors, ***a 3-point peak,*** already puts a restriction on the lateral extent of most peaks and further, realizing that noise may be riding on the signal, some vertical threshold should also be selected. With no such threshold, it is easily seen that between 1/4 and 1/3 of all points sampled will be peaks (Whitehouse and Archard, 1970). Having chosen a definition for an n-point peak, a finite-difference approximation to its curvature in terms of the n sampled heights will yield the mean peak curvature $\boldsymbol{\kappa}$, or the mean peak radius $\boldsymbol{\beta}$, the first of the 3 contact model parameters. As a parameter involving both sample height and in-plane sampling distance, it is an example of ***a hybrid parameter***.

The second parameter required, η, gives the number of asperities or ***summits*** per unit area, where a summit is the 2-D extension of a 1-D profile peak Originally, since only 1-D profiles were available, η was derived from the number of peaks per unit length: if the probability of any point being a peak in a profile sampled at interval $\boldsymbol{\delta}$ is p, then η may be estimated as $\boldsymbol{\alpha}(p/\boldsymbol{\delta})^2$, where $\boldsymbol{\alpha}$ is a factor of about 1.2, whose value for an actual surface depends on the precise definition of peaks and summits, their anisotropy and the data sampling interval (Nayak, 1971; Greenwood, 1984). This parameter measures the distribution of peaks in the plane, rather than providing direct information about their height, an example of ***a spatial parameter***.

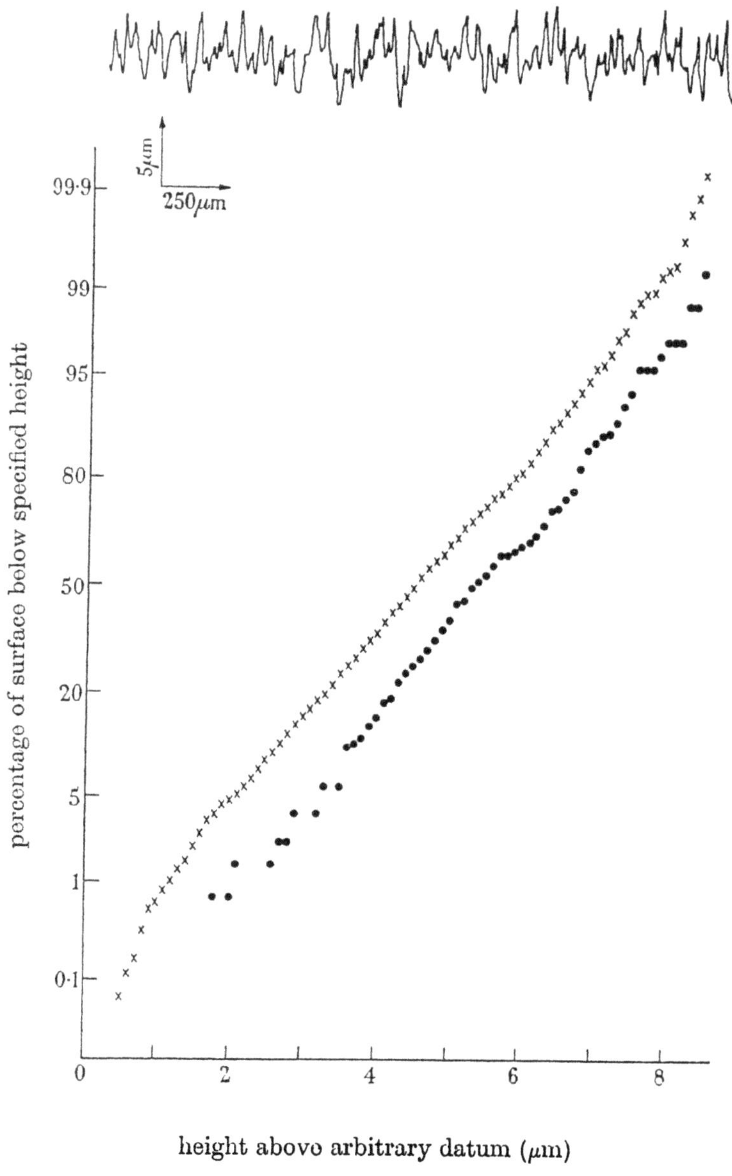

Figure 1. Cumulative height distribution of bead-blasted aluminium. Both the distributions of all heights (×) and of peak heights (●) are Gaussian, at least in the range ± 2 standard deviations. The profile of the same surface is shown in the upper diagram. From Greenwood and Williamson, (1966).

The third parameter required is the standard deviation of the peak height distribution, σ_p, giving the probability of finding a peak sufficiently high to make contact at any given load - ***an amplitude parameter*** depending only on sampled heights. Since peaks are selected by the adopted definition from the distribution of all heights, σ_p is not the same as the standard deviation of all heights, R_q, although the relationship between them is not difficult to derive for Gaussian surfaces (Bush *et al*, 1976).

This will serve as an introduction to the kinds of parameter used to characterize the many different roughness topographies realized on today's engineering surfaces. The number of such 1-D profile parameters in general use is at least 20 of which, for example, a dozen or so are defined in ISO 4287 (1997). This document also refers to the filtering which should be applied to the raw data before the profile is constructed. As shown in Sec.3, filtering often has a profound effect on the values yielded for the parameters and so must be included as part of the standard definitions. While in general, parameters in all three classes depend on the *i-j* ordering of the sampled heights, $\eta(x_i, y_j) = \eta_{ij}$, the bearing area curve and all parameters derived from it, such as R_q, are completely insensitive to ordering. Thus, except insofar as the pre-filtering depends on the ordering of the raw data, such parameters cannot describe the shape of surface features on which surface function depends.

A fourth class of surface characterization parameters is sometimes introduced. Known as ***functional parameters***, this class contains topography parameters which are expected to determine specific kinds of surface behavior. The kinds of behavior exhibited by a surface is vast, of course, with friction, lubrication and wear being only three, so that no parameter set of reasonable size could hope to characterize them all. Rather than attempting to define a universal set, the preferred approach is to construct physical models which will reveal those aspects of the topography determining each function. Algorithms for extracting the appropriate parameters from the sampled heights are then never too difficult to construct. Since parameters in the first three classes may very well also determine function, there seems little point to creating a fourth class to include these specifically defined parameters.

2.3 THE 'BIRMINGHAM 14'

In their 1993 BCR Report, Stout *et al* (1993) made an extensive study aimed at introducing a measure of standardization to 3-D surface characterization, obviously prompted by the rapid development of techniques for sampling surface height over a 2-D area rather than along a 1-D line. Much of the work was devoted to making useful recommendations for uniformity of data sampling, filtering, file format, and for selecting the nominal reference surface, $\eta_n(x, y)$. Of chief concern here, however, are the recommendations given for a set of characterizing topography parameters. From considerations of a far larger set, the so-called ***Birmingham 14*** were selected, partly on the grounds that they should give the most stable and generally useful representation of a surface and partly because, in several cases, they were easily traceable to their 1-D counterparts.

These 14, as shown in Table 1, fall into the four classes described above. In 3-D the convention has been adopted to label parameters with the letter S and a descriptive subscript, in contrast to the notation R used for 1-D profile parameters. Surface summits are here defined by the condition that a central point is higher than its 8 nearest neighbors.

Table 1: The primary 3-D parameter set

Type	Parameter Name	Description
Amplitude	S_q	RMS Deviation
	S_z	Ten Point Height
	S_{sk}	Skewness
	S_{ku}	Kurtosis
Spatial	S_{ds}	Density of Summits
	S_{tr}	Texture Aspect Ratio
	S_{td}	Texture Direction
	S_{al}	Fastest Decay Autocorrelation Length
Hybrid	$S_{\Delta q}$	RMS Slope
	S_{sc}	Mean Summit Curvature
	S_{dr}	Developed Area Ratio
Functional	S_{bi}	Surface Bearing Index
	S_{ci}	Core Fluid Retention Index
	S_{vi}	Valley Fluid Retention Index

The functional and amplitude parameters, with the exception of S_z, come directly from the Abbott-Firestone curve and as such are identical to their 1-D counterparts. It was already noted that bearing curve parameters are influenced only indirectly, if at all, by the shape of surface features. The value of S_z, on the other hand, is the mean height difference between the 5 highest summits and the 5 deepest pits and so depends on the configuration defined for summits and pits.

The hybrid slope $S_{\Delta q}$ gives the scalar magnitude of the surface slope by first evaluating its x- and y-components separately as for a 1-D profile, while similarly the mean summit curvature S_{sc} is the arithmetic mean of the x- and y-summit curvatures. Except in special cases these x- and y-components are, of course, different but when combined in this way, the directional or anisotropy information they contain is lost. The developed area ratio S_{dr} is a measure of how much the true area of the surface exceeds that of the nominal reference surface and constitutes a genuine new 3-D parameter, discussed further in Sec.4.1.

The density of summits S_{ds} is a simple extension of the Greenwood and Williamson η already introduced but with 2-D sampling is found by a simple counting of all detected 9-point summits. The remaining 3 spatial parameters are all based effectively on the 2-D area autocorrelation function, AACF, of the surface heights, a function of the vector distance (delay) between two surface points. Considering the 2-D contours in delay space of this function enclosing the central maximum of value unity at the origin, S_{al} denotes the shortest distance from the origin to the 0.2 contour, while S_{tr} is the ratio of this to the longest distance to the same contour. For an isotropic surface, the contours are all circular, yielding $S_{tr} = 1$. The texture direction S_{td} is related to the direction in which this longest correlation distance occurs but is in fact found from the direction of maximum power in the 2-D power spectral density. These 3 parameters are the only ones giving information on the directionality or anisotropy of the surface, a vital aspect of characterization in 3-D.

3. Stability of Surface Parameters

After about five years of experience with digitized roughness profiles, certain problems began to emerge, chief amongst which were the dependence on sampling conditions of the computed parameter values. The root of this problem is simple to appreciate. Assume that the distribution of sampled heights (Abbott-Firestone curve) and their AACF are known. From these the expectation of, for example, the difference in height of two points can be found as a function of their separation. In other words, the mean surface slope can be found for a given sampling interval, δ. Similarly the expected curvature required in the Greenwood and Williamson model may be found. Typical results calculated by the method of Whitehouse and Archard (1970) for a 1-D profile are given in Fig.2, showing that as the sampling interval decreases, the slope $m = \Delta q$ may be expected to increase approximately as $\delta^{-1/2}$ and the mean peak radius β to decrease as $\delta^{3/2}$. This behavior reflects the well-known experimental observation that the measured rms slope and peak curvature of a surface increase as the radius of the stylus tip decreases. Corresponding behavior is found for $S_{\Delta q}$ and S_{sc} in 2-D (Greenwood, 1984).

This result clearly challenges any claim for such parameters to describe intrinsic properties of surface geometry and poses the problem of what actual values to use in models of surface behavior. At the same time it appears to establish a new basis for these parameters since, given the Abbott-Firestone curve, essentially S_q, and the AACF, essentially S_{al} and S_{tr}, then their values can be found for any δ. In the discussion of the proposed parameters in Stout *et al* (1993), their dependence on δ is of course pointed out and recommendations for values of δ are made. It is also noted that the variation of S_q, S_{al} and S_{tr} with δ is comparatively small, so that at least the height distribution and autocorrelation might seem "intrinsic". Unfortunately, this also does not hold since their values are influenced strongly, not by the smallest features included, but by the largest, i.e. by the linear dimension L of the sample area or by the cut-off length D, reciprocal of the cut-off frequency used in high-pass filtering. Some results are given in Fig.3, showing S_q increasing approximately as $D^{1/2}$ and a correlation length, effectively S_{al}, as $D^{2/3}$. Thus, the intrinsic nature even of the bearing area ratio and the autocorrelation function are in question.

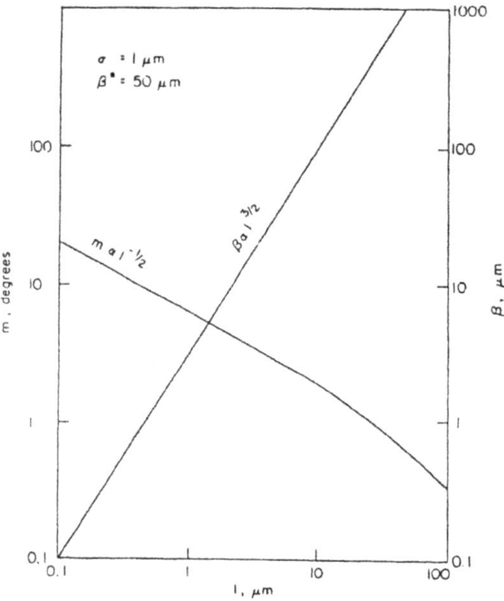

Figure 2. Variation of mean absolute slope, *m*, and peak radius of curvature, β, with sampling interval, *l*, plotted from Whitehouse and Archard's theory for a profile of rms roughness σ = 1μm and correlation length β^* = 50 μm. From Thomas and Sayles, (1978).

Turning attention next to S_{ds}, the summit density, it was mentioned earlier that for 3-point profile peaks, between $(1/2)^2$ and 1/3 of all sampled points may be expected to satisfy the peak condition if no threshold is set for the height difference between the peak and its two neighbours. In 2-D, a similar result holds for 9-point summits, where between $(1/2)^8$ and 1/9 of all sampled points are expected to be summits. Inspection of results given in Stout *et al* (1993) for a wide range of surface topographies reveals the summit density indeed falling comfortably in this range, between about 2% and 8%. Hence, S_{ds} proves to be another measure without basic physical significance.

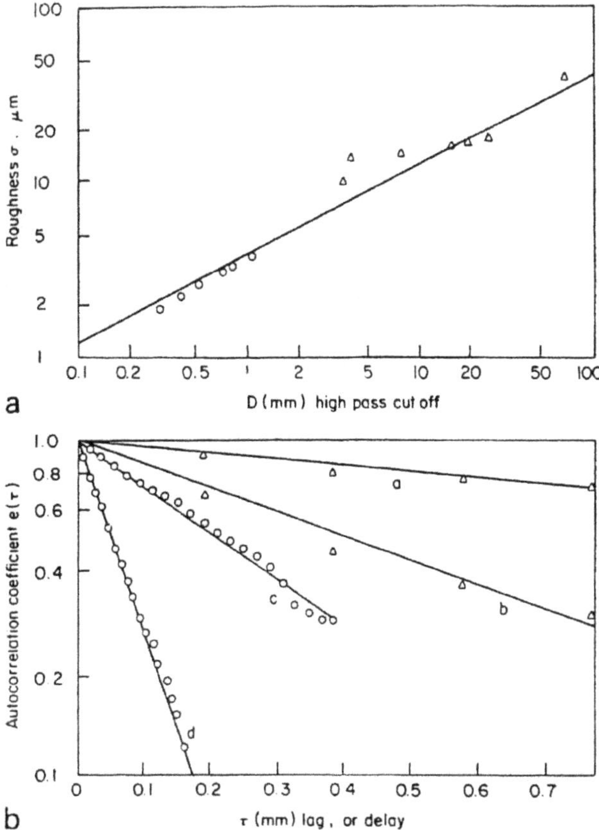

Figure 3. (a) Variation of rms roughness with high-pass cut-off for a profile from a grit-blasted surface. Solid line is best fit of slope 1/2.

(b) "Exponential" correlation functions from the same profile after high-pass filtering at cut-offs: a = 40 mm; b = 20 mm; c = 2 mm; d = 0.6 mm. From Thomas and Sayles, (1978).

At this point, without entering into detailed analysis of the remaining parameters, it may be conjectured that none of the usual roughness parameters is intrinsic to the topography, a conclusion already firmly stated by Whitehouse (1994). Once pointed out, this is not so surprising. Imagining the surface of a road as typical of any surface, the long wavelength hills and valleys clearly influence the average height deviation from a mean plane, S_q, whereas the small stones determine the slope, $S_{\Delta q}$, if the sampling interval is sufficiently small. Including both larger and smaller features, these effects will extend over many length scales.

4. A Functional Approach

4.1 FRACTAL SURFACES

Fortunately, this geometrical scaling behavior of surfaces has already received much attention under the name ***fractal geometry*** (Mandelbrot, 1982). A curve which exhibits many of the properties of a surface profile can be generated by the simple recursive construction sketched in Fig.4. Beginning with a unit line segment as ***initiator*** and a ***generator*** consisting of a line of length l_G whose middle third is replaced by two sides of an equilateral triangle, the curve is constructed by first replacing the initiator with the generator scaled to $l_G = 1$. For the second step the generator is scaled to $l_G = 1/3$ and then used to replace each line segment produced by the first process. Continuing, the initiator becomes divided into ever more but shorter segments, approaching in the limit a ***Koch curve***. In this construction, l_G represents δ, so that as the profile is sampled on ever smaller scale, the number of peaks or valleys and the average slope increase, while the local appearance of the curve remains exactly the same, just as for measured profiles. The 'length' of the Koch curve increases by the factor 4/3 at each stage and is thus actually unbounded. If the length of a unit measuring rod or tile is scaled by the factor $1/f$, then the number of rods required to cover a unit length becomes $N = 1/f$, while for a unit area, the number of tiles becomes $N = 1/f^2$. Thus, the Euclidean dimension, D, of an object is given by $D = \log N/\log(1/f)$. For the curve just constructed, this yields $D = \log 4/\log 3$, clearly a non-integral value between 1 and 2, named by Mandelbrot (1982), the ***fractal or similarity dimension***. A real surface with some degree of randomness cannot, of course, have the exact ***self-similarity*** of the Koch curve but, with different scaling factors in directions parallel and perpendicular to the surface, the geometry of the surface becomes instead ***self-affine***.

In the last 20 years, much attention has ben given to the fractal behavior of surfaces, where $2 < D < 3$. One reason was the hope that D would be a truly intrinsic, scale-invariant, property of the surface. Relevant in the present context, Majumdar and Bhushan (1990, 1991) have investigated some implications of fractal geometry for the mechanical contact of surfaces, comparing their model predictions with those of Greenwood and Williamson. A simple way to obtain D is from the power spectrum of the surface, of the form $S(\omega) \propto T^{(2D-4)}\omega^{(2D-7)}$, where ω is the spatial frequency and the ***topothesy***, T, is a scale invariant amplitude, whose magnitude is the distance between two points in the surface for which the rms slope is unity, i.e. 1 radian. Logarithmic plots of the measured spectra of many surfaces actually are piece-wise linear, so that D-values may indeed be regarded as intrinsic to the surface.

Another method of finding D, more directly related to the definition given above, is to measure the surface area as a function of δ. The hybrid parameter S_{dr} is a measure of the excess of the surface area over its projected area which, as shown in Stout *et al* (1993), increases quite rapidly as δ decreases. Brown *et al* (1994) quantify this δ-dependence by covering a sampled topographic area with triangular tiles of ever-smaller size. The resulting logarithmic plots of area against patch size are again piece-wise linear, usually with one, sometimes two, ***cross-over frequencies.*** A typical result is shown in Fig. 5. Both the slopes and the cross-over points of these lines have been found useful in understanding certain physical behavior of the surfaces.

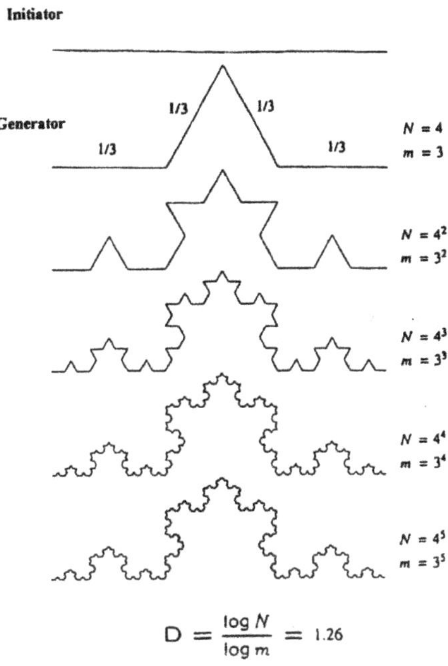

$$D = \frac{\log N}{\log m} = 1.26$$

Figure 4. Recursive construction of a Koch curve of dimension 1.26. From Majumdar and Bhushan, (1990).

4.2 FUNCTIONAL TOPOGRAPHY PARAMETERS

This raises a final point - the parameters defined and used to characterize a surface should be useful for describing some physical behavior of interest. Since it is now clear that surfaces contain features on many different scales, one of the most important considerations is to tailor the parameters to measure just those features whose scale allows them to influence the functional behavior. Such a concept, ***functional filtering,*** was introduced some years ago (Thomas and Sayles, 1978), but the present context shows more fundamentally why it is necessary. Physical modelling of the surface interaction of interest will reveal precisely what topographic features are most influential. The design of engineering surfaces with improved performance has already made great strides, so that for example, surfaces can be manufactured to have lower friction and wear, build thicker lubricant films, promote longer rolling contact fatigue life or better sealing. Each of these functions involves features of different scale, sometimes of two widely separated scales. In such cases, the topography is said to be ***patterned***, since the eye can readily recognize the set of larger features against the background of the smaller. Patterned surfaces also include those traditionally described as having ***surface lay***. For such topographies, the behavior may depend, for example, only on those heights in the vicinity of the larger valleys or summits.

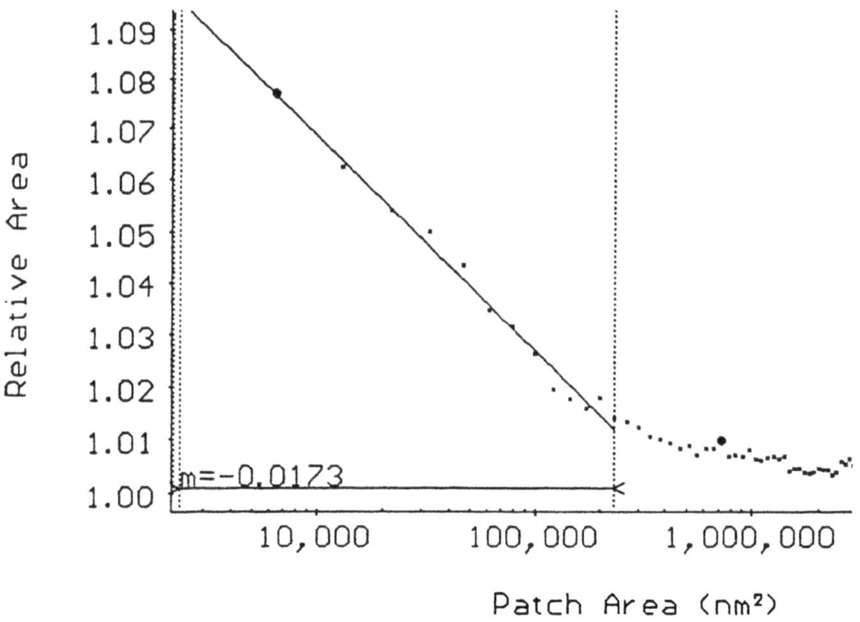

Figure 5. Scale-area plot showing relative areas calculated from a topographic data set acquired by atomic force microscopy. The data set consisted of 262,144 points about 18 nm apart on a square grid 9.44 μm x 9.44 μm. From Brown, (1994).

A good illustrative example of such a topography is the surface described by Akamatsu *et al* (1990), who used a tumbling process to impose a number of pits on otherwise smooth (by engineering standards) rolling bearing surfaces. The supposed function of these pits was to increase the elastohydrodynamic (EHD) film thickness in rolling contact. To verify this behavior, it is first necessary to build a theoretical EHD model for non-smooth surfaces and use it to find what size scales of pits affect the film. Another example is given by Vermeulen *et al* (1995) concerning electron beam textured (EBT) steel sheet. Here, pits were machined on the surface to improve the deep drawability and paintability of the sheet. To improve the drawability, the proposed function of the pits is, as for the bearing surfaces, to retain lubricant, while paintability depends on adhesion and developed surface area. The topography optimal for one of these functions is not optimal for the other. A model showing how each of these functions depends on pit size, depth, density and ***connectedness*** can then be used both to show the relevant scales for the pits and also to design the best compromise surface. In both these examples, the modelling allows the construction of algorithms to detect, count and measure just those pits of the relevant sizes, thereby imparting real physical meaning to parameters such as ***Developed Area Ratio***, S_{dr}, and ***Valley Fluid Retention Index***, S_{vi}, of Table 1 or ***Density of Pits***, S_{dp}, a counterpart to ***Density of Summits***, S_{ds}. With the feature size scale and shape specified, the algorithms return unique parameter values, provided of course that the sampling interval is always smaller than the smallest feature of interest. In general, such algorithms involve scales and thresholds for lengths both parallel and normal to the nominal surface. Gratifyingly, the software packages supplied with some commercial surface profilers have finally begun to incorporate such pattern recognition routines. If such a procedure is followed for each surface function of interest then, as these illustrations show, not only are the parameters well-defined and physically useful but also, not unexpectedly, their number tends to increase - a ***richness*** rather than a ***rash*** of parameters (Whitehouse, 1982).

'Conventional' parameters such as the 'Birmingham 14', often prove valuable in manufacturing process control, where the parameter does not control the behavior of the surface but indicates the behavior of the machine producing the topographic finish. This is a useful role in cases where the true parameters determining a surface function are not known but where, for a given finishing process, the required surface behavior is found to correlate with one or more of the conventional parameters. In such cases, the conventional parameter may be regarded as a ***representative*** for the unknown or ***hidden*** parameter. However, in changing to a different finishing process, this correlation is generally lost.

5. Conclusion

A surface, in fractal terms, lies somewhere between an area and a volume. Thus, conventional geometrical parameters used to characterize a surface all suffer from scale dependency, either from the high or low frequency cut-offs which all physical measurement processes have imposed on them. They do not characterize intrinsic properties of a surface. Such parameters may be valuable in process or quality control but for understanding the behavior and improving the performance of surfaces, greater appreciation of the nature and scale of features determining surface interaction is required. Introduction of parameters describing these features at the relevant scale means inevitably more, rather than fewer, surface topography parameters.

References

Abbott, E.J. and Firestone, F.A. (1933) Specifying surface quality, *Mech Eng* **55**, 569-572.

Akamatsu, Y., Tsushima, N., Goto, T. and Hibi, K. (1990) Improvement in oil film formation under rolling contact by controlling surface roughness pattern, *Proc Japn Intl Trib Conf (Nagoya)*, 761 - 766.

Brown, C.A. (1994) A method for concurrent engineering design of chaotic surface topographies, *J Mater Process Technol* **44**, 337-344.

Bush, A.W., Gibson, R.D. and Keogh, G.P. (1976) The limit of elastic deformation in the contact of rough surfaces, *Mech Res Commun* **3**, 169-174.

Greenwood, J.A. and Williamson, J.B.P. (1966) Contact of nominally flat surfaces, *Proc Roy Soc Lond A* **295**, 300-319.

Greenwood, J.A. (1984) A unified theory of surface roughness, *Proc Roy Soc Lond A* **393**, 133-157.

ISO 4287:1997(E/F) Geometrical product specifications (GPS) -Surface texture: Profile method - Terms, definitions and surface texture parameters.

Majumdar, A. and Bhushan, B. (1990) Role of fractal geometry in roughness characterization and contact mechanics of surfaces, *Trans ASME J Trib* **112**, 205-216.

Majumdar, A. and Bhushan, B. (1991) Fractal model of elastic-plastic contact between rough surfaces, *Trans ASME J Trib* **113**, 1-11.

Mandelbrot, B.B. (1982) *The fractal geometry of nature*, W.H. Freeman and Company, San Francisco.

Nayak, P.R. (1971) Random process model of rough surfaces , *Trans ASME J Lub Tech* **93**, 398-407.

Stout, K.J., Sullivan, P.J., Dong, W.P., Mainsah, E., Luo, N., Mathia, T. and Zahouani, H. (1993) The development of methods for the characterisation of roughness in three dimensions, Publication No. EUR 15178 EN, (Final Report) BCR, European Community, Brussels.

Thomas, T.R. and Sayles, R.S. (1978) Some problems in the tribology of rough surfaces, *Tribology International* **11**, 163-168.

Vermeulen, M., Scheers, J., de Mare, C. and de Cooman, B. (1995) 3D-characterisation of EBT-steel sheet surfaces, *Int J Mach Tools Manufact* **Vol.35**, No.2, 273-280.

Whitehouse, D.J. and Archard, J.F. (1970) The properties of random surfaces of significance in their contact, *Proc Roy Soc Lond A* **316**, 97-121.

Whitehouse, D.J. (1982) The parameter rash - is there a cure? *Wear* **83**, 75-78.

Whitehouse, D.J. (1994) *Handbook of surface metrology*, Institute of Physics Publishing Ltd, Bristol and Philadelphia.

Williamson, J.B.P., Pullen, J. and Hunt, R.T. (1970) The shape of solid surfaces, *Surface Mechanics, ASME Annual Winter Meeting, Los Angeles*, 24-35.

FRICTIONAL INSTABILITIES

N. S. Eiss
Virginia Polytechnic Institute and State University
Blacksburg, VA 24061-0238

Abstract

Some examples of the detrimental consequences of frictional instabilities such as stick slip and oscillations include annoying noises, positioning errors, poor surface roughness in machined materials, and irregular drug delivery. The causes of frictional instabilities are the negative slope of the friction-relative velocity curve, frictional coupling of two or more motions which have natural frequencies near each other, and lubricants which can crystallize. Several cures for instabilities such as stiffening the components of the system, adding damping, selecting materials with positive slopes of the friction-velocity curve, and separating the natural frequencies of the vibration modes are suggested.

1. Introduction

Friction instabilities such as stick-slip (Fig. 1) or oscillations (Fig. 2) are phenomena that occur in sliding systems. Some of the undesirable effects of frictional instabilities are noise and accelerated wear rates. There are many examples of frictional instabilities and their detrimental effects on performance and the costs that are associated with reducing and eliminating them.

The automotive industry has incurred large warranty costs because of customer complaints about noises generated by sliding surfaces in automobiles. These include brake and drive belt squeal, stabilizer bar bushing squawks, window seal itch and ticks, instrument panel squeaks, ride level sensor squeaks, and door seal pops and squeaks, Nolan et.al. (1996) These problems have arisen because the automobile manufacturers have been very successful in quieting the automobile interiors by reducing wind, engine, and road noises which used to mask these annoying noises.

Another transportation related friction-induced noise is the screech of railroad engine and car wheels on rails especially in curves in the track, Bleedhorn et.al. (1977). These noises result when the wheels roll and slide on the rails because the rigid axles force the two wheels to have the same rotational velocity on rails that have different radii of curvature. The sliding accelerates the wear of the rail thereby shortening the time between regrinding of the surfaces and ultimate rail replacement.

B. Bhushan (ed.), Tribology Issues and Opportunities in MEMS, 149-156.

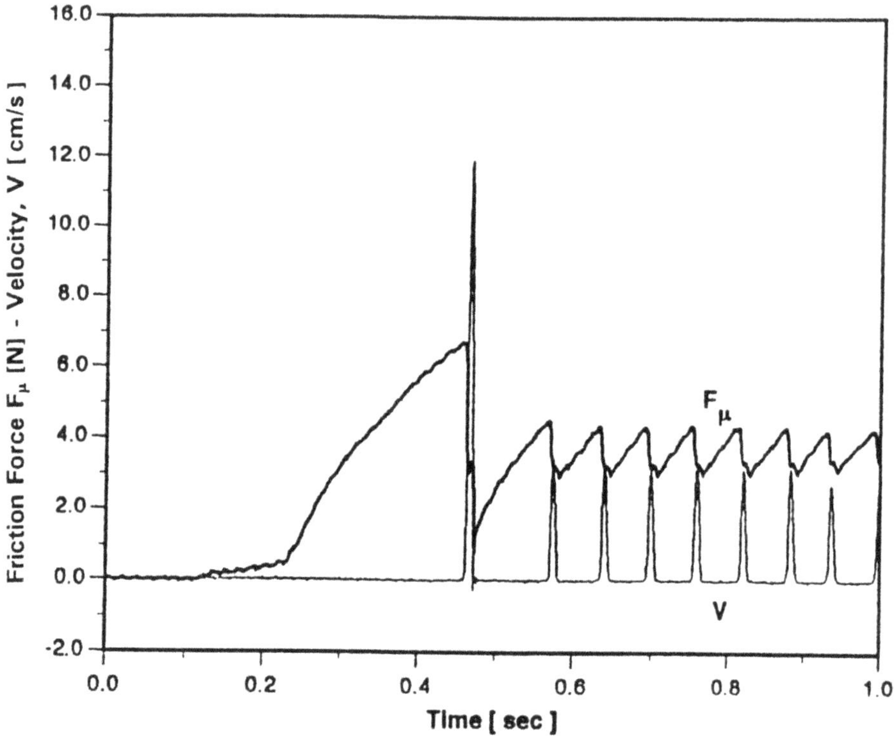

Figure 1 Example of stick-slip motion, ABS on ABS, 20.7 N normal load, 0.3 cm/s actuator speed , 3.2 N/m stiffness

Stick-slip limits the positioning accuracy of precision machine tools, robot manipulators, and large mass optical and radio telescopes. Stick-slip between the plunger seal and the cylinder causes the uneven delivery of drugs by automated syringes. The chatter of machine tools when cutting metal is caused by a friction instability. The chatter causes unsatisfactory roughness of the machined component.

Last, but not least, stick-slip causes the screech of finger nails or chalk on a slate board which has caused many a student and teacher to cringe. In this paper the causes of friction instabilities will be presented as well as some of the techniques that can be used to eliminate or minimize frictional instabilities. For those who would like an in depth review of friction instabilities, Ibrahim (1992) which has over 330 references is recommended.

2. Causes of Friction Instabilities

Friction instabilities are the unstable responses of systems which have surfaces in sliding contact to changes in friction, applied loads, displacements, or velocities. All systems have mass and components which are deflected by the applied loads. The

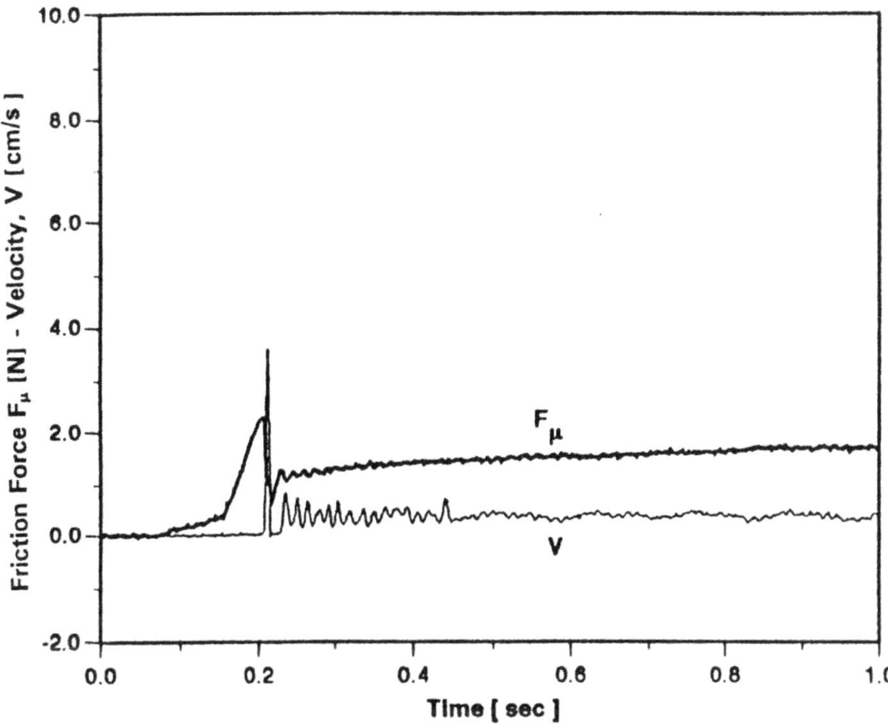

Figure 2 Example of harmonic oscillations, ABS-PC on ABS-PC, 10.7 N normal load, 0.3 cm/s actuator speed, 12.7 N/m stiffness

applied loads are necessary to initiate or maintain sliding. If the applied loads are in equilibrium with the friction then the system is at rest or is sliding with a constant relative velocity. If there is a change in friction, the system of forces are no longer in equilibrium and the net force causes the system to accelerate. The potential energy stored in the elastic deflection is converted to the kinetic energy of the accelerating masses. The resulting vibration will either decay as energy is dissipated in the system components or increase in amplitude if the system has destabilizing characteristics. If the amplitude of the oscillations is large enough to cause the relative velocity to become zero then the surfaces stick together and stick-slip motion is initiated. In the following sections characteristics of friction and the system which cause unstable motion will be discussed.

2.1 FRICTION-VELOCITY RELATIONSHIP

The stability of mechanical systems which consist of masses and elastic components is determined by the components which dissipate energy. The differential equation for the displacement, x, of a one-degree of freedom system consisting of a mass, M, damping constant C, and spring constant, K, acted on by a force, F, is:

$$M(d^2x/dt^2) + C(dx/dt) + Kx = F \tag{1}$$

If C is positive, the system is stable because the oscillations of free vibration decrease with time, t. When C is zero, free vibration amplitude is constant and when C is negative, oscillations increase in amplitude. If the force is friction which is represented by the following equation:

$$F = F_0 - \lambda\,(V - dx/dt) \tag{2}$$

where F_0 is the friction at zero relative velocity, (V - dx/dt), λ is the slope of the friction-relative velocity curve, and V is the absolute velocity of a second mass sliding on mass M. If Eq. 2 is substituted in Eq. 1 and rearranged:

$$M(d^2x/dt^2) + (C - \lambda)(dx/dt) + Kx = F_0 - \lambda V \tag{3}$$

If the slope, λ, of the friction-relative velocity, f-v, curve is greater than C, then the system will be unstable. If the system damping is small then the friction-relative velocity relationship for a pair of materials will likely determine if the system is stable or unstable.

Most material pairs have negative slopes of their f-v curves over some relative velocity range. Hard metal pairs have negative slopes over the all positive relative velocities tested. When a soft metal is one of the pair, the slope is positive at low velocities and negative at high velocities. Plastic pairs also have a maximum friction as a function of relative velocity and the value of the relative velocity at which the maximum occurs increases as the hardness of the plastic decreases, Rabinowicz (1995).

The measurement of the f-v curve used to present some problems when analog data was recorded on analog recorders. When (C - V) was negative both the friction and velocity values oscillated. Because of the low resolution of the analog recorders, the average value or the range of the friction fluctuations were plotted against the average velocity. With the advent of A-to-D interfaces and high speed sampling of the analog data, the variations of friction and velocity could be sampled simultaneously and then cross plotted to give the f-v curve. Figure 3 shows the f-v curve for the friction- and velocity-time data shown in Fig.1in which stick-slip followed the transition from static contact to relative motion.

These data were obtained on a test apparatus in which a polymer flat is slid on a polymer flat at a constant linear velocity, Eiss et.al. (1997) . The f-v curve was measured during the slip-phase of the stick slip motion. It is noted that the slope of the f-v curve is steepest at the lowest velocities and becomes less steep as the velocity increases. The double valued nature of the f-v curve is similar to that obtained on a variety of material pairs experiencing stick-slip motion under dry and lubricated conditions, Sakamoto (1987), Hess et.al. (1990), and Brockley et. al. (1970).

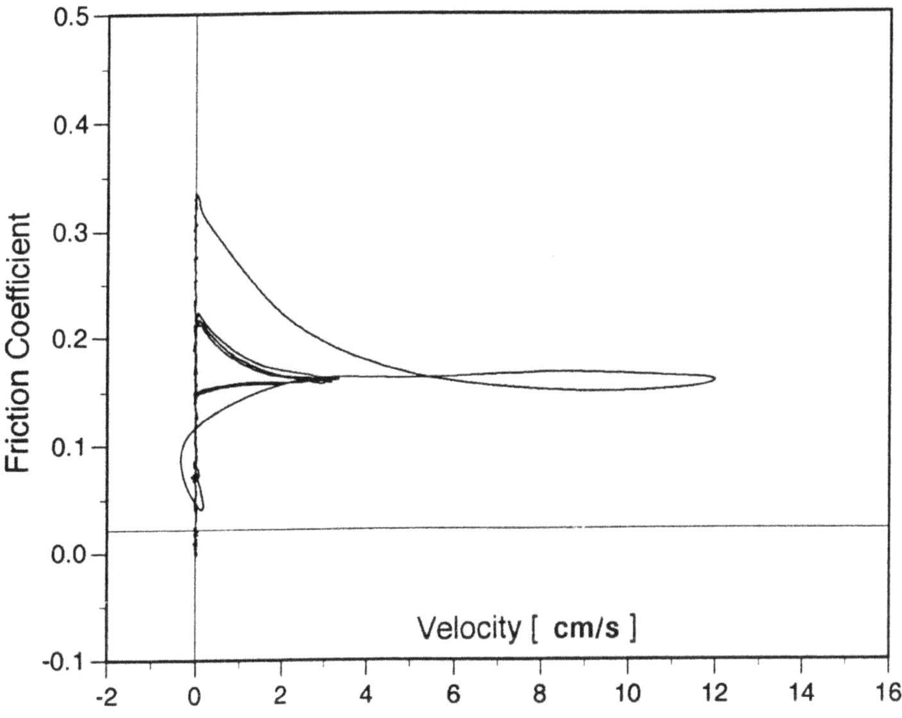

Figure 3 Friction-velocity curve for the data in Fig. 1

The explanation for the f-v relationship depends on the materials. Elastomers, such as rubber, have a positive slope of the f-v curve at low velocities and a negative slope at high velocities. This relationship is explained by the frequency dependence of the energy dissipated in the rubber when it is strained cyclically over a range of frequencies. The energy dissipation-frequency curves have the same shape as the friction-velocity curves. There is a direct correspondence between friction and energy dissipation and between sliding velocity and the rate at which the sliding materials are being strained. Thus, there is a velocity range at which friction is a maximum which corresponds to a frequency range at which energy is dissipated most rapidly in the elastomer.

The positive slopes of the f-v curve that occur at extremely low sliding velocities for soft metals and polymers are a result of diffusion and creep, respectively. During static contact or at very small motions, metallic atoms can diffuse across the contact area and cause an increase in the adhesive bonding forces at the interface. At polymer asperity contacts, the high local stresses cause creep deformation which increases the area of contact. Increases in adhesive bonds and contact area will cause friction to increase. These phenomena have also been used to explain why the static friction is often higher than kinetic friction. Once sliding commences, there is insufficient time for diffusion or creep to occur at the asperity junctions which are forming and breaking rapidly.

2.2 FRICTION COUPLING TWO OR MORE DEGREES OF FREEDOM

In this section, models are presented which indicate that frictional oscillations can occur when the slope of the f-v curve is zero, i.e. friction is independent of velocity.

Sliding systems are usually modeled with a single degree-of-freedom for motion tangential to the interface. However, measurements of motions normal to the surface during sliding by Tolstoi (1967) have spawned multidegree-of-freedom models of sliding bodies which included rotational as well as normal and tangential translation modes. In Oden et. al. (1985), a non-linear relation between normal load and asperity deflection defined the vertical stiffness and the friction was assumed independent of velocity. The model predicted an apparent coefficient of kinetic friction which was different from the interface coefficient of friction. In Tworzydlo et. al. (1992), a four degree-of-freedom model of two sliding bodies incorporated a normal motion for one body and normal, tangential and rotational motions for the second body. It was found that when the natural frequencies in the normal and rotational modes were near each other frictional coupling between these modes caused self-excited oscillations to occur.

In this and the preceding section models have been presented which predict frictional instabilities. When the slope of the f-v curve exceeds the system damping constant a departure from equilibrium will result in sustained self excited oscillations. For multidegree-of-freedom systems frictional coupling between different modes of vibration can cause self-excited vibrations when the natural frequencies are close in value.

3. Friction and Phase Changes

Both models in Sect. 2.1 and 2.2 require that a disturbance from equilibrium conditions initiate the instabilities. These disturbances could be a sudden change in friction, velocity, or displacement which cause a force imbalance or change the energy of the system. While there are many causes for sudden changes in friction, one cause which occurs in very thin liquid films on very smooth surfaces is a phase change from liquid to solid behavior. While this topic is covered in more detail in other papers in this volume, the phenomenon will be summarized here because these cyclic transitions can lead to stick-slip behavior in systems which apply forces through elastic members.

The process of lubrication by molecularly thin liquid films has been described in Israelachvili (1995). For surfaces at rest, liquid-solid surface interactions can cause the liquid molecules to become highly oriented (crystallize). The crystallized films behave as elastic solids. As tangential stresses are applied the solidified films experience elastic displacements and the two surfaces are stuck together with the film. As the applied stress reaches a critical value, the film suddenly melts (becomes disordered). The film reverts to its liquid properties and is unable to support the applied stresses. The resulting displacement relieves some of the force transmitted through the elastic components of the system as the system tries to establish an equilibrium position based on the value of the liquid viscosity. However, the inertia of the system causes it to overshoot the equilibrium position and the relative velocity goes to zero at which time the liquid refreezes which starts a new stick-slip cycle.

In Israelachvili (1995) it is noted that molecular structure has a strong influence on the ability of molecules to change phase from liquid to solid. Data indicates that there is a critical velocity above which stick slip disappears. Molecules which are spherical can rearrange very quickly and have high critical velocities and exhibit stick slip. Chain and branched chain molecules have much lower critical velocities and tend to slide smoothly. For example, cyclohexane and octane sheared between two pieces of mica surfaces at 20°C exhibit stick-slip sliding. Polydimethylsiloxane which has a molecular weight of 3700 has smooth sliding.

4. Cures for Friction Instabilities

The above causes suggest that there are several approaches to eliminate or reduce frictional instabilities which are described below.

- Increase system stiffness. Low stiffness components will store more potential energy than high stiffness components as they are deflected by the applied forces. Force imbalances cause components to accelerate and the potential energy is converted to the kinetic energy of the moving masses. Figure 4 shows that an increase in stiffness by a factor of four causes the response to change from multiple stick-slip to a single stick slip.
- Increase system damping constants. When oscillations occur and the system energy is fluctuating between potential and kinetic, energy dissipaters are needed to reduce the system energy and bring it to a stable equilibrium.
- Choose material pairs with f-v curve slopes which are positive at the operating relative sliding velocity. Negative slopes can negate the system damping and cause oscillations to be sustained.
- Separate the natural frequencies of friction-coupled modes of motion. Material pairs with friction which is independent of velocity can cause instabilities if the friction-coupled natural frequencies are close in value.
- Select lubricants with chain or branched chain structures for molecularly thin film lubrication: Lubricants with spherical shaped molecules are likely to experience cyclic solid-liquid behavior when used on atomically smooth surfaces.

5. References

Bleedhorn, T. G., Johnson, B., (1977) Steerable Steel Wheel System and Wheel Noise Suppression, IEEE Industry Application Society Annual Meeting Paper No. 37E, 957-963

Brockley, C. A., Ko, P. L., (1970) Quasiharmonic Friction-Induced Vibration, *ASME J. Lubrication Tech.*, **92**, 550-556

Eiss, N. S., Lee, E., Trapp, M. (1997) Frictional behavior of Automotive Interior Polymeric Material Pairs, Society of Automotive Engineers Paper No. 972056

Hess, D. P.,, Soom, A., (1990), Friction at a Lubricated Line Contact Operating at Oscillating Sliding Velocities, *Jl. Tribology*, **112**, 147-152

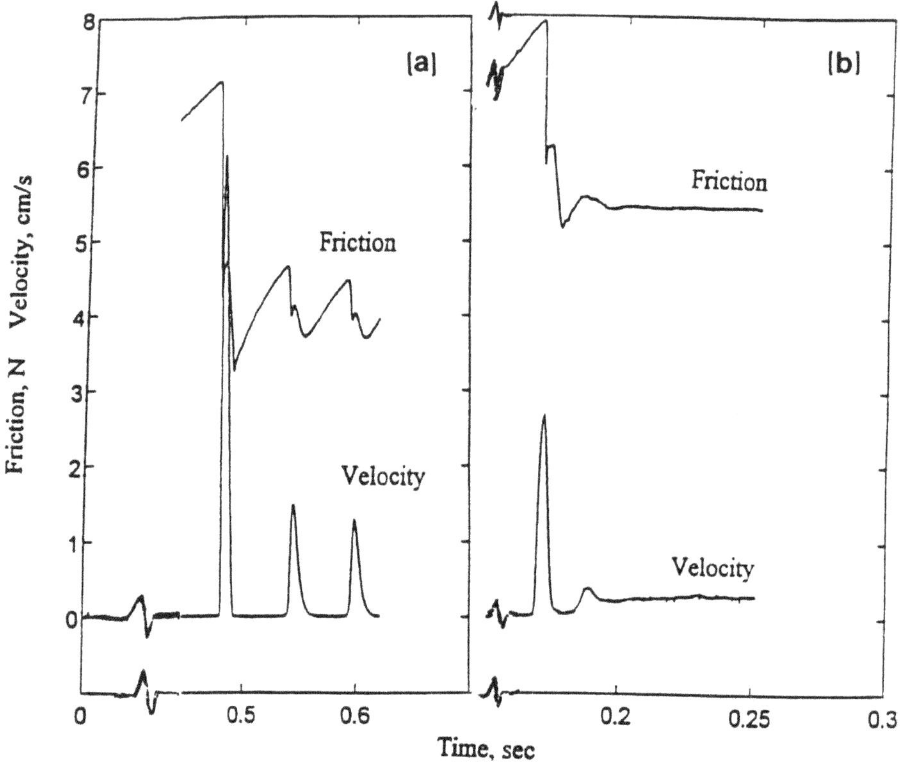

Figure 4 Friction- and Velocity-time plots for ABS on ABS, 20.7 N normal load, 0.3 cm/s actuator speed, (a) 3.2 N/m stiffness, (b) 12.7 N/m stiffness

Ibrahim, R. A., (1992), Friction-Induced Vibration, Chatter, Squeal, and Chaos, Part I - Mechanics of Friction, 107-122, Part II - Dynamics and Modeling, 123-138, in *Friction-Induced Vibration, Chatter, Squeal, and Chaos*, Ed. Ibrahim, R. A., Soom, A. ASME DE-Vol 49, ASME, NY, NY

Israelachvili, J. N., (1995), Surface Forces and Microrheology in Molecularly Thin Liquid Films, in *Handbook of Micro/Nano Tribology*, Ed. Bhushan, B., CRC Press, Boca Raton , Fl, 268-319

Nolan, S., Radiers, B., Loftus, H., Leist. T., (1996), Vehicle Squeak and Rattle Benchmarking, *Proc. 14th Inter. Modal Analysis Conf.* 12-15 Feb. 1996, Dearborn, MI, Soc. Exp. Stress Analysis, Bethel, CT, 483-489

Oden, J. T., Martins, J. A. C., (1985), Models and Computational Methods for Dynamic Friction Phenomena, *Compr. Methods and Appl. Mech and Engineering*, **52**, 527-634

Rabinowicz, E. (1995), *Friction and Wear of Materials*, John Wiley & Sons, NY, NY, 111

Sakamoto, T., (1987), Normal Displacement and Dynamic Friction Characteristics of the Stick Slip Process, *Tribology Inter.*, **20**, 25-31

Tolstoi, D. M., (1967), Significance of the Normal-Degree-of-Freedom and Natural Normal Vibration in Contact Friction, *Wear*, **10**, 199-213

Tworzydlo, W. W., Becker, E. B., Oden, J. T., (1992), Numerical Modeling of Friction-induced Vibration and Dynamic Instabilities, in *Friction-Induced Vibration, Chatter, Squeal, and Chaos*, Ed. Ibrahim, R. A., Soom, A. ASME DE-Vol 49, ASME, NY, NY, 13-32

WEAR OF CERAMICS AND METALS

T.E. FISCHER
Stevens Institute of Technology
Hoboken NJ 07030

Abstract

The principal wear modes likely to occur in MEMS will be briefly described. Wear is a material's response to the stresses occurring in a moving mechanical contact. The local contact stresses depend on the surface topography, which is usually modified by the wear processes, and by plastic or elastic deformation of contacting asperities. In metals, the contact stresses are roughly equal to the hardness of the softer material, wear occurs by plastic deformation and fatigue. The amount of material removed is roughly proportional to the contact load, sliding distance and inversely proportional to the hardness. In ceramics, three main wear modes occur, depending on the contact load: microfracture on a sub-grain scale at low loads, grain-boundary fatigue at intermediate loads and macroscopic fracture at high loads. (The latter is unlikely in MEMS). Wear is also influenced by chemical reactions of the sliding materials with the environment or with each other. Such reactions are often accelerated by friction, in which case they are called tribochemical. Tribochemical reactions can increase wear, when they increase fracture rates; they can decrease wear when they produce a smooth surface or a soft reaction product and reduce contact stresses; or they can produce a lubricating layer that decreases friction and wear. When dissimilar materials slide against each other, the softer material usually wears more and is transferred to the harder surface, except when the latter wears by tribochemical reaction. Practical examples are presented.

1. The Wear Modes

Depending on the design and the utilization of the system, one observes different mechanisms of wear. These are caused by different stress distributions and respond to different properties of the materials. Abrasive wear is caused by hard particles that penetrate the surface and remove material by scratching. It is encountered in abrasive machining, in earth-moving equipment and whenever a hard particle (from contamination or wear) is trapped between the two sliding surfaces. Fatigue wear is prominent in rolling-element bearings: it is a classic fatigue phenomenon caused by the cyclic application of the large compressive stresses in rolling contacts. Sliding wear, also called adhesive wear, occurs when two relatively smooth surfaces slide on each other. Since this is the wear mode to be expected in MEMS, we shall concentrate our attention to this phenomenon. Tribochemical wear plays a large role in the sliding of ceramics. It is the combination of mechanical and chemical attack. Chemical reactions can be vastly

B. Bhushan (ed.), Tribology Issues and Opportunities in MEMS, 157-164.

increased by simultaneous friction; these reactions can increase or decrease wear, depending on whether they accelerate crack propagation or when they remove stress concentrations on the sliding surfaces. In the following, we will describe the mechanical aspects of sliding wear first and will describe the modifications to it brought about by tribochemical phenomena.

2. Sliding Wear

Wear of a material is its response to the localized stresses caused by the normal forces pressing two surfaces together and by the friction that occurs at the contact. In order to understand the wear mechanisms and wear resistance of materials, it is necessary to have a clear view of the stresses caused by the contact load and the friction forces.

2.1. CONTACT STRESSES

The contact stresses are determined by the macroscopic forces and by the microgeometry of the contact, namely the design shape and the surface roughness of the moving parts. The compressive stresses are caused by the load and the roughness of the surfaces; the frictional stresses are determined by the adhesive interaction between the surfaces in the areas of real contact. These are modified profoundly by the chemistry of the surface. In particular, the adsorption of boundary lubricants (also called friction modifiers) reduces the friction stresses by, usually, one order of magnitude. The topography of real surfaces and their effect on contact stresses are discusses in the contributions of Tripp, of Kapoor, and of Danyluk in this volume. In order to describe the wear mechanisms, it is important to realize that these stresses are usually tri-axial and inhomogeneous. In metals, the asperities deform plastically under the load; thus the local compressive stress under the asperities is equal to the hardness and independent of the applied load. [1,2]. In ceramics, asperities deform elastically and the contact stress depends on the radius of the asperity and is again, is first approximation, large but independent of the applied load according to Greenwood and Williamson [3]. An increase of the load increases the density of contacting asperities and the real contact area [1-3].

The stresses contact stresses are illustrated schematically in figure 1. Below each asperity the stresses can be described, on a microscopic scale, by the Hamilton-Goodman theory [4] corresponding to the asperity geometry and the fraction of the load carried by the asperity (region A in Fig. 1). Between asperities, the stresses from neighboring asperities add to each other, they decrease with increasing depth below the surface. Since the distance between asperities decreases with increasing load, these stresses are load dependent in a complex fashion. This is region B in Fig. 1. At sufficiently large depth, the addition of the stresses from all the asperities amount to the macroscopically calculated stress [4] that corresponds to the design geometry and the applied forces. This is region C in Figure 1.

Wear is the response of the materials to these stresses and depends on the mechanical and chemical properties of the material in question. Adhesion between the surfaces causes friction and shear stresses underneath the asperity contact and tensile stresses behind it. These are the stresses that are responsible for the plastic deformation and fracture that constitute wear. It follows that wear is a function of friction and is reduced by lubrication.

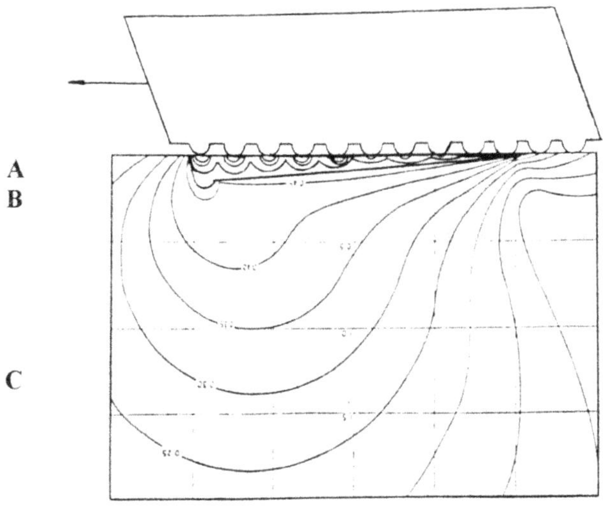

Figure 1. Schematic of the stress distribution underneath a sliding rough surface. A: surface region, B: intermediate region, C: region corresponding to continuum mechanics.

2.2. MECHANICAL WEAR OF CERAMICS

Mechanical wear of ceramics occurs by fracture. The nucleation of cracks is still poorly understood. Most theories of fracture consider the propagation velocity of a preexisting crack. It is well known that this crack propagation velocity is zero up to a threshold stress and increases very rapidly with the increase of the stress. It is a highly non-linear dependence. The above description of the stresses under a friction contact allows us to understand the observed wear of ceramics.

The large local stresses on a microscopic scale just underneath the asperities cause a large number of microcracks on a scale smaller than the grains of the ceramic. We call this phenomenon "microchipping". Since these stresses decrease rapidly with increasing depth, microchipping is shallow. The intensity of these stresses does not depend on the applied load, but the number of asperities is proportional to it. Thus the density of microcracks and, therefor, the volume of material removed by microchipping is roughly proportional to the load.

A few micrometers below the surface, stresses are smaller, but still larger than the macroscopic stress. Since they are caused by the addition of stresses from different asperities, they increase with load. These stresses are insufficient for crack nucleation; if the load is large enough, they cause grain boundary fatigue. The resultant wear mechanism is intergranular fracture and loss of entire grains. This stress depends on the density of asperities, thus on the applied load and fracture increases very rapidly with stress above a threshold. Consequently, grain boundary fatigue wear increases rapidly with load. Experimental observations [5,6] suggest that the wear rate increases with the fifth power of the load (increasing the load by a factor 4 increases the wear rate 1,000 fold).

When the load becomes large, the macroscopic stresses are large enough to cause propagation of a macroscopic crack from the surface. This causes the removal of large chunks of ceramic and constitutes "catastrophic wear" at high loads [7,8].

Wear of ceramics is thus takes on a rather complex dependence on applied load, sliding distance, contact geometry and material properties as shown in figure 2. For ceramics, the fracture toughness is a more reliable predictor of wear resistance than hardness [9], but even at similar hardness and toughness, the properties of the grain boundaries affect wear since grain boundary fracture is capable of relieving stresses and preventing the macroscopic failures.

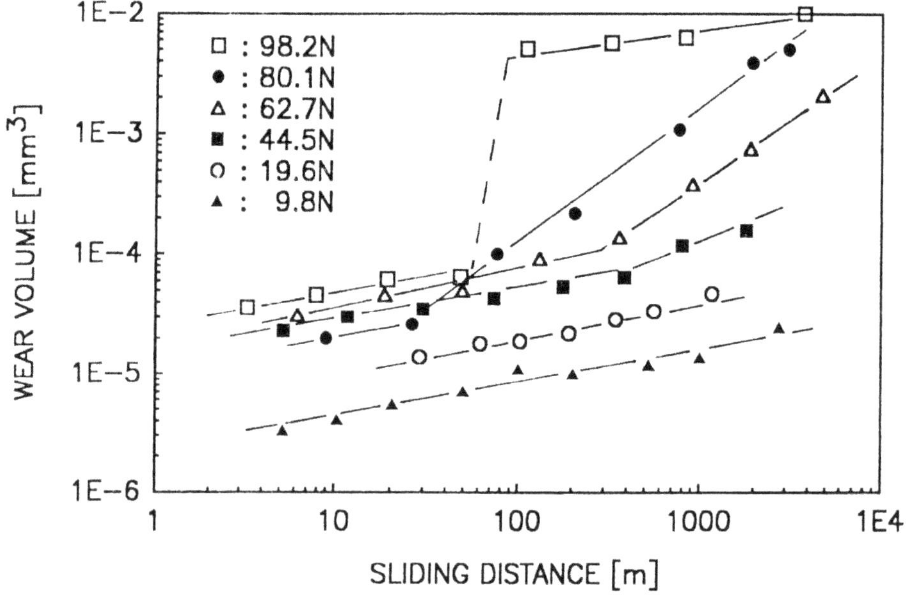

Figure 2. Wear of aluminum oxide (from Ref. 5). Mild wear with micro-chipping at 9.8 and 19 N, grain boundary fatigue 62.7 and 80.1 N, macroscopic fracture at 98.2N

2.3. MECHANICAL WEAR OF METALS

Metals respond to the shear stresses at the contact by plastic deformation. The resulting strains ε are very large. Thus work hardening and eventual fracture take place. In dry sliding, macroscopic fractures behind the contacts are usually observed. The amount of strain and work hardening is similar to those encountered in hardness testing, so that the wear resistance of metals is, in first approximation, proportional to their hardness rather than their yield stress. Most engineering alloys contain hard particles for strengthening, thus one must expect a considerable abrasive component to the wear of metals; this is commonly called plowing.

2.4. CERAMICS SLIDING ON METALS

When a ceramic slides on a metal, wear occurs mostly in the latter because it is softer. As a consequence, wear particles from the metal are transferred on the ceramic surface where they form a flat and adherent layer because of continued friction. Sliding occurs then mostly between two metal surfaces. The ceramic wears much less than when sliding against another ceramic, because the contact stresses are smaller. Wear of the ceramic is caused by fatigue due to the cyclic stresses transmitted through the metal layer.

3. Tribochemistry

Since friction, in its simplest form, involves the making and breaking of adhesive bonds between the sliding, it is almost obvious that chemical reactions of the bodies with each other, with the environment or with liquid lubricants play a major role in tribology. It is widely observed that chemical reaction rates are strongly modified, usually accelerated, by the simultaneous occurrence of friction. This phenomenon bears the name of tribochemistry and has been extensively studied in the Institute of Peter-Adolf Thiessen in Berlin [10]. This laboratory was mainly concerned with tribochemistry in ball milling, but these phenomena are equally important in sliding [11].

3.1. MECHANISMS OF TRIBOCHEMISTRY

The most obvious mechanism by which friction increases the rate of chemical reactions is frictional heat, which has been used since the earliest times for the production of fire, but one can observe or imagine various mechanisms by which friction accelerates chemical reactions, even at room temperat re{12].

Temperature effects are, of course important. One distinguishes between general increase in temperature and flash temperature [13]; the latter consists of the temperature flashes that occur at contacting asperities. The increases in reaction rates caused by frictional heat are no different from those caused by other increases in temperature.

When the sliding velocity and load are kept low to avoid frictional heating, one observes that other mechanisms operate by which friction increases reaction rates. One such mechanism is the exposure of clean surfaces. It is well known that the rates of chemical reactions are controlled by diffusion of reagents through the layer of reaction product that is formed on the surfaces. Wear exposes fresh surfaces and accelerates the reaction; in steady state, the reaction rate equals the rate of removal of the reaction product. Since wear constitutes a severe deformation of material on a small scale, it can introduce defects in the surface layer which serve as high-energy sites with increased reactivity on the surface and as diffusion paths that cause large increases in diffusion-limited reaction rates in the subsurface region.

Rubbing and fracture cause charge separations in ionic materials and set up large electrostatic potentials that can lead to discharge in the surrounding gas. With metallic bodies separated by a lubricant, electrochemical potentials are often observed. In addition, exo-electron emission is often quoted as a cause of tribochemical reactions despite the very low emission currents usually observed.

Finally, there is the possibility of direct mechanical stimulation of chemical reaction. Consider two atoms that are separated by large mechanical stresses. As the distance of these atoms increases, the energy splitting between the highest occupied molecular orbital (HOMO, usually bonding) and the lowest unoccupied molecular orbital (LUMO, usually antibonding) decreases. This diminishes the activation energy of the electron transfer taking part in a chemical reaction. Such phenomena occur only where atoms are separated; namely, in friction and in fracture. In solids, such a mechanism participates in the phenomenon of adsorption-induced fracture or stress corrosion cracking [15,16]; it is responsible for the wear increase of oxide ceramics by water and polar hydrocarbons.

3.2. TRIBOCHEMISTRY AND WEAR OF CERAMICS

The tribochemistry of ceramics takes several forms, depending on the materials, the environment and the mechanical conditions of rubbing; it can consist in chemically induced cracking that increases wear rates [14], in modifications of surface composition and topology that decrease wear [15], in a purely chemical form of wear (by dissolution in the liquid environment) [16].

The ambient humidity has a pronounced effect on the wear of silicon nitride and other ceramics, not only the amount of wear, but the wear mechanism itself is modified by humidity. Silicon nitride wears rapidly in dry argon, but if the environment contains various amounts of water vapor, the wear rate decreases by as much as two orders of magnitude[15]. Under these conditions, the wear scar is much smoother than after sliding in dry gases and is covered with an amorphous silicon oxide, which is probably strongly hydrated. The result is a reduction of the local stresses responsible for the mechanical wear. In the absence of friction, measurable oxidation rates of silicon nitride are obtained only above 1000 K [17]; they are increased one thousand fold by the presence of humidity in the air [18]. During friction, massive oxidation is obtained even at room temperature[15]. How this occurs exactly has not been determined yet. One can speculate that the reaction is accelerated because the hydroxide formed on the surface is continuously removed and a fresh surface is exposed by friction, but clear experimental evidence on the mechanisms still lacking.

In zirconia, humidity increases the wear rate by at least one order of magnitude. No reaction products are observed, but intergranular fracture is increased by the phenomenon of adsorption embrittlement[14]. The large stresses at the crack tip increase the reactivity of the bonds, a concerted dissociation reaction takes place with water molecules; consequently the cohesive bond of the material is decreased and the crack propagates at lower stresses than in dry environment.

Diamond-like carbon films likewise are known to interact with the environment that changes the friction coefficient and the wear of the material.

When silicon nitride slides in water, the tribological reaction is a dissolution (16) of the material at the contacting asperities. The resultant surfaces are so flat that hydrodynamic lubrication is obtained even in water at low sliding speeds. This modified topography of the surfaces also reduces the contact stress concentrations and therefor suppresses micro-cracking and other mechanical wear phenomena that introduce defects in the material. This phenomenon has been used to develop a novel polishing techniques that produces surfaces with a roughness of 5 Angstrom [19]. It is also the basic phenomenon that underlies the novel Chemical-Mechanical Polishing (CMP) presently in use in the integrated circuit industry.

3.3 TRIBOCHEMISTRY AND WEAR OF METALS

Oxidative wear of metals is a combination of oxidation and removal of the oxide. It has been shown that the oxidation reaction is accelerated by friction and that this creates an oxide surface film that shears preferentially to the substrate with consequent decrease in wear . A specific dependence of the wear rate on the sliding speed allows to compare the wear rate with a chemical kinetic model that provides information on the mechanisms by which friction modifies the reaction rate.[20] At high sliding velocities, the increase in surface temperature by frictional heat is considered the prime cause of oxidational wear[21]. Tribochemical phenomena also underlie the action of lubricants, especially their friction modifiers, antiwear and extreme-pressure additives.

4. Conclusion

This brief survey endeavored to show that a number of basic principles allow us to gain at least an approximate understanding of wear phenomena. Wear is an application of mechanical and chemical phenomena in response to the contact stresses and the chemical environment. From the mechanical viewpoint, it is importance to recognize the importance of the scale at which the phenomena occur. Depending on the material and the service, these phenomena occur on a microscopic scale, smaller than the grain structure, on a mesoscale where grain boundaries are important or on a macroscopic scale well described by continuum mechanics concepts. Likewise, one must consider the material properties at the different scales: the behavior of the basic material of the grains, the grain boundary properties and the combined properties that make up the macroscopic behavior. In addition, one must consider the phenomenon of, tribochemistry: during friction, chemical reactions will occur, at room temperature, in ways that they would otherwise be observed only at high temperatures. These phenomena, if well understood, can be utilized to combat wear.

Acknowledgments

This review is based on work supported by the National Science Foundation under grants #MSM-8814312 and CMS-9414976. Any opinions, findings and conclusions or recommendaions expressed in this material are those of the author and do not necessarily reflect the views of the National science Foundation.

References

1. Archard, J. F. : Contact of Rubbing Surfaces, *J. Appl. Phys.* **14**(1953), 891-98
2. Tabor, A.: A simplified account of surface topography and the contact between n solids, *Wear* 32 (1975), 269
3. Greenwood, J.A. and Williamson, J.B.P.: The contact of nominally flat surfaces, *Proc. Roy. Soc., London*, **A296** (1966), 300
4. Hamilton, G. M. and Goodman, L.E.: The stress field created by a circular sliding contact, *J. Appl. Mechanics* **33** (1966), 371-6
5. Kim H., Shin D.S. and Fischer T. E.: Mechanical and Chemical Aspects in the Wear of Alumina, *Proc. Jpn. Intl.Tribol. Conf., Nagoya* (1990), 1437-42
6. Liang, H. and Fischer, T. E.: Effect of grain boundary impurities on the mechanical and tribological properties of zirconia surfaces, *J. Amer. Ceram. Soc.* **76** (1993), 325-29
7. Hsu S.M., Lim D.S., Wang Y.S. and Munro R. G.: Ceramic wear maps: concept and method development, *Lubr. Engr.* **47** (1991), 49-55
8. Jahanmir, S. and Dong X.: Mechanism of mild to severe wear transition in alumina, *J. Tribology* **114** (1992), 403
9. Fischer T. E., Anderson M.P. and Jahanmir S.: Influence of fracture toughness on the wear resistance of yttria-doped zirconium oxide *J. Amer. Ceram. Soc.* **72** (1989), 252-57
10. Heinicke, G. *Tribochemistry,* Carl Hanser Verlag, Munich, 1984.
11. Fischer T. E.: Tribochemistry, *Ann. Rev. Mater. Sci.* **18** (1988), 303-23
12. Archard, J. F.: The temperature of rubbing surfaces, *Wear* **2** (1972), 438
13. Michalske T.A., Bunker B.C. and Freiman, S.W.: Stress corrosion of ionic and mixed ionic/covalent solids, *J. Amer. Ceram. Soc.* **66** (1983), 184-88
14. Fischer T. E., Anderson M.P., Jahanmir S. and Salher R.: Friction and wear of tough and brittle zirconia in nitrogen, air, water, hexadecane and hexadecane containing stearic acid, *Wear* **124** (19880, 133-148
15. Fischer T.E. and Tomizawa H.: Interaction of microfracture and tribochemistry in the friction and wear of silicon nitride, *Wear* **105** (1985) 29
16. Tomizawa H. and Fischer T. E. : Friction and Wear of silicon nitride and silicon carbide in water: hydrodynamic lubrication at low sliding speed obtained by tribochemical wear *ASLE Trans.* **30** (1987) 41
17. Kiehle A.J., Heung L.d. , Gielisse P.J. and Rockett T. J.: Oxidation of hot-pressed silicon nitride, *J. Amer. Ceram. Soc.* **58** (1975), 17
18. Singhal S.C.: The effect of water vapor on the oxidation of hot-pressed silicon nitride and silicon carbide, *J.Amer. Ceram. Soc.* **59** (1976), 81
19. Hah S.R. and Fischer T. E. : Tribochemical polishing, submitted to *J. Electrochem. Soc.* (1997)
20. Fischer T. E. and Sexton M. D.: The tribochemistry of oxidative wear, *Physical Chemistry of the Solid State: Applications to Metals and their Compounds,* P. Lacombe Ed. , Elsevier, 1984, p.97
21. Quinn T. F. : Theory of oxidative Wear, *Tribology Intl.* **16** (1983) 257 and 305

MECHANICS OF WEAR: FROM CONVENTIONAL COMPONENTS TO MEMS

F.J.FRANKLIN AND A.KAPOOR
The University Of Sheffield
Department Of Mechanical Engineering
Mappin Street, Sheffield, S1 3JD, UK

Abstract. Recently principle of mechanics have been used successfully to model wear from first principles. Two basic mechanisms have been considered: in one, the material fractures and produces wear debris; in the other, the material below the surface is compressed and extrudes out in the form of thin slivers leading to loss of material from the contact. While the latter mechanism is not intrinsically a fracture process, fracture occurs to separate the wear debris which, otherwise, would remain attached just outside the contact. The analyses lead to Archard-type relationships and the wear coefficient may be predicted by considering material properties, surface topography and operating conditions.

The current analyses are correct under the normal working conditions of conventional tribological components; however, in their present form, they cannot be used for MEMS. Adhesion, lubrication and material properties are vital issues which need to be addressed. Adhesion becomes very important for lightly-loaded smooth surfaces. The wear rate under such conditions is expected to be greatly under-estimated by the present theories.

1. Introduction

Wear under different combinations of load, friction coefficient and atmospheric conditions takes different forms. The mechanism of wear debris formation has been a subject of many excellent reviews [1, 2]. The common mechanisms in the wear of a mild material are: (i) fracture of a thin surface layer leading to flake like debris, as shown in Figures 1(a) and (b), and (ii) extrusion of material from below the contact into thin slivers, as shown in

B. Bhushan (ed.), Tribology Issues and Opportunities in MEMS, 165-174.

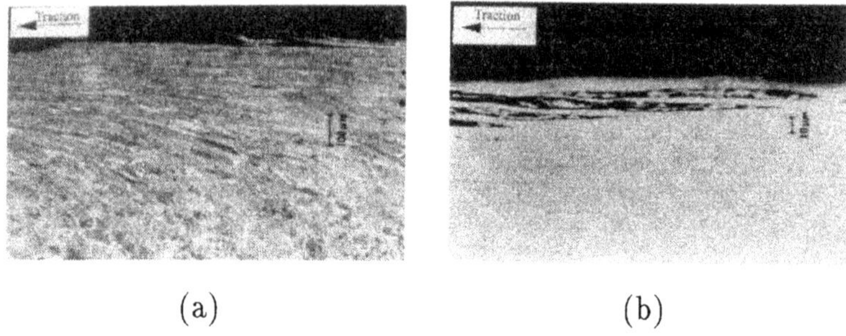

(a) (b)

Figure 1. Delamination of a thin surface layer leading to debris formation.

Figure 2. Extrusion of thin slivers from the sides of the contact, leading to wear.

Figure 2. These slivers may subsequently break off to provide debris or may remain attached to the sides of the contact.

The mechanism involving fracture of surface material has received considerable attention. Suh's group at MIT, USA worked extensively on this problem in the seventies. He suggested that flake like debris can be produced by delamination (hence the name 'delamination wear'), i.e. by propagation of a sub-surface crack and spalling of the surface [3]. The crack propagation was analysed by the research groups of Johnson at Cambridge University, UK [4, 5], Barber [6] and Keer [7] in USA and Murakami in Japan [8]. However, the LEFM based approach is faulty in: (i) it does not explain the initiation of cracks; (ii) it predicts the cracks to propagate in Mode II, but in laboratory simulations these cracks turn at an angle and propagate in Mode I; and (iii) more importantly, the crack is in a heavily plastically deformed region (as evidenced by cross-sections of worn surfaces) and so the assumption of LEFM becomes invalid.

Kragelski suggested [9] that wear is fatigue process. Cyclic stressing occurs both through repeated sliding and repeated asperity encounters. Oxley's group in Australia uses Coffin-Manson low cycle fatigue equation (1), below, to estimate the wear rate of material being slid by a wedge shaped asperity [10].

$$\left(\frac{\Delta\epsilon_p}{2}\right) N_f{}^n = C \tag{1}$$

A similar approach is used by Torrance's group in Ireland [11]. These approaches, however, have the following drawbacks: (i) for sliding by a smooth hard surface there would be no plastic deformation in the steady state (i.e. a state of elastic shakedown) and the Coffin-Manson equation would not apply; (ii) even for asperities with large slopes, the typical plastic strain cycles are not closed loops, normally associated with low cycle fatigue. Besides the reversing plastic strain acting parallel to the surface, there is accumulation of shear strain confirmed in cross-sectional views of worn surfaces. The use of Coffin-Manson equation to model wear in this case then becomes suspect.

2. A Ratchetting Failure Based Approach

Material under a sliding contact accumulates plastic strain with each load pass, a phenomenon known as ratchetting. Kapoor [12] found that a material element subjected to ratchetting fails by either low cycle fatigue, when the number of cycles to failure is given by the Coffin-Manson equation (1) or it fails after accumulating a critical strain (the so-called 'ratchetting failure'), which is comparable to strain to failure in a monotonic test. The number of cycles to failure by ratchetting is given by equation (2),

$$N_r = \epsilon_c / \Delta\epsilon_r. \tag{2}$$

He considered these two failures to be competitive so that the actual failure corresponds to the one occuring first, i.e.

$$N = \mathrm{Min}(N_f, N_r) \tag{3}$$

A wide range of tests in the literature was revealed to follow the above hypothesis, including some of Coffin's early experiments.

The above hypothesis has been used to model delamination wear [13]. Essentially, the wearing material is visualised as made up of layers – each layer fails by either ratchetting or low cycle fatigue and leads to wear debris. After the first layer has delaminated the second layer is exposed and so on.

The first layer has no accumulated strain at the start of rolling/sliding. Consider it to fail by ratchetting after N cycles as given by eq. (3). This would correspond to a certain wear rate. During failure of this layer (the first N cycles) other layers also accumulate strain. When layer 2 reaches the surface it has accumulated some strain, so the number of cycles to failure would be expected to be less than N, and the wear rate would be correspondingly higher. However, if the material strain hardens, this layer will have a higher shear yield strength, and for the same contact pressure, will undergo a smaller plastic deformation per cycle. This would lead to an increase in the number of cycles to failure and a reduction in the wear rate. The net change would be the difference of these two effects. Layer 3 would have a different wear rate. After some time all the layers reaching the surface will have the same accumulated strain and hardness as their predecessor, and so the wear rate would become constant; i.e. a steady state would be achieved.

Tyfour, Beynon and Kapoor [14] conducted wear experiments on a twin disc machine, in which a pearlitic rail steel (BS11) was run dry against a railway wheel steel (W8A) under 1500 MPa maximum contact pressure (line contact condition) and 1% slip. Wear occurred by both plastic flow and fracture. Plastic flow caused the material to accumulate in the lips to the side of the contact (as illustrated in Fig. 11 in the reference). Wear caused by fracture was separated from the total wear by weighing the disc at regular intervals. The results of the test have been reproduced in Fig. 3(a). The material can be seen to strain harden with the progress of rolling/sliding – the surface hardness nearly doubles after about 17500 cycles. The accumulated strain can also be seen to increase from zero to a constant value over the same number of cycles. Also, the wear rate can be seen to increase from a low value to a steady state value in about 17500 cycles. The increase in wear rate demonstrates that the effect of accumulation of strain, prior to the layer being exposed to the surface, is larger than that of strain hardening.

Note that the same argument applies even if the material were to fail by LCF. During the time the first layer fails, the second layer, and indeed other layers, are undergoing cyclic straining and exhausting life. When exposed to the surface, the second layer has a smaller residual life. It fails relatively quickly and leads to a higher wear rate than the first layer. As before, strain hardening will cause a reduction in the wear rate. The process continues until all the layers reaching the surface have exactly the same residual life-fraction and strain hardening as their predecessor, and then a steady state is achieved. Even though Kapoor's [12] calculations for a wedge sliding on a rigid perfectly plastic material and a cylinder sliding/rolling on copper, show that the material would fail by ratchetting, the possibility of failure by

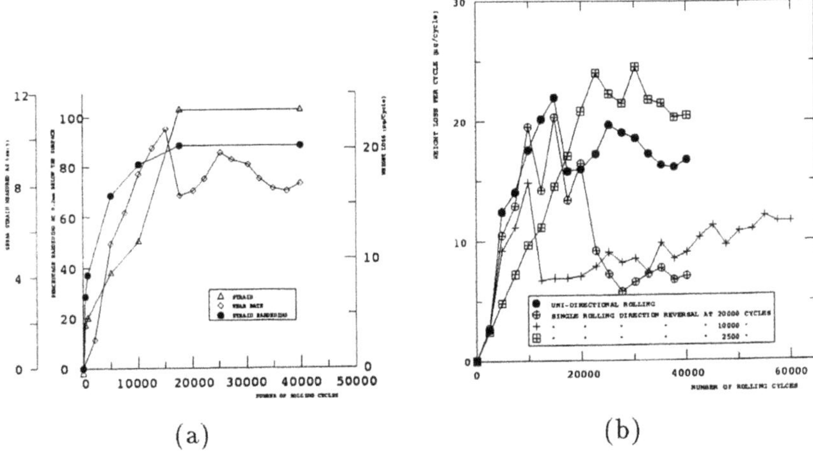

Figure 3. (a) Effect of rolling on wear rate, accumulated plastic shear strain and percentage hardening of rail steel, from Ref14. (b) Effect of reversal on wear rate of a rail steel, from Ref15.

LCF for some combination of material properties and operating conditions cannot be ruled out.

Consider next a reversal in the direction of rolling/sliding after some cycles. Under the reversed direction the amplitude of alternating plastic strain remains the same, so the Coffin-Manson relationship, eq. (1), would not distinguish between the reversed and non-reversed rolling directions. The direction of accumulating strain would, however, reverse and the material would start accumulating strain in the other direction. It would take longer to reach the critical strain to failure and the life would be longer. For failure by ratchetting (i.e. $N_r < N_f$), this would be expected to result in a drop in the wear rate. Indeed the results of experiments by Tyfour and Beynon [15], reproduced in Fig. 3(b), show that the wear rate drops at the point of reversal. If the rolling/sliding is continued in the reversed direction then all the layers with deformation in the other direction would delaminate with time and the wear rate would reach the original value. Results presented by Tyfour and Beynon do show the wear rate to rise to its steady state value in about 60000 cycles. Note that the material strain hardened irrespective of the change in the direction of rolling/sliding and achieved a steady state profile. For reversals before this steady state hardness profile was achieved, the drop in wear rate is due to both the prior straining in the opposite direction, and an increase in the shear yield strength of the material. For reversal after this steady state profile has been achieved, the drop is solely due to the prior straining. Even though in principle it is possible to separate these two effects, the general scatter in wear results makes a meaningful conclusion impossible.

3. Modelling Wear by Sliver Formation

In many situations, the material leaves the contact (extrudes out into slivers) before it can fracture. The fracture plays no part in wear and is necessary only for the extruded slivers to detach from the surface. This mechanism was analysed by Kapoor and Johnson [16], who found that it is also a result of ratchetting. As a result of repeated loading, the sub-surface material is progressively compressed and extruded out in the form of thin slivers, these subsequently separate and produce wear debris.

Real surfaces are rough and the contact is made at asperity summits, where the contact pressure is much higher than the nominal (average) pressure. Kapoor and Johnson [16] showed that if it exceeds a critical limit, termed 'ratchetting threshold' then slivers continue to extrude. They also estimated these limits for isotropic roughness (surfaces produced by bead blasting). More common surfaces are however produced by turning and grinding and have a directionality in roughness; critical limits for such roughness are also known [17]. Kapoor *et al.* [18] also estimated the total load for a nominally flat rough surface below which no asperity pressure exceeds the critical limit. This safe load limit against wear by ratchetting is easily exceeded for most components and then wear occurs.

Kapoor, Williams and Johnson [18] modelled this process by representing the wearing soft asperities as truncated wedges as in Fig. 4. A realistic roughness was considered by randomly varying the asperity height. This gave rise to a distribution of contact pressures from zero for those short asperities which just make contact to a much higher value for tall asperities. The normal load is carried by all the contacting asperities, but wear is caused only by those asperities which have contact pressures exceeding the ratchetting threshold. Let the density of such asperities be n_r and their average contact width be a_r. As sliding proceeds each transverse cross-section experiences contact loadings which are randomly distributed across the width. If, for effective pummelling, the contacts should be spaced no further than $C_1 a_r$, then a sliding distance of $1/(C_1 a_r n_r)$ will ensure that all the material elements in a cross-section have experienced a contact pressure in excess of the ratchetting threshold. After a sliding of this cycle distance the softer material in a layer of thickness h will be compressed and extruded out as shown in Fig. 4. Based on the calculations by Bower and Johnson [19], the compressive strain in the layer $\Delta\epsilon^p$ was related to the difference between the contact pressure and the ratchetting threshold, and then to the interference of the asperities. In each cycle, the layer compresses by $h\Delta\epsilon^p$

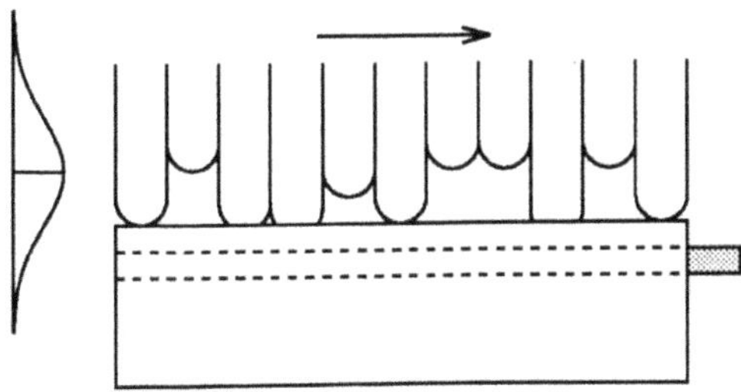

Figure 4. Hard, rough surface (exaggerated) sliding against soft, smooth surface giving rise to forward plastic extrusion.

and the volume of material displaced equals this compression multiplied by the length and the width of the wearing asperity.

The total wear volume was estimated by relating the number of times this happens to the distance slid. The resulting Archard wear coefficient k' is a function of the plasticity index in repeated sliding ψ_s. As ψ_s varies from 1.0 to 3.5, the wear coefficient increases by several orders of magnitude. Another feature is that the wear coefficient comes out to be a function of the nominal pressure. Even though it is contrary to popular belief, such dependence is known [20].

In many situations, both wear by fracture and by sliver extrusion occur simultaneously (indeed, other forms of wear may also be present). In experiments by Tyfour *et al.* [14] they were 55% and 45% of the total wear. However, this effect is geometric and so the numbers depend on component size. Kapoor [21] in a recent paper mentions that a complete switch from wear by fracture to that by sliver extrusion is possible and depends on the size.

4. The Effect Of Plasticity

The above model assumes all asperity pressures to be elastic. However, for taller asperities the contact becomes plastic. The relationship between pressure and the strain measure a/R is shown in Figure 5.

Contact behaviour has three distinct regimes: perfectly-elastic; elastic-plastic; and perfectly-plastic. One model of this behaviour is the Spherical Cavity Model which may be used for elastic-plastic indentation by a cone [22] or a wedge [23]. However, this model has been found to be inapplicable to indentation by a sphere or a cylinder [24]; an empiric model has been obtained by curve-fitting in Figure 5(a) yielding, for the elastic-plastic regime:

$$\bar{p}/\bar{p}_s - 1 \propto \ln({\psi_s}^2\delta/\sigma) \tag{4}$$

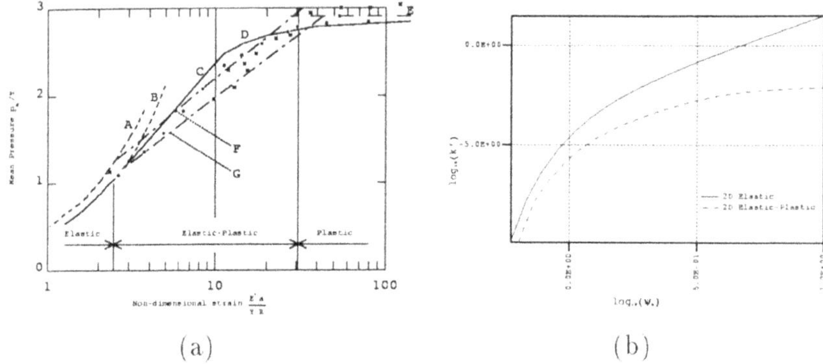

Figure 5. (a) Reproduction of Figure 6.14 from *Contact Mechanics*: the effect of plasticity — pressure as a function of semi-contact width. (b) The Archard wear coefficient (k') against the plasticity index (ψ_s). Comparison of 2D elastic model and the 2D elastic-plastic model for surface separation $d = 3\sigma$. $E^* = 115GPa$ and $H = 3GPa$.

where $\bar{p}$ is the mean pressure and δ is the overlap.

The effect of plasticity is to limit the contact pressures at the tall asperities to the fully-plastic pressure. A decrease in pressure leads to a decrease in the ratchet strain – assuming that the form of the pressure distribution is not too different from the Hertzian. A complete analysis of the wear rate requires that the ratchet rate be re-analysed for these new pressure distributions and then used in the rough surface model. A first guess has been obtained by Franklin and Kapoor [24] limiting the pressure in the analysis to the one given by the elastic-plastic model. The results from the cylindrical-asperity model are graphed in Figure 5 along with those of the earlier fully-elastic case, where:

$$\psi_s = \frac{0.7\pi}{8} \frac{E^* \sqrt{\sigma}}{\bar{p}_s \sqrt{R}}. \tag{5}$$

5. The Influence Of Adhesion

Various models for adhesion exist. The traditional model is the JKR model [25] where adhesive forces act only through the contact. Rigid body adhesion is described better by the DMT model [26]. Adhesive contact of rough surfaces is described by Maugis [27] for both these models. The more elaborate Maugis-Dugdale theory reduces to one or the other at opposite extremes of its parametric range (see Johnson [28]). All these theories deal with the elastic adhesion of a sphere to a plane. Roy Chowdhury and Pollock [29] have examined plastic adhesion of a sphere to a plane.

Adhesive contact is deceptive. For instance, when the mean contact pressure at an asperity is zero (i.e. the load supported is zero) then the pressure at the centre is given by:

$$p = \frac{5}{3}\left(\frac{3\gamma E^{*2}}{\pi^2 R}\right)^{\frac{1}{3}} \tag{6}$$

where γ is the combined surface energy. If it exceeds the ratchetting threshold, wear will occur even at zero and negative loads. For a rough surface contact, where many asperities touch, the effect of adhesion is to increase the maximum compressive pressure at all asperities. More asperities will cause wear and thus the wear rate will increase.

The incorporation of adhesion into the wear model is currently under way.

6. Discussion

Two models for wear were considered. In one the material ruptures after accumulating a critical deformation and leads to delamination wear debris. In the other material from below the contact extrudes out in the form of thin slivers. Subsequently these break off to provide debris; the fracture plays no role in wear (transport) of the material from the contact zone.

The results of the earlier, fully elastic model (see Figure 5(b)) are implausably high for large values of the plasticity index and implausibly low for small values. The introduction of plasticity into the model rectifies the former failing and introduction of adhesion is expected to rectify the latter.

References

1. A. Kjer. A lamination wear mechanism based on plastic waves. *Proc. Int. Conf. on Wear of Materials, 1987, ASME - NY*, pages 191–198.
2. Kato. *WTC*, 1997.
3. N.P. Suh. The delamination theory of wear. *Wear*, 25:111, 1973.
4. A.D. Hearle and K.L. Johnson. Mode II stress intensity factors for a crack parallel to the surface of an elastic half-space subjected to a moving point load. *Journal of Mechanics And Physics Of Solids*, 33:61, 1985.
5. A.F. Bower. The influence of crack face friction and trapped fluid on surface initiated rolling contact fatigue cracks. *Transactions ASME, Journal of Lubrication Technology*.
6. S. Shepherd, J.R. Barber, and M. Comninou. Subsurface cracks under conditions of slip, stick and separation caused by a moving compressive load. *Transactions ASME, Journal of Applied Mechanics*, E54:393, 1987.
7. L.M. Keer, T.M. Farris, and R.K. Steele. On some aspects of fatigue crack growth in rails induced by wheel/rail contact loading. *Proceedings of the Second Internation Symposium on* Contact mechanisms and wear of wheel-rail systems, *University of Rhode Island*, page 327, 1986.
8. M. Kaneta and Y. Murikami. Effects of oil pressure on surface crack growth in rolling/sliding contact. *Tribology International*, 20:210, 1987.

9. I.V. Kragelski, M.N. Dobychin, and V.S. Kombalov. *Friction and Wear: Calculation Methods.* Pergamon Press, 1982.
10. J.M. Challen, P.L.B. Oxley, and B.S. Hockenhull. Prediction of Archard's wear coefficient assuming a low cycle fatigue mechanism. *Wear*, 111:275–288, 1986.
11. P. Lacey and A.A. Torrance. The calculation of wear coefficients for plastic contacts. *Wear*, 145:367–383, 1991.
12. A. Kapoor. A re-evaluation of the life to rupture of ductile metals by cyclic plastic strain. *Fatigue And Fracture Of Engineering Materials And Structures*, 17(2):201–219, 1994.
13. J.H. Beynon and A. Kapoor. The interaction of wear and rolling contact fatigue. *in R.A. Smith (ed.) 'Reliability Assessment of Cyclic Loaded Engineering Structures', Kluwer Academic Publishers*, pages 1–26, 1997.
14. W.R. Tyfour, J.H. Beynon, and A. Kapoor. The steady state wear behaviour of pearlitic rail steel under dry rolling-sliding contact conditions. *Wear*, 180:79–89, 1995.
15. W.R. Tyfour and J.H. Beynon. The effect of rolling direction reversal on the wear rate and wear mechanism of pearlitic rail steel. *Tribology International*, 27:401–412, 1994.
16. A. Kapoor and K.L. Johnson. Plastic ratchetting as a mechanism of metallic wear. *Proceedings of the Royal Society, A*, 445:367, 1994.
17. A. Kapoor and C.F. Cocks. Wear through the plastic interaction of cylindrical asperities in sliding. *International Journal of Mechanical Science*, 36(11):1045–1059, 1994.
18. A. Kapoor, J.A. Williams, and K.L. Johnson. The steady state sliding of rough surfaces. *Wear*, 175:81–92, 1994.
19. A.F. Bower and K.L. Johnson. The influence of strain hardening on cumulative plastic deformation in rolling and sliding line contact. *Journal of Mechanics And Physics Of Solids*, 37:471–493, 1989.
20. E. Rabinowicz. *Proc. Conf. on Engineering Materials for Advanced Friction and Wear Applications, ASME, Gaithersburg*, page 202, 1988.
21. A. Kapoor. Wear by ratchetting. *Wear*, 1997.
22. K.L. Johnson. *Contact Mechanics.* Cambridge University Press, 1985.
23. K.L. Johnson. The correlation of indentation experiments. *Journal of Mechanics And Physics Of Solids*, 18:115–126, 1970.
24. F.J. Franklin and A. Kapoor. Under preparation.
25. K.L. Johnson, K. Kendall, and A.D. Roberts. Surface energy and the contact of elastic solids. *Proceedings of the Royal Society, A*, 324:301–313, 1971.
26. B.V. Derjaguin, V.M. Muller, and Toporov. Effect of contact deformations on the adhesion of particles. *Journal of Colloid and Interface Science*, 53(2):314–326, 1975.
27. D. Maugis. On the contact and adhesion of rough surfaces. *Journal of Adhesive Science And Technology*, 1995.
28. K.L. Johnson. Adhesion and friction between a smooth elastic spherical asperity and a plane surface. *Proceedings of the Royal Society, A*, 453:163–179, 1997.
29. S.K. Roy Chowdhury and H.M. Pollock. Adhesion between metal surfaces: The effect of surface roughness. *Wear*, 66:307–321, 1981.

RHEOLOGICAL MODELING OF THIN FILM LUBRICATION

JOHN A. TICHY
Department of Mechanical Engineering, Aeronautical Engineering, and Mechanics
Rensselaer Polytechnic Institute, Troy NY 12180-3590

1. Introduction

Impressive strides through molecular dynamics (MD) simulations have been made in the modeling of thin film (nanometer-scale) lubrication. This mode of lubrication is likely to exist in MEMS applications. Furthermore, there have been many recent advances in experimentation from a pure science perspective -- the ability to perform meaningful tests on nano-meter-scale films. This convergence of MD and experiment has led to an optimism in scientific circles that the mechanisms of friction and boundary lubrication, studied for centuries, can soon be understood and predicted by first principles.

From the engineering point of view, however, practical predictive ability seems far away. Conditions of MD simulations are highly idealized and the method is computationally intensive. Simulations nearly always use parallel perfectly smooth surfaces, simple well-characterized lubricating fluids, periodic boundary conditions, unrealistically high rates of shear, short time and length scales, etc.

Conventional engineering continuum based methods are thought to be of little value in describing and predicting lubrication of thin films. According to Reynolds' theory, only the viscosity of a Newtonian fluid is of importance in determining the behavior of a lubricated contact of a given geometry and kinematics. In thin film lubrication it is well known that many other factors are significant. One of the definitions of boundary lubrication is the regime where viscosity no longer affects friction.

Reasons cited for the inadequacy of continuum methods applied to thin film lubrication are, (1) the problem is so complex that any theoretical approach is doomed to failure, and (2) the film is so thin, being inherently of molecular scale, that the assumption that the material is a continuum is not valid. Chan and Horn [1], however, found that the drainage rate of a thin film of fluid between two crossed molecularly smooth mica cylinders well predicted by the Newtonian Reynolds' equation down to about 30 nanometers. At thinner gaps, good correlation with experiment is obtained by simply adding a fictitious rigid layer to the mica surfaces in the Reynolds' equation model.

B. Bhushan (ed.), Tribology Issues and Opportunities in MEMS, 175-183.

When confined between molecularly smooth surfaces of only several molecular lengths, great differences are seen between behavior in this setting and in bulk (macroscopic) rheological measurements. The apparent viscosity of the thin film can be many orders of magnitude greater than that of the bulk lubricant.

Many of these studies show that stick-slip motion, and other solid-like behavior, results from sliding of thin films [2]. The authors of such studies describe the process in terms of a molecular-scale mechanism, and often state that this kind of response indicates the futility of trying to describe such behavior from a continuum point of view. Often, viscosity alone is the only continuum property considered.

The fluid rheology classification constructed below summarizes most of the aspects of the rheology of fluids, pointing to several examples. In nearly all existing cases in tribology only monopolar purely viscous models are considered -- a small subset of the field. Many powerful aspects of fluid rheology are not brought to bear on tribology problems. Most of the terminology and perspective below is due to Bird *et al.* [3].

2. Rheological Models

2.1 NOMENCLATURE

τ_{ij} deviatoric stress tensor [N/m^2].

p_{ij} couple stress [m-N/ m^2].

v_i flow velocity vector [m/s].

$\dot{\gamma}_{ij}$ rate of strain tensor = $(\nabla v)_{ij} + (\nabla v)_{ij}^T$ [1/s].

$\dot{\gamma}$ magnitude of strain rate (2nd invariant) = $\sqrt{\dot{\gamma}_{mn}\dot{\gamma}_{mn}/2}$, sum on repeated indices.

τ magnitude of stress = $\sqrt{\tau_{mn}\tau_{mn}/2}$.

η viscosity [N-s/m^2].

λ_0 material (relaxation) time parameter [s].

β_0 couple stress material parameter [N-s].

κ_0 elastic director stress material parameter [N].

$\dot{\gamma}_{ij}^{(1)}, \tau_{ij}^{(1)}$ superscript denotes first derivative w.r.t. material coordinates rotating or deforming with the particle. Generates nonlinear coupled terms in spatial coordinates.

$\gamma_{ij}^{(0)}$ finite strain, reference frame at present time, $\gamma_{ij}^{(1)} = \dot{\gamma}_{ij}$.

n_i microstructure orientation vector (director) [-].

2.1 NOMENCLATURE (Continued)

$\dot{\Omega}_i, \omega_{ij}$ fluid rotation rate (vorticity), $\dot{\Omega}_i = (curl\, v)_i$ [1/s].
corresponding spin tensor, $\omega_{ij} = (\nabla v)_{ij} - (\nabla v)^T_{ij}$,

$\dot{\Psi}_i, \psi_{ij}$ microstructure rotation rate, corresponding spin tensor [1/s].

m, n and subscripted symbols $\beta, \dot{\gamma}, \eta, \kappa, \lambda, \tau$ refer to material parameter values.

2.2 MONOPOLAR FLUIDS

In monopolar fluids, the material point has the property of mass only. Angular momentum identically satisfied by isotropy, i.e., $\tau_{ij} = \tau_{ji}$.

2.2.1 *Purely Viscous Fluids*

For this large class of fluids, stress depends only on instantaneous local rate of strain. In the Newtonian fluid case, viscosity is independent of rate of strain: $\tau_{ij} = \eta_0\, \dot{\gamma}_{ij}$. In the so-called purely viscous or generalized Newtonian case, viscosity depends on the strain rate invariant:

$$\tau_{ij} = \eta(\dot{\gamma})\,\dot{\gamma}_{ij}. \tag{1}$$

In the field of tribology, terms such as "lubricant rheology," or "non-Newtonian lubrication" generally refer only to the effect of variable viscosity. In this discussion, x is the sliding direction and y is the cross-film direction. In a one dimensional thin film, only one component of the shear stress tensor τ_{xy} need be considered. The shear stress is a nonlinear function of shear rate $\dot{\gamma}_{xy}$. The governing rheological equation is then $\tau_{xy} = \eta(|\dot{\gamma}_{xy}|)\,\dot{\gamma}_{xy}$, where η is viscosity, a function of the absolute value of the shear rate. In the Newtonian case, viscosity is constant.

In two-dimensional thin films things get more complex -- two shear stress components (τ_{xy} and τ_{zy}) and two shear rate components ($\dot{\gamma}_{xy}$ and $\dot{\gamma}_{zy}$) are significant. To satisfy invariance considerations, viscosity must be seen as a function of the *magnitude* of the strain rate (the second invariant of the strain rate tensor), $\eta(\dot{\gamma})$, where $\dot{\gamma} = \sqrt{\dot{\gamma}^2_{xy} + \dot{\gamma}^2_{zy}}$. There are many studies which do not satisfy this formalism, which means that the same rheological model will predict different results if the coordinate system is oriented in a different direction.

Among the many examples of purely viscous models is the so-called power law model:

$$\eta = m\dot{\gamma}^{n-1}, \tag{2}$$

and the Eyring model:

$$\eta = \frac{\tau_0}{\dot{\gamma}} \arcsin(\lambda_0 \dot{\gamma}), \tag{3}$$

which exhibits a limiting shear stress τ_0. There are many other models of this type.

2.2.2 *Yield (viscoplastic) Fluids*

Intrinsic film-induced stick-slip is believed to be caused by phase transition between solid-like and liquid-like states, Israelachvili [4], when the liquid film thickness is of two or three layers of molecules. Stick-slip motion is generated due to the phase transitions which are very abrupt and stepwise. Solid-like films exhibit a yield point or critical shear stress, beyond which they behave like ductile solids undergoing plastic deformation.

The simplest yield model is the Bingham model, characterized by two material parameters: a yield shear stress and a constant viscosity. When the magnitude of the deviatoric stress tensor is less than the yield stress, the material can only move as a rigid body; when the yield stress is exceeded, the material flows in a quasi-Newtonian manner:

$$\tau \le \tau_Y : \dot{\gamma} = 0; \qquad \tau > \tau_Y : \quad \dot{\gamma}_{ij} = \frac{\tau_{ij}}{\eta_0 + \tau_Y / \dot{\gamma}}. \tag{4}$$

The Bingham model may be useful to model the flow of grease, mud, ER fluids and granular materials. However, for stick-slip to occur, it is necessary to have a region of decreasing stress with increasing rate of deformation, Jang and Tichy [5]. A transition to solid-like flow (often alluded to in basic science studies) is not sufficient. The following model is called the Bingham overshoot model:

$$\eta(\dot{\gamma}) = \eta_0 + \frac{\tau_d}{\dot{\gamma}\pi/2} \arctan\left(\frac{\dot{\gamma}}{\dot{\gamma}_d}\right) + \frac{(\tau_s - \tau_d)}{\dot{\gamma}_s} \exp\left[\frac{1}{2}\left(1 - \frac{\dot{\gamma}^2}{\dot{\gamma}_s^2}\right)\right]. \tag{5}$$

The relationship between stress and strain rate in simple shear shows a local region of decreasing stress at small strain rate, see Figure 1. The peak of the overshoot shown (analogous to static frictional stress) is controlled by the static yield stress parameter τ_s and the value of stress found when the overshoot drops off is related to the dynamic yield stress parameter τ_d. The slope of the curve at larger strain rates is the viscosity parameter η_0, while $\dot{\gamma}_s$ governs the sharpness of the overshoot, and $\dot{\gamma}_d$ governs the sharpness of the transition from the Bingham yield shear stress. The viscous yield model shows yield behavior but not the decreasing stress. The overshoot and decrease are meant to simulate the phase transition to fluid-like behavior in thin film flow. If τ_d and τ_s go to zero, the Newtonian fluid model is obtained.

Within the film, such a region of decreasing stress would be unstable, see Tanner [6], and the material in the film would quickly seek a stable region on the flow curve with increasing stress.

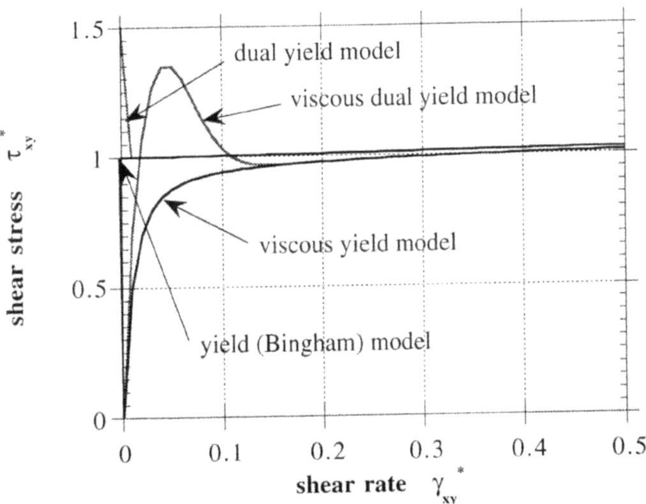

Figure 1 - Rheological Flow Curves (dimensionless form): a) Bingham Model, b) Viscous Yield Model, Dorier-Tichy [7], c) Dual Yield Model, and d) Bingham overshoot (viscous dual yield) model - Eq. (5).

2.2.3 *Viscoelastic or Time-Dependent Fluids*

One type of behavior not predicted by purely viscous models is time dependence. In purely viscous fluids, the stress must always be precisely in phase with the applied surface motion. However, it is known that many lubricants possess a characteristic (relaxation) time λ which is of order 10^{-4} - 10^{-6} s. If a lubrication process time is of this order, we can expect strong time dependent effects, i.e., the fluid cannot respond to imposed conditions. For example, in a contact with sliding speed U = 0.25 m/s and width L = 25 μm, the process time L/U is 10^{-4} s and time dependent effects are expected. The Deborah number $De = \lambda U/L$ is a measure of such time dependence. For Newtonian quasi-steady behavior Deborah number is zero. An alternative interpretation is that the fluid possesses elasticity, and the ratio of the viscosity to the elastic modulus is the relaxation time.

One of the simplest and most common types of models to account for fluid time dependence is the Maxwell model. Neighboring fluid particles are idealized as connected by an embedded differential position vector of length *ds*. The stretching of *ds* causes stress between the adjoining particles, as if induced by a viscous damper (dashpot) in series with a spring. For homogeneous flow all such *ds* elements experience the same relative stretching (strain) and the Maxwell model looks like,

$$\tau_{xy} + \lambda \frac{d\tau_{xy}}{dt} = -\mu \dot{\gamma}_{xy}. \qquad (6)$$

If relaxation time is zero, the Newtonian model is recovered

The question arises as to how to change this homogeneous Maxwell model for more general flows. The answer is that we must replace the time derivative either by one written with respect to coordinates which translate and rotate with the fluid particles (the co-rotational or Jaumann derivative), or coordinates which deform with the fluid (the co-deformational or convected derivative). The derivative in the co-rotational or co-deformational system is then transformed back to the conventional spatial coordinates by the chain rule. Unfortunately, such derivatives greatly complicate problems and are rarely used due to nonlinear coupling of stresses, e.g., the equation for τ_{xy} contains terms like $\tau_{yy}(\partial v_x/\partial y)$.

Using the partial derivative in the above equation is clearly not sufficient -- it is the rate of change of a property (velocity, stress, etc.) at a fixed location. A position vector *ds* in this system refers to *different particles* passing through the location. The material or substantial derivative is an improvement -- it is the rate of change of a property of a given particle with respect to coordinates which translate with the particle but maintain their orientation (do not rotate or deform). However, the position vector *ds* in the translating system is not embedded but connects to *different neighboring particles* which would be found at the fixed orientation. In the strictest sense, only if a co-rotational or co-deformational derivative is used will the resulting model be *admissible*, i.e., all observers will calculate the same stress state. The preference of one admissible derivative relative to another is based on the practical issue of which seems to best predict experimental results. From the standpoint of the Maxwell model, the derivatives differ as to the reference or unstressed configuration for the idealized spring.

It is important to keep in mind that viscoelastic models, even at high Deborah numbers (1) *do not exhibit stick-slip behavior*, and (2) *do not exhibit solid-like (slug) velocity profiles.* In oscillatory shear flow, at high Deborah numbers, the stress of viscoelastic fluids becomes in-phase with the strain (as for the elastic solid), rather than the strain rate (as for the viscous fluid). This is the only sense in which the flow is "solid-like," Tichy [8].

In general flows, the Maxwell model looks like:

$$\tau_{ij} + \lambda_0 \tau_{ij}^{(1)} = \eta_0 \dot{\gamma}_{ij} \tag{7}$$

With the Maxwell fluid, the time response is introduced through a rate-of-stress term thus and a coupled differential equation must be solved for the stress.

An alternative approach is the second order fluid, in which the time response arises due to a rate-of-rate-of-strain term. The second order fluid model is thus explicit in stress and is therefore easier to apply. One form is the Rivlin-Ericksen model:

$$\tau_{ij} = \eta_0 \dot{\gamma}_{ij} + \lambda_0 \eta_0 \dot{\gamma}_{ij}^{(1)} + \eta_0 \lambda_{00} \dot{\gamma}_{im} \dot{\gamma}_{mj} \tag{8}$$

Another difference between the Maxwell and ordered fluids is their conceptual origin. The Maxwell model arises from idealizing neighboring fluid particles as connected by an embedded differential position vector, Sawyer and Tichy [9]. The second order fluid, however, accounts systematically for small deviations from Newtonian behavior. The Maxwell fluid can account for large elastic effects, but relies on a specific empirical description of the behavior (springs and dashpots). The second order fluids can account, in theory, for only vanishingly small elastic effects. However, no specific idealizations are required -- all proper time dependent models asymptotically reduce to the second order fluid in the limit of small λ_0.

A third type of viscoelastic fluids are the integral models. An example is the Lodge rubberlike liquid:

$$\tau_{ij} = \int_{-\infty}^{t} \frac{\eta_0}{\lambda_0^2} \exp\left[-(t-t')/\lambda_0\right] \gamma_{ij}^{(0)}(t')dt'. \tag{9}$$

Integral models are very difficult to use, requiring the solution of integral-differential equations. On the other hand, the fading memory nature of viscoelastic materials is easy to appreciate.

2.3 MULTIPOLAR FLUIDS

For multipolar materials, the material point has microstructure, i.e., some other vector or tensor properties, in addition to the scalar property of point mass. A point can support couple (torque-like) stress as well as stress. Angular momentum equation must be solved for microstructure behavior.

The lubricant in the contact may posses solid-like characteristics near the surfaces in thin films, and fluid-like characteristics many molecular lengths away. The bulk (macroscopic) rheology may be different than that observed in the thin film. Due to the molecular orientation, the lubricant has an underlying microstructure. The material may be anisotropic and non-homogeneous. Measurements of apparent viscosity have little predictive value on friction and load.

From the continuum mechanics view, an additional dependent variable is used to describe the microstructure, a unit dimensionless vector called the director n_i. Its direction corresponds to the molecular orientation. In formulating the problem for thin films, position and time are the independent variables, there are three dependent variables: the displacement or velocity of a material point, the director, and a normal stress variable. There are three governing conservation equations for mass, momentum and moment of momentum. Thus, if another vector dependent variable is required (the director), another governing differential equation is required, namely, the moment of momentum equation. Body forces and inertia can be neglected, i.e., a mass point is in quasi-static equilibrium acted on only by the tractive forces.

The ideas and equations put forward follow directly from liquid crystal flow modeling, see Leslie [10], DeGennes [11], and Tichy [12]. Liquid crystals may be thought of as fluid materials possessing elastic microstructure. In thin films, conventional lubricants may behave as liquid crystals, a view put forward Cognard [14] and others. Coupling of viscous and elastic effects in the momentum and angular momentum equations determines the overall rheological behavior

2.3.1 *Anisotropic Fluids*

For these fluids, the stress depends on rate of strain and microstructure rotation, and couple stress depends on microstructure rotation rate:

$$\tau_{ij} = \eta_0 \dot{\gamma}_{ij} + \eta_1 \left(\psi_{ij} - \omega_{ij}\right), \qquad p_{ij} = \beta_0 \frac{\partial \dot{\Psi}_i}{\partial x_j}. \tag{10}$$

2.3.2 *Liquid Crystal or Nematic Materials*

In this case, viscous stress depends on rate of strain and microstructure orientation itself (not rotation rate), while elastic stress depends on gradient of microstructure orientation:

$$\tau_{ij} = \eta_0 \dot{\gamma}_{ij} + \eta_2 \dot{\gamma}_{im} n_m n_j, \qquad \tau_{ij}^E = \kappa_0 \frac{\partial}{\partial x_i}\left(n_j \, divn\right). \qquad (11)$$

3. Conclusions

Despite the increasing power of atomistic simulations, engineering predictive ability may remain far in the future. Continuum approaches appear to have validity down to the nanometer range. In nearly all tribology studies, coming both from the science and engineering sides, only a very small subset of the field of rheology is considered, namely, the purely viscous liquid. In particular, materials which exhibit yield behavior and time dependent behavior should be considered, as well as multipolar materials like anisotropic liquids.

4. References

[1] Chan, D.Y.C. and R.G. Horn, "The Drainage of Thin Liquid Films between Solid Surfaces," *Journal of Chemical Physics*, Vol. 83, No. 10, pp. 5311-5324, 1985.

[2] Gee, M. L., P.M. McGuiggan and J. Israelachvili, "Liquid to Solid-like Transitions of Molecularly Thin Films Under Shear," *Journal of Chemical Physics*, Vol. 93, No. 3, pp. 1895 1906, 1995.

[3] Bird, R.B., R. Armstrong and O. Hassager, *Dynamics of Polymeric Liquids, Vol. 1, Fluid Mechanics*, pp. 169-243, 351-352, 1987.

[4] Israelachvili, J., *Intermolecular & Surface Forces*, London, Academic Press, 1992.

[5] Jang, S. and J.A. Tichy, "Rheological Models for Stick-Slip Behavior," *ASME Journal of Tribology*, Vol. 119, No. 4, pp. 626-631, 1997.

[6] Tanner, R.I., *Engineering Rheology*, Clarendon Press, Oxford, pp. 387-388, 1992.

[7] Dorier, C. and J.A. Tichy, "Behavior of a Bingham-like Viscous Fluid in Lubrication Flows," *Journal of Non-Newtonian Fluid Mechanics*, Vol. 45, pp. 291-310, 1992.

[8] Tichy, J.A., "Hydrodynamic Lubrication with the Convective Maxwell Model," *ASME Journal of Tribology*, Vol. 118, No. 2, pp. 344-349, 1996.

[9] Sawyer, G.W. and J.A. Tichy, "Non-Newtonian Lubrication with the Second Order Fluid,", *ASME Journal of Tribology*, 1997.

[10] Leslie, F.M., "Theory of Flow Phenomena in Liquid Crystals," Advances in Liquid Crystals, edited by G.H. Brown, Academic Press, NY, v. 4, pp. 23, 1979.

[11] De Gennes, P.G., *The Physics of Liquids Crystals*, Oxford Press, 1974.

[12] Tichy, J.A., "Modeling of Thin Film Lubrication," *STLE Tribology Transactions*, Vol. 37, No. No. 1, pp. 24-36, 1995.

[13] Cognard, J., "Lubrication with Liquid Crystals, "Tribology and the Liquid-Crystalline State, edited by G. Biresaw, American Chemical Society, Washington, DC, pp. 1-47, 1989.

SURFACE ROUGHNESS INDUCED EFFECTS IN HYDRODYNAMIC LUBRICATION

KRISTIAN TØNDER
The Norwegian University of Science and Technology
IMM
N-7034 Trondheim, Norway

Abstract

The following deals with effects of surface roughness in hydrodynamic lubrication. Although most of the effects described apply to several types of tribological devices, the discussion is essentially related to bearings. The main reason for this is simply the suitability of bearings for illustration purposes. It is shown that roughness effects are mainly expressed as modifications of Reynolds' equation. This applies to compressible as well as incompressible fluids alike. The methods used and the difficulties encountered in the derivation of these modifications are described, as are special effects resulting from certain surface roughness structures.

1. Introduction

In classical tribological design the lubricant was a liquid and ideal lubricating film was thought to be rather thick - for even moderately large bearings several tens of microns. This was assumed to give the bearing safer operating conditions since unwanted motion such as jerks and shocks would be less likely to bring the lubricated surfaces in contact. Such contact, which can obviously be very damaging, is clearly related to surface roughness. A thick film would thus be rather more tolerant as regards surface texture excursions. In addition, in most classical applications, speeds and loads were modest. This means that even if the coefficient of friction of, say, a bearing is large, the power loss might be modest. This is because the losses are proportional to the friction force and to the relative velocity, so for a given coefficient of friction, the power lost is proportional to both speed and load. Similar arguments may be used in conection with gears.

For liquid lubricated tribological devices the designer had, and still has, a simple but very powerful tool - the Stribeck curve. It relates the coefficient of friction to some form of the Sommerfeld number, the operating quantity of which is the group, $\eta U/W$. Here η is the dynamic viscosity, U the effective relative velocity and W the load. Let us use the symbol S for this particular version of the Sommerfeld number. It can be shown that S is related to the minimum film thickness of the tribo-contact in question:

B. Bhushan (ed.), Tribology Issues and Opportunities in MEMS, 185-206.

When the film is thickness is infinite, so is the S-value; when S approaches zero, so does the film thickness.

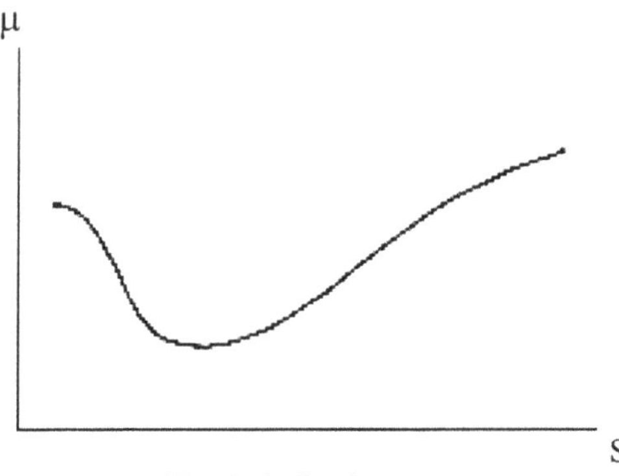

Fig. 1 Stribeck curve

Now, a bearing of a given geometry has a fixed Stribeck curve. The curve has a minimum point, to the right of which conditions of full hydrodynamic lubrication apply. Here the curve is seen to turn upwards towards the right, showing that in the fully lubricated regime, the higher the Sommerfeld number, and thereby the minimum film thickness, the higher is the coefficient of friction. This is the traditional area of operation.

There is an additional aspect to the above: when films are thick, the roughness excursions, which are always present on real surfaces, tend to be small in comparison. This means that when predicting bearing performance, one may ignore roughness and apply the Reynolds equation to the nominal film shape without appreciable error.

In recent years, however, demands on the performance of tribo-devices have gone up: higher loads and speeds, lower losses and longer life etc. are new requirements. With the exception of the longevity one, they all mean thinner films. New materials and modern surface technology have led to vastly improved surface quality. However, roughness of a certain amplitude is unavoidable. Accordingly, whatever roughness is present on tribological surfaces, it has become an increasingly more important part of the film geometry and must therefore be allowed for.

2. Reynolds' Equation

The classical equation governing hydrodynamic lubrication is that of Reynolds, which may be written, in cartesian coordinates, as follows,

$$\frac{\partial}{\partial x}\left(\frac{\rho}{\eta}\frac{\partial p}{\partial x}h^3\right)+\frac{\partial}{\partial y}\left(\frac{\rho}{\eta}\frac{\partial p}{\partial y}h^3\right)=6\rho(U_1-U_2)\frac{\partial h}{\partial x}+6h\frac{\partial}{\partial x}\{\rho(U_1+U_2)\}+12\rho\frac{\partial h}{\partial t}+12h\frac{\partial\rho}{\partial t} \quad (1)$$

Here p is pressure, h film thickness, U velocity, η viscosity and ρ density. Further, x is the coordinate in the direction of motion and t is time. The indices refer to the lower and upper surfaces respectively.

The derivation of eq. 1 is based on a number of assumptions, so the application has certain limitations. The lubricant should be Newtonian; forces due to inertia or external fields have been neglected. Further, laminar flow and essentially parallel surfaces have been assumed. Moreover, the lubricant is to be a continuum.

Clearly, several of these conditions are violated; however, good correction terms have been devised in many cases, notably for inertia, for eletromagnetic fields, for turbulent flow. A number of non-Newtonian fluid models have also been accounted for. Very thin films representing rarefaction problems have also been treated successfully by modifications to Reynolds' equation.

Any attempt to use the Reynolds equation as a basis for the calculation of roughness effects may seem to be self-contradictory, since its derivation assumes near-parallel surfaces. However, roughnesses found on real engineering surfaces have very gentle slopes of a few degrees, and very seldom exceeding some 10 to 12 degrees. Moreover, a limited number of steep slopes will not upset the permissibility of using Reynolds' equation for rough surfaces. Accordingly, most roughness effect theories are based on this equation.

3. Roughness Theories

3.1. GENERAL

In the preceding section it is argued that Reynolds' equation is applicable to real, ie. rough surfaces. Hence to find the pressure and other desired quantities of a given tribo-element we need only apply eq. 1 and solve for p; other relevant quantities are then easily derived. Accordingly, there seems to be no need for a roughness theory. Unfortunately, Reynolds' equation does not have a general analytical solution, so a numerical approach must be taken. However, the number of peaks and valleys and other surface structure characteristics is exceedingly high within the boundaries of, say, a bearing. Even when the finest details are excluded, the number might run into millions. Each surface feature would require a certain number of data points for its description. All this implies that a computer far exceeding the performance of current ones would be required. Even if such computers were available, the computation time would be formidable. (With modern concepts of self-affinities down to near-molecular levels one also faces the problem of deciding what details need to be included in the modelling.) One is therefore left with the requirement of expressing the roughness effects by some kind of average. For this to be useful we clearly also need to be able to relate these averages of effects - pressure, flow, load capacity etc. - to averages of roughness excursions and related functions.

Let us apply eq. 1 to a hydrodynamic bearing and form the averages, over the bearing area, of the various terms. We see that without further knowledge we cannot form all our averages since pressure or pressure gradients must be known. Moreover, the only information we might hope to get would be the average pressure, ie. total load divided by total area. This is not enough; we need a different kind of average - one that is representative of the local conditions. To illustrate what is required, we may consider the surface of the sea: an essentially spherical shape superimposed on which are the tidal excursions and various types of waves. Similarly, a bearing surface has a three-dimensional smooth component with shape errors, waviness and roughness superimposed. From our experience with Reynolds' equation, we know that shape excursions give rise to pressure excursions. Therefore, as with the sea or with a bearing surface, the pressure surface may be decomposed into a smooth component (3D) superimposed on which are pressure ripples or excursions. As with the sea or the metal surface, the smooth three-dimensional function is defined so as to make positive and negative excursions cancel.

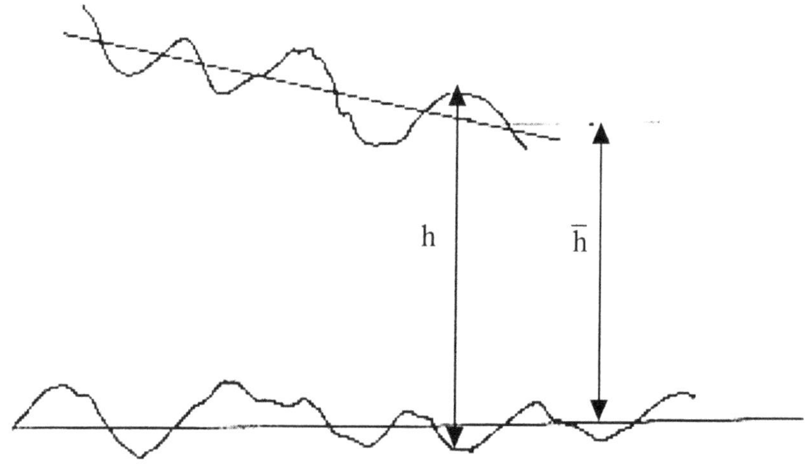

Fig. 2 Rough Surfaces

Our task is then to form equations relating this mean pressure function to known mean functions, such as averages of powers of the film thickness. Unlike a direct application of Reynolds' equation to the case at hand, such equations would lend themselves to numerical treatment since only a modest number of grid point would be required.

Having found this pressure function we may derive the load quite easily by simple integration and, as will become apparent below, also the local average of other quantities needed: shear stress, flow flux etc. Parameters such as load centre and friction force or moment are also readily computed.

3.2. SMOOTH FUNCTION CONCEPTS

There has been some discussion regarding the interpretation of the average or mean functions mentioned above. In early works by the author and H. Christensen, (1969) the term stochastic was used in connection with such averages. This term is often associated with an averaging process that in our context may be described as follows: at a given point (x,y) in a bearing, the average or expected value of a function is the mean of the function value, at that point, of a large number of statistically and nominally similar bearings (running under the same conditions). Although Christensen held that view, the author did not: all the reasoning and derivations of the early works were done based on the concept of local averages (Christensen and Tønder, 1969; Christensen, 1970; Tønder 1972). This latter concept defines the mean or expected value of a function at a point (x,y) as the area-average of that function in a small neighbourhood surrounding the point considered. This interpretation appears to be shared by most researchers in the field, such as Patir and Cheng (1978, 1979), Elrod (1973, 1979), Tripp (1983).

The underlying idea is that the roughness undulations have a very high spatial frequency. This means that even a small area-element surrounding a given point will contain a large number of roughness undulations. If we form the average value of, say, the film thickness, that value will depend only negligibly on the exact choice of area. A similar reasoning applies to other powers of the film thickness. However, a high spatial frequency of the roughness excursions must necessarily be reflected by essentially the same frequency in the pressure excursions. Hence local averaging of pressure-related functions will also generate desired expected values.

Formally we may put this as follows, for an arbitrary function, F, of the class discussed,

$$F = \frac{1}{a}\int_a F\,da + e_a(F) = \frac{1}{a}\iint_a F\,dx\,dy + e_a(F) \equiv E_a(F) + e_a(F) \qquad (2)$$

Here the E-operator generates the smooth function whilst e is just the local excursion; both $e_a(F)$ and $E_a(F)$ are functions of x and y. It follows from the above that,

$$E_a\{E_a(F)\} = E_a(F) \qquad E_a\{e_a(F)\} = 0 \qquad e_a\{E_a(F)\} = 0 \quad (3)$$

We may also define two-dimensional averaging or smoothing operators E_x and E_y thus,

$$F = E_x(F) + e_x(F) = E_y(F) + e_y(F) \tag{4}$$

Here $E_x(F)$ is smooth in the x-direction but may be oscillating in the y-direction. It is also readily seen that relations similar to those of eq. 3. apply.

We further see from the definitions that we have,

$$E_a = E_x E_y = E_y E_x \tag{5}$$

Let it also be pointed out that in roughness work it is quite customary to express expected values as an overbar, thus,

$$E_a(F) = \overline{F} \tag{6}$$

For unidirectional roughesses all three expectancies are the same, so the overbar may then be used without ambiguity.

From the above we also recognise that an integral of a smooth function is a smooth function and that the integral of an oscillating function of zero mean, such as $e_a(F)$ above, is an oscillating function of (almost) vanishing amplitude. Finally, a derivative of a smooth function is smooth.

3.3. METHODS OF APPROACH

The first modern roughness treatment was, as far as the author knows, due to Tzeng and Saibel, 1968. The approach was simply to solve Reynolds' equation for the one case that permits an exact solution, namely the two-dimensional bearing ie. one without any sideways flow. The second term of eq.1 then vanishes (expressing the y-derivative of the sideflow). Then, under steady state conditions,

$$p = 6\eta U\left(\int_0^x \frac{dx}{h^2} - h_m \int_0^x \frac{dx}{h^3}\right) + p_0 \tag{7}$$

Here p_o is the inlet pressure, usually set to zero for incompressible lubricants, so that p refers to pressure above atmospheric. The term h_m is an integration constant, determined from the outlet boundary condition.

Tzeng and Saibel averaged on both sides of the equality sign and obtained the following,

$$\overline{p} = 6\eta U\left(\int_0^x \overline{h^{-2}} dx - h_m \int_0^x \overline{h^{-3}} dx\right) \tag{8}$$

This result is correct provided the roughness structure is constant in the y-direction, ie. it must consist of ridges and grooves running across the direction of motion; otherwise it would violate the assumptions of a two-dimensional situation. This kind of roughness structure is often termed "transverse". In addition, the roughness must be located on the surface which is stationary with respect to the bearing boundaries. This is a point that will be taken up later.

If the statistics of the surface roughness is known, the average functions of eq. 8 will be available and the mean or smooth part of the pressure may be computed.

With reference to eqs. 2 to 6, eq. 8 is obtained as follows: Express p, h^{-2} and h^{-3} as sums of one smooth and one oscillating component. Collecting smooth (barred) and oscillating terms respectively we readily see that eq. 8 holds.

Eq. 7 is the only reasonably general solution to the Reynolds equation, in the sense that is valid for more than one film geometry. Another fairly general though approximate solution is the so-called short-bearing (or narrow bearing) solution. It assumes that the flow flux gradients in the y-direction are substantially larger than those in the direction of motion. Based on this, the first term of eq. 1 is neglected. This leads to the following result,

$$p = 3\eta U \frac{\partial h}{\partial x} h^{-3} (y^2 - b^2) \tag{9}$$

Here b is the half-width of the bearing. Proceeding as for the (infinitely) wide case above, we get,

$$\bar{p} = 3\eta U \frac{\partial \bar{h}}{\partial \mathrm{x}} \overline{h^{-3}} (y^2 - b^2) \tag{10}$$

We recognise that in order for this to be valid, the assumptions behind eq. 9 must hold. This means that even the roughness-induced pressure gradients must essentially vanish in the x-direction. This can only happen if the roughnesses are unchanging in this direction, which in turn requires a pattern of straight ridges and grooves running in the direction of motion. This case of striated roughness is usually referred to as “longitudinal”.

The above results, eqs. 8 and 10, led the author to a more general solution, applicable to finite bearings. Two separate cases were considered, the two ones mentioned already, transverse and longitudinal roughnesses.

We select an area element very much longer along the grooves than across them. Though the dimensions of the element are very small, it contains a large number of ridges and valleys. This is possible since roughness oscillations are asumed to have a very high spatial frequency. This element may be treated as a (nearly) infinitely wide miniature bearing. Hence the flow terms in the direction of motion must formally be the same as for the case of eq. 8. By differentiating that equation and comparing with standard expressions, we obtain for the flow flux terms,

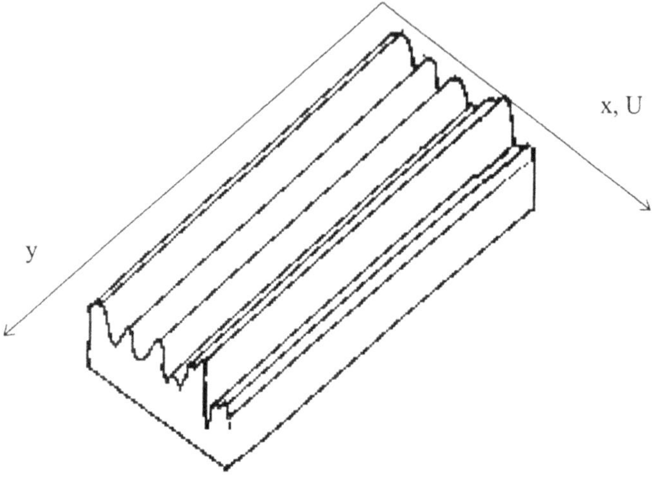

Fig. 3 Schematic of transverse roughness

$$\bar{q}_{px} = -\frac{1}{12\eta}\frac{\partial \bar{p}}{\partial x}\frac{1}{\overline{\overline{h^{-3}}}} \qquad \bar{q}_{ux} = \frac{U}{2}\frac{\overline{h^{-2}}}{\overline{\overline{h^{-3}}}} \tag{11}$$

The indices p and u refer to pressure and shear respectively; x to direction.

The flow flux in the y-direction is,

$$q_{py} = -\frac{1}{12\eta}\frac{\partial p}{\partial y}h^3 \tag{12}$$

For symmetry reasons the pressure and thereby its gradient cannot oscillate in the y-direction, ie. these functions are smooth in this direction. If this is applied to the element, the mean or expected function comes out as follows,

$$\bar{q}_{py} = -\frac{1}{12\eta}\frac{\partial \bar{p}}{\partial y}\overline{h^3} \tag{13}$$

Since the pressure gradient in the y-direction is a smooth quantity, its overbar may be omitted. However, as seen in eq. 2 it is permissible to leave it in, which has been done for consistency reasons.

We may now form a modified Reynolds equation based on eqs. 11 and 13, applicable to transverse roughness under steady-state, isoviscous conditions; constant density also having been assumed,

$$\frac{\partial}{\partial x}\left(\frac{\partial \bar{p}}{\partial x}\frac{1}{\overline{h^{-3}}}\right)+\frac{\partial}{\partial y}\left(\frac{\partial \bar{p}}{\partial y}\overline{h^{3}}\right)=6\eta U\frac{\partial}{\partial x}\left(\frac{\overline{h^{-2}}+s}{\overline{h^{-3}}}\right) \tag{14}$$

The s-term is caused by a moving roughness component and will be discussed below.

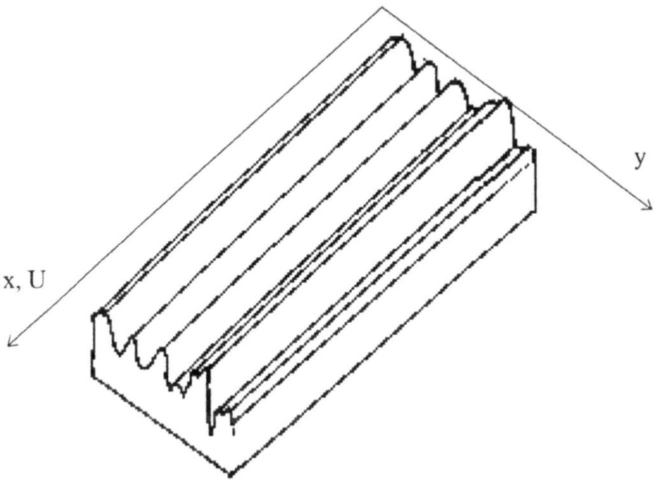

Fig. 4 Schematic of longitudinal roughness

Proceeding in a similar manner for longitudinal roughness, we select a small element, now very narrow in the y-direction. The expression for pressure flow across the ridges must be quite similar in the two cases, as must the expression for pressure (Poiseuille) flow along them. The shear flow for longitudinal roughness may be found by differentiating eq. 10 twice with respect to y and dividing by $\overline{h^{-3}}$. Comparing terms we find the shear flow flux to be $U\bar{h}/2$

Hence, under the same conditions as for eq. 14, for longitudinal roughness, the following applies,

$$\frac{\partial}{\partial x}\left(\frac{\partial \bar{p}}{\partial x}\overline{h^{3}}\right)+\frac{\partial}{\partial y}\left(\frac{\partial \bar{p}}{\partial y}\frac{1}{\overline{h^{-3}}}\right)=6\eta U\frac{\partial \bar{h}}{\partial x} \tag{15}$$

If there is roughness on both surfaces, there may be local film thickness changes which will be reflected by non-zero $\partial \bar{h}/\partial t$-terms. Clearly this will not happen for longitudinal roughness so eq. 15 also holds for two-sided longitudinal roughness. In the general case, including the one with transverse roughness, the time component must be included. For transverse striations, from geometry considerations, one has, $\partial h/\partial t = U \partial \varepsilon_m/\partial x$. Here ε_m is the roughness on the moving surface. Going through the same steps as before, but including this new term, we get an expression for s thus,

$$s = 2\overline{(\varepsilon_m/h^3)} \qquad (16)$$

The two equations, 14 and 15, may be combined to yield a single equation for unidirectional striations as a function of the angle of orientation of the grooves and ridges with respect to the direction of motion.

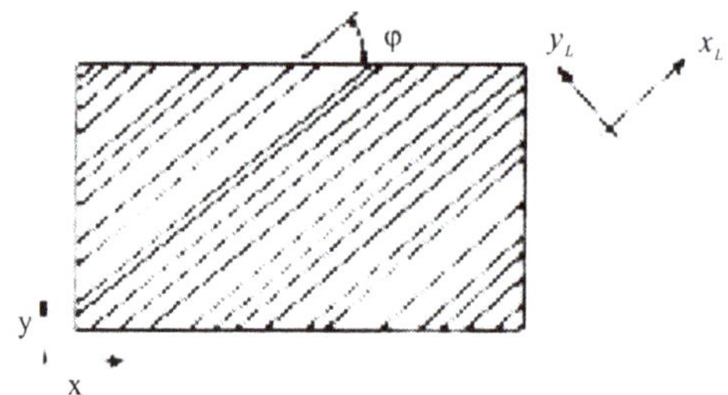

Fig. 5 Skew roughness

This was first shown by Elrod who also derived the above equations using multiple scales and perturbation in 1973. Tønder (1977a) showed that the equation may be obtained by a straightforward combination of flow components in the two directions, leading to,

$$\frac{\partial}{\partial x_L}\left(\frac{\partial \bar{p}}{\partial x_L}\overline{h^3}\right) + \frac{\partial}{\partial y_L}\left(\frac{\partial \bar{p}}{\partial y_L}\frac{1}{\overline{\overline{h^{-3}}}}\right) = 6\eta U\left\{\cos\varphi\frac{\partial \bar{h}}{\partial x_L} - \sin\varphi\frac{\partial}{\partial y_L}\left(\frac{\overline{h^{-2}}+s}{\overline{\overline{h^{-3}}}}\right)\right\} \qquad (17)$$

Here the coordinates x_L and y_L are measured along and across the striations respectively and φ is the angle between the striations and the direction of motion.

The equation may easily be transformed into one in x and y.

No other surface roughness pattern seems to yield a modified Reynolds equation in terms of analytically derived statistical parameters. The author believed isotropic roughness structures would lend themselves to an analytical treatment, but the equations found were shown to be incorrect. (Tønder, 1977b, 1980a). This led the author to look for solutions based on numerical simulation: Randomly generated but statistically similar area elements might be treated as the above-mentioned representative neighbourhood surrounding the point considered. Since only a small area is studied, fewer peaks, valleys and other special topographical features will occur. It is therefore possible to reduce the required number of grid points in the necessary numerical computations. Statistical results based on these concepts were presented by Tønder in 1972.

However, it was the works of Patir and Cheng (1978,1979) that represented a break-through. One of their main contributions was the introduction of the flow factor concept and the corresponding notation. This permitted the construction of a rather general Reynolds-type equation valid for roughness patterns having no directional bias. (For instance, striations running at an angle diiferent from 0 or 90 degrees are not covered by the Patir/Cheng notation.) Under steady-state, isoviscous conditions, the equation may now be expressed as,

$$\frac{\partial}{\partial x}(\phi_x h^3 \frac{\partial p}{\partial x}) + \frac{\partial}{\partial y}(\phi_y h^3 \frac{\partial p}{\partial y}) = 6\eta\{(U_1 + U_2)\frac{\partial h}{\partial x} + (U_1 - U_2)\sigma \frac{\partial}{\partial x}\phi_s\} \quad (18)$$

The function h is the nominal film thickness, and p is the smooth part of the pressure; ie. they correspond to $\bar{h}$ and $\bar{p}$ of the previous equations. The symbol σ is the standard deviation of the film thickness excursions.The non-dimensional flow factors (so named because they are associated with flow terms in the equation), represented by the symbol ϕ, express the appropriate average functions. As an example, $\overline{h^3} = \bar{h}^3 + 3\sigma^2 \bar{h}$. Hence, for transverse striations, $\phi_y = \overline{h^3}/\bar{h}^3 = 1 + 3(\sigma/\bar{h})^2$.

As stated already, the flow factors were determined from numerical simulations. Patir and Cheng presented their results in the form of charts in terms of standard deviation and the Kubo/Peklenik number, $\gamma = \lambda_x/\lambda_y$
where λ is the distance to a correlation value of 0.5.

Although the number of grid points is now lower, other difficulties occur, in particular those associated with the appropriate boundary conditions of the sample area. This point will be discussed later.

In 1979 Elrod generated solutions to a wider set of surface roughness structures. This was followed by a paper by Tripp, 1983, also giving solutions in terms of the Kubo/Peklenik number, γ. They both obtained results as closed form flow factors, based on sampling in the x- and y-directions and selecting a suitable autocorrelation function. Their results are in rather good agreement with those of Patir and Cheng, and are illustrated on fig. 6, where ϕ is plotted against h/σ.

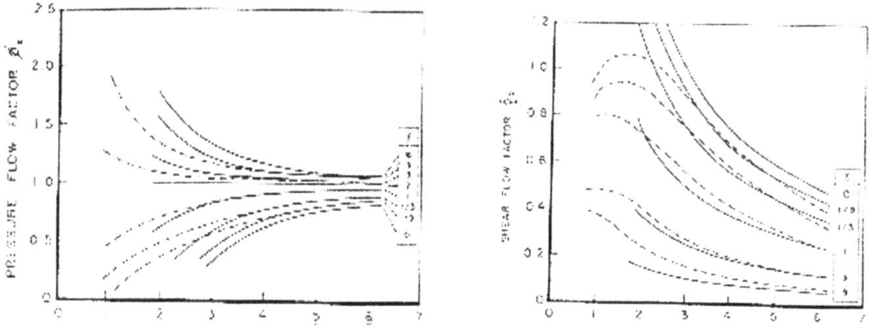

Fig. 6 Flow factors
Tripp results - solid lines ; Patir/Cheng - dotted lines

However, the γ - value alone cannot express surface roughness structure, as shown by Tønder (1984a). For instance, $\gamma = 1$ is a necessary but not sufficient condition for the roughness structure to be isotropic. This point will also be discussed later.

3.4. COMPRESSIBLE LUBRICANTS

If the lubricant is compressible, such as in the case of gases or bubbly liquids, the general assumption of constant density no longer holds. This means that ρ must remain throughout eq. 1 and will necessarily appear in a Reynolds-type equation for rough surfaces. A very important parameter in compressible lubricant situations is the bearing number, Λ, given by,

$$\Lambda = 6\eta Ul/(p_a\, h_m^2) \tag{19}$$

For low values of Λ the density may be treated as constant and our modified equations for liquids will be applicable for gas as well. However, for higher bearing numbers the density changes must be included. This makes Reynolds' equation non-linear. Typically one assumes proportionality between density and pressure, hence the symbol ρ may be replaced by p, the proportionality constant vanishing.

White (1980) gave a solution for very high bearing numbers, by employing the well-known approximation, $p = p_a\, h_i/h$, where the indices a and i refer to atmospheric and inlet respectively. From this follows White's solution,

$$\overline{p} = p_a\, h_i\, \overline{h^{-1}} \tag{20}$$

Tønder (1984b), by directly solving the full Reynolds equation for a case involving a limited number of sinusoidal oscillations, obtained a solution indicating that the same modified Reynolds equation as obtained for liquids was applicable to a compressible medium, the only difference being the appearance of $\bar{p}$, which corresponds to density, in all the terms. Thus, omitting the overbar in the notation, we may write, in a notation slightly different from that of Patir and Cheng,

$$\frac{\partial}{\partial x}(\phi_x h^3 p \frac{\partial p}{\partial x}) + \frac{\partial}{\partial y}(\phi_y h^3 p \frac{\partial p}{\partial y}) = 6\eta U \frac{\partial}{\partial x}(\phi_u ph) \tag{21}$$

In other words, the flow factors are the same as in the case of liquids. This result was later proved mathematically, Tønder 1985.

White and coworkers showed in a paper of 1986 that if the number of undulations in a transverse sinusoidal roughness structure is kept constant, solving Reynolds' equation directly will yield White's solution as the minimum film thickness approaches zero.This mathematical paradox was shown by Greengard (1989) to be caused by the limiting processes employed: If the number of undulations is constant, the limiting process in film thickness does lead to White's solution. If, however, the film thickness is kept constant, at any level, the limiting process increasing the number of oscillations leads to Tønder's result. In other words, in a sense both are right, so the mathematical paradox has been replaced by a physical one. This has prompted a closer look at the physics of the problem, and the author claims that eq. 21, with the liquid flow factors, is the better solution. This will be dealt with subsequently.

3.5. RAREFACTION

In certain situations, particularly in the case of very thin films, rarefaction becomes an important element of the lubrication of compressible media. This has been accounted for by a modification of Reynolds' equation, in the form of additional terms. This has led to the construction of roughness versions of these equations. Several researchers have been active in this area, but only a few will be mentioned here. Mitsuya (1986) and Mitsuya et al (1989) have given important contributions, as has Bhushan (Bhushan and Tønder, 1989). In essence, his modified equation is,

$$\frac{\partial}{\partial x}\{\phi_x ph^3(1+\frac{6aM}{ph})\frac{\partial p}{\partial x}\}+\frac{\partial}{\partial y}\{\phi_y ph^3(1+\frac{6aM}{ph})\}=6\eta\{U\frac{\partial}{\partial x}(\phi_u ph)+2\frac{\partial}{\partial t}(ph)\} \tag{22}$$

Here the presence of an overbar is understood. The parameter M is the Knudsen number, ie. the ratio of the mean free path to the nominal minimum film thickness, λ_a/h_m; a is a surface correction factor of order unity. In principle, it has not been shown that the previously determined flow factors apply to the rarefaction terms, but this is assumed because of the near-constancy of ph under the actual conditions. This permits computations of relevant performance data when rarefaction occurs.

In a paper by Crone et al, (1991), direct calculations under rarefaction conditions on a finite bearing having various sinusoidal roughnesses are reported. The resulting solutions deviate from those of Bhushan and Tønder (1989) as well as from those of White (1980). This will be dicussed subsequently.

4. Dynamic Effects

Surface roughness can cause several types of dynamic effects, some of which are direct ones in the sense that they are caused by the structure. Others may be termed indirect, representing an altered response to motion.

4.1. DIRECT DYNAMIC EFFECTS

When tribo-elements having roughness on both surfaces are running at low film thicknesses, asperities may come into contact and abrupt changes in the friction force may occur. Such impacts, though generally of small amplitude, may generate vibrations. There does not seem to be much literature on these phenomena but it might be a good subject for further studies.

Another effect that may have important implications is the one that is caused by the changes in local film thickness when there is roughness on both sufaces. A gas film study is due to Raad and Kuria (1989). Strong load variations were predicted. Tønder (1995) presented a work on liquid lubrication that showed that strong changes in hydrodynamic load or normal force could occur, particularly for transveres roughness of the same spatial frequency on the two surfaces. In certain situations that might occur in practice, the deviations may be of the same order of magnitude as the load itself.

In 1994 Tønder presented another example of direct dynamic effects. The situation analysed was one of surface disturbances moving past the boundaries of a bearing. If the surface features have important components of low spatial frequency, tending towards waviness rather than roughness, again strong dynamic effects will take place. As for the two-sided roughness described above, the amplitudeof the variations in the normal force generated by the bearing may be of the same order of magnitude as the load itself. Again it should be pointed out that the amplitdes of the surface disturbances that are needed for this to happen, are rather realistic.

4.2. INDIRECT DYNAMIC EFFECTS

There are at least two dynamic effects that are worth mentioning. The first of these was proposed by Kammüller (1986), by Müller and Ott (1984), Horve (1987) and others and later analysed by Salant (1992) and Tønder and Salant (1992). The mechanism proposed is the following: Transverse striations of a flexible material, such as in the case of a lip seal, will deform due to friction forces or shear stresses. The originally straight grooves and ridges will become curved. Because of the curvature the grooves will now tend to pump fluid. If the shear stresses on the flexible surface, and thereby the deformations, are symmetric about the mid-plane of the tribo-element considered, the same amount of fluid will be pumped in both directions, so no net flow will take place. If, however, the shear stresses are non-symmetric, the curving of the striations will be skew, and a net flow will result. The shear stresses are mainly caused by $\eta U/h$, so by giving the surfaces a suitable shape, one may tailor he shear stress distribution in a desired manner. Seals that are leakage free during operation are therefore possible. Indeed, it was the descovery of non-leaking seals that led to the above model. Salant and Flaherty have shown (1995) that the effect may occur for non-striated patterns too.

The other dynamic phenomenon to be mentioned is the effect of DDC (deep, disconnected cavities) on squeeze action. Tønder (1992) has shown that DDC-surfaces have superior fluid-retaining properties and that such surfaces are closely governed by Reynolds' equation, with only a modification of the viscosity expressing the presence of the cavities. The effective viscosity for a honeycomb-like surface is proportional to the remaining material, so the squeeze-time is reduced. However, after contact the cavities are quite well sealed off and a considerable part of the load is carried by hydrodynamic action. As the fluid escapes, more load is gradually taken up by the solid contact, which is being compressed in step with the descent of the roof of the cavities. It can be shown that time to reach say half the solid-to-solid frictional coefficient can be a matter of several hours in quite realistic cases, (Tønder, 1997).

Finally, the direct dynamic effects mentioned previously will lead to indirect ones, for instance in the form of varying friction forces, varying film thicknesses and, obviously, vibrations. This, however, is considered to be outside the scope of this paper

4.3. TURBULENCE

If viscosities are low and velocities high, turbulence effects may occur. In hydrodynamic bearings where roughness effects are important, the overall film thickness is small. In such cases a liquid lubricant is usually one of high viscosity; for a gas the density is very low. This all means that in the context of bearings and gears etc the Reynolds number is low, which means that turbulence will not occur. There are bearings lubricated by water, which has low viscosity; even then turbulence is often avoided because of the thin films generated. There is one related area where turbulence may become very important indeed, namely that of high-pressure seals when the fluid involved may attain high Reynolds numbers and thus cause turbulence effects. Even here surface roughness plays an important role, though one very different from that of lubrication. For this reason, this point will not be dealt with further.

5. Discussion

There are several points in hydrodynamic roughness theory that need to be cleared up. The most important of these are,

The validity of Reynolds' equation.

The satisfaction of the high spatial frequency requirement.

The boundary conditions in numerical calculations and related problems

The shortcomings of the parameter γ in expressing surface roughness structure

The gas bearing paradox

Experimental evidence

5.1. REYNOLDS EQUATION

The validity of the Reynolds equation in connection with rough surfaces has been questioned. Particularly, if the roughness gradients are large, important deviations from the solutions described will occur. The problem has been studied by Sun and Chen (1977) and by Mitsuya and Fukui (1986). The current opinion among researchers in the field seems to be that for realistic roughness values, such as observed on engineering surfaces the non-Reynolds effects - often referred to as Stokes effects - are small. Clearly also, a finite number of steep gradients or steps are acceptable to Reynolds' equation though clearly this represents approximate solutions only.

The DDC-surface has, by definition, steep gradients. Though there is a finite number of pits only, the accumulated effect of the approximation may be considerable. However, the inaccuracy in modelling the flow pattern inside a cavity is, to a large extent, to include pockets of more or less stagnant or rotating fluid. Not to include these regions thus appears to improve the modelling. This in turn may be interpreted as an effective modification of the actual shape of the cavity. Now in DDC theory the precise shape of a cavity is unimportant. This certainly indicates that the theory may be better than the its simplified derivation might warrant. The question of local turbulence may also be raised. The main consequence of that seems to be a somewhat higher viscosity. It appears unlikely that its presence would enhance leakage. Accordingly, DDC theory seems to be reasonably accurate.

5.2. SPATIAL FREQUENCY

It is difficult to predict precisely what the spatial frequency requirements are. However, numerical experiments testing this have been performed (Tønder and Christensen, 1972). It is seen that only very modest numbers of undulations of the surface texture within a bearing are required for the approximations to be of the order of per mille or less. These requirements are easily met for real surfaces.

Another aspect of rough surface undulations is the possibility of cavitation to occur in areas just following narrow passages. This is not directly included in current roughness theory. However, it can be shown that the pressure undulations tend to vanish as the spatial frequency increases. In practice, therefore, such effects will not be common outside areas of near-atmospheric pressure. In such areas it may be useful to establish the amplitudes of pressure excursions. Bounds on such values may be obtained reasonably easily from the definition of the smooth and rough quantities combined with an approximate solution to the unmodified Reynolds equation. It is, of course also possible to deal with cavitation directly in connection with numerical calculations as a function of the (assumed) pressure level.

5.3. NUMERICAL PROBLEMS

Although the concept of a representative neighbourhood has reduced the number of mesh points required, that number is still rather large, whether the method of solution is by finite differences or finite elements. Hence, the most common way of dealing with the requirement of having many undulations and representative roughness features in the modelled miniature bearing is to generate many such mini-bearings all having the same macro-geometry and the same surface statistics. They will then be different from each other but statistically equivalent. This is usually less time consuming than generating a larger area. However, a recurring problem is that of applying the correct boundary conditions. The ones chosen by Patir and Cheng, (1978, 1979) permitting no side flow at all, are obviously wrong. Tønder (1980a) tried the following approach: generate pressures using simplified boundary onditions. Use pressure trains computed inside the element as the oscillating components of the boundary values for new computations. This approach is also questionable. Newer work on the problem exists, however, by Lunde and Tønder (1994, 1995) Lunde (1996). Here correlations between boundary values and the calculated quantities were studied. It led to an approach by which a steadily smaller portion of a rather large representative surface was considered. The idea was that calculated values at points far away from the boundaries are less affected by the function values at the latter. This was seen to hold and new updated flow factors have been generated. They are not, however, very much different from the ones calculated using the simplified boundary conditions.
Realistic calculations for gas bearings are even more challenging: the same problems as for liquids occur; in addition the non-linear character of the appropriate Reynolds equation requires some kind of iteration.

5.4. THE KUBO/PEKLENIK NUMBER

As has been mentioned already, the Kubo/Peklenik number, γ has been used to characterise the surface structure. However, this parameter alone cannot express this in any detailed way. Isotropic roughness, ie. a structure having no preferred region or direction, sems to be an excellent model of textures found on many real surfaces. Accordingly, many studies have been devoted to this type of roughness, which must necessarily have a γ - value of unity. However, interpreted a little wider, the Kubo/Peklenik number may be equal to one for regular roughness patterns consisting of two-ways repetitions of a unit disturbance. Such patterns give flow factors very different from one another and from an isotropic one (Tønder, 1984). Even if we use the strict definition, a pattern consisting of unidirectional but otherwise random striations at 45 degrees to the x-direction would be defined as isotropic. It may be worth mentioning that the flow factors for this situation coincide with the values obtained by Tripp (1983) and Elrod (1979).

The shortcoming of the Kubo/Peklenik number emphasises that in order to model truly isotropic surfaces the autocorrelation function must be the same in all directions. Hu's method (1992) has proved efficient.

Bhushan (Bhushan and Tønder, 1989) devised a number, γ_B that requires the ratio be formed for all directions. This does away with some of the problems in connection with γ. However, even the Bhushan number cannot fully characterise surface feature structure.

5.5. THE GAS BEARING PARADOX

As has been pointed out already both the model of White and his coworkers and that of Tønder are mathematically correct as limiting cases. The question is, however, which (if any) is correct physically? A study of the White solution shows that the pressure and (thereby the load) is proportional to the inlet film thickness. However, particularly in the presence of roughness the definition or interpretation of what is the the inlet film thickness is rather arbitrary: it depends on the macro-geometry (where does a circular bearing have its inlet?) as well as on the micro-geometry (if there is roughness undulations in the y-direction, what is h_i ?). The same questions can be raised for each small subdivision of the bearing and the answers are as elusive. Tønder's solution avoids the situation altogether by only involving averages. Moreover, the standard solution corresponding to very large bearing numbers comes out in this analysis too; however, the proportionality factor is different.

Finally, the very high frequencies of the pressure oscillations when the roughness frequency is very high, may be limited if goverened by the inlet conditions, particularly if the relative velocity of the surfaces is high. This tends to favour a more local control of the oscillations, which again supports the Tønder solution. The problem should be looked into more closely, however.

As already mentioned, the results of Crone et al (1991) were seen to differ from both those of Tønder and those of White. However, the number of undulations considered was very low, thus violating the conditions implicit in Tønder's results. Further, since the bearing is very narrow whilst the bearing number is not extremely high, the conditions required for White's solutions are also violated.

5.6. EXPERIMENTAL EVIDENCE

Unfortunately there is not very much experimental evidence to support the above results. There are a few, though, and they do seem to support the established theories. In a paper by Tønder and Jakobsen (1992) experiments are run on an elastohydrodynamic testrig consisting of a rotating steel ball being pressed against a sapphire disc. The latter has fine grooves etched into its surface. The grooves may be oriented along or across the direction of motion and may therefore implement situations of longitudinal or transverse roughness. Intermediate positions are also possible, the orientation being adjustable during operation. The results are in accordance with theory; it is even possible to observe the reduction in ball-disc separation as the striations are rotated away from the transverse orientation.

In a test of 1984, Tønder obtained results supporting current theory by studying fluid flow across transverse ridges. Further experimental support was also reported by the author in 1976, but uncertainties in the interpretations make it a little less convincing than the one of 1984..

In her Master's thesis, Seljeseth (1994) showed experimentally that the DDC load-carrying concept is valid. The experiment consisted in dropping a small test specimen from a height of 250 μm onto a glass plate. The specimen, 20 x 20 mm^2, carrying a load of some 100 N, was provided with a hard plastic coating having a honeycomb-like structure, 1 mm thick The compression of the coating was lower than the resolution of the distance gauge used, i.e. lower than $1\,\mu m$, even when he experiment was stopped after a period of 20 hours.

Mitsuya et al, (1989) presented experiments for testing the validity of results obtained for the slip flow or rarefied regime. They made small test specimens or bearing pads having patterns in the form of striations, both transverse and longitudinal, as well as a checkered pattern. The results were in most respects in very good agreement with current theory for this regime.

6. Conclusion

We have seen from the above that progress has been made in our knowledge of roughness effects in hydrodynamic lubrication. However, there are still many aspects of the field and of related areas that are poorly understood. A deeper understanding is desirable from a scientific point of view, but may also have practical consequences. One such area that needs further study is that of the phenomena occurring when asperity contact has been established. Thicker films or the appearance of turbulence; temperature effects and non-Newtonian fluids combined with roughness are other grey areas. Other aspects requiring further investigation may be dynamic effects and extremely thin films. Moreover, the utilisation of the surface roughness effects in practical applications is just beginning to appear. This all means that the research area of roughness effects is certainly not dead.

The following basic knowledge has been established,

Reynolds' equation, in some form, appears to be well suited for handling rough surfaces likely to be found in practice.

There is strong interaction between microscopic and macroscopic effects.

We are able to model the effects of most surface structures.

Compressible lubricants may be handled, even in the rarefaction regime.

The mixed lubrication regime needs further investigation.

Very high speeds and surface roughness is also an area requiring closer studies.

7. References

Bhushan, B. and Tønder, K. (1989) Roughness-Induced Shear- and Squeeze-Film Effects in Magnetic Recording-Part I: Analysis, *Trans. ASME, J. Tribology*, **111**, 220-227

Christensen, H. (1969/70) Stochastic Models for Hydrodynamic Lubrication of Rough Surfaces, *Proc. IME*, **55**, 1013-126

Christensen, H. and Tønder, K. (1969) Tribology of Rough Surfaces. Stochastic Models of Hydrodynamic Lubrication, *SINTEF Report*, SINTEF, Trondheim, Norway, no. 18-69/10

Crone, R.M.; Jhon, M.S.; Bhushan, B. and Karis, T.E.(1991) Modeling the Flying Characteristics of a Rough Magnetic Head over a Rough Rigid-Disk Surface, *Trans. ASME, J. Tribology*, **113**, 739-749

Elrod, H.G. (1973) Thin-Film Lubrication Theory for Newtonian Fluids with Surfaces Possessing Striated Roughness or Grooving, *Trans. ASME, J. Lubrication Technology*, **95**, 484-489

Elrod, H.G. (1979) A General Theory for Laminar Lubrication with Reynolds Roughness, *Trans. ASME, J. Lubrication Technology,* **101**, 8-14

Greengard, C. (1989) Large Bearing Numbers and Stationary Reynolds Roughness, *Trans. ASME, J. Tribology,* **111**, 136-141

Horve, L. (1987) A Macroscopic View of the Sealing Phenomenon for Radial Lip Seals, *Proc. 11th Intnl. Conf. on Fluid Sealing*, BHRA, 710-731

Hu, Y. and Tønder, K. (1992) Simulation of 3-D Random Surface by 2-D Digital Filter and Fourier Analysis. *Intnl. J. Machine Tool Manufact.* **32**, 52-56

Kammüller, M. (1986) Zur Abdichtwirkung von Radial-Wellendichtringen, *Dr. Ing. Thesis*, Universität Stuttgart, Stuttgart

Lunde, L. (1996) Numerical Simulation of the Lubrication of Rough Surfaces: Methodology and Boundary Conditions, *Dr. Ing. Thesis*, Norwegian Univ. of Science and Technology, Trondheim

Lunde, L and Tønder, K. (1997) Pressure and Shear Flow in a Rough Hydrodynamic Bearing, Flow Factor Calculation, *Trans. ASME, J. Tribology*, **119**, 549-555

Mitsuya, Y. (1986) Stokes Roughness on Hydrodynamic Lubrication, Part II - Effects under Slip Flow Boundary Conditions, *Trans. ASME, J. Tribology*, **108**, 159-166

Mitsuya, Y. and Fukui, S. (1986) Stokes Roughness Effects on Hydrodynamic Lubrication, Part I - Comparison between Incompressible and Compressible Lubricating Films, *Trans. ASME, J. Tribology*, **108**, 151-158

Mitsuya, Y. and Koumura, T. (1995) Transient Response Solution Applying ADI Scheme to Boltzmann Flow-Modified Reynolds Equation Averaged with Respect to Surface Roughness, *Trans. ASME, J. Tribology*, **117**, no. 3, 430-436

Mitsuya, Y. ; Ohkubo, T. and Ota, H.(1989) Averaged Reynolds Equation Extended to Gas Lubrication Possessing Surface Roughness in Slip Flow Regime: Approximate Method and Confirmation Experiments, *Trans. ASME, J. Tribology*, **111**, no. 3 495-503

Mitsuya, Y. and Ota, H.(1991) Stiffness and Damping of Compressible Lubricating Films between Computer Flying Heads and Textured Media: Perturbation Analysis Using the Finite Element Method, *Trans. ASME, J. Tribology*, **113**, no. 4, 819-827

Müller, H.K. and Ott, G.W. (1984) Dynamic Sealing Mechanism of Rubber Rotary Shaft Seals, *Proc. 10th Intnl.Conf. on Fluid Sealing, BHRA*, 451-466

Raad, P.E. and Kuria, I.M. (1989) Two-Sided Texture Effects on Ultra-Thin Wide Wedge Gas Bearings, *Trans. ASME, J. Tribology*, **111**, 719-725

Patir, N. and Cheng, H. (1978) An Average Model for Determining Effects of Three-Dimensional Roughness on Hydrodynamic Lubrication, *Trans. ASME, J. Lubrication Technology*, **100**, 12-17

Patir, N. and Cheng, H. (1979) Application of Average Flow Model to Lubrication between Rough Sliding Surfaces, *Trans. ASME, J. Lubrication Technology*, **101**, 220-230

Salant, R.F. (1992) Numerical Analysis of the Flow Field Within Lip Seals Containing Microundulations, *Trans ASME, J. Tribology*, **114**, no. 3, 485-491

Salant, R.F. and Flaherty, A.L. (1995) Elastohydrodynamic Analysis of Reverse Pumping in Rotary Lip Seals with Microasperities, *Trans. ASME, J. Tribology*, **117**, 53-59

Seljeseth, O.A. (1994) Lubrication of Natural and Artificial Human Hip Joints: Theoretical Studies and Experimental Work. Master's thesis, Høgskolen i Narvik, Narvik, Norway.

Sun, D.C. and Chen , K.K. (1977) First Effects of Stokes Roughness on Hydrodynamic Lubrication, *Trans. ASME, J. Lubrication Technology*, **99**, 2-9

Tønder, K. (1972) Surface Distributed Waviness and Roughness. *Proc. World Conf. in Industrial Tribology*, New Delhi, India

Tønder, K. (1976) Experimental Investigation of Effects of Striated Roughness on Hydrodynamic Lubrication, *SINTEF report,* STF18 A76019, Trondheim

Tønder, K. (1977a) Mathematical Verification of the Applicability of Modified Reynolds Equations to Striated Rough Surfaces, *Wear*, **44**, no. 2

Tønder, K. (1977b) Lubrication of Surfaces Having Area-Distributed Isotropic Roughness, *Trans. ASME, J. Lubrication Technology,* **99**

Tønder, K. (1980a) Simulation of the Lubrication of Isotropically Rough Surfaces, *ASLE Trans.*, **23**, 326-333

Tønder, K. (1980b) Numerical Investigation of the Lubrication of Doubly Periodic Unit Roughneses, *Wear*, **64**, 1-14.

Tønder, K. (1984) A Numerical Assessment of the Effect of Striated Roughness on Gas Lubrication, *Trans. ASME, J. Tribology,* **106**, 315-321

Tønder, K. (1985) Theory of Effects of Striated Roughness on Gas Lubrication, *Proc. JSLE Intnl. Tribology Conference*, Tokyo, 761-766

Tønder, K. (1985) The Effects of Machining Methods on Lubricant Flow, *Tribologia*, 4, 50-68

Tønder, K. (1986) The Lubrication of Unidirectional Striated Roughness: Consequences for Some General Roughness Theories, *Trans. ASME, J. Tribology*, **108**, 167-170

Tønder, K. (1992) DDC Lubrication: A New Concept in Tribology, *Trans. ASME, J. Tribology*, **114**, no.1, 181-185

Tønder, K. (1996) Dynamics of Rough Slider Bearings: Effects of One-Sided Roughness/Waviness, *Tribology International,* **29**, 117-122

Tønder, K. (1997) Reducing Failure Risc by Surface Structure Modification: A Microstructure Model of the Lubrication of Nominally Thin Films, *Proc. ESRA 1997 Symposium*, Takamatsu, Japan

Tønder, K. and Christensen, H. (1972) Waviness and Roughness in Hydrodynamic Lubrication, *Proc. IME*, **186**

Tønder, K. and Jakobsen, J. (1992) Interferometric Studies of Effects of Striated Roughness on Lubricant Film Thickness Under Hydrodynamic Conditions, *Trans. ASME, J. Tribology*, **114**, no. 1, 52-56

Tønder, K. and Salant, R.F. (1992) Non-Leaking Lip Seals:A Roughness Effect Study, *Trans ASME, J. of Tribology,* vol. **114**, no. 3, 595-599

Tripp, J.H. (1983) Surface Roughness Effects in Hydrodynamic Lubrication: The Flow Factor Method, *Trans. ASME, J. Tribology*, **105**, 458-465

Tzeng, S.T. and Saibel, E. (1967) Surface Roughness Effect on Slider Bearing Lubrication, *ASLE Trans.*, **10**, 334-344

White, J.W, (1980) Surface Roughness Effects on the Load Carrying Capacity of Very Thin Compressible Lubricating Films, *Trans. ASME, J. Lubrication Technology*, **102**, 445-451

White, J.W.; Raad, P.E.; Tabriz, A.H.; Ketkar, S.P. and Prabhu, P.P. (1986) A Numerical Study of Surface Roughness Effects on Ultra-Thin Gas Films, *Trans. ASME, J. Tribology*, **108**, 171-177

TRANSITION FROM ELASTOHYDRODYNAMIC TO PARTIAL LUBRICATION

HSING-SEN S. HSIAO
BERNARD J. HAMROCK
Mechanical Engineering Department
The Ohio State University, Columbus, Ohio, USA

SHASHI K. SHARMA
The Ohio State University, and
Materials Directorate, Air Force Research Laboratory
Wright-Patterson Air Force Base, Ohio, USA

JOHN H. TRIPP
SKF Engineering and Research Centre B.V.
The Netherlands

1. ABSTRACT

The recognition and understanding of elastohydrodynamic lubrication (EHL) represents one of the major developments in the field of tribology in the last half of the twentieth century. The revelation of a previously unsuspected lubrication film is clearly an event of some importance in tribology. In this case it not only explained the remarkable physical action responsible for the effective lubrication of many nonconformal machine elements such as gears, rolling-element bearings, cams, and continuously variable traction drives, but also brought order to the understanding of the complete spectrum of lubrication regimes, ranging from boundary to hydrodynamic.

The transition from EHL to partial lubrication has been the focus of the last twenty years of the twentieth century. That is, if (for example) in various machine element applications the pressures are too high or the running speeds are too low, the lubricant film will be penetrated. Some contact will take place between the asperities, and partial lubrication (sometimes referred to as "mixed lubrication") will occur. The behavior of the conjunction in a partial lubrication regime is governed by a combination of boundary and fluid film effects. Interaction takes place between one or more molecular layers of boundary lubricating films. A partial fluid film lubrication action develops in the bulk of the space between the solids.

It is important to recognize that the transition from elastohydrodynamic to partial lubrication does not take place instantaneously as the severity of loading is increased, but rather a decreasing proportion of the load is carried by pressures within the fluid that fills the space between the asperities of the solids. Recent results showing the transition from elastohydrodynamic to partial lubrication while considering an individual asperity are discussed.

B. Bhushan (ed.), Tribology Issues and Opportunities in MEMS, 207-228.

2. NOMENCLATURE

A	amplitude of surface feature in z direction, m
b_x	semiaxis of Hertzian elliptical contact in x direction, m
b_y	semiaxis of Hertzian elliptical contact in y direction, m
E_a	Young's modulus of elasticity for solid a, N/m^2
E_b	Young's modulus of elasticity for solid b, N/m^2
E'	effective Young's modulus of elasticity, $2/\left[\left(1-\nu_a^2\right)/E_a+\left(1-\nu_b^2\right)/E_b\right]$, N/m^2
G	dimensionless materials parameter, $\xi E'$
h	film thickness, m
h_{min}	minimum film thickness, m
h_0	offset film thickness, $h_c-\delta_c$, m
H	dimensionless film thickness, hR_x/b_x^2
$H_{c-rough}$	dimensionless film thickness H at center of contact for case with asperity
$H_{c-smooth}$	dimensionless central film thickness for smooth surfaces, h_cR_x/b_x^2
H_0	dimensionless offset film thickness, h_0R_x/b_x^2
l_x	longitudinal base width of the surface feature, m
l_y	transverse base width of the surface feature, m
l	base diameter of the symmetrical surface feature, m
p	local pressure; gage pressure, N/m^2
P	dimensionless pressure, p/p_H
p_H	maximum Hertzian pressure, N/m^2
r_{ax}	x radius of curvature of surface a, m
r_{ay}	y radius of curvature of surface a, m
r_{bx}	x radius of curvature of surface b, m
r_{by}	y radius of curvature of surface b, m
R_x	equivalent x radius of curvature, $1/\left(1/r_{ax}+1/r_{bx}\right)$, m
R_y	equivalent y radius of curvature, $1/\left(1/r_{ay}+1/r_{by}\right)$, m
s	geometric separation, $x^2/2R_x+y^2/2R_y+\delta_{ruf}(x,y)$, m
S	dimensionless geometric separation, $X^2/2+\left(k^2/2\alpha\right)Y^2+\bar{\delta}_{ruf}(X,Y)$
T_0	ambient temperature, °C
u	fluid velocity in x direction, m/s
$\bar{u}$	mean surface velocity in x direction, $\left(u_a+u_b\right)/2$, m/s
u_a	velocity of surface a in x direction, m/s
u_b	velocity of surface b in x direction, m/s
U	dimensionless speed parameter, $\eta_0\left\|\bar{V}\right\|/E'R_x$
v	fluid velocity in y direction, m/s
v_a	velocity of surface a in y direction, m/s
v_a	dimensionless velocity of surface a in y direction, $v_a/\left\|\bar{V}\right\|$

v_b velocity of surface b in y direction, m/s

v_b^* dimensionless velocity at surface b, $v_b/|\bar{V}|$

$|V_a|$ entraining velocity at surface a, $\sqrt{u_a^2 + v_a^2}$, m/s

$|V_b|$ entraining velocity at surface b, $\sqrt{u_b^2 + v_b^2}$, m/s

$|\bar{V}|$ average entraining velocity, $(|V_a|+|V_b|)/2$, m/s

w fluid velocity in z direction, m/s

w_z applied normal load, N

W dimensionless applied normal load in point contact, $w_z/E'R_x^2$

x longitudinal coordinate, m

x_c, y_c locations of center of surface features, m

X dimensionless longitudinal coordinate, x/b_x

y transverse coordinate, m

Y dimensionless transverse coordinate, y/b_y

z film thickness coordinate, m

z_a z coordinate of surface a, m

z_b z coordinate of surface b, m

Z dimensionless film thickness coordinate, $(z - z_a)/h$

Z_0 Roelands pressure-viscosity index

α radius ratio, R_y/R_x

δ elastic deformation, m

$\bar{\delta}$ dimensionless elastic deformation, $\delta R_x/b_x^2$

δ_c elastic deformation at center of Hertzian contact for smooth surfaces, m

$\bar{\delta}_c$ dimensionless elastic deformation at center of Hertzian contact for smooth surfaces, $\delta_c R_x/b_x^2$

δ_{ruf} height of surface feature in z direction, at location (x,y) , m

$\bar{\delta}_{ruf}$ dimensionless height of the surface feature in the z direction, at location (x,y), $\delta_{ruf} R_x/b_x^2$

γ limiting-shear-strength proportionality constant

$\dot{\gamma}_{zx}$ x shear strain rate on z plane, $\partial u/\partial z$

$\dot{\gamma}_{zy}$ y shear strain rate on z plane, $\partial v/\partial z$

η fluid viscosity, N/m^2

η_0 fluid viscosity at atmospheric pressure, $\text{N}-\text{s/m}^2$

$\hat{\eta}$ dimensionless fluid viscosity, $\eta|\bar{V}|R_x/\tau_L H b_x^2$

λ dimensionless base width of surface feature, $\lambda = l/b_x$

ν_a Poisson's ratio for solid a

ν_b Poisson's ratio for solid b

ρ fluid density, kg/m^3

ρ_0 fluid density at $p = 0$, kg/m^3

σ dimensionless amplitude of surface feature, AR_x/b_x^2
τ field shear stress, N/m^2
τ^* dimensionless field shear stress, τ/τ_L
τ_0 limiting shear stress at atmospheric pressure, N/m^2
τ_L limiting-shear-strength, $\tau_0 + \gamma p$, N/m^2
$\hat{\tau}_L$ dimensionless limiting shear strength, τ_L/E'
τ_{zx}^* dimensionless X shear stress on Z plane, τ_{zx}/τ_L
τ_{zy}^* dimensionless Y shear stress on Z plane, τ_{zy}/τ_L
τ_{zxa}^* dimensionless X shear stress on surface a
τ_{zya}^* dimensionless Y shear stress on surface a
τ_{zxb}^* dimensionless X shear stress on surface b
τ_{zyb}^* dimensionless Y shear stress on surface b
ξ pressure-viscosity coefficient of lubricant, Pa^{-1}

Subscripts
a solid a or surface a
b solid b or surface b
HL hydrodynamic lubrication
HEHL hard elastohydrodynamic lubrication

3. INTRODUCTION

Fluid film lubrication occurs when opposing surfaces are completely separated by a lubricant film and no asperities are in contact. The applied load is carried by pressure generated within the fluid, and frictional resistance to motion arises entirely from the shearing of the viscous fluid.

Conformal surfaces fit snugly into each other with a high degree of geometrical conformity so that the load is carried over a relatively large area. For example, the lubrication area of a journal bearing would be 2π times the radius times the length. The load-carrying surface area remains essentially constant while the load is increased. Fluid film journal bearings (Fig. 1) and slider bearing have conformal surfaces. In journal bearings the radial clearance between the journal and the sleeve is typically one-thousandth of the journal diameter; in slider bearings the inclination of the bearing surface to the runner is typically one part in a thousand. Another example of conformal surfaces is a human hip joint.

Many fluid-film-lubricated machine elements have surfaces that do not conform to each other well. The full burden of the load must then be carried by a small lubrication area. The lubrication area of a nonconformal conjunction is typically three orders of magnitude less than that of conformal conjunction. In general, the lubrication area between nonconformal surfaces enlarges considerably with increasing load, but it is still smaller than the lubrication area between conformal surfaces. Some examples of nonconformal surfaces are mating gear teeth, cams and followers, and rolling-element bearings (Fig. 2).

Features that distinguish the three lubrication regimes present in machine elements are considered here. Hydrodynamic lubrication (HL) is generally characterized by conformal surfaces with fluid film lubrication. A positive pressure develops in a hydrodynamically lubricated journal or thrust bearing because the bearing surfaces converge and their relative motion and the viscosity of the fluid separate the surfaces. The existence of the pressure developed (usually less than 5 MPa) is not generally large enough to cause significant elastic deformation of the surfaces. The minimum film thickness in a hydrodynamically lubricated bearing is a function of normal applied load w_z, velocity u_b, lubricant absolute viscosity η_0, and geometry (R_x and R_y). Figure 3 shows some of these characteristics of hydrodynamic lubrication. Minimum film thickness $h_{\min}$ as a function of u_b and W for sliding motion is given as

$$\left(h_{\min}\right)_{\mathrm{HL}} \propto \left(\frac{u_b}{w_z}\right)^{1/2} \tag{1}$$

The minimum film thickness normally exceeds 1 μm. In hydrodynamic lubrication the films are generally thick, so that opposing solid surfaces are prevented from coming into contact. This condition is often called the ideal form of lubrication because it provides low friction and high resistance to wear. The lubrication of the solid surfaces is governed by the bulk physical properties of the lubricant, notably the viscosity; and the frictional characteristics arise purely from the shearing of the viscous lubricant.

Elastohydrodynamic lubrication is a form of hydrodynamic lubrication where elastic deformation of the lubricated surfaces becomes significant. The features important in a hydrodynamically lubricated slider bearing--converging film thickness, sliding motion, and a viscous fluid between the surfaces--are also important here. Elastohydrodynamic lubrication is normally associated with nonconformal surfaces and fluid film lubrication. There are two distinct forms of EHL, namely hard EHL and soft EHL.

Hard EHL relates to materials of high elastic modulus, such as metals. In this form of lubrication the elastic deformation and the pressure-viscosity effects are equally important. Figure 4 gives the characteristics of hard elastohydrodynamically lubricated conjunctions. The maximum pressure is typically between 0.5 and 4 Gpa; the minimum film thickness normally exceeds 0.1 μm. These conditions are dramatically different from those found in a hydrodynamically lubricated conjunction (Fig. 3). At loads normally experienced on nonconformal machine elements the elastic deformations are two orders of magnitude larger than the minimum film thickness. Furthermore, the lubricant viscosity can vary as much as 20 orders of magnitude within the lubricated conjunction. The minimum film thickness is a function of the same parameters as for hydrodynamic lubrication (Fig. 3) but with the additions of the effective elastic modulus E' and the pressure-viscosity coefficient ξ of the lubricant.

$$E' = 2 \Big/ \left[\left(1-\nu_a^2\right)/E_a + \left(1-\nu_b^2\right)/E_b\right] \tag{2}$$

where ν = Poisson's ratio

E = modulus of elasticity, Pa

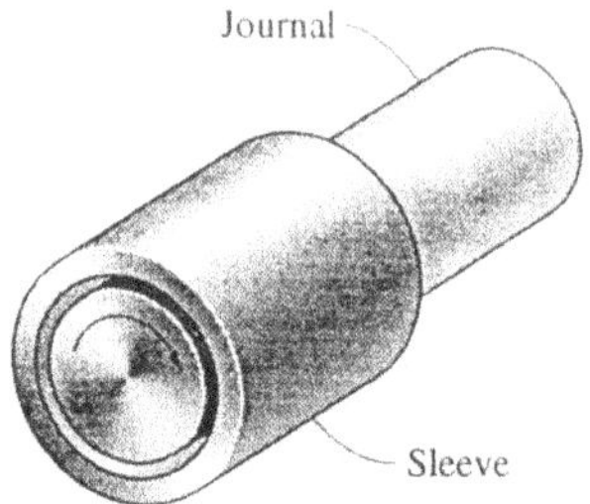

Figure 1 Conformal surfaces. [From Hamrock and Anderson (1983).]

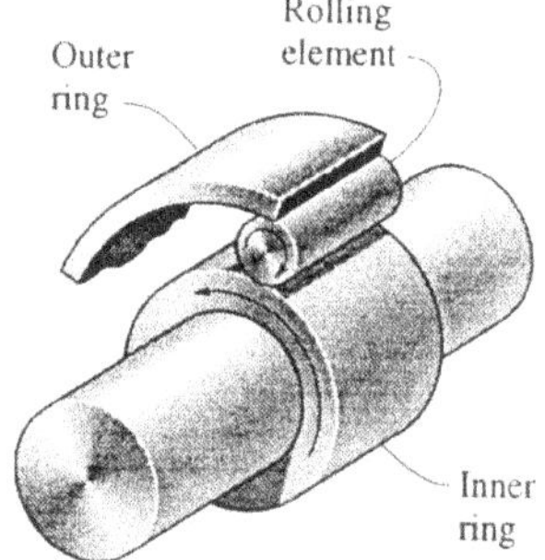

Figure 2 Nonconformal surfaces. [From Hamrock and Anderson (1983).]

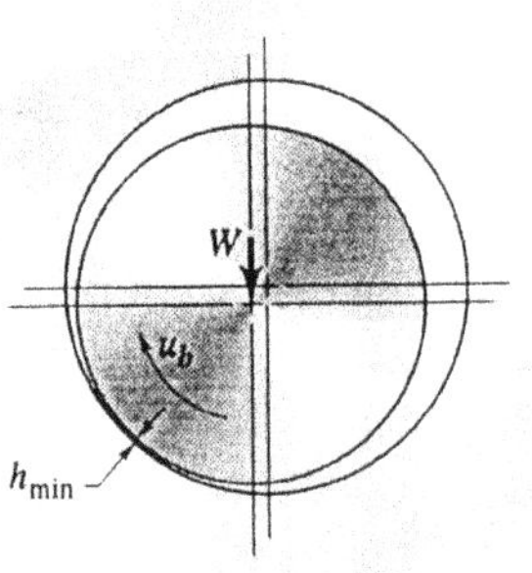

Conformal surfaces
$p_{max} \approx 5$ MPa
$h_{min} = f\left(w_z, u_b, \eta_0, R_x, R_y\right) > 1\ \mu\text{m}$
No elastic effect

Figure 3 Characteristics of hydrodynamic lubrication. [From Hamrock (1991).]

The relationship between the minimum film thickness h_{min} and the normal applied load w_z and upper surface velocity u_b for hard EHL as obtained from Hamrock and Dowson (1977) are

$$\left(h_{min}\right)_{HEHL} \propto w_z^{-0.073} \tag{3}$$

$$\left(h_{min}\right)_{HEHL} \propto u_b^{0.68} \tag{4}$$

Comparing the results for hard EHL [(Eqs. (3) and (4)] with those for hydrodynamic lubrication [Eq. (1)] yielded the following conclusions:

1. The exponent on the normal applied load is nearly seven times larger for hydrodynamic lubrication than for hard EHL. This difference implies that the film thickness is only slightly affected by load for hard EHL but significantly affected for hydrodynamic lubrication.
2. The exponent on mean velocity is slightly higher for hard EHL than for hydrodynamic lubrication.

Engineering applications in which elastohydrodynamic lubrication is important for high-elastic-modulus materials include gears, rolling-element bearings, and cams.

Soft EHL relates to materials of low elastic modulus, such as rubber. Figure 5 shows the characteristics of soft-EHL materials. In soft EHL the elastic distortions are large, even with light loads. The maximum pressure for soft EHL is 0.5 to 4 MPa (typically 1 MPa), in contrast to 0.5 to 4 GPa (typically 1 GPa) for hard EHL (Fig. 4). This low pressure has a negligible effect on the viscosity variation throughout the conjunction. The minimum film thickness is a function of the same parameters as in hydrodynamic lubrication with the addition of the effective elastic modulus. The minimum film thickness for soft EHL is typically 1 μm.

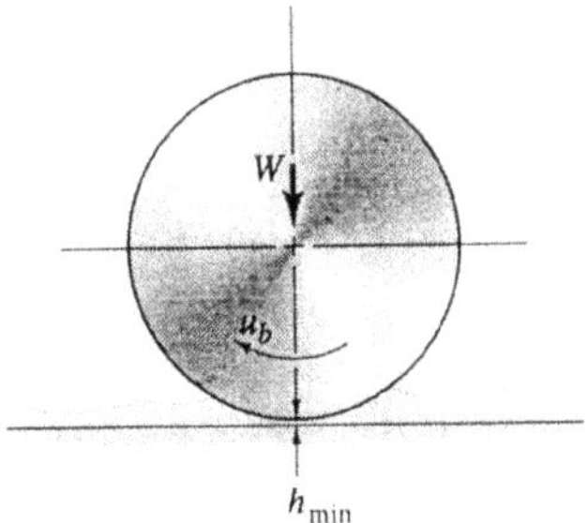

Nonconformal surfaces
High elastic modulus materials (e.g. steel)
$p_{max} \approx 0.5$ to 4 GPa
$h_{min} = f\left(w_z, u_b, \eta_0, R_x, R_y, E', \xi\right) > 0.1\ \mu m$
Elastic and viscous effects both important

Figure 4 Characteristics of hard elastohydrodynamic lubrication. [From Hamrock (1991).]

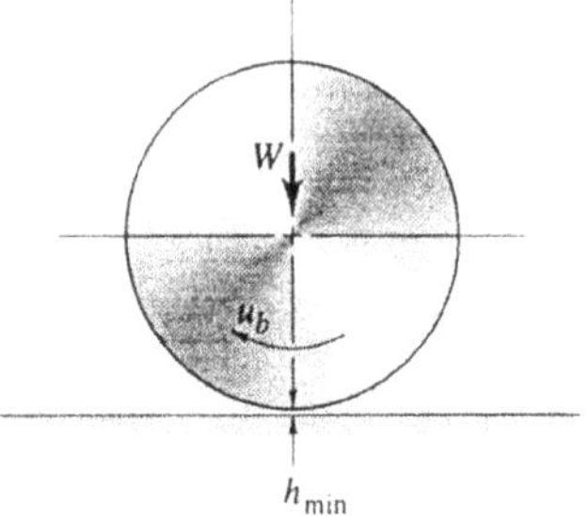

Nonconformal surfaces (e.g., nitrile rubber)
$p_{max} \approx 0.5$ to 4 MPa
$h_{min} = f(w_z, u_b, \eta_0, R_x, R_y) \approx 1\ \mu m$
Elastic effects predominate

Figure 5 Characteristics of soft elastohydrodynamic lubrication. [From Hamrock (1991).]

Engineering applications in which elastohydrodynamic lubrication is important for low-elastic-modulus materials include seals, human joints, tires, and a number of lubricated machine elements that use rubber as a material. The common feature of hard and soft EHL are that the local elastic deformation of the solids provides coherent fluid films and that asperity interaction is largely prevented. Lack of asperity interaction implies that the frictional resistance to motion is due to lubricant shearing.

If the pressures in elastohydrodynamically lubricated machine elements are too high or the speed are too low, the lubricant film will be penetrated. Some contact will take place between the asperities, and partial lubrication (sometimes called "mixed lubrication") will occur. The behavior of the conjunction in a partial lubrication regime is governed by a combination of boundary and fluid film effects. Interaction takes place between one or more molecular layers of boundary lubricating films. A partial fluid film lubrication action develops in the bulk of the space between the solids. The average film thickness in a partial lubrication conjunction is less than $1\ \mu m$ and greater than $0.01\ \mu m$. It is important to note that the transition from elastohydrodynamic to partial lubrication does not take place instantaneously as the severity of loading is increased, but rather a decreasing proportion of the load is carried by pressures within the fluid that fills the space between the opposing solids. As the load increases, a larger part of the load is supported by the contact pressure between the asperities of the solids. Under these conditions the local pressure and film thickness near the asperities are severely affected. The size and shape of the asperity, its location inside the contact, and the operating parameters affect the resulting pressures and the film thickness. Under certain conditions, an asperity could pierce through the fluid film and make contact on the opposing surface, thereby changing the lubrication regime from full EHL to partial elastohydrodynamic lubrication (PEHL).

Many researchers have studied the effect of surface roughness in EHL. Patir and Cheng (1979), Majumdar and Hamrock (1982), and Zhu et al. (1990) and others used the modified Reynolds equation to account for the surface roughness. Lubrecht et al. (1988) studied the effect of longitudinal and transverse roughness on the EHL point contact using Newtonian fluid. Chang et al. (1989) considered a single asperity or groove moving through an EHL line contact for an Eyring fluid model. Steady-state solutions under isothermal conditions were obtained at different time steps . The effects of lubricant rheology on the film thickness and the pressure profile were studied by comparing solutions obtained with Newtonian, two-slope, and Eyring models of viscosity. Elsharkawy and Hamrock (1991) considered a single asperity or groove moving through an EHL conjunction, extending an earlier work of Lee and Hamrock (1990). An isothermal, circular non-Newtonian fluid model with no side leakage and no time-variant effects was considered in solving the line-contact problem under different slide-to-roll ratios. Shieh and Hamrock (1991) incorporated streamlines into the solution of line contacts with rough surfaces. The results showed the location of stagnation points and flow reversals. Wedeven and Cusano (1979) studied the effects of a groove and a dent moving through an EHL conjunction by using optical interferometry. Dents, longitudinal grooves, and transverse grooves were studied. The study found that the surface geometry associated with the dents and grooves became intimately involved in the lubrication process itself, creating local pressure variations that deformed the local surface geometry. These results are consistent with the later theoretical work discussed earlier in the paper. For pure rolling and sliding cases of line contact, Venner et al. (1996) studied the amplitude reduction of a waviness traveling through an EHL line contact and found that the smaller wavelength asperities deformed very little and vice-versa. Venner also found that the contact operating conditions affected the amplitude reduction of the asperity and suggested a single parameter to describe the dependency of the amplitude reduction.

There is a general agreement in the literature that the asperities inside an EHL contact flatten to some degree and cause localized pressure ripples and a reduction in the film thickness. Most work on elliptical contacts with surface features has been with Newtonian fluids, whereas the fluids in an EHL contact behave in a non-Newtonian fashion. Because solving elliptical contact problem using non-Newtonian fluids is complex, some of the recent work with non-Newtonian fluids has been restricted to line contacts. The side leakage in elliptical contacts plays a major role and surface features further affect the conditions inside the contact. The predicted elastic deformation is much higher with Newtonian fluids than with non-Newtonian fluids. Thus, with non-Newtonian fluids it is more likely for an asperity to make contact with the opposing surface. An understanding of the surface feature effects in elliptical contacts using non-Newtonian fluids is needed to better understand the conditions that might lead to a transition from full-film EHL to partial EHL.

With the advent of more powerful computers and better solution methods, it is now possible to study in detail the effect of surface features in elliptical contacts. This study considered the deflection of an asperity and its effect on film thickness for elliptical contacts lubricated with non-Newtonian fluids under pure sliding. The column continuity approach to the modified Reynolds equation, and the circular non-Newtonian viscosity model is used. The finite element system approach is used to numerically solve the modified Reynolds equation. The effect of asperity width and amplitude on the pressure, and the film thickness near the asperity were studied to predict the conditions under which the transition from EHL to partial EHL could occur.

4. RELEVANT EQUATIONS

The modified Reynolds equation was derived by using the column continuity approach. Ambient conditions were assumed at the inlet and the side boundaries, whereas a Reynolds boundary was assumed at the exit. Film thickness equation was modified to account for the presence of surface asperities. The relevant equations are summarized below.

4.1 ASSUMPTIONS

The following assumptions were made:

1. The velocity in the z direction is much smaller than in the x and y directions; that is,

$$w << u, \quad w << v, \quad \frac{\partial w}{\partial x} << \frac{\partial u}{\partial z}, \text{ and } \frac{\partial w}{\partial y} << \frac{\partial v}{\partial z}$$

2. The pressure gradient in the film thickness direction is neglected; that is,

$$\frac{\partial p}{\partial z} = 0$$

3. Isothermal conditions prevail.
4. Inertia effects are negligible relative to viscous effects.
5. Steady-state conditions prevail.
6. Gravity effects are neglected.
7. No slip occurs at the bounding surfaces.

4.2. EQUATIONS OF STATE

The lubricant viscosity is highly dependent on pressure. Roelands pressure-viscosity relationship

$$\eta = \eta_0 e^{(9.67+\ln\eta_0)\left(-1+\left(1+5.1\times10^{-9}p\right)^{z_0}\right)} \tag{5}$$

was used. The fluid is considered compressible. The pressure-density relation proposed by Dowson and Higginson (1966)

$$\rho = \rho_0\left(1+\frac{0.6\times10^{-9}p}{1+1.7\times10^{-9}p}\right) \tag{6}$$

was used. The limiting shear stress of the fluid was assumed to vary linearly with pressure as

$$\tau_L = \tau_0 + \gamma p \tag{7}$$

where τ_0 is the limiting shear strength of the fluid at ambient pressure, and γ is the limiting-shear-strength proportionality constant.

4.3. NON-NEWTONIAN VISCOSITY MODEL

The circular non-Newtonian viscosity model was used in this analysis. Then, the shear-stress/shear-strain-rate equation in the x direction is

$$\frac{\partial u}{\partial z} = \frac{\tau}{\eta} \frac{1}{\sqrt{1 - \left(\frac{\tau}{\tau_L}\right)^2}} \tag{8}$$

4.4 SURFACE ASPERITY

A single asperity on the upper stationary surface, located at the center of the Hertzian contact was used in this study. Figure 6 show an asperity on the upper surface (surface b). Different sizes of the asperity were used to study how the presence of an asperity affects elastohydrodynamic lubrication under pure sliding conditions. The following equation represents the surface feature used:

$$\delta_{ruf} = A \times 10^{-10\left[(x-x_c)^2/l_x^2 + (y-y_c)^2/l_y^2\right]} \cos\left[2\pi\left\{(x-x_c)^2/l_x^2 + (y-y_c)^2/l_y^2\right\}^{1/2}\right] \tag{9}$$

where

δ_{ruf} is the height of the feature in the z direction, at location (x, y), m
A is the maximum amplitude of the feature, m
l_x is the maximum width of the feature in the x direction, m
l_y is the maximum width of the feature in the y direction, m
x_c, y_c is the location of the center of the feature, (m, m)

Since the feature is located on the upper solid (surface b), for an asperity (bump) both δ_{ruf} and A are negative.

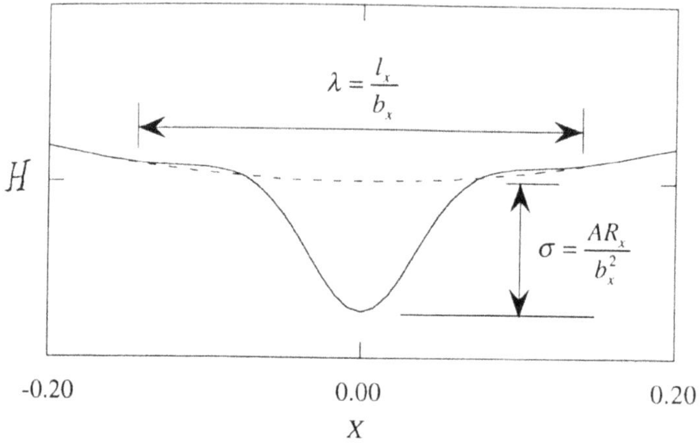

Figure 6 Asperity located at center of EHL contact

4.5 FILM THICKNESS EQUATION

The film thickness equation for the EHL conjunction with a surface feature can be written as

$$h(x,y) = h_0 + \frac{x^2}{2R_x} + \frac{y^2}{2R_y} + \delta(x,y) + \delta_{ruf}(x,y) \tag{10}$$

where the offset film thickness h_0 is given by

$$h_0 = h_c - \delta_c \tag{11}$$

The elastic deformation δ at point (x, y) under distributed pressure $p(\tilde{x}, \tilde{y})$ can be written as

$$\delta = \frac{2}{\pi E'} \iint \frac{p d\tilde{x} d\tilde{y}}{\sqrt{(x-\tilde{x})^2 + (y-\tilde{y})^2}} \tag{12}$$

4.6. LOAD EQUATION

The pressure distribution inside the EHL conjunction must be equal to the external normal load:

$$w_z = \iint p dx dy \tag{13}$$

4.7. MOMENTUM EQUATION

Using the assumptions stated earlier, the momentum equation can be written as

$$\frac{\partial p}{\partial x} = \frac{\partial \tau_{zx}}{\partial z} \tag{14}$$

$$\frac{\partial p}{\partial y} = \frac{\partial \tau_{zy}}{\partial z} \tag{15}$$

4.8. CONTINUITY EQUATION

$$\frac{\partial}{\partial x}\int_{z_a}^{z_b} \rho u dz + \frac{\partial}{\partial y}\int_{z_a}^{z_b} \rho v dz = 0 \tag{16}$$

5. RESULTS

The dimensionless width λ of the asperity was varied from 0.1 to 0.607. Asperities of widths smaller than 0.1 were not used because of limitation of computer resources. Asperities of height ranging from -0.03 to -0.097 were used in the analyses; the negative sign indicates that the surface-feature extended in the negative z direction (downward from the upper surface). With respect to the central film thickness for smooth surfaces, the asperity height ranged from 20% to 66% of the film thickness.. Higher asperity amplitudes were used for the narrower asperities. For the wider asperities when $\sigma/H_{c\text{-smooth}} < -0.5$ was used, the surface shear stress approached the limiting shear stress of the fluid.

Figures 7 to 10 show, respectively, the pressure profile, film thickness, viscosity, and shear stress on surface a, for smooth surfaces. For an asperity case $(\lambda = 0.1,\ \sigma/H_{c\text{-smooth}} = -0.66)$ the results are shown in Figures 11 to 16. After comparing these results to those from the smooth case (Figs. 7 to 10), the following observations were made:

1. The pressure increased sharply near the asperity but remained almost constant in the rest of the contact.
2. The increase in pressure resulted in a steeper increase in viscosity near the asperity.
3. The surface shear stresses increased near the asperity.
4. The asperity underwent elastic deformation and affected the local separation between the two surfaces.

The film thickness has three components (see equation (10)), the offset film thickness, the geometric separation, and the elastic deformation and can be written as

$$h(x,y) = h_0 + s(x,y) + \delta(x,y) \tag{17}$$

where the geometric separation is given by

$$s(x,y) = x^2/2 + y^2/2 + \delta_{ruf}(x,y) \tag{18}$$

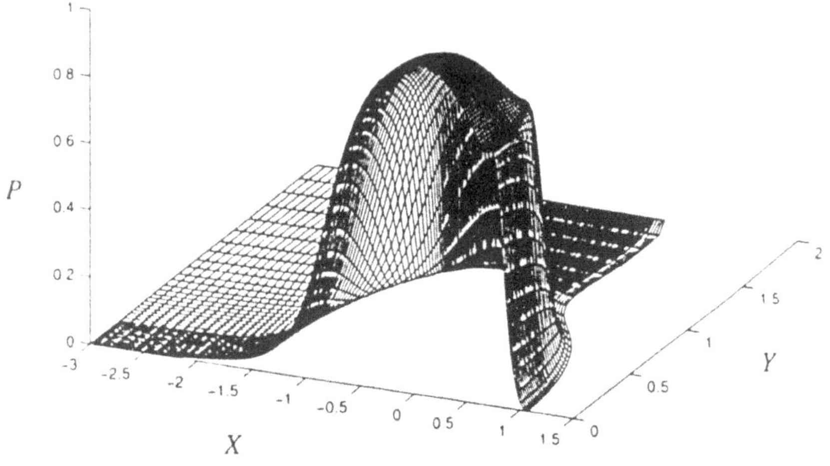

Figure 7 Pressure profile, smooth case

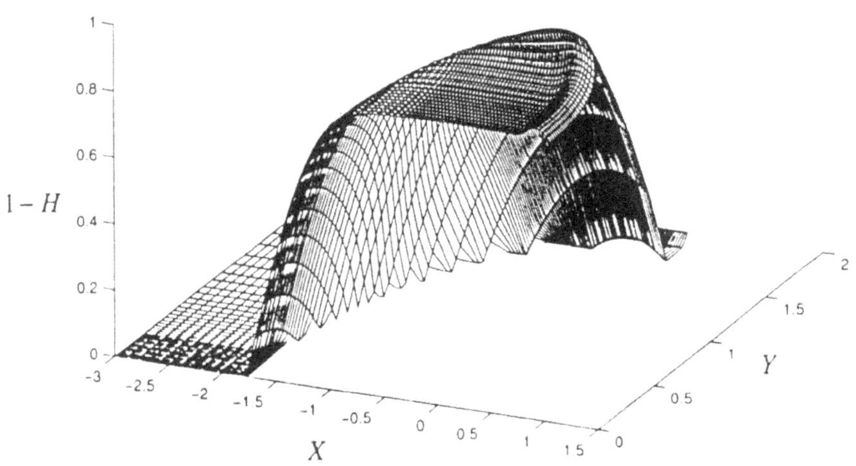

Figure 8 Film shape, smooth case

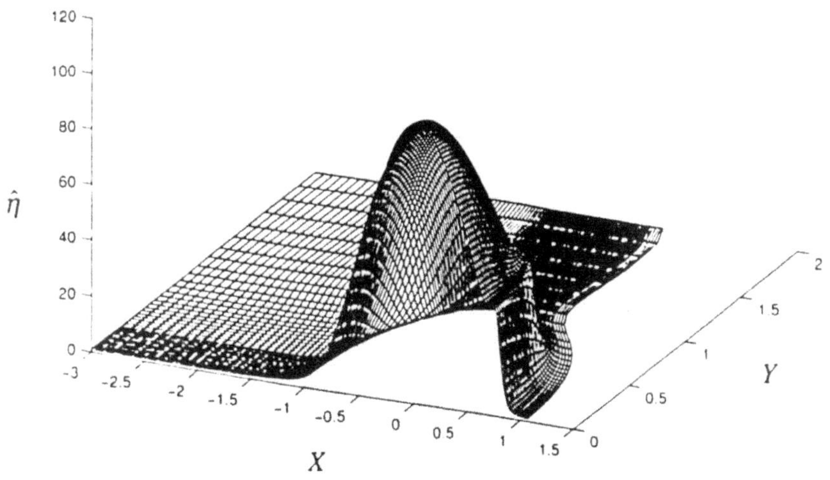

Figure 9 Viscosity, smooth case

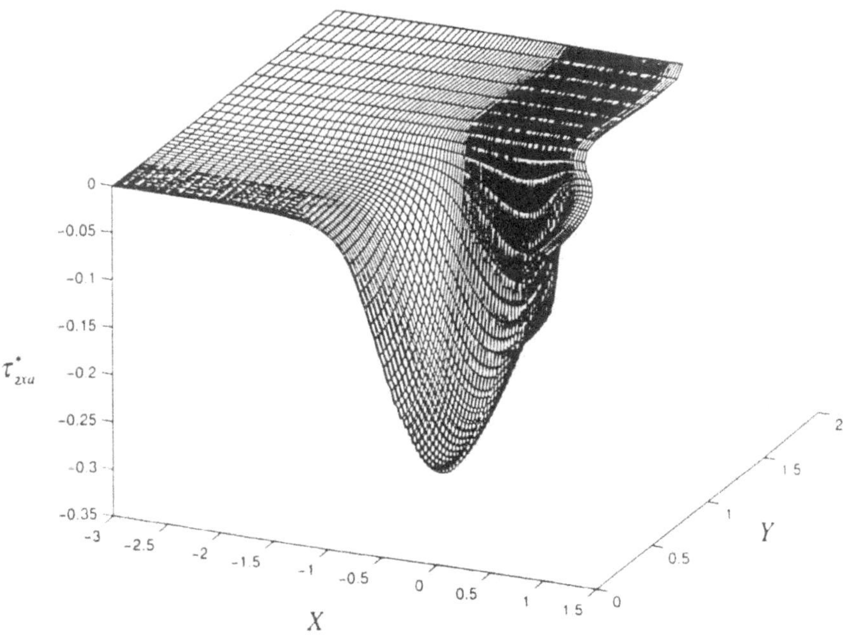

The film thickness, the geometric separation, and the elastic deformation for the smooth case and the asperity case are plotted in Fig. 11, which is enlarged in Fig. 12, for $y = 0$. The dashed lines represent the smooth case and the solid lines represent the asperity case. The offset film thicknesses for both cases were indistinguishable. The differences in the film thickness between the smooth case and the asperity case were due to the differences in the geometric separation and the elastic deformation. The elastic deflection near the asperity was less than that for the smooth surface, resulting in a decrease in the film thickness. This trend agrees with the results of Elsharkawy and Hamrock (1991) and Venner et al. (1996) for line contacts with waviness.

The effects of asperity amplitude and width on the central film thickness are shown in Figs. 17 and 18. Figure 17 reveals that, for constant asperity width, the central film thickness (in this case at the tip of the asperity) decreased almost linearly as the asperity amplitude increased. The linearity was more pronounced as the asperity became narrower. It also shows that, for constant amplitude, a wider asperity gave a larger central film thickness. This can be explained as follows: The wider asperity that had a larger projective area carried a more significant portion of the load. As the asperity carried more load, the pressure at the asperity also increased. The higher pressure on the wider asperity bearing area resulted in a larger elastic deformation. In other words, the narrower asperity that had less elastic deformation gave a smaller film thickness. It can be seen that, when the asperity dimensionless width λ became as small as 0.1, the central film thickness would become zero when the amplitude parameter $\sigma/H_{c-\text{smooth}}$ became –0.7. This indicates that the transition from EHL to partial lubrication was more likely to happen for the narrower asperities.

Figure 18 gives an alternative view for the asperity effects on the film thickness reduction. It shows that as the asperity width increased, the central film thickness asymptotically approached the smooth surface result. As the asperity width decreased, on the other hand, the central film thickness approached zero.

6. CONCLUSION

The effects of surface features such as asperities (bumps) in an elliptical EHL contact were studied under pure sliding conditions. An isothermal, circular non-Newtonian, and compressible flow was assumed. The column continuity approach to the modified Reynolds equation was used, which was solved numerically by using a finite element system approach. The effects of asperity width and height on the pressure, film thickness, and surface shear stresses were studied.

For a constant asperity width the film thickness at the center of the asperity decreases almost linearly with increasing amplitude. The film thickness reduction is more pronounced for the narrower asperities. As the asperity width increases while keeping the amplitude the same, the film thickness asymptotically approach the smooth surface result. As the asperity width decreases, the film thickness quickly approaches zero. Therefore, the narrower asperities are more likely to cause a transition from elastohydrodynamic to partial lubrication.

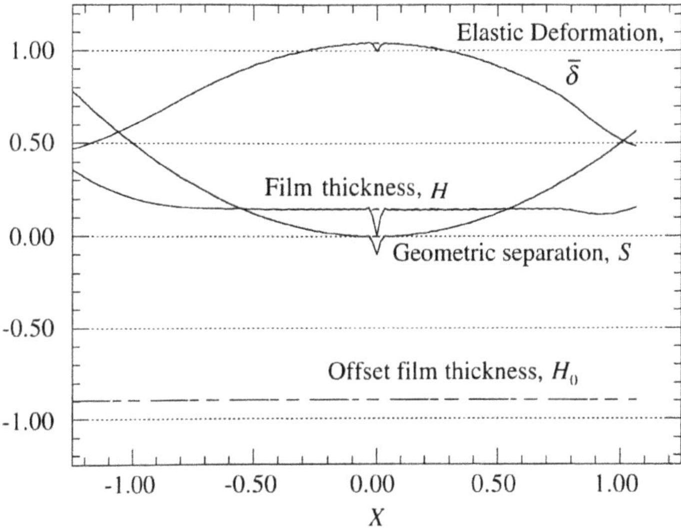

Figure 11 Elastic deformation, geometric separation, film thickness, and offset film thickness, asperity case, $-1.25 < X < 1.25$, $\lambda = 0.1$, $\sigma / H_{c\text{-smooth}} = -0.66$

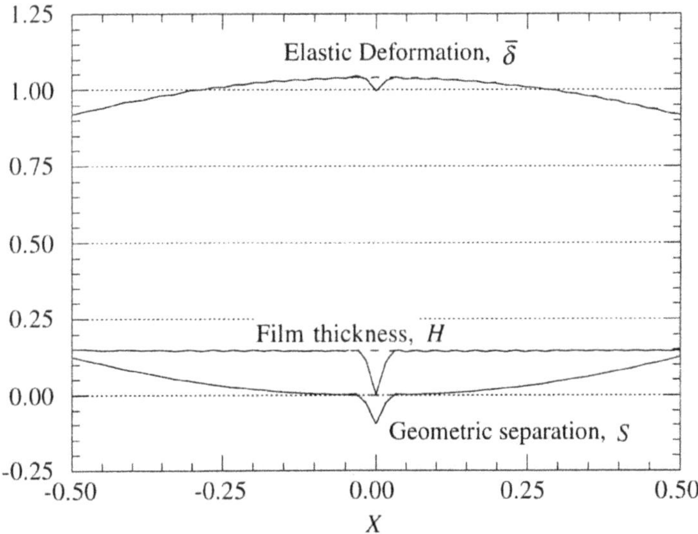

Figure 12 Elastic deformation, geometric separation, and film thickness, asperity case, $-0.5 < X < 0.5$, $\lambda = 0.1$, $\sigma / H_{c\text{-smooth}} = -0.66$

Note: Dash line--smooth case, Solid line--asperity case

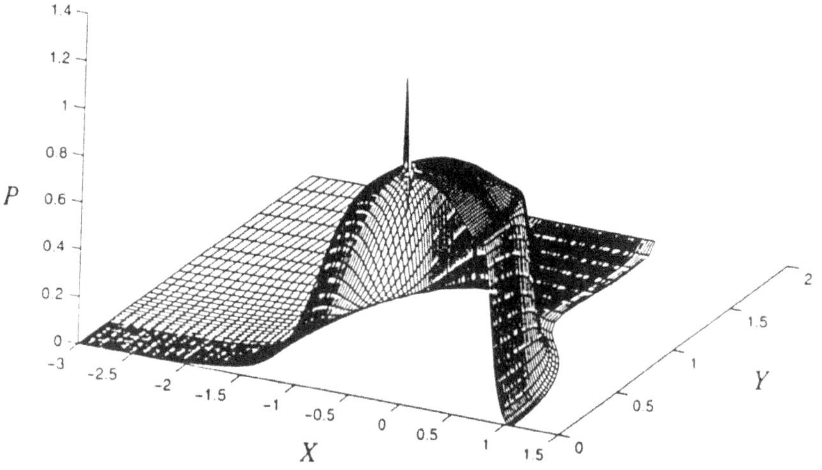

Figure 13 Pressure profile, asperity case, $\lambda = 0.1$, $\sigma / H_{c\text{-smooth}} = -0.66$

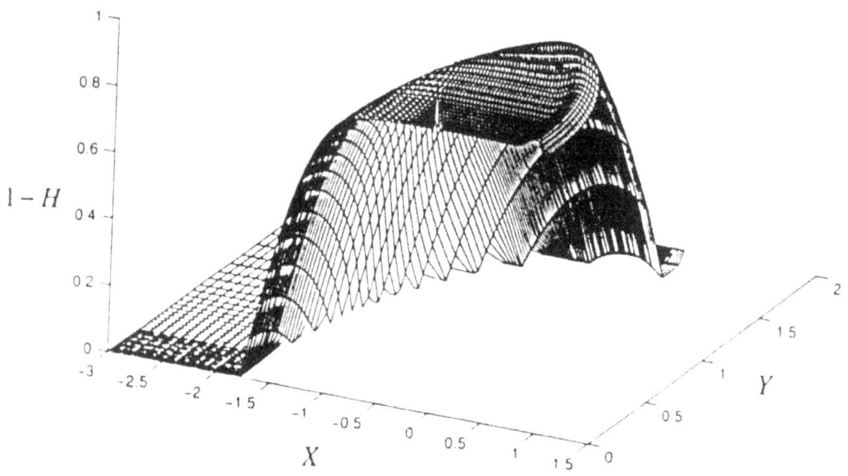

Figure 14 Film shape, asperity case, $\lambda = 0.1$, $\sigma / H_{c\text{-smooth}} = -0.66$

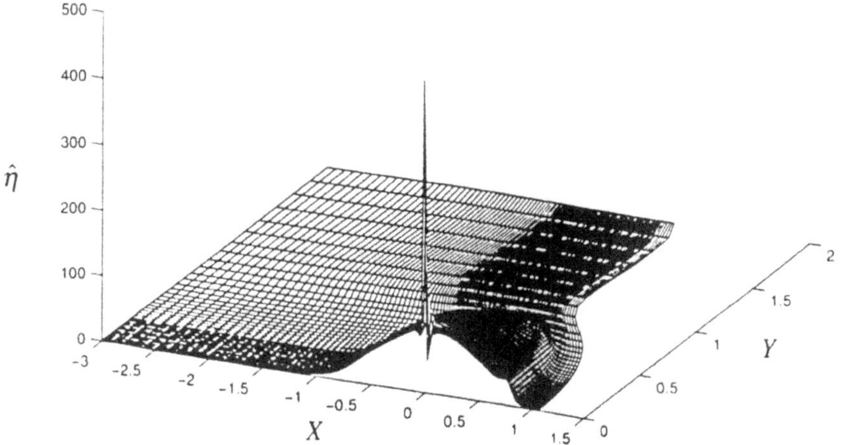

Figure 15 Viscosity, asperity case, $\lambda = 0.1$, $\sigma / H_{c\text{-smooth}} = -0.66$

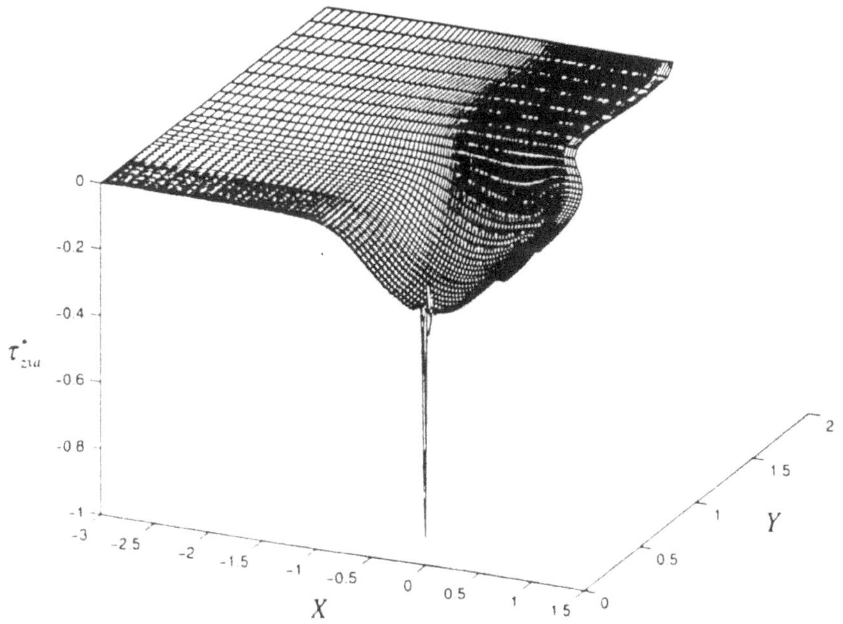

Figure 16 X shear stress on surface a, smooth case, asperity case, $\lambda = 0.1$, $\sigma / H_{c\text{-smooth}} = -0.66$

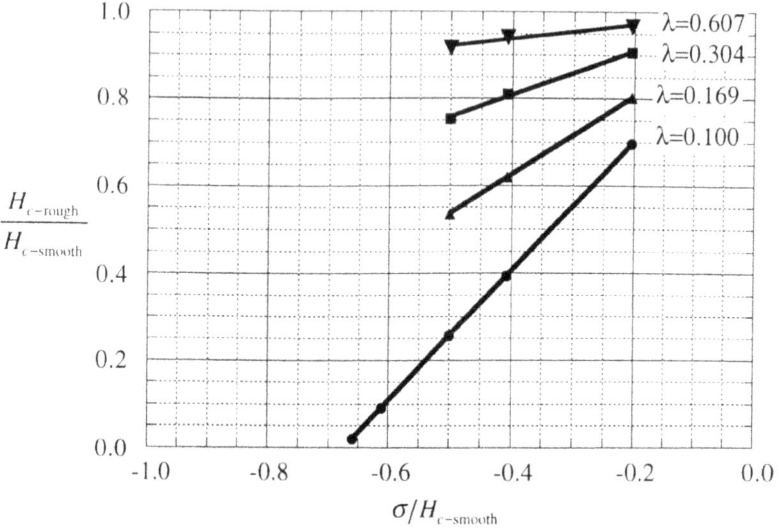

Figure 17 Effect of asperity amplitude on film thickness for $U = 1.08 \times 10^{-12}, G = 3564, W = 1.2 \times 10^{-7}, p_H = 356$ MPa

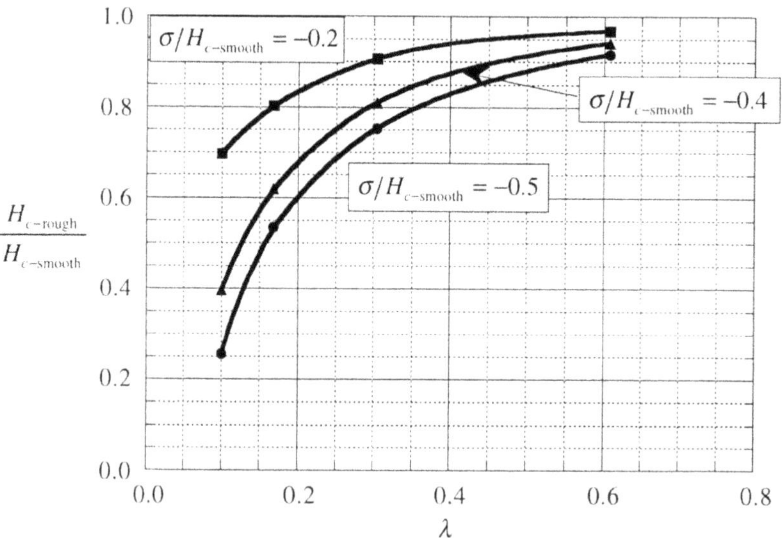

Figure 18 Effect of asperity width on film thickness for $U = 1.08 \times 10^{-12}, G = 3564, W = 1.2 \times 10^{-7}, p_H = 356$ MPa

7. REFERENCES

Chang, L., Cusano, C., and Conry, T. F. (1989): Effect of Lubricant Rheology and Kinematic Conditions on Micro-Elastohydrodynamic Lubrication, *J. Tribol.*, vol. 111, pp. 344-351.

Elsharkawy, A. A., and Hamrock, B. J. (1991): Subsurface Stresses in Micro-EHL Line Contacts, *J. Tribol.*, vol. 113, pp. 645-655.

Hamrock, B.J. (1991): *Fundamentals of Fluid Film Lubrication*, NASA Reference Publication 1255.

Hamrock, B.J., and Anderson, W.J. (1983): Rolling-Element Bearings, NASA Reference Publication 1105.

Hamrock, B.J., and Dowson, D. (1977): Isothermal Elastohydrodynamic Lubrication of Point Contacts, Part III--Fully Flooded Results, *J. Lubr. Technol.*, vol. 99, no. 2, pp. 264-276.

Lee, R. T., and Hamrock, B. J. (1990): A Circular Non-Newtonian Fluid Model-Part II-Used in Microelastohydrodynamic Lubrication, *J. Tribol.*, vol. 112, No. 3, pp. 497-505.

Lubrecht, A. A., Ten Napel, W. E., and Bosma, R. (1988): The Influence of Longitudinal and Transverse Roughness on the Elastohydrodynamic Lubrication of Circular Contacts, *J. Tribol.*, vol. 110, pp. 421-426.

Majumdar, B. C., and Hamrock, B. J. (1982): Effect of Surface Roughness on Elastohydrodynamic Line Contact, *J. Lubr. Technol.*, vol. 104, pp. 401-409.

Patir, N., and Cheng, H. S. (1979): Application of Average Flow Model to Lubrication Between Rough Sliding Surfaces, *ASME* J. *Lubr. Technol.*, vol. 101, pp. 220-227.

Patir, N., and Cheng, H. S. (1979): Effect of Surface Roughness Orientation on the Central Film Thickness in EHD Contacts, Elastohydrodynamic and Related Topics, Proceedings of the 5th Leeds-Lyon Symposium on Tribology, D. Dowson, C. M. Taylor, M. Godet, and D. Berthe, eds., Mechanical Engineering Publications for Institute of Tribology, Univ. of Leeds and Institut National des Sciences Appliquees de Lyon, pp. 15-21.

Shieh, J., and Hamrock, B. J. (1991): Film Collapse in EHL and Micro-EHL, *J. Tribol.*, vol. 113, pp. 372-377.

Venner, C.H., Couhier, F., Lubrecht, A.A., and Greenweed, J.A. (1996): Amplitude Reduction of Waviness in Transient EHL Line Contacts, Elastohydrodynamic s-96, Fundamentals and Applications in Lubrication and Traction, Proceedings of the 23rd Leeds-Lyon Symposium on Tribology, Mechanical Engineering Publications for Institute of Tribology, Univ. of Leeds, and Institut National des Sciences Appliquees de Lyon.

Wedeven, L. D., and Cusano, C. (1979): Elastohydrodynamic Film Thickness Measurements of Artificially Produced Surface Dents and Grooves, *ASLE Trans.,* vol. 22, no. 4, pp. 369-381.

Zhu, D., Cheng, H. S., and Hamrock, B. J. (1990): Effect of Surface Roughness on Pressure Spike and Film Constriction in Elastohydrodynamically Lubricated Line Contacts: *Trib. Trans.*, vol. 33, no. 2, pp. 267-273.

MICRO/NANOTRIBOLOGY: STATE OF THE ART AND ITS APPLICATIONS

BHARAT BHUSHAN
Ohio Eminent Scholar and The Howard D. Winbigler Professor
Director, Computer Microtribology and Contamination Laboratory
Department of Mechanical Engineering
The Ohio State University
Columbus, Ohio 43210-1107, U.S.A.

Abstract. Atomic force microscopy/friction force microscopy (AFM/FFM) techniques are increasingly used for tribological studies of engineering surfaces at scales, ranging from atomic and molecular to microscales. These techniques have been used to study surface roughness, adhesion, friction, scratching/wear, indentation, detection of material transfer, and boundary lubrication and for nanofabrication/nanomachining purposes. Measurement of atomic-scale friction of a freshly-cleaved highly-oriented pyrolytic graphite exhibited the same periodicity as that of corresponding topography. However, the peaks in friction and those in corresponding topography were displaced relative to each other. Variations in atomic-scale friction and the observed displacement has been explained by the variations in interatomic forces in the normal and lateral directions. Local variation in microscale friction is found to correspond to the local slope suggesting that a ratchet mechanism is responsible for this variation. Directionality in the friction is observed on both micro- and macro scales which results from the surface preparation and anisotropy in surface roughness. Microscale friction is generally found to be smaller than the macrofriction as there is less ploughing contribution in microscale measurements. Microscale friction is load dependent and friction values increase with an increase in the normal load approaching to the macrofriction at contact stresses higher than the hardness of the softer material. Wear rate for single-crystal silicon is negligible below 20 μN and is much higher and remains approximately constant at higher loads. Elastic deformation at low loads is responsible for negligible wear. Mechanism of material removal on microscale is studied. At the loads used in the study, material is removed by the ploughing mode in a brittle manner without much plastic deformation. Most of the wear debris is loose. Evolution of the wear has also been studied using AFM. Wear is found to be initiated at nano scratches. AFM has been modified to obtain load-displacement curves and for measurement of nanoindentation hardness and Young's modulus of elasticity, with depth of indentation as low as 1 nm. Hardness of ceramics on nano scales is found to be higher than that on micro scale. Ceramics exhibit significant plasticity and creep on nanoscale. Scratching and indentation on nanoscales are the powerful ways to screen for adhesion and resistance to deformation of ultrathin films. Detection of material transfer on a nanoscale is possible with AFM. Boundary lubrication studies and measurement of lubricant-film thickness with a lateral

B. Bhushan (ed.), Tribology Issues and Opportunities in MEMS, 229-246.

resolution on a nanoscale have been conducted using AFM. Self-assembled monolayers and chemically-bonded lubricant films with a mobile fraction are superior in wear resistance.

1. Introduction

The micro/nanotribological studies are needed to develop fundamental understanding of interfacial phenomena on a small scale and to study interfacial phenomena in micro- and nano structures used in magnetic storage systems, microelectromechanical systems (MEMS) and other industrial applications (Bhushan, 1995b, 1996a, 1996b; Bhushan et al., 1997a, 1997b). Friction and wear of lightly-loaded micro/nano components are highly dependent on the surface interactions (few atomic layers). These structures are generally lubricated with molecularly-thin films. Micro- and nanotribological studies are also valuable in fundamental understanding of interfacial phenomena in macrostructures to provide a bridge between science and engineering (Bowden and Tabor, 1950, 1964; Bhushan and Gupta, 1991; Bhushan, 1992, 1996a).

At most solid-solid interfaces of technological relevance, contact occurs at numerous asperities, a sharp atomic force microscope tip sliding on a surface simulates just one such contact. However, asperities on a surface exists with all shapes and sizes. To study the effect of asperity radius, tips can be produced with different radii (Bhushan and Sundararajan, 1998). Atomic force microscopy (AFM) and friction force microscopy (FFM) can be used for measurement of all engineering surfaces which may be either electrically conducting or insulating. AFM/FFM has become a popular surface profiler for topographic and friction measurements on micro to nanoscale. By using a standard or a sharp diamond tip mounted on a stiff cantilever beam, we have used AFM for scratching, wear, and measurements of elastic/plastic mechanical properties (such as load-displacement curves, indentation hardness and modulus of elasticity). Boundary lubrication studies of molecularly-thick lubricant films can be conducted using FFMs. Status of current understanding of micro/nanotribology of engineering interfaces follows.

2. Experimental Techniques

A commercial AFM/FFM commonly used to conduct studies of friction, scratching, wear, indentation, and lubrication from micro- to atomic scales and nanofabrication/nanomachining is shown in Fig. 1 (Bhushan, 1995a). Simultaneous measurements of surface roughness and friction force can be made with this instrument. In the AFM/FFM, the sample is mounted on a PZT tube scanner which consists of separate electrodes to precisely scan the sample in the X-Y plane in a raster pattern and to move the sample in the vertical (Z) direction. A sharp tip at the end of a flexible cantilever is brought in contact with the sample. Normal and frictional forces being applied at the tip-sample interface are measured using a laser beam deflection technique. A laser beam from a diode laser is directed by a prism onto the back of a cantilever near its free end, tilted downward at about 10 deg with respect to a horizontal plane. The reflected beam from the vertex of the cantilever is directed through a mirror onto a quad photodetector (split photodetector with four quadrants). The differential signal from the top and bottom photodiodes provides the AFM signal which is a sensitive measure of the cantilever vertical deflection. Topographic features of the sample cause the tip to deflect in the vertical direction as the sample is scanned under the tip. This tip deflection will change the direction of the reflected laser beam, changing the intensity difference between the top and bottom photodetector (AFM signal). In the AFM operating mode of the "height mode," for topographic imaging or for any other operation in which the applied normal force is to be kept a constant, a feedback circuit is used to modulate the voltage applied to the PZT scanner to adjust the height of the PZT, so that the cantilever vertical deflection (given by the intensity difference between the top and bottom detector) will remain almost constant during scanning. The PZT height variation is thus a direct measure of surface roughness of sample.

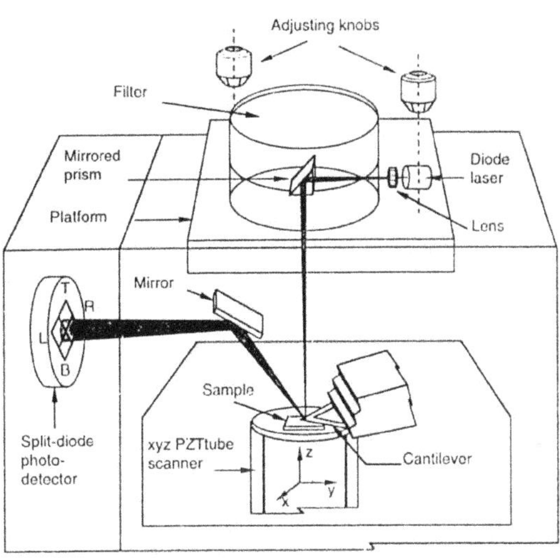

Fig. 1 Schematic of a commercial atomic force microscope/friction force microscope (AFM/FFM) using laser-beam deflection method.

For measurement of friction force being applied at the tip surface during sliding, the other two (left and right) quadrants of the photodetector (arranged horizontally) are used. In the "friction mode", the sample is scanned back and forth in a direction orthogonal to the long axis of the cantilever beam. Friction force between the sample and the tip will produce a twisting of the cantilever. As a result, the laser beam will be reflected out of the plane defined by the incident beam and the beam reflected vertically from an untwisted cantilever. This produces an intensity difference of the laser beam received in the left and right quadrants of the photodetector. The intensity difference between the left and right detectors (FFM signal) is directly related to the degree of twisting and hence to the magnitude of friction force. One problem associated with this method is that any misalignment between the laser beam and the photodetector axis would introduce error in the measurement. However, by following the procedures developed by Ruan and Bhushan (1994a), in which the average FFM signal for the sample scanned in two opposite directions, is subtracted from the friction profiles of each of the two scans to eliminate the misalignment effect. This method provides 3-D maps of friction force. By following the friction force calibration procedures developed by Ruan and Bhushan (1994a) and Bhushan (1995a), voltages corresponding to friction forces can be converted to force units.

Topographic measurements are typically made using a sharp tip on cantilever beam with normal stiffness of about 0.58 N/m at a normal load of about 10 nN and friction measurements are carried out in the load range of 10-150 nN. The tip is scanned in such a way that its trajectory on the sample forms a triangular pattern. Scanning speeds in the fast and slow scan directions depends on the scan area and scan frequency. A maximum scan size of 125 μm x 125 μm and scan rate of 122 Hz typically can be used. Higher scan rates are used for small scan lengths.

For nano-scale boundary lubrication studies, the samples are typically scanned over an area of 1 μm x 1 μm at a normal force of about 300 nN, in a direction orthogonal to the long axis of the cantilever beam. The samples are generally scanned with a scan rate of 1 Hz and the scanning speed of 2 μm/s. Coefficient of friction is monitored during scanning for a desired number of cycles. After scanning test, a larger area of 2 mm x 2 mm is scanned at a normal force of 40 nN to observe for any wear scar.

For micro-scale scratching, micro-scale wear and nano-scale indentation hardness measurements, sharp single-crystal natural diamond tip mounted on a stainless steel cantilever beam with normal stiffness of about 25 N/m is used at relatively higher loads (1 mN - 150 mN). For scratching and wear studies, the sample is generally scanned in a direction orthogonal to the long axis of the cantilever beam (typically at a rate of 0.5 Hz) so that friction can be measured during scratching and wear. The tip is mounted on the beam such that one of its edge is orthogonal to the long axis of the beam; therefore, wear during scanning along the beam axis is higher (about 2x to 3x) than that during scanning orthogonal to the beam axis. For wear studies, typically an area of 2 mm x 2 mm is scanned at various normal loads (ranging from 1 to 100 mN) for selected number of cycles.

For nanoindentation hardness measurements the scan size is set to zero and then normal load is applied to make the indents. During this procedure the diamond tip is continuously pressed against the sample surface for about two seconds at various indentation loads. Sample surface is scanned before and after the scratching, wear or indentation to obtain the initial and the final surface topography, at a low normal load of about 0.3 mN using the same diamond tip. An area larger than the indentation region is scanned to observe the indentation marks. Nanohardness is calculated by dividing the indentation load by the projected residual area of the indents. Special sensors in conjunction with AFM are also used for measurement of load-displacement data which is then used to calculate hardness and modulus of elasticity (Bhushan et al., 1996; Kulkarni and Bhushan, 1996a, b, 1997).

For measurements of surface roughness, friction force, nanoscale scratching and wear, a microfabricated square-pyramidal Si_3N_4 or silicon tip with a tip radius ranging from 10 to 50 nm (Fig. 2a) is generally used at loads ranging from 10 to 150 nN. For measurements of microscale scratching and wear and for nanoindentation hardness measurements and nanofabrication, a three-sided pyramidal single-crystal natural-diamond tip with a tip radius of about 100 nm (Fig. 2b) is generally used at relatively high loads ranging from 10 to 150 mN.

3. Friction

To study friction mechanisms on an atomic scale, a well characterized freshly-cleaved surface of highly oriented pyrolytic graphite (HOPG) has been studied by Mate et al. (1987) and Ruan and Bhushan (1994b). The atomic-scale friction force of HOPG exhibited the same periodicity same as that of corresponding topography (Fig. 3a), but the peaks in friction and those in topography were displaced relative to each other, (Fig. 3b). A Fourier expansion of the interatomic potential was used to calculate the conservative interatomic forces between atoms of the FFM tip and those of the graphite surface. Maxima in the interatomic forces in the normal and lateral directions do not occur at the same location, which explains the observed shift between the peaks in the lateral force and those in the corresponding topography. Furthermore, the observed local variations in friction force were explained by variation in the intrinsic lateral force between the sample and the FFM tip (Ruan and Bhushan, 1994b) and these variations may not necessarily occur as a result of atomic-scale stick-slip process (Mate et al., 1987), but can be due to variation in the intrinsic lateral force between the sample and the FFM tip.

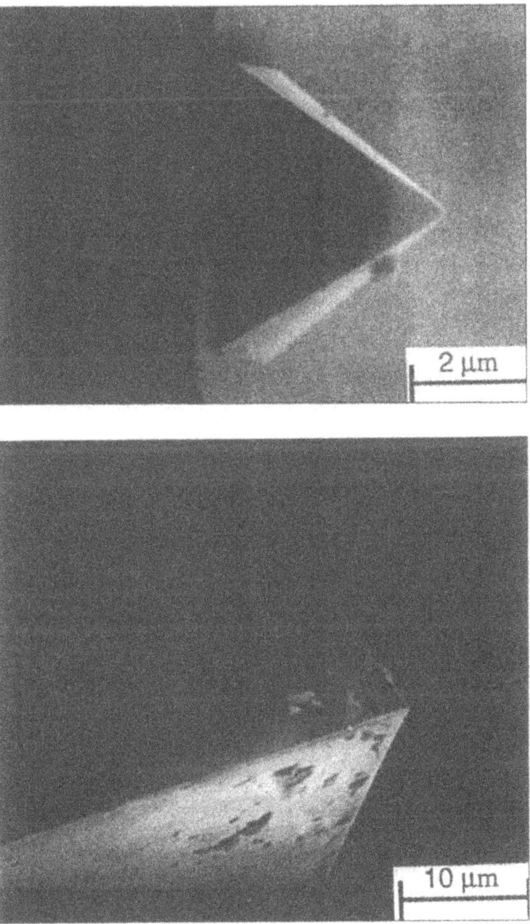

Fig. 2 SEM micrographs of a PECVD Si_3N_4 cantilever beam with tip (top) and a stainless steel cantilever beam with diamond tip (bottom).

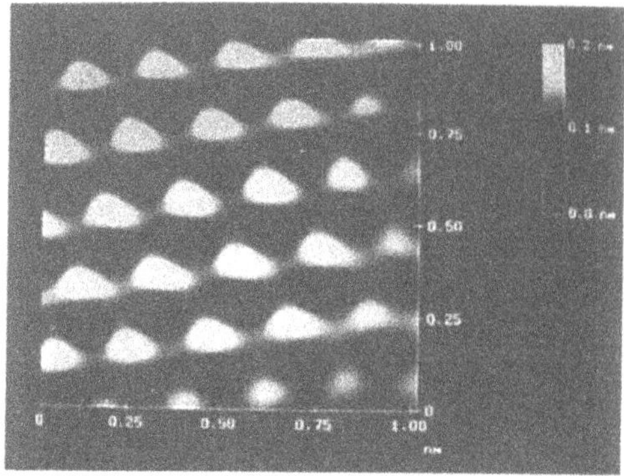

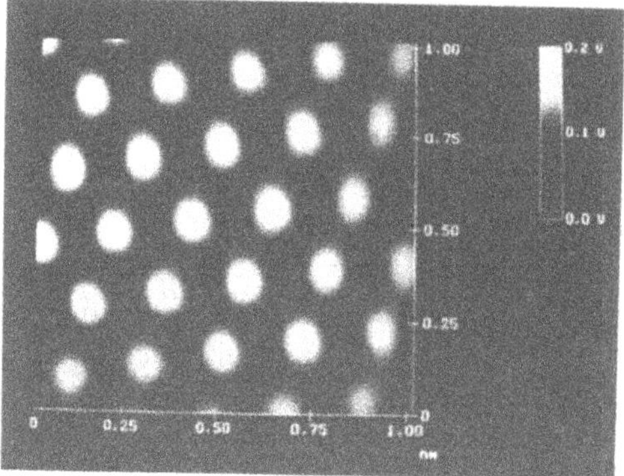

(a)

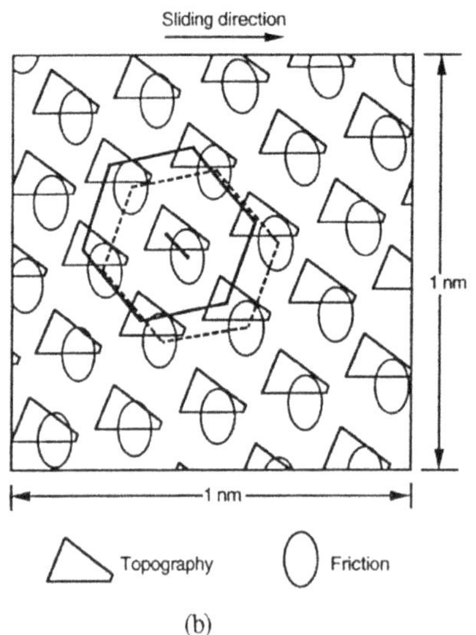

(b)

Fig.3 (**a**) Gray-scale plots of surface topography (top) and friction profiles (bottom) of a 1 nm x 1 nm area of freshly cleaved HOPG, showing the atomic-scale variation of topography and friction, (**b**) diagram of superimbosed topography and friction profiles from (a); the symbols correspond to maxima. Note the spatial shift between the two profiles (Ruan and Bhushan, 1994b).

Friction forces of HOPG have also been studied. Local variations in the microscale friction of cleaved graphite are observed, which arise from structural changes occur during the cleaving process (Ruan and Bhushan, 1994c). The cleaved HOPG surface is largely atomically smooth but exhibits line-shaped regions in which the coefficient of friction is more than order of magnitude larger. Transmission electron microscopy indicates that the line-shaped regions consist of graphite planes of different orientation, as well as of amorphous carbon. Differences in friction have also been observed for multi-phase ceramic materials (Koinkar and Bhushan, 1996c) and for organic mono- and multi-layer films (Meyer et al., 1992), which again seems to be the result of structural variations in the surfaces. These measurements suggest that the FFM can be used for structural mapping of the surfaces. FFM measurements can be used to map chemical variations, as indicated by the use of the FFM with a modified probe tip to map the spatial arrangement of chemical functional groups in mixed organic monolayer films (Frisbie et al., 1994). Here, sample regions that had stronger interactions with the functionalized probe tip exhibited larger friction.

Local variations in the microscale friction of scratched surfaces can be significant, and seen to depend on the local surface slope rather than the surface height distribution (Bhushan et al., 1994b; Bhushan, 1995a; Koinkar and Bhushan, 1997a). Directionality in friction is sometimes observed on the macroscale; on the microscale this is the norm (Bhushan et al., 1994b; Bhushan, 1995a). This is because most engineering surfaces have asymmetric surface asperities so that the interaction of the FFM tip with the surface is dependent on the direction of the tip motion. Moreover, during surface finishing processes material can be transferred preferentially onto one side of the asperities, which also causes asymmetry and directional dependence. Reduction in local variations and in directionality of frictional properties therefore requires careful optimization of surface roughness distributions and of surface-finishing processes.

Table 1 shows the coefficient of friction measured for two surface micro- and macroscales. The coefficient of friction is defined as the ratio of friction force to the normal load. The values on the microscale are much lower than those on the macroscale. Contact stresses at AFM conditions generally do not exceed the sample hardness which minimizes plastic deformation. When measured for the small contact areas and very low loads used in microscale studies, indentation hardness and modulus of elasticity are higher than at the macroscale. Lack of plastic deformation and improved mechanical properties reduce the degree of wear. In addition, the small apparent areas of contact reduce the number of particles trapped at the interface, and thus minimize the 'ploughing' contribution to the friction force.

At higher loads (with contact stresses exceeding the hardness of the softer material), however, the coefficient of friction for micro-scale measurements increases towards values comparable with those obtained from macroscale measurements, and surface damage also increases, Fig. 4 (Bhushan and Kulkarni, 1996a). Thus Amontons' law of

Table 1 Surface roughness and micro- and macro-scale coefficients of friction of various samples.

Material	R.M.S. roughness, nm	Micro-scale coefficient of friction versus Si_3N_4 tip[1]	Macro-scale coefficient of friction versus alumina ball[2]
Si (111)	0.11	0.03	0.18
C^+-implanted Si	0.33	0.02	0.18

[1]Tip radius of about 50 nm in the load range of 10-150 nN (2.5 - 6.1 GPa), a scanning speed of 5 μm/s and scan area of 1 μm x 1 μm.

[2]Ball radius of 3 mm at a normal load of 0.1 N (0.3 GPa) and average sliding speed of 0.8 mm/s.

friction, which states that the coefficient of friction is independent of apparent contact area and normal load, does not hold for microscale measurements. These findings suggest microcomponents sliding under lightly loaded conditions should experience very low friction and near-zero wear.

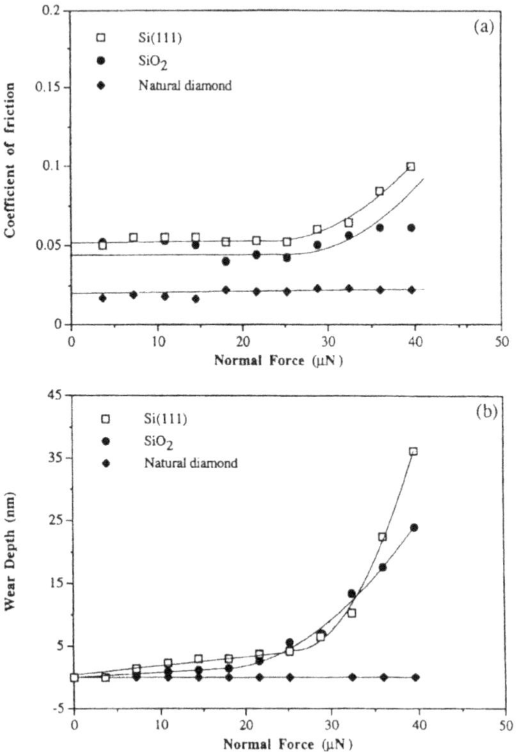

Fig. 4 (a) Coefficient of friction as a function of normal force and (b) corresponding wear depth as a function of normal force for silicon, SiO_2 coating and natural diamond. Inflections in the curves for silicon and SiO_2 correspond to the hardnesses of these materials (Bhushan and Kulkarni, 1996a).

4. Scratching, Wear and Indentation

The AFM can be used to investigate how surface materials can be moved or removed on micro- to nanoscales, for example in scratching and wear (Bhushan, 1995a) (where these things are undesirable), and nanomachining/nanofabrication (where they are desirable). The AFM can also be used for measurements of mechanical properties on micro- to nanoscales. Figure 5 shows microscratches made on Si(111) at various loads and a scan velocity of 2 μm/s after 10 cycles (Bhushan and Koinkar, 1994a). As expected, the depth of scratch increases with load. Such microscratching measurements can be used to study failure mechanisms on the microscale and to evaluate the mechanical integrity (scratch resistance) of ultra-thin films at low loads.

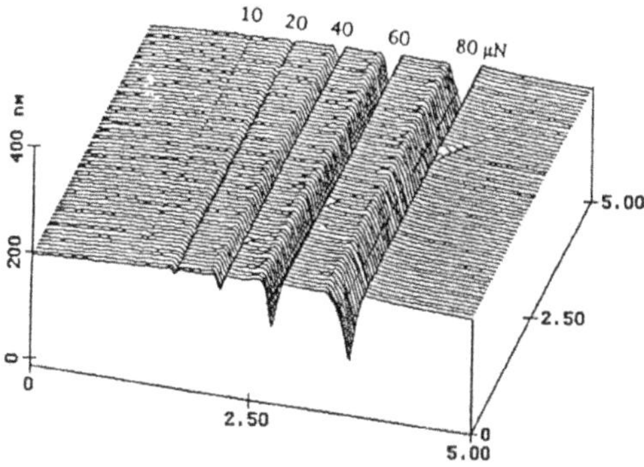

Fig. 5 **Surface plots of Si(111) scratched at various loads. Note that x and y axes are in μm and z axis is in nm (Bhushan and Koinkar, 1994a).**

By scanning the sample in two dimensions with the AFM, wear scars are generated on the surface. Figure 6 shows the effect of normal load on the wear rate. We note that wear rate is very small below 20 μN of normal force. A normal force of 20 μN corresponds to contact stresses comparable to the hardness of the silicon. Primarily, elastic deformation at loads below 20 μN is responsible for low wear. Typical wear mark generated at a normal load of 40 μN for one scan cycle and imaged using AFM at 300 nN load, is shown in Fig. 7(a). Inverted line plot of wear mark is shown in the Fig. 7(b) showing the uniform material removal at the bottom of the wear mark. Next we examine the mechanism of material removal at light loads on microscale in AFM wear experiments (Koinkar and Bhushan, 1997b). Figure 8 shows a secondary electron image of wear mark and associated wear particles. Specimen used for the SEM was not scanned after initial wear, to retain wear debris in the wear region. Wear debris is clearly observed. AFM image of the wear mark shows small debris at the edges, swiped during AFM scanning. Thus the debris is "loose" (not sticky) and can be removed during the AFM scanning. SEM micrographs show both cutting type and ribbon-like debris. TEM studies were performed to understand the material removal process. TEM micrograph of the worn region showed evidence of bend contours passing though the wear mark. The bend contours around and inside the wear mark suggests that there are some residual stresses around and inside the wear mark region. No major surface dislocation activity, phase transformation or crack formation was observed in plan view of TEM. The dislocation activity and/or cracking probably occurs at the subsurface. Observations in TEM studies suggested that for loads of 40 μN, material is mostly removed in a brittle manner or is chipped, although presence of some ribbon-like debris in SEM studies suggests some plastic deformation as well.

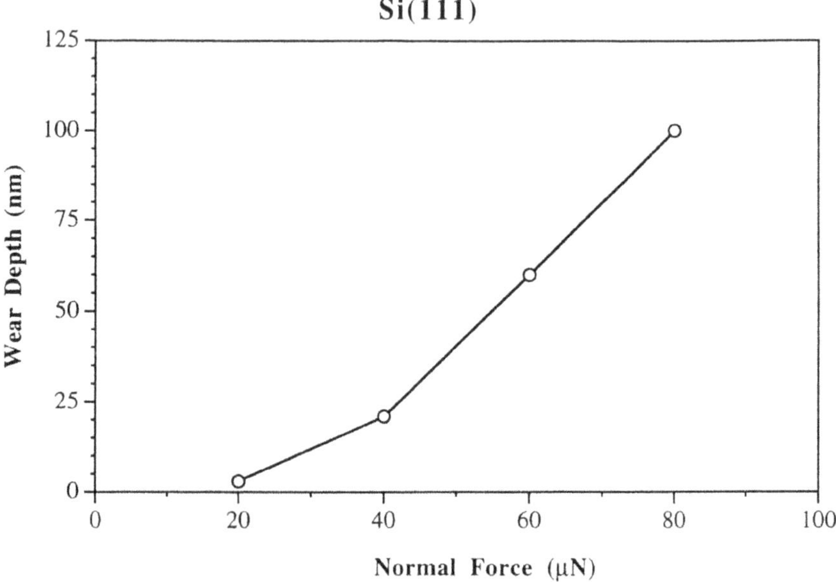

Fig. 6 Wear depth as a function of normal force for Si after one cycle (Koinkar and Bhushan, 1997b).

The evolution of wear of a diamond-like carbon coating on a polished aluminum substrate is shown in Fig. 9 which illustrates how the micro-wear profile for a load of 20 μN develops as a function of the number of scanning cycles (Bhushan et al. 1994b). Wear is not uniform, but is initiated at the nanoscratches indicating that surface defects (with high surface energy) act as initiation sites. Thus, scratch-free surfaces will be relatively resistant to wear.

Mechanical properties, such as hardness and Young's modulus of elasticity can be determined on micro- to picoscales using the AFM (Bhushan and Koinkar, 1994c; Bhushan, 1995a) and a new nano/pico indentation system used in conjunction with an AFM (Bhushan et al., 1996b; Kulkarni and Bhushan, 1996a, b, 1997; Koinkar and Bhushan, 1997c). Indentability on the scale of sub-nanometers can be studied by monitoring the slope of cantilever deflection as a function of sample traveling distance after the tip is engaged and the sample is pushed against the tip. For a rigid sample, cantilever deflection equals the sample traveling distance; but the former quantity is smaller if the tip indents the sample. Figure 10 shows the load-displacement curves at different peak loads for Si(100). Load-displacement data at residual depths as low as about 1 nm can be obtained. Loading/unloading curves are not smooth, but exhibit sharp discontinuities particularly at high loads (shown by arrows in the figure). Any discontinuities in the loading part of the curve probably results from slip. The sharp discontinuities in the unloading part of the curves are believed to be due to formation of lateral cracks which form at the base of median crack which results in the surface of the specimen being thrust upward.

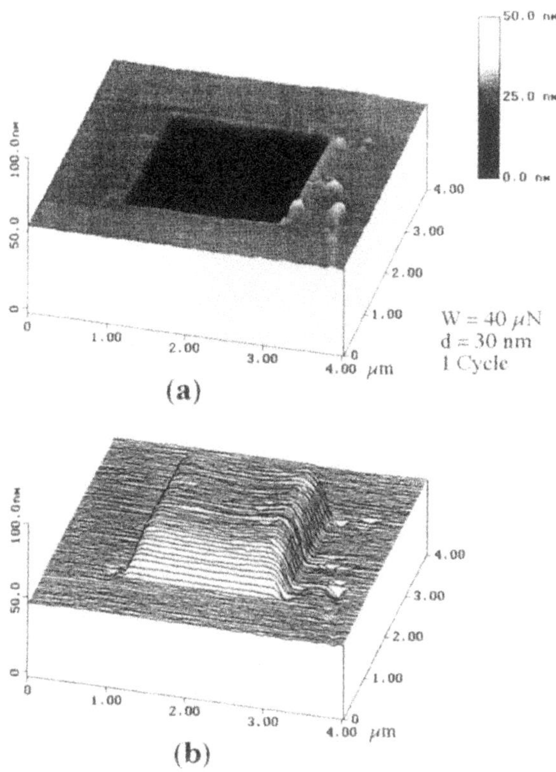

Fig. 7 (**a**) Typical gray scale and (**b**) inverted AFM images of wear mark created using a diamond tip at a normal force of 40 μN and one scan cycle on Si(111) surface (Koinkar and Bhushan, 1997b).

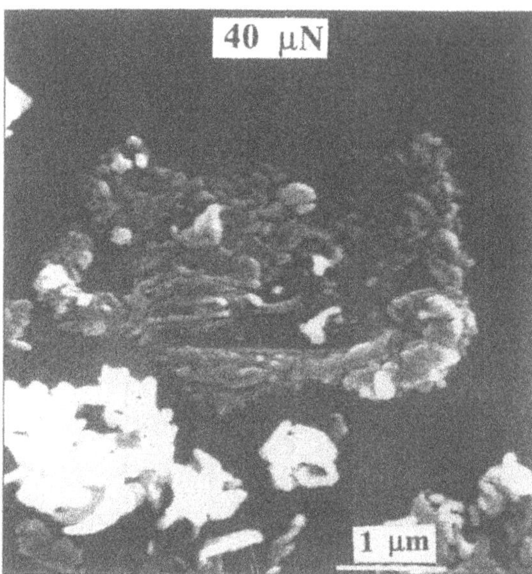

Fig. 8 Secondary electron image of wear mark and debris particles for Si produced at a normal force of 40 μN and one scan cycle (Koinkar and Bhushan, 1997b).

The indentation hardness of surface films with an indentation depth of as small as about 1 nm has been measured for Si(111) (Bhushan and Koinkar, 1994c; Bhushan et al., 1996b). Triangular indentations are observed for shallow penetration depths. The hardness of silicon on nanoscale is found to be higher than on microscale, Fig. 11. This decrease in hardness with an increase in indentation depth can be rationalized on the basis that as the volume of deformed material increases, there is a higher probability of encountering material defects. Bhushan and Koinkar (1994c) have used AFM measurements to show that ion implantation of silicon surfaces increases their hardness and thus their wear resistance. Formation of surface alloy films with improved mechanical properties by ion implantation is growing technological importance as a means of improving the mechanical properties of materials. Hardness of 20-nm thick diamond like carbon films have been measured by Kulkarni and Bhushan (1997).

The Young's modulus of elasticity is calculated from the slope of the indentation curve during unloading (Bhushan, 1995a; Bhushan et al., 1996b; Kulkarni and Bhushan, 1996a, b, 1997). Maivald et al. (1991) and DeVecchio and Bhushan (1997) used an AFM in "force modulation mode" to measure local surface elasticities. AFM tip is scanned over the modulated sample surface with the feedback loop keeping the average force constant. For the same applied force, a soft area deforms more, and thus causes less cantilever deflection, than a hard area. The ratio of modulation amplitude to the local tip deflection is then used to create a force modulation image. The force modulation mode makes it easier to identify soft areas on hard substrates.

The nano/picoindentation system has been used to study the creep and strain-rate effects of ceramics. Bhushan et al. (1996b) and Kulkarni et al. (1996b, 1997) have reported that ceramics exhibit significant plasticity and creep on nanoscale.

Detection of transfer of material on a nanoscale is possible with the AFM. Indentation of C60-rich fullerene films with an AFM tip has been shown (Ruan and Bhushan, 1993) to result in the transfer of fullerene molecules to the AFM tip, as indicated by discontinuities in the cantilever deflection as a function of sample traveling distance in subsequent indentation studies.

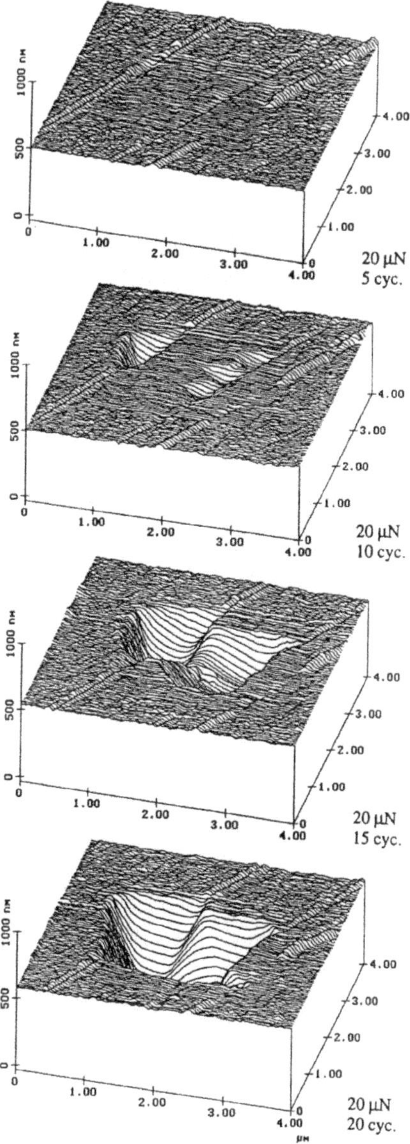

Fig. 9 **Surface plots of diamond-like carbon-coated thin-film disk showing the worn region; the normal load and number of test cycles are indicated (Bhushan et al., 1994c).**

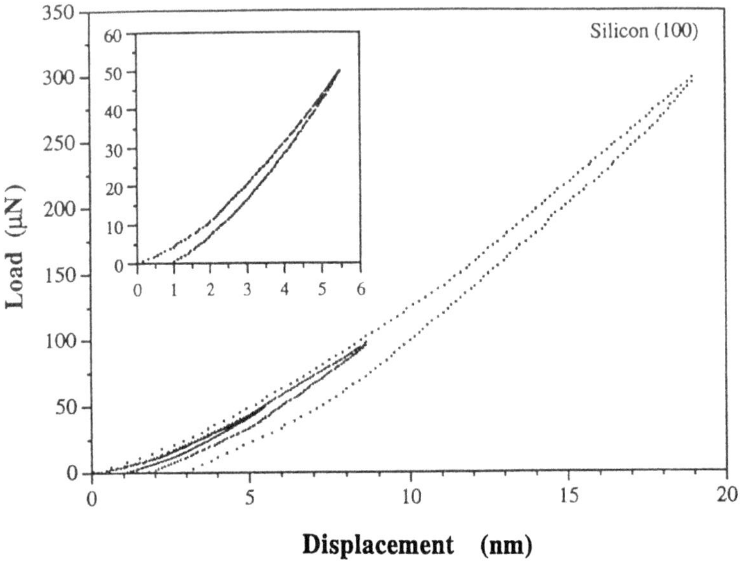

Fig. 10 Load-displacement curves at various peak loads for Si(100) (Bhushan et al., 1996b)

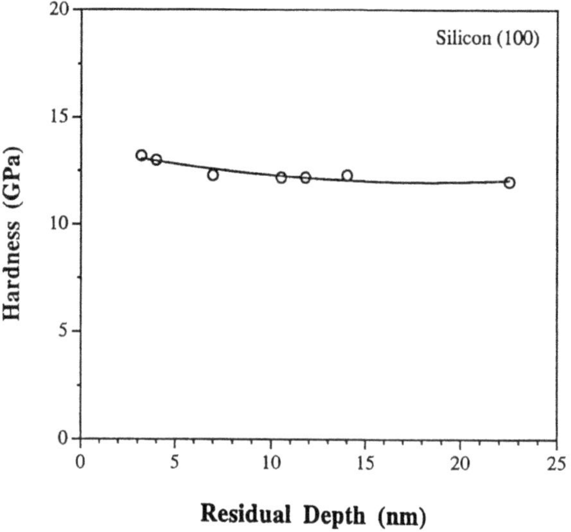

Fig. 11 Indentation hardness as a function of residual indentation depth for Si (100) (Bhushan et al., 1996b).

5. Boundary Lubrication

The classical approach to lubrication uses freely supported multimolecular layers of liquid lubricants (Bowden and Tabor, 1950, 1964; Bhushan, 1996a). The liquid lubricants are chemically bonded to improve their wear resistance (Bhushan, 1995a, 1996a). To study depletion of boundary layers, the micro-scale friction measurements were made as a function of number of cycles of virgin Si(100) surface and silicon surface lubricated with Z-15 and Z-Dol PFPE lubricants, Fig. 12 (Koinkar and Bhushan, 1996a, 1996b), Z-Dol is PFPE lubricant with hydroxyl end groups. Its lubricant film was thermally bonded at 150° C for 30 minutes (BUW- bonded, unwashed) and, in some cases, unbonded fraction was washed off with a solvent to provide a chemically bonded layer of the lubricant (BW) film. In Fig. 12(a), the unlubricated silicon sample shows a slight increase in friction force followed by a drop to a lower steady state value after 20 cycles. Depletion of native oxide and possible roughening of the silicon sample are believed to be responsible for the decrease in friction force after 20 cycles. The initial friction force for Z-15 lubricated sample is lower than that of unlubricated silicon and increase gradually to friction force value comparable to that of the silicon after 20 cycles. This suggests the depletion of the Z-15 lubricant in the wear track. In the case of the Z-Dol coated silicon sample, the friction force starts out to be very low and remains low during the 100 cycles test. It suggests that Z-Dol does not get displaced/depleted as readily as Z-15. The nanowear results for BW and BUW/Z-Dol samples with different thicknesses are shown in Fig. 12(b). The BW with thickness of 2.3 nm exhibits an initial decrease in the friction force in the first few cycles and then remains steady for more than 100 cycles. The decrease in friction force possible arises from the alignment of any free liquid lubricant present over the bonded lubricant layer. BUW with a thickness of 4.0 nm exhibits the similar behavior to BW (2.3 nm). Lubricated BW and BUW samples with thinner films exhibit a higher value of coefficient of friction. Among the BW and BUW samples, BUW samples show the lower friction because of extra unbonded fraction of the lubricant.

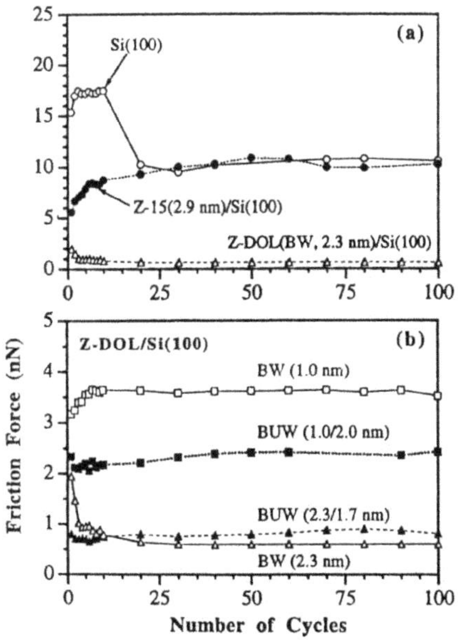

Fig. 12 Friction force as a function of number of cycles using silicon nitride tip at a normal force of 300 nN for (a) unlubricated silicon, Z-15 and bonded washed (BW) Z-Dol and (b) bonded Z-Dol before washing (BUW) and after washing (BW) Koinkar and Bhushan, 1996as).

For lubrication of microdevices, a more effective approach involves the deposition of organized, dense molecular layers of long-chain molecules on the surface contact. Such monolayers and thin films are commonly produced by Langmuir-Blodgett (L-B) deposition and by chemical grafting of molecules into self-assembled monolayers (SAMs). Based on the measurements, SAMs of octodecyl (C18) compounds based on aminosilanes on an oxidized silicon exhibited lower coefficient of friction of (0.018) and greater durability than LB films of zinc arachidate adsorbed on a gold surface coated with octadecylthiol (ODT) (coefficient of friction of 0.03) (Fig. 13) (Bhushan et al., 1995b). LB films are bonded to the substrate by weak van der Waals attraction, whereas, SAMs are chemically bound via covalent bonds. Because of the choice of chain length and terminal linking group that SAMs offer, they hold great promise for boundary lubrication of microdevices.

Measurement of ultra-thin lubricant films with nanometer lateral resolution can be made with the AFM (Bhushan, 1995a). The lubricant thickness is obtained by measuring the force on the tip as it approaches, contacts and pushes through the liquid film and ultimately contacts the substrate. The distance between the sharp 'snap-in' (owing to the formation of a liquid of meniscus between the film and the tip) at the liquid surface and the hard repulsion at the substrate surface is a measure of the liquid film thickness. This technique is now used routinely in the information-storage industry for thickness measurements (with nanoscale spatial resolution) of lubricant films, a few nanometers thick, in rigid magnetic disks.

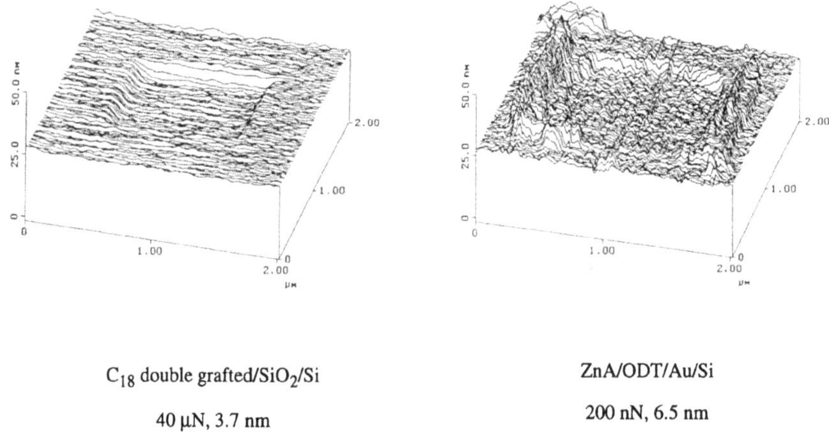

C_{18} double grafted/SiO_2/Si

40 µN, 3.7 nm

ZnA/ODT/Au/Si

200 nN, 6.5 nm

Fig. 13 **Surface plots showing the worn region after one scan cycle for self-assembled monolayers of octodecyl silanol (C18) (left) and zinc arachidate (ZnA) (right). Normal force and wear depths are indicated. Note that wear of ZnA occurs at only 200 nN as compared to 40 mN for C18 film (Bhushan et al., 1995b).**

References

Bhushan, B. (1992) *Mechanics and Reliability of Flexible Magnetic Media*, Springer-Verlag, New York.

Bhushan, B. (1995a), *Handbook of Micro/Nanotribology*, CRC Press, Boca Raton, Fl.

Bhushan, B. (1995b), *Micro/Nanotribology and its Application to Magnetic Storage Devices and MEMS*, Tribol. International **28**, 85-95.

Bhushan, B., (1996a), *Tribology and Mechanics of Magnetic Storage Devices*, second edition, Springer-Verlag, New York.

Bhushan, B. (1996b), *Nanotribology and Nanomechanics of MEMS Devices*, Proc. MEMS '96, IEEE, NY, 91-98.

Bhushan, B. (1996c), "Contact Mechanics of Rough Surfaces in Tribology: Single Asperity Contact," *Appl. Mech. Rev.* **49**, 275-298.

Bhushan, B. (1997a), *Micro/Nanotribology and its Applications*, NATO ASI Series E: Applied Sciences - Vol. **330**, Kluwer Academic Pub., Dordrecht, Netherlands.

Bhushan, B. (1997b), "Contact Mechanics of Rough Surfaces in Tribology: Multiple Asperity Contact," Trib. Lett. **4**, in press.

Bhushan, B. and Blackman, G.S. (1991), "Atomic Force Microscopy of Magnetic Rigid Disks and Sliders and its Applications to Tribology," *ASME Journal of Tribology* **113**, 452-458.

Bhushan, B. and Gupta, B.K. (1991), *Handbook of Tribology: Materials, Coatings and Surface Treatments*, McGraw-Hill, New York.

Bhushan, B., and Koinkar, V.N. (1994a), "Tribological Studies of Silicon for Magnetic Recording Applications," *J. Appl. Phys.* **75**, 5741-5746.

Bhushan, B., Koinkar, V.N. and Ruan, J. (1994b), "Microtribology of Magnetic Media," *Proc. Inst. Mech. Engrs.*, Part J: J. Eng. Tribol. **75**, 5741-5746.

Bhushan, B. and Koinkar, V.N. (1994c), "Nanoindentation Hardness Measurements Using Atomic Force Microscopy," *Appl. Phys. Lett.* **64**, 1653-1655.

Bhushan, B., Israelachvili, J.N., and Landman, U. (1995a), "Nanotribology: Friction, Wear and Lubrication at the Atomic Scale," *Nature* **374**, 607-616.

Bhushan, B., Kulkarni, A.V., Koinkar, V.N., Boehm, M., Odoni, L., Martelet, C. and Belin, M. (1995b), "Microtribological Characterization of Self-Assembled and Langmuir-Blodgett Monolayers by Atomic and Friction Force Microscopy," *Langmuir* **11**, 3189-3198.

Bhushan, B. and Kulkarni, A.V. (1996a), "Effect of Normal Load on Microscale Friction Measurements," *Thin Solid Films* **278**, 49-56.

Bhushan, B., Kulkarni, A.V., Bonin, W. and Wyrobek, J.T. (1996b), "Nano/Picoindentation Measurements Using a Capacitive Transducer System in Atomic Force Microscopy," *Philos. Mag.* **74**, 1117-1128.

Bhushan, B. and Koinkar, V.N. (1997a) "Microtribological Studies of Doped Single-Crystal Silicon and Polysilicon Films for MEMS Devices," *Sensors and Actuators* A **57**, 91-102.

Bhushan, B. and Li, X. (1997b), "Micromechanical and Tribological Characterization of Doped Single-Crystal Silicon and Polysilicon Films for Microelectromechanical Systems," *J. Mat. Res.* **12**, 54-63.

Bhushan, B. and Sundararajan, S. (1998), "Micro/Nanoscale Friction and Wear Mechanisms of Thin Films Using Atomic Force and Friction Force Microscopy", *Acta Mater.* submitted for publication.

Bowden, F.P. and Tabor, D, (1950 & 1964), *The Friction and Lubrication of Solids*, Parts I & II, Clarendon, Oxford, U.K.

DeVecchio, D. and Bhushan, B. (1997), "Localized Surface Elasticity Measurements Using an Atomic Force Microscope," *Review Sci. Instrum.* **68** (in press).

Frisbie, C.D., Rozsnyai, L.F., Noy, A., Wrighton, M.S., and Lieber, C.M. (1994), "Functional Group Imaging by Chemical Force Microscopy," *Science* **265**, 2071-2074.

Koinkar, V.N. and Bhushan, B. (1996a), "Micro/Nanoscale Studies of Boundary Layers of Liquid Lubricants for Magnetic Disks," *J. Appl. Phys.* **79**, 8071-8075.

Koinkar, V.N. and Bhushan, B. (1996b), "Microtribological Studies of Unlubricated and Lubricated Surfaces Using Atomic Force/Friction Force Microscopy," *J. Vac. Sci.* Technol. **A14**, 2378-2391.

Koinkar, V.N. and Bhushan, B. (1996c) "Microtribological Studies of Al_2O_3, Al_2O_3-TiC, Polycrystalline and Single-Crystal Mn-Zn Ferrite and SiC Head Slider Materials," Wear, **202**, 110-122.

Koinkar, V.N. and Bhushan, B. (1997a), "Effect of Scan Size and Surface Roughness on Microscale Friction Measurements," *J. Appl. Phys.* **81**, 2472-2479.

Koinkar, V.N. and Bhushan, B. (1997b), "Scanning and Transmission Electron Microscopies of Single-Crystal Silicon Microworn/machined Using Atomic Force Microscopy," *J. Mat. Res.* in press.

Koinkar, V.N. and Bhushan, B. (1997c), "Microtribological Properties of Hard Amorphous Carbon Protective Coatings for Thin Film Magnetic Disks and Heads," *Proc. Instn Mech. Engrs. Part J: J. Eng. Trib.* (submitted for publication)

Kulkarni, A.V. and Bhushan, B. (1996a), "Nanoscale Mechanical Property Measurements Using Modified Atomic Force Microscopy," *Thin Solid Films* **290-291**, 206-210.

Kulkarni, A.V. and Bhushan, B. (1996b), "Nano/picoindentation Measurements on Single-Crystal Aluminum Using Modified Atomic Force Microscopy," *Materials Letters* **29**, 221-227.

Kulkarni, A.V. and Bhushan, B. (1997), "Nano/picoindentation Measurement of Amorphous Carbon Coatings Using a Capacitive Transducer System in Atomic Force Microscopy," *J. Mat. Res.* (submitted for publication).

Maivald, P., Butt, H.J., Gould, S.A.C., Prater, C.B., Drake, B., Gurley, J.A., Elings, V.B., and Hansma, P.K. (1991), "Using Force Modulation to Image Surface Elasticities with the Atomic Force Microscope," *Nanotechnology* **2**, 103-106.

Mate, C.M., McClelland, G.M., Erlandsson, R., and Chiang, S. (1987), "Atomic-Scale Friction of a Tungsten Tip on a Graphite Surface," *Phys. Rev. Lett.* **59**, 1942-1945.

Meyer, E., Overney, R., Luthi, R., Brodbeck, D., Howald, L., Frommer, J., Guntherodt, H.-J, Wolter, O., Fujihira, M., Takano, T., and Gotoh, Y. (1992), "Friction Force Microscopy of Mixed Langmuir-Blodgett Films," *Thin Solid Films* **220**, 132-137.

Oden, P.I., Majumdar, A., Bhushan, B., Padmanabhan, A. and Graham, J.J. (1992), "AFM Imaging, Roughness Analysis and Contact Mechanics of Magnetic Tape and Head Surfaces," *ASME Journal of Tribology* **114**, 666-674.

Poon, C.Y. and Bhushan, B. (1995), "Comparison of Surface Roughness Measurements by Stylus Profiler, AFM and Non-contact Optical Profiler," *Wear* **190**, 76-88.

Ruan, J. and Bhushan, B. (1993), "Nanoindentation Studies of Fullerene Films Using Atomic Force Microscopy," *J. Mat. Res.* **8**, 3019-3022.

Ruan, J. and Bhushan, B. (1994a), "Atomic-Scale Friction Measurements Using Friction Force Microscopy: Part I- General Principles and New Measurement Techniques," *ASME Journal of Tribology* **116**, 378-388.

Ruan, J. and Bhushan, B. (1994b), "Atomic-Scale and Microscale Friction of Graphite and Diamond Using Friction Force Microscopy," *J. Appl. Phys.* **76**, 5022-5035.

Ruan, J. and Bhushan, B. (1994c), "Frictional Behavior of Highly Oriented Pyrolytic Graphite," *J. Appl. Phys.* **76**, 8117-8120.

Pulsed Force Mode: A new method for characterizing thin silane films by adhesive force measurements

S. Hild, U. Krotil, O. Marti
Experimental Physics, University of Ulm
D89069 Ulm, Germany

1. Introduction

The Scanning Force Microscope (SFM) has become a useful tool for the investigation of isolating surfaces with respect to their surface properties. Methods have been developed that enable the simultaneous mapping of topographic features and material properties like local stiffness, adhesive force or friction on nanometer scale [1-6]. During the last few years the characterizing of by chemical contrast, so called Chemical Force Microscopy (CFM) has become a field of great interest. First, the CFM has been performed using Friction Force Microscopy (FFM) [7-9]. Because FFM is performed in contact mode for soft sample often occurs the problem of destructive lateral forces. This complicates the investigation of delicate samples like polymers or organic materials.

To overcome the deficiency of contact mode imaging non- or intermittent contact modes have been introduced [10,11]. Here, the cantilever is oscillated at his resonance frequency which means in the 100 kHz-range with amplitudes up to 150 nm. For topography images the SFM is operated in constant amplitude mode. An additional material contrast can be determined by the phase shift between free oscillation and contact or near to contact of the cantilever. It is known that the phase shift reflects adhesive and viscoelastic properties of the sample up, but to now a quantitative determination of these properties has not been possible [12].

Another method of CFM without lateral forces is taking force vs. distance curves [13]. Here the specific interactions between molecules are deter-

B. Bhushan (ed.), Tribology Issues and Opportunities in MEMS, 247-260.

mined as a mechanical force acting between tip and surface, the so-called "pull-off force" which is correlated to the adhesion properties of the system. To get a two dimensional image force vs. distance curves are recorded sequentially in a line [15]. Although this method is commonly is used it does not appear to be convenient for imaging surface properties with lateral resolution. On the one hand, the recording frequency of force vs. distance curves is limited at about 100 Hz. This is due to the fact that a triangular modulation of the piezo is used, which can excite disturbing resonances far above the modulation frequency working with higher frequencies. On the other hand, taking a whole array of force vs. distance curves results in a high amount of data.For data reduction only sections of the force vs. distance curves are taken at definite distance to the sample surface. Whereas elastic properties of the surface can be investigated by this method, the determination of the exact adhesive force is not possible because of missing the correct jump-off point.

To overcome these disadvantages a new operation system called "Pulsed Force Mode" has been developed [16], where a sinusoidal z-modulation is used to perform force vs. distance curves. The amount of data is reduced without sacrificing the essential information due to the recording of a few characteristic points of the curve by Sample&Hold circuits. With this set up the topography of a sample can be imaged without shear forces at high scan speed. The imaging of a sample takes times comparable to performing the image in contact mode. Additionally, surface properties like adhesion but also local elastic behavior and surfaces charges can be recorded with high lateral resolution.

A wide range of applications for the PFM can be found in the mapping of adhesive or mechanical properties of organic surfaces [16,17]. In this study, we would like to show the suitability of the PFM for characterizing surfaces with chemical sensitivity by determine the adhesive force. As example silicon surface laterally structured with silane molecules are used. Such surfaces can easily prepared by Micro Contact Printing (μCP) [18]. Goal of μCP is to influence the function of the surface by intention. Therefore small organic molecules are transferred to an inorganic surface using an microstructured stamp consisting of silicone polymer. These stamps are replicas of microstructured silicon masters with structural dimensions from several μm down to several 100 nm. A schematic of this μCP process is shown Figure 1.

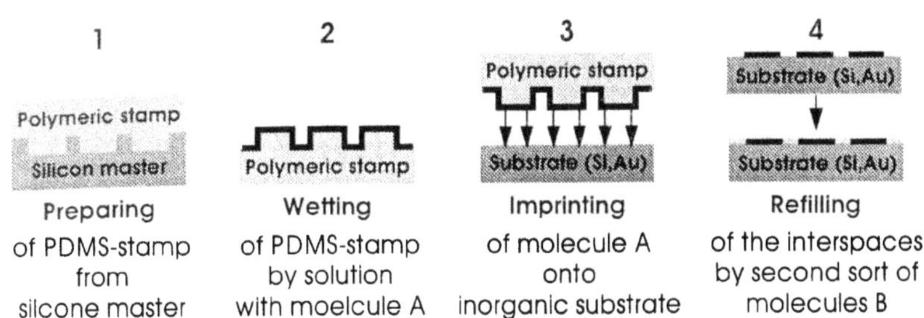

Fig. 1: Schematic model of surface structuring of inorganic using Micro Contact Printing (μCP). By additional adsorption of molecules after the imprintin process at least bifunctionalized surface can be created.

When the stamp is getting into contact to the surface molecules are transferred at the contact areas. The areas between them remain empty. They can be functionalized with a second sort of molecule that case that a laterally structured surface with at least two functionalities is created. Up to now the system thiole-gold is preferred for investigations. Because thioles spontaneous chemisorb [19] to gold surfaces and form self-assembled monolayers with high orientation. Presumption for a successful imprinting process are freshly prepared gold substrates [18]. Although this method enable to microstructure silicon surfaces the process is complicated [20]. An alternative system for direct perform silicon surface microstructuring is to imprint silane molecules. Silanes are available in a wide range of derivatives consisting of mono-. bi- or trifunctionalized Si-atom with alkyl chains having different endgroups. Silanes are chemically bond to silicondioxide surfaces or also polymeric surfaces [21,22]. Although this advantages μCP with silane molecules is poorly investigated [23,24]. Maybe because silane molecules are more difficult to handle then thioles.

2. Instrumentation

The PFM has been performed as an external box which simulates a contact mode experiment. This box can be added to the feedback circuit of usual force microscopes [25] if the system allows to interrupt the signal of the force detector and to modulate the piezo in z-direction. The modulation of the piezos is done by applying a sine voltage to the the z-offset at a variable frequency ranging from 100 Hz up to 3 kHz. This low modulation frequency enables to control the force acting to the surface, because you work far away from resonance frequency of the cantilever. The z-modulation can be chosen to have peak to peak amplitudes between 10 nm and 1 μm. At least an amplitude has to be used that the cantilever is retracted from the surface. The maximum amplitude is limited by the maximum z-range of the used piezo. The modulation results in a force vs. time curve schematically shown in Figure 2. During one period the tip contacts the surface only for a short time between 10^{-3} and 10^{-4} seconds. Because of this small contact times the shear forces are reduced and even delicate because soft samples can be imaged.

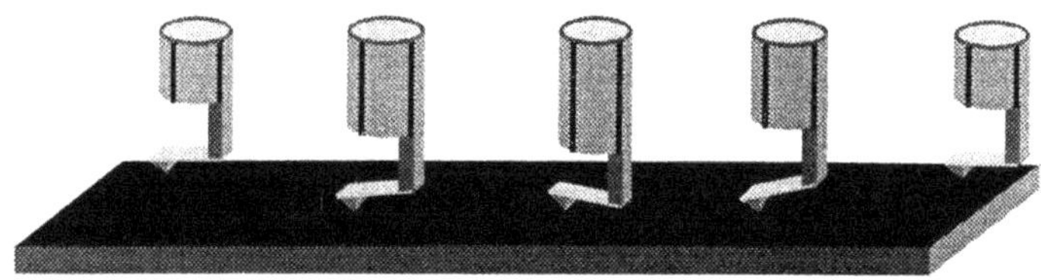

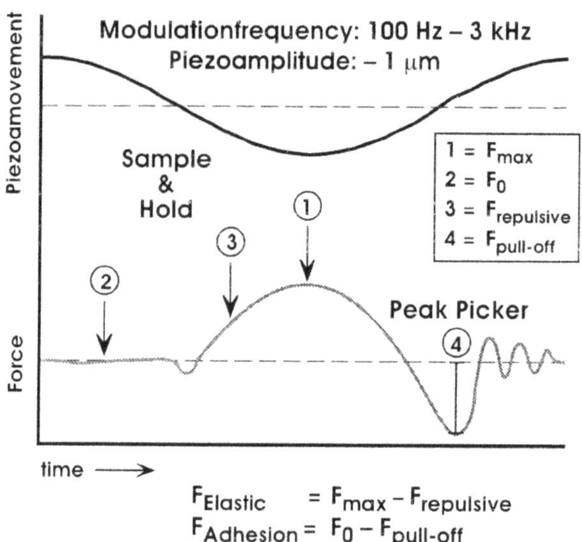

Fig. 2: Schematic of PulsedForceMode. (A) Shows the movement of the piezo (B) gives a schematic of the resulting force vs. time curve. Numbers shows the points where specific values are detected by sample&hold (S&H) or peakpicker circuits.

To image the topography a constant force mode is performed, where the point of maximum repulsive force (S&H1) is used for the feedback control. The deflection of the free cantilever is stored by S&H2 for zero calibration. To determine the elastic response of the sample the slope of the repulsive region is recorded by S&H3. The compliance of the sample is given by a function of the difference between S&H1 and S&H3. The pull-off force is detected by a peak picker circuit and stored in S&H4. The recorded adhesive force is determined by the difference S&H2 – S&H4. Filtering is necessary because of the cantilever oscillates with it resonance frequency after the jump-off.

3. Experimental

As samples silicon substrate structured by µCP are used. The necessary µCP for stamps are made from polydimethylsilane (PDMS, Silgard 184, DOW Chemicals) like described in [18]. In Fig. 3 the topography image of a PDMS stamp is shown. For the quality of the imprinted structure the elevated areas should be plain.Because the stamp is mechanically deformed during the printing process, the aspect ratio - the ratio between the distance of two elevations and the depth of the structure - has to be a maximum value of. Otherwise the deeper areas will get into contact to the silicon surface, too.

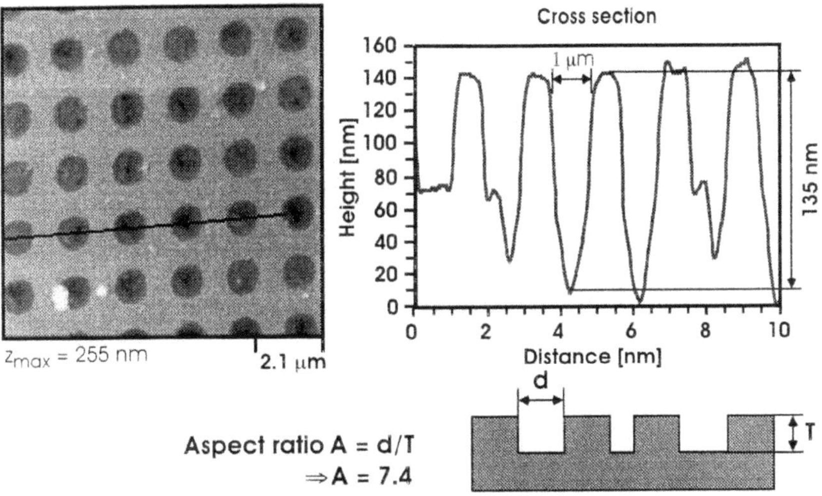

Fig. 3: SFM image of a Polydimethylsiloxane stamp. To transfere definite structures the aspect ratio A has to be smaller then 5, because mechanical stress will deform the stamp.

As molecular ink solutions of 2% Dichlorodimethylsilane (DCS) in chloroform and 10 mM Octadecyltrichlorosilane (OTS) in hexane are used. Additionally for gasphase modification Heptadecafluoro-1,1,2,2-(tetrahydrodecyl)trichlorosilane (FTS) or Monochlorotrimethylsilane (MCS) are taken. For the gasphase reaction ether the stamps or the sample are exposed to a saturated atmosphere of the silanes for 10 min with reduced pressure of 1mbar. The imprinting process is performed under ambient conditions.

For the PFM measurements Si_3N_4-cantilever with spring constant of 0.1 N/m and resonance frequencies of 50 kHz are used. The chosen modulation frequency has been 1 khz and the modulation amplitude where about 200 nm. The measurement has performed under ambient conditions.

4. Results

The behavior of structure development during the transfer of silane molecules solvent in chloroform or toluene could be visualized by scanning force microscope images. In Fig. 4 the surface of a silicon substrate is shown that was imprinted by DCS solution. After a short contact of 2 min between stamp and substrate aligned point structures at the surface with a height of about 13 nm (light points in Fig. 4) can be seen. They should consist out of collections of silane molecules.

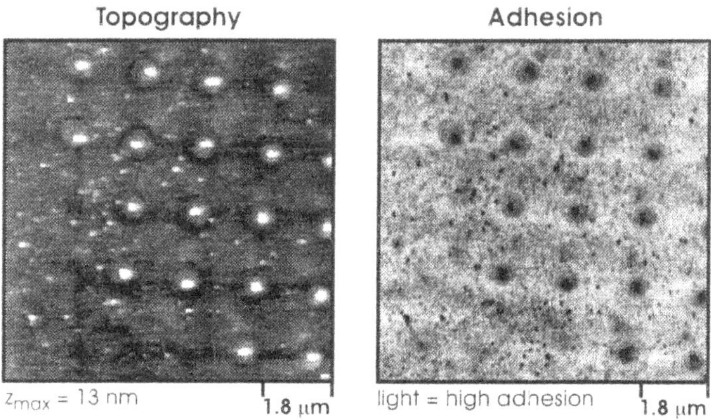

Fig. 4: SFM-image of a transfered Dichlordimethylsilane structure after 2 min contact time solution of toluene. Molecules are transfered as cluster which are correlated to the center of interspaces of the stamp. The adhesion in the areas where molecules are transfered is reduced.

If the contact period is increased up to 30 min the complete stamp structure will be observed (Fig. 5). The silanized areas seemed to slightly corrugated. The thickness of the generated layer is 5 to 10 nm. This high values and the corrugation seems to indicate that amultilayer is a multilayer formed.

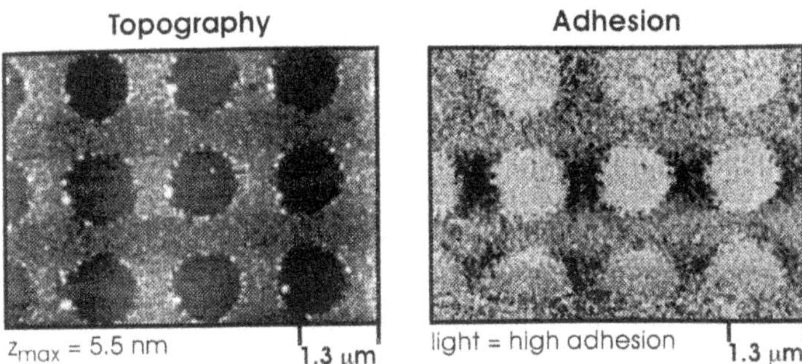

Fig. 5: SFM-image of a transfered Dichlordimethylsilane structure after 30 min contact time solution of toluene. Whole structure ist transfered. The adhesion in the areas where molecules are transfered is reduced.

The time related structure development behavior can be explained by the model shown in Fig. 6: Some solvents, e. g. chloroform or toluene, swell the PDMS-stamp (Fig. 3.6, 2) If such a swollen silicone stamp is used for imprinting (Fig. 3.6, 3), there is only a first contact between the exposed areas of stamp and substrate. Later, solvent molecules will diffuse out of the stamp, so that it reduces to its former shape. The contact area between stamp and silicone surface reaches a maximum. During evaporating the solvent, solved silane molecules can be transported so that they can react with "non-silanized" silicon surface areas.

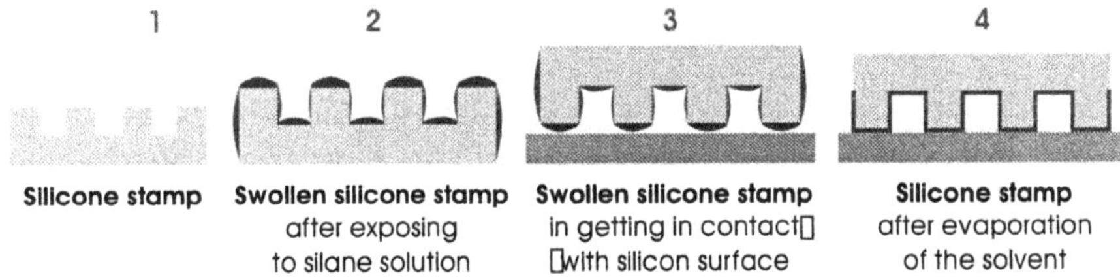

Fig. 6: Schematic model for the structural changes of a silicone stamp during µCP. The solvent will swell the polymeric material (2). When the deformed stamp print to the surface first elevated areas get into contact (3). After evaporating the solvent the whole structure the origninal strucutre is rebuild (4).

Simultaneously to the topography the adhesive force is mapped like shown in Figure 4 and 5. This images shows that the adhesive force that have been measured between the tip and the silicon substrate is lower than the value determined for the regions where molecules have been transmitted. In this areas are at the surface. That means the adhesion between methyl groups and the tip is lower than the adhesion between silicon surface and the tip. These effect further be used to characterize the transmitted structures additional with respect to the chemical contrast.

An alternative method to "wet" the stamp without swelling it, is to absorb the silane molecules out of a saturated gasphase. Then the stamp is brought into contact with the silicon surface. The SFM examinations of the functionalized silicon surfaces show that very thin silane layers are formed which cannot be characterized by their topographic changes (Fig.7). However, in the adhesive force image different areas can be distinguished: dark areas which are silanized and non-silanized light areas.

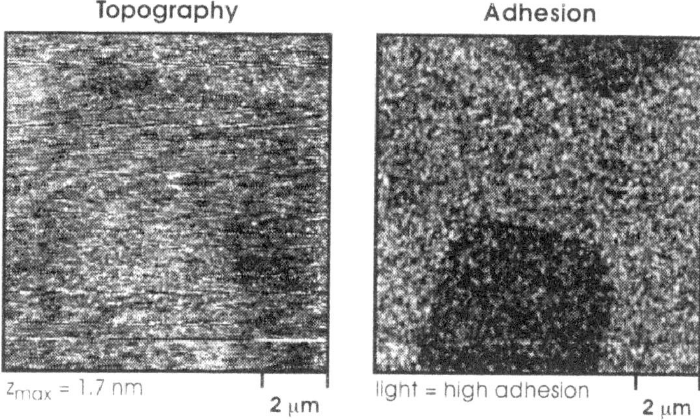

Fig. 7: SFM images of the opography and adhesive force of Trimethylchlorsilane attached to silicon surface. The stamp has been wetted by gasphase reaktion. Thin films maybe amonolayer is transfered. No molecules in the intespace are visible.

As last example we would study a surface consisting of two different modified area formed by μKD and additional functionalization of the free non-silanized areas (compare to Fig. 1). In the first step large functional areas of about 5 μm were created by stamping 10 mM OTS solved in hexane on a silicon surface. The resulting structure is shown in figure 8. On the one hand a significant increase of the topography in the silanized areas can be detected, on the other hand the adhesion force slows down (Fig. 9). The adhesion according the non-modified silicon surface is significant lowered by the functionalization with OTS molecules. This can be explained due to the fact that the surface energy of the silanized areas is reduced compared to the pure silicon surface because of the methyl groups of the OTS. Opposed to the example shown before the OTS molecules performed no closed layers but a cluster-like structure which clearly can be seen in figure 9.

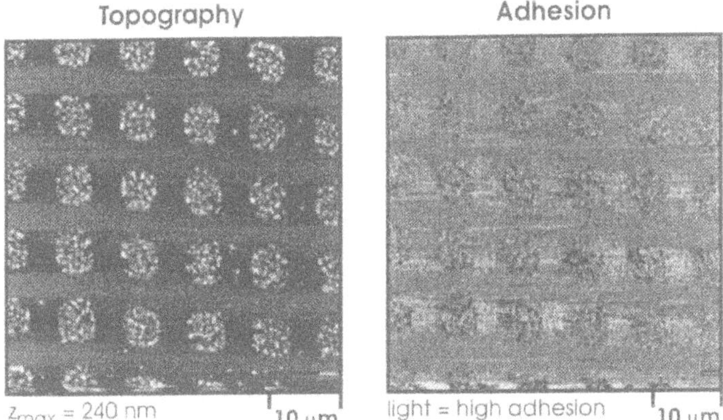

Fig. 8: SFM image performed in PFM of silicon surface where OTS stamped from 2% solvent in hexan. In the adhesion picture on the right side a significant decreasement of the adhesion force (dark areas) in the silanized locations can be seen.

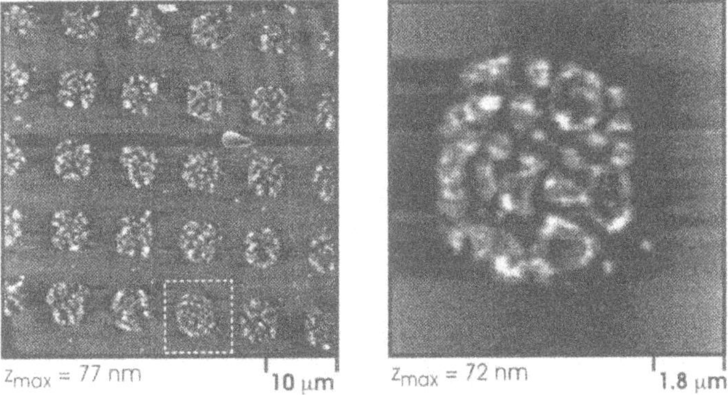

Fig. 9: SFM-Image of a transfered Octatrymethylsilane molecules. Imprinting was performed from wurde mit 2% solution in hexene. No closed layer can be observed. The inset shows clusterd structure because of transfered polymerized molecules.

The formation of the aggregated structure can be explained by the fact that both bi- and tri-functional silanes like to polymerize under the influence of moisture and form silanol oligomers or at least polymers. Because all experiments are performed under common ambient conditions it might be possible that the given humidity accelerates polymerization. A thin water layer on the silicone stamp can be cause the polymerization reaction of some of the silane molecules during the wetting of the stamp. These polymerized molecules will be transmitted during stamping. Therefore the sub-structure of the stamped silanes will be consist of a collection of more than 100 nm up to 1mm large aggregates. A schematic of the polymeriation reaction taking place during μCP is shown in figure 10.To avoid the polymerization monofunctional silanes, like monochlorotrimethylsilane (MTS) should be preferred for further modifications.

1. Wetting of the stamp using silane solution under ambient conditions

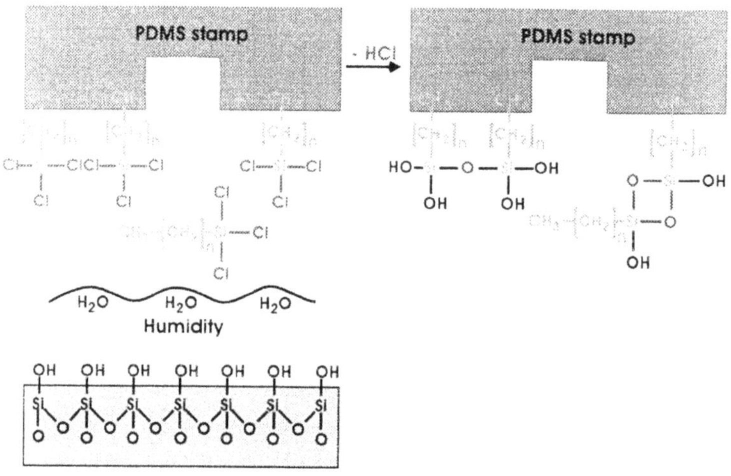

2. Contacting wetted stamp and silicone surface

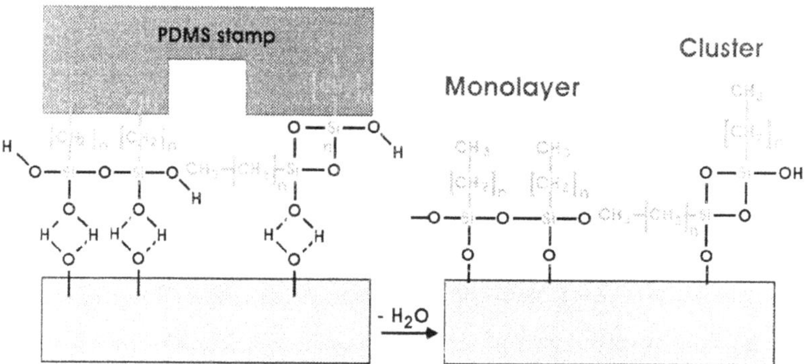

Fig. 10 Schematic model for the silanization reaction during mCP. Due to the humidity the bi- or trifuctional silane molecules can polymerize. Because of this aggregates are transferred to the silicon surface

Besides the formation of a corrugated areas no contaminations can be found, so that these regions are effective for further functionalization with another type of molecules. To functionalize the remaining effective silicon surface the microstructured sample was exposed within a saturated atmosphere of FTS. Afterwards the surface was imaged by SFM in PFM using a non-modified Si_3N_4-cantilever which supposed to show chemical neutral behavior. The gasphase reaction is performed under reduced pressure but not under nitrogen atmosphere so that during the process of functionalization the fluorochlorosilanes can polymerize because of humidity. These polymerized molecules are bond the remaining free areas of the surface. That is the reason why the FTS functionalized intermediate regions seem to be higher than the OTS structured one. Although a significant difference in topographic height can be observed, no adhesion contrast can be detected with a chemical neutral tip.

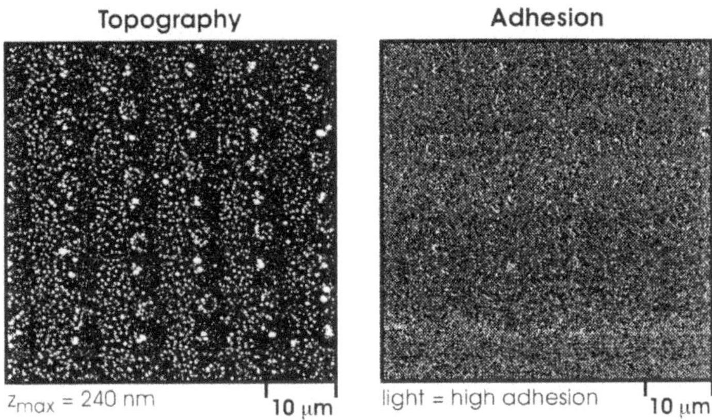

Fig. 11: SFM-image of surface functionalized first with OTS by μCP. In e second step the interspaces are functionalized by FTS. The adhesion detection using a chemical neutral tip shows no adhesive contrast. This can be explained if the chemistry of the surface are examinated closer.

This can be explained by the small differences in surface tension between the areas modified by OTS molecules and these ones modified by FTS molecules. Whereas for a surface covered with CF_3 molecules a surface tension of 19 nN/m [26] is proposed, a methylsiloxane surface will have can have of about 20 nN/m [26]. Therefore the chemical contrast determined by surface tension should only be low. As a matter of fact the usage of functionalized tips improves the chemical contrast [26-29]. Generally, chemical similar or equal materials cause larger interactions between tip and surface than chemical different materials. To enlarge the contrast we functionalized the tip with organic molecules that case that OTS molecules cover the tip. The functionalization was done by dipping the silicon cantilever in a 10 mM OTS solvent in hexane for 5 minutes. This modified tip was used to examine the surface like, shown in fig. 12. The functionalization of the silicon tip increases its chemical sensibility evidently. Chemical different regions can be visualized with better contrast. In the current case the adhesion force measured in OTS areas is increased that it it appears light in adhesion picture.

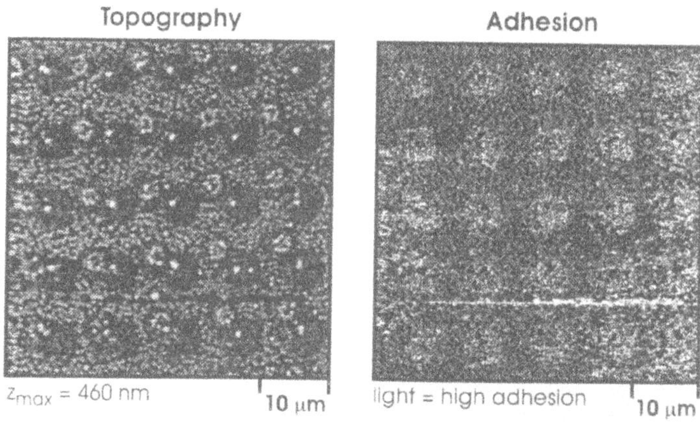

Fig. 12: Similar surface than shown in fig. 11, but imaged with silicon tip functionalized with OTS molecules. Now a adhesive force contrast can be observed. In contrast to the pure OTS structure on silicon the adhsion contrast is inverted which means higher interaction between tip ans surface for similar materials.

5. Conclusion

The PulsedForceMode is a new operation principle which combines the advantage of force vs. distance curves and intermittent contact mode. At a speed of conventional contact SFM measurements samples can be imaged without lateral shear forces. Simultaneous mapping of topography and adhesive force enables the characterization of heterogenous surfaces by chemical contrast. The sensitivity of the adhesive force measurements can increased by using cantilevers where the tip is modified with organic molecules.

The investigation of silicon substrates modified with a lateral structure by μCP using the PFM shows that alkylchlorosilanes on hydrated silicon surfaces behaves complementary to thiole - gold system: (a) silanes have no instant self-organization; (b) silane layers appears significant less structured than thiole layers; (c) during stamping of bi- or trifunctional chlor- or alkylsilanes creates multilayer or even clustered structures.

6. Acknowledgements

We would like to thank G. Volswinkler for building and improving the PFM control electronics. We thank the AFM-group, Th. Stifter, H. Waschipky, B. Zink, L. Weiss, A. Rosa, E. Weilandt- for their help, ideas and fruitful discussions. R. Brunner, J. Barenz and F. Rothe are also acknowledged for their help for this paper. This work has been supported by the government of Baden-Würtemberg and the Sonderforschungsbereich 239.

7. References

[1] M. Radmacher, R. W. Tillmann, H. E. Gaub, Biophys. J., 64, 735, 1993
[2] R. M. Overney, H. Takano, M. Fujihira, Europhys. Lett. 26, 443, 1994
[3] R. M. Overney, E. Meyer, J. Frommer, H.-J. Güntherodt, Langmuir 10, 1281, 1994
[4] O. Marti, J. Colchero, J. Mlynek, Nanotechnology 1, 141, 1990
[5] O. Marti, J. Colchero, Physikalische Blätter, 12, 1007, 1992
[6] E. Meyer, R. Lüthi, L. Howald, H.-J. Güntherodt, Fores in Scanning Probe Methods, 285, 1985, Kluwer Academic Publishers
[7] R. M. Overney, E. Meyer, J. Frommer, H.-J. Güntherodt, Langmuir 10, 1281, 1994
[8] S. Akari, D. Horn, H. Keller, Adv. Materials, 7, 594, 1995
[9] C. D. Frisbie, L. F. Rosznay, A. Noy, M. S. Wrighton, C. M. Lieber, Science, 265, 2071, 1994
[10] D. Sarid, Scanning Force Microscopy, revised ed. (Oxford University Press, New York, 1994)
[11] D. Zhong, D. Innis, K. Kjoller, V.B. Ellings, Surface Science 290, 688, 1993
[12] J. Spatz, S. Sheiko. M. Möller, R. Winkler, P. Reinecker, O. Marti, Nanotechnology 6 40 1995
[13] H. A. Mizes, K.-G. Loh, R. J. D. Miller, S. K. Ahuja, E. F. Grabowski, Appl. Phys. Lett. 59, 2901, 1991
[14] M. Radmacher, M. Fritz, J. P. Cleveland, D.A. Walters, P. K. Hansma, Langmuir 10 3809, 1992
[15] K. O. van der Werf, C. A. Putman, B. G. de Grooth, J. Greve, Appl. Phys. Letter, 65, 1195, 1994
[16] A. Rosa, E. Weilandt, S. Hild, O. Marti, Meas. Sci. and Technol., 8, 1, 1997
[17] A. Rosa, S. Hild, O. Marti, Polymer Preprints, 37, 616, 1996
[18] A. Kumar, G. M. Whitesides, Appl. Phys. Lett., No. 14, 63, 1993
[19] M. D. Porter, T. B. Bright, D. L. Allara, C. E. D. Chidsey, J. Am. Chem. Soc. 109, 3559, 19987
[20] N. Abbot, A. Kumar, G. M. Whitesides, Chem. Mat. 6, 596, 1994
[21] W. S. Gutowsky, D. Y. Wu, S. Li, J. Adhesion, 43, 139, 1993
[22] G. S. Ferguson, M. K. Chaudhury, H.A. Buybuck, G. M. Whitesides, Macromolecules, 26, 5870, 1993
[23] Y. Xia, M. Mrksich, E. Kim, G. M. Whitesides, J. Am. Soc., 117, 9576, 1995
[24] S. Akari, D. Horn, H. Keller, Adv. Materials, 7, 594, 1995
[25] S. Hild, A. Rosa, G. Volswinkler, O. Marti, TopoMetrix Newsletter, 7, 5, 1997

[26] J. Brandrup, E. H. Immergut, Polymer Handbook, 3. ed. (Wiley-Interscience publications, New York, 1994)

[27] C. D. Frisbie, L. F. Rosznay, A. Noy, M. S. Wrighton, C. M. Lieber, Science, 265, 2071, 1994

[28] T. Nakagawa, K. Ogawa, T. Kurumizawa, S. Ozaki, Jpn. J. Appl. Phys, 32, 294, 1993

[29] V. V. Tsukruk, V. N. Bliznyuk, J. Wu, D. W. Visser, Polym. Prepr. 37 (2), 575, 1996

[30] S. K. Sinniah, A. B. Steel, C. J. Miller, J. E. Reutt-Robey, J. Am. Chem. Soc. 118, 8925, 1996

FORMATION OF NANOMETER-SCALE CONTACTS TO VISCOELASTIC MATERIALS

Implications for MEMS

K.J. WAHL
Code 6170
Naval Research Laboratory
Washington, D.C. 20375-5342

W.N. UNERTL
Laboratory for Surface Science & Technology
University of Maine
Orono, ME 04469

Abstract

The making and breaking of nanometer-scale contacts is an essential operation in MEMS devices with moving parts. The behavior of contacts in this size range is not well understood, especially if viscoelastic materials are involved. This article describes shear modulation spectroscopy, a new scanning force microscope technique especially well suited for quantitative studies of nanometer-scale contacts to viscoelastic materials such as lubricants and some polymers. The technique is illustrated by measurements and analysis of contacts to poly(vinylethylene).

1 Introduction

The operation of MEMS devices with moving parts requires the making and breaking of contacts with typical dimensions of 10-100 nm and loads in the nanoNewton to milliNewton range. In general, these contacts may be subjected to both normal and shear loading. They may, or may not, be lubricated and the lubricant film may be as thin as a single monolayer. Recent theoretical [1,2] and experimental studies [3] suggest that continuum theories of contact mechanics [4] should be applicable to MEMS size contacts. However, factors such as adhesion [2,5] and capillarity [6] contribute more significantly compared to macroscopic contacts. In addition, other important factors, such as microslip [7] and viscoelastic response [8], may be altered when the contacts have dimensions comparable to the microstructure or molecular dimensions. A major experimental difficulty in the study of small contacts is the accurate measurement of the contact area.

In this article, we describe shear modulation spectroscopy (SMS). SMS is implemented with a scanning force microscope and provides quantitative information about nanometer-scale contacts including changes during loading and unloading. Emphasis is on contacts to viscoelastic materials. We also discuss the application of contact mechanics models that include viscoelastic response. The article concludes with a few remarks about the implications for MEMS technology.

B. Bhushan (ed.), Tribology Issues and Opportunities in MEMS, 261-271.

2 Basics of Shear Modulation Spectroscopy

Shear modulation spectroscopy [7] is an extension of the well-developed mechanical modulation techniques used to measure the viscoelastic and rheological properties of bulk polymers [9] and has been used previously to study elastic behavior in scanning force microscope contacts [10-13]. Similar measurements on much larger area contacts have been made with the surface forces apparatus [14,15] and microscopic sphere-on-flat contacts [16]. SMS is also closely related to friction loop methods [17] and to force modulation spectroscopy [18,19].

Both SMS, which uses shear modulation, and force modulation spectroscopy, which modulates the contact normal to the surface, provide information about the viscoelastic properties of the contact. However, SMS has two significant advantages over force modulation spectroscopy. First, since the normal load is constant in SMS, the contact area does not change during a modulation cycle. This simplifies the analysis. Second, during normal modulation, the end of the cantilever moves up and down along a circular arc. Thus, the contacting tip always has a small component of shear motion parallel to the surface. As shown recently by Mazeran and Loubet [20], this combined normal and lateral motion significantly complicates quantitative interpretation of the force modulation spectroscopy data.

2.1 Shear Modulation Spectroscopy

Figure 1a illustrates how SMS is carried out with a scanning force microscope. The scanning force microscope tip (height H) is placed in contact with a viscoelastic sample above point O under applied load F_N. The substrate is then displaced parallel to its surface by distance x_o to O'. This causes a displacement of the tip's contact point by $x_t = x_o - x_c$ where x_c is the distortion of the sample surface relative to point O'. This distortion results from the shear force $F_s = \kappa_\Theta x_t$, where κ_Θ is the torsional stiffness of the cantilever. For a viscoelastic material and oscillatory sample displacement $x_o(t) = X_o\exp(i\omega t)$, the contact of the tip on the surface is displaced by $x_t = X_t\exp[i(\omega t - \alpha)]$ which has a phase lag α. The experimentally measured quantities are the amplitude and phase of the tilt angle Θ of the tip; $\Theta = x_t/H = \Theta_o\exp[i(\omega t - \alpha)]$ where $\Theta_o \equiv X_t/H$ and X_t is the amplitude of the tip motion at the contact. The shear stiffness of the contact is $\kappa_c \equiv dF_s/dx_c$. For the case of an elastic Hertzian contact with no slip, $\kappa_c = 8G^*a$ where a is the effective radius of the contact.

Figure 1b shows the relationships between the amplitudes and phases of the various motions. Starting from the condition $x_t = x_o - x_c$, the amplitude and phase of the contact displacement can be expressed in terms of the measurable quantities as

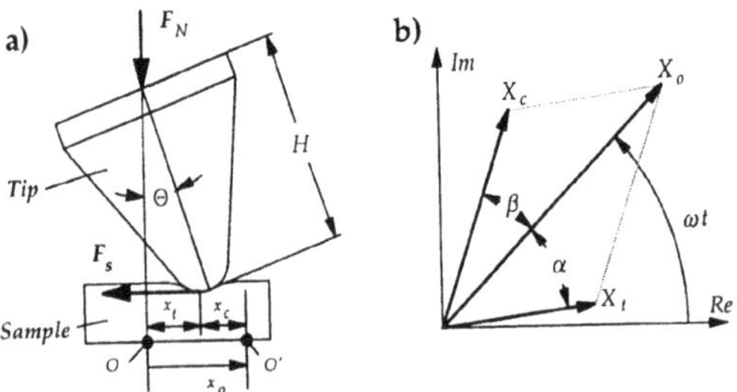

Figure 1. (a) Schematic diagram of the scanning force microscope shear measurement. (b) Relationship between the amplitudes and phases.

$$\frac{X_c}{X_o} = \sqrt{1+\left(\frac{X_t}{X_o}\right)^2 - 2\left(\frac{X_t}{X_o}\right)\cos\alpha} \tag{1}$$

and

$$\sin\beta = \frac{\left(\frac{X_t}{X_o}\right)\sin\alpha}{\sqrt{1+\left(\frac{X_t}{X_o}\right)^2 - 2\left(\frac{X_t}{X_o}\right)\cos\alpha}}. \tag{2}$$

The phase angle β is the more robust quantity. It requires only a relative measurement of the displacement amplitude of the tip whereas determination of X_c requires absolute knowledge of X_o, which is typically only a few tenths of a nanometer.

2.2 Experimental ASPECTS OF sms

SMS can be implemented on any scanning force microscope capable of making quantitative measurements of the torsional and bending motions of the cantilever force detector. We used a scanning force microscope that has been described previously [21] and is based on the optical lever method. The sample is oscillated laterally with respect to the cantilever by applying a small sinusoidal modulation, typically ≈ 0.1-0.2 nm, to the piezoelectric scan tube. This amplitude is well below the value required to initiate sliding, except as the contact initially forms and just prior to release [7]. The photodiode detector output is analyzed with a lock-in amplifier to obtain the amplitude X_t and phase α of the torsional response of the scanning force microscope cantilever. The amplitude of the shear force acting on the tip can be obtained from $|F_s| = \kappa_\Theta X_t$ if the torsional force constant κ_Θ is known. For the cantilevers we used $\kappa_\Theta \approx 67H$ where $H \approx 4$ µm is the tip height [22].

The poly(vinylethylene) (PVE) samples discussed below were made from the same 96% 1,2-polybutadiene material with 134,000 number-average molecular weight used in a previous study of the bulk mechanical properties [23]. From this study, the measured bulk relaxation time at 289 K is $\tau_b \approx 60$ µs. Young's modulus was measured by indentation to be in the range 2-4 MPa. Films of PVE were cast onto glass slides from toluene solution, dried in air for about 10-15 min, followed by vacuum drying for another 10-20 min. Since the surface properties of the PVE samples were found to vary slowly for several days following preparation [7], only freshly prepared samples were used to obtain the data presented below.

2.3 SMS MEASUREMENTS ON A VISCOELASTIC MATERIAL

Figure 2 shows the viscoelastic shear response measured on a freshly prepared PVE sample during a force-distance curve measurement [7]; i.e., during the formation, loading, unloading, and rupture of a contact [24]. The origin for the time axis is taken as the point of maximum applied load (*C*). The shear modulation amplitude was $X_o \approx 0.15$ nm. The lower graph displays the force-distance curve. The sample was moved toward the tip at constant speed starting out of contact. Jump-to-contact occurs at point *A* (t = -36 s). The approach is continued until the predetermined maximum load at *C* (t = 0 s). Then

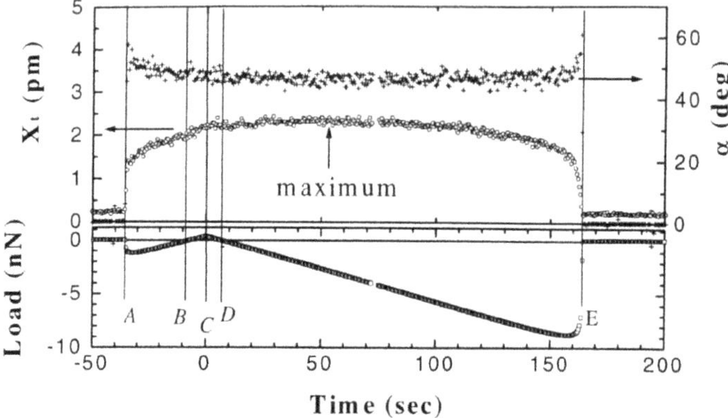

Figure 2. Shear response of a freshly prepared PVE sample during a loading and unloading cycle. The modulation frequency is 1.0 kHz and the modulation amplitude is about 0.13 nm. Lower panel: force-distance curve. Upper panel: amplitude X_t and phase lag α of tip motion on the surface. (From [7].)

the direction of motion is reversed until the tip and sample separate at the pull-off point E (t = 164 s). The rounding just prior to pull-off is typical of both macroscopic and scanning force microscope-scale viscoelastic contacts [25,26]. The magnitude of the pull-off force and the degree of rounding depend on the speed, again indicative of the viscoelastic nature of the contact. For times between points B and D, the contact was under compressive load; otherwise, the load was tensile. The force-distance curve was identical with and without shear modulation indicating that the applied shear modulation does not significantly modify the evolution of the contact.

The upper part of Fig. 2 shows the measured displacement X_t and phase α of the scanning force microscope tip. At jump-to-contact (A), X_t increases suddenly to about 1.3 pm, then more slowly to about 2.4 pm at maximum load (C). X_t continues to increase slightly until $t_{max} \approx 55$ s, long after maximum load was reached, then gradually decreases. Even at its maximum $X_t << X_o$. After the maximum, X_t decreases with the rate of decrease becoming large just as the force-distance curve begins to round significantly prior to pull-off (E). Phase response mirrors the amplitude response: α increases rapidly to about 60°, then decreases smoothly, reaching a minimum of about 43° at the same time X_t reaches t_{max}. For both amplitude and phase response, the rate of change during jump-to-contact (at A) and pull-off (at E) is limited by the overall response of the measuring system. The large decrease in phase just after A and the increase just prior to E may indicate that sliding begins just prior to pull-off.

The maximum value of X_t and the corresponding minimum in phase lag α always occur *after* the contact was under tension (i.e., $F_N < 0$), even for the slowest speeds (50 pm/s). This delayed maximum seems to be a characteristic of viscoelastic materials and was not observed for any of the elastic materials we have also studied, e.g., diamond, mica, silicone [27].

Measurements on PVE were carried out for frequencies over the range 50 Hz $\leq f \leq$ 1.2 kHz and were all qualitatively similar to the results for 1.0 kHz (Fig. 2) except that X_t and α varied with frequency as expected for a viscoelastic character of the contact [9].

3 Interpretation of Shear Modulation Spectroscopy Data

This section begins with a discussion of how the contact radius can be determined from the SMS data. The variation of the contact area during loading and unloading is then explained qualitatively in terms of creep using the formalism for contact of viscoelastic bodies developed by Ting [28]. Finally, a more quantitative description that includes interfacial adhesion is evaluated.

3.1 DETERMINATION OF THE CONTACT RADIUS

Perhaps the simplest way to extract information about the materials properties of the contact from the data like that in Fig. 2 is to model the contact using linear springs and dashpots. The simplest mechanical models of this type are the Maxwell and Voigt models [9,29]. We discuss the latter here. In the Voigt model, the viscoelastic component of the contact is described by a linear spring of stiffness κ_c in parallel with a dashpot with damping coefficient r_c. The cantilever is modeled by a linear spring of stiffness κ_Θ connected in series with the Voigt model. The stiffness of the contact is given in terms of the measurable quantities by

$$\kappa_c = \kappa_\Theta \frac{\left(\frac{X_t}{X_o}\right)\cos\alpha - \left(\frac{X_t}{X_o}\right)^2}{1+\left(\frac{X_t}{X_o}\right)^2 - 2\left(\frac{X_t}{X_o}\right)\cos\alpha} . \tag{3}$$

This result was used previously to determine the stiffness of the PVE contact and its increase to the bulk value as the sample aged [7]. The stiffness of the contact is also related to the effective storage shear modulus G^* by [4]

$$\kappa_c = 8G^* a . \tag{4}$$

Since, as pointed out above, $X_t/X_o << 1$, eqns. (3) and (4) can be combined to yield

$$a \cong \left(\frac{\kappa_\Theta}{8G^*}\right)\left(\frac{X_t}{X_o}\right)\cos\alpha . \tag{5}$$

Assuming constant G^*, eqn. (5) shows that the torsional response of the force sensor is a measure of the contact radius a. This has been pointed out previously for elastic materials [11-13]. This is an important result because the contact radius is one of the most difficult characteristics of a nanometer-scale contact to determine experimentally. Figure 3 is a plot of $(X_t/X_o)\cos\alpha$ calculated from the data in Fig. 2 for times between jump-to-contact and pull-off. The maximum load of 0.4 nN occurs at about 36 s while the maximum in $(X_t/X_o)\cos\alpha$ is at about 85 s. The shaded areas mark regions where the contact may be sliding. The solid curve is a ninth order polynomial fit to the data. Figure 3 shows that the delayed maximum in X_t implies that a reaches its maximum value well after the maximum load has been applied.

3.2 VISCOELACTICITY AND CONTACT AREA

A qualitative understanding of the origin of the delayed maximum in contact area is provided by the work of Ting [28,30]. Following Lee and Radok [31], Ting showed that

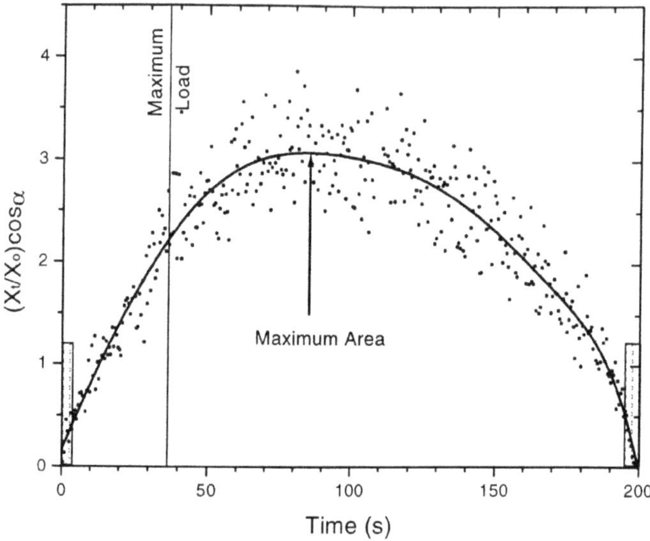

Figure 3. Area function vs. contact time. The maximum load occurs at 36 s and the maximum contact area at about 85 s. The shaded areas indicate time periods where the contact may be sliding.

the solutions of the Hertzian contact problem can be extended to linear viscoelastic materials if the elastic constants are replaced by corresponding integral operators obtained from the viscoelastic stress-strain relations. For example, if the time evolution of the contact force $F_N(t)$ is known, the contact radius is given by

$$a(t)^3 = \frac{3}{8} R \int_0^t \Phi(t-t') \frac{\partial F_n(t')}{\partial t'} dt' \tag{6}$$

where R is the radius of the contacting body and the effective elastic shear modulus G^* has been replaced by an integral operator involving the creep compliance $\Phi(t)$. In contrast to the elastic case, the contact radius now depends on the history of loading through $\partial F_N(t)/\partial t$. As discussed by Ting, care must be taken in determining the upper limit of integration in Eqn. (6) once the area has begun to decrease. Figure 4 gives an example for the case of a rigid spherical indenter on a Maxwell solid. Similar results are obtained for a Voigt model. In this example, the load (Fig. 4a) is a half-sinusoid; $F_N(t) = F_0 \cdot \sin(t/\tau)$ where τ is the relaxation time of the contact. The contact size (Fig. 4b) continues to grow by creep even after the load has begun to decrease. The time t_{max} at which the maximum contact radius occurs depends on the ratio $\Delta t/\tau$ where Δt is the total contact time. Fig. 4c shows this dependence. The general trend is for t_{max} to occur closer to the time of maximum load as Δt is increased. As Δt decreases, t_{max} approaches Δt. In Fig. 3, the maximum occurs roughly one-third of the way between the time of maximum load and pull-off. From Fig. 4c, we see that this implies that $\tau \approx \Delta t$.

The theory of Ting is too simple to provide a quantitative description of Fig. 3. It neglects both adhesion, which is responsible for the pull-off force, and dispersion forces,

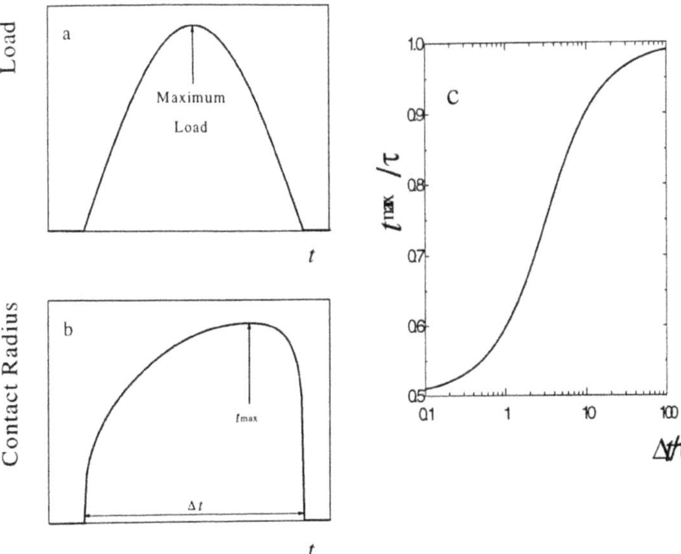

Figure 4. Contact of a rigid sphere with a Maxwell solid. (a) Sinusoidal load applied for time Δt. (b) Contact radius a as a function of contact time. (c) t_{max} as a function of the total contact time.

which cause the jump-to-contact. Additionally, freshly prepared PVE is so compliant that, even for the small forces used here, the usual approximation that the diameter of the contact is much smaller than the probe is not satisfied. Appropriate expressions for an elastic indentation by a spherical punch are given by Sneddon [32] and Maugis [33]. We know of no treatment for the case of shear. The contact radius for F_N can be estimated from a Sneddon analysis of the contact and shows that the contact radius at maximum load in Fig. 2 is approximately equal to the nominal tip radius.

Johnson, Kendall and Roberts (JKR) [34] and others [35,36] have shown how adhesion can be incorporated into the theory of elastic contact. Tirrell and co-workers [37] have recently attempted to extend JKR theory to include the effects of linear viscoelasticity. They show that the analog of eqn.(6) is

$$\frac{8}{3R}\int_0^t \Psi(t-t')\frac{\partial a^3(t')}{\partial t'}dt' = F_N(t) + 3\pi RW + \sqrt{6\pi RWF_N(t) + (3\pi RW)^2} \tag{7}$$

where W is the work of adhesion and $\Psi(t)$ is the stress relaxation function. Equation (7) is only valid as long as the contact radius is increasing; i.e., $t < t_{max}$. Equation (7) was found to give a good description of contacts between 0.7-1.2 mm diameter spheres of diblock copolymers of poly(ethylene)-poly(ethylene-propylene) [37].

We used the data for PVE to evaluate eqn. (7) for the case of nanometer-scale contacts. $F_N(t)$ was taken directly from Fig. 2. The product RW was treated as a parameter for the evaluation of the right hand side (RHS) of eqn. (7). The optimum value ($RW \approx$ 1.85 N) was taken as that which produces a discontinuity in the RHS at the measured value of the pull-off force. The nominal R of the tips used to obtain the data in Fig. 2 is expected to be in the range 20-50 nm. Thus, $RW \approx 1.85$ nN corresponds to $W \approx 40$-90

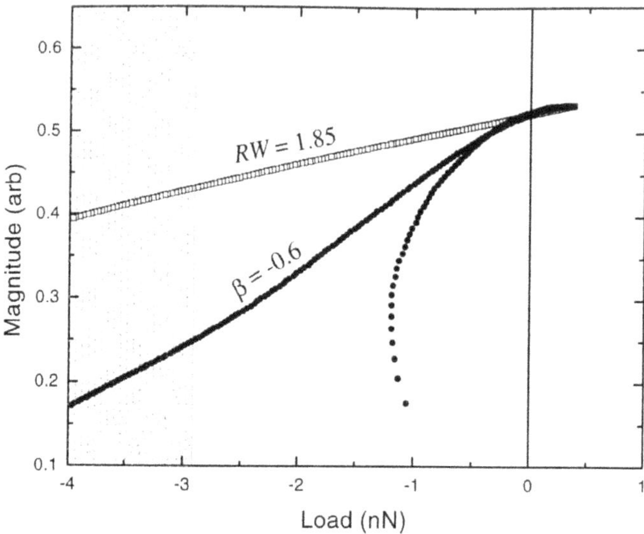

Figure 5. LHS (•'s) and RHS (□'s) of eqn. (7) as a function of the applied load. The shaded region marks the load range for which the contact area is decreasing.

mJ/m^2 which is a reasonable value [5]. The resulting RHS is plotted as the squares in Fig. 5 for the range of $F_N(t)$ for which the contact area (Fig. 3) is increasing.

The left hand side (LHS) of eqn. (7) was evaluated using the ninth order polynomial fit shown in Fig. 2 and assuming that the stress relaxation function has a power law time dependence of the form

$$\Psi(t) = G_o^* t^{\beta} \tag{8}$$

where G_o^* is zero-frequency value of the effective shear modulus and β is the power law exponent. The LHS was evaluated analytically for various values of β. The best value of β was taken as that for which the contributions to the LHS for the loading ($t < 0$ s in Fig. 2) and unloading ($t > 0$ s) parts of the force-distance curve were identical over the widest range. The resulting LHS is plotted as the dots in Fig. 5 for β = -0.6. The LHS and RHS were normalized at $F_N = 0$ N by adjusting G_o^*. The resulting $G_o^* \approx 2$ MPa is consistent with the measured bulk storage modulus assuming a Poisson ratio of 0.5.

It is certainly encouraging that physically reasonable values of the parameters result from applying eqn. (7). However, as can be seen in Fig. 5, the LHS and RHS have significantly different slopes after the maximum load. There is also a large difference in their functional forms during the initial phases of the contact formation. Clearly, the model expressed in eqn. (7) is incomplete. The analysis was also carried out using the stress relaxation functions of the Maxwell and Voigt model but no significant improvement was obtained. What is missing? First, the model assumes that both the interfacial and bulk viscoelasticity properties are identical. Since the polymer molecules are expected to take up different configurations near the surface [38,39], this may not be a good assumption, particularly since we know that the properties of the PVE samples were continuing to change with sample age [7]. Second, the interfacial adhesion has been assumed to be constant throughout. This will not be true in general, especially when the contact area is changing rapidly. Barquins [25], Barquins and Maugis [40], and

Greenwood and Johnson [41] have discussed this from the viewpoint of fracture mechanics but neglected any effects due to creep. These models have also been applied to qualitative interpretation of scanning force microscope force distance curves [26]. Whether or not the conditions required by these theories can be realized experimentally has been questioned [37]. Third, the full range of data in Fig. 3 cannot be analyzed because eqn. (7) does not apply when the contact area is decreasing. Fourth, the analysis given above is only valid if the contact radius a is much smaller than the radius R of the tip. Maugis [33] has shown how to extend JKR theory to this case for elastic, but not viscoelastic, materials. It is not known at present whether all of these effects can be brought together in a single comprehensive theory of viscoelastic contacts with adhesion.

4 Implications for MEMS

The results described in this paper provide a basis for describing the response of MEMS-scale contacts during their formation and separation. The behavior can be complex and rate dependent if adhesion and/or viscoelastic materials, such as lubricants, are involved. For viscoelastic materials, creep can cause the maximum contact area to continue to increase long after the maximum load is reached, even when adhesion is neglected. Adhesion increases the force required to break the contact and, when combined with creep, can substantially increase this pull-off force. The behavior can be complex and is not quantitatively understood at present.

The new technique of Shear Modulation Spectroscopy, described in this paper, offers the possibility to substantially improve our understanding of the mechanics of nanometer-scale contacts.

5 Acknowledgments

The authors thank P. Kleban, R. J. Colton, and K. L. Johnson for useful discussions and encouragement. We are grateful to S. V. Stepnowski for assistance with the measurements, C. M. Roland for providing the PVE samples, and M. Tirrell for providing a preprint prior to publication. The authors thank the Office of Naval Research for support. WNU also acknowledges support from the Department of Energy and the Maine Science and Technology Foundation.

6 References

1. Landman, U., Luedtke, W.D., Burnham, N.A., Colton, R.J. (1990) Atomistic Mechanisms and Dynamics of Adhesion, Nanoindentation, and Fracture, *Science* **248**, 454-461.
2. Johnson, K.L. (1997) Adhesion and friction between a smooth elastic spherical asperity and a plane surface, Proc. Roy. Soc. London A 453, 163-179.
3. Burnham, N.A. and Colton, R.J. (1993) Force Microscopy, in D.A. Bonnell (ed.), *Scanning Tunneling Microscopy and Spectroscopy*, VCH Publishers, New York, pp.191-249.
4. Johnson, K.L. (1985) *Contact Mechanics*, Cambridge University Press, Cambridge.
5. Chaudhury, M.K. (1996) Interfacial interaction between low energy surfaces, *Matl. Sci. Eng. R16*, 97-158.

6. Maboudian, R. and Howe, R. T. (1997) Critical Review: Adhesion in surface micromechanical structures, *J. Vac. Sci. Technol. B* **15**, 1-20.
7. Wahl, K.J., Stepnowski, S.V. and Unertl, W.N. (1997) Viscoelastic effects in nanometer-scale contacts under shear, *Tribology Lett.* (submitted).
8. Bhushan, B., Israelachvili, J.N., and Landman, U. (1995) Nanotribology: friction, wear and lubrication at the atomic scale, Nature 374, 607-616.
9. Ferry, J.D. (1980) *Viscoelastic Properties of Polymers*, John Wiley, New York.
10. Yamanaka, K. and Tomita, E. (1995) Lateral Force Modulation Atomic Force Microscope for Selective Imaging of Friction Forces, *Jpn. J. Appl. Phys.* **34**, 2879-2882.
11. Carpick, R.W., Ogletree, D. F. and Salmeron, M. (1997) Lateral stiffness: a new nanomechanical measurement for the determination of shear strengths with friction force microscopy, *Appl. Phys. Lett.* **70**, 1548-1550.
12. Lantz, M.A., O'Shea, A. C., Hoole, F. and Welland, M.E. (1997) Lateral stiffness of the tip and tip-sample contact in frictional force microscopy, *Appl. Phys. Lett.* **70**, 970-972.
13. Lantz, M.A., O'Shea, A. C., Welland, M.E. and K. L. Johnson (1997) Atomic-force-microscope study of contact area and friction on $NbSe_2$, *Phys. Rev. B* **55**, 10776-10785.
14. Luengo, G., Schmitt, F.J., Hill, R., and Israelachvili, J. (1997) Thin film Rheology and Tribology of Confined Polymer Melts: Contrasts with Bulk Properties, *Macromolecules* **30**, 2482-2494.
15. Granick, S. and Hu, H.W. (1994) Nanorheology of Confined Polymer Melts. 1. Linear Shear Response at strongly Adsorbing surfaces, *Langmuir* **10**, 3857-3866.
16. Georges, J. M., Tonck, A., Loubet, J.L., Mazuyer, D., Georges, E. and Sidoroff, F. (1996) Rheology and Friction of Compressed Polymer Layers Adsorbed on Solid surfaces, *J. Phys. II France* **6**, 57-76.
17. Cohen, S.R., Neubauer, G., and McClelland, GM, (1990) Nanomechanics of a Au-Ir contact using a bidirectional atomic force microscope, *J. Vac. Sci. Technol. A* **8**, 3449.
18. Burnham, N.A., Gremaud, G., Kulik, A.J., Gallo, P.J., and Oulevey, F. (1996) Materials' properties measurements: Choosing the optimum scanning probe microscope configuration, *J. Vac. Sci. Technol. B* **14**, 1308-1312.
19. Tanaka, K., Taura, A., Ge, S.R., Takahara, A., and Kajiyama, T. (1996) Molecular Weight Dependence of Surface Dynamic Viscoelastic Properties for the Monodisperse Polystyrene Film, *Macromolecules* **29**, 3040-3042.
20. Mazeran, P.E. and Loubet, J.L. (1997) Force modulation with a scanning force microscope: an analysis, *Tribology Lett.* **3**, 125-132.
21. Koleske, D.D., Lee, G.U, Gans, B.I., Lee, K.P., DiLella, D.P., Wahl, K.J., Barger, W.R., Whitman, L.J. and Colton, R.J. (1995) ,Design and calibration of a scanning force microscope for friction, adhesion, and contact potential studies, *Rev. Sci. Instrum.* **66** 4566-4574.
22. Ogletree, D.F., Carpick, R.W. and Salmeron, M. (1996) Calibration of frictional forces in atomic force microscopy, *Rev. Sci. Instrum.* **67**, 3298-3306.
23. Roland, C. M. (1994) Constraints on Local Segmental Motion in Poly(vinylethylene) Networks, *Macromolecules* **27**, 4242-4247.
24. Burnham, N.A., Colton, R.J., and Pollock, H.M. (1993) Interpretation of force curves in force microscopy, *Nanotechnology* **4**, 64-80.
25. Barquins, M. (1982) Influence of Dwell Time on the Adherence of Elastomers, *J. Adhesion* **14**, 63-82.
26. Aimé, J.P., Elkaakour, Z., Odin, C., Bouhacina, T., Michel, T., Curély, J. and Dautant, A. (1994) Comments on the use of the force mode in atomic force microscopy for polymer films, *J. Appl. Phys.* 76, 754-762.

27. K. J. Wahl, S. V. Stepnowski and W. N. Unertl (to be published).
28. Ting, T. C. T. (1966) The Contact Stresses Between a Rigid Indenter and a Viscoelastic Half-Space, *J. Appl. Mech.* **33**, 845-854.
29. Harris, C.M. (ed) (1996) *Shock and Vibration Handbook*, 4th edition, McGraw-Hill, New York, p. 2.5.
30. Ting, T.C.T. (1968) Contact Problems in the Linear Theory of Viscoelasticity, *J. Appl. Mech.* **35**, 248-254.
31. Lee, E.H. and Radok, J.R.M. (1960) The Contact Problem for Viscoelastic Bodies, *J. Appl. Mech.* **27**, 438-444.
32. Sneddon, J. N. (1965) *Int. J. Eng.* **3**, 47.
33. Maugis, D. (1995) Extension of the Johnson-Kendall-Roberts Theory of the Elastic Contact of Spheres to Large Contact Radii, *Langmuir* **11**, 679-682.
34. Johnson, K.L., Kendall, K. and Roberts, A.D. (1971) Surface energy and the contact of elastic solids, *Proc. R. Soc. London A* **324**, 301-313.
35. Derjaguin, B.V., Muller, V.M. and Toporov, Y.P. (1975) Effect of Contact Deformations on the Adhesion of Particles, *J. Colloid Interface Sci.* **53**, 314-326.
36. Greenwood, J. A. (1997) Adhesion of Elastic Spheres, *Proc. R. Soc. London* **A** 453, 1277-1297.
37. Falsafi, A., Deprez, P., Bates, F. S. and Tirrell, M. (1997) Direct Measurement of adhesion between viscoelastic polymers: A contact mechanical approach, *J. Rheology* (submitted).
38. Tanaka, K., Takahara, A., and Kajiyama, T. (1995) Surface molecular motion in thin films of poly(styrene-block-methyl methacrylate) diblock copolymer, *Acta Polymer* **46**, 476-482.
39. Mayes, A.M. (1994) Glass Transition of Amorphous Polymer Surfaces, *Macromolecules* **27**, 3114-3115.
40. Barquins, M. and Maugis, D. (1981) Tackiness of Elastomers, *J. Adhesion* **13**, 53-65.
41. Greenwood, J.A. and Johnson, K.L. (1981) The mechanics of adhesion of viscous solids, *Phil Mag. A* **43**, 697-711.

NANOTRIBOLOGY OF VAPOR-PHASE LUBRICANTS

J. KRIM AND M. ABDELMAKSOUD
Physics Department, Northeastern University
Boston, MA 02115

Abstract

Vapor-phase lubricants have traditionally been studied within the context of macroscopic system performance. Nonetheless, they may well prove to be of critical importance to the tribological performance of MEMS devices as well since the vapor phase may ultimately prove to be the most effective, if not only, means to deliver and/or replenish a lubricant in certain submicron scale devices. This work discusses how a Scanning Probe Microscope (SPM) can be used in combination with a Quartz Crystal Microbalance (QCM) to investigate the performance of vapor phase lubricants at nanometer length scales.

1. Introduction

Micro-Electro-Mechanical Systems (MEMS) are an emerging, cutting-edge technology which relies on the microfabrication of small scale mechanical components (switches, sensors, mirrors, etc.) and the integration of those components with on-board electronic processing. There is widespread belief that the future is likely to be a world dominated by MEMS and/or NEMS (Nano-Electro-Mechanical Systems) devices, which will be used in such diverse applications as gas and pressure sensors, accelerometers, chemical analytic "microlaboratories", and airborne "nanosatellites" . [1] The MEMS of today rely heavily on silicon-based materials and/or fabrication processes which were originally developed for the microelectronics industry. While such processes and materials have yielded working devices, the materials choices are largely historical and may not have resulted in optimal performance and mechanical reliability.

Because MEMS devices must react to mechanical signals, many employ construction topologies that require physical motion. Suspended plates and beams which are fabricated a few microns away from their supporting sub-

B. Bhushan (ed.), Tribology Issues and Opportunities in MEMS, 273-284.

strates are in common use, and these structures typically have relatively large areas and very small stiffnesses. These combined characteristics makes MEMS devices highly susceptible to surface forces which can cause the suspended member to deflect towards the substrate, collapse and/or adhere permanently to the substrate. This can happen during device fabrication as well as during normal operation. [2] A number of surface treatments have met with varying degrees of success for alleviating such stiction/adhesion problems, and in these cases friction and wear become the limiting factors to device operational lifetimes. With the current impetus towards device components extending well into the nanometer regime, friction/wear/adhesion complications currently encountered in MEMS devices are expected to be even more severe in NEMS devices due to the enhancement of surface effects as length scales diminish. There is thus a growing need for concomitant studies of the mechanical and tribological properties of sub-micron to molecular-scale systems, and the need to carry out new studies relevant to the development of optimal, or even operational submicron-scale mechanical systems, becomes increasingly pressing.

We discuss here how a Scanning Probe Microscope (SPM) can be used in combination with a Quartz Crystal Microbalance (QCM) for nanometer-scale investigations of the tribological performance of vapor phase lubricants under consideration for MEMS/NEMS devices. Our discussion begins by describing the role of non-lubricated solid-solid contacts in selected MEMS device failures. We then present examples of how a combined SPM/QCM can be used to probe both non-lubricated and vapor-lubricated solid-solid contacts in geometries analogous to those encountered in the actual devices. (Experimental details of the construction of the SPM/QCM apparatus itself are contained in [10]).

2. Failure of MEMS through unlubricated solid-solid contact.

When solid surfaces touch each other, the contact between them occurs predominantly at the summits of the surface roughness. [3] Because each of these contact areas is small, perhaps a few atoms in extent, both the topology and the mechanics of the contacting asperities must be investigated in order to gain basic understanding of the system's overall behavior. If the system itself is of mesoscopic or molecular extent, then an understanding of the nature of such contact zones becomes essential. Solid surface nanocontacts abound among MEMS devices, and a myriad of device complications and failures are associated with them. Here are two representative examples:

The *Digital Micromirror Device*, or DMD, is a large-scale integrated spatial light modulator which consists of an array of rotatable aluminum mirrors fabricated using semiconductor-like processing. [4] The array can consist of a a million or more independently controlled reflective, digital light switches and, like most other MEMS devices, has contacting surfaces in relative motion. The aluminum-aluminum contacts exhibit friction and wear behavior, and need to be lubricated to prevent their adhesion and permanent bonding. Working devices may be required to operate for up to 10 years and 50,000 operational hours without failure of a single mirror after device shipment! [4] While a wide variety of lubricants have been investigated for use with the DMD, little or no research has been performed concerning the nature of the metal-metal contact itself. Even though aluminum-on-aluminum represents a poor choice of materials in terms of avoiding adhesion, insufficient fundamental knowledge is available to select alternative materials in an informed manner.

An *electrostatically actuated micromechanical switch* [5]can have superior combinations of parasitic resistance and capacitance as compared with a semiconductor device, even in this day of advanced microelectronics. As such, microfabricated switches and relays could replace both conventional switches and electronic devices in many applications where signal fidelity is crucial and switching speeds are moderate. Although the literature on the performance of these devices is relatively sparse, a number of groups are actively working to fabricate them on account of the anticipated benefits. [6] With the exception of a capacitively coupled microwave switch, [7] the performance of all of these devices is limited by the performance of the contact. [8] The most common failure mechanisms are stiction at the contact point, and progressive degradation of the electrical properties due to unspecified chemical or morphological changes at the contacting interface. Traditional macroscopic approaches to contact improvement have proven irrelevant, since the physical laws which govern the performance at mesoscopic length scales are quite distinct from those of the macroscopic scale. [9] Indeed, it has become evident that the physical nature of the contact must be characterized (which is estimated to consist of approximately ten distinct nanometer-scale asperities [8]) and understood at a fundamental level in order for large device improvements to be achieved.

The metal-metal point-contact problems associated with long-term reliable operation of the DMD and the microswitch are representative of those encountered in a variety of micromechanical devices: A basic understanding of such contacts would be of great use for overall device performance optimization. The adhesion of atomic-scale metal-metal point contacts can be measured in a controlled environment with a scanning probe microscope (SPM) and a quartz crystal microbalance operating in combination [10]. The SPM allows a single asperity contact to be formed, and opens up the possibility of applying and testing existing theories of continuum contact mechanics with the properties of a nm-scale single-asperity. [11] The microbalance, whose surface is oscillating horizontally, can be employed to measure shear forces associated with the breaking of formed metal contacts, and to monitor the uptake and mechanical properties of adsorbed lubricant species. The shear strength and the work of adhesion are thus two important parameters that can be measured by this technique. The shear strength corresponds to the shear force per unit area required to shear the interface. The work of adhesion corresponds to the energy per unit area required to pull apart the interface.

While it is clear from the preceeding discussion that a QCM/SPM can provide powerful information concerning the nature of an unlubricated contact, the major advantages of combining the two techniques are realized upon introduction of a lubricant at the contact. In this case, the QCM can be employed to probe both static (thickness, density, etc...) and dynamic (viscosity, interfacial energy dissipation, etc..) properties of the lubricant that otherwise would be unattainable.

In the sections which follow, we begin with a description of how a QCM acting alone is employed to probe nanotribological properties of potential vapor-phase lubricants. Three examples of joint SPM/QCM capabilities are described in subsequent sections.

3. The Quartz Crystal Microbalance as a Probe of Atomic-scale Mechanical Properties

Studies of the fundamental origins of friction have undergone rapid progress in recent years with the development of new experimental and computational techniques for measuring and simulating friction at atomic length and time scales. [9] Employing established technologies, such as ultra-high vacuum, for the preparation of crystalline samples, nanotribologists have been gathering information in situations where the nature of the contacting surfaces is determined in advance of the measurement. They have collectively measured friction forces per unit true contact area which span twelve orders of magnitude,[9]

[12] If the precise nature of the contacting asperities between macroscopic objects in sliding contact could be determined (such studies do in fact represent an area of high research activity within the tribological community), then the results of nanotribological studies could begin to be directly implemented into mainstream tribological considerations. Meanwhile, the results can be immediately applied to solid-vapor or solid-liquid interfaces, where the complicating factors associated with asperity contacts are less of an issue, and to MEMS/NEMS related issues, where machine components with astoundingly small dimensions are rapidly approaching

the length scales routinely probed by the nanotribological community.

The concept of lubricating high temperature bearing surfaces with organic vapors has a much longer history. It has existed for at least forty years, with substantial efforts beginning in the 1980's and continuing on to the present day. [13] Vapor phase lubrication occurs via three distinct forms: (a) organic films which are intentionally reacted with a surface to form a lubricating film, (b) vapors which condense to form a lubricating liquid film on the surface of interest, and (c) light weight hydrocarbon vapors deposited onto hot catalytic nickel surfaces. Vapor phase lubricants are advantageous for use at high temperature, (meaning that either the ambient temperature of the entire system is elevated, or the local surface "flash point" temperature due to frictional heating is elevated) and in situations where the vapor can be used as a reservoir for replenishment of areas where the lubricant has been depleted in the course of the bearing lifetime.

Several organic vapors have been identified which exhibit desirable tribological properties, [14] but a detailed and fundamental understanding of their surface chemical reactions in the tribological processes of interest is far from complete. In particular, an understanding of the particular surface reactions which occur, and how they are affected by tribological conditions such as temperature and/ or lubricant concentration level in the carrier gas remains inadequate. Ultra-high vacuum and modern nanotribological techniques can be brought to bear on these issues by examining in detail the properties of a known (macroscopic-scale) vapor-phase lubricant, and the knowledge gained (if not the lubricant itself) is extremely likely to be applicable to NEMS/MEMS operations as well. Indeed, the vapor phase may ultimately prove to be the most effective, if not only, means to deliver and/or replenish a lubricant in the case of a submicron scale device. (Alternate approaches which are in development for lubrication of such devices, for example the attachment of a self-assembled monolayers to contacting surfaces, may be more difficult to initially apply, and may be difficult or impossible to replenish if damaged while the device is in operation. Such is not the case for a lubricant applied from the vapor phase, which can reattach itself continuously.)

The Quartz Crystal Microbalance (QCM) is a technique which is particularly amenable to studies of lubricant surface films and/or metal-metal contacting asperities on account of its sensitivity to both the amount of material present as well as the mechanical properties of that material. It has been used for decades for microweighing purposes,[15] and was adapted for atomic-scale friction measurements in 1986-88 by Widom and Krim.[16] [17] A QCM consists of a single crystal of quartz which oscillates in transverse shear motion with a quality factor Q near 10^5. The driving force is kept at constant magnitude at the series resonant frequency of the oscillator, typically in the range $f = 4$ to 10 MHz. Each major face is polished to better than optical flatness, and then plated with a metal electrode which is prepared by deposition in ultra-high vacuum (UHV) and can be characterized *in situ* with Auger electron spectroscopy. In general, the (111) surfaces of metals are studied, and Scanning Tunneling Microscopy (STM) measurements are performed on the surface electrodes to select suitable samples for the measurements of interest.

The quartz crystal microbalance can be an extremely sensitive probe of the mechanical properties of films which may form on its metal electrodes, and/or probe tips which it may come in contact with. In general, all parameters except for the interfacial slippage and viscosity are accounted for in the research discussed here. Measurements of potential lubricant films are carried out *in situ* by transferring the sample, within the ultra-high vacuum chamber, to a tip which can be cooled to 2K, or alternatively heated to 450 C. Vapor is then admitted to the chamber, producing a change in the resonant frequency of the microbalance which is proportional to the fraction of the mass of the vapor particles which attach to the surface and are able to track its oscillatory motion. (Sensitivities of tenths to hundreths of a monolayer are typical.)

Film adsorption onto the microbalance produces shifts in the frequency of vibration and also the amplitude of vibration if slippage of the adsorbed material is occurring. Amplitude shifts are due to frictional shear forces exerted on the surface electrode by the adsorbed film, or alternatively by a contacting metal asperity. The microbalance in its present arrangement can detect shear forces in excess of 2.5×10^{-7} N. Characteristic slip times τ, and friction coefficients (i.e. shear stresses per unit velocity) η, are determined via the relations:[16]

$$\delta(Q^{-1}) = 4\pi\tau\delta f_o \qquad \eta = \frac{\rho_2}{\tau} \tag{1}$$

where ρ_2 is the mass per unit area of the adsorbate, and τ corresponds to the time for an object's speed to fall to $1/e$ of its original value, assuming that it has been pushed at constant speed and then released, allowing frictional forces to bring it to a stop. (Amplitude shifts are converted to quality factor shifts $\delta(Q^{-1})$ through calibration with a gas which does not condense at the temperature of interest.) The "object" of interest is the film which is adsorbed on the oscillatory substrate, and τ represents an average over all film particles. Substantial decoupling effects occur whenever the slip time exceeds the period of oscillation. Partial decoupling is also quite evident for values of $\omega\tau = 2\pi f_o \tau \geq .5$ where the percent of film mass which effectively couples to the oscillation is 80% or less.

If the adsorbed material is more than molecularly thick, then the oscillating surface can in principle generate plane-parallel motion in the contacting material. The response of the oscillator in such conditions depends critically on the internal motion of the film adjacent to the electrode, as well as on whether slippage is occurring at the interface with the substrate. A variety of models have been developed to predict the oscillator response in a such an environment, incorporating the possibility of, among other things, liquid-like viscosity, multilayered structures and interfacial slip. [18] [19] [20] In the event that any of the the lubricant films to be studies exhibits such slippage/viscoelastic effects, the behaviors will be readily observable.

4. Example 1: Molecular-scale investigations of vapors which are known to lubricate at the macroscopic scale

As mentioned earlier, a variety of organic vapors have been identified which exhibit desirable tribological properties,[14] but a detailed and fundamental understanding of their surface chemical reactions in the tribological processes of interest is far from complete. In particular, an understanding of the particular surface reactions which occur, and how they are affected by tribological conditions such as temperature and/ or lubricant concentration level in the carrier gas remains inadequate. In an entirely analogous manner, lubricant materials are being utilized on current MEMS devices based primarily on performance, with incomplete understanding of the manner in which the lubricant functions at a molecular level.

As an example, we discuss here how nanotribological characterizations of a particular (macroscopic) lubricant, phenyl phosphate-based TBPP, could aid tribological performance of both macroscopic and nanometer-scale devices. This vapor-phase lubricant consists of a blend of tertiary-butyl phenyl phosphate (TBPP) molecules,[21] whose atomic constituents are carbon, hydrogen, oxygen and phosphorus. It demonstrates high quality performance at elevated temperatures [13] and exhibits oxidation inhibiting characteristics as well as a number of other desirable tribological properties, particularly when iron is present. While the precise mechanisms for its beneficial properties are uncertain, it is believed that after reacting with the surface, the phosphate contained in the original lubricant molecule acts as a binder for graphitic carbon, which in turn may be the actual lubricant. The tribological properties of TBPP are highly sensitive to its concentration levels in the carrier gas material, presumably due to a balance between the reaction rate of the vapor with the surface and the rate at which the lubricant is depleted in a particular setting.

The oxidation inhibiting properties of TBPP are presumably due to preferential adsorption of TBPP molecules over oxygen molecules, but this has not been established in a controlled environment. Assuming the phosphate binding mechanism to be valid, then a phosphate-free lubricant should exhibit markedly different tribological properties. Moreover, since the presence of iron is known to enhance the lubricating properties of TPBB, its behavior on a nonferrous substrate is expected to be distinct.

In order to examine these issues, detailed examinations of TBPP could be performed in a very straightforward manner with a joint QCM/SPM apparatus, where the electrodes of the QCM and the tip of the SPM are composed of the materials of relevance to the device performance:

(1) Measurements of the uptake rate of the lubricant as a function of exposure rate and substrate temperature (a QCM can be successfully operated up to 500 C) [22] to aid in our understanding of the sensitivity of the lubricant to concentration rates.

(2) Measurements of the oxygen uptake rate of the metal surface, to examine whether the lubricant is indeed preferentially bound to the metal surface.

(3) Auger electron spectroscopy measurements of the surface at various stages of uptake of the reactive lubricant, for monitoring of the chemical composition of the surface, for example its graphitic carbon content.

(4) Measurements of the mechanical properties of the lubricating film, in the event that slippage and/or film viscosity can be detected in response to the oscillatory motion of the QCM electrode (as described in an earlier section).

(5) SPM measurements of the lubricant surface, to determining both molecular structure and larger scale surface morphology as a function of overall film thickness, and in particular to determine whether graphitic carbon is present at the surface.

(6) SPM/QCM measurements of the drag force of the SPM tip on the oscillating QCM surface electrode.

We note that analogous studies could be performed on vapors which are known to lubricate MEMS devices to understand the lubricating mechanisms involved. Moreover, there may exist lubricants which exhibit desirable tribological characteristics for both macroscopic *and* microscopic device applications.

5. Example 2: Nanomechanical properties of potential lubricants

In addition to probing the molecular-scale properties of vapors which are known lubricants, a combined SPM/QCM is itself a simple nanometer-scale mechanical system whose response to vapor-phase adsorbents can be monitored for nanotribological performance. In particular, the apparatus can be employed to examine the effect of the rubbing action of the scan tip on the QCM can induce tribochemical reactions which result in the formation of reaction products with desirable tribological properties. As an example, we assume graphitic carbon to have desirable tribological properties [23] and discuss the particular case of ethylene/platinum.

The tribological properties of ethylene/Pt are unsubstantiated. Nonetheless, its temperature dependence and structure has been thoroughly characterized by the surface science community. The adsorption and reaction products of ethylene adsorbed in ultra high vacuum conditions (UHV) on Pt(111) have been characterized by a wide variety of surface science techniques. [24] These studies have determined that ethylene (C_2H_4) bonds molecularly to the Pt surface at temperatures up to about 200K. Beginning at about 230K, ethylidyne ($C\text{-}CH_3$) is formed by loss of hydrogen and this species is stable up to 430K. Upon annealing above 430 K, further dehydrogenation occurs until, above 700K, only carbon is left on the surface. Annealing this carbon covered surface to temperatures above 800 K results in the formation of graphite islands on the Pt surface. The entire sequence can be imaged and identified by STM. Although the formation of graphitic carbon on an open Pt surface occurs at elevated temperatures, it may well occur at much reduced temperatures in the presence of a sliding contact, where "flash point" temperatures are frequently much higher than ambient.

The following studies could be performed on this potential lubricant, where the SPM is taken as a scanning tunneling microscope employed for atomic-scale surface characterizations of morphology or chemical species identification. The QCM is utilized for simultaneous measurements of the dynamical properties of the adsorbed species and to measure the drag force which the SPM tip exerts on the surface of the QCM surface when the QCM is oscillating.

A QCM with Pt(111) electrodes can be dosed with ethylene in UHV conditions and the frequency and quality factor shifts of the microbalance upon uptake of the gas will be monitored to determine the ethylene film's mass per unit area and dynamical properties. The QCM oscillatory motion can then be halted, to allow the surface to be imaged with the STM in order to identify the structure of the chemical species which is present at the tip-substrate interface. The QCM then resumes its (horizontal) oscillation with a STM tip in (tunneling) contact with the surface, allowing tip-surface sliding speeds to be achieved (up to 10 cm/s) which are as much as five orders of magnitude greater than those typical of STM scans on stationary surfaces. The QCM frequency and amplitude during this process are monitored to measure the drag force of the STM tip along the surface, a parameter which is likely to be chemically sensitive. The oscillation will then be halted and the surface reimaged investigate whether the same chemical species is still present or whether a tribologically induced reaction has occurred, the most likely reaction product being graphitic carbon.

By performing such experiments at varying tip-substrate sliding speeds, sliding times and temperatures, a sliding speed vs. interfacial contact temperature vs. sliding time phase diagram can be mapped out. Repeating the measurements with a friction force microscope rather than an STM would in addition allow for investigations of the lateral force as a function of the normal force applied to the scan tip.

6. Example 3: Sliding friction/molecular transport

The increased recent interest in the fundamental origins of friction has sparked a variety of discussions and debates concerning the nature of the atomic-scale mechanisms which dominate the dissipative process by which mechanical energy is transformed into heat. While much recent attention has focused on phonon contributions to friction on account of the relative ease with which atomic lattice vibrations can be modelled by molecular dynamics simulations, [25] Electronic mechanisms for friction due to sliding-induced excitations of conduction electrons at metallic interfaces have also been put forward. Such mechanisms are quite distinct from the friction associated with static charge buildup, a theoretical problem which, from Coulomb's original work on friction until the present time, remains unsolved.

We have recently observed direct experimental evidence for electronic contributions to friction [26] by investigating the friction associated with sliding of an arbitrary material (nitrogen) along an arbitrary superconductor (lead) above and below the metal's superconducting transition temperature. We observed the friction associated with nitrogen sliding over lead in its superconducting state to be approximately half that observed for its normal state. The reduction in friction is best explained in terms of an abrupt drop in conduction electron contributions to the friction when the lead enters the superconducting state, leaving phonon excitations as the sole probable mechanism for energy dissipation. The result implies that a variety of intriguing applications may be possible, in those systems where electronic contributions are found to be significant. For example, systems characterized by substantial enough electronic friction forces would allow surface transport of adsorbed atoms or molecules via the drag forces exerted by an electrical current. The surface transport might be employed for delivery of lubricant molecules, or for transport of molecule which have been chemically modified by a SPM to a specified collection point.

These particular applications are perhaps years away. Nonetheless, they demonstrate how our current knowledge of friction at the atomic scale can allow us to be able to predict, at the fundamental level, which interfaces will be characterized by reduced friction levels.

Acknowledgments

This work has been supported by NSF grants DMR9625921 and CMS9634374

References

[1]Spencer, N.D. , Tysoe, W.T. and Maboudian, R. (1997),*Tribology Letters*, **3**,preface

[2]Maboudian, R. and Howe, R.T.(1997), Stiction reduction processes for surface micromachines, *Tribology Letters*, **3**,215-221

[3]Greenwood, J.A. (1992), *in Fundamentals of Friction: Macroscopic and Microscopic Processes.* Singer, I.L. and Pollock, H.M. ,eds. (Kluwer, Dordrecht).

[4]Henck, S.A. (1997), *Tribology Letters*,**3**,239

[5]Zavracky, P.M. , Majumder, S. and McGruer, N.E., (1997), Micromechanical switches fabricated using nickel surface micromachining, *J. Electromechanical Systems*, **6**, 3

[6]Sakata, M. (1989), *Proc. IEEE MEMS Workshop '89 (Salt Lake City, UT),149*

[7]Randall, J.N. , Goldsmith, C. , Denniston, DT.-H. (1996) fabrication of micromechanical switches for routing radio frequency signals, *J. Vac. Sci. Technol.* ,**14B**,3692-3696

[8]Majumder, S. , McGruer, N.E. ,P. , Adams, G.G. , Morrison, R.H. and Krim, J. (in press) Measurement and modeling of surface micromachined, electrostatically actuated microswitches , *Transducers '97*

[9]Krim, J. (1996) Friction at the atomic scale, *Scientific American* , **275**,74-80

[10]Krim, J. , Dayo, A. and Daly, C. (Plenum Press, New York, 1994) Combined scanning tunntling micoscope and quartz microbalanc study of molecularly thin water layers , *Atomic Force Microscopy/Scanning Tunneling Microscopy , edited by S.H. Cohen* 211-215

[11]Agrait, N. , Rubio, G. and Vieira, S. (1996) Plastic deformation in nanometer scale contacts , *Langmuir*, **12**,4505-4509

[12]Krim, J. (1995) Progress in Nanotribology: Experimental Probes of Atomic Scale Friction,*Comments in Condensed Matter Physics*,**17**,263-280 (Invited Review)

[13]Forster, N.H. and Trivedi, H.K. (1997) Rolling contact testing of vapor phase lubricants - part I: material evaluation,*Trib. Trans.*,**40**,421-428

[14]Faut, O.D. and Wheeler, D.R. (1983) On the mechanism of lubrication by Tricresylphosphate (TCP)-The coefficient of friction as a function of temperature for TCP on M-SO steal ,*A.S.L.E. Trans.*,**26**,344-350

[15]Lu, C. and Czanderna, A. (1984),*Applications of Piezoelecic Quartz Crystal Microbalances*,eds. (Elsevier,Amesterdam)

[16]Krim J. and Widom, A. (1988) Damping of a crystal oscillator by an adsorbed monolayer and its relation to interfacial viscosity,*Phys. Rev.* , **38B**,12184-12189; Widom, A. and Krim, J. (1986) Q factors of quartz oscillator modes as a probe of submonolayer-film dynamics,*Phys. Rev.* ,**34B**,1403-1404

[17]Watts, E.T. , Krim, J. and Widom, A. (1990) Experimental observation of interfacial slippage at the boundary of molecularly thin films with gold substrates, *Phys. Rev.*,**41B**,3466-3472

[18]Lea, M.J. and Fozooni, P. (May,1985) The transverse acoustic impedance of an inhomogeneous viscous fluid, *Ultrasonics*,**23**,133-137

[19]Duncan-Hewitt, W.C. and Thompson, M. (1992) Four-layer theory for the acoustic shear wave sensor in liquids incorporating interfacial slip and liquid structure,*Anal. Chem.*,**64**,94-105

[20]Yang, M. , Thompson, M. and Duncan-Hewitt, W.C. (1993) Interfacial properties and the response of the thickness-shear-mode acoustic wave sensor in liquids ,*Langmuir* ,**9**,802-811

[21]Marino, M.P. and Placek, D.G. (CRC Press, Boca Raton, 1994),*CRC Handbook of Lubrication and Tribology Vol.III,R.R. Booser, ed.*, 269-286

[22]Mecea, V.M. , Carlsson, J.O. , Heszler, P. and Bartan, M. (1995) Development and testing of a high temperature quartz crystal microbalance, *Vacuum*,**46**,691-694
[23]We emphasize that this is an assumption, and not an established atomic-scale result.
[24]Land, T.A. , Michely, T. , Behm, R.J. , Menninger, J.C. and Comsa, G. (1991) STM investigation of the adsorption and temperature dependent reactions of ethylene on Pt(111) ,*Appl. Physics A, Solids and Surfaces*, **53**,414-417; Land, T.A. , Michely, T. , Behm, R.J. , Hemminger, J.C. and G. Compsa (1992) Direct observation of surface reactions by scanning tunneling microscopy : Ethylene—ethylidyne—carbon particles—grphite on Pt(111), *J. Chem. Phys.* ,**97**,6774-6783
[25]]Cieplak, M. , Smith, E.D. and Robbins, M.O. (1994) Molecular origins of frction: The force on adsorbed layers, *Science*,**265**, 1209-1212
[26]Dayo, A. , Alnasrallah, W. and Krim, J. (in press) Superconducting-dependent sliding friction,*Phys. Rev. Lett.*

The Tribology of Hydrocarbon Surfaces Investigated using Molecular Dynamics

JUDITH A. HARRISON AND STEVEN J. STUART
United States Naval Academy
Chemistry Department
Annapolis, MD 21402

MARTIN D. PERRY
Arkansas Technical University
School of Physical and Life Sciences
Russellville, AR 72801

Abstract

Diamond's potential usefulness as a material for MEMS has renewed interest in the atomic-scale friction and wear properties of diamond. With this in mind, molecular dynamics simulations have been used to examine the tribology of diamond surfaces in contact under a number of experimentally relevant conditions. Friction was examined as a function of applied load, in systems with hydrocarbon chains chemisorbed to one diamond surface, and in systems with molecules trapped between the contacting surfaces. The origin of friction was shown to be directly related to the amount of energy dissipated during sliding. Differences in the friction were correlated to the differences in simulation systems and atomic-scale sliding mechanisms. Conditions that lead to the wear of the diamond surfaces and specific tribochemistry of the trapped molecules were elucidated.

1. Introduction

The advent of new technology, capable of investigating friction and wear on the atomic scale, has shifted the focus of the experimental examination of friction from the macroscopic to the microscopic scale. The atomic force microscope (AFM) (Carpick *et al.*, 1997; Mate *et al.*, 1987; Sheehan *et al.*, 1996), the surface force apparatus (SFA) (Israelachvili *et al.*, 1988; Van Alsten *et al.*, 1988), and the quartz crystal microbalance (Krim *et al.*, 1991) have all been used to examine atomic-scale friction in a number of different systems. Advances in computer technology have allowed for complementary theoretical investigations of atomic-scale friction and wear (Harrison *et al.*, 1995a; Hirano *et al.*, 1991; Sokoloff, 1984, 1996; Zhong *et al.*, 1990; Cieplak *et al.*, 1994; Smith *et al.*, 1996; Sorensen *et al.*, 1996). In some cases, theory and experiment agree quite well. For example, the wearless atomic-scale friction of diamond has been investigated using both the AFM (Germann *et al.*, 1993) and molecular dynamics (MD) simulations (Perry *et al.*, 1995b). The MD simulations showed that the friction for sliding on the (100) face of diamond is approximately the same as it is for sliding on the (111) face of diamond, in agreement with experimental observations.

Because diamond possesses unique friction and wear properties, the friction of diamond has historically been of great interest (Tabor *et al.*, 1992). The development of chemical vapor deposition (CVD) technology has renewed interest in the atomic-scale frictional properties of diamond and carbon coatings. More recently, there has been a rapid growth in the field of micromachining (Bhushan, 1996a). Because of the advanced

B. Bhushan (ed.), Tribology Issues and Opportunities in MEMS, 285-299.

microfabrication technology developed for use in the production of silicon microprocessors, the fabrication of microelectromechanical systems (MEMS) has been almost exclusively limited to silicon and silicon-based compounds. However, the applicability of MEMS made of silicon is limited due to the physical characteristics of silicon (Hunn *et al.*, 1994). Diamond's low friction, high thermal conductivity, and resistance to abrasion make it a physically superior material for MEMS (Jansen *et al.*, 1996). In the past, the major roadblock to the use of diamond in MEMS was the difficulty involved in generating microstructures of diamond. It is now possible to pattern and then selectively etch diamond crystals, thereby generating microstructures of diamond (Hunn *et al.*, 1994). Thus, the use of diamond for MEMS applications is becoming a more likely prospect. Because these devices are small, on the order of hundreds of microns, the atomic-scale mechanisms of friction and wear in the MEMS materials must be understood.

With these things in mind, we have been using MD simulations to examine the atomic-scale tribology of diamond under a number of different conditions. The atomic-scale friction and wear between two diamond (111) surfaces has been investigated as a function of crystallographic sliding direction, applied load, and sliding speed. In addition, we have repeated these simulations with hydrocarbon groups chemisorbed to one of the diamond surfaces or with third-body molecules trapped between the surfaces. These simulations have yielded new insight into the way energy is dissipated during sliding and into specific tribochemical reactions that occur during sliding. This knowledge might ultimately aid in the design of components for MEMS or diamond coatings with tailored friction and wear properties.

2. Methods and Procedures

The MD sliding simulations are briefly described below, the details having been given elsewhere (Perry *et al.*, 1997). A typical starting configuration showing two hydrogen-terminated diamond (111) surfaces is shown in the left panel of Figure 1. Systems containing chemisorbed groups are generated by replacing one or more of the hydrogen atoms on the upper diamond surface with methyl, ethyl, or *n*-propyl groups. (These systems will be referred to as the methyl-terminated, ethyl-terminated, and *n*-propyl-terminated systems, respectively.) Systems containing third-body molecules are generated by adding methane, ethane, or isobutane molecules between the two diamond surfaces. (These systems will be referred to as the methane, ethane, and isobutane systems, respectively.) Friction is simulated by moving the rigid-layer atoms (topmost two or three layers) in the chosen sliding direction. The equations of motion are not integrated for these layers or the bottom two layers of the lower surface. A thermostat (Berendsen *et al.*, 1984) is applied to the middle five layers of each lattice to maintain the temperature of the system at 300 K. All other atoms are allowed to dynamically evolve in time according to Newton's equations of motion. The forces governing the motion of the atoms are derived from a reactive empirical bond-order potential (Brenner, 1990; Brenner *et al.*, 1991). Because the outermost layers of the lattice are held rigid, their relative positions are used to define the distance between the two surfaces. The applied (normal) load is increased by decrementing the distance between the rigid layers. Unless otherwise indicated, the sliding speed and direction are 100 m/s and the $[11\bar{2}]$ crystallographic direction, respectively. Periodic boundary conditions were applied in the plane of the contacting surfaces to simulate an infinite interface.

3. Friction and Energy Dissipation of Diamond

Diamond surfaces typically used in experiments are rough on the atomic scale. With this in mind, sliding simulations are performed with hydrocarbon groups chemisorbed to one of the contacting diamond surfaces. These groups may model hydrocarbon groups that remain on the surface of diamond films grown by CVD. In addition, experiments have shown that debris present between diamond surfaces during sliding has a profound effect on the observed friction. Thus, the friction between diamond surfaces in the presence of debris or third-body molecules is also examined.

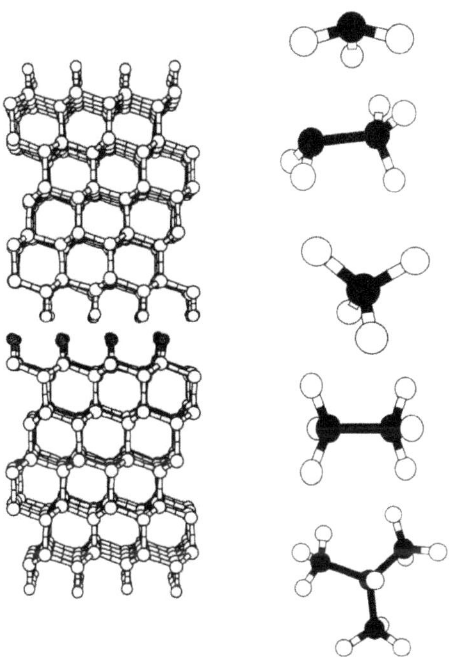

Figure 1.

Representation of two hydrogen-terminated diamond (111) surfaces in contact viewed along the [$\bar{1}$10] direction (left panel). This system is composed of 24 layers of 16 atoms each. The right panel shows representations of some of the chemisorbed groups and third-body molecules that are examined. From the top these are, methyl groups, ethyl groups, methane molecules, ethane molecules, and isobutane molecules. Large and small spheres represent carbon and hydrogen atoms, respectively. The carbon atoms in the chemisorbed groups and the third-body molecules are colored dark gray. Hydrogen atoms from one of the diamond surfaces are colored light gray so that they may be tracked throughout the course of the simulations.

For all the systems examined, the average frictional force as a function of average normal load is shown in Figure 2. In general, the frictional force increases as a function of normal load for all these systems. However, as the load is increased the presence of both the larger chemisorbed groups and the third-body molecules markedly reduces the average frictional force compared to the atomically flat hydrogen-terminated system. In addition, the average frictional force of all the third-body molecule systems is lower than the chemisorbed-molecule systems. These observations can be explained by examining the atomic-scale sliding mechanisms and the associated pathways of energy dissipation in these systems.

3.1 HYDROGEN-TERMINATED DIAMOND

When two hydrogen-terminated diamond lattices are in sliding contact the hydrogen atoms on opposing surfaces interact with one another. For example, when the hydrogen atoms on opposing diamond surfaces are aligned, movement in the sliding direction will bring them into head-on contact with one another. Analysis of video sequences of the dynamics reveals that as the hydrogen atoms on opposing surfaces first begin to interact (or "collide") they "revolve" around each other in the sliding plane. This motion causes oscillations, with the periodicity of the unit cell, in the normal and frictional forces as a function of sliding distance. It becomes increasingly difficult for the hydrogen atoms to "revolve" or "pass by" one another as the load is increased. Thus, the amplitude of the oscillations in the frictional and normal forces increases (Harrison *et al.*, 1992).

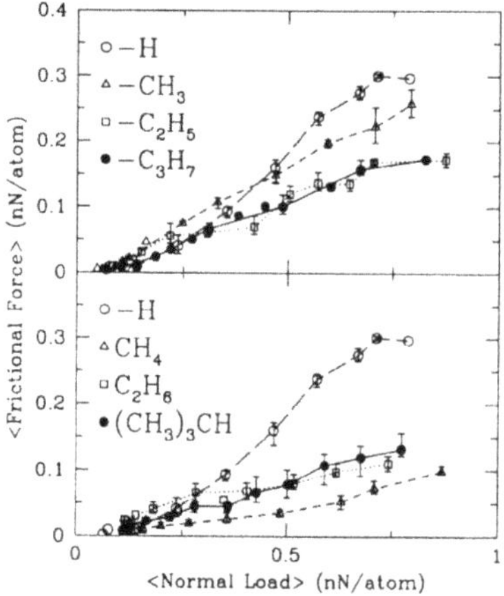

Figure 2.

Average frictional force per rigid-layer atom as a function of average normal load per rigid-layer atom. Data for the chemisorbed groups and third-body molecule systems are shown in the upper and lower panels, respectively. The hydrogen-terminated system contains 48 rigid-layer atoms, all other systems contain 32.The error bars on selected data points represent the range of average frictional force values obtained from five simulations with independent starting configurations (Perry et al., 1997). Lines have been drawn to aid the eye.

Animated sequences of the sliding simulations illustrate that the "collisions" suffered by the interfacial hydrogen atoms result in mechanical excitation of both the hydrogen atoms and subsurface atoms. Computation of the vibrational energy between the hydrogen layer and the first carbon layer of the diamond lattices as a function of sliding distance (Fig. 3) confirms this behavior. Peaks in the vibrational energy occur directly after the hydrogen atoms revolve around one another. (At high loads, the opposing diamond surfaces become "stuck" then "slip" past one another a distance of one unit cell. During the slip, the hydrogen atoms on opposing surfaces move past one another very quickly.) The vibrational energy is transferred to the rest of the lattice and eventually dissipated by the thermostat (Harrison *et al.*, 1995b). This conversion of the work of sliding into heat dissipated by the bulk is the essence of wearless atomic-scale friction. Because more work is required for the hydrogen atoms to pass by one another as the load is increased, the amount of vibrational excitation in the diamond lattices increases with increasing load. Thus, more energy is dissipated to the bulk in the form of friction.

3.2 CHEMISORBED GROUPS

Because the average frictional force as a function of load is approximately the same in the methyl-terminated system as it is in the hydrogen-terminated system (Fig. 2), approximately the same amount of energy must be dissipated by the lattice. That is, sliding must initiate approximately the same amount of vibrational excitation in both systems for a given load. Examination of the vibrational energy between the interface carbon-hydrogen bonds as a function of sliding distance for these two systems, at similar loads, confirms this conclusion (Perry *et al.*, 1995a). The chemisorbed methyl groups initiate approximately the same amount of mechanical excitation in the diamond lattices because their small size allows them to act in much the same way as the hydrogen atoms they replaced. That is, the methyl groups collide with hydrogen atoms on the opposing surface. In addition, because methyl groups are small, and fit well into the valleys on the (111) surface, they do not reduce the interaction between the two diamond surfaces significantly (Fig. 4). In addition to exciting the carbon-hydrogen bonds in the diamond lattice, the methyl groups are able to rotate like turnstiles as they interact with the hydrogen atoms on the opposing surface. This motion introduces a novel mechanism of energy dissipation, although one that appears to have little influence on the friction.

Lengthening the chemisorbed hydrocarbon chain to two carbon atoms (ethyl) or three carbon atoms (*n*-propyl) reduces the average frictional force, particularly at higher loads (Fig. 2). The size of these groups allows them to act as a boundary layer between the hydrogen atoms on opposing diamond surfaces. The plot of the average frictional force versus interface separation (Fig. 4) clearly shows that the larger hydrocarbon groups force the interface layers of the diamond further apart at a given value of the normal force. Thus, the majority of the frictional force must arise from the interaction of the chemisorbed group with the opposing diamond surface. Because the chemisorbed groups do not cover the entire surface, the total frictional force is lower in these systems. In addition, the flexibility of the longer hydrocarbon chains, in conjunction with the conformations they adopt while sliding, allows these groups to interact less with the opposing diamond surface at higher loads than the smaller methyl groups. For instance, the carbon-carbon bond of a chemisorbed ethyl group, or the chain's backbone, is not usually parallel to the sliding direction at the start of the simulation. However, when the load is small, sliding induces alignment of the chain's backbone in the sliding direction. The tail of the ethyl group then interacts directly with the hydrogen atoms on the opposing surface. Therefore, for average normal load values of approximately 0.4 nN/atom or less the average frictional force is approximately the same for the hydrogen-terminated, methyl-terminated, and ethyl-terminated systems.

At loads above approximately 0.4 nN/atom, however, the backbone of the ethyl group does not align itself in the sliding direction during the course of the simulation. The potential energy associated with passing directly over hydrogen atoms is too great, and the tails of the chemisorbed ethyl and *n*-propyl groups realign to occupy the regions between hydrogen atoms on the lower diamond surface during sliding. Because the tail portion of the chemisorbed group interacts less with the opposing diamond surface, less mechanical excitation in the diamond is induced; therefore, the average frictional force is lower than in the methyl- and hydrogen-terminated systems (Harrison *et al.*, 1995b; Harrison *et al.*, 1993).

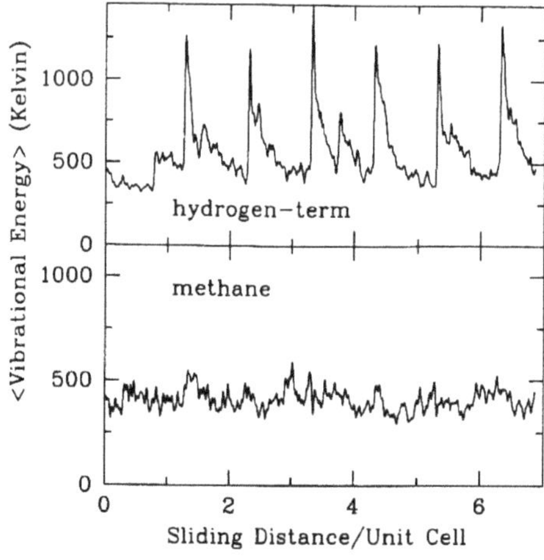

Figure 3.

Average vibrational energy in the carbon-hydrogen bonds of the upper (sliding) diamond surface as a function of sliding distance. (The unit-cell distance in the [11$\bar{2}$] direction is 4.357 Å.) Data for the hydrogen-terminated system and the methane system are shown in the upper and lower panels, respectively. For the hydrogen-terminated system, the average normal load and the average frictional force are 0.79 nN/atom and 0.32 nN/atom, respectively. For the methane system, the average normal load and average frictional force are 0.85 nN/atom and 0.094 nN/atom, respectively (Perry et. al., 1996b).

3.3 THIRD-BODY MOLECULES

In general, the systems with third-body molecules also reduce friction by acting as a boundary layer between the opposing diamond surfaces. At a given load, the separation of the hydrogen atoms on opposing diamond surfaces is always significantly larger when the third-body molecules are present compared to the separation in the absence of these molecules (Fig. 4). As a result, the direct interaction of the hydrogen atoms on opposing diamond surfaces is reduced to an almost negligible amount. Therefore, the observed friction must arise from the interaction of the third-body molecules with the diamond surfaces. Because the frictional force as a function of load is approximately the same in the ethane and isobutane containing systems (Fig. 2), approximately the same amount of energy must be dissipated while sliding. In contrast, the lower frictional force in the methane system is the result of less energy being dissipated during sliding. Thus, the methane molecules must interact to a lesser extent with the confining diamond surfaces than the other third-body molecules during sliding.

The motion of the third-body molecules and the extent of their interactions with the confining diamond surfaces are dictated by the molecule's size, its shape, and the relative positioning of the diamond surfaces. The motion of the third-body molecules during an individual sliding simulation is ascertained by analyzing animated sequences of individual sliding simulations, plots of frictional force versus sliding distance (friction traces), and calculated various quantities related to the third-body molecule itself. In this way, the motion of the third-body molecules, the extent of their interaction with the confining diamond surfaces, and the effects of confinement and sliding on the third-body molecules can be fully characterized.

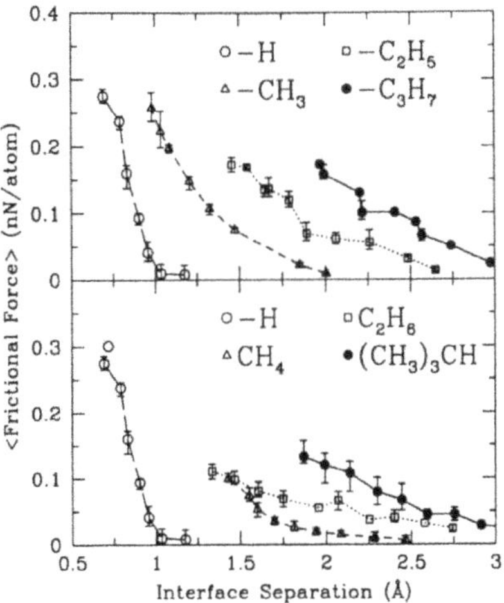

Figure 4.

Average frictional force per rigid-layer atom as a function of interface separation. These data for the chemisorbed groups and third-body molecule systems are shown in the upper and lower panels, respectively. The hydrogen-terminated system contains 48 rigid-layer atoms, all other systems contain 32. Error bars on selected data points represent the range of average frictional force values obtained from five simulations that have independent starting configurations (Perry et al., 1997). Lines have been drawn to aid the eye.

The spherical shape and small size of the trapped methane molecules cause the motion of the molecule during sliding to be similar regardless of the load or diamond surface alignment. The frictional force as a function of sliding distance is periodic. The spacing of hydrogen atoms on the confining surfaces allows the small methane molecule to avoid strong interactions with hydrogen atoms on the diamond by traveling in the regions between the hydrogen atoms (Perry *et al.*, 1996b). Instead of moving parallel to the sliding direction as attached groups are required to do, the methane molecules move at a 45° angle to the sliding direction in the potential energy valleys between rows of hydrogen atoms on the (111) diamond surface. Virtually no vibrational excitation is initiated in the diamond during sliding (Fig. 3) because strong interactions with the hydrogen atoms on the diamond surfaces are avoided. As a result, the average frictional force is very small at all the loads examined.

In contrast, in the ethane and isobutane systems, the shape and periodicity of the friction traces varies with load, alignment of the surfaces, and the orientations these molecules adopt while sliding. Several initial orientations of these molecules, relative to the sliding direction, are possible because ethane and isobutane are nonspherical molecules. For example, we use the angle between the carbon-carbon bond in ethane and a line parallel to the sliding direction to define the ethane molecule's orientation. Both ethane and isobutane are too large to move between hydrogen atoms on the confined diamond surfaces as the methane molecules do. As a result, more mechanical excitation of the diamond lattices is initiated, and so more energy is dissipated by the larger third-body molecules than when the methane molecule is present. The magnitude of the excitation depends on the molecule's trajectory, which depends somewhat on its starting configuration, and the alignment of the opposing surfaces.

One way to characterize the fate of a confined third-body molecule is to calculate the molecule's potential energy, vibrational energy, rotational energy, and translational energy during the sliding. These "signature" quantities are shown in Fig. 5 for a simulation in which ethane is the third-body molecule and the upper diamond surface is moved in the [110] crystallographic direction. Several important aspects of this trajectory are apparent from these data. First, because all the signature quantities are periodic, repeating once every unit cell length, the motion of the confined molecule must be regular. In addition, the maxima in the translational energy plot indicate that the molecule does not move smoothly between the two confining surfaces during sliding. Rather, the molecule is stationary for a certain amount of time, with stresses building up in the molecule (an increasing potential energy trace) as the upper surface moves, until the molecule is able to move past the obstacles (hydrogen atoms) in its path. The peaks in the translational energy correspond to the movement of the ethane molecule. Peaks in the plot of vibrational energy of the molecule indicate that the third-body molecule is being excited vibrationally as a result of the sliding. This excess energy is then transferred to the confining diamond surfaces and eventually dissipated by the thermostat.

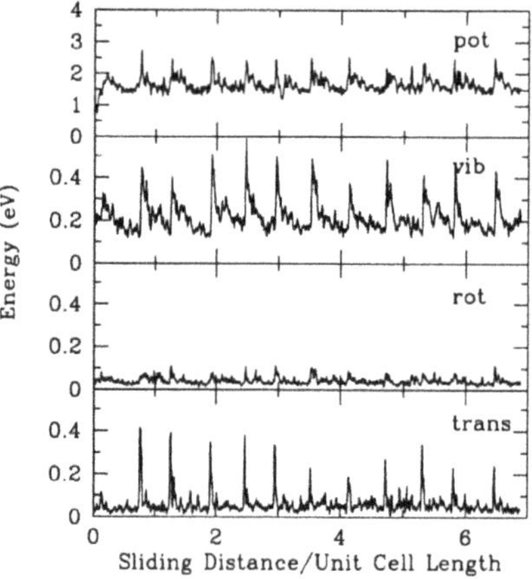

Figure 5.

Energy of the trapped ethane molecule as a function of sliding distance for sliding the upper diamond surface in the [1$\bar{1}$0] crystallographic direction. The average normal force and frictional force values are 0.68 nN/atom and 0.15 nN/atom, respectively. The unit cell length in the [1$\bar{1}$0] direction is 2.5157 Å.

4. Wear and Tribochemistry

4.1 THIRD-BODY MOLECULES

Analysis of the potential energy of third-body molecules during the particular run described by Figure 5, shows that while the ethane molecule is slightly distorted during sliding (as evidenced by the maxima in the potential energy signature), it remains intact throughout the course of the simulation. This is not always the case. Occasionally the stress on the ethane molecule caused by the sliding is large enough to break a chemical bond within the molecule. Bond breaking is usually observed for loads greater than approximately 0.65 nN/atom in this system. For example, a second simulation was run using the same starting configuration that yielded the data shown in Figure 5. However, in this case, the upper diamond surface was slid in the [11$\bar{2}$] crystallographic direction. The "signature" data for the ethane molecule in this simulation are shown in Figure 6. These data are not as periodic and regular as they are when sliding in the [1$\bar{1}$0] direction. That is, while the molecule is swept along by the sliding motion, the molecule becomes stuck for variable periods of time causing the peaks in the translational energy plot to be spaced at irregular intervals. The periodic peaks in the molecule's vibrational energy indicate that the molecule still dissipates energy to the contacting diamond surfaces. Lastly, the potential energy trace indicates that the ethane molecule is deformed by the sliding. In fact, after sliding approximately 5 unit cells, the ethane molecule undergoes a chemical reaction. This is apparent from the very large value of potential energy.

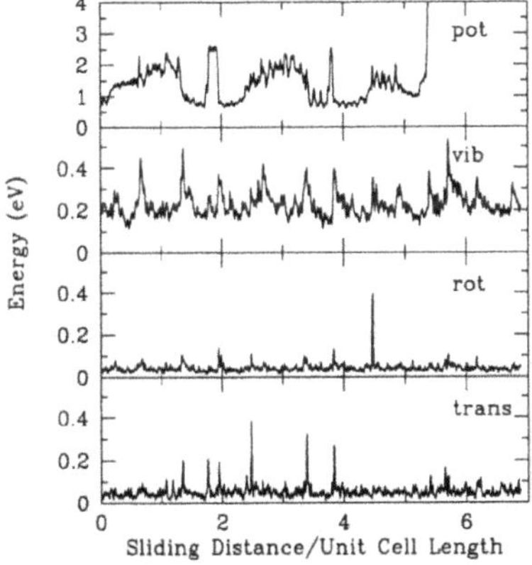

Figure 6.

Energy of the trapped ethane molecule as a function of sliding distance for sliding the upper diamond surface in the [11$\bar{2}$] crystallographic direction. The average normal force and frictional force values are 0.65 nN/atom and 0.091 nN/atom, respectively. The unit cell length in the [11$\bar{2}$] direction is 4.357 Å.

As the upper diamond surface slides in the [112] direction, significant stresses build up on the trapped ethane molecule (Fig. 7, upper left panel). In this case, the stress on a carbon-hydrogen bond of the ethane molecule becomes large enough to break the chemical bond. That is, shear-induced bond scission occurs (Fig. 7, upper right panel). This bond-breaking event results in a hydrogen atom and a C_2H_5 radical trapped at the sliding interface. These two species are very reactive and initiate a number of other tribochemical reactions.

First, the trapped hydrogen atom comes into close proximity with a hydrogen atom on the lower diamond surface. The hydrogen atom on the diamond surface is abstracted to form a hydrogen molecule, H_2, and a surface radical site (Fig. 7, lower left panel). Because the hydrogen molecule is stable, and small, it avoids subsequent chemical reactions during the remainder of the simulation. This is not the case for the reactive C_2H_5 radical. The sliding process causes this radical to be in close proximity to the radical on the diamond surface that was created by the hydrogen atom abstraction. The C_2H_5 radical then forms a bond with the lower diamond surface (Fig. 7, lower right panel). The system is now identical to the ethyl-terminated systems (Fig. 1) except for the hydrogen molecule trapped at the interface. It should also be noted that because the sliding results in different behavior when sliding in different crystallographic directions in the ethane system, the friction and wear are anisotropic.

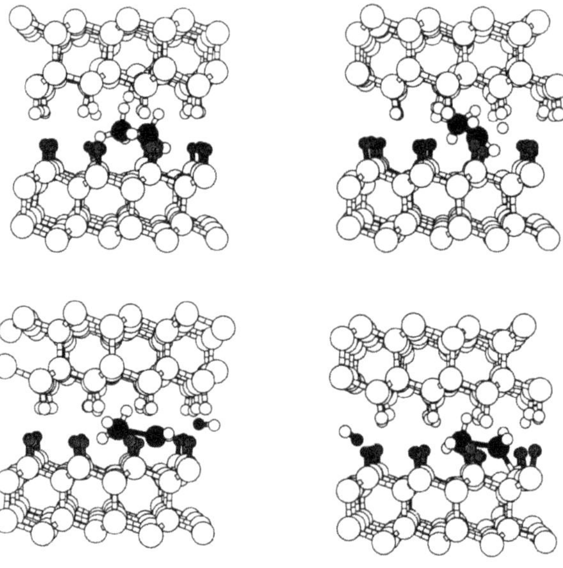

Figure 7.

Simulation configurations at different times for a system with an ethane molecule at the interface. These configurations are from the same simulation that yielded the data shown in Figure 6. In the upper left panel the ethane molecule is visible. In the upper right panel a hydrogen atom has been worn (sheared) from the ethane molecule (visible at the right of the panel). In the lower left panel the free hydrogen atom has abstracted a hydrogen atom from diamond to form H_2 (visible at the right of the panel). In the lower right panel the C_2H_5 radical has added to the diamond surface to form a system much like the ethyl-terminated systems described earlier. Color coding is the same as in Figure 1.

Additional tribochemical reactions were observed in the ethane system and also in the isobutane system but not in the methane system.

4.2 CHEMISORBED GROUPS

When the load is high, tribochemical reactions are also observed during sliding simulations that begin with ethyl groups chemisorbed to one of the diamond surfaces

(Harrison *et al.*, 1994). In contrast, no tribochemistry is observed in the hydrogen-terminated or the methyl-terminated systems at any of the loads examined. When ethyl groups are present, the simulations show wear that is, in the majority of cases, initiated by the shearing of hydrogen atoms from the tail (-CH_3) portion of the ethyl groups (Fig. 8, upper left panel). Once the hydrogen atom has been sheared from the tail of the ethyl group, it is free to react at the interface in a number of ways. These atoms can recombine with an existing radical site (including the site from which they were abstracted), they can abstract a hydrogen atom from either diamond surface, or they can abstract a hydrogen atom from the tail of another chemisorbed chain to produce an H_2 molecule.

Once the shearing of hydrogen from the tail of the ethyl group (or initiation) occurs, the reactive radicals left on the tails of the adsorbed molecules can also react in a number of ways. These radicals can be eliminated by recombination with free hydrogen atoms, they can abstract hydrogen atoms from the opposing surface, or they can form bonds with existing radical sites on the opposing surface (Fig. 8, upper right panel). The formation of chemical bonds between the two surfaces results in adhesion of the two surfaces. This type of adhesion is consistent with conclusions inferred but not directly observed in experimental studies of the friction and wear of diamond. Continued sliding causes these adhesive bonds between the diamond surfaces to break simultaneously, leaving molecular wear debris (C_2H_4) trapped at the interface (Fig. 8, lower panel). Macroscopic friction experiments that have examined diamond sliding on diamond in ultrahigh vacuum have shown that friction is initially low and increases dramatically as the sliding progresses. This is thought to be due to adhesive contacts between the nascent radical sites formed by the sliding. Wear debris is then thought to form from the shearing of the newly formed chemical bonds during sliding. The wear debris is opaque and powdery and is composed of mostly unsaturated hydrocarbons with small amounts of hydrocarbon and graphite. The molecular wear debris formed in our simulations is consistent with the debris formed in these experiments.

Wear rates observed when diamond is in sliding contact with diamond in ultrahigh vacuum have been shown to depend upon the direction of sliding. Our MD simulations also show that the sequence of tribochemical reactions, and the products of these reactions, can differ depending on the sliding direction. Beginning with the same starting configuration that led to the tribochemical reactions outlined in Fig. 8 but sliding the upper diamond surface in the $[1\bar{1}0]$ direction yields very different tribochemistry. The rows of hydrogen atoms perpendicular to the $[1\bar{1}0]$ sliding direction are closer together than the rows perpendicular to the $[11\bar{2}]$ sliding direction. Therefore, the tail portion of the ethyl groups becomes more tightly lodged during sliding and eventually results in the breakage of the carbon-carbon bond in the ethyl group. These different reaction mechanisms imply different wear rates for each direction, consistent with experimental observations (Feng *et al.*, 1991; Wilks *et al.*, 1979).

5. Summary

Molecular dynamics simulations have been used to examine the atomic-scale friction between contacting hydrogen-terminated diamond surfaces in sliding motion. Simulations have examined the effects of many different variables on the average frictional force. These variables are: applied load (normal force), crystallographic sliding direction, the presence of chemisorbed groups on one diamond surface, and the presence of third-body molecules between the contacting surfaces.

For the hydrogen-terminated system, the frictional force increases with increasing applied load and no wear of the surfaces is observed at any load. The friction arises from the mechanical (vibrational) excitation of the interface layers of the diamond lattices during sliding. Once the interface region of the surfaces is excited, the excitation propagates to the rest of the lattice, and is eventually dissipated as heat. This process is the essence of wearless atomic-scale friction and is consistent with a number of analytic models (McClelland *et al.*, 1992; Frenkel *et al.*, 1938; Tomlinson *et al.*, 1929). Chemisorption of short hydrocarbon chains to the diamond surfaces causes the friction to be reduced at larger loads compared to the hydrogen-terminated system. This is due to the flexibility of the groups, which allows them to travel a trajectory that leads to less mechanical excitation of the contacting diamond surfaces.

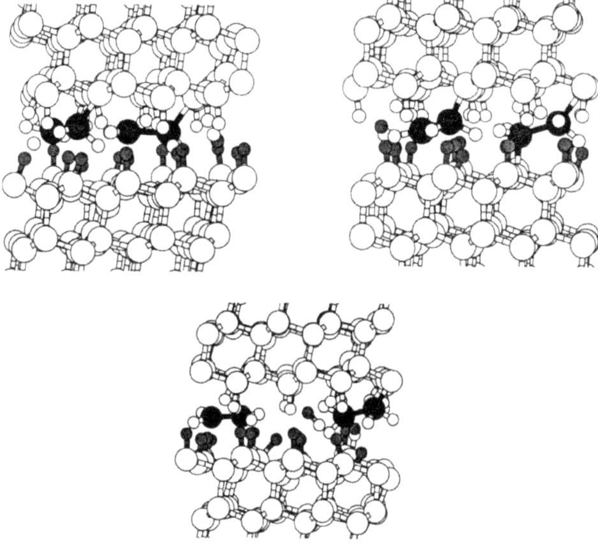

Figure 8.

Simulation configurations at different times for a system with chemisorbed ethyl groups at the interface. In the upper left panel hydrogen atoms have been sheared from the tail region of both ethyl groups. In the upper right panel adhesion of the two surfaces has occurred via formation of a carbon-carbon bond from the ethyl group to the lower diamond surface. In the lower panel the adhesive bonds have been broken and the resulting ethene molecule is visible at the right of the panel. Color coding is the same as in Figure 1.

The friction is also reduced at high loads when hydrocarbon molecules are confined between the hydrogen-terminated diamond surfaces. In these systems, the third-body molecules act as a boundary layer between the two diamond surfaces. As a result, the movement of one of the surfaces relative to the other does not initiate much vibrational excitation at the interface. Thus, the observed friction is lower than when the third-body molecules are absent. The friction that is observed in these systems is due to the interaction of the third-body molecule with the contacting surfaces. The larger the third-body molecule, the stronger are the interactions it has with the confining surfaces while sliding, so vibrational excitation of the lattice and the third-body molecule increase. In this case, the diamond lattice and the third-body molecules both dissipate energy and contribute to the observed friction.

There are relatively few experiments that have examined the atomic-scale friction of diamond (Germann *et al.*, 1993; Mate, 1993; Bhushan *et al.*, 1996b, van den Oetalaar *et al.*, 1997). An AFM has been used to investigate the atomic-scale friction of polished natural diamond in air using both a microfabricated Si_3N_4 tip and a single-crystal natural diamond tip (Bhushan *et al.*, 1996b). In both cases, the frictional force increased as a function of load. Comparison of these data to our simulated data is difficult because of the differences in the systems examined. In addition, because these experiments were conducted in air, the surfaces were probably contaminated. Despite the obvious differences between these experiments and the simulations, it is encouraging that similar trends in frictional force with load were found.

The experiments that most closely resemble our simulation conditions are the ultrahigh-vacuum AFM experiments of Germann *et al.* (1993) and van den Oetelaar *et al.* (1997). The measured frictional force between a CVD diamond tip and a hydrogen-terminated diamond (111) surface was found to be independent of load (Germann *et al.*, 1993). However, by comparing the shape and magnitude of the attractive force upon the tip-sample retraction and approach, Germann *et al.* conclude that the sample or tip might be covered by some physisorbed or chemisorbed species. The simulations described here show that chemisorbed species or trapped molecules (perhaps resembling physisorbed species) can have a profound effect on the frictional force.

Many experiments have examined the friction of diamond on the macroscopic scale (Tabor *et al.*, 1992; Ruan *et al.*, 1994; Hayward *et al.*, 1987; Hayward, 1991). These experiments indicate that lubricating hydrocarbon debris is formed when diamond is in sliding contact with diamond. It is not clear whether this debris originates from small fragments of diamond abraded from one surface or from the polymerization of adsorbed hydrocarbons. However, it is clear that this debris changes the friction significantly. The most recent evidence suggests that the lubricating hydrocarbon debris lowers the friction (Hayward *et al.*, 1987; Hayward, 1991). Simulations described here confirm that third-body molecules, that could perhaps model debris, do indeed lower the friction between diamond surfaces.

The simulations also show that, under certain conditions, tribochemical reactions do occur. The debris formed as a result of these reactions is consistent with debris observed in experiments. Depending on the starting state of the system, these reactions can lead to wear of the diamond surfaces, degradation of the third-body molecule, and chemisorption of new groups to the contacting diamond surfaces. All of these things change the nature of the contacting surfaces and may ultimately affect the observed friction.

6. Acknowledgments

This work was supported by T*he Office of Naval Research* (ONR) under Contract N00014-98-WR-20003. The authors would like to thank Donald W. Brenner, Carter T. White, and Richard J. Colton for many helpful discussions throughout the course of this work. Some of the figures were generated with the program XMol (XMol, version 1.3.1, Minnesota Supercomputer Center, Inc., Minneapolis, MN, 1993).

7. References

Berendsen, H. J. C., Postman, J. P. M., Van Gunsteren, W. F., DiNola, A., and Haak, J. R. (1984), "Molecular Dynamics with Coupling to an External Bath", *J. Chem. Phys.* **81**, 3684-3690.

Bhushan, B. (1996a), "Nanotribology and Nanomechanics of MEMS Devices", *Ninth Annual Workshop on Microelectromechanical Systems*, Feb 11-15, 91-98.

Bhushan, B. and Kulkarni, A. V. (1996b), "Effect of Normal Load on Microscale Friction", *Thin Solid Films* **278**, 49-56.

Brenner, D. W. (1990), "Empirical Potential for Hydrocarbons for use in Simulating Chemical Vapor Deposition of Diamond Films", *Phys. Rev. B* **42**, 9458-9471.

Brenner, D. W., Harrison, J. A., White, C. T., and Colton, R. J. (1991), "Molecular Dynamics Simulations of the Nanometer-Scale Properties of Compressed Buckminsterfullerene", *Thin Solid Films* **206**, 220-223.

Carpick, R. W. and Salmeron, M. (1997) "Scratching the Surface: Fundamental Investigations of Tribology with Atomic Force Microscopy", *Chem. Rev.* **97**, 1163-1194.

Cieplak, M., Smith, E. D., and Robbins, M. O. (1994) "Molecular Origins of Friction: The Force on Adsorbed Monolayers", *Science* **265**, 1209-1212.

Feng, Z. and Field, J. E. (1991), " Friction of Diamond on Diamond and Chemical Vapor Deposition Diamond Coatings", *Surf. Coatings Technol.* 47, 631-645.

Frenkel, F. C. and Kontorova, T. (1938), *Zh. Eksp. Teor. Fiz.* **8**, 1340.

Germann, G. J., Cohen, S. R., Neubauer, G., McClelland, G. M., Seki, H., and Coulman, D. (1993), "Atomic Scale Friction of a Diamond Tip on Diamond (100) and (111) Surfaces", *J. Appl. Phys.* **73**, 163-167.

Hayward, I. P. and Field, J. E. (1987), Proc. Inst. Mech. Eng., IMechE Conf. 1, 205.

Hayward, I. P. (1991), "Friction and Wear Properties of Diamonds and Diamond Coatings", *Surf. Coatings Technol.* **49**, 554-559 and references therein.

Harrison, J. A. and Brenner, D. B. (1995a) , "Atomic-Scale Simulation of Tribological and Related Phenomena", in *The Handbook of Micro/Nanotribology*, CRC Press, Boca Raton, FL, pp. 397-439 and references therein.

Harrison, J. A., White, C. T., Colton, R. J., and Brenner, D. W. (1995b), "Investigation of the Atomic-Scale Friction and Energy Dissipation in Diamond using Molecular Dynamics", *Thin Solid films* **260**, 205-211

Harrison, J. A. and Brenner, D. W. (1994), "Simulated Tribochemistry: An Atomic-Scale View of the Wear of Diamond ", *J. Am. Chem. Soc.* **116**, 10399-10402.

Harrison, J. A., White, C. T., Colton, R. J., and Brenner, D. W. (1993), "Effects of Chemically-Bound, Flexible Hydrocarbon Species on the Frictional Properties of Diamond Surfaces", *J. Phys. Chem.* **97**, 6573-6576.

Harrison, J. A., Colton, R. J., White, C. T., and Brenner, D. W. (1992), "Molecular Dynamics Simulations of Atomic-Scale Friction on Diamond Surfaces", *Phys. Rev. B* **46**, 9700-9708.

Hirano, M. and Shinjo, K. (1991), "Anisotropy of Frictional Forces in Muscovite Mica", *Phys. Rev. Lett.* **67**, 2642-2645.

Hunn, J. D., Withrow, S. P., White, C. W., Clausing, R. E., and Heatherly, L. (1994), "Fabrication of Single-Crystal Diamond Microcomponents", *Appl. Phys. Lett.* **65**, 3072-3074.

Israelachvili, J. N., Holmola, A. M., and McGuiggan, P. M. (1988), "Dynamic Properties of Molecularly Thin Liquid Films", *Science* **240**, 189-191.

Jansen, E., Dorsch, O., Obermeier, E., and Kulisch, W. (1996), "Thermal Conductivity Measurements on Diamond Films Based on Micromechanical Devices", *Diamond and Related Materials* **5**, 644-648.

Krim, J., Solina, D. H., Chiarello, R. (1991), "Nanotribology of a Kr Monolayer: A Quartz-Crystal Microbalance Study of Atomic-Scale Friction", *Phys. Rev. Lett.* **66**, 181-184.

Mate, C. M. , McClelland, G. M., Erlandsson, R., and Chiang, S. (1987) "Atomic-Scale Friction of a Tungsten Tip on a Graphite Surface", *Phys. Rev. Lett.* **59**, 1942-1945.

McClelland, G. M. and Glosli, J. N. (1992), "Friction at the Atomic Scale", in *Fundamentals of Friction: Macroscopic and Microscopic Processes*, (Singer, I. L. and Pollock, H. M., eds) Kluwer Academic Publishers, Dordrecht, pp. 405-422.

Perry, M. D. and Harrison, J. A. (1997), "Friction between Diamond Surfaces in the Presence of Small Third-Body Molecules", *J. Phys. Chem. B* **101**, 1364-1373.

Perry, M. D. and Harrison, J. A. (1996a), "Molecular Dynamics Studies of the Frictional Properties of Hydrocarbon Materials", *Langmuir* **12**, 4552-4556.

Perry, M. D. and Harrison, J. A. (1996b), "Molecular Dynamics Investigations of the Effects of Debris Molecules on the Friction and Wear of Diamond", *Thin Solid Films* **290-291**, 211-215.

Perry, M. D. and Harrison, J. A. (1995a), "The Effects of Methane-Debris Molecules on the Frictional Properties of Diamond", *Tribology Letters* **1**, 109-119.

Perry, M. D. and Harrison, J. A. (1995b), "Universal Aspects of the Atomic-Scale Friction of Diamond Surfaces", *J. Phys. Chem.* **99**, 9960-9965.

Ruan, J.-A. and Bhushan, B. (1994), "Atomic-Scale and Microscale Friction of Diamond Surfaces", *J. Appl. Phys.* 76, 5022-5035.

Sheehan, P.E. and Lieber, C. M. (1996) "Nanotribology and Nanofabrication of MoO_3 Structures by Atomic Force Microscopy", *Science* **272**, 1158-1161.

Smith, E. D., Robbins, M. O., and Cieplak, M. (1996), "The Friction on Adsorbed Monolayers", *Phys. Rev. B* **54**, 8252-8260.

Sokoloff, J. (1984), "Theory of Dynamical Friction between Idealized Sliding Surfaces", *Phys. Rev. B* **42**, 760-765.

Sokoloff, J. (1996), "Theory of Electron and Phonon Contributions to Sliding Friction", in *Physics of Sliding* (Persson, B. N. J. and Tosatti, E., eds.), Kluwer Academic Publishers, Dordrecht, pp. 217-229.

Sorensen, M. R., Jacobsen, K. W., and Stoltze, P. (1996), "Simulations of Atomic-Scale Sliding Friction", *Phys. Rev. B.* **53**, 2101-2113.

Tabor D. and Field, J. E. (1992), *The Properties of Natural and Synthetic Diamond*, (Field, J. E. Ed.), Academic Press, London, pp. 547-571 and references therein.

Tomlinson, G. A. (1929), "A Molecular Theory of Friction", *Philos. Mag. Ser.* **7**, 905-939.

Van Alsten, J. and Granick, S. (1988), "Molecular Tribometry of Ultrathin Liquid Films", *Phys. Rev. Lett.* **61**, 2570-2573.

van den Oetelaar, R. J. A. and Flipse, C. F. J. (1997), "Atomic-Scale Friction on Diamond (111) Studied by Ultra-High Vacuum Atomic Force Microscopy", *Surf. Sci.* **384**, L828-L835.

Wilks, J. and Wilks, E. M. (1979), *The Properties of Diamond*, (Field, J. E. Ed.), Academic Press, London, pp. 351-381.

Zhong, W. and Tomanek, D. (1990), "First-Principles Theory of Atomic-Scale Friction", *Phys. Rev. Lett.* **64**, 3054-3057.

DUAL-AXIS PIEZORESISTIVE AFM CANTILEVER FOR INDEPENDENT DETECTION OF VERTICAL AND LATERAL FORCES

B.W. CHUI, T.W. KENNY

Stanford University
Terman 551, Stanford, CA 94305-4021, USA

H.J. MAMIN, B.D. TERRIS, D. RUGAR

IBM Almaden Research Center
650 Harry Road, San Jose, CA 95120-6099, USA

1. Abstract

A novel dual-axis atomic force microscope (AFM) cantilever has been fabricated with independent piezoresistive sensors for simultaneous detection of lateral and vertical forces. The cantilever consists of a flat, triangular probe connected to a base via several tall, narrow "ribs." The triangular probe, which is predominantly compliant in the vertical direction, is implanted with piezoresistors to form a vertical deflection sensor. The ribs, which are predominantly compliant in the lateral direction, have a separate set of sidewall-embedded piezoresistive sensors that form a lateral deflection sensor. A unique ion implant at approximately 45° to the vertical is used to produce the piezoresistive sensors and associated electrical interconnects on the cantilever structure. The device has been used to obtain vertical-force and lateral-force AFM images and to measure microscale friction.

2. Introduction

Some areas of atomic force microscopy (AFM) [1-2], such as nanotribology [3] and magnetic imaging [4], require the detection of lateral forces [5] as well as vertical forces with micromachined cantilevers. Existing detection methods are primarily based on the use of conventional planar cantilevers with optical [6] or piezoresistive [7-8] means of deflection sensing. The vertical deflection mode of the cantilever is used to generate information about the topography of the sample, while the torsional deflection mode is used to measure lateral forces that correspond for example to friction between the tip and the sample. In particular, the piezoresistive detection approach typically involves the fabrication of piezoresistors on the surface of a flat, U-shaped cantilever [9] with one piezoresistor on each leg. External electronic circuitry is used to generate the sum of the two resistances as well as their difference. The vertical deflection signal is derived from the summed value,

B. Bhushan (ed.), Tribology Issues and Opportunities in MEMS, 301-312.

while the torsional deflection signal is derived from the difference. Since the same mechanical element is used in two deflection modes, signal intermixing may occur. In addition, the vertical stiffness of the cantilever has to be balanced against its torsional stiffness in the determination of cantilever dimensions. These issues may cause complications in cantilever design as well as signal analysis.

This paper describes a novel AFM cantilever with two orthogonal axes of compliance, each with an independent piezoresistive sensor. This makes it possible for vertical and lateral forces to be measured with a minimum of external circuitry, and also allows the cantilever geometry to be separately optimized for each detection mode.

3. Cantilever Design and Principle of Operation

The cantilever consists of a vertical force sensor and a lateral force sensor. The vertical force sensor is in the form of a flat, triangular cantilever with a piezoresistive boron layer (fig. 1a). This design is similar to piezoresistive cantilevers developed previously in our group [10] and elsewhere [2]. Vertical deflection of the cantilever tip causes a change in resistance that can be converted to a vertical deflection signal by means of an external Wheatstone bridge. The triangular probe is connected to the base via four high-aspect ratio "ribs" which give the structure lateral compliance. The inner ribs have embedded piezoresistive layers on the sidewalls, forming a lateral force sensor that produces a lateral deflection signal (fig. 1b). During operation, forces at the cantilever tip are mechanically resolved into vertical and lateral components that are separately detected by the corresponding sensors, giving rise to synchronized vertical and lateral force output signals. This represents the first use of piezoresistive sensors fabricated on vertical and horizontal surfaces of a microstructure to achieve simultaneous force sensing in orthogonal directions.

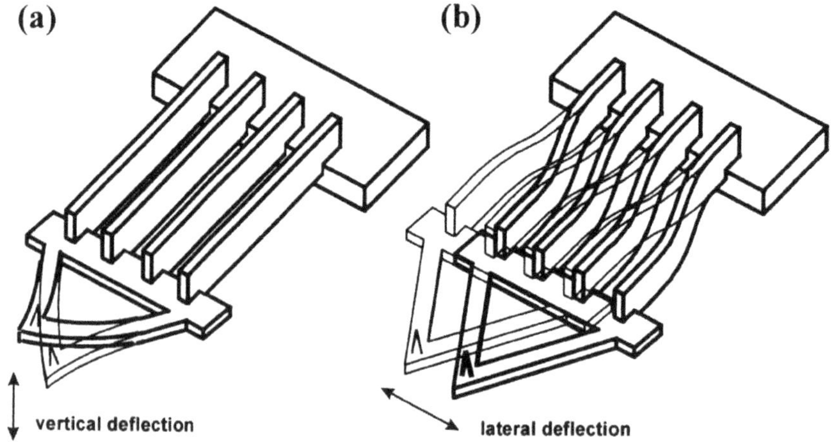

Figure 1: Novel dual-axis AFM cantilever with orthogonal axes of compliance: (a) vertical deflection mode (i.e. bending of the flat triangular probe); (b) lateral deflection mode (i.e. bending of the high-aspect-ratio ribs).

Figure 2a shows the electrical path along the vertical force sensor. The two outer ribs are heavily doped with boron (i.e. p-doped) to provide high conductivity, allowing current to flow from the base to the triangular probe, through the two piezoresistors, and back. Figure 2b shows the electrical path along the lateral force sensor. Current flows from the base onto one of the two inner ribs with its sidewall-mounted piezoresistor, then to the crossbar, then doubles back and returns to the base via the other inner rib. To prevent electrical crosstalk between the vertical force sensing circuit and the lateral force sensing circuit, back-to-back p-n-p diodes are fabricated on the crossbar at the locations shown in fig. 2b.

The electrical paths described above require current to flow in a vertical plane in some parts of the structure and in a horizontal plane in others, as well as turn corners between the two planes. A novel ion implant technique was developed to make this possible, i.e. to produce electrically conducting vertical sidewalls and horizontal surfaces at the same time. In this technique, dopant ions are implanted at approximately 45 degrees to the vertical, allowing vertical sidewalls and horizontal surfaces to be implanted simultaneously and ensuring electrical continuity at the interfaces.

Oblique ion implantation on high-aspect-ratio micromachined devices is a powerful fabrication technique that can be extended to a general variety of piezoresistive lateral force sensors, such as accelerometers, fluidic sensors and magnetic sensors. It offers an integrated low-cost, high-performance alternative to other sensing methods such as capacitive and optical methods.

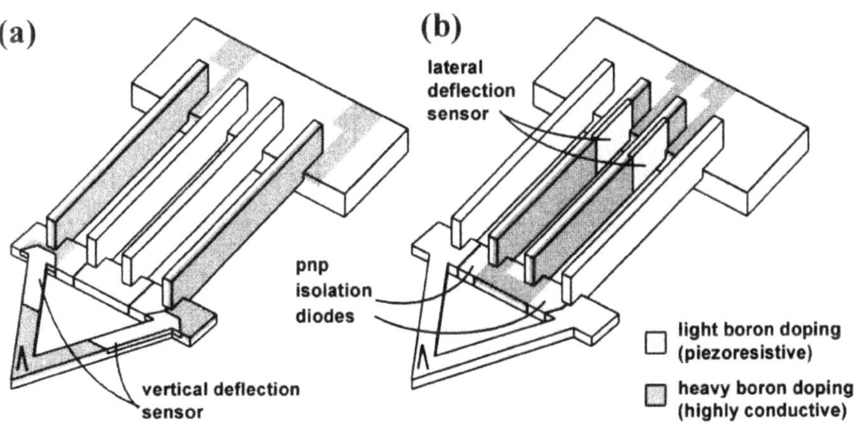

Figure 2: (a) Vertical and (b) lateral piezoresistive deflection sensing circuits. The heavily doped regions (dark shaded regions in the figure) on the structure form electrical connections to the piezoresistive sensors (light shaded regions). Electrical isolation between the two circuits is achieved by means of two back-to-back p-n-p junctions on the crossbar. The n-doped region simply results from the original n-type starting material. (The two sensing circuits actually co-exist on the same cantilever, but are drawn separately here for the sake of clarity.)

After the implant, a 10-second rapid thermal anneal at 1000 °C is performed to activate the boron dopants while causing minimal diffusion (thus preserving the shallow piezoresistors), followed by a 40-minute furnace anneal at 800 °C to repair implant-related damage in the silicon. Contacts vias are etched through the 1000 Å oxide layer, aluminum is sputtered on and patterned to form metal leads to the piezoresistive sensors, and a forming gas anneal is performed. Finally, tetramethyl ammonium hydroxide (TMAH) solution (25% wt. in H_20) is used to etch away the backside silicon underneath the cantilever, and buffered HF is used to remove the buried oxide layer leaving a released structure (fig. 3d).

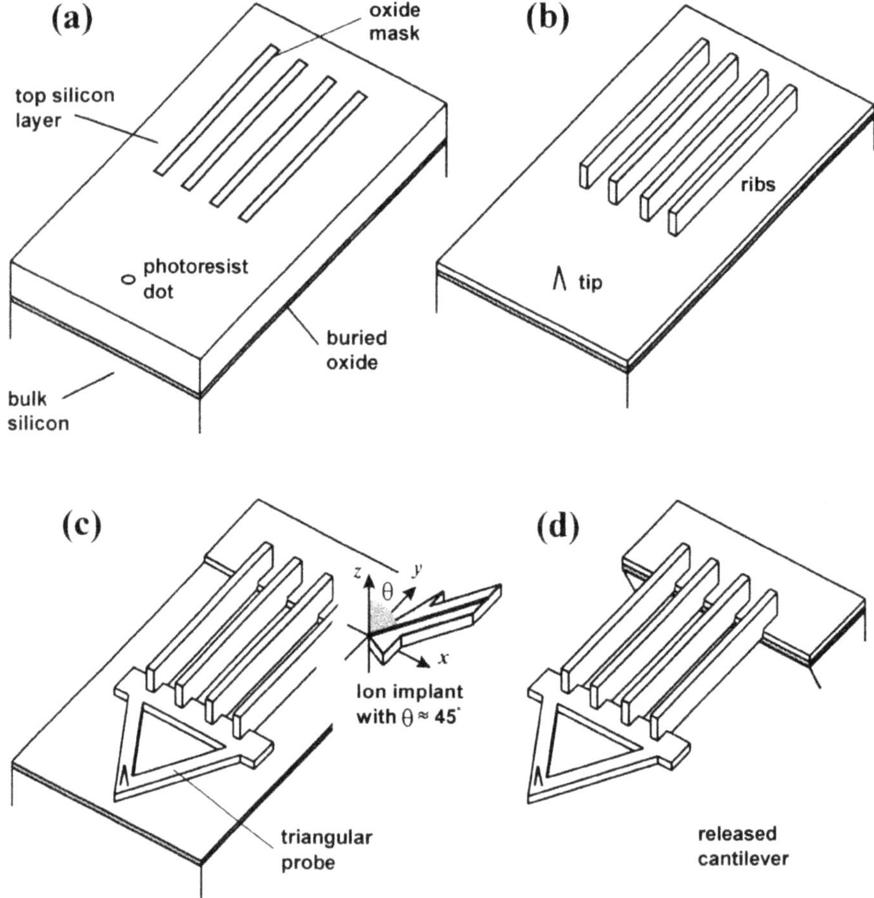

Figure 3: Fabrication process for dual-axis cantilever. Note the ion implantation step illustrated in part (c) which is performed at a large angle to the vertical, enabling the fabrication of electrical elements on vertical and horizontal surfaces at the same time, while ensuring electrical continuity at the interfaces.

4. Cantilever Fabrication

The entire fabrication process of the cantilever is shown in fig. 3. The starting material is a silicon-on-insulator (SOI) wafer (fig 3a) with a 10 μm thick, n-type, single-crystalline top silicon layer. An oxide mask is used to define the position of the ribs, while a circular photoresist dot is used to define the position of the tip. In fig. 3b, the high-aspect-ratio ribs are formed using a high-density plasma etching system with a Cl_2/HBr plasma [11]. This highly anisotropic etch produces essentially vertical sidewalls. The etch is halted when the top silicon layer has been thinned down to the desired thickness of the triangular probe. Careful monitoring of the etch rate and precise timing is essential for this step.

The tip is also formed in the same etching step. Because the photoresist dot is relatively less resilient to the etch compared with the oxide mask, it is eroded from the sides at an appreciable rate and gradually decreases in diameter during the etch. The silicon underneath is therefore etched to form a tapering conical tip, with a radius of curvature as small as 200 Å. The chemistry of the plasma etch also helps produce a self-sharpening effect. In addition, the photoresist dot is subjected to a 90-second reflow on a 150 °C hot plate before the etch in order to smooth out the photoresist profile. It was experimentally found that gently sloping photoresist sidewalls assisted in the formation of a conical silicon tip profile.

It should be emphasized that the one-step plasma etch described above is capable of producing two different etch profiles (rectangular and conical) with the help of two different masking materials (oxide and photoresist). This approach is advantageous in that it eliminates the need to devote separate steps towards the fabrication of the conical tip and the high-aspect-ratio ribs.

In fig. 3c, the triangular probe itself is patterned and etched. Photoresist thicker than the height of the ribs is used in this step to protect the already formed ribs. A 1000 Å thermal oxide layer is then grown on the surface of the entire cantilever structure for electrical passivation, in preparation for the ion implant step.

A basic ion implant process for producing flat piezoresistive cantilevers has been developed by Tortonese *et al* [9], and the technique has since been extended to the fabrication of shallow piezoresistive layers in thin cantilevers of thickness 1 μm and below [10,12]. For the present dual-axis cantilever structure, boron is implanted at 55 keV with a dose of $5 \times 10^{14}/cm^2$ at approximately 45 degrees to the vertical in the x-z plane, forming piezoresistive layers on vertical and horizontal surfaces at the same time (fig. 3c). (In the case of the ribs, only one sidewall of each rib receives the implant by design, since the other sidewall in the shadow of the ion beam.) A second, heavier boron implant (energy 55 keV, dosage $5 \times 10^{15}/cm^2$) is then performed at the same angle, but with the piezoresistive regions masked off with thick photoresist, in order to form electrical connections to the piezoresistors. The central regions of the back-to-back p-n-p isolation diodes on the crossbar (see fig. 2b) are masked off with thick photoresist during both the light and the heavy implants, and those regions thus remain n-type.

Care has to be taken during the cantilever design process to ensure that any tall structures, e.g. ribs, are spaced sufficiently far apart so that they do not shadow each other during the oblique ion implants. The same precaution applies to the placement of the photoresist implant masks, since the thick photoresist presents a rather tall profile as well.

Figure 4 shows an SEM image of a finished cantilever, made of single-crystal silicon. The triangular probe is 1.3 μm thick, 90 μm long and each leg is 8 μm wide; the ribs are 1.3 μm wide, 110 μm long and 10 μm tall. One of the advantages of the dual-axis cantilever design is that the stiffness of the triangular probe and the ribs can be independently adjusted by changing their dimensions, unlike the case of the simple planar cantilever. These adjustments can also be used to indirectly achieve a desired level of vertical or lateral piezoresistive sensitivity for a given application.

Other design enhancements can be seen in fig. 4. The extremities of the ribs have been made wider in order to form more rigid perpendicular joints with the triangular probe and the base. A mass has been added to the mid-section crossbar of the cantilever for mechanical reinforcement and for vibrational decoupling between the triangular probe and the ribs.

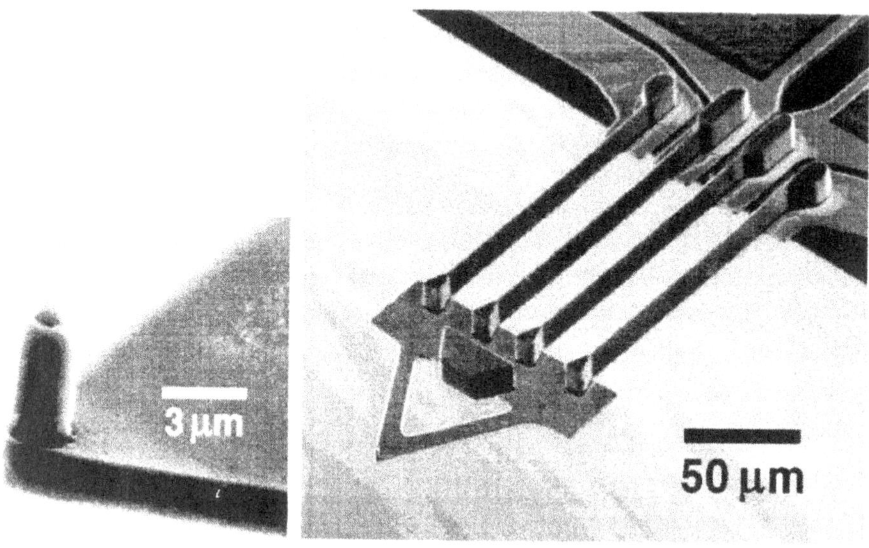

Figure 4: SEM image of dual-axis cantilever. The device shown has a 1.3 μm, 70 μm long thick triangular probe which is connected to the base by four 1.3 μm wide, 110 μm long, 10 μm tall high-aspect-ratio ribs. On the left is a close-up view of the tip, formed in a single plasma etch.

5. Cantilever Characterization

5.1. PIEZORESISTIVE SENSOR PERFORMANCE

The performance of the piezoresistive deflection sensors was measured with Wheatstone bridge circuits based on the low-noise Burr-Brown INA103 instrumentation amplifier configured with a gain of 1000. The bridge voltage was 2.0 V (i.e. 1.0 V across each resistor). The vertical and lateral deflection sensors on the cantilever showed piezoresistive sensitivities $\Delta R/R$ of 0.25 ppm and 1.0 ppm per Å respectively. The noise characteristics were favorable as well: the noise floor for both the vertical and the lateral deflection sensors was less than 20% above the theoretical Johnson noise floor. (Both sensors had resistances of approximately 10 kΩ.) In a 10 kHz bandwidth, the minimum detectable deflection was 20 Å in the vertical direction and 5 Å in the lateral direction.

The calculated mechanical stiffness of the triangular probe is 2 N/m in the vertical direction, and that of the four ribs combined is 11 N/m in the lateral direction. Because the cantilever has been designed so that the triangular probe and the ribs each has only one predominant direction of compliance, mechanical crosstalk between the two types of elements can be minimized. In one experiment, the cantilever was mounted in an atomic force microscope and the tip was scanned laterally back and forth across a polycarbonate surface. While the lateral deflection sensor output showed clear reversals of polarity as the scan direction was reversed (due to frictional drag on the tip), the vertical deflection sensor showed no observable change in output signal at any point in time.

5.2. DUAL-CHANNEL AFM IMAGING

AFM imaging capability has been demonstrated with the dual-axis cantilever. During each scan, the vertical and lateral force signals from the cantilever were processed in parallel using two sets of Wheatstone bridges and amplifiers. Figure 5 shows a pair of AFM images obtained using the cantilever in z-servo mode (i.e. constant vertical loading force). Figure 5a is the vertical force image corresponding to bending of the triangular probe, while figure 5b is the lateral force image corresponding to bending of the ribs. The sample is a silicon sample with a grid of ridges 1600 Å in height and 2 μm in width, with a pitch of 10 μm. The vertical force image shows the overall topography of the sample, while the lateral force image displays correlated information about the lateral displacement of the tip as it is scanned from left to right. The bright north-south bands in the lateral force image correspond to an increase in lateral tip displacement. The location of these bands coincide with the left edges of the ridges in the vertical force image, suggesting that the tip was "tripped" momentarily by the step before climbing it. If this is true, the apparent width of the north-south oriented ridges in the vertical force image would be smaller than that of the east-west oriented ones, since the tip would be spending a disproportionately small amount of time on top of the north-south oriented ridges. Indeed, this is observed to be so in the vertical force image.

Figure 6 shows a pair of more detailed AFM images taken over a smaller area of the sample covering only one ridge intersection. These images can be examined in conjunction with a pair of derived line scans (fig. 7). The line scans corroborate our interpretation of the behavior of the cantilever tip upon encountering abrupt steps. The "tripping" of the tip at the leading edge of the ridge is visible as a positive spike on the lateral deflection signal; this spike indeed coincides with a positive step on the vertical deflection signal. Only a weak perturbation in the lateral deflection signal is observed at the negative step on the vertical deflection signal; this is expected since there should be no "tripping" that occurs when the tip falls off the trailing edge of the ridge.

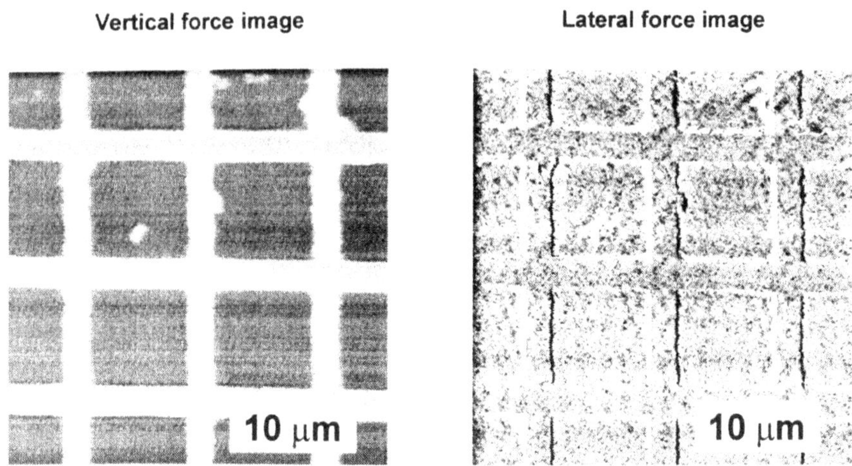

Figure 5: Vertical force (left) and lateral force (right) AFM images taken in parallel during one scan with a dual-axis cantilever. The substrate is a silicon sample with a grid of ridges 1600 Å in height, 2 μm in width and 10 μm in pitch.

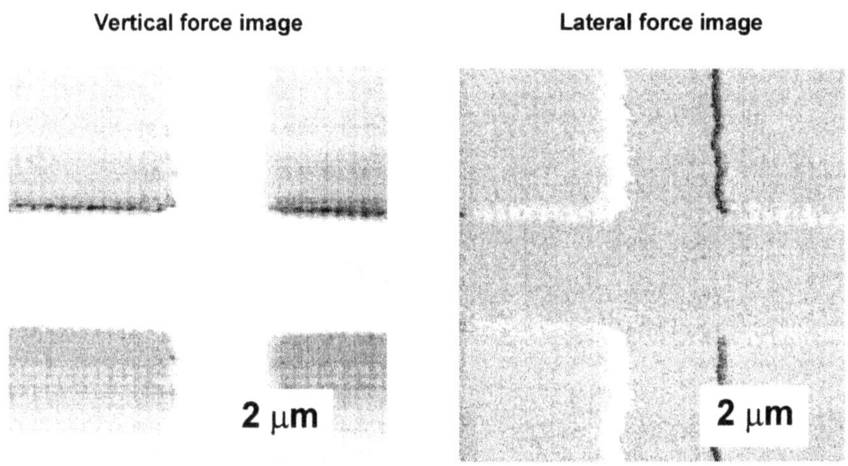

Figure 6: Close-up AFM images of a single ridge in each direction, obtained on the sample used in fig. 5. On the lateral force image, note the bright north-south oriented bands along the left edge of the north-south oriented ridge, corresponding to increased lateral deflection of the probe tip as it is "tripped" at the positive step edge before climbing it. The scan direction is from left to right.

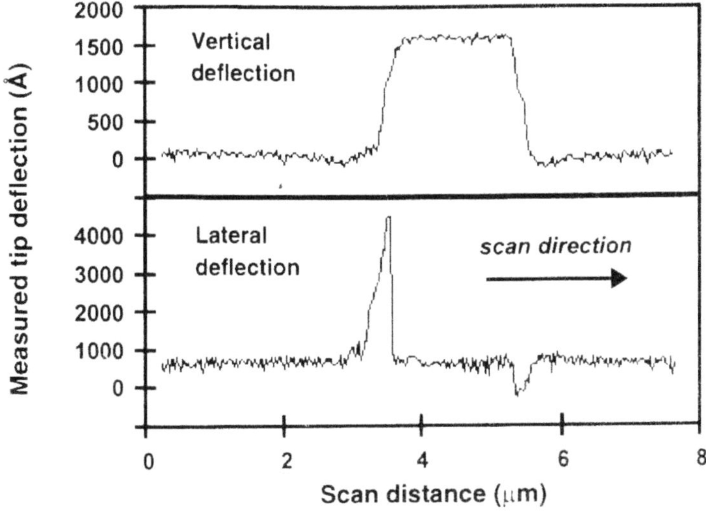

Figure 7: Line scans across a ridge on the sample of figures 5 and 6. Note the positive spike on the lateral deflection signal which occurs at the same location as the positive step on the vertical deflection signal, suggesting that the cantilever tip was momentarily "tripped" at the edge of the step before climbing it. Such tripping led to increased lateral tip deflection which is observable as a positive signal.

5.3. MICROSCALE FRICTION MEASUREMENTS

Experiments have also been performed with the dual-axis cantilever to measure microscale friction. In one experiment performed in air, the silicon tip of the cantilever was scanned across a polycarbonate substrate at various tip velocities (10 μm/s to 320 μm/s) and various loading forces (0.6 μN to 2.2 μN). The lateral deflection of the cantilever was measured. Based on a knowledge of the lateral piezoresistive sensitivity and the calculated lateral stiffness of the cantilever, the lateral frictional force acting on the probe tip can be calculated. The resulting data points for measured lateral force vs. applied loading force are plotted in fig. 8.

It is observed that lateral deflection of the cantilever increased generally linearly with loading force for any given tip velocity. When straight lines are fitted to the data points, the slope of such lines can give an indication of the coefficient of friction of a silicon tip on a polycarbonate surface under the conditions of the particular experiment. In this case, the coefficient of friction is observed to be approximately 0.4 for low tip velocities around 10 μm/s, increasing to a maximum of approximately 0.6 for tip velocities of 40 μm/s and above.

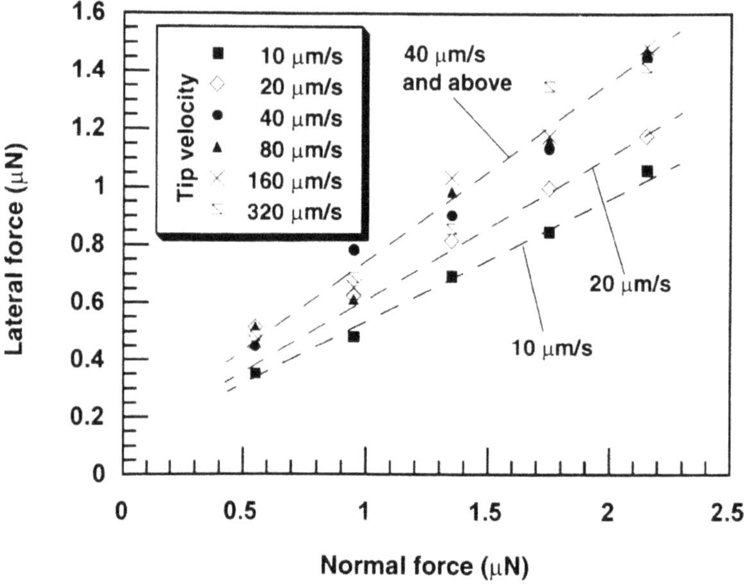

Figure 8: Measured lateral force vs. applied loading force for silicon tip scanned across polycarbonate surface in air at various tip velocities.

6. Conclusion

A novel AFM cantilever has been designed and fabricated that allows simultaneous, independent detection of vertical and lateral forces, as well as individual optimization of its mechanical characteristics in two orthogonal directions. The device exhibits favorable piezoresistive sensor performance, and its utility in dual-channel AFM imaging and microscale friction analysis has been demonstrated.

A special ion implant process involving a large oblique angle of incidence was developed for producing electrically conducting elements (including piezoresistive elements) on the sidewalls of high-aspect-ratio micromachined beams. With this method, electrically conductive layers, including piezoresistors, can be fabricated both on vertical sidewalls and on horizontal surfaces of microstructures using a minimum of implant steps. Electrical continuity can also be guaranteed between the two planes. This fabrication approach is not limited to AFM cantilevers and can be applied to lateral force sensors in general. For example, accelerometers can be fabricated with high-aspect ratio beams implanted at an oblique angle to form piezoresistive lateral deflection sensors, and these sensor elements can be duplicated along additional axes of sensitivity.

The dual-axis AFM cantilever described in this paper has potential uses in a wide variety of applications. Examples include nanotribology (for frictional analysis), tip-based data storage (for concurrent servo control and data readback) [10] and magnetic imaging (for simultaneous imaging of vertical and lateral fields at domain interfaces).

7. Acknowledgments

We are grateful to C.F. Quate, S.C. Minne, H.T. Soh, T.D. Stowe, J. Harley, J. McVittie and J.S. Kim of Stanford for their assistance. We would also like to thank R.P. Ried and L.S. Fan of IBM Almaden Research Center for helpful discussions. The staff of the Stanford Nanofabrication Facility Center for Integrated Systems provided invaluable assistance during the device fabrication process.

In addition, we would like to acknowledge the support of an IBM Cooperative Fellowship, the National Science Foundation CAREER Program (ECS-9502046), the National Science Foundation Instrumentation for Materials Research Program (DMR-9504099), the Charles Lee Powell Foundation, and DARPA.

The authors may be contacted via electronic mail at *chui@leland.stanford.edu*, *kenny@cdr.stanford.edu*, and *mamin@almaden.ibm.com*.

8. References

1. Binnig, G., Quate, C.F., and Gerber, Ch., "Atomic Force Microscope," *Phys. Rev. Lett.***56**, pp. 930-3 (1986).
2. Rugar, D., and Hansma, P.K., "Atomic Force Microscopy," *Physics Today* **43**, pp. 23-30 (1990).
3. Bhushan, B., ed., *Micro/Nanotribology and its Applications*, NATO Advanced Science Institutes Series, Kluwer Academic Publishers, Boston (1997).
4. Gruetter, P., Mamin, H.J. and Rugar, D., "Magnetic Force Microscopy (MFM)" in *Scanning Tunneling Microscopy II: Further applications and related scanning techniques*, 2nd ed., R. Wiesendanger and H.-J. Guntherodt (eds.), Springer-Verlag, Berlin, pp. 151-207 (1995).
5. Mate, C.M., McClelland, G.M., Erlandsson, R., and Chiang, S., *Phys. Rev. Lett.* **59**, p.1942 (1987).
6. Kawakatsu, H., and Saito, T., "An atomic force microscope with two optical levers for detection of the position of the tip end with three degrees of freedom" in ref. [3], pp. 55-60.
7. Kassing, R., and Oesterschulze, E., "Sensors for scanning probe microscopy" in ref.[3], pp.35-54.
8. Brugger, J., Burger, J., Binggeli, M., Imura, R., and de Rooij, N.F., "Lateral force measurements in a scanning force microscope with piezoresistive sensors," Proceedings of the 8th Int'l Conf. on Solid State Sensors and Actuators, Stockholm, 1995, pp. 636-9.
9. Tortonese, M., Yamada, H., Barrett, R.C., and Quate, C.F., "Atomic force microscopy using a piezoresistive cantilever," Proceedings of the 6th Int'l Conf. on Solid State Sensors and Actuators, San Francisco, 1991, pp. 448-51.

10. Chui, B.W., Stowe, T.D., Kenny, T.W., Mamin, H.J., Terris, B.D., and Rugar, D., "Low-stiffness silicon cantilevers for thermal writing and piezoresistive readback with the atomic force microscope," *Appl. Phys. Lett.* **69,** pp. 2767-9 (1996).

11. LAM TCP 9400 plasma etcher, Lam Research Corp., Fremont, California, USA.

12. Ried, R. P., Mamin, H. J., Terris, B. D., Fan, L. S., and Rugar, D., "5 MHz, 2 N/m Piezoresistive AFM Cantilevers with INCISIVE Tips," Proceedings of 9th Int'l Conf. on Solid State Sensors and Actuators, Chicago, 1997, pp. 447-50.

STATISTICAL THERMODYNAMIC TREATMENT OF THE AFM TIP IN LIQUID

Kenichiro Koga and X.C. Zeng
Department of Chemistry, University of Nebraska-Lincoln, Lincoln, NE 68588

Dennis J. Diestler
Department of Agronomy, University of Nebraska-Lincoln, Lincoln, NE 68583

Abstract

A statistical thermodynamic analysis of the atomic force microscope (AFM) is presented and applied to a single-atom tip moving quasistatically over a rigid substrate immersed in water. The RISM integral equation technique is used to compute the normal force (load) F_z on the tip as a function of its height above the substrate. F_z is found to be oscillatory, which implies that multiple scanning trajectories of the tip are possible under constant load. The unique trajectory along which the system is thermodynamically stable is revealed. This study shows that the tip may undergo hopping motion even over a defect-free substrate, due to layering of water molecules between the tip and substrate.

1 Introduction

There are great advantages to performing atomic force microscope (AFM) measurements in liquids: first, the elimination of capillary forces makes true atomic resolution possible at very small separation between the probe tip and the substrate[1]; second, biological samples can be probed in their native environments [2-6]. To develop the applications of AFM in liquids, it is of critical importance to understand effects of liquids on the probe tip movements. O'Shea, Welland and Rayment[2] observed oscillations in the force curves (normal forces on an AFM tip versus tip-substrate separation) in the AFM measurements in octamethylcyclotetrasiloxane liquid. Cleveland, Schaffer, and Hansma[3] also found oscillatory forces acting on the AFM tips over hydrophilic cleavage planes of ionic crystals in water. Moreover,

B. Bhushan (ed.), Tribology Issues and Opportunities in MEMS, 313-323.

recent computer simulations of the AFM in liquids confirmed the oscillatory forces[7, 8]. These studies demonstrated an important effect of liquids on the force acting on the AFM tips; at the same time they leave an unanswered question: How are the scans of the AFM tip affected when the system is immersed in liquid environment?

Scanning motions of AFM tips in air over various surfaces have been the subject of several reviews[9, 10]. The purpose of this paper is to investigate distinctive features of scanning motions of an AFM tip over a rigid monolayer substrate in liquid water. Integral equation techniques are used to calculate the force on the tip when scanning over a rigid substrate. A unique scanning trajectory is obtained from thermodynamic analyses. This trajectory shows that the tip can undergo hopping motion in scanning over a defect-free substrate, because of layering of the water molecules between the tip and the substrate.

2 Thermodynamic Analysis of the AFM in Liquid

In figure 1 we schematize the AFM probe tip immersed in a pure liquid in contact with a solid substrate. From a strict thermodynamic perspective we take the *system* to consist of the finite portion of liquid contained within the rectangular prism indicated in Fig. 1. The tip and substrate belong to the environment, although they influence the system *via* the external fields that they impose upon the latter. The walls of the prism that confine the system are assumed to be very far from the tip and to remain stationary. Thus, temperature (T) and density, or T and pressure, or T and chemical potential (μ) are fixed. The only means by which mechanical work can be performed by the system is through movements of the probe tip.

We assume that reversible transformations of the system are described quantitatively by Gibbs' fundamental relation

$$d\Omega = -SdT - Nd\mu - \mathbf{F} \cdot d\mathbf{r} \tag{1}$$

where the symbols Ω, S and N respectively stand for the grand potential, the entropy and the number of molecules. The displacement of the probe tip $d\mathbf{r} = \mathbf{e}_x dx + \mathbf{e}_y dy + \mathbf{e}_z dz$ against an *applied force* (*e.g.*, due to cantilever)

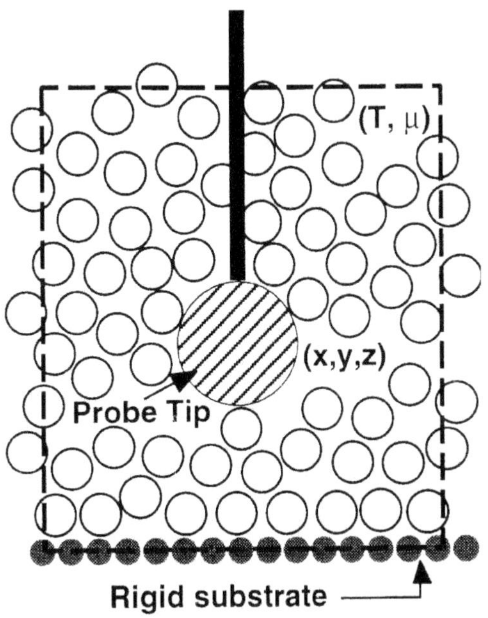

Figure 1: Scheme of the system consisting of finite portion of liquid contained within the rectangular prism denoted by the dotted line. Note that the probe tip and the substrate belong to the environment.

$-\mathbf{F} = -\mathbf{e}_x F_x - \mathbf{e}_y F_y - \mathbf{e}_z F_z$ results in the system's doing work $dW = \mathbf{F} \cdot d\mathbf{r}$. We assume that the tip's position can be specified by the single vector $\mathbf{r}$. In essence we treat the tip as a rigid body having a fixed orientation. The grand potential can then be regarded as a function of T, μ, and $\mathbf{r}$, the force $\mathbf{F}$ being given by

$$\mathbf{F} = -(\nabla_{\mathbf{r}}\Omega)_{T,\mu}. \tag{2}$$

AFM measurements are carried out extremely slowly on the molecular scale. It is therefore reasonable to view the scanning motions of the probe tip as quasistatic and properly describable by means of equilibrium thermodynamics and statistical mechanics. AFM scans are generally performed at constant T and μ in either of two mechanical modes: (1) constant height of the probe tip; (2) constant load on the probe tip. In the first mode the position of the tip is varied quasistatically and the force is measured. For example, Figure 2 shows plots of the normal component (load) F_z as a function of z for several fixed lateral positions (x, y). An interesting feature of these plots is their oscillatory character, which reflects the layering of liquid in the gap between the tip and the substrate. Depending upon the conditions, several heights of the tip may correspond to the same force. As z becomes very large, F_z must of course tend to zero.

In the second (i.e. constant load) mode of operation of the AFM, F_z is held fixed at F_z' and the probe tip is moved quasistatically along a lateral trajectory prescribed by the parametric equations

$$x = x(t), \quad y = y(t) \tag{3}$$

where t represents "time". The lateral trajectory is simply the projection of the three-dimensional trajectory $[\mathbf{r}(t)]$ onto the plane of the substrate. The description of the behavior of the system in this mode is facilitated by introducing the auxiliary energy

$$\Psi = \Omega + F_z z. \tag{4}$$

Combining Eqs. (1) and (4), we obtain

$$d\Psi = -SdT - Nd\mu - F_x dx - F_y dy + zdF_z. \tag{5}$$

At a given point (x, y) on the lateral trajectory, any of several values of the height may satisfy the requirement $F_z = F_z'$. These several values $\{z_i\}$ can be obtained by computing the curve F_z versus z. Doing this for a selection of points (x, y), we generate a set of possible vertical $z(t)$. However, some of these are either unstable or metastable. If the system is truly in thermodynamic equilibrium at all "times" (i.e., at all points on the quasistatic trajectory), then the tip must follow the stable trajectory, for which the energy Ψ is minimum under the conditions of fixed T, μ, and F_z.

Now to find the (relative) minimum of Ψ at each point (x, y), we compute the difference in Ψ between each of the heights z_i satisfying the condition $F_z = F_z'$ and a reference point z_0 sufficiently far from the substrate that the force acting on the tip is zero, that is

$$\begin{aligned} \Delta\Psi &= \Psi(z_i) - \Psi(z_0) && (6) \\ &= \Delta\Omega + F_z' z_i. && (7) \end{aligned}$$

The change in the grand potential is calculated from the supermolecule approach described in 3.2. The height which gives the smallest $\Delta\Psi$ is the truly stable position of the tip. We can also examine local stability of the system at each of the height z_i. If the system is locally stable,

$$\frac{\partial^2 \Delta\Psi}{\partial z^2} > 0 \tag{8}$$

whereas if the system is unstable,

$$\frac{\partial^2 \Delta\Psi}{\partial z^2} < 0. \tag{9}$$

3 Model and Computational Methods

3.1 Model for the AFM in liquid

We consider the *system* consisting of water molecules and the *environment*, including a single-atom tip and a rigid hexagonal monolayer substrate. The modified TIPS site-site potential function[11] is employed for the intermolecular interaction of water. The tip-substrate interaction in vacuum, $U^{\mathrm{vac}}(\mathbf{r})$, is taken to be a pairwise sum, $U^{\mathrm{vac}}(\mathbf{r}) = \sum_j u_{\mathrm{ts}}(r_j)$, where u_{ts} is the Lennard-Jones (LJ) potential: $u_{\mathrm{ts}}(r) = 4\epsilon_{\mathrm{s}}[(\sigma_{\mathrm{s}}/r)^{12} - (\sigma_{\mathrm{s}}/r)^6]$ and r_j denotes the distance between the tip atom and j-th atom of the substrate; ϵ_{s} and σ_{s} are the LJ energy and size parameters. We use parameters for a site-site model of naphthalene[12] in the calculations, $\epsilon_{\mathrm{s}}/k = 67.1\mathrm{K}$, $\sigma_{\mathrm{s}} = 3.395\text{Å}$, where k_B denotes the Boltzmann constant. The water-tip and water-substrate intermolecular interactions are described by LJ potentials with the parameters determined by combining rules, $\epsilon_{\mathrm{ws}} = \sqrt{\epsilon_{\mathrm{w}}\epsilon_{\mathrm{s}}}$ and $\sigma_{\mathrm{ws}} = (\sigma_{\mathrm{w}} + \sigma_{\mathrm{s}})/2$, where ϵ_{w} and σ_{w} are the LJ energy and size parameters for oxygen site-site interaction of water molecules. The lattice constant l_a of the hexagonal substrate is set to be $2^{1/6}\sigma_{\mathrm{s}}$. Water temperature is set at 298.15K and density at $1\mathrm{g/cm}^3$.

3.2 Integral equation techniques

To calculate the force curve, F_z versus z at each point (x, y), we use the supermolecule approach[12, 13] based on the extended reference interaction site model (RISM) integral equation techniques[14, 15, 16].

The extended RISM theory is an approximate integral-equation theory that has been used in the calculation of site-site distribution functions for intermolecular potential models including long-range interactions. The key equation of the RISM theory is an Ornstein-Zernike-like relation, given in reciprocal (Fourier) space by

$$\hat{\mathbf{h}} = \hat{\mathbf{w}}\hat{\mathbf{c}}\hat{\mathbf{w}} + \hat{\mathbf{w}}\hat{\mathbf{c}}\boldsymbol{\rho}\hat{\mathbf{h}}, \tag{10}$$

where the matrix elements $\hat{h}_{\alpha\beta}(k)$ and $\hat{c}_{\alpha\beta}(k)$ are the site-site pair correlation and direct correlation functions, respectively. $\boldsymbol{\rho}$ and $\mathbf{w}$ are respectively a density matrix and an intramolecular correlation matrix, which are given parameters or functions. The RISM relation, coupled with the site-site hypernetted chain analogue closure relation among $h_{\alpha\beta}$, $c_{\alpha\beta}$ and the site-site pair potential functions $u_{\alpha\beta}$, can be solved numerically for $h_{\alpha\beta}$ and $c_{\alpha\beta}$.

When the probe tip is moved from a reference point $\mathbf{r}_0$ to a point $\mathbf{r}$, the work done by the system, $(\Delta\Omega)_{T,\mu}$, can be written as the sum of two parts: one is the work done to overcome the tip-substrate interaction in vacuum, $\Delta\Omega^{\mathrm{vac}} = U^{\mathrm{vac}}(\mathbf{r}) - U^{\mathrm{vac}}(\mathbf{r}_0)$, and the other is the work required to alter the liquid structure, $\Delta\Omega^{\mathrm{ex}}$. The latter is equivalent to a difference between the excess chemical potentials $\mu^{\mathrm{ex}}(\mathbf{r})$ and $\mu^{\mathrm{ex}}(\mathbf{r}_0)$ of a *supermolecule* consisting of the tip and the substrate with the two *fixed* relative configurations ($\mathbf{r}$ and $\mathbf{r}_0$). Thus, the work can be expressed as

$$\begin{aligned} \Delta\Omega &= \Delta\Omega^{\mathrm{vac}} + \Delta\Omega^{\mathrm{ex}} & (11) \\ &= U^{\mathrm{vac}}(\mathbf{r}) - U^{\mathrm{vac}}(\mathbf{r}_0) + \mu^{\mathrm{ex}}(\mathbf{r}) - \mu^{\mathrm{ex}}(\mathbf{r}_0). & (12) \end{aligned}$$

The excess chemical potential of a supermolecule with a fixed relative configuration is obtained from the relation[16]

$$\beta\mu^{\mathrm{ex}} = 4\pi\rho\sum_{ij}\int drr^2[\frac{1}{2}(h_{ij}^{\mathrm{vu}}(r))^2 - c_{ij}^{\mathrm{vu}}(r) - \frac{1}{2}h_{ij}^{\mathrm{vu}}(r)c_{ij}^{\mathrm{vu}}(r)], \tag{13}$$

where ρ is the density of the solvent and the superscripts vu denote the site-site pair and direct correlation functions between solvent and the tip molecules. To treat the supermolecule consisting of an infinitely-large number of atoms of the substrate, we extend the RISM relation by introducing a two-dimensional periodicity for the supermolecule. This is essentially the same idea as those made by other workers to study polymers in solvents[17]. This method is discussed more fully elsewhere[18].

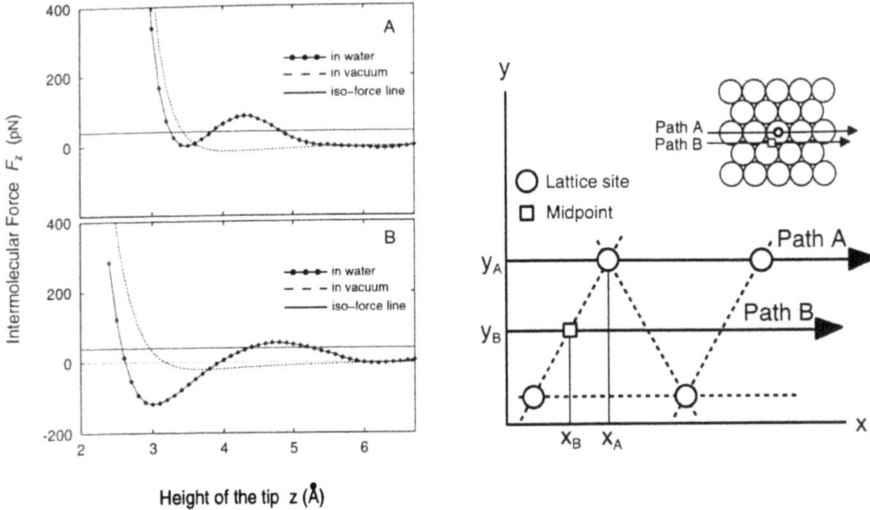

Figure 2: [Left] Force curves F_z of the single-atom tip as a function of the tip height z above the substrate: (A) on top of a hexagonal lattice site and (B) on top of a midpoint between two lattice sites (see Fig. 3). The horizontal solid line denotes an iso-force line (40 pN).

Figure 3: [Right] The two pathways of the probe tip over the hexagonal close-packed substrate. Coordinates (x_A, y_A) and (x_B, y_B) are the initial positions of the tip on paths A and B.

4 Results

Figure 2 displays the calculated force curves, *i.e.*, F_z versus the tip height z. The curves are in general oscillatory, regardless of the lateral position of the tip above the lattice [see Figs. 2(a) and (b)]. The atomic-scale force oscillations are attributed to the layering of liquid molecules near solid surfaces[19]. The intersections of the iso-force line with the force curve correspond to the heights z_i satisfying the condition $F_z = F_z'$. Consequently, there exist a number of trajectories for the tip under the same load. In contrast, there exists only one trajectory for the tip in vacuum[20].

We examine two scanning paths [i.e., two lateral trajectories (x(t),y(t))] shown in Fig. 3. Figure 4 shows the stable trajectories along the path A under three different loads[21]. The dotted curves indicate all possible metastable

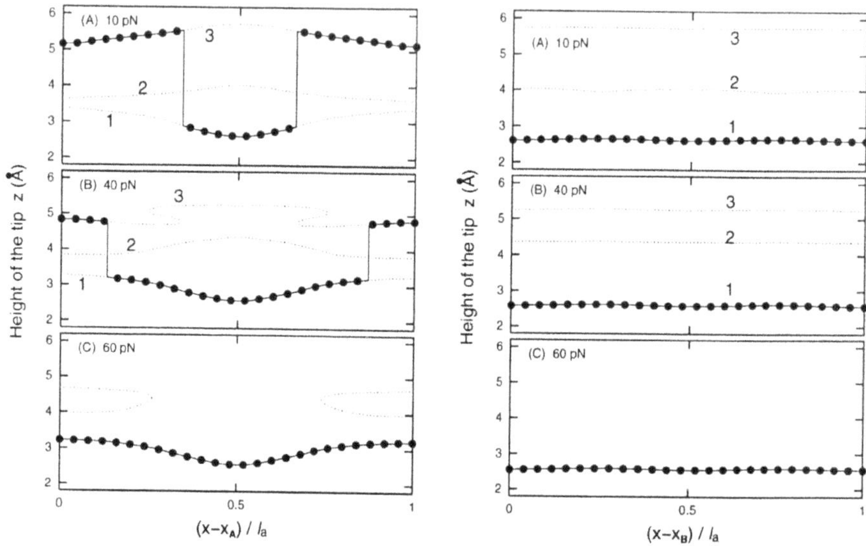

Figure 4: [Left] Solid lines (with heavy dots) delineate the stable trajectory of the tip when moving along the path A under three different constant loads: (A) 10 pN, (B) 40 pN, and (C) 60 pN. Light dotted lines delineate all possible metastable and unstable trajectories of the tip. l_a is the lattice constant of the monolayer substrate.

Figure 5: [Right] The same as Fig. 4 except for the tip moving along the path B.

and unstable trajectories of the tip along the path A. Under the smallest load ($F_z' = 10$ pN) three continuous trajectories are found [denoted by 1,2, and 3 in Figs. 4(a) and (b)]. Along trajectory 1 the tip is always in contact with the substrate ($z \sim 3$Å), whereas along trajectory 3 the tip is separated from the substrate by one hydration layer ($z \sim 5$ Å). Since the tip is unstable on trajectory 2, the stable trajectory can only be portions of either trajectory 1 or 3, depending upon the tip's position. We found that trajectory 3 is more stable when the x coordinate of tip is within $0.3 l_a$ from a lattice site, whereas trajectory 1 is more stable otherwise. This is because when the tip is on top of a lattice site the potential $\Delta\Omega$ is higher for trajectory 1 than that for trajectory 3. As the tip moves away from the lattice site $\Delta\Omega$ gradually decreases for trajectory 1 and eventually becomes lower than that for trajectory

3. Here $\Delta\Omega^{ex}$ plays a key role. When the tip is on top of a lattice site $\Delta\Omega^{ex}$ is about 2.4 k_BT for trajectory 1, which is higher than that for trajectory 3. However, when the tip is on top of the midpoint $\Delta\Omega^{ex}$ is about 1.5 k_BT for trajectory 1, which is lower than that for trajectory 3. Consequently, the stable trajectory exhibits a *step-wise* feature, indicating that the tip may hop back and forth between trajectory 1 to 3 during the quasi-static scan along the path A[22]. If the load is increased, more portions of trajectory 1 become stable [Fig. 4(b)]; under very high load the entire trajectory 1 becomes stable [Fig. 4(c)]. It is of interest to calculate the potential barrier $\Delta^{\ddagger}\Omega$ separating two trajectories 1 and 3. Note that trajectory 2 corresponds to the ridge of the potential barrier. The barrier, which we take to be equal to the difference between Ω on trajectories 2 and 3 at the $x-$coordinate where the tip jumps, is $1.2k_BT$ under the load 10 pN and $0.6k_BT$ under the load of 40 pN, which is comparable to the thermal energy. Similar conclusion has been made from experiment[3].

Figure 5 display the stable trajectories along the path B under three loads. Although three trajectories are also found under the lowest as well as the medium loads [Fig. 5(a) and (b)], the trajectory 1 is always the stable one throughout the path B. Under the highest load no multiple trajectories are found; the sole trajectory is the stable one [Fig. 5(c)] on which the tip is always in close contact with the substrate.

5 Conclusions

This study demonstrates that a single-atom tip in liquid water may undergo hopping motion even over a defect-free substrate, if the probe tip always follows the thermodynamically stable trajectory. One may question the extent to which this hopping motion affects a sample's image quality. It is our hope that this theoretical work will inspire parallel AFM experiments to answer this question. Since the resolution achieved by most AFM measurements in liquid environments so far is about one nanometer[4, 5], this hopping motion may be too small to affect the AFM image. However if the AFM is capable of true atomic resolution in liquids, the hopping motion may complicate the image analysis. The newly developed carbon nanotube AFM by Dai et al[23] seems very promising for detecting this hopping motion because the AFM tip is "sharp" on the atomic scale.

Acknowledgements

We are grateful to Dr. H. Tanaka for providing an original code of the RISM equation program. XCZ and DJD thank the National Science Foundation and the Office of Naval Research for support of this work.

References

[1] F. Ohnesorge and G. Binnig, *Science* **260**, 1451 (1993).

[2] S. J. O'Shea, M. E. Welland and T. Rayment, *Appl. Phys. Lett.* **60**, 2356 (1992).

[3] J. P. Cleveland, T. E. Schäffer and P. K. Hansma, *Phys. Rev. B.* **52**, R8692 (1995).

[4] C. Bustamante and D. Keller, *Physics Today* **48**, 32 (1995).

[5] J. Yang, L.K. Tamm, A.P. Somlyo, and Z. Shao, *J. Microsc.* **171**, 183 (1993).

[6] Y.L. Lyubchenko, P.I. Oden, D. Lampner, S.M. Lindsay, and K.A. Dunker, *Nucl. Acids Res.* **21**, 1117 (1993).

[7] U. Landman, W. D. Luedtke, N. A. Burnham and R. J. Colton, *Science* **248**, 454 (1990).

[8] L. D. Gelb and R. M. Lynden-Bell, *Chem. Phys. Lett.* **211**, 328 (1993).

[9] *Forces in Scanning Probe Methods*, NATO ASI **E286**, edited by H.-J. Güntherodt, D. Anselmetti, and E. Meyer (Kluwer Academic Publishers, Dordrecht, 1995).

[10] B. Bhushan, J.N. Israelachvili, and U. Landman, *Nature* **374**, 607 (1995).

[11] W. Jorgensen, *J. Am. Chem. Soc.* **103**, 335 (1981); B. M. Pettitt and P.J. Rossky, *J. Chem. Phys.* **77**, 1451 (1982).

[12] K. Koga, H. Tanaka, and X.C. Zeng, *J. Phys. Chem.* **100**, 16711 (1996).

[13] S. W. Chen and P.J. Rossky, *J. Phys. Chem.* **97**, 6078 (1993); M. Matsumoto, H. Tanaka, and K. Nakanishi, *J. Chem. Phys.* **99**, 6935 (1993).

[14] D. Chandler and H. C. Andersen, *J. Chem. Phys.* **57**, 1930 (1972).

[15] F. Hirata, B. M. Pettitt, and P. J. Rossky, *J. Chem. Phys.* **77**, 509 (1982).

[16] S. J. Singer and D. Chandler, *Mol. Phys.* **55**, 621 (1985).

[17] F. Hirata and R. M. Levy, *Chem. Phys. Lett.* **136**, 267 (1987); K. S. Schweizer and J. G. Curro, *Phys. Rev. Lett.* **58**, 246 (1987).

[18] K. Koga and X.C. Zeng, *Phys. Rev. Lett.* **79**, 853 (1997).

[19] R.G. Horn and J.N. Israelachvili, *J. Chem. Phys.* **75**, 1400 (1981).

[20] D. J. Diestler, E. Rajasekaren, and X. C. Zeng, *J. Phys. Chem.* **101**, 4992 (1997).

[21] The loads ($\sim 10^{-11}$N) used here are comparable to typical load per atom in laboratory AFM experiments.

[22] Similar behavior has also been found in computer simulation of liquids between two parallel solid surfaces, where the surface may take hopping motion normal to the surface due to the drainage or imbibition transition of the confined liquids. See M. Schoen, D.J. Diestler and J.H. Cushman, *J. Chem. Phys.* **100**, 7707 (1994).

[23] H. Dai, J.H. Hafner, A.G. Rinzler, D.T. Colbert, and R.E. Smalley, *Nature* **384**, 147 (1996); D.J. Keller, *Nature* **384**, 111 (1996).

TRIBOLOGICAL ISSUES OF POLYSILICON SURFACE-MICROMACHINING

J. J. Sniegowski
Sandia National Laboratories
Albuquerque, NM 87185-1080

1. Abstract

Polysilicon surface-micromachining is a Micro-Electro-Mechanical Systems (MEMS) manufacturing technology where the infrastructure for manufacturing silicon integrated circuits is used to fabricate micro-miniature mechanical devices. This presentation describes a multi-level mechanical polysilicon surface-micromachining technology and includes a discussion of the issues which affect device manufacture and performance. The multi-level technology was developed and is employed primarily to fabricate microactuated mechanisms. The intricate and complex motion offered by these devices is naturally accompanied by various forms of friction and wear in addition to the classical stiction phenomena associated with micromechanical device fabrication and usage.

2. Introduction

Fabrication by polysilicon surface micromachining can provide both sensors and actuators. However, microactuators with associated micromechanisms typically present the most severe cases of tribological surface contact and thus the greatest challenge to tribologists. Microsensors, on the other hand, typically fail due to adhesive effects which occur after contact between various surfaces of the device and the supporting substrate. These adhesive effects, often referred to as stiction by micromachinists, is a very significant subset to the general subject of tribology in these devices. Mastrangelo recently [1] presented an in-depth discussion of adhesion related failures and discussed several practical methods for the elimination of the forces which induce contact between surfaces and for the elimination of the intersolid adhesion mechanisms. Briefly summarized, adhesion related failures for structures which are elastically supported above the substrate can be circumvented by design using normalized elastic member dimension bounds. This prevents the collapse or pinning of the various elements of a device structure by assuring their mechanical stiffness exceeds the contact forces. If this is not

B. Bhushan (ed.), Tribology Issues and Opportunities in MEMS, 325-340.

possible, as with the design of highly compliant structures needed for sensitive accelerometers, Mastrangelo reviewed the use of a wide variety of process, surface treatment, and physical schemes. Further, Henck [2] provides another recent and excellent summary of Texas Instruments Incorporated extensive work on adhesion related failure and lubrication of their Digital Micromirror Device™ (DMD™). This is probably the most comprehensive work to date that addresses the tribological issues related to a micromechanical device. Another case, where no elastic supports are present to separate potentially contacting surfaces, is common to many microactuated mechanisms. In this case, surface-to-surface contacts are inevitable.

In this paper, we present a general description of Sandia's polysilicon surface-micromachining technology, the level of device complexity accomplished, and the technology's tribological issues. Examples of geared actuators coupled to other geared micromechanisms are used to illustrate the tribology requirements of this technology. The latter portion of this paper presents several tribology related challenges yet to be solved. The detailed nature of the device surfaces and their processing history affects the mechanical and tribological behavior of surface-micromachined devices often in a more dramatic and peculiar fashion than analogous macro-scale devices. An overview of the types of tribological surfaces, as determined by the fabrication process, are described. These surfaces usually exhibit different surface topography which may have significant impact at these dimensions. For example, the surface topography can be generated by dry plasma etch, or by the surface of a deposited film. Another challenge is the quantitative investigation into the details of the surface condition and topography is hindered by the "micro" size of the features.

3. Polysilicon Surface-Micromachining Technology

Polysilicon surface-micromachining has a number of very desirable manufacturing attributes resulting from its close relation to the infrastructure used in the manufacture of microelectronic integrated circuits (ICs). However, the technology does not readily lend itself to producing devices with tribologically preferred materials nor to producing devices readily amenable to classical tribological treatments such as the use of liquid lubricants or hydrodynamic bearings. New solutions to tribological issues must be developed for surface-micromachining to reach its fullest potential.

3.1. WHY USE POLYSILICON SURFACE MICROMACHINING

The key advantages to polysilicon surface-micromachining are that the process is a batch-fabrication technology which does not require piece-part assembly, and that it utilizes the IC infrastructure which is capable of large-scale, low-cost (per unit) production. Polysilicon micromachining, relative to other micromachining technologies, utilizes the greatest cross-section of IC tools and process technology. The IC manufacturing infrastructure has the potential to produce tremendous quantities of fully-assembled, ready-to-operate, micromechanical devices at very low per-unit costs as long as the device fabrication remains within the constraints of the technology. The aspect of pre-assembled parts is very significant. These parts are in the fully-assembled state after the final release step of the process and are ready to be operated. Without this integral method of assembly and batch-fabrication, intricate techniques for mass assembly would need to be developed.

Secondly, considering the physical size of the devices, there are not many means of accurately defining such small geometries. The more conventional means of machining, although capable of holding tolerances to micron dimensions, are not capable of defining complex geometries with such dimensions. Thus, producing a gear with dimensional control to a few microns with a computer controlled milling machine is readily accomplished. But, this same tool cannot produce a gear of 100 micron diameter with 10 micron gear teeth. It is the photolithographic tools developed for the definition of the IC electronic devices that readily allow the definition of such microscopic gears. A quick review of the MEMS literature clearly illustrates that nearly all forms of micromachining, from bulk micromachining to LIGA, utilize the IC photolithographic tool set to define their devices.

Because of its small size, polysilicon surface-micromachining technology presents unique tribology challenges. Their IC-like fabrication constrains the choice of construction materials. Polysilicon, although a very reasonable mechanical material, does not have optimal tribological properties. In general, the technology presents like-material contacts and polysilicon is poor with regards to wear load. Consideration of more desirable tribological material in the form of non-standard IC films pose contamination issues for IC foundries or are often incompatible with standards processes. Finally, the use of thin film technology and small geometries make the devices highly susceptible to surface forces. Fortunately, multiple film technology enables extensive innovation regarding design solutions to tribology related limitations. For example, mechanical stiffening of structures significantly reduces their susceptibility to external forces. In addition, multiple layers are allowing us to produce much higher force actuators which greatly improve device reliability in regards to overcoming stiction. Recent results hold promise that we will be able to move out of the regime where stiction is very significant.

3.2. THE FABRICATION TECHNOLOGY

Surface micromachining manufactures micromechanical devices utilizing the planar fabrication techniques common to the microelectronic circuit fabrication industry. Constructing devices in this technology consists of adding and selectively defining layers of thin films onto a flat silicon substrate. The complexity of the technology is defined by the number of polysilicon layers. Therefore, a N-level technology refers to N layers of polysilicon (poly) films deposited including both non-structural and structural films. Most often, the first layer is a thin (≤3000Å) electrically conductive film of polysilicon which is used for electrical interconnect and for defining voltage reference planes. Subsequent polysilicon layers, typically 0.5-4.0 μm thick, provide the structural layers from which the mechanical elements are constructed. Interspersed between the polysilicon layers are layers of a sacrificial film, usually silicon dioxide, which provides the spacing between the polysilicon layers and acts as a mold to hold the layers of polysilicon during the fabrication process. Therefore, a N-level polysilicon surface-micromachining technology has, as its basic process module, the repetitive cycle of deposition and definition of two primary films, a sacrificial silicon dioxide film and a structural polycrystalline silicon film. The deposition, photolithography, and etch processes are based on those commonly used for integrated circuit (IC) fabrication, but modified for thicker, mechanically-optimized films. For example, the polysilicon and silicon dioxide films are typically deposited by low-pressure chemical vapor deposition (LPCVD), while reactive ion etch techniques are used for film definition. A 4-level process in essence repeats this base sequence 4 times.

At the completion of the process, the sacrificial layers, as their name suggests, are selectively etched away in hydrofluoric acid (HF), which does not attack the polysilicon layers. The result is a construction system which can be used to form mechanical elements ranging from simple cantilevered beams to complex systems of springs, linkages, mass elements and joints. Because the entire process is based on standard integrated-circuit fabrication technology, hundreds to thousands of devices are batch-fabricated in a ready-to-operate form on a single six-inch silicon substrate.

Typically, structures constructed with one level of polysilicon are only capable of restricted movement because they are attached to the substrate by elastic members. Although the degree of mechanical complexity possible with a single level process is limited, it can nevertheless produce very useful and commercially viable devices, particularly in sensor applications. One such example is Analog Device's surface-micromachined accelerometer [3] which is similar to the simple comb drive pictured in Figure 1a. Extension to a double-level process (Figure 1b) begins to allow considerably greater mechanical design flexibility, particularly with regard to rotating elements. As seen in Figure 1b, a free-spinning gear attached to the substrate with a free-spinning pin at some radius from its center can be produced. However, a third level of polysilicon is needed to couple energy to and from this gear. Figure 1c illustrates this ability to interconnect elements with absolute, hard linkages made possible through the use of three levels. Note also that any or all of the mechanical layers can be made electrically conductive, thus providing additional layers for electrical interconnect or electrodes.

To illustrate actual fabrication steps in relation to producing functional elements, Figure 2a shows the gear in Figure 1b in cross-section. Figure 2b schematically indicates the gear, the hub which anchors the gear to the substrate while allowing rotation, and a pin joint at the outer radius for connection to a link to drive the gear. This represents the functional elements of the structure, while Figure 2c illustrates the same cross-section but with respect to the order of polysilicon film depositions used to form the structure.

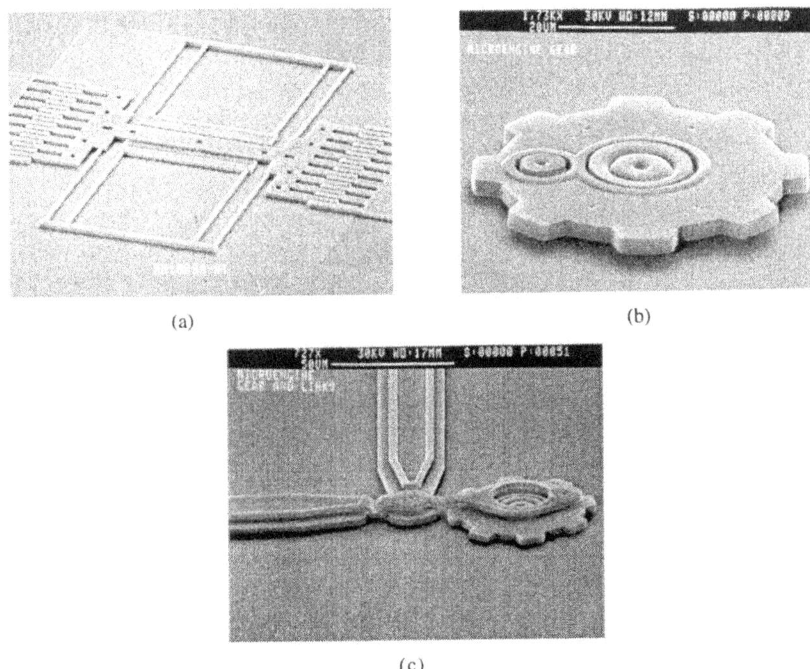

Figure 1. a) Simple, yet very useful structures, particularly for sensor applications, can be fabricated using a two-level process with one layer of interconnect poly and one layer of mechanical polysilicon. b) A three-level process produces movable mechanical elements. However, connection to these structures is limited. Here a gear with a central hub attached to the substrate and a free-spinning pin along its radius is shown. However, connection to the radial pin is not possible without a fourth layer of polysilicon. c) A four-level process allows the fabrication of complex, interconnected, interactive mechanisms with actuators. That is, the gear in Figure 1b is now connected to a linkage element and can be actuated through that element.

From the perspective of physical embodiment, this technology allows the construction of quite complex machinery (see Figures 3-7). However, micromechanical machinery have not yet seen the wide-spread industrial use that micromechanical sensors have achieved. Several issues historically limiting their widespread application have been low force/torque levels, difficulty in coupling tools to engines (for technologies with few mechanical layers), and susceptibility to surface effects such as stiction, friction, and wear. The four-level polysilicon micromachining process developed at Sandia has aided in producing higher-force actuators and the ability to couple tools to microengines and produce useful work. The surface issues remain as an active area of research.

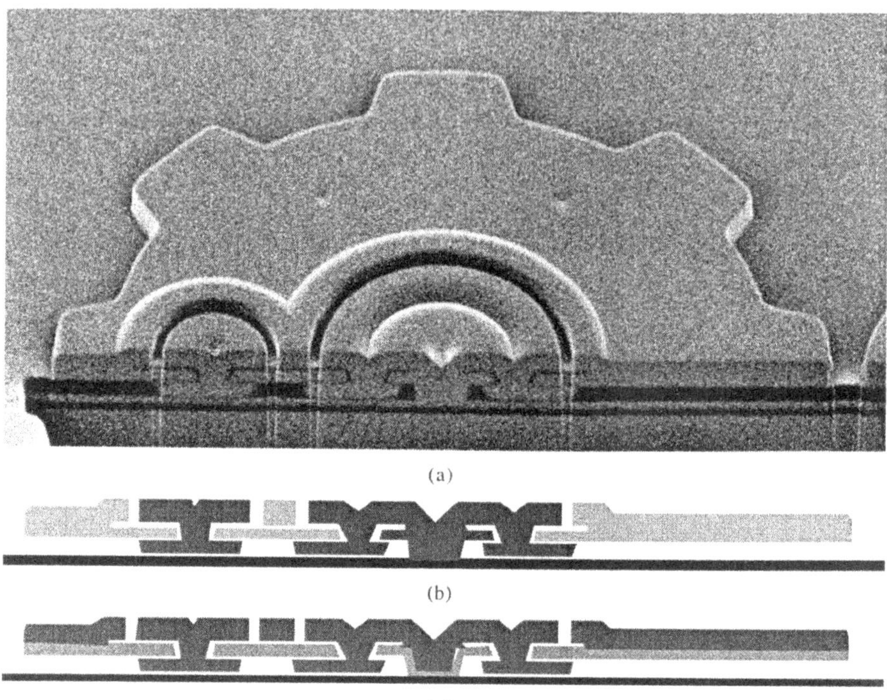

Figure 2. The top image a) is of the cross-section through a microgear generated by focused ion beam (FIB) milling . The schematic b) below it indicates parts of the gear by functional design, while c) illustrates the order of the film depositions during fabrication.

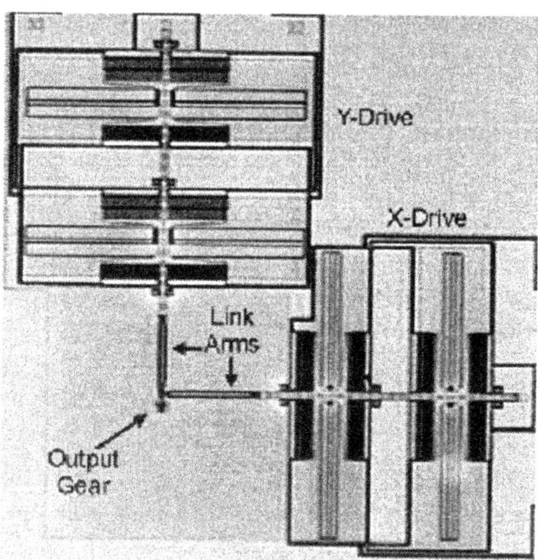

Figure 3. The microengine converts orthogonal movement from the X and Y comb drives into rotational movement of the output gear.

3.3. DEVICE EXAMPLES

The micromachine [4,5] has evolved as one of the primary actuation mechanisms for MEMS at Sandia. The device consists of two orthogonal linear drivers and a pair of linkages coupled to an output gear. This is essentially a 2-D version of pistons, connecting rods, and a crankshaft. Our standard form of linear drives are electrostatically controlled comb drives (Figure 3).

The microengine, in turn, can be coupled to a gear cluster capable of increasing the available torque (Figure 4). An example of a system powered by the microengine and transmission is the fold-up mirror used to redirect an optical signal (Figure 5).

Figure 4. SEM of the gear cluster unit and linear rack. The gear ration is approximately 10:1 in this case. The linear speed of the rack is approximately one tenth that of the linear tooth velocity of the drive gear. The rack can be used to drive a folding mirror, for example.

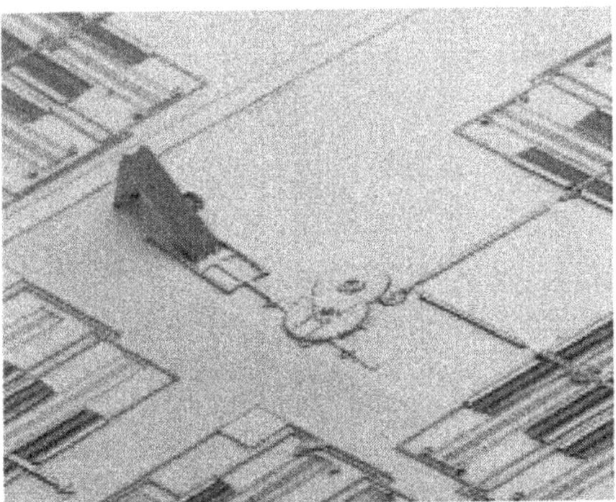

Figure 5. SEM of the gear cluster unit and linear rack being used to fold up a pop-up micromirror.

As a final example, Figure 6 illustrates the combination of several functions which form the basic elements of a combined micro-lock and micro-optical system. The entire operation of unlocking the lock and switching the position of the optical shutter have been demonstrated and can be accomplished in under 100 milliseconds. Several of the images illustrate surfaces which have sliding contact (Figures 6a,b,c for example) or normal contact (Figure 6g). Therefore, the issue of adhesion and friction are critical to the assembly's operation.

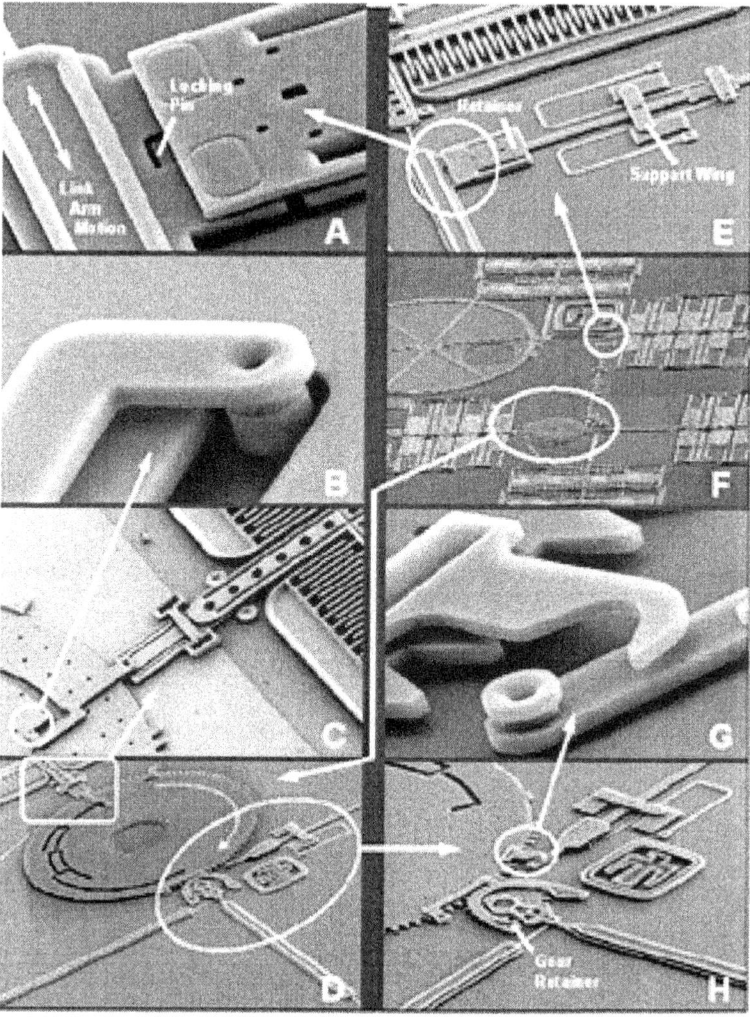

Figure 6. Several images are taken of a micromechanical assembly which combines the functions of a lock and optical shutter. Figure 6f is the overview of the entire pin-in-maze lock, center bottom, and the shutter wheel, upper-left. The other images are close-up of the various elements, such as a latch and hook (images d,h,g).

3.4. STATUS OF THE TECHNOLOGY

The current level of device complexity may be best illustrated by figure 7 where a gear box with a three million-to-one gear ratio has been fabricated in the 4-level process. The device is being used to demonstrate gear cluster complexity and to study the coupling efficiency of geared elements. It has already been demonstrated that gear clusters such as this can multiply torque to a level where buckling pop-up mirrors (Figure 7) from the substrate plane is reliably accomplished. Results indicate that force multiplication places very high loads on elements and may be potentially capable of fracturing the material. This capability is currently being quantified. This may have application to materials testing in addition to a broad range of other applications.

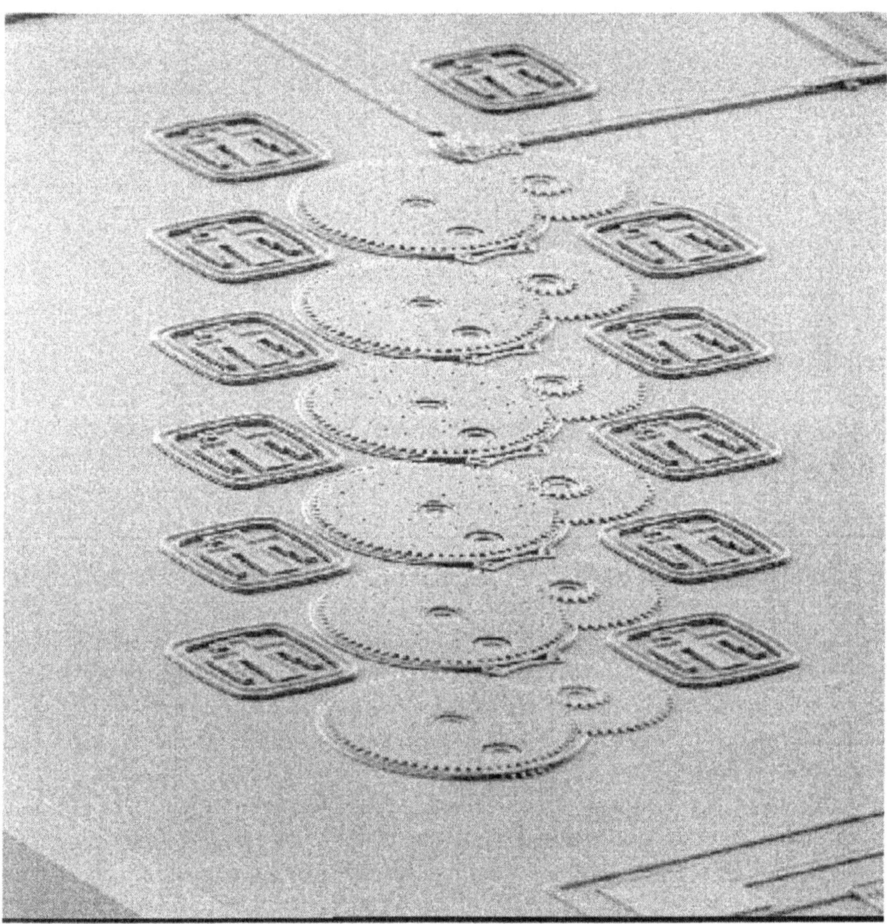

Figure 7. A perspective view of a 3,000,000:1 gear cluster illustrates the complexity available with a 4-level poly process.

Analogous to the IC world, the addition of independent layers permits greater design freedom and complexity. Thus, extension to a 5-level technology is being pursued [6]. From the design perspective, the utility of a 5-level polysilicon/oxide stack is schematically depicted in Figure 8. This technology allows a designer to consider stacking complex, moving structures on top of movable structures. This, in essence, allows micromechanisms to be conveyed about the substrate surface to interact with each other or with the environment. From the manufacturing perspective, this adds another level of challenge for the tribologist.

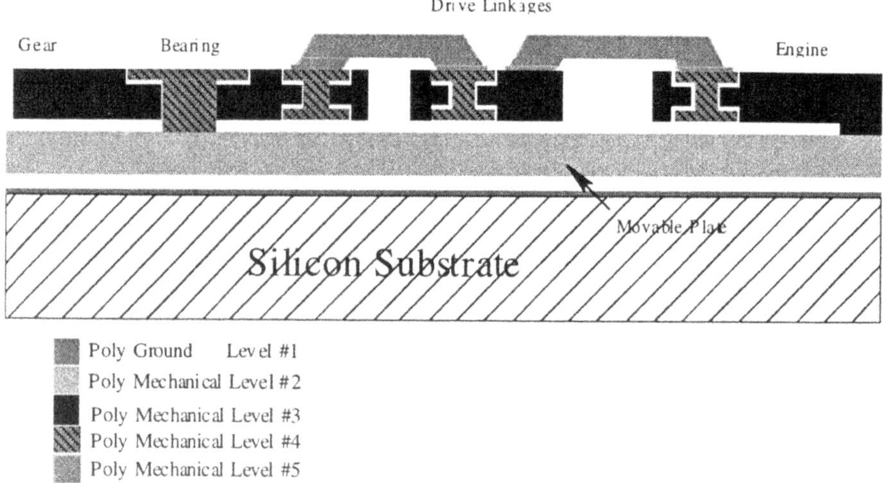

Figure 8. Schematic depiction of the utility of a 5-level technology. Structures capable of movement and manipulation can be placed on plates which can convey them to other locations on a substrate.

4. The Tribological Issues

This section presents examples of the contacting/sliding surfaces encountered with polysilicon surface-micromachined devices. A major aspect of this technology is that the devices, upon release, are fully-assembled. Thus, parts cannot be lubricated and then assembled. Even if separated, it would be extremely difficult to handle the individual elements for coating.

The support springs for the microengine and travel limiters in Figure 9 illustrate a set of "open" contacting surfaces which could be treated with a line-of-sight application method. However, the application of a lubricant will not, in most cases, be line-of-sight due to the fully-assembled nature of the technology. Rather, the lubricant will need to penetrate under surfaces and into relatively small apertures ranging from as small as 0.1 micron to a couple of microns. For example, the cross-section of the gear in Figure 2 showing the gear teeth, the hub, and pin joint is much more typically of surfaces which need lubrication. The teeth are a vertical surface and are line-of-sight.

However, the contacting surfaces of the pin joint and hub present covered surfaces and apertures on the order of 0.5 micron. The effective coverage of coatings, such as the self-assembled monolayers, on these surfaces is under investigation.

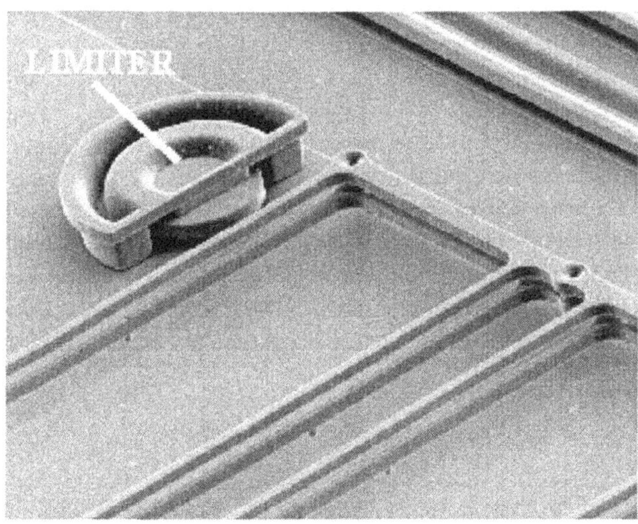

Figure 9. SEM close-up of multi-level support springs and a travel limiter which significantly increase the robustness and reliability.

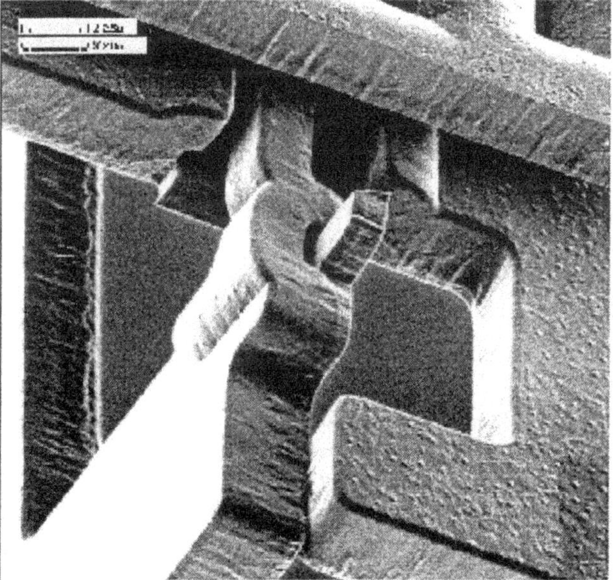

Figure 10. SEM close-up of a hinge used to connect and allow the fold-up of the mirrors show in Figure 5. Note that all three layers of mechanical poly are easily discernible.

The joint shown in Figure 10 allows the folding of the large plates which comprise the micromirrors in Figure 5. Several aspects of the fabrication technology are illustrated by the figure. The 2-D nature of the design with the third dimension arising from the stacking of the films is apparent. The joint does not have the cylindrical appearance of a typical journal bearing. This results from the journal being defined in the thin film by a plasma etch which produces vertical walls relative to the surface of the film. True 3-D machining is beyond the capability of the technology at this time. The surfaces can be inspected and illustrate two types of topography. One is the surface produced by the deposition of the film itself. These are the top and bottom surfaces of the films. This surface typically has RMS roughness on the order of tens of nanometers. The other surface encountered are the sidewalls of the films defined by the plasma etch technique. They typically have a characteristic striation pattern with the striations running vertically. Measurement of the sidewall roughness is much more difficult. The impact of these surface roughness on friction and wear is an active area of research. Surface roughness is known to affect stiction and is discussed below.

5. The Critical Issues: Stiction, Friction, and Wear

Although several material issues continue to be important to polysilicon surface-micromachining, e.g., the measurement of thin film materials properties, and thin film residual stress measurement and control, they are not hindering the use of the devices to the extent of the surface related phenomena. Experience with the microengine, geared micromechanisms, and sensor structures has convinced us that many devices can be adequately designed and fabricated with the desired mechanical functions, but that stiction, friction, and wear present the greatest impediment to common usage. Using the basic microengine with the analysis of Miller et al [7], we are extracting information on the frictional behavior. Under many conditions the devices display 'macroscopic' friction behavior with a friction coefficient of approximately 0.5.

5.1. STICTION (ADHESION-RELATED FAILURE)

Post-release chemical treatments and wafer handling greatly affect the possible outcome of the devices. In particular, the phenomena of stiction affects both sensors and actuators during the final release stage of fabrication and during use of the devices (in-use stiction). The issues of friction and wear primarily applies to actuators. As a MEMS fabrication technology, polysilicon surface-micromachined structures tend to be the most sensitive to stiction. This is chiefly due to their surface-to-volume ratio and the scaling behavior of various surface effects to these small dimensions, e.g. liquid-vapor surface tension. In particular, liquid-vapor surface tension forces (typically water - air menisci) which occur during the drying of parts after the final liquid HF release etch are responsible for bringing surfaces into contact. Recall that the final removal of the sacrificial films to produce free-standing structures is done with a liquid hydrofluoric acid etch. It is during the subsequent drying process that the effects of stiction affect the structures. A dominant force being the surface tension of the liquid-vapor surfaces which form as the device dries. The references by Alley et al [8], Legtenberg, et al [9] provide a review of stiction forces. Typical surface micromachine device fabrication also exacerbates the problem. Nominally, they are constructed close to the supporting substrate (≤3μm) and are comprised of relatively thin films (≤4μm). Thus their dimensions enhance the attractive forces which induce them to adhere and often do not provide sufficient mechanical stiffness to prevent it.

By their very nature, actuators such as the microengine will always have some parts of the structures in contact with each other, e.g., in the joints. Therefore, chemical modification of active surfaces is desirable to reduce adhesion. It has been known for some time that physically roughening the surfaces make them less susceptible to adhesion [10]. Yee et al [11] used this approach to reduce sticking of microstructures. However, it is often undesirable, for mechanical or optical reasons, to have rough surfaces. Maboudian and Howe [12], Mastrangelo [1], and Henck [2] recently contributed reviews which detail the methods currently available for either circumventing or eliminating stiction or adhesion-related failure.

5.2. FRICTION AND WEAR

Beyond adhesion-related failure, friction and wear obviously impact the performance of the microengine and associated structures. The issue of friction is closely tied to the stiction problem. Once stiction is overcome, the devices behave quite well exhibiting 'normal' coefficients of friction. Our experience to date with various surface treatments, including SAMs, indicates this as a promising path. Although we do not have conclusive data regarding the reduction of friction and wear with the application of SAMs, the indication of the preliminary test data clearly dictates a thorough examination.

Wear will probably be the last issue to be attended to directly. However, with the strong relationship between these phenomena and a little serendipity, solutions to stiction and friction may also provide a solution for long lifetime. Clearly though, the order of solution must be first, to move structures (solve stiction), second, to move them smoothly and easily (solve friction), and thirdly, with the first two solved, to move them for a long time (solve wear).

5.2.1 *Liquid Lubricants*

With the application of liquid lubricants, the formation of menisci with the accompanying surface tension forces will disallow a liquid lubricant to be localized to the moving joint or slide areas only. Liquid lubrication will require total immersion of the devices. If total immersion is acceptable in the application of the device, the research of Deng et al [13] with their micromotors and our own microengine [14] results with a lubricant such as silicone oil predicts very acceptable results. A specific result is that operation in a liquid tends to make the device behavior more repeatable and uniform. Unfortunately, the speed of operation in these fluids is greatly reduced relative to operation in air due to viscous drag. When considering that typical lifecycles are in the millions to billions of cycles for operation in air, to repeat this number of cycles at the reduced speed implies a nearly infinite lifetime. On the other hand, other failure mechanisms such as slow drifting of contaminants in the fluid due to electrostatic attractive forces exist.

5.2.2 *Solid Lubricants*

The use of materials of dissimilar hardness in bushing applications for rotating devices is common practice in the macro-world. This concept was reasonably extrapolated into the micro-world. Therefore, films such as diamond-like-carbon, silicon carbide, and silicon nitride for example have been proposed and tried in microstructures. Constraints on the film include compatibility with the process and process tool set and whether it can be deposited on the proper surfaces. Silicon nitride is a film which readily qualifies on both issues. The friction-reduction benefit was demonstrated with some of the early micromotors [15] and friction test structures [16]. Several researchers since that time have seen similar results. This area continues to show potential in dealing with friction and wear, and new films are continually being explored.

5.2.3 *Self Assembled Monolayers as Lubricants*

The micromotor work of Deng et al [17] and our microengine research suggests that both friction and wear can be decreased with the application of SAMs. However, the type of micromotor they investigated operates in a mode with rolling friction as opposed to the sliding friction that occurs in actuators such as the microengine. Whether the effects of SAMs are beneficial to friction and wear needs to be determined for the class of devices with sliding friction such as the microengine. The question whether the SAMs will readily wear off has been partially answered by Henck [2]. His results are consistent with our observations of the behavior of the microengine. There is initial reduction in the coefficient of friction (cof), followed by slightly longer lifetime than untreated surfaces but the SAMs appear to wear off and the device begins to behave as an untreated device. Senft and Dugger [18] have reported initial results on the study of SAMs using surface micromachined tribological test devices.

5.2.4 *Fluorocarbon coatings*

The work of Man et al [19] regarding the deposition of conformal fluorocarbon (Teflon-like) films onto released structures suggests that these films may also eliminate stiction and reduce wear. Their results indicate a tough, very stable film which conformally coats the released structures. We have also initiated an effort to investigate these films, but using commercially available plasma tools and standard IC process gases (e.g. CHF_3) [20]. Initial results are promising and research is continuing. The cof is reduced to less than 0.1, comparable to bulk Teflon™, however these films also indicate wear-off behavior. Therefore, lifetime has not been extended dramatically.

6. The Future

Polysilicon surface micromachining is a very promising technology that Sandia has exploited for a variety of micromechanical sensors and actuators. The general process available provides four levels of polysilicon for mechanical constructions ranging from simple doubly clamped beams for sensing to intricate interconnected linkages and joints for actuation. The basic process has been enhanced by the inclusion of chemical-mechanical polishing to allow easier mechanical design and processing while offering the extension to additional levels of polysilicon and a novel approach to the monolithic integration of CMOS microelectronics with surface micromechanics

Complex devices have been shown to operate with full functionality and to operate at speeds up to several hundred thousand revolutions per minute while accumulating several billions of cycles. While major improvements have been made, the issues of stiction, friction, and wear are not yet fully resolved. They must be further addressed before we will be able to use the microengine as a reliable driver in applications which have requirements from start-up after long dormancy to continuous high-speed operation. Fortunately, surface treatments with self-assembled monolayers, possibly followed by deposition of thin Teflon-like films, appear to be very promising approaches. Recent successes demonstrated by several researchers in these areas suggests these surface treatments to be broadly applicable to polysilicon surface micromachining, including multi-level devices such as the microengine.

7. Acknowledgment

This paper discusses the work of many people at the Sandia National Laboratories' Microelectronics Development Laboratory including G. LaVigne (Microengine testing), P. J. McWhorter (Integrated Micromechanics, Microsensors, & CMOS Technology Department manager), S. Miller (Microengine modeling/testing), M. S. Rodgers (Device design/layout), and B. K. Smith (Plasma coatings), J. Sniegowski (Actuators, four/five-level processes). The process development engineers, operators, and technicians of the Microelectronics Development Laboratory under the direction of J. F. Jakubczak and H. Stewart are acknowledged for their contributions to the process development and fabrication of these devices.

Sandia is a multiprogram laboratory operated by Sandia Corporation, a Lockheed Martin Company, for the United States Department of Energy under contract DE-AC04-94AL85000.

References

1. Mastrangelo, C.H., Adhesion-related failure mechanisms in micromechanical devices, *Tribology Letters* 3 (1997) 223-238.
2. Henck, S.A., Lubrication of digital micromirror devices™, *Tribology Letters* 3 (1997) 239-247.
3. Kuehnel, W., and Sherman, S., 1994, "A surface micromachined silicon accelerometer with on-chip detection circuitry," *Sensors and Actuators A, Vol. 45, no. 1,* 7-16

4. Sniegowski, J. J., Miller, S. L., LaVigne, G., Rodgers, M. S., and McWhorter, P. J., 1996, "Monolithic geared-mechanisms driven by a polysilicon surface-micromachined on-chip electrostatic microengine," Proceedings, *Solid-State Sensor and Actuator Workshop*, Hilton Head Is., SC, USA, 178-182

5. Garcia, E. J., and Sniegowski, J. J., 1995, "Surface micromachined microengine", *Sensors and Actuators A, Vol. 48*, 203-214.

6. Sniegowski, J. J. and Rodgers, M. S., Multi-layer enhancement to polysilicon surface-micromachining technology, (in press) Proc. Intl. Electron Device Meeting (IEDM97), Dec. 7-10, 1997 Washington DC.

7. Miller, S. L., Sniegowski, J. J., LaVigne, G., and McWhorter, P. J., 1996, "Friction in surface micromachined microengine," Proceedings, *SPIE Smart Electronics and MEMS, vol. 2722,* San Diego, CA, 197-204.

8. Alley, R. L. Cuan, G. J., Howe, R. T., and Komvopoulos, K., 1992, "The effect of release-etch processing on surface microstructure stiction," Proceedings, IEEE Solid-State Sensor and Actuator Workshop, Hilton Head Is., SC, 202-207.

9. Legtenberg, R., Tilmans, H. A. C., Elders, J., Elwenspoek, M., 1994, "Stiction of surface microachined structures after rinsing and drying: model and investigation of adhesion mechanism," *Sensors and Actuators A*, vol. 43, 230-238.

10. Guckel, H., Sniegowski, J. J., Christenson, T. R., Mohney, S., and Kelly, T. F., 1989, "Fabrication of micromechanical devices from polysilicon films with smooth surfaces," *Sensors and Actuators, Vol. 20,* 117-122.

11. Yee, Y., Chun, K., and Lee, J. D., 1995, "Polysilicon surface modification technique to reduce sticking of microstructures," Proceedings, *8th International Conference on Solid-State Sensors and Actuators, and Eurosensors IX*, vol. 1, Stockholm, Sweden, 206-209.

12. Maboudian, R., and Howe, R.T., Stiction reduction processes for surface micromachines, *Tribology Letters* 3 (1997) 215-221.

13. Deng, K., Dhuler, V. R., and Mehregany, M., 1993, "Measurement of micromotor dynamics in lubricating fluids," Proceedings, *Micro Electro Mechanical Systems '93,* Fort Lauderdale, FL, USA, 260-264.

14. Miller, S. L., Sniegowski, J. J., LaVigne, G., and McWhorter, P. J., 1996, "Performance trade-offs for a surface micromachined microengine," Proceedings, *SPIE Symposium on Micromachining and Microfabrication*, vol. 2879, Austin, TX.

15. Fan, L-S., Tai, Y-C. and Muller, R. S., 1989, "IC-processed electrostatic micromotors", *Sensors and Actuators, Vol. 20,* 41-47.

16. Lim, M. G., Chang, J. C., Schultz, D. P., Howe, R. T., and White, R. M., 1990, "Polysilicon microstructures to characterize static friction," Proceedings, *Micro Electro Mechanical Systems '90,* Napa Valley, CA, USA, 82-88.

17. Deng, K., Collins, R. J., Mehregany, M., and Sukenik, C. N., 1995, "Performance impact of monolayer coating of polysilicon micromotors," *J. Electrochem. Soc.*, vol. 142, no. 4, 1278-1285.

18. Senft, D. and Dugger, M. T., Friction and wear in surface micromachined tribological test devices, Proceedings, *SPIE Symposium on Micromachined Devices and Components III*, vol. 3224, Austin, TX, 31-38.

19. Man, P. F., Gogoi, B. P., and Mastrangelo, C. H., 1996, "Elimination of post-release adhesion in microstructures using thin conformal fluorocarbon films," Proceedings, *Micro Electro Mechanical Systems '96,* San Diego, CA, USA, 55-60.

20. Smith, B. K. , Sniegowski, J. J., and LaVigne, G., Thin Teflon-like films for eliminating adhesion in released polysilicon microstructures, *TRANSDUCERS97*, Chicago, June 16-19, 1997, 245-248.

ADVANTAGES AND LIMITATIONS OF SILICON AS A BEARING MATERIAL FOR MEMS APPLICATIONS

M. N. GARDOS
Hughes Aircraft Company, Components & Materials Laboratory
El Segundo, CA 90245

Abstract

The useful life of rotating-sliding Si MEMS moving mechanical assembly (MEMS-MMA) machine elements is short. This is attributed to high wear exacerbated by high coefficients of friction (COF) measured during actual micromachine operation *and* during bench-top tribometry, even under the relatively mild environmental stresses of room-ambient conditions. A summary of recent SEM tribometric experiments is given to show that high friction and wear are caused by a variety of factors related to the low cohesive energy density and surface chemistry of poly-Si, Si(100) and Si(111). Compared to polycrystalline diamond (PCD) films, Si performs poorly as a MEMS bearing material. In particular, the 1.8-times strength of the C-C bond in PCD as opposed to the Si-Si bond in bulk silicon translates into more than *10,000 times* lower PCD wear rates under thermal ramping to 850°C, in vacuum. This major difference in behavior has been ascribed to the way Si and diamond interact with the environment. The respective material removal rates are controlled by shear-induced surface cracking superimposed by the adhesive interaction caused by the incipient linkage of dangling bonds between the sliding counterfaces, generated by heating in vacuum above the desorption temperature of adsorbates. The magnitude of the COF is material-specific, generally defined by the number and distribution of dangling (high friction), reconstructed (reduced friction) or adsorbate-passivated (low friction) surface bonds, as a function of temperature and the atmosphere. Even in the benign atmospheric environment of a 0.2 Torr partial pressure of hydrogen gas (P_{H_2}), the COF of the various Si crystallinities at the 15 g (low) SEM tribometer load is as high or higher than in vacuum and significantly higher than those of the PCD under a 28 g load. There is, however, a remarkably repeatable reduction in COF_{Si} within a narrow thermal range at that low load, just below the desorption temperature of hydrogen, indicating the same type of heating- and sliding-catalyzed dissociative chemisorption of H_2 previously observed with PCD. In contrast with PCD, deconstructed (*dangling*) Si bonds do not seem to react with H_2 on cooling, leaving the COF at high values. Only the Si(100) benefited somewhat from P_{H_2} in terms of some wear rate reduction. The other crystallinities exhibited essentially the same wear rates in vacuum and P_{H_2} on thermal ramping to 850°C (near 10^{-12} $m^3/N{\cdot}m$, 10^4-times more than PCD), although the rates were reduced during testing at room temperature by one to two orders-of-magnitude. These values could be further diminished by keeping the COF_{Si} the lowest by staying within the 250°C to 450°C temperature range of the tribocatalytic reaction region. The results indicate that operating in the right thermal-atmospheric environment may extend the life of Si MEMS-MMAs sufficiently for a variety of practical applications.

B. Bhushan (ed.), Tribology Issues and Opportunities in MEMS, 341-365.

1. Introduction

The production processes used for the fabrication of Si-based integrated circuits have also been employed with increasing finesse to make micron-sized Si mechanical structures combined with electronic circuitry on the same piece of semiconductor. As summarized in [1-5] and reiterated here, the useful life of rotating Si micromachines and other MEMS-sized moving mechanical assemblies (MEMS-MMAs) is not nearly long enough for practical applications. They are considered as demonstration hardware only due to excessive wear of Si exacerbated by high coefficients of friction (COF). These unsuitable parameters were measured during both actual micromachine operation *and* bench-top tribometry, even under the relatively mild environmental stresses of room-ambient conditions.

This paper also contains a brief review and critique of the friction and wear testing techniques most often used to characterize and screen MEMS-MMA bearing materials. Atomic force microscopy, friction force microscopy or molecular tribometry are deemed insufficient to approximate the actual contact conditions of most MEMS-MMAs. To present data more representative of the conditions and degradation found in micromechanisms, the results of recent scanning electron microscope (SEM) tribometric experiments are summarized. It is demonstrated that high friction and wear are caused by a variety of factors related to the low cohesive energy density and surface chemistry of Si crystallinities such as undoped Si(100), Si(111) and polycrystalline silicon (poly-Si). Compared to polycrystalline diamond (PCD) films, all Si crystallinities perform poorly as MEMS bearing materials. In particular, the 1.8-times strength of the C-C bond in PCD as opposed to the Si-Si bond in bulk silicon translates into more than *10,000 times* lower PCD wear rates under thermal ramping to 850°C, in vacuum. The extraordinarily low wear of PCD is surprising in view of its flexural strength being only 20-times, the hardness 10-times and the fracture toughness about 5-times higher than those of Si.

This major difference in behavior has been ascribed to the way Si and PCD interact with the environment. The respective material removal rates are controlled by shear-induced microcracking of the relatively low tensile strength surfaces. The dominant failure mode is the drastically increased adhesive interaction via the incipient linkage of dangling σ bonds between the sliding counterfaces, especially above the desorption temperature of adsorbates in vacuum. The magnitude of the COF is material-specific, generally defined by the number of dangling (high friction), reconstructed (reduced friction) or adsorbate-passivated (low friction) surface bonds as a function of temperature and test atmosphere.

New SEM-tribometric results are also given in this paper, indicating that Si micromachine elements operated in the right thermal-atmospheric environment may result in much longer MEMS-MMA lifetimes, without having to apply self-assembled monolayers as lubricants. A remarkably repeatable reduction in COF and wear of all examined Si crystallinities, in a narrow temperature range just below the thermal desorption temperature in a low partial pressure (0.2 Torr) of hydrogen (P_{H_2}), appears to be advantageous. It is hypothesized that in this limited regime, heating- and rubbing-catalyzed dissociative chemisorption of molecular hydrogen turns H_2 into an atomic-level lubricant.

2. Current State of Si MEMS-MMA Improvements

To date, measures aimed at mitigating the high COF and wear of Si for MEMS applications included (a) implantation with carbon [6], boron [7,8] or phosphorus [8] ions, (b) dry oxidative generation or CVD deposition of SiO_2 [8], and (c) coating Si with thin layers of metals such as Ag, Cu, Sn or Zn [10]. These methods have been offered as partial solutions to the tribological misbehavior of Si MEMS-MMAs, despite the fact that oxidized Si surfaces are hydrophilic. Hydrated surfaces and capillary water cause high "stiction" (break-away friction force) in micromotors [11]. Replacement of Si with bare-metallic bearing surfaces (e.g., LIGA-processed Ni coatings deposited on the microelements) also resulted in extremely high friction and wear [12].

Some promise has been shown by recently fabricated electrostatic micromotors spun up to >100,000 rpm, using metals and polymers as bearings. However, lifetimes greater than 6 months could be achieved only when the speeds were reduced to 10,000 rpm [13]. Yet, in many applications, sufficiently high drive torques can only be provided by much higher rotational velocities to compensate for the small rotor mass. By appropriately designing the involute profiles of Si microgears, Sniegowski [14] reported $>2.3\times10^9$ revolutions with 66,300 start/stop cycles for gear trains driven at 200,000 rpm, in air. If such high speeds were a design requirement for the performance of a MEMS motor and gear train assembly, the time of continuous operation before failure would be only 8 days. This still-impressive achievement represented the best results among several attempts, with other gear trains failing after much shorter periods of time. No adequate explanation was given for the lack of repeatability.

At first glance, a simple solution is to hydrogenate all the Si surface bonds before use [15]. This technique might be useful for static release applications in the absence of prolonged contact with water vapor, but it cannot provide repassivation of the wear-generated dangling bonds throughout the tribological process. Another way of lengthening the life of Si MEMS-MMAs appears to be the application of hydrophobic, self-assembled (low surface energy) monolayers on the bearing surfaces [11,16-19]. Again, for static release applications this may be fine, but in the sliding or rolling mode the tribooxidative stability of these thin (mostly hydrocarbon-based) lubricating films, especially with their degradation catalyzed by active surface sites at elevated flash (and environmental) temperatures, is in question. Also, as enumerated in [19], some of these layers are incompatible with certain MEMS materials of construction. The presence of SiO_2 is tribologically undesirable, because "stiction" is caused by hydrogen bonded or capillary water trapped within the bearing interfaces. The formation of surface silica is exacerbated by enhanced tribooxidation of Si when water vapor is there.

The presence of SiO_2 is ubiquitous. It is either an integral part of the MEMS fabrication process or an unavoidable byproduct generated during handling, storage or operation of Si-based MEMS-MMAs, even if only in room temperature air and humidity. The rubbing-enhanced reaction of Si with water vapor to generate ≡Si-OH groups at the interface changes both the chemistry and the morphology of the surfaces. It is the precursor of oxide-hydroxide wear debris formation. In spite of the "lubricating" effects of high partial pressures of water previously shown to provide low friction via the mechanism of Si $\rightarrow$ SiOH $\rightarrow$ SiO_2 $\rightarrow$ hydrated SiO_2 during bench-top tribometry at RT [20], the MEMS literature is awash with reports of the harmful effects of ordered molecular layers of hydrogen-bonded H_2O or worse, capillary (bulk) water and grease-like hydrated silica trapped within the exceedingly small clearances of Si MEMS-MMAs.

Friction-induced energy dissipation on the atomic-molecular-scale and the lubrication of solids with low shear strength surfactant monolayers in benign environments are reasonably well understood [21,22]. Nevertheless, the application of this knowledge towards reducing the high stiction/friction of Si-based MEMS-MMA devices is scarce, even for applications at RT. Instead, more attention has been paid to mitigating the problem by reducing the real area of contact (A_r) through a controlled increase of surface microroughness [23-30]. This methodology is presently used with computer hard discs. Microscopic texture marks are purposely stamped on the disc to reduce A_r between it and the head, and thus mitigate stiction during head lift-off and touchdown, i.e., start-up and shut-down of the computer [23]. Nanoscale scribing of Si microbearing surfaces to reduce A_r with some control (similar to efforts with computer discs) has also been attempted [28]. In both macroscopic [27] and microscopic [4] tribosystems, surface pores and asperity valleys act as debris traps. The depressions keep abrasive debris away from the load-carrying A_r.

Even though surface roughness is an important parameter, the stiction/friction-increasing or reducing effects of any interfacial layer (unavoidably generated by tribochemical reactions with atmospheric adsorbates or purposely preapplied as a lubricant) cannot be considered separately from the roughness [30,31]. The main reason is embodied in the simple and well-known equation $F_k = \tau_s \cdot A_r$, where F_k = friction force, and τ_s = shear strength of the surface layer (A_r as previously defined). Both τ_s and A_r must be controlled simultaneously. Electrostatic effects, which also have to be taken into account as a contributing factor to the residual adhesion and/or repulsion of MEMS surfaces in contact, are closely related to asperity height and shape as well as charge bleed-off [29,31]. The interaction of all these parameters is complex. Even the most thoroughly constructed predictive models can deal with well-defined, statically contacting surfaces only [31]. The initial conditions change radically with progressive wear, rendering the practicality of such models for operating MEMS-MMAs questionable.

These problems stand unresolved. To make matters worse, current research on advanced MEMS still use Si almost exclusively for MMA construction [13,14,32]. The troubling status of MEMS bearing materials is well-represented by the tacit admission in [32]: "....Silicon was selected as the bearing material for its micromachinability, although it is not necessarily the ultimate material of choice for micromachine bearing components." The problem of rapid wear-out and high friction remains pervasive, because the vast majority of these MEMS-MMA devices are still based on Si due to the technological know-how accumulated on manipulating, machining and manufacturing this widely used microelectronics material. The fact is that very few of the practical MEMS incorporate sliding or rolling parts under load, because their wear lives are not long enough. Yet, in many cases, dynamic (tribological) assemblies such as a single micromotor can do the same job better than a large array of static devices [14, 33-38].

3. Tribometric Characterization of MEMS-MMA Bearing Materials

At the present time, atomic force microscopy/friction force microscopy (AFM/FFM) [39] and molecular tribometry (MT) [40-43] are most often used or recommended for approximating MEMS-MMA conditions by bench-testing in the lab. Even a cursory review of the systems aspects of this approach indicates that AFM/FFM and/or MT may be necessary but are not sufficient for the task. The surface behavior of those MEMS microelements which move (but just bend or twist) or even if they touch a counterface (e.g., in sensors, membrane pumps or microtweezers) is vastly different from those that slide-roll under surprisingly high Hertzian stresses (e.g., in micromotors and actuators). Tribometers must be able to reveal the adhesive and friction forces as well as the realistic surface damage (wear) resulting from these particular contact conditions. The essential differences lie in the presence or absence of shear under load and the associated roughening of the worn bearing surfaces and the formation of wear debris, all in the presence of chemical bonds ranging from the weak van der Waals to strong covalent bonding affected by the environment. The combined macroscopic effects of hydration, capillary and electrostatic forces are superimposed on the compaction, plowing and rejection of wear debris from the contacts and the attendant loss of tolerances by wear.

In nearly all MEMS-MMAs the interaction is within area contacts, under a variety of normal loads. As an example, the 16% wear-induced increase in just 5 minutes of the originally 28 μm inside (bearing hole) diameter of an air-driven, Si microturbine wheel driving nothing but itself at 150,000 rpm [44] is not approximated by the action of a sharp AFM/FFM Si, diamond or any other tip brought close to a highly flat Si surface. On the other hand, the hydrogen-bonding-induced hydration forces causing MEMS stiction failures such as Si microtweezers not being able to completely release sapphire microbeads [29] or other similar adhesion problems related to handling of microparts [45] can be quantified by AFM/FFM [46]. Significant differences have been shown between thin hydrocarbon or fluorocarbon lubricant layers serving as stiction-reducing surface films [47]. A properly instrumented AFM/FFM tip can help determine (a) the shear strength of an MoO_3 island displaced laterally on a basal plane of MoS_2 or those of (b) a Nb_2O_3 island similarly sheared off its own disulfide basal plane [48,49], or (c) a C_{60} buckyball island sheared on the basal plane of graphite [47,50].

As to present-day MT apparatus, their crossed-cylinder contacts invariably consist of mica. Although the Hertzian contact area and the related unit stress are controlled better than those of an AFM tip occasionally penetrating (scratching) the counterface on purpose, mica is not a MEMS-MMA bearing material. Molecular tribometry is of little use, unless the surface energy-related friction of ultra-thin fluid lubricant layers deposited on mica are the data of interest [43].

Therefore, at interfaces where the degree of adhesion ("stiction" or repulsion) of essentially statically contacting surfaces is the critical value for MEMS operation, AFM/FFM is useful. Even MT might be applicable, provided the specimen materials were changed from mica to more realistic candidates (e.g., Si). It is essential to recognize that the performance of macroscopic sliding-rolling interfaces is controlled by adhesion, friction and wear in a load range where Amonton's Law is followed. In nanotribometry, the friction force is *not* linearly proportional to the normal load, however [51]. Approximating MEMS-MMA bearing surface behavior at a bench top level needs more suitable test machines.

The scanning electron microscope (SEM) tribometer schematically described in Fig. 1a was specially designed and constructed to fill the gap between an AFM/FFM or MT and a benchtop-type, conventional friction and wear tester. Its primary purpose has been to help reveal tribological effects of bearing material surface chemistry influenced by elevated temperatures and various gas atmospheres, under realistic engineering Hertzian stresses. Transcending the microscopic region into the macroscopic regime, a small (7 mm x 5 mm x 2 mm) flat is oscillated against a hemispherically tipped and dead-weight-loaded pin (2 mm dia.), either in the vacuum of the SEM column (~1 x 10^{-5} Torr; over 90% water vapor in the residual gas atmosphere at room temperature) or in a lidded, Knud-

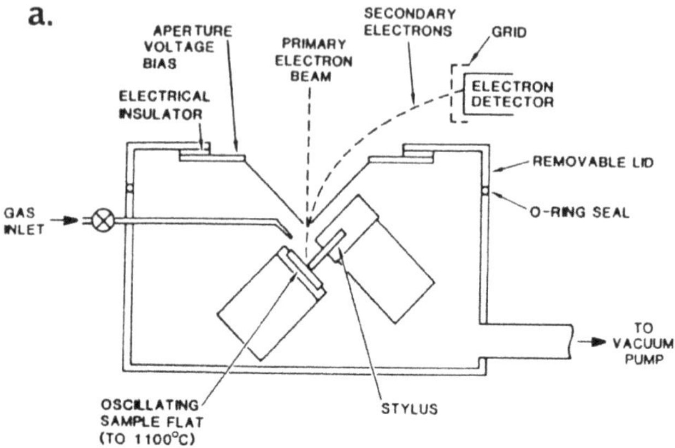

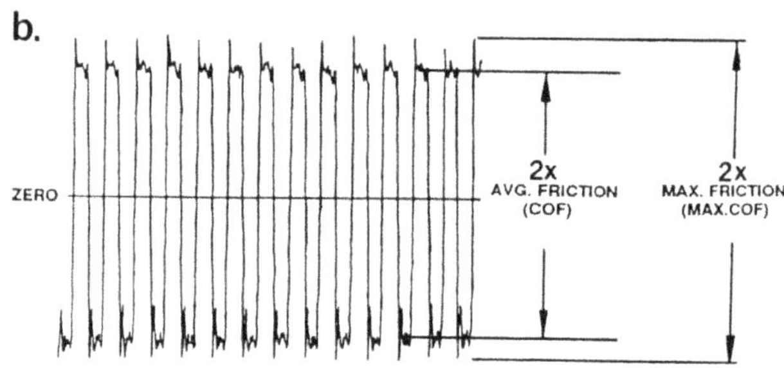

Figure 1. The SEM tribometer: (a) tester schematic, and (b) an idealized friction trace depicting twice the COF and MAX.COF about the zero friction baseline during oscillatory sliding. MAX.COF values are not reported in this paper.

sen cell-like subchamber backfilled and purged with low partial pressures of selected gases such as H_2, He, N_2 and O_2. Depending on the modulus, Poisson's ratio and wear rate of the specimen materials, the Hertzian contact stresses range from GPa to MPa (from many thousands to hundreds of psi) values. The flat may be held at RT or heated to as high as ~1100°C during tribometry. The normal load-force and the average friction force are logged and converted to the average coefficient of kinetic friction (designated as COF in the present paper, see Fig. 1b) by a tabletop computer using commercial data logging and analysis software. This apparatus has been employed to test SiC and intercalated graphites [51], substoichiometric rutile-based lubricious oxides [51-56], MoS_2 [57,58], carbon-graphite seal materials [59,60], as well as Si and PCD for MEMS-MMA bearing material applications [1-5].

4. SEM Tribometry of Si in Vacuum and Hydrogen

SEM tribometric data are presented here on poly-Si, Si(100) and Si(111) (a) heated to 850°C, then cooled to room temperature (RT) at a relatively rapid (40°C/min) thermal ramping rate in ~1.33 x 10^{-3} Pa = 1 x 10^{-5} Torr vacuum (over 90% residual H_2O), followed by similar experimentation in 26 Pa (0. 2 Torr) partial pressures of 99.999%-pure H_2 (P_{H_2}), (b) slowly thermal-ramped to adsorbate (hydrogen) desorption temperatures in P_{H_2}, and (c) kept at RT during sliding in P_{H_2}. Mainly, we wanted to examine the possibility of using molecular hydrogen as an atomic level lubricant, chemisorbed on the dangling bonds generated on rubbed Si surfaces.

The most promising analogous experiments to date have shown that H_2 became dissociatively chemisorbed on PCD in certain temperature ranges, providing the lowest COF measured with PCD sliding against itself [4]. The wear rates in vacuum and in P_{H_2} were in the low 10^{-16} $m^3/N{\cdot}m$ regime, about 10^4-times less than those of the various Si crystallinities examined in vacuum. According to the basic hypothesis governing the fundamental tribological behavior of diamond and Si described in [1-4], the degree of wear is controlled by high COF-induced surface (tensile) cracking directly resulting from the extent of the adhesive interaction caused by the incipient linkage of dangling bonds between the sliding counterfaces.

The magnitude of adhesive COF of PCD and Si are essentially defined by the number of dangling (high friction), reconstructed (reduced friction) or adsorbate-passivated (low friction) surface bonds. Incipient linking of the sliding counterfaces by unsaturated bonds on heating to sufficiently high temperatures, and their passivation by benign adsorbates on cooling, have been suggested as the main causes of radically increased and reduced adhesion and friction, respectively. Strong circumstantial evidence has also been repeatedly given for a trough-like "bathtub" curve dip in the COF at the highest temperatures, attributed to surface re(de)construction. Continued heating of the progressively degased and worn PCD and Si surfaces in vacuum appears to dimerize the dangling bonds to lower the surface energy, unhindered by the ongoing tribological action. However, once the heating stops and the rubbed surfaces cool below a certain temperature, the loss of activation energy needed to keep the bonds reconstructed causes a significant rise from the deepest part of the COF trough due to deconstruction. This rise culminates in another COF peak, where the surface adsorbates rapidly passivate the reappearing dangling bonds to lower the COF once more on further cooling to RT. The best examples of such ideal stepfunction-with-trough COF signatures to date have been generated during SEM tribometry of the mechanically polished and hot chromic acid-cleaned, C(100)-textured PCD films [4]. These layers were sliding against each other under thermally ramped conditions, both in ~1x10^{-5} Torr vacuum and in low P_{H_2} (Fig. 2). The offset of the peaks between COF trough (the first forms at a higher temperature than the second) was attributed to the known thermal-atmospheric stability of the reconstructed surfaces.

During SEM tribometry to ~800°C and cryogenic wear testing of well-characterized amorphous and turbostratic carbons sliding against α-SiC in vacuum, P_{H_2} and P_{He} also showed that a 0.2 Torr P_{H_2} environment was particularly able to reduce the room temperature COF and wide temperature range wear rate of these materials [59-61]. This reduction was attributed to their low sp^2/sp^3 ratio and the passivation of a significant number of dangling *carbon* σ (and π) bonds on the wear-torn surfaces.

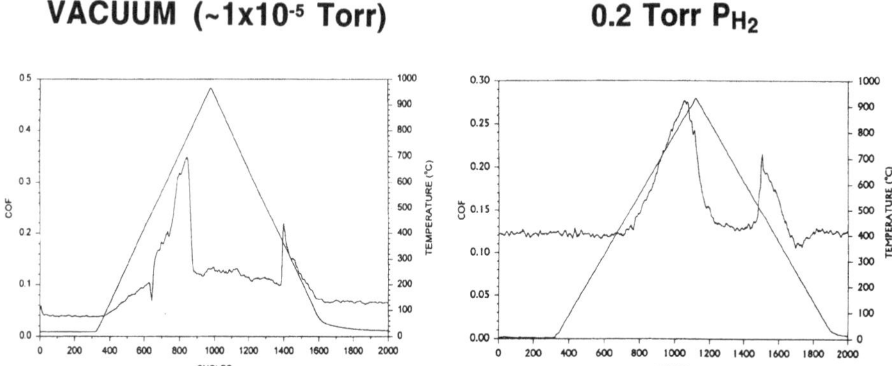

Figure 2. Typical stepfunction-with-re(de)construction trough COF_{PCD} signatures generated by SEM tribometry in vacuum and partial pressure of hydrogen; from [4].

If the dangling (but less energetic) Si σ bonds generated on the sliding counterfaces were also able to break the stable H-H bonds to create the reactive hydride intermediates and hydrogen radicals needed for surface passivation, both the high COF and the frictional tensile stress-induced cracking (i.e., incipient wear) could be mitigated. This way, artificial generation of the reactive hydrogen species in the vicinity of MEMS-MMAs e.g., with hot filaments or UV lamps (convoluted and potential harmful techniques not readily tailorable to the task at hand [62]) would be avoided. MEMS-MMAs fabricated from Si might then be transplanted from the narrow realm of demonstration mechanisms into the ranks of practical hardware.

As a secondary benefit, operating rapidly spinning MEMS components in low P_{H_2} could reduce the extent of wind resistance they normally suffer in atmospheric air or high P_{gas}. Instrument (ball) bearing-operated gyroscopes are routinely rotated in low P_{He} to mitigate such parasitic viscous losses [63]. Inasmuch as the diffusivity of molecular hydrogen through poly- and single-crystal Si is low [64-67], hermetic enclosure of ultrahigh speed microrotors in low P_{H_2} is feasible.

4.1. TEST METHODOLOGY

During the low-load (15 g = 0.15 N) Si vacuum experiments more thoroughly described in [1,2], each test was started at RT, followed by heating to 850°C at a 40°C/min ramping rate. From there, the flat was cooled at the same rate to near RT. Each test lasted for 2000 cycles in vacuum (~1 x 10^{-5} Torr of the SEM column). A pin was tested four times: twice against one flat on progressively fresh test paths starting at a 15 g (the lowest reliable) normal load, in two pumpdown and backfill cycles. After the first test, the flat was moved sideways with a second, built-in x-y stage housing the heater to expose a fresh wear path for the second test. The first two tests yielded usable wear rate ($\mathcal{W}$) data, but only unreliable COF signatures due to the high initial wear of the pin. These experiments were followed by (a) disassembly, photomicrography and HF+H_2O etch/rinse cleaning of the used pin and a new flat (also to simultaneously achieve complete hydrogenation of the surface bonds), and (b) two additional tests with the used pin against fresh wear tracks in separate pumpdown cycles. Repeated testing with the same pin but several successive wear paths on two flats were found necessary to enlarge the wear scar and to mitigate the effects of excessive wear debris formation and plowing of the compacted debris into furrows. These approaches were needed to provide the largest and most debris-free apparent and real areas of contact (A_{app} and A_r) for clearer COF signatures that are footprints of the varying magnitudes of adhesion at the atomic level.

During the following three rounds of P_{H_2} tests, we tried several techniques to eliminate the need for the two starting experiments with each Si specimen pair. These initial tests were only used to enlarge the wear scars. During the fourth round, slow thermal ramping was performed to the temperature signalling the mass desorption of hydrogen to confirm a remarkably repeatable reduction in COF found during the first three rounds. The lowest COF_{Si} ever measured appeared in a narrow thermal range immediately preceding this temperature of sudden COF increase. The $\mathcal{W}$ within this beneficial region was also measured and compared to the $\mathcal{W}$ obtained during (a) conventional high temperature (850°C) thermal ramping in vacuum and in P_{H_2}, and (b) the RT "grinds" in P_{H_2}. Since it is unlikely that silicon would serve in MEMS anywhere near 850°C in any atmosphere, the SEM-tribometric $\mathcal{W}$ values determined at RT are deemed more representative of the $\mathcal{W}$ found with conventional Si MEMS-MMAs.

4.1.1. First Round (Room Temperature "Grind" and Standard Thermal Ramp in P_{H_2})

Here, scars equivalent to some minimum size were prepolished on the tip in an attempt to shorten the overall procedure. The prepolished pin and its mating flat were ($HF+H_2O$)-rinsed to hydrogenate the surface bonds before testing. Next, the plane-parallelism of each artificial "scar" mated to its flat still had to be assured by a 1000-cycle "grind" under the 15 g = 0.15 N load, before the first real test was performed under the same load in an unbroken P_{H_2} atmosphere. As explained before, two marginally useful "grinding" tests had to be performed first during the previous vacuum tests before any valid COF results could be generated.

The seating "grind" in P_{H_2} was done only after several pumpdown and backfill cycles with 99.999%-pure H_2 issued from a lecture bottle, following the procedure delineated in [4]. The lidded tribometer subchamber, with its own separate mechanical vacuum pump continually removing the admitted H_2, was repeatedly purged at gas pressures low enough so that the turbomolecular pump of the SEM column could handle the spill-over H_2 load emanating through the pinhole of the subchamber lid. The thorough H_2 purge cycles removed as much residual air and moisture as possible mainly by momentum transfer, without having to bake the subchamber. After "grinding", the flat was moved sideways with a second motorized stage to expose a new wear path for the first 15 g test, without disturbing the P_{H_2} atmosphere. As in [4], SEM imaging was not done in P_{H_2} to preclude any H-H bond-breaking or desorption assist by the e-beam.

It was found after Round 1 with all three Si crystallinities that the slight shift of each wear scar on the flat that suddenly developed due to the equally slight (but nevertheless detrimental) wobbling of the oscillating sample stage. No sharp scar boundaries were on the flat. Since both the "grind" and the following test were done in an unbroken P_{H_2} atmosphere with no disassembly and photography in-between, well-defined boundaries were essential for accurate estimation of $\mathcal{W}$. As an additional problem, the 1000-

cycle "grind" did not render the artificial scar sufficiently plane-parallel with the flat, despite of the otherwise good alignment of the pin secured in its well-fitting holder. For these reasons, the Round 1 tests generated valid COF results, but $\mathcal{W}$ could not be calculated with confidence.

4.1.2. *Second Round (Room Temperature "Grind" and Standard Thermal Ramp in P_{H_2})*

The observations during Round 2 and photography of the worn specimens indicated again that prepolishing the starting pin tip scars did not work as well as originally anticipated. Furthermore, relubricating the main stage jackscrew could not completely eliminate the stage wobble problem. Therefore, the estimation of reliable $\mathcal{W}$ data was again prevented, although some good COF curves (not reported here), similar to those in Round 1, were generated.

4.1.3. *Third Round (Room Temperature "Grind" and Standard Thermal Ramp in P_{H_2})*

"Grinding" at RT, but only for 2000 cycles (2000 cycles less than in vacuum) was tried again in the hope that less (and less harmful) debris would be caught between the counterfaces in P_{H_2} than in vacuum. This procedure did provide a starting pin tip scar large and debris-free enough (although not as large as those worn in vacuum or the prepolished scars during the first two rounds in P_{H_2}) to yield meaningful COF data for the immediately following thermal-ramp test. We also developed an accurate scar width measurement procedure after the "grind" to allow the estimation, for the first time, of $\mathcal{W}$ in RT P_{H_2}. Knowing the "ground" scar size on the pin tip further permitted the calculation of $\mathcal{W}$ during the thermal ramp test that followed. It was discovered that the pin could be removed and reassembled from/into the friction force transfer arm for photography after "grind" without losing conformity between the pin scar and the flat. The close tolerances of the pin holder's tightly-fitting dovetail that slides securely into the end of the arm may have been responsible for this serendipitous find. However, disassembly had to be followed by another pump-and-purge cycle, introducing an added time penalty to the new procedure.

The wear scar measurement problem was resolved by this revised procedure. However, the slight but persistent stage wobble and the resulting smearing of the wear scars on the flat to a width greater than the width of the pin tip scar were still not eliminated. Additional testing revealed that the wobble of the oscillating stage was not caused by reduced lubrication or increasing wear of its jackscrew. It was the activation of the second stage, used to move the flat sideways between tests without breaking atmosphere, that imparted the skewed motion of the flat during the next experiment. An inexplicably developed backlash of the drive gears and/or the jackscrew of the secondary stage seemed to have affected the behavior of the primary stage. Even though this problem no longer prevented the measurement of the pin tip $\mathcal{W}$ due to the newly devised tip removal and photographic technique after "grinding", the wobble was nevertheless circumvented by manually moving the flat sideways to expose a fresh track for the next test on the same flat. The second motorized stage originally designed for the job was simply not activated. Inasmuch as the pin had to be removed for post-"grind" photography anyway, repositioning the flat in its holder and tightening it down against the heater strip simply became part of the overall test procedure. As a result, the pin scar width could be reconciled with the scar widths on the flat, and Round 3 yielded the best Si friction and wear results generated in P_{H_2} to date. It also gave the first set of reliable RT $\mathcal{W}$ values measured during the initial "grind" cycles, in P_{H_2}.

4.1.4. Fourth Round (Room Temperature "Grind" and Slow Thermal Ramp in P_{H_2})

Using ad-hoc thermal ramping procedures tailored to the general friction behavior of each Si crystallinity, the temperature of the flats was slowly raised to the point of drastic COF increase. We aimed to duplicate the unusual COF reduction in this apparently beneficial thermal-atmospheric regime, and show a possible parallel reduction in $\mathcal{W}$. The RT "grind" cycles preceding the slow-ramp procedures were also meant to yield additional RT $\mathcal{W}$ values, complementing those measured during the Round 3 "grinds".

5. Test Results

5.1. FRICTION DATA

The COF of the Si crystallinities measured in vacuum and those obtained during the Round 1 P_{H_2} tests are compared in Fig. 3. The Round 1 and 3 P_{H_2} experiments are juxtapositioned in Fig. 4 to show the extent of COF repeatability in light of the differences in scar size: those in Round 1 were somewhat larger than during Round 3. Although the stepfunction-with-re(de)construction trough COF signatures previously also observed with PCD in vacuum and P_{H_2} (Fig. 2) manifest themselves with Si as well, there are some important modifications to the curve shapes:

a. There is a remarkable and repeatable reduction in the COF_{Si} in P_{H_2} at the estimated surface temperatures of 250°C to 450°C. The relatively high starting RT COF (as high as 0.45) is reduced there to near 0.1, as low as 0.02. This low friction region occurs just below the temperature of drastic COF increase attributed mainly to the removal of surface hydrogen by thermal desorption. The same type of COF reduction below the thermal desorption temperature of hydrides was observed with PCD in P_{H_2}, but not in vacuum [4]. Apart from this beneficial region, the COF_{Si} at all other temperatures are generally higher in P_{H_2} than in vacuum.

b. There was some reduction in COF_{Si} on cooling in vacuum, but the final values are not as low as those at the onset at RT. This was attributed to the considerable increase in both A_{app} and A_r due to the high $\mathcal{W}$ of Si producing larger wear scars (see the low average Hertzian stresses associated with each test, given in brackets within the respective COF curves). In contrast, the final COF values in P_{H_2} came down only in the case of poly-Si (Round 1). For all other tests, the final RT COF in P_{H_2} was significantly higher than in vacuum, in spite of the smaller scars (i.e., smaller A_{app} or A_r) producing larger Hertzian stresses.

c. The true surface temperature of the flat is heavily influenced by the thermal conductivity of the test atmosphere [60]. Considering the ~200°C ΔT between the heater thermocouple and the estimated surface temperature of the flat in vacuum, but only a ~100°C ΔT in 0.2 Torr H_2 [4,60], the presence of P_{H_2} increases the approximate hydrogen desorption temperatures of Si from about 300° to 400°C in vacuum to 400° to 500°C in P_{H_2}. This phenomenon was observed with PCD as well [4].

d. The thermal range of the re(de)construction-attributed troughs observed in vacuum shifted to higher values during the Round 1 Si(111) P_{H_2} test only. However, the troughs invariably shifted in the same direction during *all* tests of Round 3. The final COF remained characteristically high in P_{H_2}, with no reduction on cooling.

The COF charts of the Round 4 slow-ramp experiments are presented in Fig. 5. The data reconfirm the unusual thermal-atmospheric region where COF is significantly reduced, especially in the case of poly-Si. This crystallinity also yielded the lowest COF in the beneficial region during both Rounds 1 and 3. The adsorbate desorption temperatures rank qualitatively with the strengths of the respective surface hydrides-to-Si bonds: SiH_2-on-Si(100)<SiH-on-Si(111), with the poly-Si desorption temperature in-between.

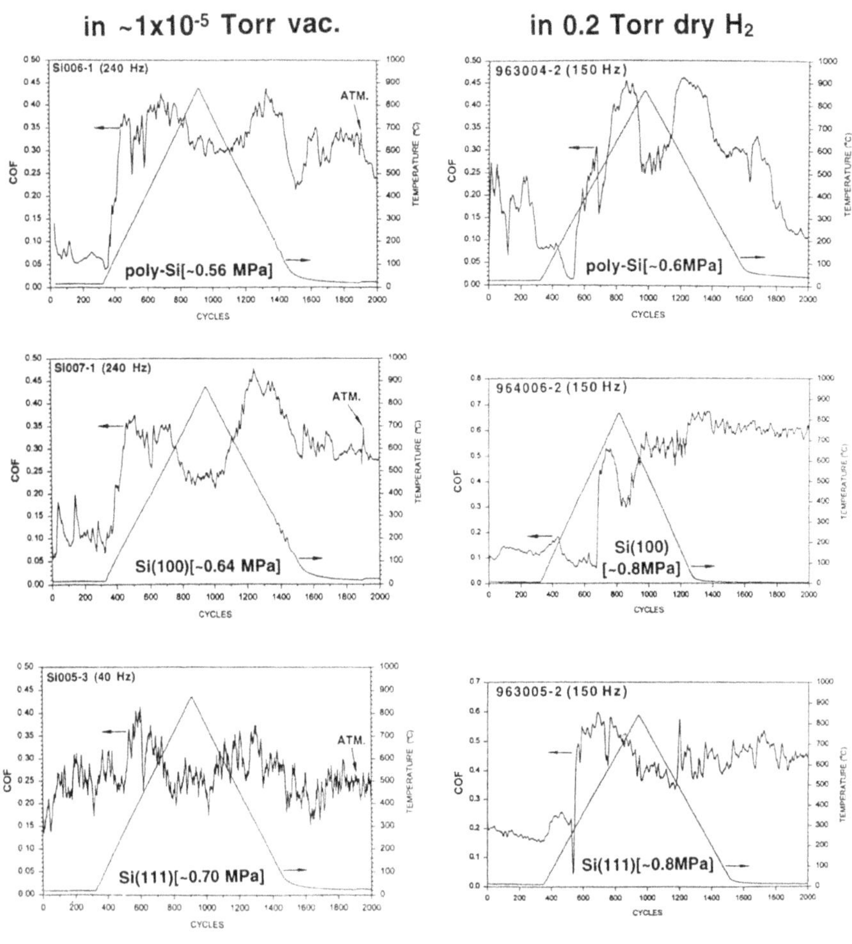

Figure 3. COF of the various Si crystallinities in vacuum and P_{H_2} (Round 1) under standard thermal ramping (Hertzian stresses calculated by wear scar diameters).

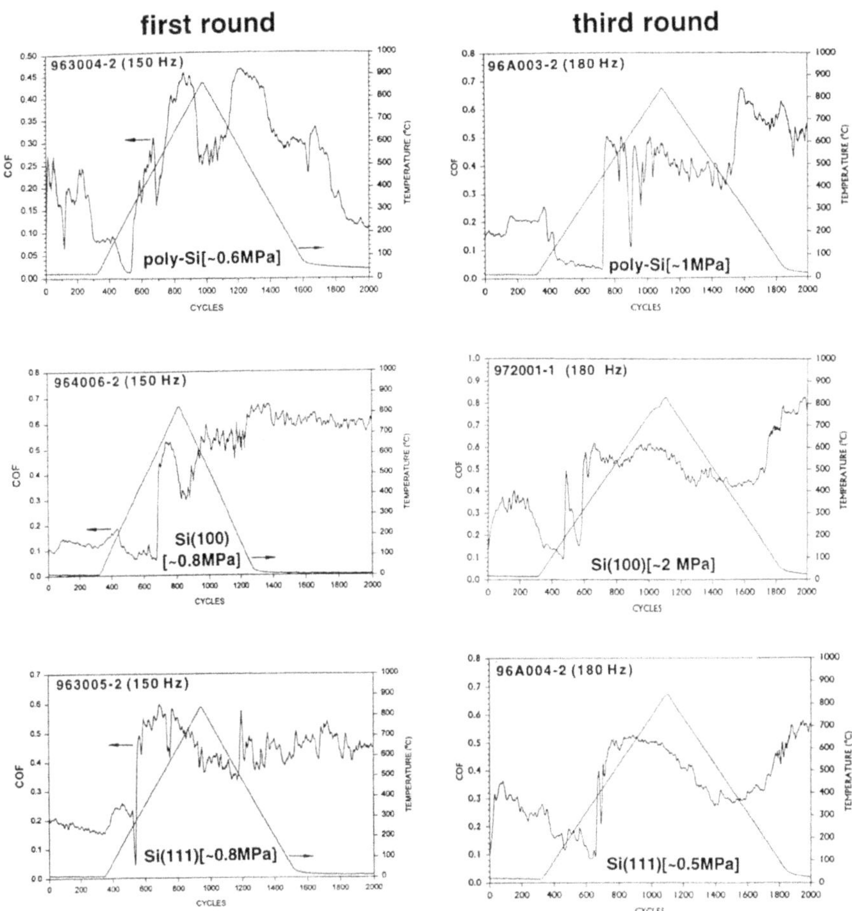

Figure 4. Comparison of COF_{Si} During Round 1 and 2 in P_{H_2} under standard thermal ramping (Hertzian stresses calculated by wear scar diameters).

The low desorption temperature of the Si(100) in Fig. 5 is the same as that indicated by the first desorption peak of the Si(100) doublet in Fig. 4, but does not agree with the Si(100) singlet temperature in Fig. 3.

Since these experiments were not duplicated, the origin and cause of the particular COF variations in Fig. 5, such as the unusually repeatable oscillations with Si(111) or the appearance of multiple desorption peaks are not speculated upon at this time.

0.2 Torr P_{H_2}

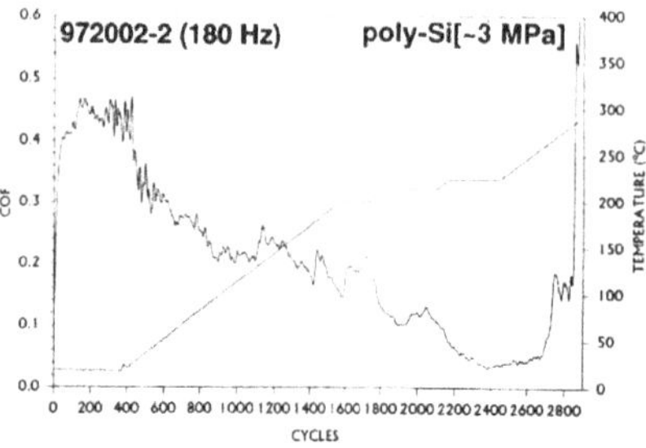

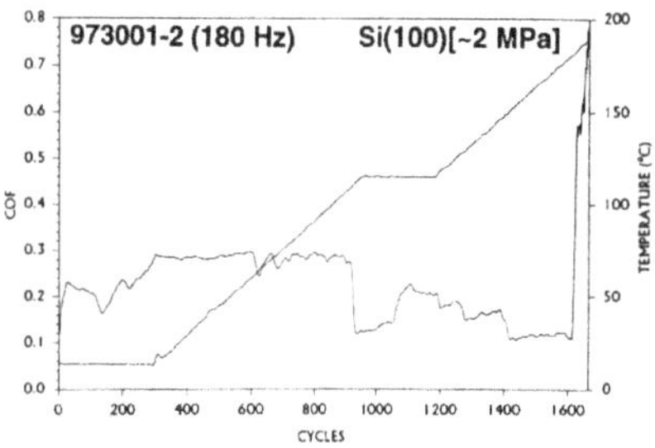

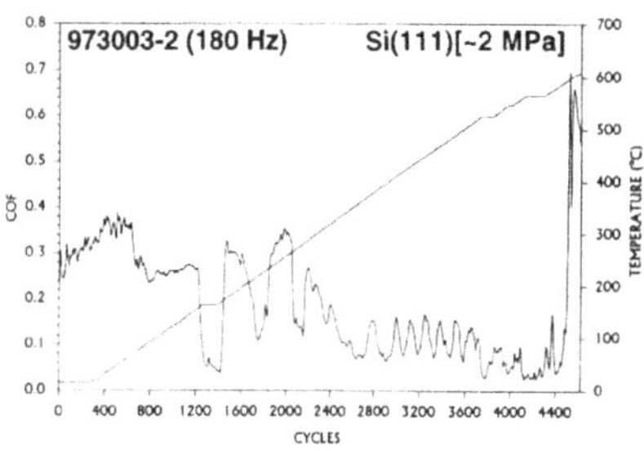

Figure 5. COF of the various Si crystallinities in P_{H_2} under *slow* thermal ramping (Hertzian stresses calulated by wear scar diameters).

5.2. WEAR RATE ($\mathcal{W}$) DATA

The $\mathcal{W}$ of the sliding combinations thermally ramped to/from 850°C in vacuum and in P_{H_2} (designated as TR), "ground" at room temperature in P_{H_2} (RT) and tribocatalytically slow-ramped (TCSR) to hydrogen desorption temperatures are presented in Table 1. The results indicate that during the TR tests, only the Si(100) benefited from the atmosphere of rarified hydrogen. The other two crystallinities exhibited essentially the same $\mathcal{W}$ both in vacuum and P_{H_2}. In P_{H_2}, sliding at RT does reduce $\mathcal{W}$ from the TR level of $\sim 10^{-12}$ m^3/N·m by one to two orders-of-magnitude. *Operating in the P_{H_2} TCSR regime yields the lowest friction and wear in each case.* The poor repeatability of the duplicated RT Si(100) $\mathcal{W}$ cannot be explained until (a) more tests are run, and (b) the SEM photomicrographs depicting the appearance of the scars on the pins and flats are examined.

Even in the absence of these data, it is known that the precision and accuracy of the $\mathcal{W}$ associated with the RT and TCSR experiments is low due to the small size of the pin scar and the error involved in measuring the small scar diameter from SEM photos.

Table 1. The wear rate ($\mathcal{W}$) of the various silicon crystallinities in vacuum and PH_2 under thermally ramped to/from 850°C (TR), room temperature (RT) "grinding" and tribocatalytically slow-ramped (TCSR) conditions.

Pin/Flat	Atm./Test Type	Stress (MPa) Start	Stress (MPa) End	Pin Wear Rate* (m^3/N·m)	No. of cycles
Si(100)	Vac./TR	0.8	0.5	2.82×10^{-12}	2000
	dry PH_2/TR	3.3	0.8	6.2×10^{-13}	2000
	dry PH_2/RT	430	3.3	3×10^{-14}	2000
		430	108	(3×10^{-17})*	2000
	dry PH_2/TCSR	108	57	(9×10^{-17})*	1625
Si(111)	Vac./TR	1.0	0.8	9.7×10^{-13}	2000
	dry PH_2/TR	2.5	0.9	1.45×10^{-12}	2000
	dry PH_2/RT	430	2.5	2.9×10^{-13}	2000
		430	2.8	8×10^{-14}	2000
	dry PH_2/TCSR	2.8	2.0	(2×10^{-15})*	4600
poly-Si	Vac./TR	0.8	0.6	1.62×10^{-12}	2000
	dry PH_2/TR	132	1.1	1.27×10^{-12}	2000
	dry PH_2/RT	430	132	1×10^{-14}	2000
		430	3.5	3×10^{-14}	2000
	dry PH_2/TCSR	3.5	2.9	(9×10^{-15})*	2850

* values in parentheses are reported to a larger number of significant figures due to the extremely small wear rates; precision/accuracy are poorer.

6. Discussion

6.1. DISSOCIATIVE CHEMISORPTION OF MOLECULAR HYDROGEN ON Si

The general thermodynamic concepts of hydrogen de(re)sorption off/on Si appear simple (Fig. 6), but they are not. The activation energy of the respective sorption paths vary widely and the hypothesized reactive hydride intermediates interacting with the Si surface during either direction have not been clearly elucidated [2,69]. The desorption kinetics themselves vary from one cleavage plane to another: first-order kinetics were observed for Si(100), but second order for Si(111) [69].

Notwithstanding the variation in activation energies and the differences in the kinetic paths that involve hydride intermediates, it is easier to desorb the surface hydrides during heating where sufficient thermal energy is available than to re(chemi)sorb molecular hydrogen during cooling. It is empirically known that H_2 molecules have very low sticking probability on clean Si(100)-2x1. Theoretical calculations [68] yielded a ~46 kcal/mol

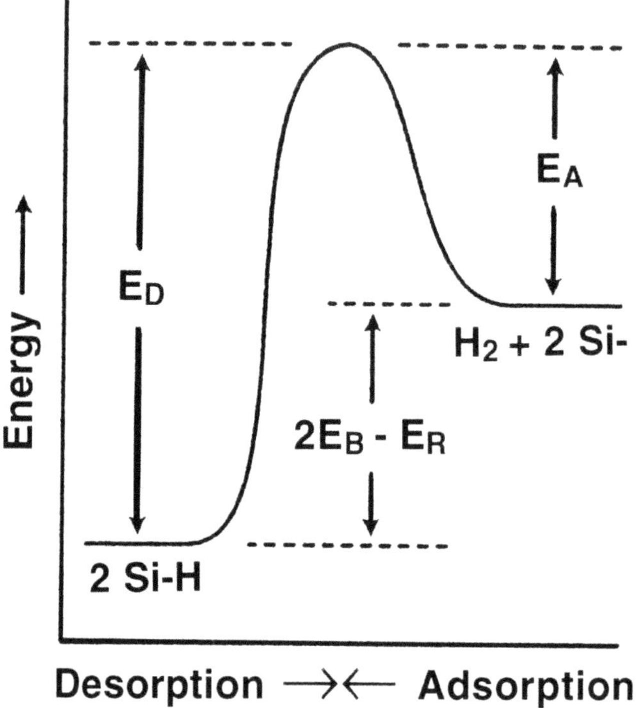

Figure 6. Bond and activation energies of hydrogen desorption and adsorption off/on Si surfaces (E_D = activation energy of desorption, E_A = activation energy of adsorption, E_R = H-H bond energy and E_B = Si-H bond energy); after [68].

activation energy for dissociative adsorption of H_2 onto Si(111), and between 15 and 19 kcal/mol for Si(100) [71,72]. In practice, Si surface bonds are capped with hydrogen mostly by HF rinses or using atomic H generated via plasmas or hot filaments [68,70]. Instead of all surface atoms becoming fully hydrogen-capped, Si(100) flats reconstruct to (2x1) upon baking in H_2 at high (≥1100°C) temperatures [73]. Reconstruction is maintained even in air and atmospheric moisture for surprisingly long periods of time [73].

However, the high stability of the (2x1) structure was not always observed [74], especially not in progressively higher relative humidities [75]. In ultrahigh vacuum (UHV), molecular dynamics simulations predicted and experiments showed that Si(100) wafers pressed together statically (no sliding) spontaneously bonded covalently at RT [76]. These wafers were previously dehydrogenated in UHV and exhibited a (2x1) reconstruction pattern before contact. As early as in 1967, a paper dealing with the passivation of dangling Si bond defects at the Si/SiO_2 interface demonstrated that baking in H_2 at temperatures between 1000° and 1200 °C negated the effects of electron trapping at the interface [77]. A patent was granted along the same lines in 1979, teaching that passivation could be achieved between 650° and 950°C already [78]. Others have adapted this method of reducing the number of charge carrier traps at the Si/SiO_2 interface [79,80]. Annealing of Si solar cells in H_2 similarly improved device performance [81,82]. The ability of both *atomic* and *molecular* hydrogen to passivate dangling bonds at the grain boundaries of poly-Si was similarly evaluated. The H_2 deactivated the free radicals during elevated temperature annealing, but not as effectively as H [82].

Epitaxial growth studies of Si [83] also deal with the dissociative chemisorption of H_2 [83]. Even Ref. 73, showing stable reconstruction of Si on hydrogen-annealing at elevated temperatures, contains a passing reference to some H_2 instability in terms of possible chemisorption: "...During the high temperature stage of the H_2 anneal the small steps on the silicon surface become unstable and by fast surface diffusion between the step edges larger flat terraces are formed so that the step density is minimized. Simultaneously or during the cooling down the surface reconstructs to the 2x1 structure, hydrogen is dissociatively absorbed [*sic*] on this surface during the cooling in the H_2 atmosphere." Heating of lapping-damaged Si surfaces in molecular hydrogen to temperatures as low as 380°C already started passivating dangling bonds, as attributed to either H_2 entering strained Si-Si bonds [81], or simply passivating the $Si_3{\equiv}Si$- dangling bonds on the surface [79,80]. Just plain heating Si to 810°C in vacuum alone was shown to deplete Si interstitials from the surface region and thus reduce dangling-bond-containing point defects [81].

Although these data show some ability on the part of molecular hydrogen to chemisorb on Si, they do not indicate unambiguously whether thermodynamic or kinetic factors control the decomposition of H_2, especially in the presence of the highly active dangling bonds continuously generated by wear.

As to *chemically induced* surface attrition (wear), annealing Si in H_2 at high temperatures (a) removes any oxide that formed during boule growth, originating from residual oxygen and moisture in the reactor [85], or (b) strips the thin native oxide that invariably forms on Si in air [86]. Further etching of the bare Si by hydrogen via the generation of volatile SiH_4 or by plain volatilization of the Si atoms is negligibly small below ~1000°C, but is observable above this temperature [86]. This is confirmed by other etching and Si evaporation data [87,88], where etching of the Si(100) surface was predicted to be nearly three-times faster than that of Si(111). The intermediate hydride products are apparently more readily stabilized on the Si(100) surface [87]. In contrast, surface cracking induced *mechanical* wear of Si should be higher in H_2 than in vacuum. Cracks forming in the wake of a high friction tribocontact act as important precursors to the wear of Si and PCD both in vacuum and P_{H_2}. These cracks propagate faster in P_{H_2} (at least in the case of Si), because hydrogen passivates the dangling bonds that would otherwise hold the crack walls together [89].

6.2. RECONCILIATION WITH THE SEM TRIBOMETRIC DATA

The COF results are in line with the sorption data in the literature. The heating- and sliding-catalyzed dissociative chemisorption of H_2 in the thermal regime just before the mass desorption of the surface hydrides appears to be responsible for the lowest COF measured for all Si crystallinities in P_{H_2}. The presence of hydrogen also skewed the surface desorption and the reconstruction temperatures to values higher than those observed in vacuum, in spite of more effective heat transfer between the heater strip and the flat in P_{H_2}. The increase in the desorption temperatures may be explained by the LeChâtelier

Principle, a basic law of chemical equilibrium adapted to the present case: if a reaction (here, the desorption of an adsorbate) results in a gas (H_2) as a product, increasing the partial pressure of this gas in the reaction chamber will retard the reaction (i.e., desorption will occur at higher temperatures). A similar argument may also be applicable to the skewed re(de)construction troughs in P_{H_2}. Gas-phase reactions at the highest temperatures may have also hindered the formation and eventual dimerization of dangling bonds.

The COF remained high on cooling in P_{H_2}, because the chemisorption energy hill could not be overcome in the absence of sufficient thermal energy, the sliding induced flash temperatures and the presence of wear-generated dangling bonds notwithstanding. Therefore, on the first order, the tribocatalytic activation in the absence of external heating should be equal to or less than 46 kcal/mol on Si(111) and 15 to 19 kcal/mol on Si(100).

Yet, there was a passivating reaction with the small amount of residual water in the moderate SEM column *vacuum* during the cooling cycles (Fig. 3). This indicates that dissociated water vapor may also be a good atomic level lubricant for Si, provided the amount present in the test environment is below a certain partial pressure. The small amount of residual H_2O (on the order of 10^{-5} Torr) remaining in the SEM column readily sorbs onto the rubbed Si surface, in contrast with the severely hindered adsorption of the <1 ppm of H_2O remaining in the 99.999%-pure P_{H_2}. The residual H_2O molecules in P_{H_2} are unable to compete effectively with the much larger number of H_2 molecules for the surface. This concept of competitive sorption was previously explained in [58]. Therefore, the low starting and final RT COF in the SEM column vacuum may indicate some lubricating usefulness of even very small (~10^{-5} Torr) P_{H_2O} interacting with the rubbed Si surfaces, provided that the thermal environment is below the desorption temperature of water. This hypothesis is now being tested for all three Si crystallinities by introducing a small P_{H_2O} into the 0.2 Torr H_2 atmosphere of the SEM tribometer subchamber during standard thermal ramping (TR) to/from 850°C.

The fact that only Si(100) has benefited some from the P_{H_2} atmosphere in terms of reduced $\mathcal{W}$ during TR may be associated with the lowest surface energy of a fully H-terminated Si(100) [2,15], maintained longer to some extent by the effects of the LeChâtelier Principle. With the rest of the crystallinities, any wear reduction under the benign conditions of the tribocatalytic hydrogen chemisorption region is probably counteracted by the otherwise generally high COF and more extensive surface microcracking in P_{H_2}. As a combined result, the $\mathcal{W}_{TR}$ appeared to be the same for poly-Si and Si(111) both in vacuum and P_{H_2}. Examination of the photomicrographic appearance of the worn pin and flat scars as well as the morphology, amount and distribution of the wear debris may shed additional light on the differences and/or similarities in the respective wear mechanisms.

7. Conclusions

A review of the tribometric techniques used for the characterization of MEMS-MMA bearing materials (mainly Si) indicates that at interfaces where the proper degree of adhesion ("stiction" or repulsion) of essentially statically contacting surfaces is the critical value for MEMS operation, AFM/FFM is useful. Even MT might be made acceptable, if the specimen materials were changed from mica to more realistic candidates. However, where the performance of the sliding-rolling interfaces is controlled by adhesion, friction and wear, more appropriate test machines are needed.

A summary of recent SEM tribometric experiments more suitable to the task is also given to show that high friction and wear are caused by a variety of factors related to the surface chemistry and low cohesive energy density of poly-Si, Si(100) and Si(111). Compared to PCD films, Si does not perform nearly as well as a MEMS bearing material. This major difference in behavior has been ascribed not just to the large difference in cohesive energy density, but also to the way Si and diamond interact with the environment. The respective $\mathcal{W}$ are controlled by shear-induced surface cracking superimposed by the adhesive interaction caused by the incipient linkage of dangling bonds between the sliding counterfaces. These bonds are generated by heating in vacuum or in P_{H_2} above the desorption temperature of the surface adsorbates and by wear. The validity of this hypothesis will be further scrutinized by the photographic examination of the P_{H_2}-worn scars.

The magnitude of the COF is material-specific, generally defined by the number and distribution of dangling (high friction), reconstructed (reduced friction) or adsorbate-passivated (low friction) surface bonds, as a function of temperature and the atmosphere. In P_{H_2}, the COF of the various Si crystallinities is as high or higher than in vacuum and significantly higher than those of the PCD. However, there is a remarkably repeatable reduction in COF_{Si} in a narrow temperature range just below the thermal desorption temperature of hydrogen, indicating the same type of heating- and sliding-catalyzed dissociative chemisorption of H_2 previously observed with PCD. Unlike PCD, however, the deconstructed (dangling) Si bonds do not react with H_2 on cooling, leaving the final COF at very high values.

Only the Si(100) appears to benefit, to a limited extent, from P_{H_2} in terms of some reduction in $\mathcal{W}$. The other crystallinities exhibited essentially the same $\mathcal{W}$ in vacuum and P_{H_2} on standard thermal ramping to 850°C (near 10^{-12} $m^3/N{\cdot}m$, 10^4-times more than PCD), although the wear rates were reduced during RT testing by about a factor of 10 to 100. *The wear rates could be further lowered by keeping COF the smallest by staying within the temperature range of the tribocatalytic reaction region.* The results indicate that operating Si MEMS-MMAs in the right thermal-atmospheric environment may lengthen their useful lifetimes sufficiently for a variety of practical applications, provided the MEMS electronics on the same chip can withstand the same temperatures.

Acknowledgments

The author would like to thank Messrs. S. Gabelich, L. Melton, B. Buller and G. Meldrum of Hughes Aircraft Company for performing the SEM tribotests with consummate skill and reducing the data to publishable form. Prof. Jiri Jonas (Beckmann Inst., U. of Illinois Champaign-Urbana) provided invaluable references essential to one of the main themes of the present paper. This research was completed under AFOSR Contract No. F49620-95-C-0002 managed by Maj. Hugh C. De Long at AFOSR/NL.

References

1. Gardos, M.N. (1996) Surface chemistry-controlled tribological behavior of Si and diamond, *Tribol. Lett.*, **2**, 173-187.
2. Gardos, M.N. (1996) Tribological behavior of polycrystalline and single-crystal silicon, *Tribol. Lett.*, **2**, 355-373.
3. Gardos, M.N. (1997) Tribological behavior of polycrystalline diamond films, in *"Protective Coatings and Thin Films," Proc. NATO Adv. Res. Workshop, May 30 - June 5, 1996, NATO ARW Series,* Y. Pauleau and P.B. Barna (eds.), Kluwer Academic Publishers, Dordrecht, The Netherlands, pp.185-196.
4. Gardos, M.N. (in press) Re(de)construction-induced friction signatures of polished polycrystalline diamond films in vacuum and hydrogen, *Tribol. Lett.*.
5. Gardos, M.N. (in press) Silicon and diamond for MEMS bearing applications in extreme environments, in *Proc. Fifth Int. Symp. on Diamond Materials*, 192nd Joint International Meeting of the Electrochem. Soc., Aug. 31-Sept. 5, 1997, Paris, France.
6. Gupta, B.K., Chevallier, J., and Bhushan, B., (1993) Tribology of ion bombarded silicon for micromechanical applications, *Trans. ASME, J. Tribol.*, **115**, 392-399.
7. Gupta, B.K., Bhushan, B., and Chevallier, J. (1994) Modification of tribological properties of silicon by boron ion implantation, *Tribol. Trans.*, **37**, 601-607.
8. Bhushan, B., and Li, X. (1997) Micromechanical and tribological characterization of doped single-crystal silicon and polysilicon films for microelectromechanical systems devices, *J. Mat. Res.*, **12**, 54-63.
9. Bhushan, B., and Koinkar, V.N. (1994) Tribological studies of silicon for magnetic recording applications, *J. Appl. Phys.*, **75**, 5741-5746.
10. Jang, D.-S., and Kim, D.-E. (1996) Optimum film thickness of thin metallic coatings on silicon substrates for low load sliding applications, *Tribol. Int.*, **29**, 345-356.
11. Alley, G.R.L., Cyuan, J., Howe, R.T., and Komvopoulos, K. (1992) The effect of release etch processing on surface microstructure stiction, in *Tech. Digest, 1992 IEEE Solid-State Sensor and Actuator Workshop* (June 21-25, 1992, Hilton Head Island, NC), pp. 202-207.
12. Mathieson, D., Beerschwinger, U., Yang, S.J., Reuben, R.L., Taghizadeh, M., Eckert, S., and Wallrabe, U. (1996) Effect of progressive wear on the friction characteristics of nickel LIGA processed rotors, *Wear*, **192**, 199-207.
13. Anon.(1996) Micromotor's lifetime exceeds six months, *Micromachine Devices (R&D Magazine Newsletter)*, **1(3)**, Nov. 1996, p.2.
14. Sniegowski, J.J. (1996) Shrinking microdesigns in microdevices: surface micromachined gearboxes, locks and mirrors, in *Proc. NAS/AVS/JPL Micromachining Workshop III,* Sept. 25-26, 1996, Anaheim, CA.; Miller, S.L., Sniegowski, J.J., LaVigne, G., and McWhorter, P.J. (1996) Performance tradeoffs for a surface micromachined microengine, *Micromachined Devices and Components II*, SPIE **Proc. Vol. 2882**, 182-191.
15. Houston, M.R., Howe, R.T., and Maboudian, R. (1997) Effect of hydrogen termination on the work of adhesion between rough surfaces, *J. Appl. Phys.*, **81**, 3474-3483.

16. Tsukruk, V.T., Bliznyuk, V. N., Visser, D., and Hazel, J. (1996) Reconstruction of fluid Langmuir monolayers under shear forces, *Tribol. Lett.*, **2**, 71-80; Tsukruk, V.V., Bliznyuk, VB., and Everson, M.P. (1996) Nanotribology of composite molecular layers, *Polymer Prepr.*, **37**, 557-558.
17. Wang, D., Thomas, S.G., Wang, K.L., Xia, Y., and Whitesides, G.M. (1997) Nanometer scale patterning and pattern transfer on amorphous Si, crystalline Si, and SiO_2 surfaces using self-assembled monolayers, *Appl. Phys. Lett.*, **70**, 1593-1595.
18. Abeln, G.C., Lee, S.Y., Lyding, J.W., Thompson, D.S., and Moore, J.S. (1997) Nanopatterning organic monolayers on Si(100) by selective chemisorption of norbornadiene, *Appl. Phys. Lett.*, **70**, 2747-2749.
19. Maboudian, R., and Howe, R.T. (1997) Stiction reduction processes for surface micromachines, *Tribol. Lett.*, **3**, 215-221.
20. Zanoria, E.S., Danyluk, S., and McNallan, M.J. (1995) Formation of cylindrical sliding-wear debris on silicon in humid conditions and elevated temperatures, *Tribol. Trans.*, **38**, 721-727.
21. Yoshizawa, H., Chen, Y.-L., and Israelaschvili, J. (1993) Recent advances in molecularl evel understanding of adhesion, friction and lubrication, *Wear*, **168**, 161-166.
22. Klein, J., Kumacheva, E., Mahalu, D., Perahla, D., and Fetters, L.J. (1994) Reducing the frictional forces between solid surfaces bearing polymer brushes, *Nature*, **370**, 634-636.
23. Elings, V.B., and Gurley, J.A. Scanning tunneling microscopy in research and development, *Res. & Dev.*, Feb. 1989, pp. 126-129.
24. Alley, R.L., Mai, P., Komvopoulos, K., and Howe, R.T. (1993) Surface roughness modification of interfacial contacts in polysilicon microstructures, in *Proc. 7th Int. Conf. Solid-State Sensors and Actuators, TRANSDUCERS '93*, June 7-10, 1993, Yohohama, Japan.
25. Suh, N.P., Moshleh, M., and Howard, P. S. (1994) Control of friction, *Wear,* **175**, 151-158.
26. Kotwal, C.A., and Bhushan, B. (1996) Contact analysis of non-Gaussian surfaces for minimum static and kinetic friction and wear, *Tribol. Trans.*, **39**, 890-898.
27. Dubrujeaud, B., Vardavoulias, M., and Jeandin, M. (1994) The role of porosity in the dry sliding wear of a sintered ferrous alloy, *Wear*, **174**, 155-161.
28. Ando, J., Ino, Y., Ishikawa, Y., and Kitahara, T. (1996) Friction and pull-off force on silicon surface modified by FIB, in *Proc. Ninth Ann. Int. Workshop on Micro Electro Mechanical Systems*, San Diego, CA Feb. 11-15, 1996, IEEE Cat. No. 96CH35856, ISSN 1084-6999, pp. 349-353.
29. Arai, F., Andou, D., and Fukuda, T. (1996) Adhesion forces reduction for micro manipulation based on micro physics, in *Proc. Ninth Ann. Int. Workshop on Micro Electro Mechanical Systems,* San Diego, CA Feb. 11-15, 1996, IEEE Cat. No. 96CH35856, ISSN 1084-6999, pp. 354-359.
30. Stanley, H.M., Etsion , I., and Bogy, D.B. (1990) Adhesion of contacting rough surfaces in the presence of sub-boundary lubrication, *Trans. ASME, J. Tribol.*, **112**, 98-104; Jeng, Y.-R., (1996) Impact of Plateaued Surfaces on Tribological Performance, *Tribol. Trans.*, **39**, 354-361.
31. Komvopoulos, K., and Yan, Y. (1997) A fractal analysis of stiction in micromechanical systems, *Trans. ASME, J. Tribol.*, **119**, 391-400.
32. Anon. (1996) Kona conference raises 4 issues about commercialization of MEMS, *Micromachine Devices (R&D Magazine Newsletter)*, **1(2)**, Oct. 1996, p. 2.

33. Helvajian , H., and Robinson, E.Y. (eds.) (1993) Micro- and nanotechnology for space systems: an initial evaluation, *ATR-93(8349)-1*, The Aerospace Corp., El Segundo, CA, 1993; Scott, W.B. Micromachines Hold Promise for Aerospace, *Aviation Week*, March 1, 1993, p. 36.
34. Helvajian, H. (ed.) (1995) Microengineering technology for space systems, *ATR-95(8168)-2*, The Aerospace Corp., El Segundo, CA, 30 Sept. 1995.
35. Hawkey, T.J., and Torti, R.P. (1993) Integrated microgyroscope, in *Proc. Workshop on Microtechnologies and Applications to Space Systems*, JPL Publ. 93-8, NASA CR-195688, June 15, 1993, pp.257-264.
36. Yates, R., Williams, C., Shearwood, C., and Mellor, P. (1996) A Micromachined rotating yaw rate sensor, *Micromachined devices and Components II*, SPIE **Proc. Vol. 2882**, 161-168.
37. Legtenberg, R., Berenschot, E., Elwenspoek, M., and Fluitman, J. (1996) Electrostatic microactuators with integrated gear linkages for mechanical power transmission, in *Proc. Ninth Ann. Int. Workshop on Mico Electro Mechanical Systems*, San Diego, CA Feb. 11-15, 1996, IEEE Cat. No. 96CH35856, ISSN 1084-6999, pp. 204-209.
38. Bright, V.M., Reid, J.R., and Comtois, J.H. (1996) Polysilicon thermal actuator arrays and applications, in *Proc. NAS/AVS/JPL Micromachining Workshop III*, Sept. 25-26, 1996, Anaheim, CA.
39. Bhushan, B. (1996) Nanotribology and nanomechanics of MEMS devices, in *Proc. Ninth Ann. Int. Workshop on Micro Electro Mechanical Systems*, San Diego, CA Feb. 11-15, 1996, IEEE Cat. No. 96CH35856, ISSN 1084-6999, pp. 91-98.
40. Israelaschvili, J.N., and McGuiggan, P.M. (1990) Adhesion and short-range forces between surfaces. Part I: New apparatus for surface force measurements, *J. Mater. Res.*, **5**, 2223-2231; ibid, Part II: Effects of surface lattice mismatch, 2232-2243.
41. Van Alsten, J., and Granick, S. (1989) Friction measured with a surface forces apparatus, *Tribol. Trans.*, **32**, 246-250.
42. Hirano, M., Shinjo, K., Kaneko, R., and Murata, Y. (1991) Anisotropy of frictional forces in muscovite mica, *Phys. Rev. Lett.*, **67**, 2642-2645.
43. Matsuoka, H., and Kato, T. (1996) Discrete nature of ultrathin lubrication film between mica surfaces, *Trans. ASME, J. Tribol.*, **118**, 832-838.
44. Gabriel, K.J., Behi, F., Mahadevan, R., and Mehregany, M. (1990) In-situ friction and wear measurements in integrated polysilicon mechanisms, *Sensors and Actuators*, **A21-A23**, 184-188.
45. Fearing, R.S. (1995) Survey of sticking effects for micro parts handling, in *Proc. 1995 IEEE/RSJ Int. Conf. on Intelligent Robotics and Systems,* **Vol. 2**, 212-217.
46. Noy, A., Frisbie, C.D., Rozsnyai, L.F., Wrighton , M.S., and Lieber, C.M. (1995) Chemical force microscopy: exploiting chemically-modified tips to quantify adhesion, friction, and functional group distributions in molecular assemblies, *J. Am. Chem. Soc.*, **117**, 7943-7951; Thomas, R.C., Houston, J.E., Crooks, R.M., Kim, T., and Michalske, T.A. (1995) Probing adhesion forces at the molecular scale, *J. Am. Chem. Soc.*, **117**, 3830-3834.
47. Meyer, E., Lüthi, R., Howald, L., Gutmannsbauer, W., Haefke, H., and Güntherodt, H.-J. (1996) Friction force microscopy on well-defined surfaces, *Nanotehnology,* **7,** 340-344.
48. Kim, Y., Huang, J.-L., and Lieber, C.M. (1991) Characterization of nanometer scale wear and oxidation of transition metal dichalcogenide lubricants by atomic force microscopy, *Appl. Phys. Lett.*, **59**, 3404-3406.

49. Lieber , C.M., and Kim, Y. (1991) Characterization of the structural, electronic and tribological properties of metal dichalcogenides by scanning probe microscopies, Thin Solid Films, **206**, 355-3590.
50. Lüthi, R., Meyer, E., Haefke, H., Howald, L., Gutmanns Bauer, W., and Güntherodt, H.-J. (1994) Sled-type motion on the nanometer scale: cohesive energies of C_{60}, *Science*, **266**, 1979-1981.
51. Feldman, K., Fritz, M., Hahner, G., Marti, A., and Spencer, N.D. (1997) Surface forces, surface chemistry, and tribology," invited paper presented at the First World Tribology Congress, Sept. 8-12, 1997, London, England, in *New Directions in Tribology*, Mech. Eng. Publ. Ltd., London, England, 1997, pp. 347-353.
52. Gardos, M.N. (1990) Determination of the tribological fundamentals of solid lubricated ceramics, Volume 1: Summary, *WRDC-TR-90-4096*, Nov. 1990, Hughes Aircraft Co., El Segundo, CA, USA.
53. Gardos, M.N. (1988) The effect of anion vacancies on the tribological properties of rutile (TiO_{2-x}), *Tribol. Trans.*, **31**, 427-436; also see discussions on this paper (1989) *Tribol. Trans.*, **32**, 30-31 ; Gardos, M.N., Hong, H.-S., and Winer, W.O. (1990) The Effect of anion vacancies on the tribological properties of rutile (TiO_{2-x}), Part II: Experimental evidence, *Tribol. Trans.*, **32**, 209-220.
54. Gardos, M.N. (1989) The Tribooxidative behavior of rutile-forming substrates, in *Mat. Res. Soc. Symp. Proc., New Materials Approaches to Tribology: Theory and Applications,* L. E. Pope, L. L. Fehrenbacher and W. O. Winer, (eds.) **Vol. 140**, 325-338.
55. Gardos, M.N. (1993) The Effect of Magnéli phases on the tribological properties of polycrystalline rutile (TiO_{2-x}), in *Proc. 6th Int. Congr. on Tribology, Eurotrib '93,* M. Kozma (ed.), **Vol. 3**, Aug. 30 - Sept. 2, 1993, Budapest, Hungary, pp. 201-206.
56. Gardos, M.N. (1994) Determination of the tribological fundamentals of solid lubricated ceramics, Part III: Molecular engineeering of rutile (TiO_{2-x}) as a lubricious oxide, *WL-TR-94-4108*, Hughes Aircraft Company, El Segundo, CA, USA, Oct. 1994.
57. Didziulis, S.V., Fleischauer, P.D., Soriano, B.L., and Gardos, M.N. (1990) Chemical and tribological studies of MoS_2 films on SiC substrates, *Surf. Coat. Technol.*, **43/44**, 652-662.
58. Gardos, M.N. (1995) Anomalous wear behavior of MoS_2 Films in moderate vacuum and dry nitrogen, *Tribol. Lett.*, **1**, 67-85.
59. Gardos, M.N., Adams, P.M., Barrie, J.D., and Hilton, M.R. (1997) Crystal-structure-controlled tribological behavior of carbon-graphite seal materials in partial pressures of helium and hydrogen. I. Specimen characterization and fundamental considerations, *Tribol. Lett.*, **3**, 175-184.
60. Gardos, M.N., Davis, P.S., and Meldrum, G.R. (1997) Crystal-structure-controlled tribological behavior of carbon-graphite seal materials in partial pressures of helium and hydrogen. II. SEM tribometry, *Tribol. Lett.*, **3**, 185-198.
61. Gardos, M.N., and Buller, B.W. (1997) Crystal-structure-controlled tribological behavior of carbon-graphite seal materials in partial pressures of helium and hydrogen. III. Cryogenic (100 K) wear tests, *Tribol. Lett.*, **3**, 199-204.
62. Uchida, Y., Deane, S.C., and Milne, W.I. (1992) Posthydrogenation of low-pressure chemical vapor deposited amorphous silicon using a novel internal lamp system and its application to thin-film transistor fabrication, *J. Appl. Phys.*, **72**, 3150-3154.

63. Gardos, M.N., and Meeks, C.R. (1984) Solid lubricated rolling element bearings, Part I: Gyro bearings and the associated solid lubricants research, *Vol.1: Summary, AFWAL-TR-83-4129*, Final Report, Hughes Aircraft Company, El Segundo, CA, February 1984.
64. Abrefah, J., Olander, D.R., and Balooch, M. (1990) Hydrogen dissolution in and release from nonmetals. II. Crystalline silicon, *J. Appl. Phys.*, **67**, 3302-3310.
65. Beyer,W. (1991) Hydrogen effusion: a probe for surface desorption and diffusion, *Physica B*, **170**, 105-114.
66. Stutzman, M., Beyer, W., Tapfer, L., and Herrero, C.P. (1991) States of hydrogen in crystalline silicon, *Physica B*, **170**, 240-244.
67. Jackson, W.B., Johnson, N.M., Tsai, C.C., Wu, I.-W., Chiang, A., and Smith, D. (1992) Hydrogen diffusion in polycrystalline silicon thin films, *Appl. Phys. Lett.*, **61**, 1670-1672.
68. Myers, S.M., Follstaedt, D.M., Stein, H.J., and Wampler, W.R. (1992) Hydrogen interaction with cavities in helium-implanted silicon, *Phys. Rev. B.*, **47**, 13,380-13,394.
69. Wu, C.J., Ionova,I.V., and Carter, E.A. (1994) First-principles-derived rate constants for H ad-atom surface diffusion on Si(100)-2x1, *Phys. Rev. B*, **49**, 13,488-13,500.
70. Wu, C.J., Ionova,I.V., and Carter, E.A. (1993) Ab initio H_2 desorption pathways for H/Si(100): the role of $SiH_{2(a)}$, *Surf. Sci.*, **295**, 64-78.
71. Sinniah, K., Sherman, M.G., Lewis, L.B., Weinberg, W.H., Yates, J.T. Jr., and Janda, K.C. (1989) New mechanism for hydrogen desorption from covalent surfaces: the monohydride phase on Si(100), *Phys. Rev. Lett.*, **62**, 567-570.
72. Sinniah, K., Sherman, M.G., Lewis, L.B., Weinberg, W.H., Yates, J.T. Jr., and Janda, K.C. (1990) Hydrogen desorption from the monohydride phase on Si(100), *J. Chem. Phys.*, **92**, 5700-5711.
73. Bender, H., Verhaverbeke, S., Caymax, M., Vatel, O., and Heynes, M.N. (1994) Surface reconstruction of hydrogen annealed (100) Silicon, *J. Appl. Phys.*, **75**, 1207-1209.
74. Ibach, H., Wagner, H., and Bruchman, D. (1982) Dissociative chemisorption of H_2O on Si(100) and Si(111) - a vibrational study, *Solid Sate Commun.*, **42**, 457-459.
75. Sakuraba, M., Murota, J, and Ono, S. (1994) Stability of the dimer structure formed on Si(100) by ultraclean low-pressure chemical-vapor deposition, *J. Appl. Phys.*, **75**, 3701-3703.
76. Gösele, U., Stenzel, H., Martioni, T., Steinkirchner, J., Conrad, D. and Schnee-schmidt, K. (1995) Self-propagating room-temperature silicon wafer bonding in ultrahigh vacuum, *Appl. Phys. Lett.*, **67**, 3614-3616.
77. Hofstein, S.R. (1967) Stabilization of MOS devices, *Solid State Electronics*, **10**, 657-670.
78. Levinstein, H.J., and Sinha, A.K. (1979) Hydrogen annealing process for stabilizing metal-oxide-semiconductor structures, *U.S. Patent No. 4,151,007*, Apr. 24, 1979.
79. Bower, K.L., and Myers, S.M. (1990) Chemical kinetics of hydrogen and (111)Si-SiO_2 interface defects, *Appl. Phys. Lett.*, **57**, 162-164.
80. Mrstik, B.J., and Rendell, R.W. (1991) Model for Si-SiO_2 interface state formation during irradiation and during post-irradiation exposure to hydrogen environment, *Appl. Phys Lett.*, **59**, 3012-3014.

81. Babsail, L.S., and Morrison, S.R. (1991) The Effect of hydrogen treatment on damaged and undamaged metal-insulator-semiconductor solar cells, *J. Appl. Phys.*, **70**, 259-265.
82. Kazmerski, L.L. (1988) Investigation of impurity neutralization and defect passivation in polycrystalline silicon solar cells, in *Mat. Res. Soc. Symp. Proc.*, **Vol. 106**, 199-211.
83. Wolff, S.H., and Wagner, S. (1989) Hydrogen surface coverage: raising the silicon epitaxial growth temperature, *Appl. Phys. Lett.*, **55**, 2017-2019.
84. Gossman, H.-J., Rafferty, C.S., Unterwald, F.C., Boone, T., Mogi, T.K., Thompson, M.O., and Luftman, H.S. (1995) Behavior of intrinsic Si point defects during annealing in vacuum, *Appl. Phys. Lett.*, **67**, 1558-1560.
85. Yanase, Y., Horie, H., Oka, Y., Sano, M., Sumita , S., and Shigematsu, T. (1994) Atomic force microscopy observation of Si(100) surface after hydrogen annealing, *J. Electrochem. Soc.*, **141**, 3259-3263.
86. Gallois, B.M., Besmann, T.M., and Stutt, M.W. (1994) Chemical etching of Silicon(100) by hydrogen, *J. Am. Ceram. Soc.*,**77**, 2949-2952.
87. Wei, L. Li, Y., and Tsong, I.S.T. (1995) Etching of Si(111)-(7x7) and Si(100)-(2x1) surfaces by atomic hydrogen, *Appl. Phys. Lett.*, **66**, 1818-1820.
88. Zhong, L., Fujimori, H., Shimbo, M., Kashima, K., Matsushita, Y., Aiba, Y., Hayashi, K., Takeda, R., Shirai, H., Saito, H., Matsushita, J.-I., and Yoshikawa, J. (1995) Determination of Si evaporation rate at 1200°C in hydrogen, *Appl. Phys. Lett.*, **67**, 3951-3953.
89. Thouless, M.D., and Cook, R.F. (1990) Stress-corrosion cracking in silicon, *Appl. Phys. Lett.*, **56**, 1962-1964; Lu, X., Cheung, N.W., Strathman, M.D., Chu, P.K., and Doyle, B. (1997) Hydrogen induced silicon surface layer cleavage, *Appl. Phys. Lett.*, **71**, 1804-1806.

SURFACE FORCE INDUCED FAILURES IN MICROELECTROMECHANICAL SYSTEMS

C.H. MASTRANGELO
Department of Electrical Engineering and Computer Science
Center for Integrated Sensors and Circuits
University of Michigan
Ann Arbor, MI 48109-2122, USA
email carlosm@eecs.umich.edu

Abstract. Over the past decade much work has been devoted to the realization of practical microelectromechanical systems (MEMS). These systems combine electronic circuits with microfabricated mechanical transducers on a miniature substrate to perform a wide range of sensing and actuation functions. MEMS are currently used to process acceleration, pressure, heat, sound, and chemical signals in a generalized manner. Because mechanical structures used in MEMS are made of thin films suspended a few micrometers off the substrate, these structures are highly susceptible to surface forces which can substantially and sometimes catastrophically degrade their manufacturing yield and reliability. This paper discusses some of the common failure mechanisms present in these devices due to adhesion forces. In particular, adhesion-related failures occurs in MEMS when suspended elastic members unexpectedly collapse and stick to their substrates. This type of device failure develops during device fabrication and operation hence is one of the dominant sources of yield loss in MEMS. The physical mechanisms responsible for the failure are discussed, and normalized elastic member dimension bounds for prevention of collapse and sticking as well as other practical failure prevention schemes are presented.

1. Introduction

Microlectromechanical systems are micrometer-size systems capable of interfacing electrical and mechanical forces. Because these devices must react to mechanical signals, many of these use construction topologies that require physical motion. For example, an accelerometer translates the motion

B. Bhushan (ed.), Tribology Issues and Opportunities in MEMS, 367-395.

of a suspended proof mass into capacitance which is later converted to a voltage output. Suspended microstructures such as plates and beams are commonly used in the manufacturing of pressure (Burns 1988, Kung and Lee 1991) and acceleration sensors (Ristic, Gutteridge, Dunn, Mietus and Bennett 1992). These structures are typically made by forming a layer of the plate or beam material on top of a sacrificial layer of another material and etching the sacrificial layer (Guckel and Burns 1989, Lysko, Stolarski and Jachowicz 1991).

Typically these suspended elastic structures have large areas, yet are constructed a few microns off their substrate. The presence of these very narrow gaps makes these structures very susceptible to surface induced phenomena such as capillarity and adhesion forces. Nathanson (Nathanson and Guldberg 1975) first reported that small electroformed elastic structures are influenced by surface tension forces. If these forces were sufficiently high the elastic member could collapse and permanently adhere to the substrate causing a device failure.

Surface forces can be present both during the device manufacture as well as during their normal operation hence their path to failure must be carefully studied. During processing, adhesion can occur when the suspended member is exposed to an aqueous rinse and dry cycle. Guckel (Guckel and Burns 1989) observed that when microscopic elastic plates are rinsed and dried, the capillary forces acting on these are large enough to bring them in contact with their underlying substrate. They also observed that, after complete drying, some of them remain pinned to the substrate held by attractive forces. Similar results have been observed by numerous researchers (Guckel, Sniegoeski and Christenson 1989, Scheeper, Voorthuyzen and Bergveld 1990, Orpana and Korhonen 1991, Takeshima, Gabriel, Ozaki, Takashashi, Horiguchi and Fujita 1991). Capillarity induced collapse can also develop when the device is under normal operation if it is exposed to high humidity conditions leading to capillary condensation (Israelachvili 1985) and the formation of a water droplet in the gap.

The route to sticking failure is hence divided into two phases. In the collapse phase the structure is brought in contact with the substrate, and in the adhesion phase, the structure is affixed to the substrate. If any of these two phases is eliminated, the adhesion failure does not occur. Either phase can be eliminated if the stiffness of the microstructure is sufficiently high yielding a characteristic stiffness threshold for the onset of failure. This threshold is observed experimentally in an array of progressively weaker suspended elements. Figure 1 shows a photograph of an array of micromachined polysilicon cantilever beams showing clearly the onset of failure at a particular beam length. The same phenomena develops in other types of suspended micromachined structures such as doubly supported beams and

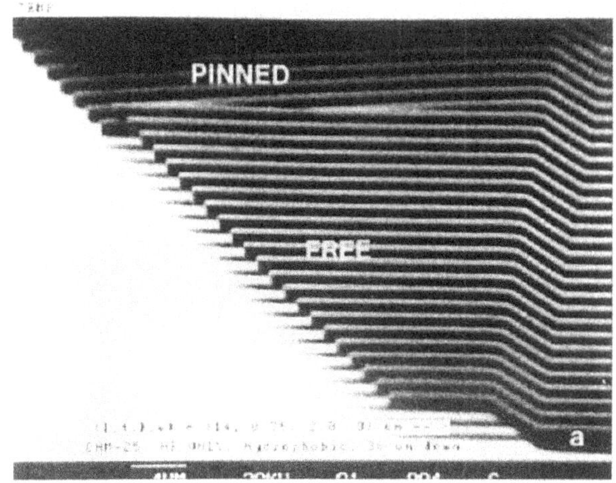

Figure 1. SEM of micromachined polysilicon cantilever beams of increasing length. The photograph shows the onset of pinning for beams longer than 34 μm.

plates.

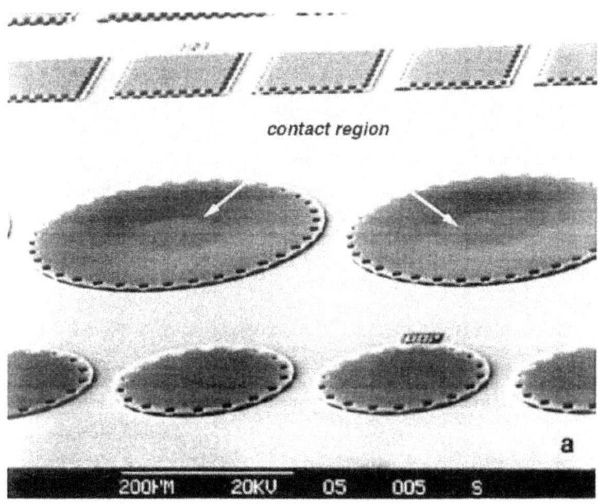

Figure 2. SEM of an array of suspended polysilicon circular plates. The two large plates have collapsed.

A third yet possible adhesion failure mode develops if the suspended member is placed in contact with another surface by an external force or by accidental shock. This is especially important in deformable mirror devices (Hornbeck 1995) and other dynamical systems where mechanical contact is intrinsic. Recently, intermittent sticking failures have been observed in microengines (Tanner, Smith, Bowman, Eaton and Peterson 1997, Miller, Vigne, Rodgers, Sniegowski, Walters and McWorther 1997) and surface micromachined accelerometers. Figure 3 shows a photograph of the Sandia microengine (Garcia and Sniegowski 1995). This device consists of a small rotating polysilicon pinion gear driven by two perpendicular linear comb drives (not shown) moving 90° out of phase. In the figure the gear drives a larger diameter rotational mechanical safety lock. During the engine opera-

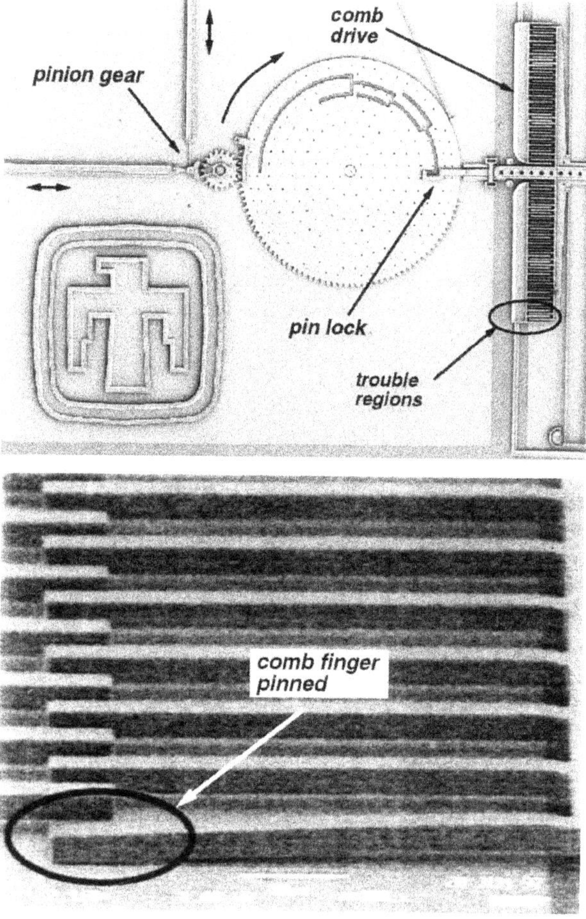

Figure 3. SEM of Sandia's microengine showing dynamic sticking failure. While in operation one of the fingers of the driving comb drives collapsed.

tion, adhesion to the substrate can occur at the pin assembly joints (Tanner et al. 1997) and at the comb drive fingers causing failure (Figure 3(b)).

This paper discusses the conditions necessary for the onset of the adhesion failure and several techniques used to eliminate it. The material for the manuscript was extracted from several of my papers in the subject as well as a recent longer review discussing this topic (Mastrangelo 1997). Other important related reviews are found in (Maboudian and Howe 1997, Komvopoulos 1996, Bhushan 1997, Kaneko 1991).

2. Collapse by Capillary Forces

The behavior of elastic structures under capillary forces has been studied in (Mastrangelo and Hsu 1993). Here, as an illustrative example, we analyze the lumped elastic structure of Figure 4. This simple structure contains all the general phenomena present in continuous elastic microstructures under a capillary pull. The plate represents the suspended member surface, the spring its stiffness, and the gap the distance between the member and the substrate.

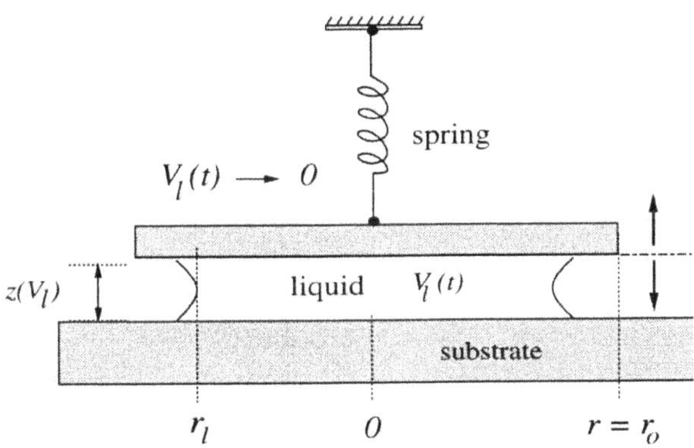

Figure 4. Deflection of a rigid plate attached to a spring by capillary forces

In Figure 4, a rigid circular plate of radius r_o is suspended above a substrate. The substrate is fixed, and the top plate is attached to a spring of constant κ. If the weight of the plate is negligible, the original plate separation is $z = h$ when the spring is relaxed. A liquid bridge of volume V_l is trapped between the plate and substrate. In order to simulate the drying process, the equilibrium positions assumed by the plate as the liquid is gradually removed are considered. Initially, the liquid spreads to a radius r_l with a volume $V_l = \pi\, r_l^2\, z$. The maximum volume that the liquid assumes without overflowing is $V_o = \pi\, r_o^2\, h$. The surface energy U_S of the spring-plate-liquid system is

$$U_S = \begin{cases} U_{S_o} + 2\pi\gamma_l \cos\theta_c \left(r_o^2 - V_l/\pi z \right) & z \geq z^* \\ U_{S_o} + \pi\gamma_l (\cos\theta_c - 1) \left(r_o^2 - V_l/\pi z \right) & z < z^* \end{cases}, \tag{1}$$

where U_{S_o} is a constant, θ_c is the contact angle, γ_l is the liquid surface tension, and $z^* = V_l/\pi r_o^2$ is the plate separation assumed when the liquid completely wets the surface of the top plate. This equation neglects the complicated nature of the small liquid-air meniscus area (Fortes 1982, Carter 1988, Finn 1986, Padday, Pitt and Pashley 1974, Matijevic 1969) since the liquid-air area is small. At $z = z^*$, U_S has a breakpoint, and for $z < z^*$, the liquid overflows. Initially, the total energy is

$$\begin{aligned} U_T &= U_S + U_E \\ &= U_{S_o} + 2\pi\gamma_l \cos\theta_c \left(r_o^2 - \frac{V_l}{\pi z} \right) + \frac{1}{2} \kappa \left(h - z \right)^2 \end{aligned} \tag{2}$$

The equilibrium plate spacing minimizes U_T. Figure 5 shows plots of U_T for different liquid volumes. The curve has one or two minima. One of

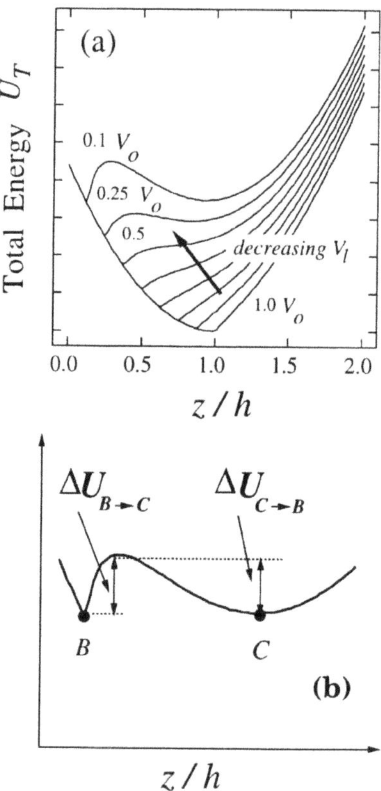

Figure 5. (a) Total energy of the two-plate system. (b) Activation energy separating the minima

these develops at the breakpoint of U_S implying that the equilibrium liquid radius is r_o. The other results when $dU_T/dz = 0$ or from Eq. (2) along the curve

$$z^2(z-h) - \frac{2\gamma_l \cos\theta_c \, V_l}{\kappa} = 0 \; . \tag{3}$$

The path traced by the reachable minimum of U_T as the liquid volume decreases from $V_l = V_o$ to $V_l = 0$ determines the equilibrium position and final state of the plate. This path is known as the equilibrium trajectory, and a reachable minimum of a potential system is that at which the system rests from a known starting state. Conceptually, this minimum is determined by placing an imaginary golf ball on the energy curve at any starting position and letting it roll to the most favorable energy well. As internal control parameters change, the shape of the curve and its minima shift making the ball slide toward one side or the other (see Figure 6). Sudden changes or

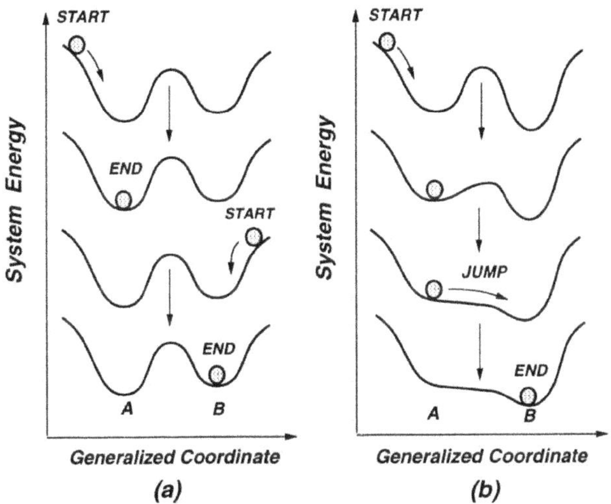

Figure 6. (a) Reachable minima of a potential system. (b) Catastrophe.

catastrophes may occur in the equilibrium for those values of the parameters at which a local minimum disappears when it merges with a local maximum (Arnold 1986, Saunders 1990, Poston and Stewart 1978, El Naschie 1990).

The behavior of the spring-plate-liquid system during the drying cycle is evident when the branches of the local extrema of U_T are plotted as a function of the control parameter V_l in a branching diagram. Letting $\xi = V_l/V_o$, $\lambda = z/h$ and $V_o = \pi r_o^2 h$, the first branch $\mathcal{B}_1$ associated with an extremum in U_S is

$$\mathcal{B}_1 : \;\; V_l = \pi r_o^2 z \; , \text{ or } \; \xi = \lambda \; . \tag{4}$$

The second branch $\mathcal{B}_2$ is found from Eq. (3)

$$\mathcal{B}_2: \; \xi = \left(\frac{\kappa\, h^2}{2\pi\gamma_l \cos\theta_c\, r_o^2}\right)(1-\lambda)\lambda^2 \;=\; N_C\,(1-\lambda)\lambda^2\,, \tag{5}$$

where the non-dimensional number $N_C = \kappa\, h^2/2\pi\gamma_l \cos\theta_c\, r_o^2$. Branch $\mathcal{B}_2$ is a minimum of U_T for $\lambda > 2/3$ and a maximum for $\lambda < 2/3$. Figure 7 shows three branching diagrams for fixed N_C. The arrows show the direction of the equilibrium trajectories.

In Figure 7(a), the equilibrium trajectory begins at A with $(\xi, \lambda) = (1,1)$. If N_C is low, branches $\mathcal{B}_1$ and $\mathcal{B}_2$ do not intersect. As V_l decreases, the trajectory follows branch $\mathcal{B}_1$ from points A to B. At $\xi(B)$ a second minimum C from $\mathcal{B}_2$ emerges in the energy curve. This new minimum is not reachable since an energy barrier $\Delta U_{B\to C}$ is present between the two minima as shown in Figure 5(b). In the limit as $V_l \to 0$, the trajectory ends at D with $(\xi, \lambda) = (0,0)$. The final plate separation is $z = 0$, and the plate remains pinned to the substrate.

N_C can be adjusted such that $\mathcal{B}_1$ and $\mathcal{B}_2$ intersect in a segment as in Figure 7(b). Setting $\mathcal{B}_1(\lambda) = \mathcal{B}_2(\lambda)$ one finds

$$\xi \;=\; (\lambda_1, \lambda_2) \;=\; \frac{1}{2} \pm \left(\frac{1}{4} - \frac{1}{N_C}\right)^{1/2}. \tag{6}$$

When λ_1 and λ_2 are real, the segments of $\mathcal{B}_1$ and $\mathcal{B}_2$ in $\lambda_1 \leq \lambda \leq \lambda_2$ combine thus disappearing from the extremum set. The equilibrium trajectory starts at A following branch $\mathcal{B}_1$ to B. At $\xi(B)$, the trajectory jumps abruptly to the now global minimum at point C and continues along $\mathcal{B}_2$ to point D where the minimum E from $\mathcal{B}_2$ reappears. Point E is is not reachable because of the barrier $\Delta U_{D\to E}$; thus, for lower V_l, the trajectory follows $\mathcal{B}_2$ ending at F with $(\xi, \lambda) = (0,1)$. The final plate separation is $z = h$, and the plate is free.

There is a threshold value in N_C, defined as N_T, that determines the final state. Trajectories with $N_C < 4$ follow $\mathcal{B}_1$ throughout the cycle, yielding pinned plates. Trajectories with N_C slightly larger than 4 are routed to branch $\mathcal{B}_2$ after a catastrophe yielding free plates. For $N_C \geq 9/2$ and $\lambda_2 \geq 2/3$, the trajectory is a smooth curve (Figure 7(c)) yielding free plates; thus $N_T = 4$.

It is convenient to define an elastocapillary number, N_{EC}, such that the plate is free for $N_{EC} > 1$ and pinned for $N_{EC} < 1$. Thus

$$N_{EC} \;=\; \frac{N_C}{N_T} \;=\; \left(\frac{\kappa\, h^2}{8\pi\;\gamma_l \cos\theta_c\, r_o^2}\right). \tag{7}$$

The analysis of the spring-plate-liquid system shows a bifurcation of the equilibrium trajectory, and a non dimensional number which determines

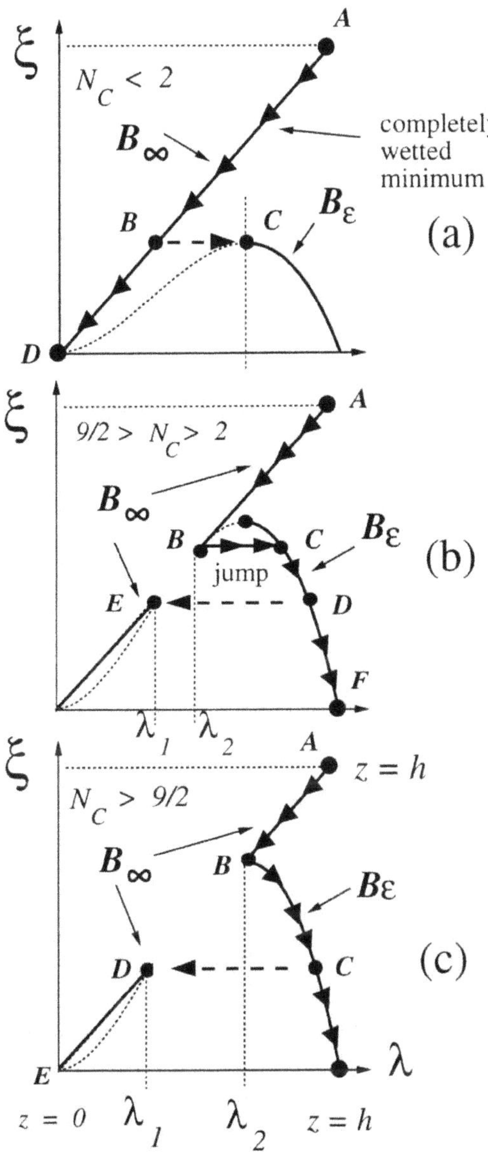

Figure 7. Branching diagrams for the spring-plate-liquid system

its final state. These characteristics are also present in continuous elastic structures. In (Mastrangelo and Hsu 1993), expressions for the elastocapillary numbers for continuous elastic beams and plates have been calculated using a variational energy approach that includes the balance of elastic, po-

Structure	*Approximate Elastocapillary Numbers*
cantilever beam	$\frac{2\,Eh^2t^3}{9\,\gamma_l\cos\theta_c\,l^4\,(1+t/w)}$
doubly supported beam	$\frac{128\,Eh^2t^3}{15\,\gamma_l\cos\theta_c\,l^4\,(1+t/w)}\left[1+\frac{2\,\sigma_R\,l^2}{7\,E\,t^2}+\frac{108\,h^2}{245\,t^2}\right]$
circular plate	$\frac{10}{9}(\frac{5}{2})^{2/3}(\frac{Eh^2t^3}{\gamma_l\cos\theta_c(1-\nu^2)\,r_o^4})\left[1+\frac{3(1-\nu^2)}{4}\frac{\sigma_R\,r_o^2}{E\,t^2}+\frac{2187}{2560}(\frac{2}{5})^{2/3}(\frac{h^2}{t^2})\right]$
square plate	$(\frac{25\,E\,h^2\,t^3}{\gamma_l\cos\theta_c(1-\nu^2)\,w^4})\left[1+\frac{2(1-\nu^2)}{9}\frac{\sigma_R\,w^2}{E\,t^2}+\frac{5}{12}\frac{h^2}{t^2}\right]$

tential, and surface energies. The results for different microstructures are shown in Table I. where E is the Young modulus, t the member thickness, h the gap, ν is Poisson's ratio, σ_R the member residual tensile stress, and l, w, r_o are the member length, width, and radius.

3. Adhesion by Intersolid Contact Forces

The second the phenomena in the failure path is a strong intersolid adhesion force between the microstructure and the substrate overcomes the elastic restoring force of the deflected suspension member. The intersolid adhesion force is a consequence of the change in the energy stored at the contact area with respect to the member deformation. The magnitude this the adhesion energy depends on the nature of the interface and the presence of surface contamination. In pure crystalline solids this energy is high (500-2000 mJm^{-2}), and in soft polymers it is much lower (5-100 mJm^{-2}).

It has been reported that the adhesion between two microstructure surfaces is also dependent on the roughness of the surface (Komvopoulos 1996). In practice, the proper treatment of this parameter is difficult as the interfacial roughness for a microstructure is by no means a-priori certain, and its influence on the onset of sticking failure is largely not understood. Therefore for all calculations that follow this energy represents an effective empirical quantity.

In this section we consider the equilibrium between these two opposing forces using an energy function formulation that includes the adhesion and elastic energy of the deformed suspension member.

The simplest problem is that of the peeling of an elastic cantilever beam from an adhesive surface. This problem is related to the peeling of sticky tapes (Kendall 1971) and the cleavage of crystals (Maszara, Goetz and McKitterick 1988, Gillis and Gilman 1964, Gilman 1960). Figure 8 shows a

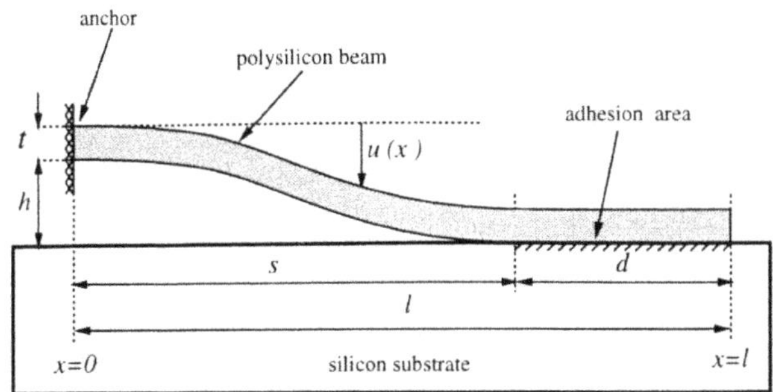

Figure 8. Schematic of cantilever beam adhering to the substrate

cross section of a cantilever beam of length l, width w, thickness t, height h, and Young's modulus E. The beam is adhering to its substrate a distance $d = (l - s)$ from its tip. The stored elastic energy of the beam in the segment $0 \leq x \leq s$ induces a restoring force that tends to peel the beam from the adhering substrate. The energy of adhesion stored in the segment $s \leq x \leq l$ induces another force that holds the beam in contact with the substrate. The equilibrium peel distance s^* is determined by the balance of these two energies. At equilibrium, s^* minimizes the the total energy of the system (bending plus adhesion energies) (Kendall 1971).

Since there are no external forces acting on the beam for $0 \leq x \leq s$, its deflection $u(x)$ is the solution of

$$E\,I\,\frac{d^4\,u}{d\,x^4} = 0 \quad , \quad I = \frac{w\,t^3}{12} \; , \tag{8}$$

where I is the moment of inertia of the beam respect to the z axis. Equation (8) is solved subject to the boundary conditions

$$\left.\frac{du}{dx}\right|_0 = \left.\frac{du}{dx}\right|_s = 0 \; , \quad u(0) = 0 \quad , \quad u(s) = h \; . \tag{9}$$

which ignores the effects of finite compliance at the step-up posts (Mullen, Mehrengany, Omar and Ko 1991, Meng, Mehregany and Mullen 1993). Equation (8) has the solution

$$u = h\,f(\eta) = h\,\eta^2(\,3 - 2\,\eta\,) \quad , \quad \eta = \frac{x}{s} \; . \tag{10}$$

The bending energy stored in the beam is

$$U_E = \frac{EI}{2}\int_0^l \left(\frac{d^2u}{dx^2}\right)^2 dx = \frac{6EIh^2}{s^3} . \tag{11}$$

The interfacial energy stored in $s \le x \le l$ is simply the surface energy per unit area of the bond γ_s times the area of contact

$$U_S = -\gamma_s w(l - s) . \tag{12}$$

The parameter γ_s has units of J m^{-2} or N m^{-1}. The sign of U_S is negative because it is a binding energy. The total energy (or free energy) of the system is the sum of the elastic plus surface energies

$$U_T = U_E + U_S = \frac{6EIh^2}{s^3} - \gamma_s w(l - s) . \tag{13}$$

Figure 9 shows a typical curve of $U_T(s)$. This curve has a single minimum corresponding to the equilibrium s^*. This is found by setting $dU_T/ds = 0$

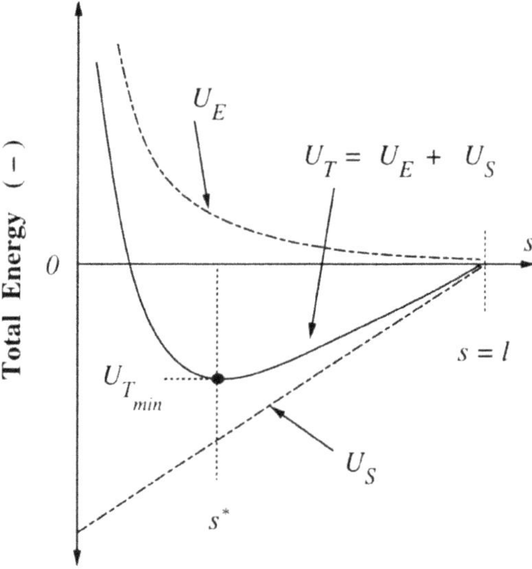

Figure 9. Typical total energy curve for cantilever beam adhering to the substrate.

to obtain

$$s^* = \left(\frac{3}{2}\frac{E\,t^3\,h^2}{\gamma_s}\right)^{1/4} . \tag{14}$$

The energy curve has a single equilibrium point if $s^* < l$ and no equilibrium point if $s^* > l$. Thus the beam is pinned to the substrate if $s^* < l$, and it is free if $s^* > l$.

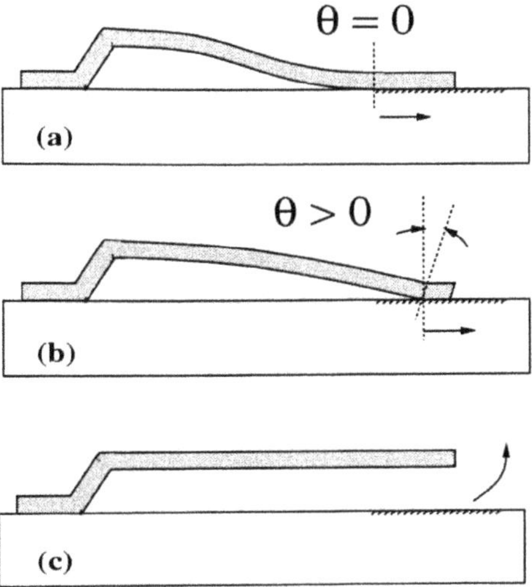

Figure 10. Pivoting of cantilever tip near detachment.

The slope boundary condition $du/dx = 0$ at $x = s$ does not allow for shear deformations of the tip of the beam. Shear deformations are particularly important for $s \to l$ since when $d = (l - s)$ is very small, the tip of the cantilever "pivots" changing its elastic energy substantially just before detachment as shown in Figure 10. This effect is taken into consideration by dividing the beam into two regions. The region of the beam for $0 < x < s$ has no external forces acting on it, so Eq. (8) is solved subject to the boundary conditions of Eq. (9) and the modified slope condition at $x = s$,

$$\left.\frac{du}{dx}\right|_s = \theta = \frac{hm}{s} \; . \tag{15}$$

where θ is the shear angle of the tip, and m is a non dimensional number. The deflection $u(x)$ and elastic energy of the beam segment are

$$u = h\eta^2((m-2)\eta + (3-m)) \; , \tag{16}$$

$$U_E = \frac{6EIh^2}{s^3}(1 - m + \frac{1}{3}m^2) \; .$$

Note that U_E decreases with increasing m (and θ) for $0 \leq m \leq 3/2$. The short segment $s \leq x \leq l$ corresponding to the beam tip experiences shear deformations which determine m. The solution of the coupled problem is

Structure	*Approximate Peel Numbers*
cantilever beam	$\frac{3}{8}\frac{E\,t^3\,h^2}{\gamma_s\,l^4}$
doubly supported beam	$\left(\frac{128Eh^2t^3}{5\gamma_s l^4}\right)\left[1+\frac{4\,\sigma_R\,l^2}{21\,E\,t^2}+\frac{256}{2205}(\frac{h}{t})^2\right]$
circular plate	$\frac{40}{3}\frac{E\,h^2\,t^3}{(1-\nu^2)\,\gamma_s\,r_o^4}\left[1+\frac{51(1-\nu^2)}{160}\frac{\sigma_R\,r_o^2}{E\,t^2}+\frac{63}{200}\frac{h^2}{t^2}\right]$
square plate	$\frac{186\,E\,h^2\,t^3}{(1-\nu^2)\,\gamma_s\,w^4}\left[1+\frac{27(1-\nu^2)}{310}\frac{\sigma_R\,w^2}{E\,t^2}+\frac{12}{31}\frac{h^2}{t^2}\right]$

given in reference (Mastrangelo and Hsu 1992) yielding

$$m(s) = \frac{\frac{16}{5}\left(\frac{t}{d}\right)^3\left(\frac{t}{s}\right)\left[1+\frac{15}{32}\left(\frac{d}{t}\right)^2\frac{E}{G}\right]}{1+\frac{32}{15}\left(\frac{t}{d}\right)^3\left(\frac{t}{s}\right)\left[1+\frac{15}{32}\left(\frac{d}{t}\right)^2\frac{E}{G}\right]} \tag{17}$$

where $G = E/2(1+\nu)$ is the beam shear modulus.

The beam detaches from its substrate when

$$l = s^* = \left(\frac{3}{8}\frac{E\,t^3\,h^2}{\gamma_s}\right)^{1/4} \tag{18}$$

therefore we can define a *peeling bound*, N_P such that the beam remains in contact with the substrate for $N_P < 1$ and free for $N_P > 1$. Thus

$$N_P = \frac{3}{8}\frac{E\,t^3\,h^2}{\gamma_s\,l^4} \tag{19}$$

Using the same techniques, peel bounds for doubly-supported beams and plates have been calculated. These bounds are shown in Table II.

Figures 11-12 shows experimental plots of the effectiveness of some of the bounds for different structures. The best fit occurs for cantilever beams because of the lack of residual and elongation stresses. The surface energies have been determined from the slope of the plot. Surface energies ranging

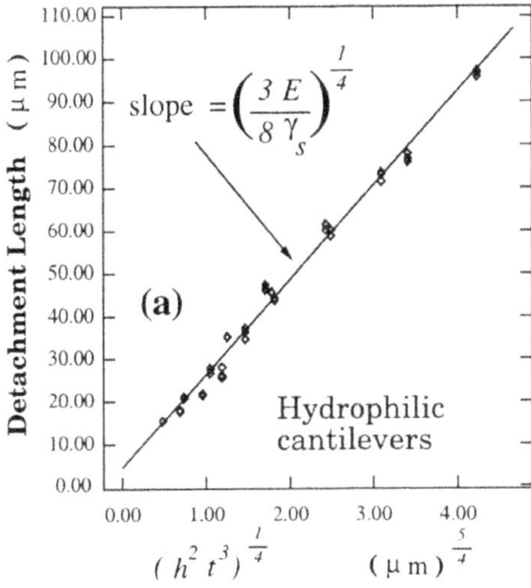

Figure 11. Peel bound for polysilicon cantilever beams

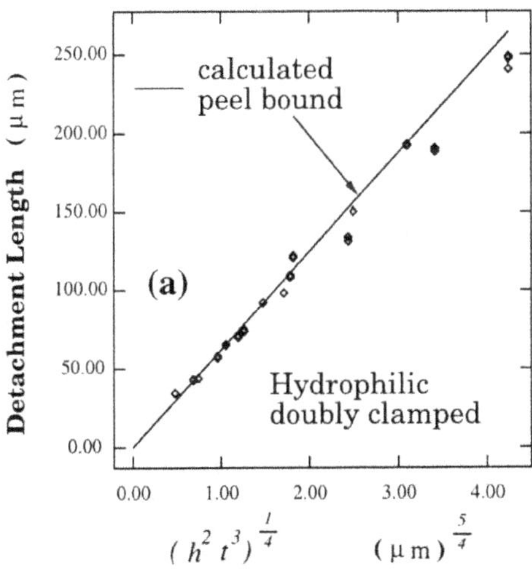

Figure 12. Peel bound for doubly supported beams

between 100-300 mJm^{-2} have been measured for polysilicon microstructures (Mastrangelo and Hsu 1993).

4. Some Failure Prevention Methods

While these bounds are useful for detecting the onset of adhesion failures in MEMS devices, many practical devices require dimensions that fall within the failure range. A number of schemes have been developed to eliminate the failure both during the fabrication and device operation. Roughly, these schemes fall within two categories. Physical schemes aim at modifying the shape of the structure to minimize the collapsing or contact force. Chemical schemes aim toward the chemical modification of the contact surfaces through the use of low-energy coatings.

4.1. ELIMINATION OF COLLAPSING FORCE

A large number of adhesion failures occur during the fabrication of the devices themselves. These failures are related to the capillary force that develops when the structures are released by wet chemical etches.

4.1.1. *Freeze-Drying Methods:*

The capillary pull force can be eliminated if the liquid phase is not present. The rinse solution can be removed by freezing the solution and sublimation. This freeze-drying (Flosdorf 1982, Holler 1979, Mellor 1978) technique can be used to eliminate the capillary pull by first freezing the sample and then exposing it to a heated vacuum environment. This idea was first applied to the release of MEMS devices by Guckel and Burns (Guckel and Burns 1989). A well known disadvantage of the freeze-drying method is the fact that the rinse solution can undergo a significant volume change. This volume change can create a stress sufficient to destroy the sample. To alleviate this problem, Guckel used a rinse solution consisting of a mixture of methanol and water that minimized the volume change. A similar process was developed by Takeshima et al. (Takeshima et al. 1991) who replaced the rinse solution through a gradual series of dilutions with t-butyl alcohol. This compound freezes at 25.6 C° and is much softer than Guckel's icy mixture; however the sublimated vapors are highly toxic.

4.1.2. *Supercritical Drying:*

Another method of eliminating the capillary pull is by the use of supercritical drying techniques. In this technique, the rinse solution is gradually replaced by liquid CO_2 at elevated pressures inside a high-pressure chamber. The sample is then taken to the critical point of CO_2 where the interface between the liquid and gas does not exists. The technique is highly successful with nearly 100% yields (Mulhern, Soane and Howe 1993). Commercial supercritical drying equipment has been available for more than 20 years (Dawes 1988) for small samples, and large scale equipment construction has been planned. The main difficulty with the technique resides in the safety considerations because of the very high ($\approx$ 72 Atm) pressures required to take the samples to the critical point. This technique is more widely accepted than the freeze-drying method.

4.1.3. *Dry Etching:*
A number of schemes have been proposed which are based on dry etching of the sacrificial layer. This operation is often difficult if the sacrificial layer is silicon based (such as silicon dioxide) because it requires strong etchants that do not have good selectivities with respect to the suspended element material. Vapor-phase HF etching (Watanabe, Ohnishi, Honma, Kitajima, Ono, WIlhelm and Sophie 1995) at elevated temperatures (Wong, Moslehi and Bowling 1993, Ruzillo, Torek, Daffron, Grant and Novak 1993) has been often used for this purpose with a high rate of attack of silicon nitride films. The structure can also be easily released by plasma etching if the sacrificial layer is silicon (Hirano, Furuhata and Fujita n.d.).

A more successful series of techniques are based upon the replacement of the sacrificial layer with a weaker solid that can be later removed by a harmless dry etching. These schemes involve primarily the replacement of sacrificial layers with plastics that can be then be etched using O_2 plasma or ozone which are harmless to silicon-based materials. Techniques based on conformal vapor deposited polymer supports (p-xylylene) (Figures 13-15) (Mastrangelo and Saloka n.d.) have been successful in releasing polysilicon plates that as large as $3000 \times 3000 \times 1\ \mu m^3$. A technique that is based

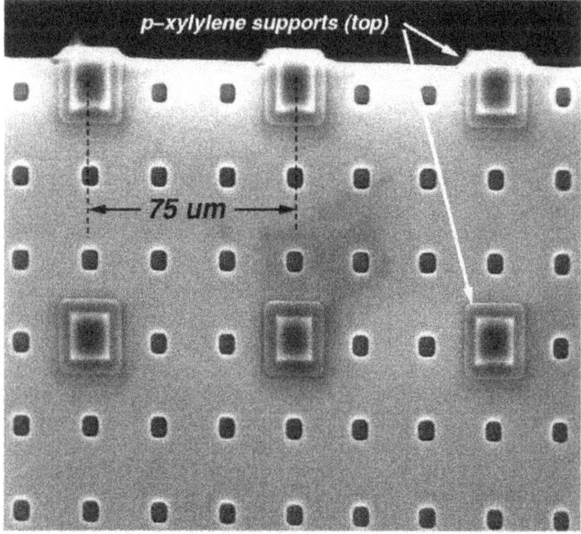

Figure 13. Top view of polymer support feet on a polysilicon plate

upon the replacement of the rinse solution with resist through a series of dilutions was developed by Orpana and Korhonen (Orpana and Korhonen 1991). Other techniques include the replacement of the rinse solution with

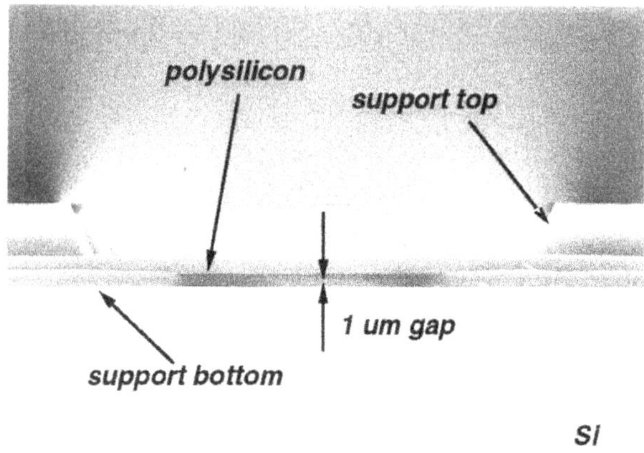

Figure 14. Cross section of polymer support

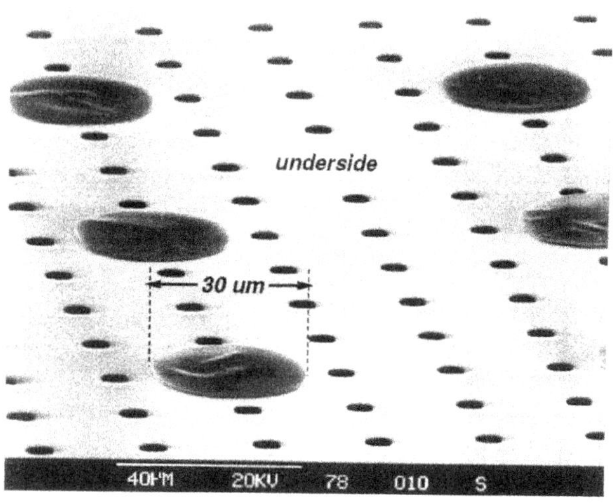

Figure 15. SEM of underside of sacrificial p-xylylene supports

monomers that polymerize in a short time as well as plasma-polymerized fluorinated plastics (Kozlowski, Lindmair, Scheiter, Hierold and Lang 1995, Jansen, Gardeniers, Elders, Tilmans and Elwenspoek 1994).

4.1.4. *Hydrophobic Coatings:*

The capillary pull can become a push if the contact angle is made larger than 90 °. Oxidized silicon surfaces are hydrophilic with contact angles ranging from $0-39°$ depending on the type of oxide and surface treatment.

Bare silicon surfaces obtained by the removal of native oxide using HF are hydrophobic. The hydrophobicity has been attributed to possible hydrogen groups (Graf, Grunder, Schulz and Muhlhoff 1990) attached to the silicon. Contact angles (Gould and Irene 1988) as high as 78° have been reported; however, the native oxide is known to regrow in both water and when exposed to air (Watanabe, Hamano and Harazono 1989, Morita, Ohmi, Hagesawa, Kawakami and Suma 1989); thus complete hydrophobicity is difficult to achieve. In order to obtain a higher contact angle, a chemical change in the surface of the suspended member is required. Recently, a number of well-known (Ulman 1991) hydrophobic self-assembled monolayers (SAM) have been grown on silicon surfaces with successful results (Alley, Howe and Komvopoulos 1992, Houston, Maboudian and Howe 1996). These layers are based on the silanization of silicon surfaces with organic groups by treating the surface with octadecyltrichlorosilane (OTS) precursor molecule ($C_{18}H_{37}SiCl_3$). The procedure first involves the replacement of the aqueous rinse with an organic solvent though a series of dilution steps. The SAM is next grown, and then the solvent is replaced by water through the reverse series of dilutions. Because the silanized surfaces have a very high contact angle ($\approx 114°$), at this stage the water recedes from the surfaces resulting in a sample that emerges dry from the rinse. Cantilever beams at least as long as 1000 μm long were successfully released using this technique with beams as long as 400 μm passing the "in-use" stiction contact test (Figure 16). More recently, self-assembled fluorinated layers have been used for

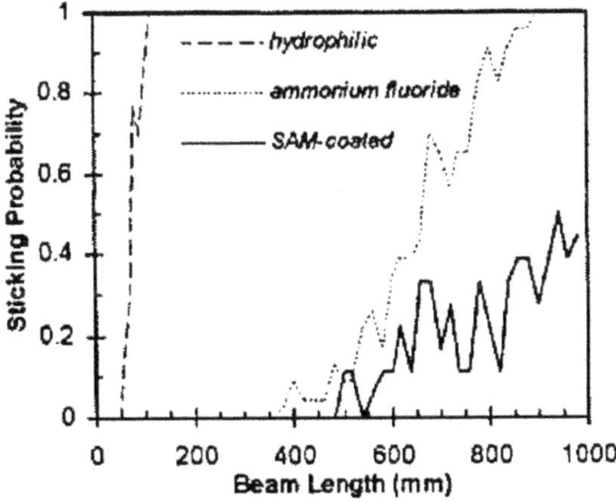

Figure 16. Sticking probability for cantilever beams that were given different treatments.

the same purpose with improved results (Srinivasan, Houston, Howe and Maboudian 1997).

Most of these techniques (with the exception of coatings) can only eliminate the capillary pull or its effects during its fabrication process. After the sample is packaged, if it is exposed to a high-humidity environment or a shock, the pinning may return hence becoming a reliability issue. Therefore these techniques are insufficient to effectively eliminate the adhesion failure but rather delay it. A permanent solution to the problem also requires additional treatments that attempt to reduce the intersolid adhesion.

4.2. REDUCTION OF INTERSOLID ADHESION

The permanent elimination of the adhesion failure requires the reduction of the intersolid surface adhesion. Several techniques have been developed toward this goal. The main ones are listed below.

4.2.1. *Textured Surfaces:*

The adhesion force can be decreased if the contact area between the elastic member and the substrate is reduced. This can be accomplished by texturing the contact surface through the generation of a rough interface (Alley, Mai, Komvopulos and Howe 1993, Yee, Chun and Lee 1995). Yee (Yee et al. 1995) reported an increase in the detachment length of approximately two by the oxidation and dry etching of a polysilicon supporting surface.

Texturing can also be introduced by constructing a periodic array of small supporting post, commonly known as "dimples". These supports are constructed by etching small indentations into the sacrificial layer before the deposition of the suspended member as shown in Figures 17-18. This method was first used by L. S. Fan (Fan 1990) for the construction of a micromotor, and later applied to comb drives (Tang, Nguyen and Howe n.d.). The force required to detach a semi-spherical dimple of radius R is (Johnson, Kendall and Roberts 1971, Johnson 1987)

$$F = (3/2)\,\pi\,\gamma_s\,R \tag{20}$$

hence it becomes zero as $R \to 0$. This technique is very simple but it suffers from two drawbacks. First, the dimple method cannot be applied for microstructures with flat surfaces as required in many transducer designs. Second, the adhesion of the dimple is often larger than that anticipated by Eq. (20) when drying out the sample. This is attributed to the formation of a solid bridge of SiO_2 as water evaporates (Watanabe et al. 1989).

4.2.2. *Low-Energy Monolayer Coatings:*

These coatings are the same hydrogenated and OTS-type monolayers discussed above. If their intersolid adhesion energy is sufficiently low, this

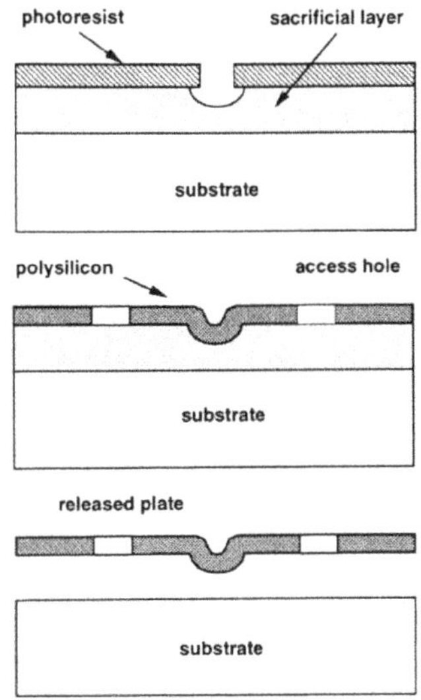

Figure 17. Construction of dimple posts.

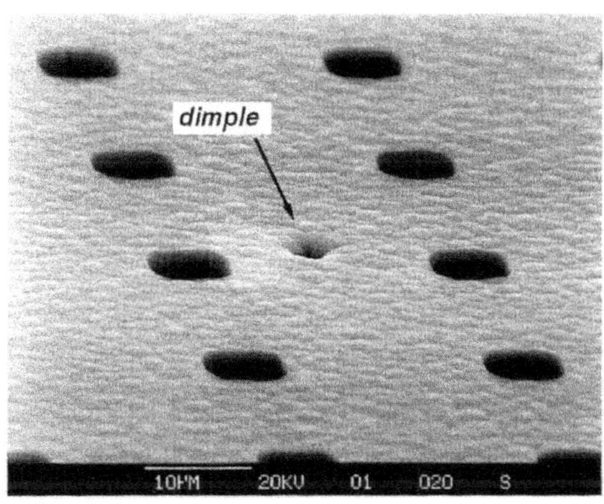

Figure 18. SEM photograph of a dimple on a polysilicon plate.

option is very attractive since it eliminates both failure originating mechanisms. Houston et al. (Houston et al. 1996) performed contact tests in silicon microstructures with surfaces that were terminated with hydrogen bonds, and coated with OTS self-assembled monolayers. Preliminary tests suggested that cantilever beams as long as 1000 μm-long exhibited about 50 % sticking probability on OTS-coated samples after they are brought in contact with the substrate (Figure 16). These films have been reported to show an extremely low effective adhesion energy (3 μJm^{-2}). More recently, coatings of perfluorinated alkylthrichlorosilane or FDTS ($C_{10}H_4F_{17}SiCl_3$) have been used for elimination of adhesion (Srinivasan et al. 1997).

The main difficulty associated with many of these films is their fragility. Because the SAM films are extremely thin ($\approx$2.5 nm), they have a high tendency to degrade at elevated temperatures, as shown in Figure 19. (Houston et al. 1996). The aging behavior of hydrogen terminated surfaces has also been observed by Houston et al. (Houston, Maboudian and Howe 1995) indicating that these surfaces degrade over time. FDTS films are less reactive

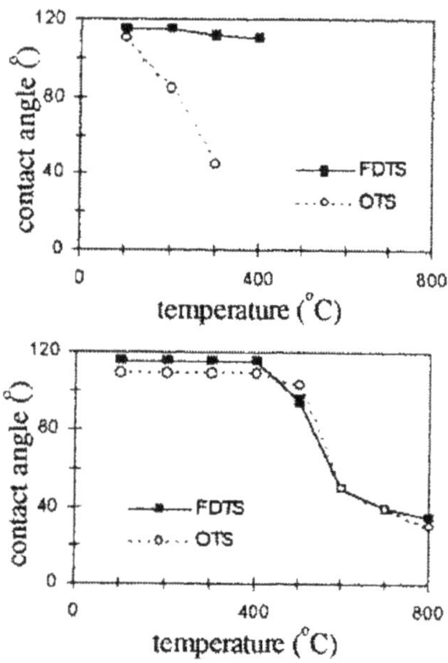

Figure 19. Degradation of the contact angle of self-assembled OTS and FDTS coatings exposed 5 minutes in air (top) and nitrogen environments.

than OTS films hence can withstand much longer periods at atmospheric conditions and at elevated temperatures. Figure 19 shows the comparison of water contact angles on the film surfaces when exposed to elevated temperatures for a brief time.

Recent experiments by Kluth *et. al.* (Kluth, Sunag and Maboudian 1997) suggest that these films are also susceptible to thermal desorption; hence their long term effectiveness remains an open question. These films however will remain effective inside sealed, controlled packaged environments at thermal equilibrium where degradation reactions cannot take place, and any vaporization is henceforth followed by a redeposition. Similar monolayer films are currently used in sealed deformable mirror displays under this strictly controlled environment (Hornbeck 1995).

4.2.3. *Fluorinated Coatings:*
Fluorinated hydrocarbon coatings consisting of CF_x chains have very low surface energies; hence they are likely candidates coatings for the reduction of intersolid adhesion. PTFE-like layers can be grown in many different ways (Licari 1970). A common way to grow these films is by plasma polymerization (Yasuda and Hsu 1977, Yasuda and Hsu 1978, Yasuda 1985). Often the growth of these films is highly directional with most of the growth occurring in the surfaces exposed to the plasma (Jansen et al. 1994). In order for these films to be effective, the film must grow under a covered surface.

Fluorinated films can also be grown at a much reduced rate in zones which are not directly exposed to the plasma (Buzzard 1978). In this regime, the deposition rate is a diffusion-limited process which is not subject to the rapid decay of the plasma density resulting in a much more uniform deposition. This idea has been recently exploited by Man et al. (Man, Gogoi and Mastrangelo n.d., Man, Gogoi and Mastrangelo March 1997) to deposit PTFE-like films on silicon microstructures with much improved conformality In this technique, the deposition takes place inside a Faraday cage that prevents the plasma from directly reaching the sample.

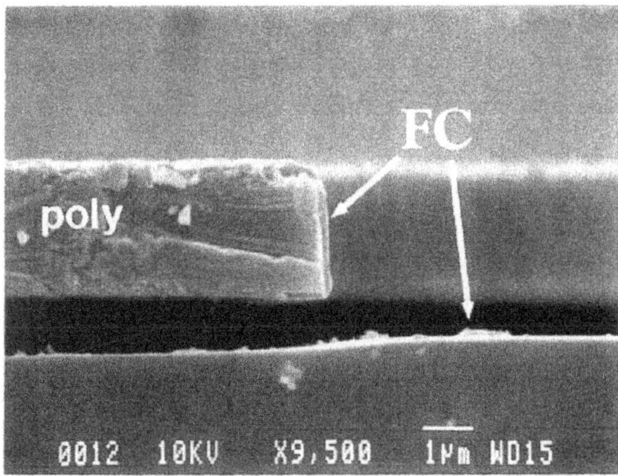

Figure 20. Cross section of poly plate showing PTFE film coverage

The fluorinated films can be quite thick (> 20 nm) hence durable and have a contact angle of about 108 °.

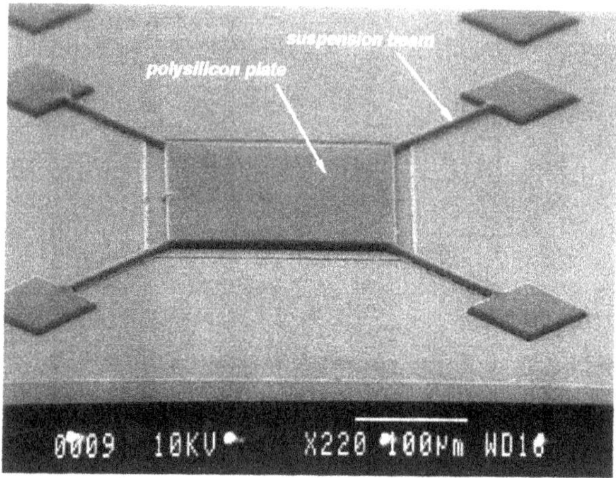

Figure 21. Example polysilicon plate coated with PTFE

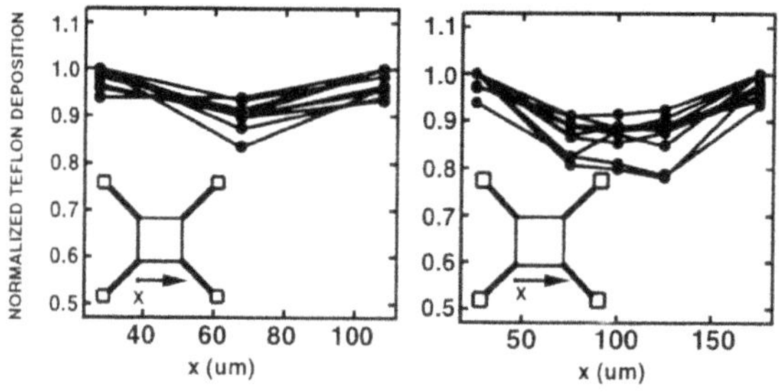

Figure 22. Uniformity of TFE-like films on the underside of suspended plates. The front of the plate has approximately twice the thickness as the underside.

Periodic wear and temperature tests were performed in these films as shown in Figure 25. The results predict a projected mean time to failure (determined by a loss in the hydrophobicity) greater than 10 years at 150 °C in atmospheric conditions. The films remain hydrophobic even at very high temperatures and display little wear even after 10^7 contact cycles.

Of all the above adhesion prevention methods, low-energy coatings are the most effective and reliable. In addition, both SAM (Deng 1994) and thicker fluorinated coatings have proved effective in the reduction of wear and friction in dynamic microstructures (Smith, Sniegowski and LaVigne 1997). The monolayer technique is very attractive since it solves both capillary pull and adhesion problems with only one application, but it requires a sealed package to maintain the long term monolayer integrity. Thicker films are more robust and can survive atmospheric conditions for

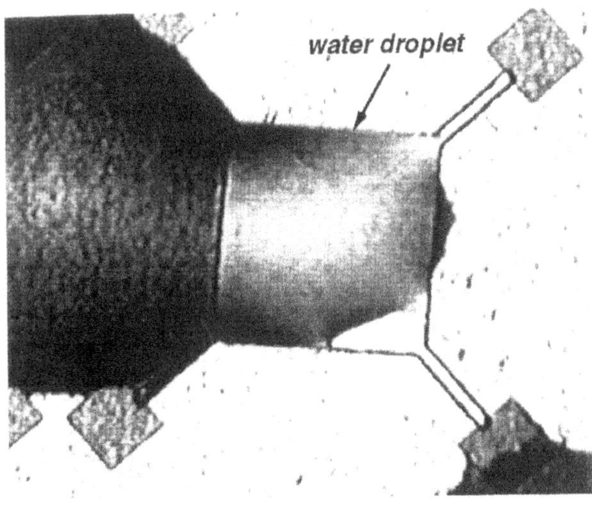

Figure 23. Wetting of underside of polysilicon plate without PTFE film

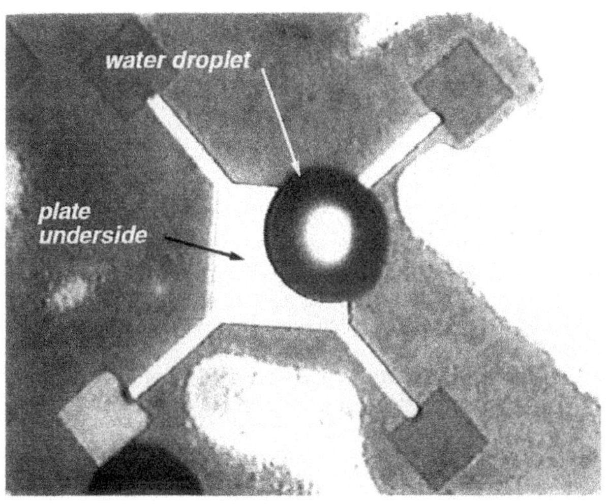

Figure 24. Wetting of underside with PTFE film

many years but these require the use of a complementary technique to eliminate the capillary pull during the device fabrication. So far the best solution seems to be a combination of both techniques.

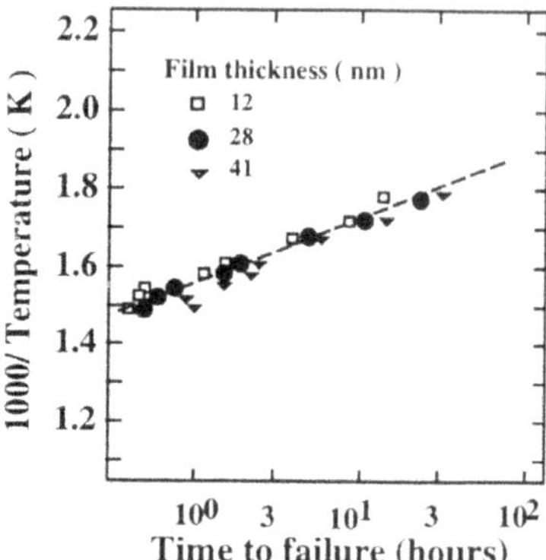

Figure 25. Degradation of TFE-like plasma polymerized coatings under atmospheric conditions at different temperatures. These films remain on the sample for many years.

5. Conclusion

This paper reviews the causes and solutions for surface force induced sticking failures in microelectromechanical systems. The failures originate from the extreme susceptibility of suspended MEMS devices to surface forces due to their close proximity to their substrates. The failure route requires the presence of a collapsing force followed by a strong intersolid adhesion. The behavior of the microstructure under these two forces has been examined and several practical techniques for the elimination of failure are discussed.

References

Alley, R., Howe, R. T. and Komvopoulos, K.: 1992, The effect of release-etch processing on surface microstructure stiction, *International Workshop on Solid-State Sensors and Actuators (Hilton Head' 92)*, pp. 202–207.

Alley, R. L., Mai, P., Komvopulos, K. and Howe, R. T.: 1993, Surface roughness modification of interfacial contacts in polysilicon microstructures, *Transducers' 93*, pp. 288–291.

Arnold, V. I.: 1986, *Catastrophe Theory*, Springer-Verlag, New York.

Bhushan, B.: 1997, Micro/nanotribology and its applications to magnetic storage devices and mems, *Tribology International* **28**, 85–96.

Burns, D. W.: 1988, *Micromechanics of Integrated Sensors and the Planar Processed Pressure Transducer*, PhD thesis, University of Winsconsin, Madison.

Buzzard, P. D.: 1978, *Plasma Polymerization of Tetrafluoroethylene in a Field Free Zone*, PhD thesis, University of California, Berkeley.

Carter, W. C.: 1988, The forces and behavior of fluids constrained by solids, *Acta Metallurgica* **36**, 2283–2292.

Dawes, C. J.: 1988, *Introduction to Biological Electron Microscopy: Theory and Techniques*, Ladd Research Industries, Burlington, VT.

Deng, K.: 1994, PhD thesis, Case Western Reserve University, Cleveland, OH.

El Naschie, M. S.: 1990, *Stress. Stability, and Chaos in Structural Engineering: An Energy Approach*, McGraw-Hill, New York.

Fan, L.-S.: 1990, *Integrated Micromachinery – Moving Structures in Silicon Chips*, PhD thesis, University of California, Berkeley.

Finn, R.: 1986, *Equilibrium Capillary Surfaces*, Springer-Verlag, New York.

Flosdorf, E. W.: 1982, *Freeze Drying: Drying by Sublimation*, University Microfilms International, Ann Arbor, MI.

Fortes, M. A.: 1982, Axisymmetric liquid bridges between parallel plates, *J. Colloid and Interface Science* **88**, 338–352.

Garcia, E. J. and Sniegowski, J. J.: 1995, Surface micromachined microengine, *Sensors and Actuators* **48**, 203–214.

Gillis, P. P. and Gilman, J. J.: 1964, Double-cantilever cleavage mode of crack propagation, *J. Appl. Phys* **35(3)**, 647–658.

Gilman, J. J.: 1960, Direct measurements of the surface energies of crystals, *J. Appl. Phys* **31(12)**, 2208–2218.

Gould, G. and Irene, E. A.: 1988, An in-situ study of aqueous HF treatment of silicon by contact angle measurement and ellipsometry, *J. Electrochem. Soc.* **135**, 1535–1539.

Graf, D., Grunder, M., Schulz, R. and Muhlhoff, L.: 1990, Oxidation of hf-treated si wafer surfaces in air, *J. Applied Phys.* **68**, 5155–5161.

Guckel, H. and Burns, D. W.: 1989, Fabrication of micromechanical devices from polysilicon films with smooth surfaces, *Sensors and Actuators* **20**, 117–122.

Guckel, H., Sniegoeski, J. J. and Christenson, T. R.: 1989, Advances in processing techniques for silicon micromechanical devices with smooth surfaces, *International Workshop on Micro Electromechanical Systems (MEMS'89)*, pp. 71–75.

Hirano, T., Furuhata, T. and Fujita, H.: n.d., Dry releasing of electroplated rotational and overhanging structures, *International Conference on Micro Electromechanical Systems (MEMS 93)*, Fort Lauderdale, FL, U.S.A. Feb. 1993, pp. 278–283.

Holler, R. O.: 1979, *Freeze-Drying Biological Specimens*, Smithsonian Institution Press, Washington, DC.

Hornbeck, L. J.: 1995, Digital light processing and MEMS: timely convergence for a bright future, *Proc. SPIE Micromachining and Microfabrication Process Technology Conference*, Vol. SPIE 2639, pp. 1–21.

Houston, M., Maboudian, R. and Howe, R. T.: 1996, Self-assembled monolayer films as durable anti-stiction coatings for polysilicon microstructures, *International Workshop on Solid-State Sensors and Actuators (Hilton Head' 96)*, pp. 42–47.

Houston, M. R., Maboudian, R. and Howe, R. T.: 1995, Ammonium fluoride anti-stiction treatments for polysilicon microstructures, *Transducers'95* pp. 210–213.

Israelachvili, J. N.: 1985, *Intermolecular and Surface Forces*, Academic Press, New York.

Jansen, H. V., Gardeniers, J. G. E., Elders, J., Tilmans, H. A. C. and Elwenspoek, M.: 1994, Applications of fluorocarbon polymers in micromechanics and micromachining, *Sensors and Actuators* **A41-42**, 136–140.

Johnson, K. L.: 1987, *Contact Mechanics*, Cambridge University Press, New York.

Johnson, K. L., Kendall, K. and Roberts, A. D.: 1971, Surface energy and the contact of elastic solids, *Proc. Royal Soc. London A* **324**, 301–313.

Kaneko, R.: 1991, Microtribology related to mems–concept, measurements, applications, *International Workshop on Micro Electromechanical Systems (MEMS'91)*, pp. 1–8.

Kendall, K.: 1971, The adhesion and surface energy of elastic solids, *J. Phys. D: Appl. Phys.* **4**, 1186–1195.

Kluth, G. J., Sunag, M. M. and Maboudian, R.: 1997, Thermal behavior of alkylsiloxane self-assembled monolayers on the oxidized si(100) surface, *Langmuir* **13**, 3775–3780.

Komvopoulos, K.: 1996, Surface engineering and microtribology for microelectromechanical systems, *Wear* **200**, 305–327.

Kozlowski, F., Lindmair, N., Scheiter, T., Hierold, C. and Lang, W.: 1995, A novel method to avoid sticking of surface micromachined structures, *Transducers'95* pp. 220–223.

Kung, J. T. and Lee, H.-S.: 1991, An integrated air-gap capacitor process for sensor applications, *Transducers' 91*, pp. 312–314.

Licari, J. J.: 1970, *Plastic Coatings for Electronics*, McGraw-Hill, New York.

Lysko, J. M., Stolarski, E. and Jachowicz, R. S.: 1991, Capacitive silicon pressure sensor based on the one-side wafer processing, *Transducers' 91*, pp. 685–688.

Maboudian, R. and Howe, R. T.: 1997, Adhesion in surface micromechanical structures, *J. Vac. Sci. Technol.* **B15**, 1–20.

Man, P. F., Gogoi, B. P. and Mastrangelo, C. H.: March 1997, Elimination of post-release adhesion in microstructures using conformal fluorocarbon films, *Journal of Microelectromechanical Systems* **5**.

Man, P. F., Gogoi, B. P. and Mastrangelo, C. H.: n.d., Elimination of post-release adhesion in microstructures using thin conformal fluorocarbon films, *Proc. IEEE Micro Electro Mech. Syst. Workshop,* San Diego, CA, U.S.A. Feb. 1996 pp. 55–60.

Mastrangelo, C. H.: 1997, Adhesion-related failure mechanisms in micromechanical devices, *Trib. Lett.* **3**, 223–238.

Mastrangelo, C. H. and Hsu, C. H.: 1992, A simple experimental technique for the measurement of the work of adhesion of microstructures, *International Workshop on Solid-State Sensors and Actuators (Hilton Head' 92)*, pp. 208–212.

Mastrangelo, C. H. and Hsu, C. H.: 1993, Mechanical stability and adhesion of microstructures under capillary forces– parts I and II, *Journal of Microelectromechanical Systems* **2**, 33–55.

Mastrangelo, C. and Saloka, G.: n.d., A dry-release method based on polymer columns for microstructure fabrication, *Proc. IEEE Micro Electro Mech. Syst. Workshop,* Fort Lauderdale, FL, U.S.A. Feb. 1993 pp. 77–81.

Maszara, W. P., Goetz, G. and McKitterick, J. B.: 1988, Bonding of silicon wafers for silicon-on-insulator, *J. Appl. Phys* **64(10)**, 4943–4950.

Matijevic, E.: 1969, *Surface and Colloid Science*, Vol. 2, Wiley, New York.

Mellor, J. D.: 1978, *Fundamentals of Freeze Drying,* Academic Press, New York.

Meng, Q., Mehregany, M. and Mullen, R. L.: 1993, Theoretical modeling of microfabricated beams with elastically restrained supports, *IEEE J. Microelectromechanical Sys.* pp. 128–137.

Miller, S. L., Vigne, G. L., Rodgers, M. S., Sniegowski, J. J., Walters, J. P. and McWorther, P. J.: 1997, Routes to failure in rotating MEMS devices experiencing sliding friction, *Proc. SPIE Micromachined Devices and Components Conference*, Vol. SPIE 3224, pp. 24–30.

Morita, M., Ohmi, T., Hagesawa, E., Kawakami, M. and Suma, K.: 1989, Control factor of native oxide growth on silicon in air or in ultrapure water, *Applied Physics Letters* **55**, 562–564.

Mulhern, G. T., Soane, D. and Howe, R. T.: 1993, Supercritical carbon dioxide drying for microstructures, *Transducers' 93*, pp. 296–299.

Mullen, R. L., Mehrengany, N., Omar, M. P. and Ko, W. H.: 1991, Theoretical modeling of boundary conditions in microfabricated beams, *International Workshop on Micro Electromechanical Systems (MEMS'91)*, pp. 154–159.

Nathanson, H. C. and Guldberg, J.: 1975, Topologically structured thin films in semiconductor device operation, *Physics of Thin Films* **8**, 251–298.

Orpana, M. and Korhonen, A. O.: 1991, Control of residual stress of polysilicon thin films by heavy doping in surface micromachining, *Transducers' 91*, pp. 957–960.

Padday, J. F., Pitt, A. R. and Pashley, R. M.: 1974, Menisci at a free liquid surface: surface tension from the maximum pull on a rod, *J. Chem. Soc. Faraday Trans. 1* **10**, 1919–1931.

Poston, T. and Stewart, I.: 1978, *Catastrophe Theory and its Applications*, Pitman, London.

Ristic, L., Gutteridge, R., Dunn, B., Mietus, D. and Bennett, P.: 1992, Surface micromachined polysilicon accelerometer, *International Workshop on Solid-State Sensors and Actuators (Hilton Head' 92)*, pp. 118–121.

Ruzillo, J., Torek, K., Daffron, C., Grant, R. and Novak, R.: 1993, Etching of thermal oxides in low pressure anhydrous HF/CH_3OH gas mixture at elevated temperature, *J. Electrochem. Soc.* **140**, p. L64–L66.

Saunders, P. T.: 1990, *An Introduction to Catastrophe Theory*, Cambridge University Press, Cambridge.

Scheeper, P. R., Voorthuyzen, J. R. and Bergveld, P.: 1990, Surface forces in micromachined structures, *Micromechanics Europe 1990 (MME '90)*, pp. 26–31.

Smith, B. K., Sniegowski, J. J. and LaVigne, G.: 1997, Thin teflon-like films for eliminating adhesion in released polysilicon microstructures, *Transducers' 97*, pp. 245–248.

Srinivasan, U., Houston, M. R., Howe, R. T. and Maboudian, R.: 1997, Self-assembled fluorocarbon films for enhanced stiction reduction, *Transducers' 97*, pp. 1399–1402.

Takeshima, N., Gabriel, K. J., Ozaki, M., Takashashi, J., Horiguchi, H. and Fujita, H.: 1991, Electrostatic parallelogram actuators, *Transducers' 91*, pp. 63–66.

Tang, W. C., Nguyen, T.-C. H. and Howe, R. T.: n.d., Laterally driven polysilicon resonant microstructures, *Proc. IEEE Micro Electro Mech. Syst. Workshop,* Salt Lake City, Utah, U.S.A., Feb. 1989 pp. 53–59.

Tanner, D. M., Smith, N. F., Bowman, D. J., Eaton, W. P. and Peterson, K. A.: 1997, First reliability test of a surface micromachined microengine using SHiMMeR, *Proc. SPIE Micromachined Devices and Components Conference*, Vol. SPIE 3224, pp. 14–21.

Ulman, A.: 1991, *An Introduction to Ultrathin Organic Films: From Langmuir-Blodgett to Self-Assembly*, Academic Press, Boston.

Watanabe, H., Hamano, M. and Harazono, M.: 1989, The role of atmospheric oxygen and water in the generation of water marks on the silicon surface in cleaning processes, *Mat. Sci. Eng.* **B4**, 401–405.

Watanabe, H., Ohnishi, S., Honma, I., Kitajima, H., Ono, H., WIlhelm, R. and Sophie, A.: 1995, Selective etching of phosphosilicate glass with low pressure vapor hf, *J. Electrochem. Soc.* **142**, 237–243.

Wong, M., Moslehi, M. and Bowling, R.: 1993, Wafer temperature dependence of the vapor-phase hf oxide etch, *J. Electrochem. Soc.* **140**, 205–208.

Yasuda, H.: 1985, *Plasma Polymerization*, Academic Press, New York.

Yasuda, H. and Hsu, T.: 1977, Some aspects of plasma polymerization of fluorine containing organic compounds, *J. Polym. Sci. Polym Chem. Ed.* **15**, 2411–2424.

Yasuda, H. and Hsu, T.: 1978, Some aspects of plasma polymerization of fluorine containing organic compounds. ii. comparison of ethylene and tetrafluoroethylene, *J. Polymer Sci.* **16**, 415–425.

Yee, Y., Chun, K. and Lee, J. D.: 1995, Polysilicon surface modification technique to reduce sticking of microstructures, *Transducers' 95*, pp. 206–209.

DEVELOPMENT, FABRICATION AND TESTING OF A MULTI-STAGE MICRO GEAR SYSTEM

C. THÜRIGEN, W. EHRFELD, B. HAGEMANN, H. LEHR, F. MICHEL
Institut für Mikrotechnik Mainz GmbH, Carl-Zeiss-Strasse 18-20, D-55129 Mainz, Germany

1 Introduction

There is a general trend towards miniaturization of technical systems. Over the last few decades the development of microelectronics has led to the integration of more and more functions into compact appliances. The same is now happening in microtechnology, i. e. the miniaturization and integration of mechanical systems such as drives. For many applications e. g. in medicine, drives with an outer diameter in the millimeter range and torques of more than 0.1 mNm are required. Drives in this context are a combination of motor and gearbox.

Until recently the smallest drives available on the market had a diameter of 8 mm. Last year drives with a diameter of 5 mm and 3 mm were introduced [5]. Now the company Faulhaber Motoren and the Institut für Mikrotechnik Mainz GmbH (IMM) have developed together an electromagnetic micromotor with an outer diameter of only 1.9 mm [2]. In order to adapt the high motor speeds of some 100,000 rpm and torques of 7.5 μNm to the requirements of most applications, a micro gear system with the same diameter was also implemented.

2 Design of the gear system

With a total diameter of 1.9 mm the gears have tip diameters of about 0.5 mm. Typical working conditions for micro gear systems are high input speeds and high gear ratios. The task of the design engineer is to find applicable tooth profiles and gear system designs. This is decisively influenced by the microtechnological manufacturing of the components. Spur gears, for examble, can well be produced using these processes. Therefore involute, cycloidal and circular arc tooth profiles were inspected. The involute profile is the most suitable because of its constant velocity ratio during the meshing and its indifference to center distance variations. Manufacturing by means of microtechnological processes is independent of cutting tools so that the tooth profile

B. Bhushan (ed.), Tribology Issues and Opportunities in MEMS, 397-402.

can be adapted to the special conditions in micro gear systems. This is supported by the very good mathematical description of the involute tooth profile [3].

In precision mechanics, gear systems are classified into stationary transmissions (Figure 1), planetary gear systems (Figure 3 and Figure 5), and special types with high gear ratios (Figure 2) [4]. In the most suitable design of stationary types for micro gear systems, combinations of wheels and pinions are mounted on at least two rigid traversing axles which have to be placed in the housing with high precision. The total gear ratio is the product of the ratios of the single stages, which is up to 3.5 in this design. In planetary gear systems, higher ratios in one stage can be realized, and the torque is distributed over three or four planetary wheels. They also center the frame so that no additional bearings are required. In this gear type typically external and internal toothed wheels are arranged. Gear systems with high ratios are special types of planetary gears. To gain the high ratios, there is only a small difference in the number of teeth of planetary and internal toothed wheels. The eccentric rotation of the planetary wheel has to be centered (Figure 2). The complicated designs of these gear systems are difficult to miniaturize. The advantages of planetary gear systems led to the decision to implement them as a micro system.

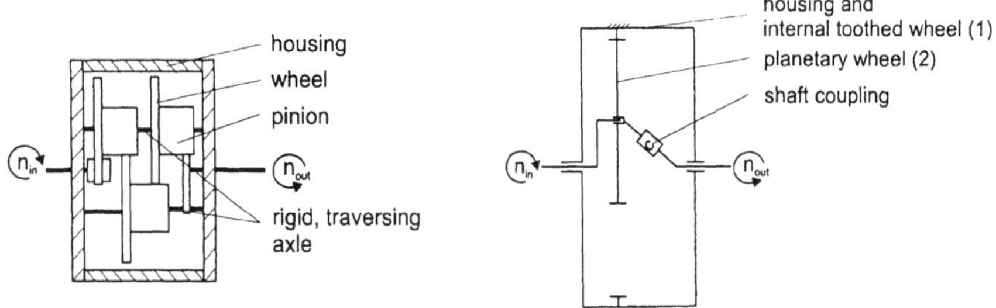

Figure 1. Stationary transmission

Figure 2. Gear system with high ratio

Two different types were implemented. At first a gear system of the Wolfrom type was designed to demonstrate the possibilities of microtechnological manufacturing. This two-stage gear system is driven by the sun wheel while the second internal toothed wheel is the output. Three planetary gears are used. So the numbers of teeth of the fixed and the output toothed wheels differ by three. With an adapted profile offset both wheels can mesh with the same planetary gear. Because the planetary wheels are supported by the sun wheel, a frame to arrange their bearing is not necessary (Figures 3 and 4). With this compact design high gear ratios for two-stage planetary gears are reached.

The single-layer toothed wheels can very well be produced with microtechnological manufacturing. The housing, input and output jewel bearings are made by precision mechanics. The modulus of the implemented gear system of the Wolfrom type is 38 μm. The external toothed wheels have a width of tooth face of 250 μm, while the tooth face of the internal toothed wheels is 500 μm wide. The gear ratio is 45. The disadvantages of this design are the inflexibility of gear ratios and a lower efficiency than standard multi-stage types.

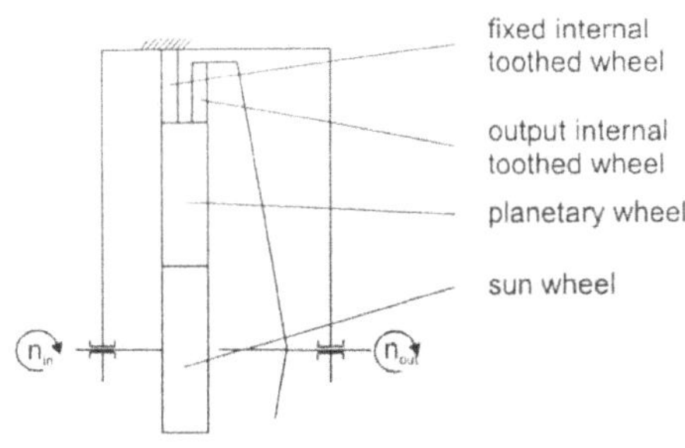

Figure 3. Planetary gear system of Wolfrom type

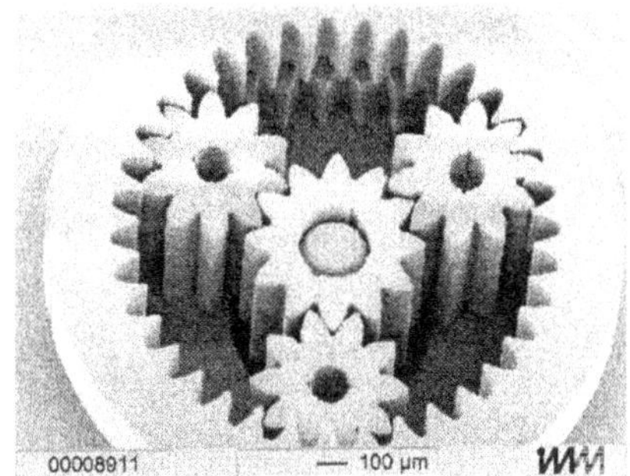

Figure 4. Assembled micro gear system of Wolfrom type

Because of these disadvantages a second micro gear system was implemented, more suited for commercialisation. This time a multi-stage gear type was chosen, whose components can be mass produced by micro injection moulding. In this design a sun wheel mounted on the shaft of the micromotor is the input. The planetary wheels with tip diameters of 560 µm (Figure 9) rotate on axles, that are fixed in the frame and only have diameters of 180 µm. The width of the tooth face is 300 µm. For a maximum output torque of 250 µNm the axles have to be fixed on both sides. Therefore, the frame has to be divided into two parts (Figure 10). Rigid connecting elements are placed between the planetary wheels. The sun wheel of the following stage (B and Figure 10) or the output shaft (C) are fixed to the upper part of the frame. To reduce friction, the bore of the bearing is a polygon with 10 edges, so that wear particles are deposited in the appearing pockets (Figure 9). In porous bearings this shape of bore in combination with a cylindrical shaft particullary reduces static friction and extends the life time [3].

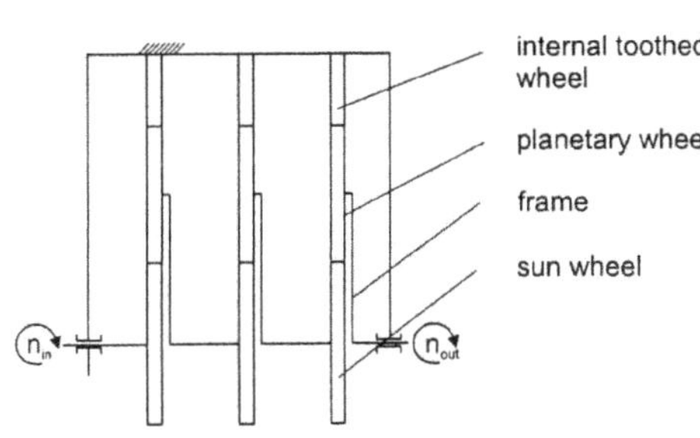

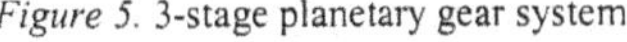
Figure 5. 3-stage planetary gear system

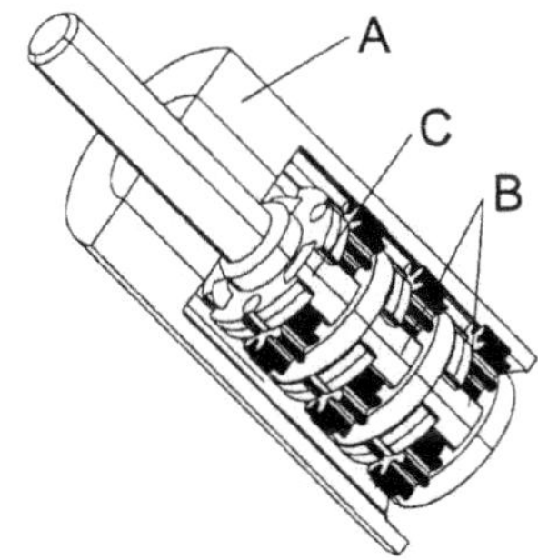

Figure 6. Design of implemented 3 stage planetary gear system

Stages with a modulus of 55 µm and gear ratios of 3.6, 4.71, and 5.33 were designed. They all use the same internal toothed wheel. It has a width of tooth face of 2,300 µm and is integrated in the housing. The free combination of up to five stages allows a great number of different ratios.

3 Fabrication and microassembly

For the production of gears, microtechnological processes like LIGA and wire EDM were compared to precision mechanics techniques such as self-generating milling in the watch industry. It was found only microtechnological processes reach the required manufacturing tolerances for micro gears. The implemented planetary gear systems have minimum radii of less than 15 μm which cannot be realized with 30 μm diameter wires in EDM technology. By means of the LIGA process [1] the gears and their moulds can be made with tooth faces which are exactly parallel to the cylinder axis. The surface roughness is about $h_{rms} = 40$ nm.

The toothed wheels of the Wolfrom type gear system are made in metal by micro electroplating. To build up the moulds for the multi-layer frame parts, LIGA-made sheets are stacked in a special support (Figure 7). For the housing a resist with a thickness of 2,300 μm has to be irradiated. The mould of gears is also made by this process. The electro plating process is continued until a solid backplane is grown up which fixes the rod for the bore (Figure 8). During micro injection moulding, the smooth side walls of LIGA-made micro structures allow a proper removal from the mould.

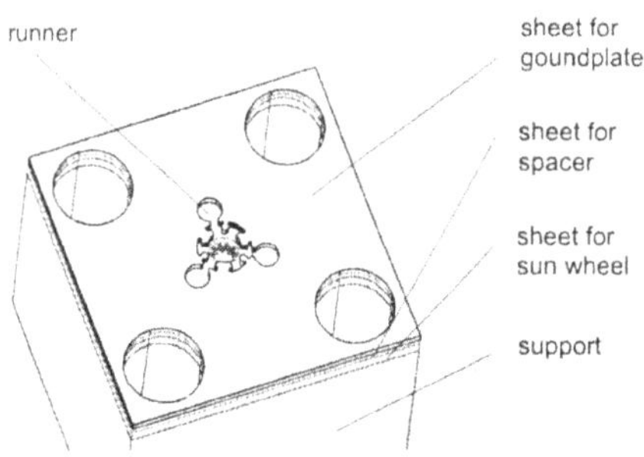

Figure 7. 3-dimensional design of mould insert of the upper frame part

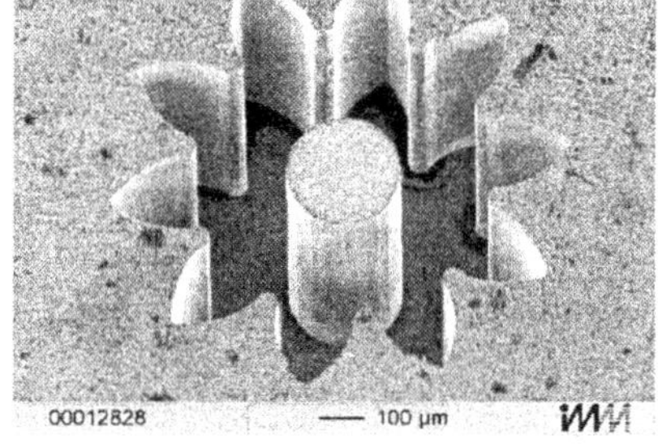

Figure 8. Mould insert for micro gear

For the assembly which has to take place in a clean room environment of class 100 to 1000, manual techniques were developed: Special tools and supports were designed and furthermore several tweezers and vacuum pipettes are used. For the observation, stereo microscopes with a variable magnification in the range of 15 to 60 are used. An assembled frame with planetary gears is shown in Figure 10. The manual asembly is suitable to produce small series. Mass production will only be possible with automated assembly. Its development will strongly be influenced by the experiences made with manual assembly.

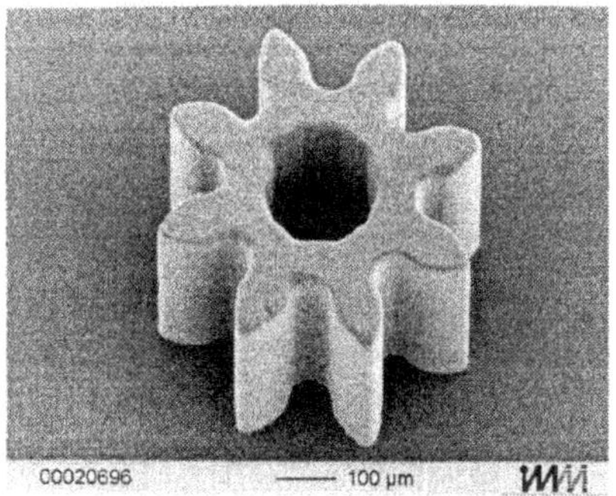

Figure 9. Micro injection moulded gear

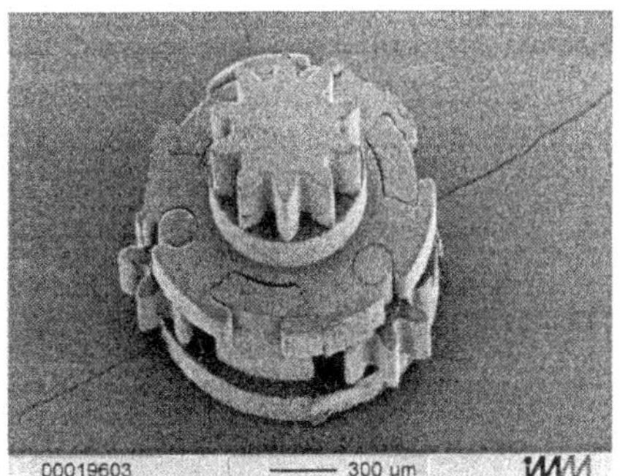

Figure 10. Assembled frame of planetary gear system

4 Testing

For applications it is important to determine the output torque of the micro drive, while the efficiency rate is an important value to qualify the gear systems. Therefore, input and output torques have to be determined. In tests standard DC-micro motors were used to generate the load while the micro motors, implemented in the project, generated the input torque. The output torque of electromagnetic motors can be calculated by the magnetic flux and the motor current. The magnetic flux, in turn, can be determined by measuring the induced voltage related to the rotation of the motor while friction losses are determined via the no-load currents. In tests the micro drives with a three-stage gear system reached output torques of 180 µNm and a maximum efficiency of about 57%. Graphs of the efficiency depending on input speed and torque are shown in Figure 11.

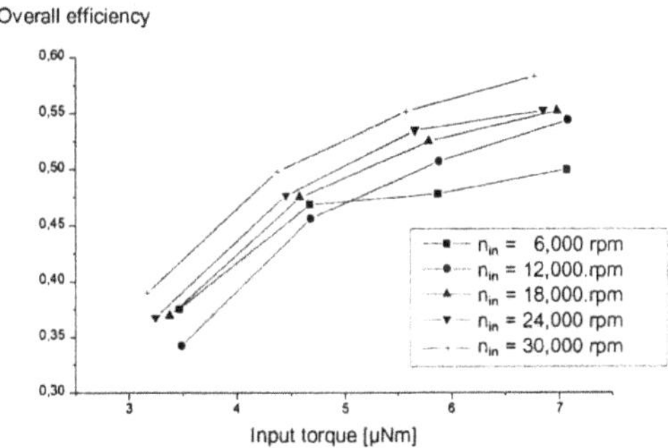

Figure 11. Efficiency rate of three-stage micro gear system with ratio of 47

In these tests a fully synthetic oil with a kinematic viscosity of $\nu = 95\ \text{mm}^2/\text{s}$ at 20°C and a precision grease lubricant were used. The basic oil of this grease lubricant had a kinematic viscosity of $\nu = 70\ \text{mm}^2/\text{s}$ at 20°C, and a PTFE-powder was used as thickener. No significant difference between oil and grease lubricant were detected in the tests.

In life time tests the micro gear systems were run with a load of 150 µNm. With a sufficient amount of lubricant they ran for up to 1500 h. In these tests an oil with a kinematic viscosity of $\nu = 320\ \mathrm{mm^2/s}$ at 20°C was used. Figures 12 and 13 show the wear on one of the axles in the frame for the planetary gear system and the sun wheel of the output stage after a working time of 1250 hours. They show that higher working times than demonstrated are possible.

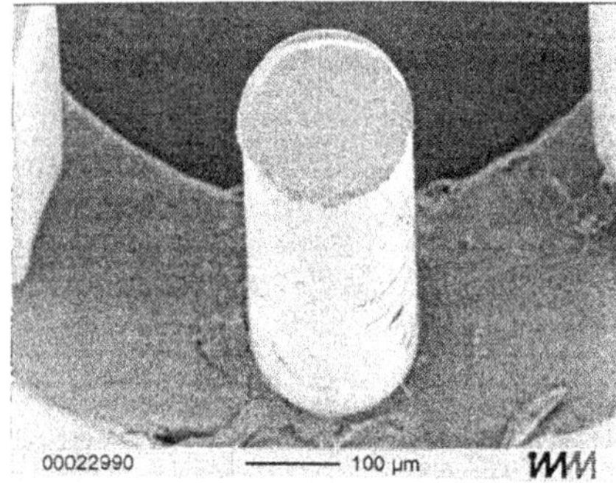

Figure 12. Wear on the axle in frame of the 3rd stage

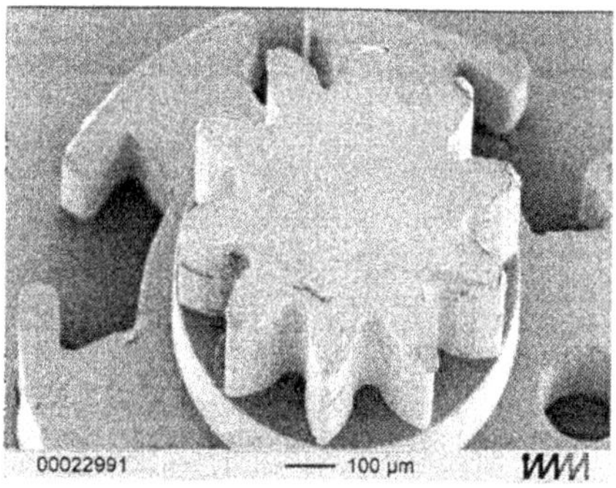

Figure 13. Wear on sun wheel of the 3rd stage

5 Conclusion

In close cooperation between a medium sized company and a research institute a new micro gear motor was developed and introduced on the market. A combination of advanced precision mechanics and microtechnological manufacturing created a new powerful product. With a diameter of 1.9 mm the motor rotates at 100,000 rpm, and with an additional gear system output torques of up to 150 µNm are reached. Lifetimes of up to 1,500 hours were demonstrated.

6 References

1. W. Ehrfeld, H. Lehr: *Deep X-Ray Lithographie for the Production of three-dimensional Microstructures from Metals, Polymers and Ceramics,* Radiation Physics and Chemistry, 1995
2. B. Hagemann: *Entwicklung von Permanentmagnet-Mikromotoren mit Luftspaltwicklung,* theses, university of Hannover, Germany, to be published
3. W. Krause: *Konstruktionselemente der Feinwerktechnik,* Carl Hanser Verlag, München, Wien, 1993
4. J. Legler: *Leistungsverhalten von hochübersetzenden Stirnradgetrieben und Räderketten der Gerätetechnik*, theses, TU Dresden 1989.
5. F. A. Stadler: *Spritzenleistungen in der klassischen Mikrobearbeitung - eine Frage der Geisteshaltung*, Tagungsband Micro-Engineering 96, Stuttgart, 1996

ANALYSIS OF GEAR TOOTH PERFORMANCE OF MECHANICALLY-COUPLED, OUTER-ROTOR POLYSILICON MICROMOTORS

K.C. Stark, M. Mehregany, and S.M. Phillips

Microfabrication Laboratory
Case Western Reserve University, Cleveland, OH 44106

1. Introduction

This paper focuses on specific design of gear teeth for micromotors, as well as performance evaluation of the coupled micromotors both in air and using different lubricants. Two basic types of gear teeth, as described in Tables 1 and 2, have been studied to evaluate their appropriateness for mechanical coupling of outer-rotor micromotors: involute shaped teeth and cycloidal shaped teeth. These tooth profiles have been chosen since the contact between teeth of mating gears has pure rolling motion at the pitch line (involute shaped teeth) or pitch circle (cycloidal shaped teeth) [1]. Figure 1 shows SEM photos of two pairs of coupled 75 μm micromotors, and Figure 2 shows SEM photos of typical involute and cycloidal gear teeth after fabrication and release.

2. Testing of mechanically-coupled micromotors

Outer-rotor wobble micromotors of radii 75, 100, 125, and 150 μm were tested for up to 30 minutes at excitation voltages of 100 V for up to 30 minutes at motor speeds up to 12 RPM. During operation, the micromotors were observed and mechanically probed when necessary to maintain constant rotation. After testing, the minimum voltage required for coupled actuation was measured. Table 3 summarizes the results of this testing and shows the minimum actuation voltage required for actuation when loaded. These results have shown that coupled micromotors with involute shaped teeth can be driven over a wider range of voltages (as low as 28 V for 150 μm radii motors), and operate more consistently and more smoothly than micromotors with cycloidal teeth. SEM photos showed that there was no change in the roughness of the tooth face after this test (see Figure 3).

3. High speed operation

From the results of the 30 minute test, it was determined that the 150 μm micromotors performed the most reliably at higher speeds. The 150 μm radius motors were then operated at an excitation speed of 47000 RPM (motor speed of 120 RPM) and at 100 V for up to 2 hours to determine if there was a noticable difference in the wear characteristics of the gear teeth. The experiment was stopped at 2 hours because the micromotor required excessive mechanical probing. Figure 4 shows an SEM picture of a Type 1 tooth after 2 hours of operation; this tooth is from the same gear as the one in Figure 3. Comparing the tooth in Figures 3 (30 minutes of operation) and the tooth in Figure 4 (2 hours of operation), it can be seen that the inherent roughness of the tooth profile associated with the reactive ion etch process during fabrication has started to smooth out by the frictional contact between the gear teeth. Polysilicon wear has been previously been observed and studied in other harmonic side-drive micromotors [2]. However, it is important to note that in the case of mechanically-coupled micromotors, wear can occur between the rotor and the bearing as well as between the teeth.

B. Bhushan (ed.), Tribology Issues and Opportunities in MEMS, 403-406.

4. Lubrication of coupled micromotors

Because performance may be seriously affected by the sidewall roughness of the gear teeth, operation of mechanically-coupled micromotors is investigated with different lubricants. Unlike metal, macroscopic gears which are cut and can be polished, the inherent roughness of the micromotor tooth profile associated with the RIE process during fabrication of micromotors may require lubricants for successful operation. Self-assembled monolayers (SAMs) and silicone oil are investigated as possible lubricants for micromotor operation.

SAMs have shown promise in coating polysilicon surfaces to provide lubrication for micromotor operation [3]; the monolayer used is Octadecyltrichlorosilane (OTS). Testing of SAM coating micromotors did not show improvement, which can be explained by the fact that the coefficient of friction between polysilicon surfaces, such as meshed gear teeth, is increased from 0.36 for uncoated polysilicon to 0.55 for OTS coated micromotors [3].

Silicone oil was next investigated as a possible. A set of motors was tested for 15 minutes first in room air, and then for another 15 minutes after being immersed in silicone oil with a viscosity of 20 cs. The motors immersed in silicone oil were placed in a vacuum dessicator and pumped down in order to ensure that the surfaces and gaps of the micromotors were covered with silicone oil. Both tests were performed at room temperature (25°C) and at a nominal operating voltages between 100 V and 135 V. The excitation frequency was 5000 RPM for an approximate motor speed of 10 RPM. These experiments showed that a dramatic improvement of performance was observed for the motors after immersion in silicon oil. In many cases, motors that did not operate consistently in room air operated with minimal mechanical probing after immersion in silicone oil.

As a comparison of the effects on frictional wear of the sidewalls of the polysilicon teeth, a pair of 150 μm pitch radius micromotors was operated at a speed of 12 RPM and an excitation voltage of 100 V for 20 hours without requiring mechanical probing. Due to the viscous oil that the motors were immersed in, the maximum speed attainable was 12 RPM. The time of 20 hours was specifically chosen in order to compare the frictional wear of the teeth with the test in Section 3. These motors were therefore operated for the same number of motor rotations: 20 hours at 12 RPM in silicone oil is the same number of rotations as the 2 hours test at 120 RPM test room air described in Section 3. Figure 5 shows an SEM photo of the sidewall profile for the teeth for the test in silicone oil. As seen here, there is no evidence of frictional wear on the tooth profile, the picture is the same as that in Figure 3. Therefore, silicone oil not only helps improve micromotor reliability, but also helps to prevent adverse wear in coupled gear teeth.

(a)

(b)

Figure 1: SEM pictures of a typical pair of 75 μm radii coupled micromotors with: (a) cycloidal shaped teeth; and (b) involute shaped teeth with a 20° pressure angle.

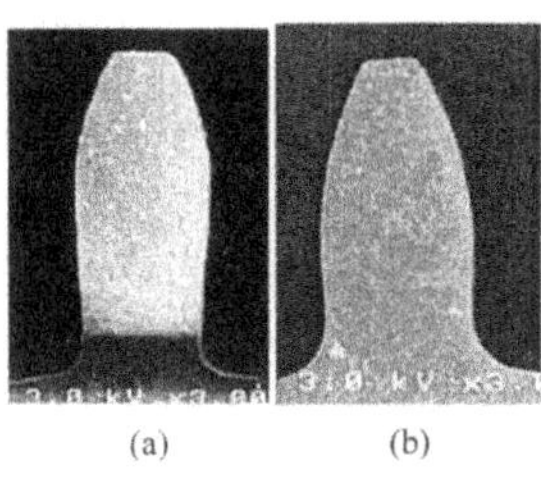

(a) (b)

Figure 2: SEM picture close-up of: (a) a cycloidal tooth (Type 3); and (b) an involute tooth (Type 2). Each are about 10 μm wide in the middle.

Coupled pair type	Tooth Profile	Pressure Angle
1	Involute	20°
2	Involute	14.5°

Table 1: Involute tooth designs types.

Coupled pair type	Tooth Profile	Circle radius for addendum (in μm)		Circle radius for dedendum (in μm)	
		Gear "a"	Gear "b"	Gear "a"	Gear "b"
3	Cycloidal	40	40	40	40
4	Cycloidal	40	20	20	40

Table 2: Cycloidal tooth designs types.

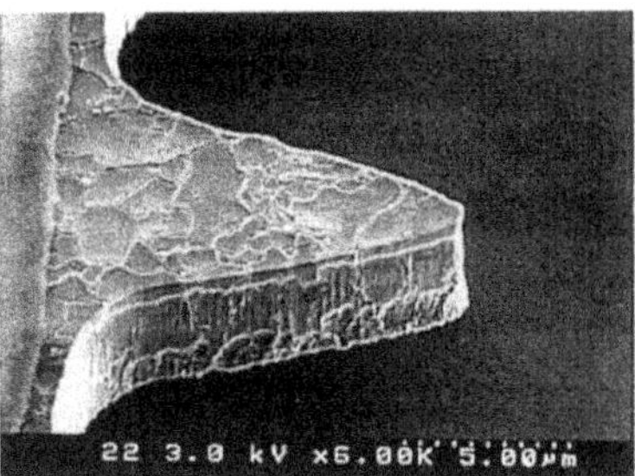

Figure 3: SEM picture close-up of tooth profile after 30 minutes of operation; no wear is evident.

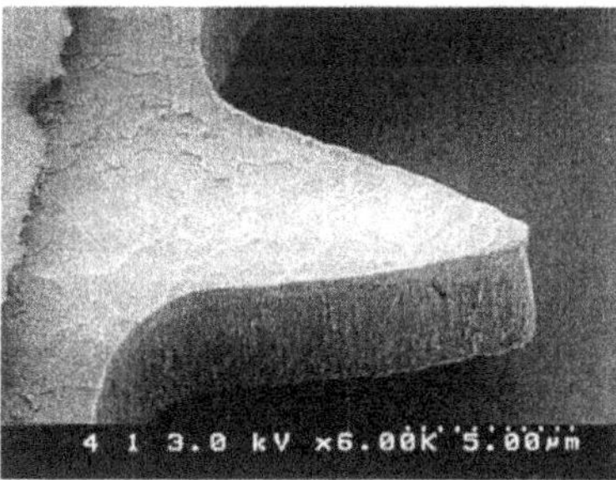

Figure 4: SEM picture close-up of tooth profile after 2 hours of operation. The tooth profile is beginning to be smoothed out from the frictional wear.

Pair Type (see Tables 1,2)	Motor radii (μm)	Minimum voltage for actuation after testing (V)	Frequency of probing necessary to maintain continuous rotation
1	150	50	every 5 minutes
2	150	68	every 2 minutes
3	150	116	1-2 times per minute
4	150	52	every 2 minutes
1	100	75	every 2 minutes
2	100	60	every 3 minutes
3	100	76	every 30 seconds
4	100	77	every 20 seconds

Table 3: Results of testing for 150 μm radii and 100 μm radii coupled micromotors.

Figure 5: SEM picture close-up of tooth profile after 20 hours of operation in silicone oil. The tooth profile does not show signs of frictional wear.

[1] J. E. Shigley. *Kinematic Analysis of Mechanisms*. McGraw-Hill, Inc., New York, 1969.

[2] M. Mehregany, S.D. Senturia, and J.H. Lang, "Measurement of wear in polysilicon micromotors," IEEE Trans. on Electron Devices, ED-39, Sept. 1992, pp. 2060-2069.

[3] K. Deng, R. J. Collins, and C. N. Sukenik. Performance impact of monolayer coating of polysilicon micromotors. *Journal of the Electrochemical Society*, 142(4):1278, April 1995.

MICRO/NANOTRIBOLOGICAL STUDIES OF SINGLE-CRYSTAL SILICON AND POLYSILICON AND SIC FILMS FOR USE IN MEMS DEVICES

Bharat Bhushan, Sriram Sundararajan and Xiaodong Li
Computer Microtribology and Contamination Laboratory,
Department of Mechanical Engineering,
The Ohio State University, Columbus, OH 43210-1107, USA

Christian A. Zorman and Mehran Mehregany
Department of Electrical Engineering and Applied Physics
Case Western Reserve University
Cleveland, Ohio 44106, USA

Abstract MEMS devices are currently made from single-crystal silicon, LPCVD polysilicon films and other ceramic films. For high temperature applications, SiC films are being developed to replace polysilicon films. Tribology in MEMS devices is of importance. Atomic force/friction force microscopy (AFM/FFM) and nanoindentation techniques have been used for tribological studies on microscale on materials of interest. These techniques have been used to study surface roughness, adhesion, friction, scratching/wear, indentation and boundary lubrication of bulk and treated silicon, polysilicon films and SiC films. Measurements of microscale and macroscale friction show that friction values on both scales of all the silicon samples are about the same and higher than that of SiC. The microscale values are lower than the macroscale values as there is less ploughing contribution in the microscale measurements. Surface roughness has an effect on friction. Microscratching, microwear and nanoindentation studies

B. Bhushan (ed.), Tribology Issues and Opportunities in MEMS, 407-430.

indicate that SiC films are superior to coated/treated silicon followed by bare silicon. Chemically bonded lubricants appear to be suitable for MEMS devices.

1. Introduction

The advances in silicon process technology over the last three decades has led to the development of microcomponents known as microelectromechanical systems or MEMS. Researchers have fabricated a wide variety of sensors, actuators, valves, gear trains, turbines, nozzles and pumps (for a collection of work see Trimmer, 1997) with dimensions in the range of a couple to a few hundred microns. MEMS research is still in its infancy and the emphasis to date has been on the fabrication and laboratory demonstration of individual components. MEMS devices have begun to be commercially used, particularly in the automotive industry. Silicon-based acceleration sensors are used in airbag-deployment (Bryzek et al., 1994), in anti-skid braking systems and four-wheel drives. Today, acceleration sensor technology, for example, is on the order of a billion dollars-a-year industry dominated by Lucas NovaSensor and Analog Devices. Another application of MEMS is the use of silicon micromachined pressure sensors for in-cylinder pressure monitoring in auto engines, in medical instrumentation, cockpit instrumentation and many hydraulic and pneumatic consumer products (Fujimasa, 1996). Another field where MEMS devices are being pursued is in magnetic storage systems (Bhushan, 1996a), where they are being developed for super-compact and ultra-high-recording-density magnetic disk drives. Horizontal thin-film heads with a single-crystal silicon substrate, referred to as silicon planar head sliders are under development and can be mass-produced through IC technology (Lazarri and Deroux-Dauphin, 1989; Bhushan et al., 1992). Several integrated head/suspension microdevices have been fabricated for contact recording applications (Ohwe et al., 1993; Hamilton, 1991). High bandwidth servocontrolled microactuators have been fabricated for ultra-high-track-density applications (Miu and Tai, 1995; Fan et al., 1995). Millimeter-sized wobble motors and actuators for tip-based recording schemes have also been fabricated (Fan and Woodman, 1995). In some cases, MEMS devices are used primarily for their miniature size, while in others, as in the case of the air bags, because of their high reliability and low-cost manufacturing techniques. This latter fact has been possible since semiconductor-processing costs have reduced drastically over the last decade, allowing the use of MEMS in many previously impractical fields.

The fabrication techniques for MEMS devices employ photolithography and fall into four basic categories: bulk micromachining, surface micromachining, wafer bonding and LIGA (a German acronym for Lithographie Galvanoformung Abformung), a German term for lithography, electroforming and plastic molding. The first two approaches, bulk and surface micromachining, use planar photolithographic fabrication processes developed for semiconductor devices in producing two-dimensional (2D) structures (Jaeger, 1988). Bulk micromachining employs anisotropic etching to remove sections through the thickness of a single-crystal silicon wafer, typically 250 to 500 μm thick. Bulk micromachining is a proven high-volume production process and is routinely used to fabricate microstructures such as pressure and acceleration sensors and magnetic heads. Surface micromachining is based on depositing and etching structural and sacrificial films to produce a freestanding structure. These films are typically made of low-pressure chemical vapor deposition (LPCVD) polysilicon film with 2 to 20 μm thickness. Surface micromachining is used to produce surprisingly complex micromechanical devices such as grippers, gears and motors. Wafer bonding is a technique used to join two silicon substrates in order to provide mechanical support, thermal isolation, heat sinking or electrical connection to a microfabricated device. Wafer bonding is used to create sealed chambers in bulk-micromachined microvalves and micropumps. It is also used to fabricate silicon-on-insulator substrates for microelectronics. LIGA is used to produce high-aspect ratio MEMS devices that are as thick as they are wide or long. The LIGA process yields very sturdy 3D structures due to their increased thickness and has been used to produce gears, motors, actuators and connectors.

In MEMS devices, various forces associated with the device scale down with the size. When the length of the machine decreases from 1 mm to 1 μm, the area decreases by a millionth and the volume decreases by a billionth. The resistive forces such as friction, viscous drag and surface tension that are proportional to the area, increase a thousand times more than the forces proportional to the volume, such as inertial and electromagnetic forces. These forces lead to tribological concerns, which become critical because friction/stiction (static friction), wear and surface contamination affect device performance and in some cases, can even prevent devices from working. For example, in micromotors, the intermittent contact at the rotor-stator interface and physical contact at the rotor-hub interface result in wear issues and also necessitates low friction/stiction between the contacting surfaces. In addition, there is a need for development of bearing/bushing materials that are both compatible with MEMS fabrication processes and which provide superior friction and wear performance. Studies have been conducted to measure the friction/stiction in micromotors (Tai and Muller, 1990), turbines and gear structures (Gabriel et al., 1990) and polysilicon microstructures (Lim et al., 1990) to understand friction mechanisms. Several studies have been conducted to develop solid and liquid lubricant and hard films to minimize friction and wear (Henck, 1997; Bhushan, 1996b; Bhushan et al., 1995b; Deng et al., 1995; Beerschwinger et al., 1995; Koinkar and Bhushan, 1996a,b). In microvalves used for flow control, the mating valve surfaces should be smooth enough to seal while maintaining a minimum roughness to ensure low friction/stiction. There are tribological issues in the fabrication processes as well. For example, in surface micromachining, the suspended structures can sometimes collapse and permanently adhere to the underlying substrate (Guckel and Burns, 1989). The mechanism of such adhesion phenomena needs to be understood (Mastrangelo, 1997). Only a few studies have been conducted on the tribology of bulk silicon and polysilicon films used in MEMS devices (Bhushan and Venkatesan, 1993a,b; Gupta et al., 1993; Venkatesan and Bhushan, 1993,1994; Gupta and Bhushan, 1994; Bhushan and Koinkar, 1994; Bhushan, 1996b) and mechanical properties of polysilicon films are not well characterized (Mehregany et al., 1987; Ericson and Schweitz, 1990; Guckel et al., 1992; Schweitz, 1991; Fang and Wickert, 1995).

The advent of atomic force/friction force microscopy (AFM/FFM) (Bhushan, 1995; Bhushan et al, 1995a), has allowed the study of surface topography, adhesion, friction, wear, lubrication and measurement of mechanical properties, all on a micro- to nanometer scale. Recently, microtribological studies were conducted using the AFM/FFM on doped silicon and polysilicon films that are used in MEMS devices (Bhushan and Koinkar, 1994,1997; Bhushan and Li, 1997).

Although silicon based MEMS devices find wide uses today, they lack high temperature capabilities with respect to both mechanical and electrical properties.

TABLE 1 Selected bulk properties[a] of 3C (β- or cubic) SiC and Si(100).

Sample	Density (kg/m^3)	Hardness (GPa)	Elastic Modulus (GPa)	Fracture Toughness ($MPa\ m^{1/2}$)	Thermal Conductivity[b] (W/m °K)	Coeff. of Thermal Expansion[b] ($\times 10^{-6}$ / °C)	Melting Point (°C)	Bandgap (eV)
β-SiC	3210	23.5-26.5	440	4.6	85-260	4.5 - 6	2830	2.3
Si(100)	2330	9-10	130	0.95	155	2 – 4.5	1410	1.1

[a] Unless stated otherwise, data shown were obtained from Bhushan and Gupta (1997).

[b] Obtained from Shackelford et al. (1994).

Recently, researchers have been pursuing SiC as material for high-temperature microsensor and microactuator applications (Tong et al., 1992; Shor et al., 1993). SiC is a likely candidate for such applications since it has long been used in high-temperature electronics, high-frequency and high-power devices, such as SiC metal-semiconductor field effect transistors (MESFETS) that operate at frequencies up to 12.6 GHz (Spencer et al., 1994) and inversion-mode metal-oxide-semiconductor field effect transistors (MOSFETS). Many other SiC devices have also been fabricated including UV detectors, SiC memories and SiC/Si solar cells. SiC has also been used in microstructures such as speaker diaphragms and X-ray masks. For a summary of SiC devices and applications, see Harris, 1995. Table 1 compares selected bulk properties of SiC and Si(100). Because of SiC's large bandgap, almost all devices fabricated from SiC have good high-temperature properties. This high-temperature capability of SiC combined with its excellent mechanical properties, thermal dissipative characteristics, chemical inertness and optical transparency makes SiC an ideal choice for complementing polysilicon (polysilicon melts at 1400 °C) in MEMS devices. Since MEMS devices need to be of low cost to be viable in most applications, researchers have found low-cost techniques of producing single-crystal 3C-SiC (cubic or β-SiC) films via epitaxial growth on large area silicon substrates (Zorman et al., 1995). This technique allows high-volume batch processing and has the advantage of having silicon as the substrate, an inexpensive material for which microfabrication and micromachining technologies are well established. It is believed that these films will be well suited for MEMS devices. Micro/nanotribological characterization of these SiC films and their comparison to the materials currently used in such small-scale devices is of critical importance. This paper presents the results of these studies conducted for the first time on 3C-SiC films and compares these films to silicon, and doped and undoped polysilicon films.

2. Experimental Technique

2.1 DESCRIPTION OF APPARATUS AND TEST PROCEDURES

A modified atomic force/friction force microscope (AFM/FFM) (Nanoscope III, Digital Instruments, Santa Barbara, California), was used for the micro/nanotribological studies. Surface roughness and microscale friction measurements were simultaneously made over a scan size of 10 μm x 10 μm with a Si_3N_4 tip (tip radius ~ 50 nm, cantilever stiffness ~ 0.6 N/m) sliding over the sample surface orthogonal to the long axis of the cantilever at 25 μm/s. A coefficient of friction and conversion factors for converting the friction signal voltage to force units (nN) were obtained through the methods developed previously by Bhushan and co-workers (Bhushan, 1995). The normal loads used in the friction measurements varied between 50 to 300 nN. The reported values are each an average of six separate measurements.

For the scratch and wear tests, specially fabricated microtips were used. These microtips consisted of single-crystal natural diamond, ground to the shape of a three-sided pyramid, with an apex angle of 60° and tip radius of about 70 nm, mounted on a platinum-coated stainless steel cantilever beam whose stiffness was 50 N/m. Samples were scanned orthogonal to the long axis of the cantilever with loads ranging from 20 to 100 μN to generate scratch/wear marks. Scratch tests consisted of generating scratches in a reciprocatory mode at a given load for 10 cycles over a scan length (stroke length) of 5 μm at 10 μm/s. Wear marks were generated over a scan area of 2 μm x 2 μm at 4 μm/s and the wear marks were observed by scanning a larger 4 μm x 4 μm area with the wear mark at the center. Imaging scans of both scratch and wear tests were done at a low normal load of 0.5. The reported scratch/wear depths are an average of three runs at separate instances. All measurements were performed in an ambient environment (21 ± 1 °C, 45 ± 5% RH).

Hardness and elastic modulus were calculated from load-displacement data obtained by nanoindentation using a commercially available nanoindenter (Bhushan,

1995). The instrument monitored and recorded dynamic load and displacement of a three-sided pyramidal diamond (Berkovich) indenter with a force resolution of about 75 nN and displacement resolution of about 0.1 nm. Multiple loading and unloading are performed to examine reversibility of the deformation and thereby ensuring that the regime was elastic. Macroscale friction measurements were conducted using a ball-on-flat tribometer under reciprocating motion (Bhushan, 1996a).

2.2 TEST SAMPLES

Studies were conducted on various types of virgin silicon samples: undoped (lightly doped) single-crystal Si(100), Si(111) and Si(110) and the following types of treated/coated silicon samples: PECVD-oxide coated Si(111), dry-oxidized, wet-oxidized and C^+ -implanted Si(111). Studies were also conducted on heavily doped (p^+ -type) single-crystal Si(100), undoped polysilicon film, heavily doped (n^+ -type) polysilicon film and 3C-SiC (cubic or β-SiC) film. A 10 mm x 10 mm coupon of each sample was ultrasonically cleaned in methanol for 20 minutes and dried with a blast of dry air prior to measurements. The undoped Si(100) was a p-type material grown by the CZ process. It had a boron concentration of 1.7×10^{15} ions/cm^3 from intrinsic doping during the manufacturing process. The doped wafer (p^+-type single-crystal silicon) was heavily doped with boron ions (from a solid source of oxide of boron) with concentration of 7×10^{19} ions/cm^3 down to a depth of 5.5 μm using thermal diffusion. Grain size of polysilicon wafer was about 5 mm. The polysilicon film was produced as follows: (i) The substrate used was thermally oxidized Si(100) wafers with the oxide layer grown using a standard wet oxidation recipe to a nominal thickness of about 100 nm; (ii) the polysilicon film was grown on the substrate using an LPCVD process (deposition temperature – 610 °C, silane flow rate – 285 sccm, deposition temperature – 230 mTorr), using the thermal decomposition of silane vapor. The films were about 3 μm thick, with columnar grains and a grain size of about 750 nm. X-ray diffraction and transmission electron microscope characterization showed the film to be highly oriented (110). The n^+ -doped polysilicon film was obtained by doping the polysilicon film with phosphorus ions from a solid source of P_2O_5 by thermal diffusion at 875 °C for 90 minutes. The 3C-SiC films were grown through an atmospheric pressure chemical vapor deposition

(APCVD) process on a Si(100) substrate. To grow the SiC film, the wafer is placed on an SiC-coated graphite susceptor, which is induction-heated by an RF-generator. Prior to growth, the reaction chamber is evacuated, then filled with hydrogen at a flow rate of 25 sccm and the wafer is heated to 1000 °C for 5 minutes in the presence of hydrogen, which etches the native oxide and other metallic and organic contaminants from the wafer surface. The wafer surface is then carbonized by heating the susceptor to a growth temperature of 1280 °C under a stable flow of 15% propane in hydrogen at a flow rate of 84 sccm. Following carbonization, the propane flow rate is decreased to 26 sccm, and silane (5% in hydrogen) is introduced at 102 sccm at 1 atmosphere. Flow rates are maintained until films with the desired thicknesses are grown. A more detailed explanation of the growth and characterization of these SiC films is given in Zorman et al. (1995). The films obtained were about 2 μm thick. Both as-deposited and polished versions of the undoped polysilicon and SiC films were studied. The polysilicon film was chemical-mechanically polished in a Struers Planopol-3 polishing machine using 100 ml colloidal silica dispersion (Rippey Corporation, particle size of 30 – 100 nm) mixed in 2000 ml de-ionized water at a force of 210 N for 15 min, with the pads running at 150 rpm in the same direction. The SiC film was polished in a Buehler Ecomet-3 polishing machine with diamond slurry (General Electric Company, particle size of 100 – 500 nm) for 30 min at a force of 50 N, with the pads running at 10 rpm in the same direction. Doped polysilicon sample was polished using Fuji film lapping tape, LT-2, the main lapping agent being 37 μm-sized Cr_2O_3 particles. Boundary lubrication studies were conducted on silicon samples coated with perflouropolyether lubricants and Langmuir-Blodgett and chemically grafted self-assembled monolayer films.

3. Results and Discussion

Three studies are presented in this section. The first study compares micro/nanotribological properties of various forms of virgin, coated and treated silicon samples. The second study is composed of similar studies conducted for the first time on SiC film and compares this material to other materials currently used in MEMS devices. The third study discusses various forms of boundary lubrication that may be suitable for MEMS devices.

3.1 COMPARISON OF VARIOUS VIRGIN, COATED AND TREATED SILICON SAMPLES

Table 2 summarizes the results of the studies conducted on various silicon samples. Microscale coefficient of friction values of all the samples are about the same. Table 3 compares macroscale and microscale friction values for two of the samples. When measured for small contact areas and very low loads used in microscale studies, indentation hardness and elastic modulus are higher than that at the macroscale. This reduces wear. This added to the effect of small apparent area of contact reducing number of trapped particles on the interface, results in less ploughing contribution in the case of microscale friction measurements. Figure 1 and Table 2 show microscale scratch data for the various silicon samples. These samples could be scratched at 10 μN load. Scratch depth increased with normal load. Crystalline orientation of silicon has little influence on scratch resistance. PECVD-oxide samples showed the best scratch resistance followed

TABLE 2 Rms, microfriction, microscratching/microwear and nanoindentation hardness data for various virgin, coated and treated silicon samples.

Material	Rms roughness[a] (nm)	Microscale Coefficient of friction[b]	Scratch depth[c] at 40 μN (nm)	Wear depth[c] at 40 μN (nm)	Hardness[c] at 100 μN (GPa)
Si(111)	0.11	0.03	20	27	11.7
Si(110)	0.09	0.04	20		
Si(100)	0.12	0.03	25		
Polysilicon	1.07	0.04	18		
Polysilicon (lapped)	0.16	0.05	18	25	12.5
PECVD-oxide coated Si(111)	1.50	0.01	8	5	18.0
Dry-oxidized Si(111)	0.11	0.04	16	14	17.0
Wet-oxidized Si(111)	0.25	0.04	17	18	14.4
C^+ -implanted Si(111)	0.33	0.02	20	23	18.6

[a] Scan size of 500 nm x 500 nm

[b] versus Si_3N_4 ball, ball radius of 3 mm at a normal load of 0.1 N (0.3 GPa) at an average sliding speed of 0.8 mm/s

[c] measured using an AFM with a diamond tip of radius of 100 nm

TABLE 3 Surface roughness and micro- and macroscale coefficients of friction of selected samples

Material	Rms roughness (nm)	Coefficient of microscale friction[a]	Coefficient of macroscale friction[b]
Si(111)	0.11	0.03	0.18
C+ -implanted Si(111)	0.33	0.02	0.18

[a] versus Si_3N_4 tip, tip radius of 50 nm in the load range of 10–150 nN (2.5 – 6.1 GPa) at a scanning speed of 5 μm/s over a scan area of 1 μm x 1 μm

[b] versus Si_3N_4 ball, ball radius of 3 mm at a normal load of 0.1 N (0.3 GPa) at an average sliding speed of 0.8 mm/s

by dry-oxidized, wet-oxidized and ion-implanted samples. Ion implantation does not appear to improve scratch resistance. Such microscratching experiments can be used to study failure mechanisms on the microscale and to evaluate mechanical integrity (scratch resistance) of ultra-thin films at low loads.

Wear data on the silicon samples are presented in Table 2. PECVD-oxide samples showed superior wear resistance followed by dry-oxidized, wet-oxidized and ion-implanted samples. This agrees with the trends seen in scratch resistance. In PECVD, ion bombardment during the deposition improves the coating properties such as suppression of columnar growth, freedom from pin-hole, decrease in crystalline size, and increase in density, hardness and substrate-coating adhesion. These effects may help in improving mechanical integrity of the sample surface.

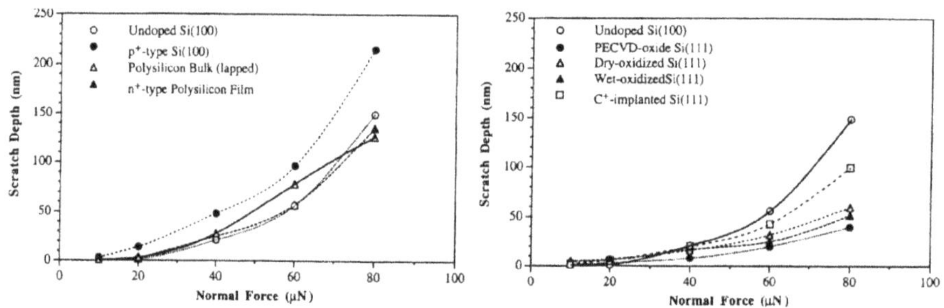

FIGURE 1 Scratch depth as a function of normal load after 10 cycles for various silicon samples, virgin, treated and coated.

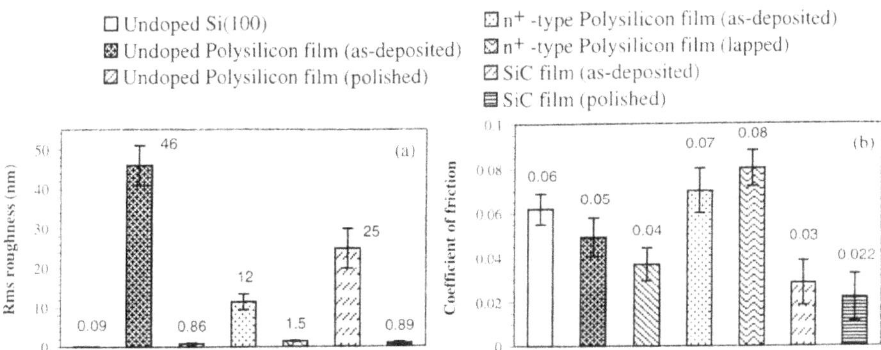

FIGURE 2 Comparisons of (a) Rms surface roughness and (b) microscale coefficient of friction values for various samples

3.2 COMPARISON OF SIC FILMS TO SINGLE-CRYSTAL SILICON AND DOPED AND UNDOPED POLYSILCION FILMS

3.2.1 Surface roughness and friction

The surface roughness of various samples obtained with the AFM can be compared in Fig. 2(a). Polishing of the as-deposited polysilicon and SiC films drastically affect the roughness as the values reduce by two orders of magnitude. Si(100) appears to be the smoothest followed by polished undoped polysilicon and SiC films, which have comparable roughness. The doped polysilicon film shows higher roughness than the undoped sample, which is attributed to the doping process. The microscale coefficients of friction of the various samples are shown in Fig. 2(b). Despite being the smoothest sample, Si(100) shows higher friction than the other samples. Doped polysilicon also shows high friction, which is due to the presence of grain boundaries and high surface roughness. In the case of the undoped polysilicon and SiC films, polishing did not affect the friction values by much. From the data, the polished SiC film shows the lowest friction followed by polished, undoped polysilicon film, which strongly supports the candidacy of SiC films for use in MEMS devices. Gray scale top-view images of surface height and corresponding friction force of selected samples obtained with the AFM/FFM are shown in Fig. 3. Brighter regions indicate higher height than darker regions in the case of the surface height images, while brighter regions indicate higher friction force

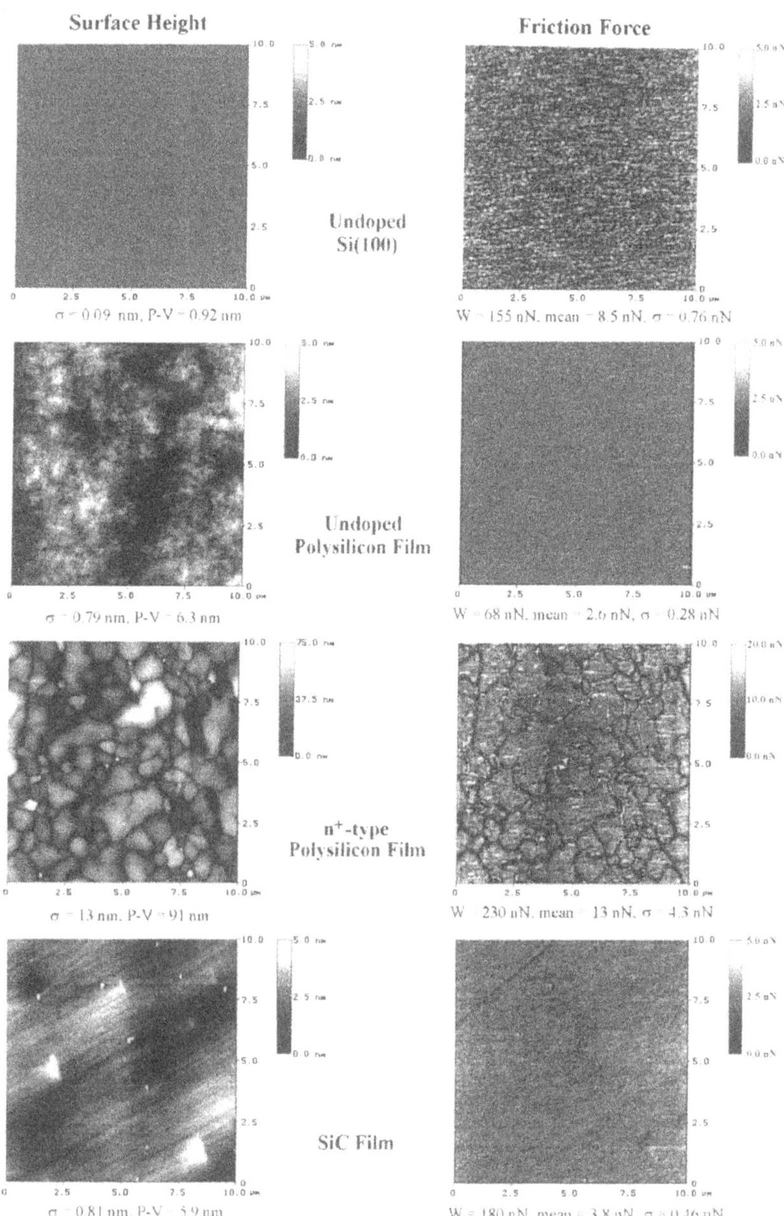

FIGURE 3 Gray scale maps of surface height and corresponding friction force maps of Si(100), undoped polysilicon film (polished), n+ -type polysilicon film and SiC film (polished) samples.

experienced by the tip than the darker regions. The doped polysilicon film shows a large number of grain boundaries. It appears that regions of high and low surface height do not necessarily correspond to regions of high and low friction.

The low microscale friction exhibited by SiC compared to the other materials agrees with the fact that many ceramic-ceramic interfaces generally show low friction on the macroscale. In this study, the Si_3N_4 tip – SiC film interface also shows the same trend. In the case of ceramic materials, formation of tribochemical films on the surface due to sliding results in low values of friction. SiC can react with water vapor to form silicon hydroxide and oxide, which improve the frictional behavior (Xu and Bhushan, in press). From this discussion, it can be seen that an obvious drawback with SiC is its readiness to form tribochemical films, which is environment-dependent. Another factor that affects friction in ceramic materials is fracture, which leads to high friction because of the ploughing contribution. But in the case of SiC film, fracture toughness is believed to follow the same trend as that of the bulk material, which is higher than that of silicon (see Table 1), thereby resulting in lower ploughing contribution and hence lower friction. Macroscale friction measurements indicate that SiC film exhibits one of the lowest friction values as compared to the other samples. Doped polysilicon sample shows low friction on the macroscale as compared to the undoped polysilicon sample due to a combination of the doping effect as well as the higher roughness of the doped film.

3.2.2 Scratch/wear tests and hardness and elastic modulus measurements

As explained earlier, the scratch tests consisted of making scratches for 10 cycles with varying loads. Figure 4(a) shows a plot of scratch depth vs. normal load for various samples. Error bars are given for the Si(100) and SiC data. The variation was typically about 12%. Scratch depth increases with increasing normal load. Si(100) and the doped and undoped polysilicon film show similar scratch resistance. From the data, it is clear that the SiC film is much more scratch-resistant than the other samples. The increase in scratch depth with normal load is very small and all depths are less than 30 nm, while the Si(100) and polysilicon films reach depths in excess of 150 nm. Figure 5 shows 3D images of the scratch marks.

Wear tests were conducted on the samples for 1 cycle at normal loads ranging from 20 to 80 μN. The resulting wear depths are plotted against the normal loads in Fig. 4(b). Again the wear depth increases with increasing normal load. Similar to the scratch resistance data, Si(100) and the polysilicon samples behave similarly in terms of wear resistance also. The SiC film starts out showing comparable wear depth at 20 μN to all the other samples, but at higher loads, SiC film shows superior wear resistance. Also there is hardly an increase in wear depth with increasing normal load in the case of SiC film. The evolution of wear on the various samples was also studied. This test was performed by wearing the same region for 20 cycles at a normal load of 20 μN, while observing wear depths at different intervals (1,2,10,15 and 20 cycles). This would give information as to the progression of wear of the material. The wear depths observed are plotted against number of cycles in Fig. 4(b). For all the materials, the wear depth increases almost linearly with increasing number of cycles. This suggests that the material is removed layer by layer in all the materials. Here also, SiC film exhibits lower wear depths than the other samples. Doped polysilicon film wears less than the undoped film, which evolution of undoped polysilicon film SiC films. Not much debris particles are seen in the figures. This is due to the fact that the debris particles created are loose and are pushed outside the imaging area by the sliding tip during scanning. It can also be seen from the figure that the wear marks are uniform at the bottom, indicating that uniform wear has occurred with particle pile-up on the edges

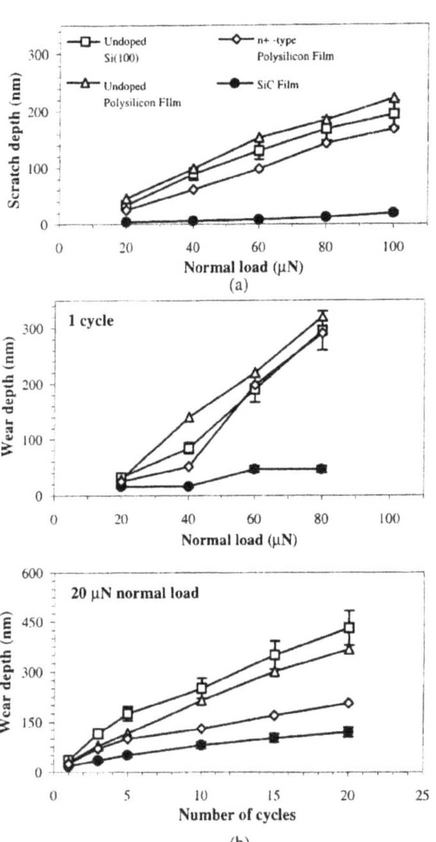

FIGURE 4 (a) Scratch depths for 10 cycles as a function of normal load and (b) wear depths as a function of normal load and as a function of number of cycles for various samples.

of the wear mark.

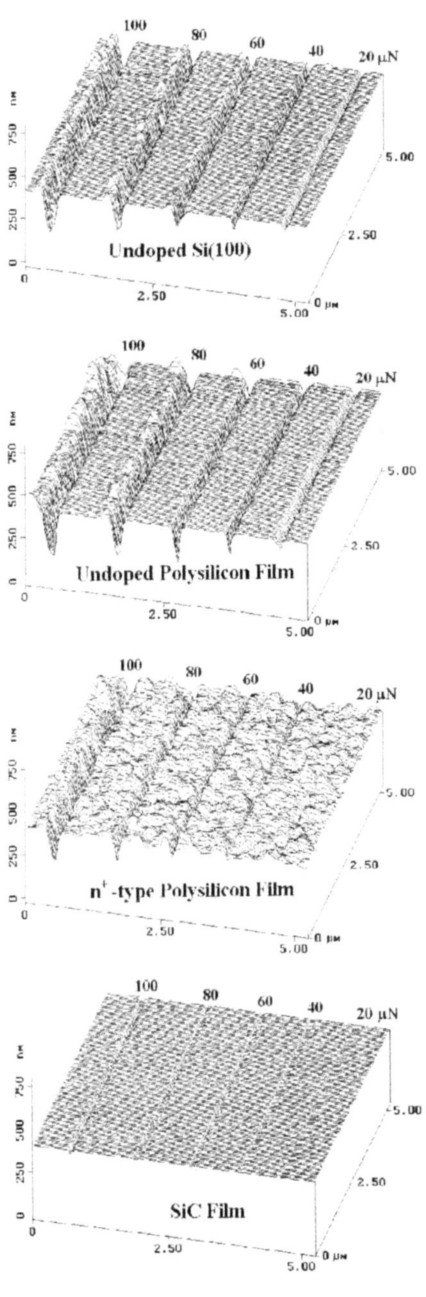

FIGURE 5 Scratch profiles for various samples.

The wear tests clearly show that SiC film possesses an extremely wear resistant surface compared to the other samples and together with the scratch tests results, indicate that SiC film has better surface mechanical properties than Si(100) and the polysilicon films. Table 4 shows a summary of the various properties of the samples measured in this study. The superior scratch/wear resistance of the SiC film is consistent with the hardness values near the surface measured with the nanoindenter. Higher hardness of SiC film is primarily responsible for its better scratch/wear resistance. Wear in ceramic materials in the case of asperity contacts occurs due to brittle fracture. As the AFM tip slides over the material surface, Hertzian cone cracks can occur when the normal stress exceeds a critical value (Hutchings, 1992). Friction (tangential) forces during sliding reduce this critical value. High fracture toughness and low coefficient of friction of SiC film help in reducing the chances of brittle fracture, which results in low wear. Therefore abrasive wear due to plastic deformation and fracture govern the wear process. In the case of all samples, surface reactions result in the formation of tribochemical films that are

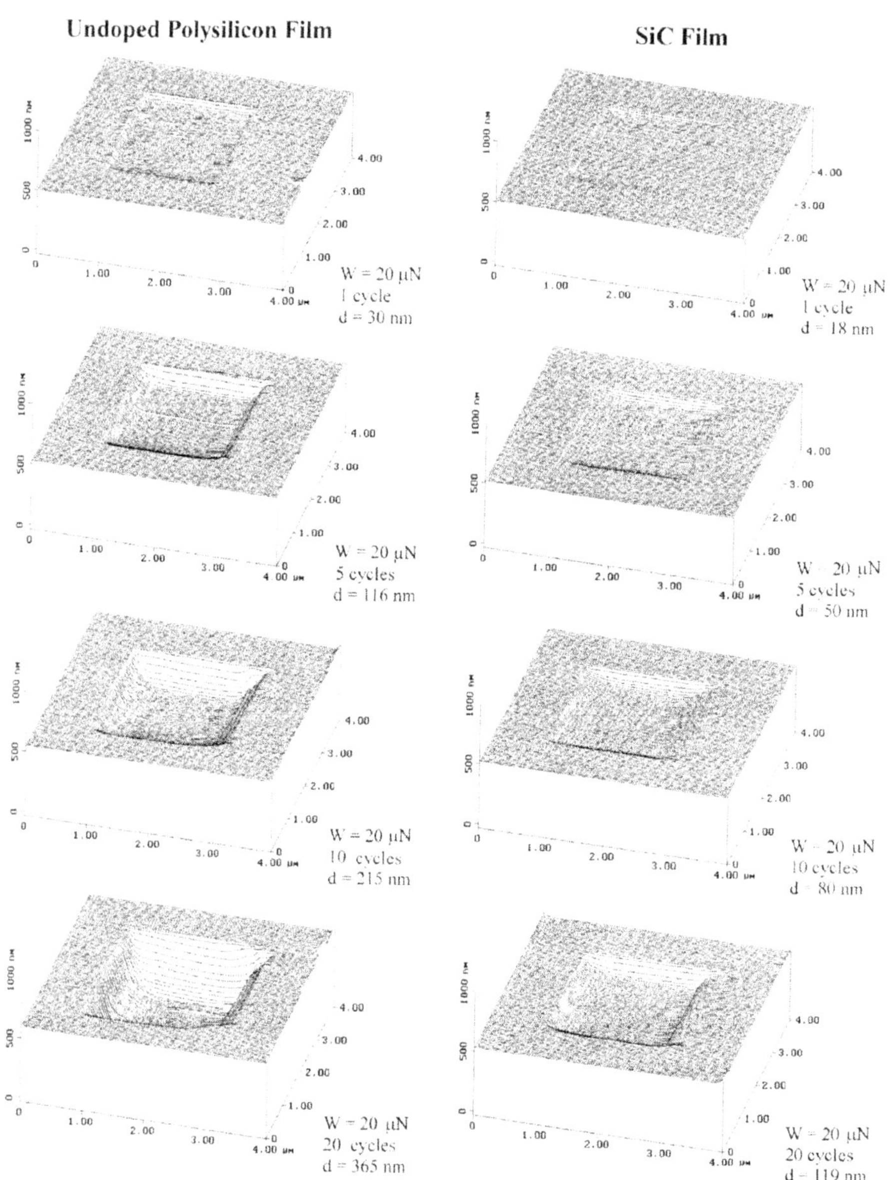

FIGURE 6 3D wear profiles of undoped polysilicon and SiC films (polished) after 1, 5 10 and 15 wear cycles at 20 μN normal load.

different in nature than the underlying material. Therefore, at low loads and at the onset of wear, all samples would show similar wear depths as they all have comparable interfacial shear strengths and attack angles (attack angle is the included angle between the leading face of the asperity and the contact plane at the point of contact). This is seen in the case of the reported wear depths at 20 μN for 1 cycle (Fig. 4b). In the case of the scratch test data (Fig. 4a), SiC shows slightly lower scratch depth at 20 μN since the data is for 10 cycles, during which time it is possible that the bulk properties of the material start coming into play in addition to that of the tribochemical films. This is consistent with the wear depths at 20 μN for 10 cycles and more (Fig. 4c). As the normal load increases, Si(100) and polysilicon films show sharp increase in degree of penetration (ratio of groove depth to radius of contact) and attack angle, which leads to higher wear (Hokkirigawa and Kato, 1988; Koinkar and Bhushan, 1997). Higher fracture toughness of SiC as compared to Si(100) is responsible for its lower wear. Also the higher thermal conductivity of SiC (see Table 1) as compared to the other materials leads to lower interface temperatures which generally results in less degradation of the surface (Bhushan, 1996). Doping of the polysilicon does not affect the scratch/wear resistance and hardness much. The measurements made on the doped sample are affected by the presence of grain boundaries.

Representative load-displacement curves of indentations made at 0.2 and 15 mN peak indentation loads on various samples are shown in Fig. 7. The obtained hardness and elastic modulus values are given in Table 4. SiC film shows higher hardness and elastic modulus than the other samples. Higher hardness of SiC is believed to be responsible for its superior scratch/wear resistance compared to the other samples.

3.3 BOUNDARY LUBRICATION

The classical approach to lubrication uses freely supported multimolecular layers of liquid lubricants (Bhushan, 1995). The liquid lubricants are chemically bonded to improve their wear resistance. To study depletion of boundary layers, the micro-scale friction measurements were made as a function of number of cycles on virgin Si(100) surface and silicon surface lubricated with about 2 nm-thick Z-15 and Z-DOL perfluoro-polyether (PFPE) lubricants, the data of which is shown in Fig. 8 (Koinkar and Bhushan,

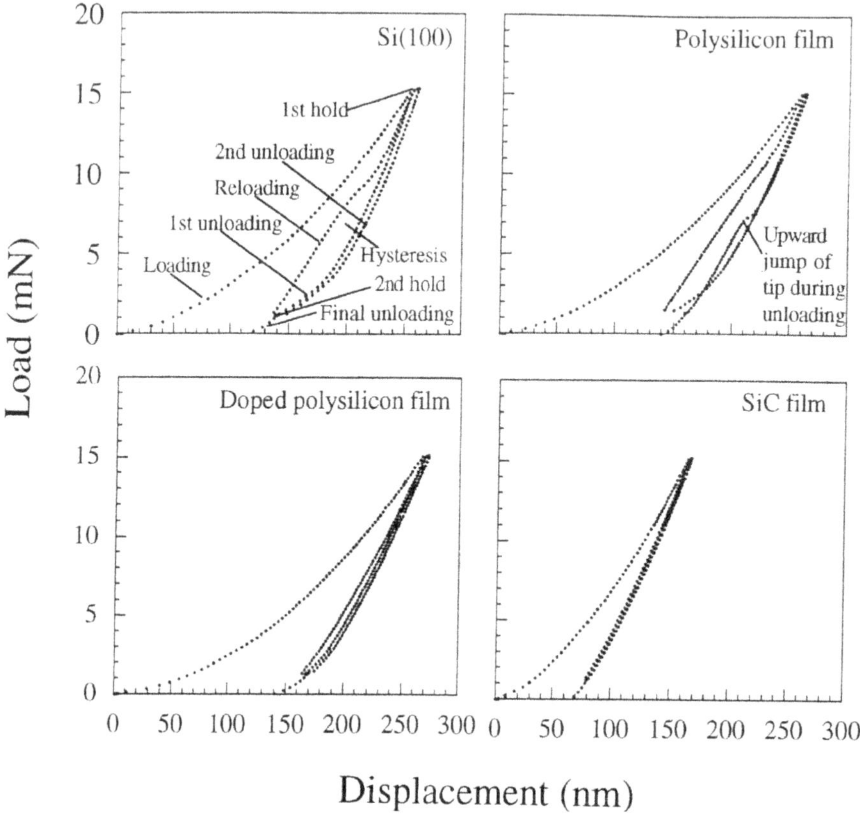

FIGURE 7 Representative load displacement curves for various samples.

1996a). Z-DOL is a PFPE lubricant with hydroxyl end groups. Its lubricant film was thermally bonded at 150° C for 30 minutes and washed off with a solvent to provide a chemically bonded layer for the lubricant film. In Fig. 8, the unlubricated silicon sample showed a slight increase in friction force followed by a drop to a lower steady state value after 20 cycles. Depletion of native oxide and possible roughening of the silicon sample are believed to be responsible for the decrease in friction force after 20 cycles. The initial friction force for Z-15 lubricated sample is lower than that of unlubricated silicon and increases gradually to a friction force value comparable to that of unlubricated silicon after 20 cycles. This suggests depletion of the Z-15 lubricant in the wear track. In the case of the Z-DOL coated sample, the friction force starts out to be low and remains low

TABLE 4 Summary of micro/nanotribological properties of the sample materials.

Sample	Rms roughness[a] (nm)	P-V Distance[a] (nm)	Coefficient of Friction		Scratch depth[d] (nm)	Wear depth[e] (nm)	Nano-hardness[f] (GPa)	Young's modulus[f] (GPa)
			Micro[b]	Macro[c]				
Undoped Si(100)	0.09	0.9	0.06	0.33	89	84	12.5	165
Undoped polysilicon film (as deposited)	46	340	0.05					
Undoped polysilicon film (polished)	0.86	6	0.04	0.46	99	140	12.3	167
n^+ -type polysilicon film (as deposited)	12	91	0.07	0.16	61	51	11.5	201
n^+ -type polysilicon film (polished)	1.45	11	0.08					
SiC film (as deposited)	25	150	0.03					
SiC film (polished)	0.89	6	0.02	0.20	6	16	24.5	395

[a] measured using AFM over a scan size of 10 μm x 10 μm

[b] measured using AFM/FFM over a scan size of 10 μm x 10 μm

[c] obtained using a 3-mm diameter sapphire ball in a reciprocating mode at a normal load of 10 mN and average sliding speed of 1 mm/s after 4 m sliding distance.

[d] measured using AFM at a normal load of 40 μN for 10 cycles, scan length of 5 μm

[e] measured using AFM at normal load of 40 μN for 1 cycle, wear area of 2 μm x 2 μm

[f] measured using Nanoindenter at a normal load of 200 μN.

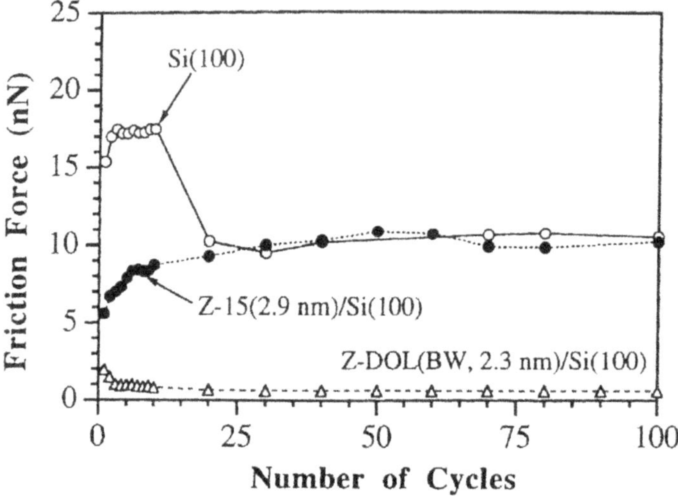

FIGURE 8 Friction force as a function of number of cycles using a Si_3N_4 tip at 300 nN normal load for unlubricated and lubricated silicon samples.

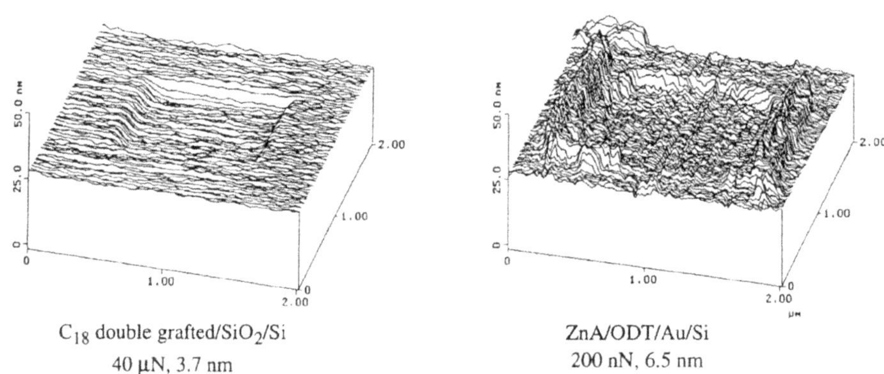

FIGURE 9 Surface profiles showing worn region after one scan cycle for self-assembled monolayers of octodecyl silanol (C_{18}) (left) and zinc arachidate (ZnA) (right). Normal force and wear depths are indicated.

during the cycle of 100 tests. It suggests that Z-DOL does not get displaced or depleted as readily as Z-15. Additional studies of freely supported liquid lubricants showed that either increasing the film thickness or chemically bonding the molecules to the substrate with a mobile fraction improves the lubrication performance (Koinkar and Bhushan, 1996a,b).

For lubrication of microdevices, a more effective approach involves the deposition of organized, dense molecular layers of long-chain molecules on the surface contact (Bhushan et al., 1995b). Such monolayers and thin films are commonly produced by Langmuir-Blodgett (L-B) deposition and by chemical grafting of molecules into self-assembled monolayers (SAMs). Based on measurements, SAMs of octodecyl (C_{18}) compounds based on aminosilanes on oxidized silicon exhibited lower coefficient of friction (0.018) an greater durability (Fig. 9) than L-B films of zinc arachidate adsorbed on a gold surface coated with octadecylthiol (ODT) (coefficient of friction 0.03) (Bhushan et al., 1995b). L-B films are bonded to the substrate by weak van der Waals attraction, whereas SAMs are chemically bound via covalent bonds. Because of the choice of the chain length and terminal linking groups that SAMs offer, they hold great promise for boundary lubrication of microdevices.

4. Conclusions

The micro/nanotribological studies conducted on the various materials on the scale pertinent to MEMS devices show that 3C-SiC film shows lower friction and superior scratch/wear resistance when compared to the other materials currently used in MEMS devices. Higher hardness and fracture toughness of the SiC film is believed to be responsible for its superior mechanical integrity and lower friction. Polysilicon films and undoped single-crystal silicon showed similar characteristics. Doping of polysilicon film does not affect its tribological properties much. Boron-doped silicon showed poor wear resistance compared to the undoped silicon, which is attributed to its lower hardness on the nanoscale. The These characteristics of 3C-SiC film added to the availability of a low-cost technique of producing the material make it an exceptional choice as a material for high-temperature MEMS applications. It appears that chemically grafted self-assembled monolayers lubricants show promising performance for boundary lubrication in MEMS devices.

Acknowledgements

Financial support for this study was provided by the Office of Naval Research, Department of the Navy (Contract No. N00014-96-1-10292). The information herein does not necessarily reflect the position or policy of the government and no official endorsement should be inferred.

REFERENCES

Beerschwinger, U., Albrecht, T., Mathieson, D., Rueben, R.L., Yang, S.J. and Taghizadeh, M. (1995), "Wear at Microscopic Scales and Light Loads for MEMS Applications", *Wear* **181-183**, 426-435.

Bhushan, B. (1992), *Mechanics and Reliability of Flexible Magnetic Media*, Springer-Verlag, New York.

Bhushan, B. (1995), *Handbook of Micro/nanotribology*, CRC, Boca Raton, FL.

Bhushan, B. (1996a), *Tribology and Mechanics of Magnetic Storage Devices, 2nd edition*, Springer, New York.

Bhushan, B. (1996b), "Nanotribology and Nanomechanics of MEMS Devices", *Proc. Ninth Annual Workshop on Micro Electro Mechanical Systems*, pp. 91-98, IEEE, New York.

Bhushan, B., Dominiak, M. and Lazarri, J.P. (1992), "Contact Start-stop Studies with Silicon Planar Head Sliders Against Thin-film Disks", *IEEE Trans. Mag.* **28**, 2874-2876.

Bhushan, B. and Venkatesan, S. (1993a), "Mechanical and Tribological Properties of Silicon for Micromechanical Applications: A Review", *Adv. Info. Storage Sys.* **5**, 211-239.

Bhushan, B. and Venkatesan, S. (1993b), "Friction and Wear Studies of Silicon in Sliding Contact with Thin-film Magnetic Rigid Disks", *J. Mater. Res.* **8**, 1611-1628.

Bhushan B. and Koinkar, V.N. (1994), "Tribological Studies of Silicon for Magnetic Recording Applications", *J. Appl. Phys.* **75**, 5741-5746.

Bhushan, B., Israelachvili, J.N. and Landman, U. (1995a), "Nanotribology: Friction, Wear and Lubrication at the Atomic Scale", *Nature* **374**, 607-616.

Bhushan, B., Kulkarni, A.V., Koinkar, V.N., Boehm, M., Odoni, L., Martelet, C. and Belin, M. (1995b), "Microtribological Characterization of Self-assembled and Langmuir–Blodgett Monolayers by Atomic Force and Friction Force Microscopy", *Langmuir* **11**, 3189-3198.

Bhushan, B. and Gupta, B.K. (1997), *Handbook of Tribology: Materials, Coatings and Surface Treatments*, Krieger, Malabar, FL.

Bhushan, B. and Koinkar, V.N. (1997), "Microtribological Studies of Doped Single-crystal Silicon and Polysilicon Films for MEMS Devices", *Sensors and Actuators* **A 57**, 91-102.

Bhushan, B. and Li, X. (1997), "Micromechanical and Tribological Characterization of Doped Single-crystal Silicon and Polysilicon Films for Microelectromechanical Systems", *J. Mater. Res.* **12**, 1-10.

Bryzek, J., Peterson, K. and McCulley, W. (1994), "Micromachines on the March", *IEEE Spectrum* May, 20-31.

Deng, K., Collins, R.J., Mehregany, M. and Sukenik, C.N. (1995), "Performance Impact of Monolayer Coatings of Polysilicon Micromotors", *Proc. MEMS 95*, Amsterdam, Netherlands, Jan-Feb.

Ericson, F. and Schweitz, J.A (1990), "Micromechanical Fracture Strength of Silicon", *J. Appl. Phys.* **68,** 5840-5844.

Fan, L.S., Ottesen, H.H., Reiley, T.C. and Wood, R.W. (1995), "Magnetic Recording Head Positioning at Very High Track Densities Using a Microactuator-based, Two-stage Servo System", *IEEE Trans. Industr. Electron.* **42,** 222-233.

Fan, L.S. and Woodman, S. (1995), "Batch Fabrication of Mechanical Platforms for High-density Data Storage", *8th Int. Conf. Solid State Sensors and Actuators (Transducers '95)/Eurosensors IX*, Stockholm, Sweden, 25-29 June, 434-437.

Fang, W. and Wickert, J.A. (1995), "Comments on Measuring Thin-film Stresses Using Bi-layer Micromachined Beams", *J. Micromech. Microeng.* **5**, 276-281.

Fujimasa, I. (1996), *Micromachines: A New Era in Mechanical Engineering*, Oxford University Press, Oxford, U.K.

Gabriel, K.J., Behi, F., Mahadevan, R. and Mehregany, M. (1990), "In Situ Friction and Wear Measurement in Integrated Polysilicon Mechanisms", *Sensors and Actuator*, **A 21-23**, 184-188.

Guckel, H. and Burns, D.W. (1989), "Fabrication of Micromechanical Devices from Polysilicon Films with Smooth Surfaces", *Sensors and Actuators* **20,** 117-122.

Guckel, H., Burns, D., Rutigliano, C., Lovell, E. and Choi, B. (1992), "Diagnostic Microstructures for the Measurement of Intrinsic Strain in Thin Films", *J. Micromech. Microeng.* **2,** 86-95.

Gupta, B.K., Chevallier, J. and Bhushan, B. (1993), "Tribology of Ion Bombarded Silicon for Micromechanical Applications", *ASME J. Tribol.* **115,** 392-399.

Gupta, B.K. and Bhushan, B. (1994), "Modification of Tribological Properties of Silicon by Boron Ion Implantation", *Tribol. Trans.* **37,** 601-607.

Hamilton, H. (1991), "Contact Recording on Perpendicular Rigid Media", *J. Mag. Soc. Jpn.* **15** (Supp. 52) 483-481

Harris, G.L. (1995) (ed.), *Properties of Silicon Carbide*, Instn. of Elect. Eng., London.

Henck, S.A. (1997), "Lubrication of Digital Micromirror Devices", *Trib. Lett.* **3,** 239-247.

Hokkirigawa, K. and Kato, K. (1988), "An Experimental and Theoretical Investigation of Ploughing, Cutting and Wedge Forming During Abrasive Wear", *Trib. Int.* **21,** 51-57.

Hutchings, I.M. (1992), *Tribology: Friction and Wear of Engineering Materials*, CRC Press, Boca Raton, FL.

Jaeger, R.C. (1988), *Introduction to Microelectronic Fabrication, Vol. 5*, Addison-Wesley, reading, MA.

Koinkar, V.N. and Bhushan, B. (1996a), "Micro/nanoscale Studies of Boundary Layers of Liquid Lubricants for Magnetic Disks", *J. Appl. Phys.* **79,** 8071-8075.

Koinkar, V.N. and Bhushan, B. (1996b), "Microtribological Studies of Unlubricated and Lubricated Surfaces Using Atomic Force/Friction Force Microscopy", *J. Vac. Sci. Technol.* **A14,** 2378-2391.

Koinkar, V.N. and Bhushan, B. (1997), "Scanning and Transmission Electron Microscopies of Single-crystal Silicon Microworn/Micromachined Using Atomic Force Microscopy", *J. Mater. Res.* (in press).

Lazarri, J.P. and Deroux-Dauphin, P. (1989), "A New Thin Film Head Generation IC Head", *IEEE Trans. Mag.* **25,** 3190-3193.

Lim, M.G., Chang, J.C., Schultz, D.P., Howe, R.T. and White, R.M. (1990), "Polysilicon Microstructures to Characterize Static Friction", *Proc. IEEE Micro Electro Mechanical Systems,* 82-88.

Mastrangelo, C.H. (1997), "Adhesion-related Failure Mechanisms in Micromechanical Devices", *Trib. Lett.* **3,** 233-238.

Mehregany, M., Howe, R.T. and Senturia, S.D. (1987), "Novel Microstructures for the In-situ Measurement of Mechanical Properties of Thin Films", *J. Appl. Phys.* **62,** 3579-3584.

Mehregany, M., Gabriel, K.J. and Trimmer, W.S.N. (1988), "Integrated Fabrication of Polysilicon Mechanisms", *IEEE Trans. Electron Dev.* **35** (6), 719-723.

Miu, D.K. and Tai, Y.C. (1995), "Silicon Micromachined Scaled Technology". *IEEE Trans. Industr. Electron.* **42**, 234-239.

Ohwe, T., Mizoshita, Y. and Yonoeka, S. (1993), "Development of Integrated Suspension System for a Nanoslider with an MR Head Transducer", *IEEE Trans. Mag.* **29,** 3924-3926.

Schweitz, J.A. (1991), "A New and Simple Approach to the Stress-strain Characteristics of Thin Coatings", *J. Micromech. Microeng.* **1,** 10-15.

Shackelford, J.F., Alexander, W. and Park, J.S. (eds.) (1994), *CRC Material Science and Engineering Handbook, 2nd edition*, CRC PRess, Boca Raton, FL.

Shor, J.S., Goldstein, D. and Kurtz, A.D. (1993), "Characterization of n-type β-SiC as a Piezoresistor", *IEEE Trans. Electron Devices* **40,** 1093-1099.

Spencer, M.G., Devaty, R.P., Edmond, J.A., Khan, M.A., Kaplan, R. and Rahman, M. (1994) (eds.), *SiC and Related Materials, Proc. Fifth Conf.*, Inst. of Physics Publishing Ltd., Bristol, U.K.

Tai, Y.C., Fan, L.S. and Muller, R.S. (1989), "IC-processed Micro-motors: Design, Technology and Testing", *Proc. IEEE Micro Electro Mechanical Systems*, 1-6.

Tai, Y.C. and Muller, R.S. (1990), "Frictional Study of IC Processed Micromotors", *Sensors and Actuators* **A 21-23**, 180-183.

Tong, L., Mehregany, M. and Matus, L.G. (1992), "Mechanical Properties of 3C Silicon Carbide", *Appl. Phys. Lett.* **60**, 2992-2994.

Trimmer, W.S. (ed.) (1997), *Micromachines and MEMS, Classic and Seminal Papers to 1990*, IEEE Press, New York.

Venkatesan, S. and Bhushan, B. (1993), "The Role of Environment in the Friction and Wear of Single-crystal Silicon in Sliding Contact with Thin-film Magnetic Rigid Disks", *Adv. Info Storage Sys.* **5**, 241-257.

Venkatesan, S. and Bhushan, B. (1994), "The Sliding Friction and Wear Behavior of Single-crystal, Polycrystalline and Oxidized Silicon, *Wear* **171**, 25-32.

Xu, J. and Bhushan, B., "Friction and Durability of Ceramic Slider Materials in Contact with Lubricated Thin-film Rigid Disks", *Proc. Instn. Mech. Engrs, Part J: J. Eng. Trib.* (in press).

Zorman, C.A., Fleischmann, A.J., Dewa, A.S., Mehregany, M., Jacob, C., Nishino, S. and Pirouz, P. (1995), "Epitaxial Growth of 3C-SiC Films on 4 in. Diam Si(100) Silicon Wafers by Atmospheric Pressure Chemical Vapor Deposition", *J. Appl. Phys.* **78**, 5136-5138.

PHASE TRANSFORMATIONS IN SEMICONDUCTORS UNDER CONTACT LOADING

YURY GOGOTSI, MICHAEL S. ROSENBERG

University of Illinois at Chicago, Department of Mechanical Engineering
842 West Taylor Street, M/C 251, Chicago, IL 60607
YGogotsi@uic.edu

ANDREAS KAILER, KLAUS G. NICKEL

Universität Tübingen, Institut für Mineralogie, Petrologie und Geochemie, Wilhelmstr. 56, D-72074 Tübingen, Germany

ABSTRACT

Phase transformations occurring in materials under contact loading are important for a wide range of problems in materials science and engineering. We studied solid-state phase transformation, including metallization and amorphization of semiconductors under high non-hydrostatic pressures using a combination of hardness indentation tests, scratch tests, Raman spectroscopy, FTIR and various microscopic techniques. Our experiments demonstrated metallization due to closing of the band gap and consequent formation of metastable phases upon decompression in silicon, germanium and several other semiconductors. For the first time high-pressure, these phases were unambiguously observed in hardness impressions, scratches and machining debris, and for some of these phases Raman spectra have not been published before.

The data obtained confirm that the hardness level of many semiconductors depends on the stress (deformation) needed to initiate the transformation. It may help to choose and/or optimize conditions of ductile-regime machining of semiconductors, as well as give new insights into their surface properties. Information on the phase transformations in the surface layer of materials resulting from contact interactions is important for understanding the mechanisms of wear, friction and erosion.

1. INTRODUCTION AND OBJECTIVES

Contact loading is one of the most common mechanical impacts that materials can experience during processing or application. Examples are cutting, polishing, indentation testing, wear and friction. This kind of loading has a very significant nonhydrostatic component of stress that leads to dramatic changes in the materials structure, such as amorphization [1] and phase transformations [2,3]. Simultaneously, processes of plastic deformation, fracture and interactions with the environment and counterbody can occur. The latter ones have been studied fairly well, but the processes of phase transformations at the sharp contact have been studied only for a select few materials and the data obtained so far can be only considered as preliminary.

B. Bhushan (ed.), Tribology Issues and Opportunities in MEMS, 431-442.

The recent observation of phase transformations under indentation using micro-Raman spectroscopy [4] gives a new tool for studying pressure induced phase transformations. The combination of indentation tests with micro-Raman spectroscopy (Fig. 1) provides a powerful and fast tool for monitoring of pressure induced phase transformations in semiconductors [4]. Raman spectra can be taken within seconds at any point of the impression. An example is shown in Fig. 2. By varying the shape and elastic compliance of the indenter, minimizing load-displacement hysteresis loops, and using statistical distributions of numerous measurements to minimize errors, reproducible values of transition pressures and deformations can be obtained. Studies of reversible transformations that occur upon unloading or heating of samples by the laser beam are also feasible. Studies of pressure induced metallization can improve our understanding of wear mechanisms and lead to advances in machining of semiconductors.

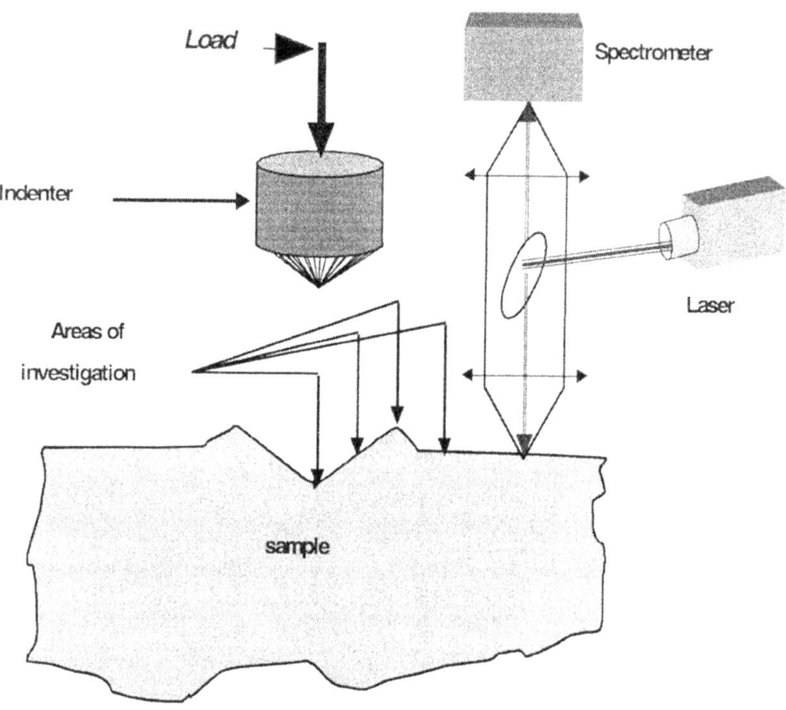

Figure 1. Schematic representation of the Raman/indentation tests. The analysis can be performed with a 1μm spot size and 1μm steps.

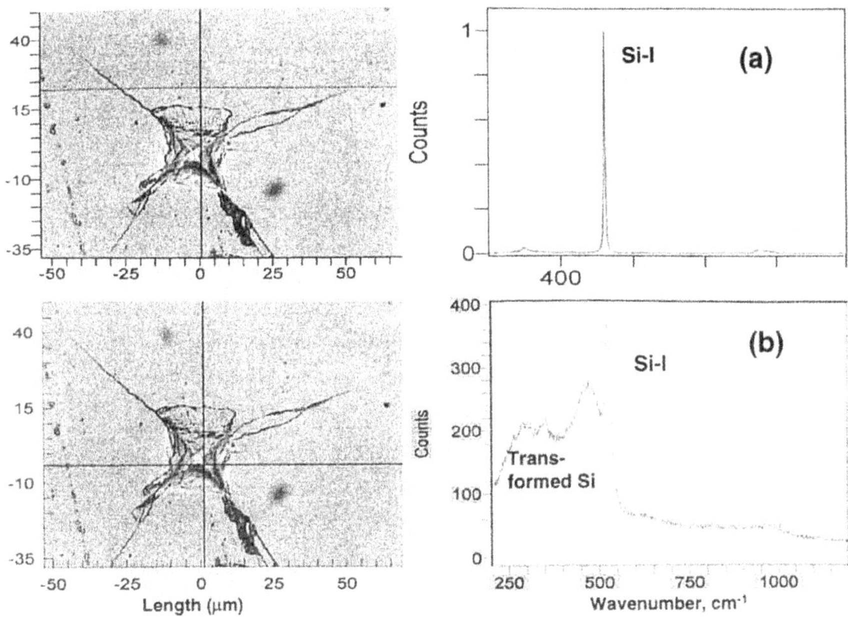

Figure 2. Optical micrographs of a Vickers impression on a silicon wafer and micro-Raman spectra taken from the wafer surface (a) and the Vickers impression (b. Crosshair shows the analysis point.

2. CURRENT STATE OF KNOWLEDGE IN THE FIELD

It is known that the indentation of materials with diamond indenters creates high stresses (hydrostatic and deviatoric) under the indenter that can cause phase transformations [5-9]. It has also been shown that the hardness correlates with metallization pressures for a number of semiconductors [10]. Indirect monitoring of phase transformations during indentation through conductivity measurements [5-7] or sudden volume changes (known as "pop-out" events [9,11]) occurring during unloading were used in most of the indentation experiments. However, the conductivity measurements can be only used to detect metallic-nonmetallic transitions. The pop-out events alone can only suggest that a transformation may have occurred if the volume change is large enough. Cracking around the impression and difficulties of analyzing the indentation force/displacement data may also lead to erroneous conclusions. Direct studies of the phase transformations induced by indentation, scratching or machining have been conducted using transmission electron microscopy (TEM) [5,12-15]. However, TEM examination is possible on quenched products only and is complicated by the difficulty in handling small and stressed samples, and the instability of metastable phases during dimpling, ion thinning, and under the electron beam. Taking into account the successful employment of Raman spectroscopy for monitoring phase transformation in conventional pressurization experiments [16-18] and for mapping internal stresses in semiconductors [19-23], it is natural to use it in indentation experiments.

3. EXPERIMENTAL APPROACH: INDENTATION/RAMAN TESTS

The experiments were conducted using LabRam II micro-Raman spectrometer (Dilor, France) and Ramascope 1000 micro-Raman spectrometer (Renishaw, UK). The use of several excitation wavelengths (Ar ion laser at 514.5 nm, He-Ne laser at 632.8 nm and a diode laser at 780 nm) allows us to distinguish between Raman bands and bands of other origin. The accuracy of the band position is ± 3cm^{-1}, even stronger shifts are possible due to strong residual stresses. In addition to Raman analysis and light microscopy, scanning electron microscopy (SEM) was used. We used standard Vickers and Knoop pyramidal indenters, Rockwell spherical indenter and a sharpened to a point (90 degree angle) diamond cone (Lunzer, USA) to produce indentations and scratches on the surface of commercial wafers.

4. MATERIALS

We investigated silicon, which is probably the most intensively studied material in the scientific literature and for which dozens of studies on high-pressure transformations have been published [15,24-27], Ge, GaAs and several other semiconductors.

5. RESULTS

5.1. Silicon Results

For silicon, 12 different phases are known from calculations and pressure cell experiments [26,28]. Most of these belong to very high-pressure regimes. Of interest for indentation studies are amorphous Si and the lower pressure types Si-I (conventional diamond structure), Si-II (metallic β-tin structure), Si-III (bc8, body-centered cubic structure), Si-IV (hd, hexagonal diamond structure) and Si-XII (r8, rombohedral distortion of bc8 [18,29]).

Although only amorphous silicon was observed in previous investigations when TEM and other techniques were used [12,14,15], micro-Raman spectroscopy revealed the metastable phases. Up to five phases can be found around a single Rockwell impression (Fig. 3) on a silicon wafer. The dark-red regions on the surface produce a Raman band at 500-510 cm^{-1} which can be ascribed to Si-IV. Si-III was primarily present in the subsurface zones of Rockwell impressions, where the highest shear stress exists. Using the indentation/Raman technique we have demonstrated that the Raman spectrum that was previously ascribed to a metastable Si-III phase [30] originates from two different high-pressure phases of Si: Si-III and Si-XII. Formation of nanocrystalline Si, that was also observed by TEM [13] can result in the down shift of the Si-I band from the 520 cm^{-1} and produce additional features on the spectrum with the position close to that of hexagonal Si-IV. Although, the ratio between the shear and compression stress depends on the indenter used, very similar results have been obtained using different indenters (Fig. 4). This shows that metallization occurs under various loading conditions. All phases described above have been observed in indentations or scratches produced using three different indenters. The shape of the indenter determines the distribution of phases within the indentation and, probably, the shape and size of the transformation zone.

The unloading rate, not the shape of the indenter, appeared to be the most important factor that determines the composition of the indentation zone. Only amorphous Si is formed under fast unloading. Slow unloading allows for transformation of Si-II to Si-XII and later to Si-III (Fig. 5).

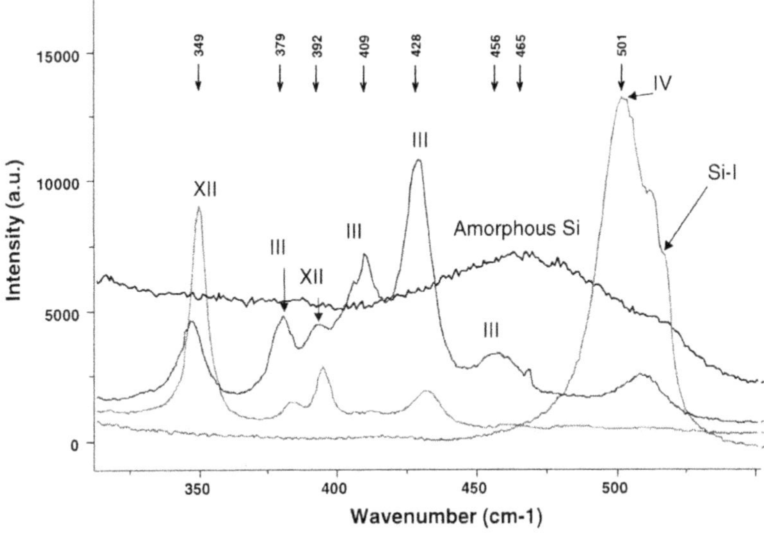

Figure 3. Raman Spectra of silicon from different regions of a Rockwell indentation showing the presence of regions of amorphous Si, Si-I, Si-III, Si-IV and Si-XII

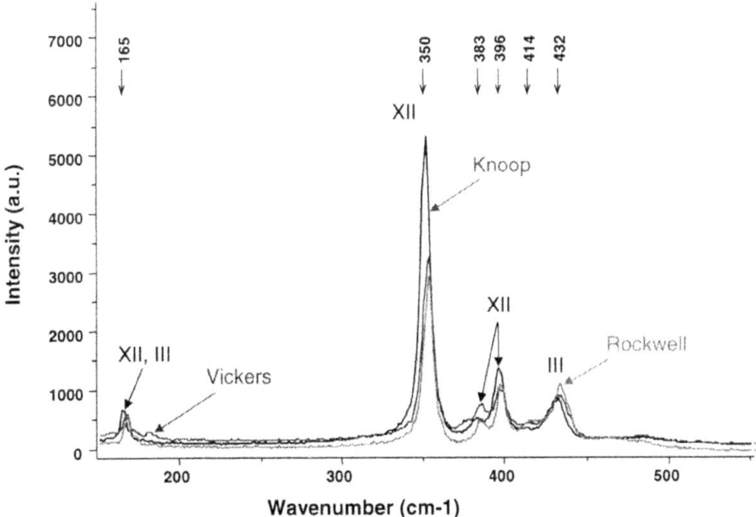

Figure 4. Raman spectra of silicon after indentation with with three different types of indenters; Knoop, Rockwell and Vickers

Quantification of conditions for phase transformations can be done using depth-sensing indentation. Comparison of the indentation diagrams (Fig. 6) with the results of Raman analysis allows us to identify the transition points from pop-in and pop-out events, as well as from changes in the slope of the diagram. Studies of reversible transformations that occur upon unloading of samples are also feasible, as can be seen from Fig. 6. It is necessary to note that distinct pop-ins or pop-outs can be registered in a certain rate interval. Too fast loading does not reveal these events. Very slow loading results in a sluggish transformation and changes in the slope of the diagram, sometimes hardly visible, are observed instead of sudden jumps.

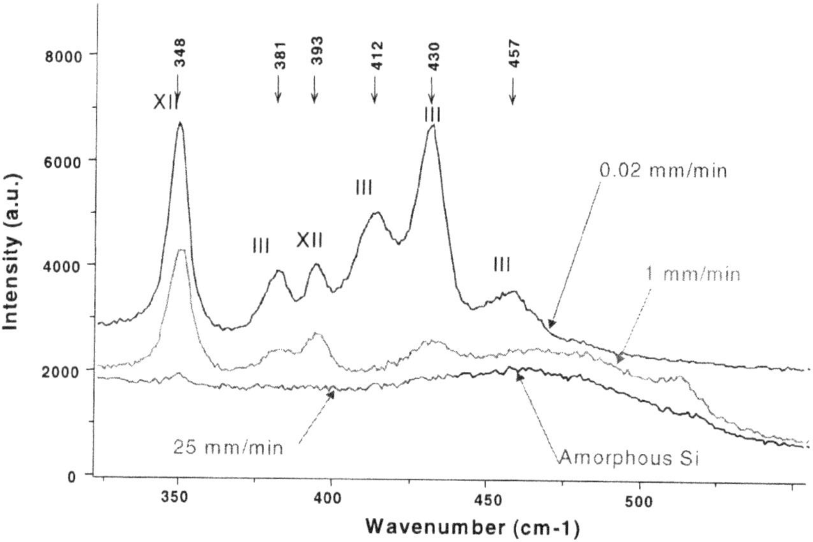

Figure 5. Raman spectra of Vickers impressions in silicon obtained with different loading rates all at a load of 1 kgf

Using micro-Raman analyses, we investigated the full cycle of transformations in Si under loading, decompression and annealing (Fig. 7). Thus, even for such a well-studied material as Si, principally new data had been obtained by using the combination of indentation and micro-Raman spectroscopy. More details on indentation induced phase transformations in Si are given in [31].

5.2. Germanium Results

In the case of Ge, there is a transition to the metallic β-Sn (Ge-II) structure at 8-11 GPa [7,30,32,33]. There are also several metastable phases known from experiments in high-pressure cells, which form during unloading from the metallic state [34]. Additionally, there exists the hd phase and a simple tetragonal phase (st12 or Ge-III). By fast pressure release the bc8 (Ge-IV) phase may be produced. We observed, in addition to the original Ge-I, the hd phase and further Raman bands, which can be assigned to at least two different polymorphs due to their changing intensity ratios (Fig. 8). Even though further detailed analysis of Raman spectra is required, the transformation is evident. It is important to note that metastable phases of Ge disappear after aging at ambient conditions within a day or less.

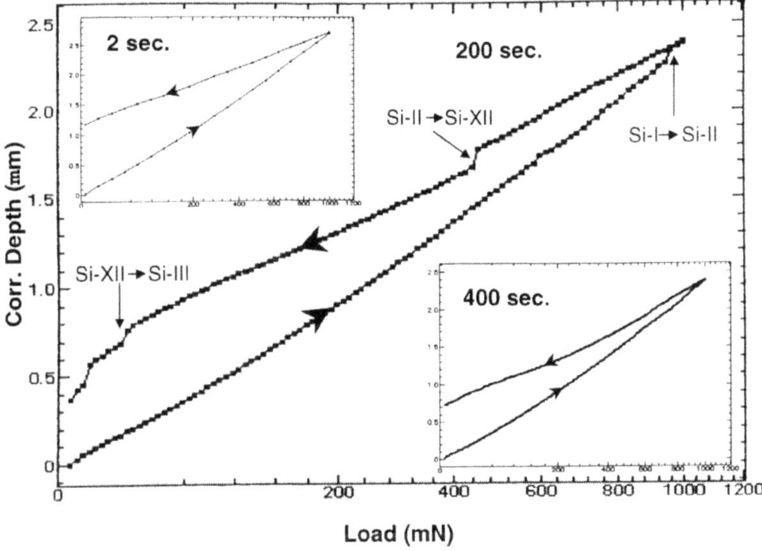

Figure 6. Load-displacement curves obtained on a Si single crystal (100) in a 4 s, 400 s and 800 s cycle of loading to 1 N. Arrows show pop-in and pop-out events caused by phase transformations. The suggested assignment is also given.

5.3. GaAs Results

Indentation tests on GaAs wafers at room temperature demonstrated an irreversible increase in relative intensity of the TO phonon peak relative to the LO phonon peak, as well as broadening and weakening of GaAs Raman bands. The breakdown of the TO selection rule suggests that the post-indentation material is composed of zincblende polycrystallites [35]. This suggests significant structural changes even after the RT experiments. However, according to the temperature dependence of the hardness, GaAs shows a plateau below the room temperature [6]. Thus, its hardness may be controlled by the transformation to a metallic phase at low temperatures. Scratching of GaAs samples at liquid nitrogen temperature clearly demonstrated the ductile behavior (Fig. 9) that can be only explained by transformation to metallic phase. The possibility of the formation of metastable phases of GaAs after decompression has been considered in the literature, but no experimental confirmation has been provided yet.

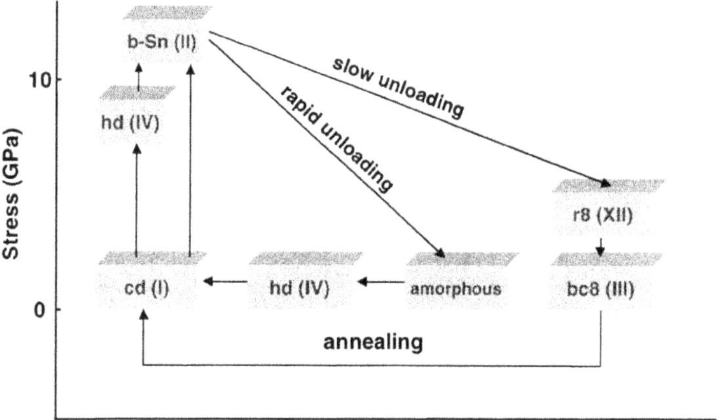

Figure 7. Cycle of the phase transformations that occur during hardness indentations and post-treatment in silicon. During the indentation experiment Si transforms from the original diamond cubic Si-I structure to the hexagonal diamond Si-IV or the metallic β-Sn phase (Si-II) depending on local stress conditions. Upon unloading other polymorphs form, including the amorphous phase, Si-XII and Si-III. During annealing there is a transition to Si- IV or, at higher temperatures, to Si-I [31].

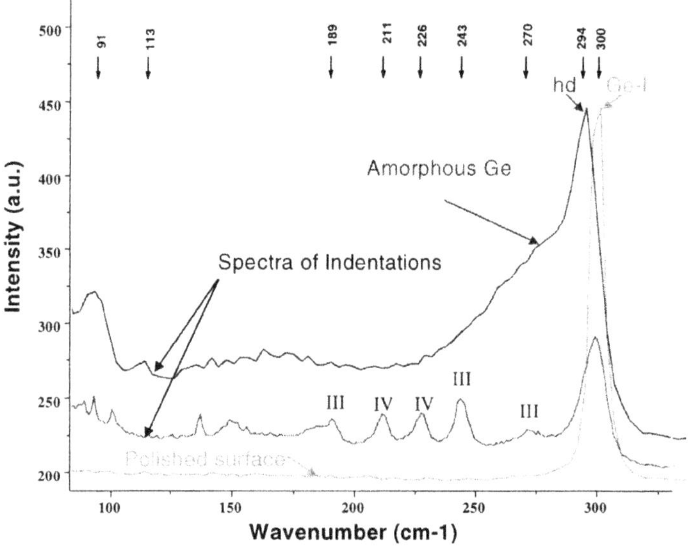

Figure 8. Micro-Raman spectra of different polymorphs of Ge in a Vickers impression. Ge transforms from the diamond structure (Ge-I, ~300 cm^{-1}) to either hexagonal diamond, hd, phase (294) or a mixture of two crystalline phases, probably the bc8 (Ge-IV) and tetragonal st12 (Ge-III, 270, 243, 189 cm^{-1} and eventually other bands).

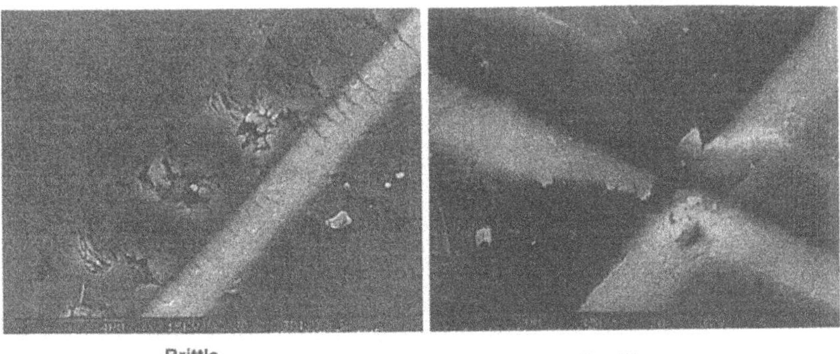

Figure 9. Typical SEM microgrphs of brittle and ductile cuts on GaAs surfaces

6. ADVANTAGES AND LIMITATIONS OF THE INDENTATION/RAMAN TECHNIQUE

In summary, the *advantages* of the indentation/micro-Raman tests include:

- fastest and most efficient of all methods available (1-5 min for the whole analysis);
- small spot size, down to 1 μm
- high shear stresses favor amorphisation and phase transformations;
- simplicity of sample preparation and measurements;
- possibility of in-situ studies;
- phase transformations upon loading and unloading, and reverse transformations upon heating;
- effects of the crystal orientation on the transformation can be studied;
- determination of both amorphous and crystalline phases;
- a very small amount of material is sufficient (e.g., a thin film of a few square micrometers in size);
- variety of easily replaceable indenters allows the use of a wide range of loading conditions;
- statistical treatment of numerous measurements minimizes errors and leads to reproducible results.

We also understand that this technique has certain *limitations:*

- difficulties with the quantification of the transformation conditions;
- experiments with soft materials require the use of thin films on hard substrates;
- scarce databanks of Raman data compared to that for the diffraction data;
- difficulties in calculation of structure parameters of new phases on the basis of the Raman spectra only;
- metals cannot be studied, but a transition to metallic state can be detected.

7. POTENTIAL APPLICATIONS

Ductile regime machining of ceramics and semiconductors is one key advance in the development and improvement of machining technology in the past decade. A significant amount of work on ductile-regime machining has been published during past seven years [37,38]. The very existence of a ductile regime implies extremely high plasticity for these normally brittle materials. There is an on-going debate about how to model ductile-regime machining of silicon, germanium and other semiconductors. In our opinion, the metallization mechanism of ductility is much more likely to operate compared to thermal melting, dislocation movement and other suggested models. However, it is hardly possible to expect that this hypothesis will be accepted by the engineering community before sufficient and direct evidence of metallization during ductile-regime machining is given. Until now, only indirect evidence was obtained using TEM. Recent TEM studies suggest that the ductile regime can be mediated by phase transformations.[14] However, until now no convincing structural evidence of the latter in machining has been presented.

Our preliminary Raman spectroscopy studies clearly demonstrated metallization due to closing of the band gap under indentation of silicon and germanium with sharp diamond indenters. Since the conditions under the cutting tool result in pressures consistent with those under indenters [39], we assume that similar transformations can occur during ductile-regime machining. Fig. 9 gives an example of metallization leading to a ductile cut. Micro-Raman spectroscopy coupled with indentation, scratch and turning experiments can be used to study metallization of semiconductors under the conditions of single-point turning and/or milling.

Understanding of the underlying mechanisms of ductile regime machining is required for choice and optimization of the machining parameters. If we understand, model, and accurately predict under what conditions the ductile regime occurs, then we can command the proper desired motion condition to the machine tool controller to realize them. Optimization of the single-point turning conditions based upon the obtained information about the mechanisms of processes under the tool, and prediction of regimes of ductile machining for new classes of semiconductors may become possible. A better understanding of the surface changes (phase composition and residual stresses) can be obtained, and a better quality of the surface finishing can be achieved. Further applications in the field of tribology (wear and erosion mechanisms), physical measurements (spreading resistance probes, pressure sensors) and surface patterning for semiconductors and electronic ceramics are also possible.

8. CONCLUSIONS

Our experimental results are in excellent agreement with predictions of J.J. Gilman [3,10,36]. The main hurdle was to demonstrate unambiguously that high-pressure metallization occurs in a variety of contact loading situations. We now have a tool to do this directly and even in-situ. The feasibility of the proposed technique has been clearly demonstrated on two of the most typical semiconductors. For the first time metastable phases in the hardness impressions were clearly shown. **Since these phases can be formed only via metallic Si and Ge, these experiments produced direct evidence of the metallization of semiconductors under contact loading**. The obtained phase compositions for Si and Ge are in agreement with the anvil work, and additionally Raman spectra of some of the metastable phases were recorded for the first time. These findings be used to choose and optimize conditions of the ductile regime machining for semiconductors.

REFERENCES

1. Brazhkin, V.V. and A.G. Lyapin, Lattice Instability Approach to the Problem of High-Pressure Solid-State Amorphization, *High Pressure Research*, **15**, 9-30 (1996).
2. Cahn, R.W., Metallic Solid Silicon, *Nature*, 357, 645-646 (1992).
3. Gilman, J.J. Mechanism of Shear-Induced Metallization, *Czech J. Phys*. **45**, 913-919 (1995).
4. Gogotsi, Y.G., A. Kailer, K.G. Nickel, Phase Transformations in Materials Studied by Micro-Raman Spectroscopy of Indentations, *Materials Research Innovations*, **1** (1) 3-9 (1997).
5. Clarke, D.R., M.C. Kroll, P.D. Kirchner, R.F. Cook, B.J. Hockey, Amorphization and Conductivity of Silicon and Germanium Induced by Indentation, *Phys. Rev. Lett*. **60**, 2156-2159 (1988).
6. Gridneva, I.V., Y.V. Milman, and V.I. Trefilov, Phase Transition in Diamond-Structure Crystals During Hardness Measurements, *Phys. Stat. Solidi* **9**(14), 177 (1972).
7. Pharr, G.M., W. C. Oliver, R. F. Cook, P. D. Kirchner, M. C. Kroll, T. R. Dinger, and D. R. Clarke, Electrical resistance of metallic contacts on silicon and germanium during indentation, *J. Mater. Res*. **7**, 961 (1992).
8. Pharr, G.M., W.C. Oliver, D.S. Harding, New Evidence for a Pressure-Induced Transformation During the Indentation of Silicon, *J. Mater. Res*. **6**, 1129-1130 (1991).
9. Weppelmann, E.R., J.S. Field and M.V. Swain, Influence of spherical indentor radius on the indentation-induced transformation behavior of silicon, *J. Mater. Sci*. **30**, 2455 (1995).
10. Gilman, J.J. Insulator-metal Transitions at Microindentations, *J. Mater. Res*. **7**, 535-538 (1992).
11. Weppelmann, E.R., J.S. Field and M.V. Swain, Observation, Analysis and Simulation of the Hysteresis of Silicon Using Ultra-micro-indentation with Spherical Indenters, *J. Mater. Res*. **8**, 830-840 (1993).
12. Callahan, D.L. and J.C. Morris, Extent of Phase Transformation in Silicon Hardness Indentations, *J. Mater. Res*., **7** 1614-1617 (1992).
13. Morris, J. C., D. L. Callahan, The microstructure of indentation (hardness) impressions in silicon and germanium, in *Microstructure of Materials*, Ed. K. M. Krishnan, San Francisco Press, San Francisco, CA, p. 104 (1992).
14. Morris, J.C., D.L. Callahan, J. Kulik, J.A. Patten, R.O., Scattergood, Origins of Ductile Regime in Single-Point Diamond Turning of Semiconductors, *J. Am. Ceram. Soc*. **78**, 2015-2020 (1995).
15. Morris, J.C., D.L. Callahan, Origins of Microplasticity in Low-Load Scratching of Silicon, *J. Mater. Res*. **9**, 2907-2913 (1994).
16. Ferraro, J.R., Vibrational Spectroscopy at High External Pressures: The diamond anvil cell, Academic Press, Orlando, 1984.
17. Jayaraman, A., Diamond Anvil Cell and High-Pressure Physical Investigations, *Rev. Modern Phys*. **55**, 65-108 (1983).
18. Crain, J., G.J. Ackland, J.R. Maclean, R.O. Piltz, P.D. Hatton, G.S. Pawley, Reversible Pressure-Induced Structural Transitions Between Metastable Phases Of Silicon, *Phys. Rev. B*, **50**, 13043-46 (1994)
19. Di Gregorio, J.F., T.E. Furtak, Analysis of Residual Stress in 6H-SiC Particles within Al_2O_3/SiC Composites through Raman Spectroscopy, *J.Am. Ceram. Soc*. **75**, 1854-1857 (1992):
20. Lucazeau, G., L. Abello, Raman Spectroscopy in Solid State Physics and Material Science. Theory, Techniques and Applications, *Analusis*, **23**, 301-311 (1995).
21. Sparks R.G. and M.A. Raesler, Micro-Raman Analysis of Stress in Machined Silicon and Germanium, *Prec. Eng*. **10**, 191 (1988).

22. Shen, H. and F. Pollack, Raman Study of Polish-Induced Strain in <100> and <111> GaAs and InP, *J.Appl. Phys.* **64**, 3233 (1988).
23. Lucazeau, G., L. Abello, Micro-Raman Analysis of Residual Stresses and Phase Transformations in Crystalline Silicon under Micro-Indentation, *J.Mater.Res.*, **12**, 2262-2273 (1997).
24. Jameison, J.C., Crystal Structures at High Pressures of Metallic Modifications of Silicon and Germanium, *Science*, **139,** 762 (1963).
25. Minomura, S. and H.G. Drickamer, Pressure-Induced Phase Transformations in Silicon, Germanium, and Some III-V Compounds, *J. Phys. Chem. Solids*, **23** 451 (1962).
26. Needs, R.J. and A. Mujica, First-principles pseudopotential study of the structural phases of silicon, *Phys. Rev.* **B 51**, 9652-60 (1995).
27. Wentorf, R.H. and J.S. Kasper, Two New Forms of Silicon, *Science*, **139** 338 (1963).
28. Clarysse, P., P. De Wolf, H. Bender, and W. Vandervorst, Recent Insights Into The Physical Modeling of The Spreading Resistance Point Contact, *J. Vac. Sci. Technol. B* **14**, 358-68 (1996).
29. Piltz, R.O., J.R. Maclean, S.J. Clark, G.J. Ackland, P.D. Hatton, J. Crain, Structure and properties of silicon XII: A complex tetrahedrally bonded phase *Phys. Rev. B* **52**, 4072-85 (1995).
30. Hanfland, M. and K. Syassen, Raman Modes of Metastable Phases of Si and Ge, *High Pressure Res.* **3**, 242-44 (1990).
31. Kailer, A., Y.G. Gogotsi, K.G. Nickel, Phase Transformations of Silicon Caused by Contact Loading, *J. Appl. Phys.* **81** (7) (1997).
32. Bates, C.H., F. Dachille, R. Roy, High-Pressure Transitions of Germanium and a New High-Pressure Form of Germanium, *Science*, **147**, 860-862 (1965).
33. Kasper, J.S., S.H. Richards, The Crystal Structures of New Forms of Silicon and Germanium, *Acta Crystallographica*, 1964, **17**, 752-755.
34. Nelmes, R.J., M.I. McMahon, N.G. Wright, D.R. Allan, J.S. Loveday, Stability and Crystal Structure of BC8 Germanium, *Phys. Rev. B*, **48**, 9883-86 (1993).
35. Nagata, K., Webb, S.J., R.A. Stralding, Raman Spectra of $InAs_{1-x}Sb_x$ Alloys and $InAs_{0.58}Sb_{0.42}$/InSb Strained Layer Superlattice under High Pressure, *Phys. Stat. Sol.* (b) **198**, 527-532 (1996).
36. Gilman, J.J. Metallization at Microindentations, in *Mat. Res. Soc. Symp. Proc.*, **276**, 191-196, MRS (1992).
37. Biffano, T.G., T.A. Dow, R.O. Scattergood, Ductile-Regime Grinding: a New Technology for Machining Brittle Materials, *J. Engng. for Industry*, **113**, 184-189 (1991).
38. Blake, P.N. and R.O. Scattergood, Ductile-Regime Machining of Germanium and Silicon, *J. Am. Ceram. Soc.* **73**, 949-957 (1990).
39. Lucca, D.A., Y.W. Seo, Effect of Tool-Edge Geometry on Energy Dissipation in Ultraprecision Machining, *CIRP Ann.*, **42** (1) 83 (1993).

INFLUENCE OF WATER ADSORPTION ON MICROTRIBOLOGY OF MICROMACHINES

N. OHMAE
Faculty of Engineering, Osaka University,
2-1 Yamadaoka, Suita, Osaka 565, JAPAN

1. Introduction

In an ordinary atmosphere, for example, at a relative humidity of 50%, 4 or 5 layers of water molecules adsorb on solid surface. Strong stiction due to adsorbed water onto micromachine (or MEMS) surfaces often causes enough resistance to stop relative motions, and therefore adsorption of water molecules has become one of the most serious problems in micromachines [1]. We have studied adsorption of water molecules onto micromachine surface and its effect on microtribology. Adsorption behavior of water molecules was detected by quartz crystal microbalance (QCM), and adhesion at each coverage of water molecules was measured by tip-flat contacts. The existence of vicinal water on the surfaces of Ni and Cu films prepared by the LIGA (Lithographie-Galvanoformung und Abformung) process has been found at water coverages of approximately three [8]. The dependence of adhesion force on pull-off velocity as well as the difference in electrical contact resistance supports this existence. Greater adhesion resulted at coverages over 2.0-3.5 was due primarily to capillary force. Contact geometry, microroughness, micropore and surface functional groups are all found to be influential on the formation of vicinal water and its tribological properties.

We have reformed our experimental apparatus after the NATO ASI, organized by Professor Bhushan, in Sesimbra, Portugal. The new apparatus basically is AFM/FFM which works in an atmosphere with backfilled water vapor. The same tendency of the dependence of adhesion force on coverage has been found. Pull-on force, which is an attractive force acting before an actual tip-flat contact has also showed the coverage dependence. The generation of capillary force depends on coverage, and thus on the structure of adsorbed water. In this paper, the threshold values at which capillary force predominates were discussed from the structure of adsorbed water molecules.

2. Analytical Procedures

2.1. QCM

Changes in the resonant frequency of QCM due to mass gain are expressed by

B. Bhushan (ed.), Tribology Issues and Opportunities in MEMS, 443-454.

$$\frac{\Delta f}{f} = -\frac{\Delta d}{d} = -\frac{\Delta m}{\rho_Q A D} \qquad (1),$$

where f is resonant frequency, d the thickness of quartz, m mass change equivalent to thickness Δd, while ρ_Q and A are density and area of quartz. For AT cut quartz,

$$\Delta f = -2.26 \times 10^{-6} \frac{f^2 \Delta m}{A} \qquad (2).$$

For QCM with a resonant frequency of 5MHz, 0.1Hz corresponds to 2ng. Thus the detection of adsorption species with a resolution of ng is possible by using QCM.

2.2. ADSORPTION ISOTHERM AND BET PLOT

By adopting the BET plot, adsorption of adsorbate is written as

$$\frac{N}{N_m} = \frac{cx}{(1-x)[1-(1-c)x]} \qquad (3),$$

where N is the number of adsorbed molecules, N_m the number of adsorption sites of adsorbent, which equals to the number of adsorbed molecules of the first layer, x the relative pressure in the form of p/p_0 and c the constant representing the interaction between adsorbate and adsorbent. By rewriting eq. (3), we obtain

$$\frac{x}{(1-x)N} = \frac{1}{cN_m} + \frac{(c-1)x}{cN_m} \qquad (4).$$

Since N and x are obtained by QCM measurements, the plot of the left-hand term of eq. (4) against x shows a linear relation; $(c-1)/cN_m$ being the slope and $1/cN_m$ the intersect. The numbers of adsorbed molecules on the second layer, the third layer $\cdots$ are considered as $2N_m$, $3N_m \cdots$.

2.3. PORE DISTRIBUTION

The distribution of pores is obtained by Kelvin equation of

$$\ln \frac{p}{p_0} = \frac{-2V_m \gamma}{rRT} \cos \alpha \qquad (5),$$

where V_m is molecular volume of adsorbate, γ surface tension of adsorbate, α an angle between liquid and a wall of pore, r radius of pore, R gas constant and T temperature. r

is obtained by the adsorption isotherm where p/p_0 is a variable. Generally, the distribution of pore radius is presented in the form of dv/dr versus r.

2.4. ISOSTERIC HEAT OF ADSORPTION

The isosteric heat of adsorption, q_{st}, is expressed by

$$q_{st} = RT^2\left(\frac{\partial \ln p}{\partial T}\right)_{V=const} \quad (6).$$

$q_{st} = q + RT$, but RT is small as ~2.4J/mole. In an actual calculation, q_{st} is given by

$$\frac{d \ln p}{d(1/T)} = -\frac{q_{st}}{R} \quad (7).$$

For more detailed analytical procedures, readers can refer to ref. 8.

3. Structure of Adsorbed Water Molecules

Based upon experimental evidences in ref. 8, the model of adsorbed water at the vicinity

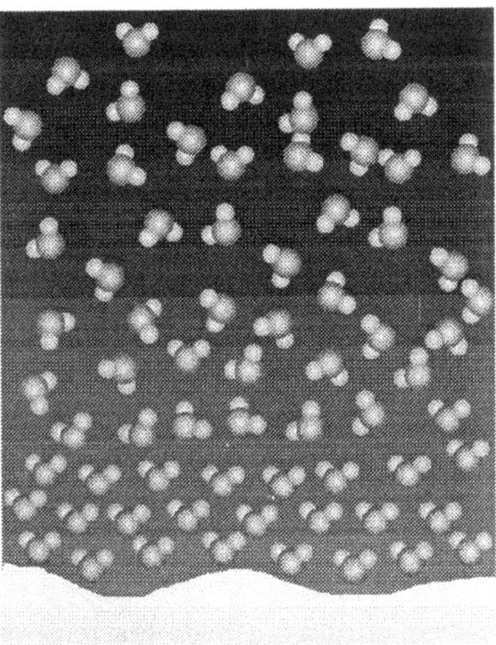

Figure 1. A model showing the existence of vicinal water.

of surface was proposed, as shown in Figure 1. Near the surface, the water molecules are in a contiguous arrangement and possess a long-range order, while the water molecules separated at a greater distance from the surface have a bulk water arrangement with short-range order. An interlayer between the vicinal water and the bulk water is a disordered region. Drost-Hansen [5], Derjaguin [4], and Peschel et al. [10] studied surface vicinal water in their work on solid-liquid interfaces. Our experimental work does not include such bulk water but solely adsorbed water molecules. It is quite interesting to note that the vicinal water exists not only at solid-water interface but also in adsorbed water.

4. Experimental Apparatus

Our apparatus previously used for measuring adhesion and friction forces made use of capacitance sensors. However, this type of displacement sensor required so much skill for adjustment. Moreover, it was difficult to precisely adjust simultaneously two capacitance sensors which could measure adhesion and friction.

To overcome this technical problem, we have designed new apparatus using a laser diode (wavelength 685nm, power 20mW) and a quadrant PSD, the schematic drawing of which is shown in Figure 2. As is noticed, our new apparatus basically is the AFM/FFM which operates in vacuum and in the atmosphere with backfilled water vapor. Typical range of force measurement is a few tens of nano Newtons to a few tens

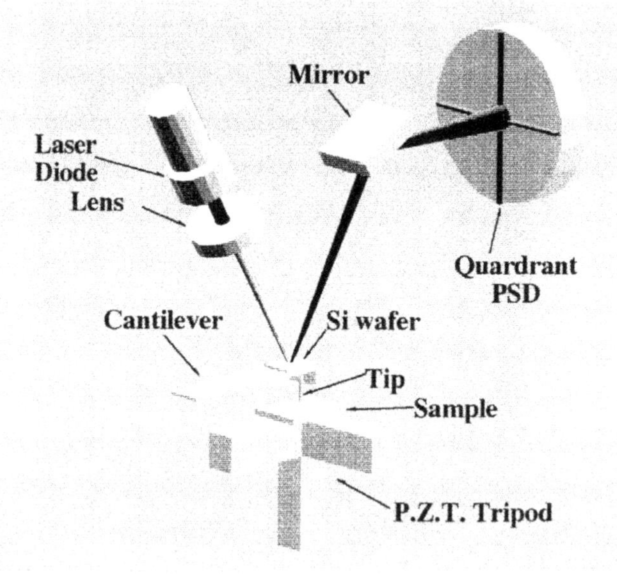

Figure 2. Schematic drawing of microtribometer.

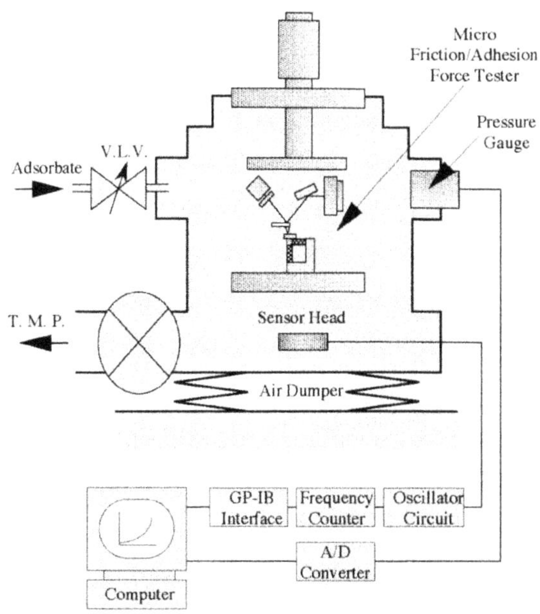

Figure 3. Entire view of microtribometer installed in a vacuum chamber.

of micro Newtons. By using a stiff tip, it is possible to detect the forces on the order of mili Newtons. The PZT configurations are almost the same as the previous one [8]. This apparatus is installed in a vacuum chamber (ultimate pressure 10^{-6} Torr) with an air dumper. To measure the pressure of water vapor and the adsorption isotherms, the same measuring systems as the previous apparatus were used. An entire view of the apparatus is shown schematically in Figure 3.

Adhesion experiments were carried out using a Ni-tip with a radius of curvature of about 40μm and a LIGA-Ni film with varied coverages of water molecules. An applied force was approximately 10μN, and the pull-off velocity was 2μm/s.

5. Results and Discussion

When the tip surface comes close enough to interact with a countersurface, in this case LIGA-Ni film with adsorbed water, an attractive force, or sometimes referred to "pull-on" force, generates. Figure 4 presents force curves for contacts at varied coverages of 1.8, 3.6 and 8.3. In such experiments that use atomically smooth surface as mica, Si wafer etc., the coverage analyzed by QCM may give a constant value over the specimen surface. The LIGA film, however, produces very rough surface on the order of a few tens of nanometer in Ra. The coverage of water molecules at the tops and the bottoms of LIGA film can vary, though an average coverage is given by a constant value.

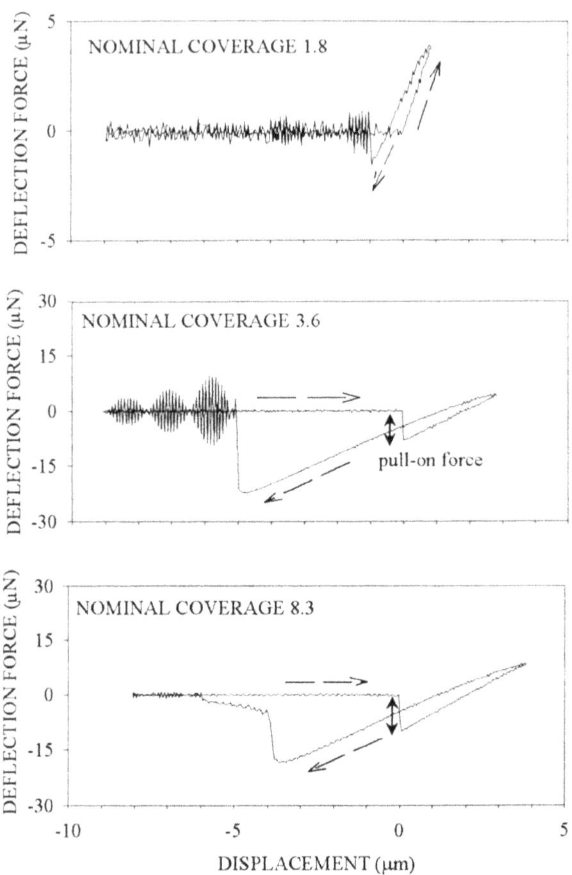

Figure 4. Force curves for different coverages of water molecules.

Therefore, Figure 4 depicts water adsorption as "nominal" coverage. This has been pointed out by helpful discussion with Professor Jean-Marie Georges at Ecole Centrale de Lyon. The pull-on forces are indicated by the both-sided arrow. The pull-on force is negligibly small at a coverage of 1.8, whereas they become high at coverage of 3.6 and 8.3. The dependence on coverage of water molecules thus is found for pull-on forces.

Figure 5 compares adhesion forces and pull-on forces. A rapid increase in adhesion force initiates at a coverage of 2.0-2.5, while an increase in pull-on force starts at a coverage of roughly 3. Figure 5 signifies that the formation of micro-menisci play a critical role in surface interaction. The onset of increased pull-on forces appears to be at a coverage of 3, slightly higher than that of adhesion. The reason for this has not yet been fully understood, but the contact processes, i.e., the processes of separation, approach and contact, are in great effect, since adhesion force results after deformation

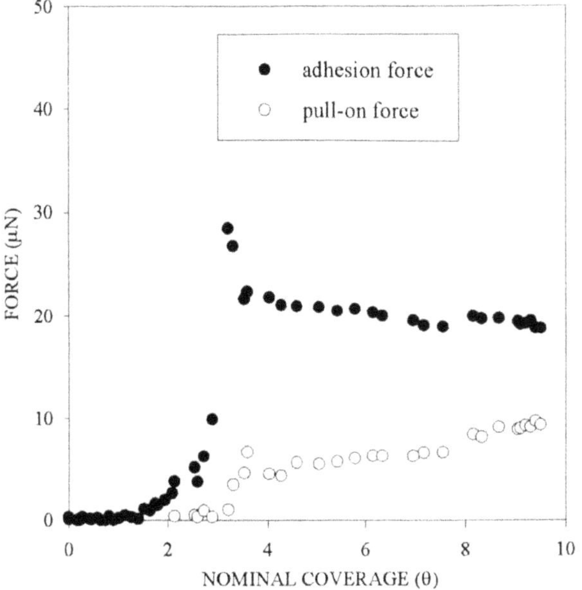

Figure 5. Adhesion and pull-on forces at different coverages.

and rearrangements of water molecules as well as of surface atoms.

Measured adhesion forces for different surface roughnesses of LIGA-Ni film are shown in Figure 6. Adhesion forces appear to increase at a coverage of roughly 3, but in the more precise expression, 2.0-2.5 for rough film, 2.5 for medium roughness, and 3.5 for smooth film. This slight but very important difference in the coverage in which high adhesion predominates will be discussed later. Once adhesion goes up at coverages higher than the threshold value, a decreased adhesion results for rough surface, while increasingly high one for smooth surface. This can be explained by the small number of contacting points for rough surfaces, the periphery of which is surrounded by water molecules. Water molecules thus formed, though over the threshold coverage, produce capillary force.

As described earlier, for rough surfaces the coverage of water molecules is an averaged value over the entire specimen surface. The adhesion forces shown in Figure 6 are replotted versus relative pressure of water vapor in Figure 7, in order to discuss the menisci formation. Rapid increase in adhesion is evident at a relative pressure of 0.7, and this transition can remarkably be seen on this replotted figure. The scatter of threshold values in Figure 6, plotted versus coverage, has been in order when viewed as a function of relative pressure. This implies that the meniscus formed at a relative pressure of 0.7 has the structure of bulk water which gives rise to capillary force. By putting the relative pressure of 0.7 into Kelvin equation, the radius of meniscus of water molecules at this pressure is estimated at 2.9nm. According to Christenson, the radius of

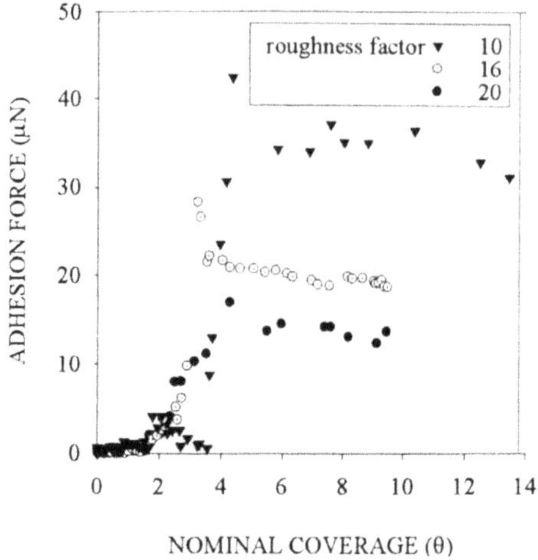

Figure 6. Adhesion of LIGA-Ni films with different roughnesses, plotted against coverage water.

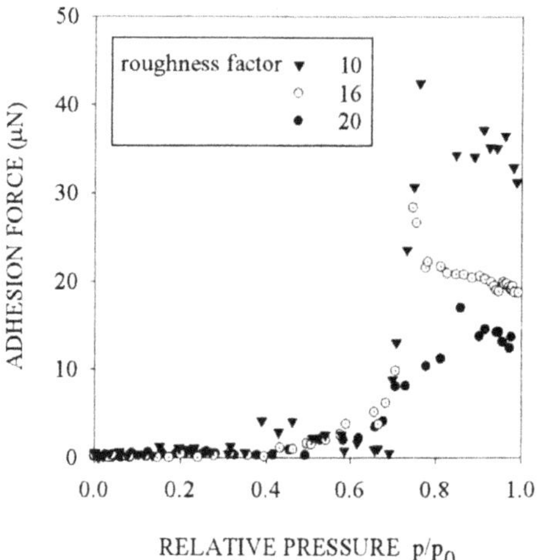

Figure 7. Replot of Figure 6 versus relative pressure, showing the same startpoint of increase despite of different roughnesses of LIGA-Ni films.

meniscus where strong capillary force of water predominates is larger than 2nm [3]. Fisher and Israelachivili have indicated that, due to structural change of adsorbed water, adhesion increase when the meniscus radius exceeds 5nm [6]. Buffey suggested that about 20 molecules are required for water to have bulk properties [2]. A calculation of

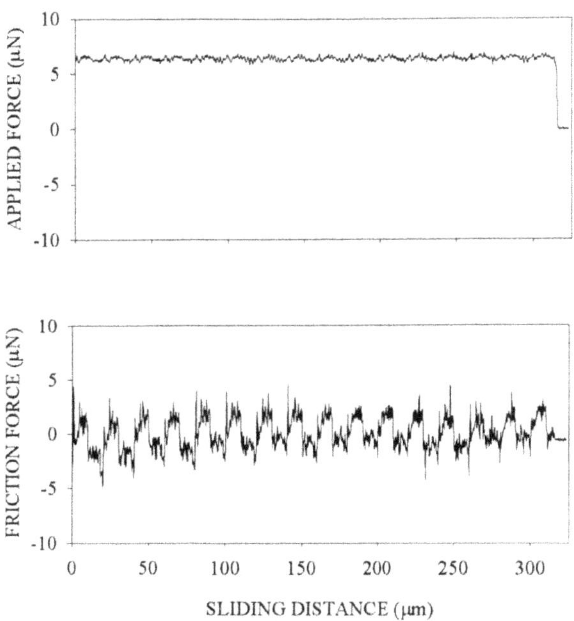

Figure 8. Traces of friction (lower figure) obtained from reciprocating slidings; applied force 6.5μN, $p/p_0 = 0.58$.

diameter of the hemisphere, in which 20 molecules of water are contained, leads to about 1.5nm. The specimens used in the present experiments are Ni-films fabricated by the LIGA process. Comparing with the experiments or theoretical calculation in refs. 2,3 and 6, there exists a difference in micro-roughness. Nevertheless, a good agreement is obtained in the threshold value which gives rise to capillary forces. Therefore, strong adhesion over a coverage of about three is thus explained by the meniscus formation of bulk water.

In our previous studies, we only conducted experiments on adhesion with adsorbed water molecules [7,8,9] because of the limitation of measuring apparatus. Our new micro-tribometer detected friction as shown in Figure 8, where friction experiments were carried out using a Ni-tip with a radius of curvature of 10μm at different coverages of water molecules. An applied load, which is servo-controlled, was typically 6-8μN. The sliding speed, or the scan speed of the tip, was 2.0μm/s, and 30 reciprocating slidings were performed on the fixed friction track. Since friction did not vary so much over 30 scans, an average value of 30 scans was plotted in Figure 9. Adhesion force also was measured after the last scan of the Ni-tip.

Figure 9 indicates that friction coefficient is high at 0.6 for dry surfaces, although a slight amount of water vapor before evacuating the test chamber may exist. Friction coefficient remains low at p/p_0 of 0.2-0.6, while a remarkable increase is observed at a p/p_0 of 0.7. Friction coefficient was high at 0.8 at p/p_0 of 0.7-0.9. A rapid increase in

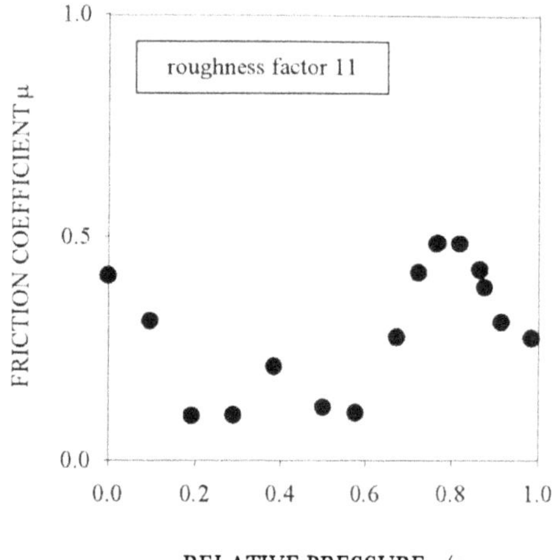

Figure 9. Friction properties with adsorbed water molecules

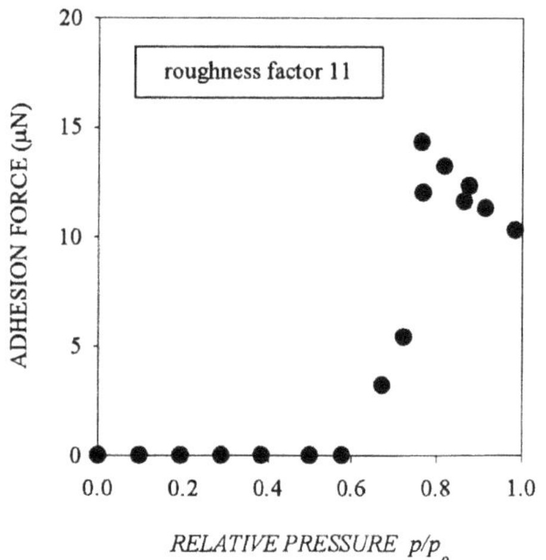

Figure 10. Adhesion measured after slidings.

friction at $p/p_0 = 0.7$ agrees very well with adhesion properties shown in Figure 7. Therefore, both adhesion and friction depend greatly on the structure of adsorbed water molecules. A tendency in the decrease in friction over $p/p_0 = 0.9$ is obtained. Low friction at very high humidity culminating in a number of studies may explain this decrease, but the adsorbed water is so thin that this mechanism is unclear.

Adhesion measured on the friction track after finishing reciprocating slidings shows almost the same tendency that is shown in Figures 7 and 9. The rapid increase in adhesion at p/p_0 = 0.7 is clearly observed. Adhesion after slidings at p/p_0 of 0-0.6 is almost zero. This implies that the structure of vicinal water is stable after slidings. Now that both friction and adhesion are low at p/p_0 of 0.2-0.6, it is advantageous to maintain low humid atmosphere for the operation of micromachines or to design hydrophobic surface. In either case, the vicinal water plays a key role.

6. Concluding Remarks

A rapid increase in adhesion at water coverages of 2.0-3.5 reported previously was explained by the formation of bulk water when replotted against relative pressure of water vapor.

Since the LIGA surface is rough, uniform coverage as in the case of atomically flat surface is not applicable, in a strict sense. In the present study, we need the term "nominal coverage" which may give an aveage value. Our new setup for measuring adhesion and friction worked well, and high friction at p/p_0 = 0.7-0.8 as well at dry conditions resulted. The rapid increase in friction at p/p_0 = 0.7 lead to high adhesion at p/p_0 = 0.7 when adhesion was measured even after sliding. These findings suggests the important role of surface vicinal water.

Acknowledgments

I would like to thank Yasuyuki Baba for carrying out many of the experimental work in this study. Support by the research contract from the Micromachine Center is acknowledged. This work was also supported in part by the Grant-in-Aid for Scientific Research from the Ministry of Education, Science, Sports and Culture, Japan.

REFERENCES

1. Bhushan, B. (1994) Tribology of magnetic storage systems, in Booser, E.R. (ed.) *Handbook of Lubrication and Tribology Vol. III*, CRC Press, Boca Raton, pp. 349-355.
2. Buffey, I.P. and Brown, W.B. (1990) A theoretical study of the infrated adsorption spectra of large water clusters, *J. Chem. Soc. Faraday Trans.* **86**, 2357-2360.
3. Christenson, H.K. (1988) Adhesion between surfaces in undersaturated vapors —A reexamination of the influence of meniscus curvature and surface farces, *J. Colloid Interface Sci.*.**121**, 170-178.
4. Derjaguin, B.V. (1966) Effect of lyophile surfaces on the properties of boundary liquid films, *Inst. Physic. Chemistry*, 109-119.
5. Drost-Hansen, W. (1969) Structure of water near solid film, *Ind. Eng. Chem.* **61**, 10-47.
6. Fisher, L.R. and Israelachvili, J.N. (1981) Direct measurement of the effect of meniscus forces on adhesion: A study of the applicability of macroscopic thermodynamics to microscopic liquid interfaces, *Colloid and Surfaces*, **3**, 303-319.
7. Ohmae, N. (1995) Recent work on tribology of micromachines, *Proceeding of the 1st International Micromachine Symposium, Japan*, 47-53.
8. Ohmae, N. (1997) Water adsorption on MEMS surface studied by quartz crystal microbalance, in Bhushan, B. (ed.), *Micro/ Nanotribology and its Applications*, Kluwer, Dordrecht, pp. 629-645.
9. Ohmae, N., Baba, Y., Umeno, M. and Tagawa, M. (1997) Structure and tribology adsorbed water on

micromachine surfaces fabricated by the LIGA Process, *Proceeding of the 2nd International Symposium on Tribochemistry, Janowice, Poland*, in press.

10. Peschel, G. and Adlfinger, K.H. (1970) Viscosity anomalies in liquid surface zones III. The experimental method, *Ber. Bunsen physik. Chem.* **74**, 351-357.

TRIBOLOGICAL EFFECTS OF SURFACE ROUGHNESS IN MEMS DEVICES

LIWEI LIN
Center for Integrated Sensors and Circuits
Department of Mechanical Engineering and Applied Mechanics
University of Michigan, Ann Arbor, MI 48109-2125, USA
Phone: (1) 313-647-6907, Fax: (1) 313-647-3170
e-mail: lwlin@engin.umich.edu

Abstract

This paper presents tribological effects of surface roughness in the emerging field of MEMS (microelectromechanical systems) where the various characteristic lengths of microscale devices are comparable in magnitude with surface imperfections. Identification and characterization of surface roughness in MEMS devices encounter scientific challenges in different regimes. A review of MEMS micromachining processes describing key manufacturing steps is followed by practical engineering examples. Tribological issues are illustrated in various areas including microscale heat transfer, solid mechanics and optics.

1. Introduction

MEMS has emerged as an interdisciplinary field in recent years [1] which encompasses a wide range of disciplines such as electrical engineering, mechanical engineering, materials science, physics and chemistry. Tribological issues dealing with problems of adhesion, abrasion, corrosion and erosion [2] are generally encountered in macroscopic machinery and are inevitable for microscopic MEMS devices. The technological and economical impacts of tribological issues in MEMS demand fundamental understanding and characterization in materials development, design, processing and testing of microstructures. Surface roughness is the center of these tribological issues. Micro devices which are fabricated by IC (Integrated Circuit) manufacturing processes [3] represent very different surface features than those macro structures fabricated by conventional mechanical manufacturing processes. It is important to characterize the surface roughness of microstructures and investigate the tribological effects in the microscale.

This paper discusses tribological effects of surface roughness for MEMS devices fabricated by surface-micromachining, bulk-micromachining and micro hot embossing processes. Three types of micro surfaces are discussed: polysilicon resistive microheaters for microfluidics applications [4,5,6]; suspended micro beams in micromechanical structures [7,8] and hot embossed plastic microstructures for optical applications [9,10]. These microstructures are fabricated by various processes and

B. Bhushan (ed.), Tribology Issues and Opportunities in MEMS, 455-462.

surface roughness is characterized independently. Tribological effects in several key applications regimes such as heat transfer, solid mechanics and optics are addressed in different sections.

2. Polysilicon Micro Resistive Heaters

MEMS devices have been using micro resistive heaters as heating elements for many years. For example, tiny micro bubbles have been nucleated to eject ink droplets in bubble-jet printers [11]. Thermophysical phenomena of bubble nucleation are interested processes not only for the microscale resistive heaters but also for macroscale boiling experiments. It has been concluded that phase change processes in the macro scale boiling depend strongly upon residual air which may be trapped inside surface imperfections of heated surfaces [12]. The surface imperfections become active sites for bubble nucleation and the phenomenon is characterized as heterogeneous nucleation [13]. In the microscale boiling experiments, microheaters are fabricated by micromachining processes which are unlikely to create big surface imperfections. Furthermore, the dimensions of microheaters are generally smaller than the sizes of imperfections which are discovered on the surfaces made by macro scale manufacturing processes. Therefore, the cavity theory which has been widely used for bubble nucleation for macroscale devices may not be applicable in the microscale. In order to investigate the microscale bubble nucleation mechanism, surface roughness on these microheaters should be characterized.

Figure 1 shows the SEM microphoto of a fabricated microheater. The fabrication process includes thermal oxidation, LPCVD (Low Pressure Chemical Vapor Deposition) polysilicon deposition, patterning and etching of the microheaters [4]. This microheaters have two parts. One is 10 x 2 x 0.5 μm^3 at the left hand side and the other is 5 x 2 x 0.5 μm^3 at the right hand side. The portion at the right hand is expected to have a high temperature if an electrical current is passed through the microheater since it is thinner than the portion at the left hand side. When the microheater is immersed in FC 43 (an electronic inert liquid provided by 3M company) and electrically self-heated, a tiny microbubble is generated as seen in Fig. 2. This microbubble has a diameter of about 5 μm. The bubble nucleation

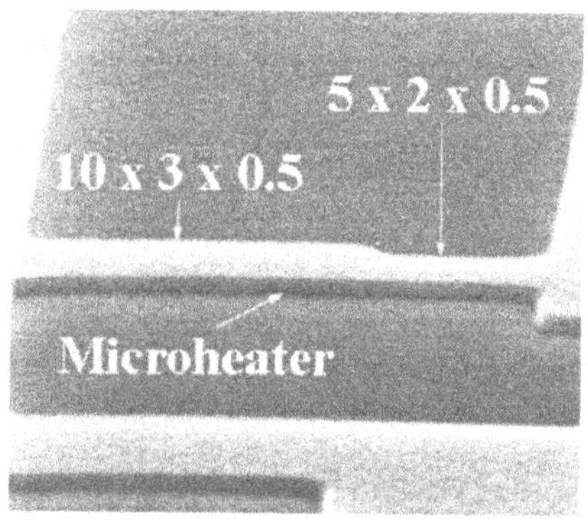

Fig. 1. A SEM microphoto showing a polysilicon resistive heater [5].

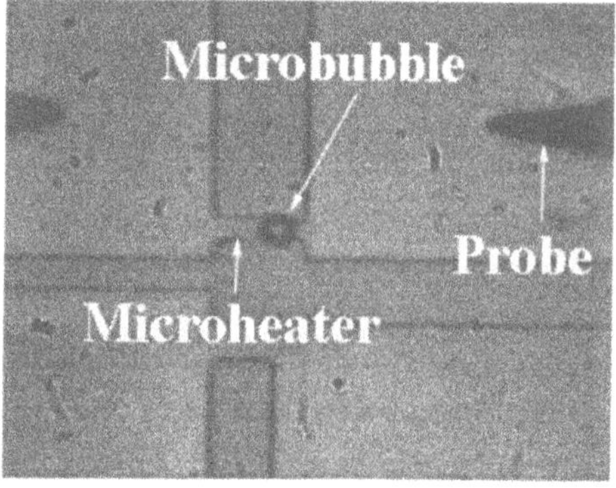

Fig. 2. A micrograph showing a 5 μm in diameter thermal bubble is nucleated.

temperature is found to be close to the critical temperature of the FC 43 liquid and it is much higher than its boiling temperature [4]. The tribological effects of surface roughness are investigated.

The characterization of surface roughness starts from the fabrication process. The original surfaces of prime silicon wafers are very smooth. For (100) type silicon wafers, measured RMS (root mean square) surface roughness is about 0.12 nm [14]. The subsequent microfabrication processes of thermal oxidation and polysilicon deposition are not expected to cause dramatic change in terms of surface roughness. Figure 3 is an AFM scan result (0.5 x 0.6 μm^2) on top of the microheater. The measured RMS surface roughness is about 6.5 nm. The bottom of Fig. 3 is one single horizontal scan which passes through the center and shows a "cavity" like shape at the right hand side. The measured peak-to-valley roughness of this cavity is about 10.7 nm [15]. The necessary bubble nucleation temperature for such a cavity size based on heterogeneous bubble nucleation mechanisms is calculated according to a theory developed by Y.Y. Hsu [16]. It is found that the bubble nucleation temperature should be about 80°C higher than the normal boiling temperature. Imperfections with larger sizes can actually reduce the necessary superheat temperature for bubble nucleation [16]. Since experimental measurements suggest a much higher superheat temperature is required for bubble nucleation, it is concluded that homogeneous nucleation is the one which occurs in this micro boiling tests [15]. Nevertheless, the role of surface roughness for microscale bubble nucleation applications can be clearly exemplified and should be taken into design consideration for bubble powered microfluidic devices.

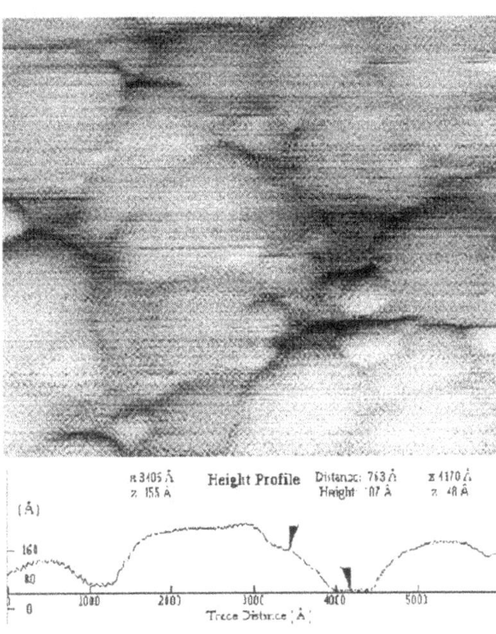

Fig. 3. (Top) AFM surface scan of microheaters. (Bottom) A single horizontal scan across the center.

3. Suspended Micro Beams

Surface roughness of beams or columns of macroscopic devices may be negligible when compared with the overall geometry. However, geometrical variations in MEMS devices may be important in solid mechanics analyses and can not be neglected. Suspended micro beams are particularly interested since they are basic components in mechanical structures such as micro accelerometers [17]; micro resonators [8]; and electromechanical filters [18]. These suspended beams are commonly fabricated by surface micromachining processes with polysilicon as the structural layer [8]. The basic process starts with clean silicon wafers. A layer of silicon nitride is deposited for electrical insulation. It is followed by a thin polysilicon layer deposition. This first polysilicon layer is patterned and etched to define the ground electrodes and ground planes. A thick (about 2 μm) sacrificial PSG layer is then deposited. Anchors are patterned and etched. The structural polysilicon layer of about 2 μm is now deposited, patterned and etched. Finally, the sacrificial layer is etched away and the microstructure is released to complete the process.

During the structural layer definition, dry plasma etching processes are widely used and they generally create uneven side walls. The surface variations of side walls may become an important factor in device functionality. For example, in the design and operation of a passive micro strain gauge [19], a tiny slope beam with geometry of 20 x 2 x 1.2 μm^3 determines the output performance of the device. The width of the slope beam is only 1.2 μm as shown in the SEM microphoto of Fig. 4. The side walls of the beam have uneven surfaces due to the dry silicon etching process. These surface imperfections may become important factors in characterizing spring constant, crack initiation and other solid mechanics properties. Figure 5 shows the schematic diagram of such a beam with uneven side walls. A study based on stochastic finite element methods has been conducted to predict the probabilistic behavior and response of microfabricated polysilicon beam structures [20]. It is found that the theoretical coefficient of variation (COV), or the uncertainty of the device readout is about 1.56% if the beam width variation is 6%. The same analysis may be applied to devices made by similar processes such as a micro electromechanical filter as shown in Fig. 6 [18]. Filters require precise frequency outputs and small dimensional variations on beam structures may greatly affect the manufacturing controllability. Improvements of dry etching process, scientific studies on the surface roughness

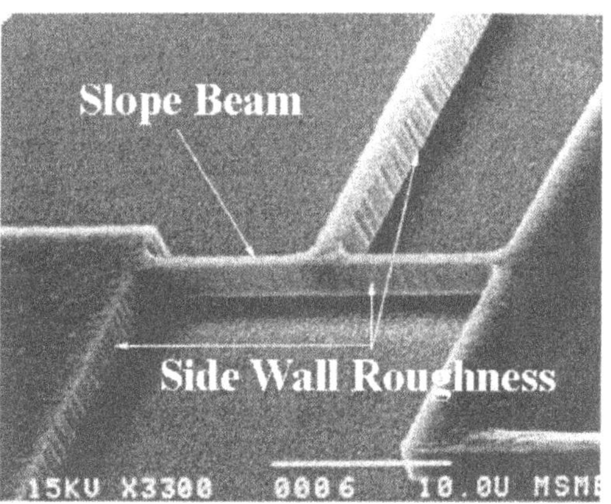

Fig. 4. SEM microphoto showing suspended slope beam with rough surfaces on side walls [7].

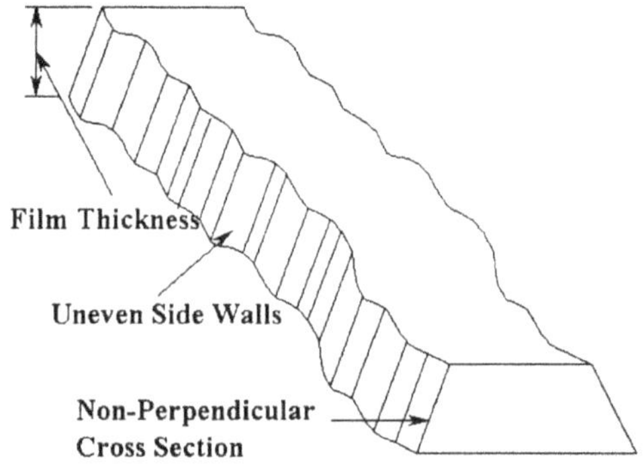

Fig. 5. Schematic diagram of irregularities at side walls.

effects and experimental characterizations are required to help solving the tribological effects of these MEMS devices.

4. Hot Embossed Plastic Microstructures

The third micromachined surfaces to be discussed in this paper are surfaces of structures fabricated by LIGA or LIGA-like processes [21]. Molding processes are generally the final steps in these processes and they determine the surface roughness. Intermediate steps such as synchrotron radiation, electroplating have little effects. Two types of surface fabricated by laboratory and commercial molding processes are discussed in this paper [22]. Figure 7 shows the fabrication process of the laboratory process which is named SMIHE (Silicon Mold Insert Hot Embossing) process [9]. The process starts with thermal oxidation and patterning of silicon dioxide which functions as the mask in the later silicon wet etching step. Anisotropic silicon etching is followed to define the micro pyramids in the silicon wafer. The fabricated silicon mold inserts are directly used in the hot embossing process to make plastic Polymethyl Methacrylate (PMMA) micro pyramids. These optically transparent pyramids are put on battery powered LCD displays to enhance brightness within viewing angles [10]. Apparently, surface roughness of molded surfaces is an important element for optical applications and should be characterized.

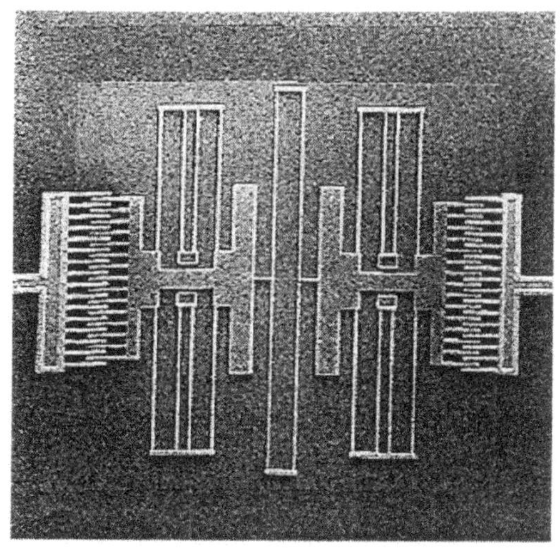

Fig. 6. SEM microphoto of a two-resonator micro electromechanical filter [18].

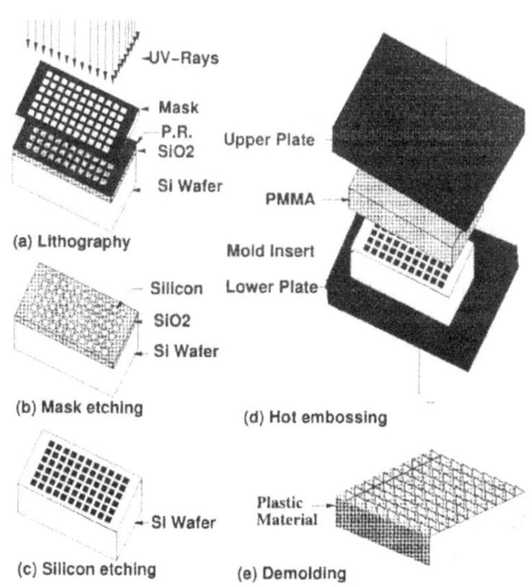

Fig. 7. The SMIHE (Silicon Mold Insert Hot Embossing) process [9].

In a separate effort, a commercial hot embossing process has also been applied to make micro pyramids. The commercial process starts with silicon mold inserts. The electroplating process is used to duplicate these silicon mold inserts with nickel plates. Commercial hot embossing processes use nickel mold inserts to manufacture Polyvinyl Chloride (PVC) films. PVC is generally used in commercial process since it has a lower glass transition temperature than PMMA. The two molding processes used different mold inserts and different plastic materials such that surface characteristics are expected to be different. First, Figure 8 shows SEM microphoto of hot embossed PMMA micro pyramids. These pyramids have a base width of about 30 μm and height of about 21 μm. Under this SEM photo, the surfaces are extremely flat without any defects. However, AFM scan results of (111) surfaces show voids on the PMMA micro pyramid as shown in Fig. 9 which gives detail examination on a 2 x 2 μm^2 area. On the other hand, bumps have been shown on (100) surfaces of PVC films as illustrated in Fig. 10 which has a scan area of 2 x 2 μm^2. The influence of these surface defects of voids and bumps to the optical performance is not clear. However, it is important to have a full understanding of surface roughness on these microfabricated structures. Table 1 list measured surface roughness from silicon, PMMA and PVC microstructures. It is found that the surface roughness of new silicon wafers is extremely small. Both hot embossed PMMA films and PVC films are rougher than the silicon mold inserts. Detail characterization and experiments are underway to investigate both the caused of the differences and their effects in optical devices.

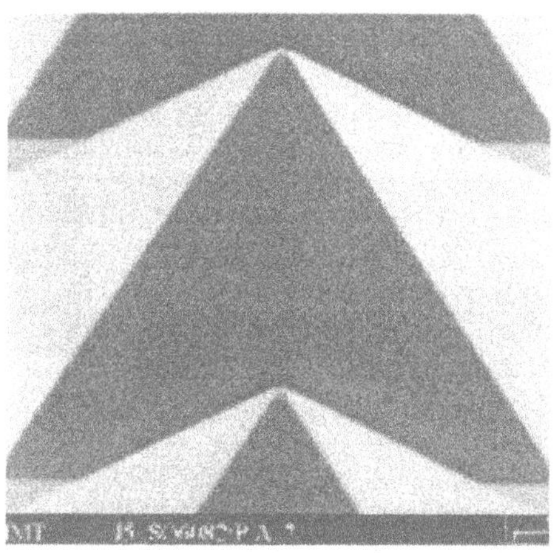

Fig. 8. SEM microphoto showing hot embossed micro pyramids made of PMMA [9].

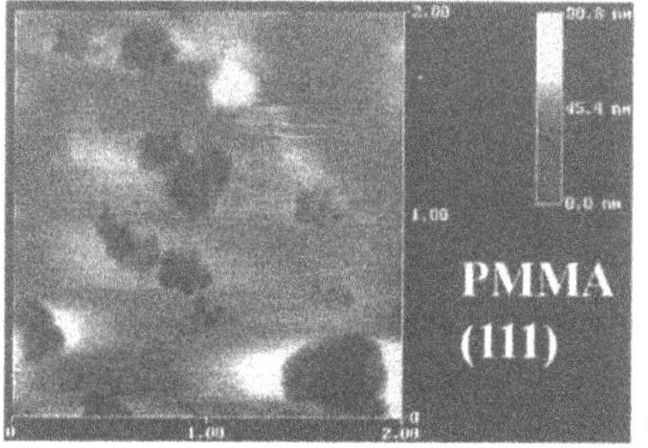

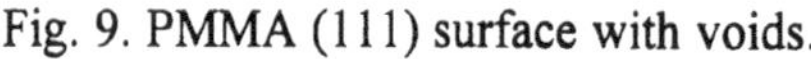

Fig. 9. PMMA (111) surface with voids.

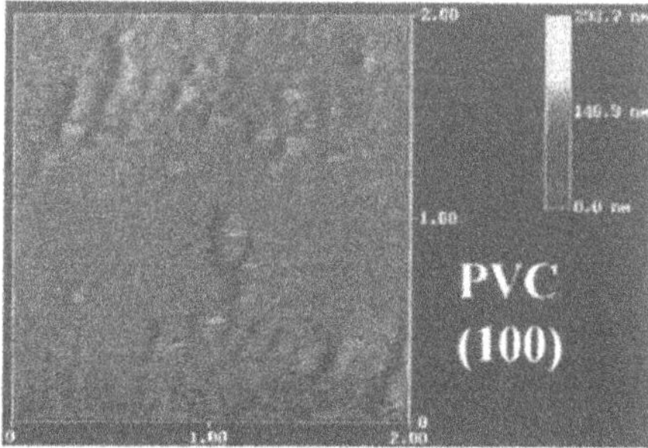

Fig. 10. PVC (100) surface with bumps.

TABLE 1. Measured surface roughness

Sample ID	Mean (nm)	Peak - valley (nm)	Remarks
Si (100)	0.06	2.4	New wafer
Si (111)	2.2	24	Mold insert
PMMA (100)	3.9	78	Laboratory process
PMMA (111)	4.2	91	Laboratory process
PVC (100)	15	380	Commercial process
PVC (111)	10	140	Commercial process

5. Conclusions

Tribological issues of surface roughness exist in the theory, experiments and engineering applications in MEMS devices. This study summarizes some of these challenges and recommends specific directions for future research. Surface roughness in three kinds of microstructures has been identified, analyzed including microheaters for bubble nucleation applications, micro beams for micro mechanical structures and hot embossed micro pyramids for optical applications. Surface imperfections of microheaters affect the bubble nucleation mechanism and should be taken into design considerations for bubble powered microfluidic devices. Side wall roughness of micromachined beams due to plasma dry etching processes may alter the mechanical characteristics of micro devices and should be carefully characterized. Finally, surface roughness of hot embossed plastic microstructure may have bumps or voids which affect the optical properties of micro devices.

6. Acknowledgments

The author would like to thank Mr. R. Alley and Y.T. Cheng for conducting AFM analyses on microheaters and hot embossed microstructures, respectively.

7. References

[1] K.E. Peterson, "Silicon as a Mechanical Material," *Proc. of the IEEE,* Vol. 70, pp. 420-457, 1982.

[2] M.J. Neale, "*Tribology Handbook*," New York, Wiley, 1973.

[3] R. C. Jaeger, "*Introduction to Microelectronic Fabrication*," Addison-Wesley, 1988.

[4] Liwei Lin and A.P. Pisano, "Thermal Bubble Powered Microactuators", *Microsystem Technologies*, Vol. 1, pp. 51-58, 1994.

[5] Liwei Lin and A.P. Pisano, "Bubble Forming on a Micro Line Heater", *ASME Winter Annual Meeting*, DSC-Vol.32, pp. 147-163, Atlanta, 1991

[6] Liwei Lin, K.S. Udell and A.P. Pisano, "Phase Change Phenomena on a Heated Polysilicon Micro Heater in Confined and Unconfined Micro Channels", *Thermal Sciences and Engineering*, Vol. 2, pp. 52-59, 1994.

[7] Liwei Lin, R.T. Howe and A.P. Pisano, "A Passive in-situ Micro Strain Gauge", *1993 Micro Electro Mechanical Systems Workshop*. pp. 206-210, 1993.

[8] W.C. Tang, C.T.-C. Nguyen and R.T. Howe (1989), "Laterally Driven Polysilicon Resonant Microstructures", *Sensors and Actuators*, Vol. A20, pp. 25-32.

[9] Liwei Lin, C.-J. Chiu, W. Bacher and M. Heckele, "Microfabrication Using Silicon Mold Inserts and Hot Embossing", *Seventh International Symposium on Micro Machine and Human Science*, pp. 67-71, October, 1996.

[10] Liwei Lin, T.K. Shia and C.-J. Chiu, "Fabrication and Characterization of IC-Processed Brightness Enhancement Films", *9th Int. Conference on Solid State Sensors and Actuators, Transducers'97*, Chicago, June, pp. 1427-1430, 1997.

[11] N.J. Nielsen, "History of ThinkJet Printerhead Development," *HP Journal*, pp. 4-10, 1985.

[12] H.B. Clark, P.S. Strenge and J.W. Westwater, "Active Sites for Nucleate Boiling", *Chem. Eng. Prog. Symposium*, Vol. 55, pp.103-110, 1959.

[13] V.P. Carey, "*Liquid-Vapor Phase-Change Phenomena*," Hemisphere Pub. Corp., Washington, D.C..

[14] B. Bhushan, "Nanotribology and Nanomechanics of MEMS Devices," *Proceedings of IEEE Workshop on Micro Electro Mechanical Systems (MEMS96)*, pp. 91-98, 1996.

[15] Liwei Lin, "*Selective Encapsulations of MEMS: Micro Channels, Needles, Resonators and Electromechanical Filters*", Ph.D dissertation, Department of Mechanical Engineering, University of California at Berkeley, 1993.

[16] Y.Y. Hsu, "On the Size Range of Active Nucleation Cavities on a Heating Surface," *J. of Heat Transfer*, Vol. 84C, pp. 207-216, 1962.

[17] T.A. Core, W.K. Tsang, and S. Sherman, "Fabrication Technology for an Integrated Surface-Micromachined Sensor", *Solid State Technology*, October, pp. 39-47, 1993.

[18] Liwei Lin, C.T.-C. Nguyen, R.T. Howe and A.P. Pisano, "Micro Electromechanical Filters for Signal Processing", *1992 Micro Electro Mechanical Systems Workshop*. pp. 226-231, 1992.

[19] Liwei Lin, A.P. Pisano and R.T. Howe, "A Micro Strain Gauge with Mechanical Amplifier,", *J. of Microelectromechanical Systems*, Vol. 6, No. 4, Dec. 1997.

[20] D. Mirfendereski, Liwei Lin, A. D. Kirueghian and A. P. Pisano, "Probabilistic Response of Micro-Fabricated Polysilicon Beam Structures: Comparison of Analysis and Experiments, " *in Proc. ASME Winter Annu. Meet., Symp. Mmicromechanical Syst.*, Vol. DSC-46, pp. 77-80, 1993.

[21] W. Bacher, W. Menz and J. Mohr, "The LIGA Technique and its potential for Microsystems - a Survey", *IEEE Trans. on Industrial Electronics*, Vol. 42, pp. 431-441, 1995.

[22] Liwei Lin, Y.T. Cheng and C.-J. Chiu, "Comparative Study of Hot Embossed Microstructures by Laboratory and Commercial Environments", *Microsystem Technologies Journal*, 1997.

Active Friction Control Using Ultrasonic Vibration

V. Scherer, W. Arnold and B. Bhushan*

Fraunhofer Institute for Nondestructive Testing (IZFP)
University, Bldg. 37, D - 66123 Saarbrücken, Germany

** Computer Microtribology and Contamination Laboratory*
The Ohio State University, 206 W. 18th Ave.
Columbus, OH 43210-1107, U.S.A.

email: vscherer@izfp.fhg.de

Abstract

In this paper, the mechanism of reduction of friction by ultrasonic vibration has been studied. Experiments have been conducted by vibrating the specimen surfaces at ultrasonic frequencies. We found that vibration of the surfaces with amplitudes as low as a few nanometers evokes a super-lubricating effect when the frequency of vibration is high enough, here greater than 500 kHz. With a Pin-on-Flat (POF) test apparatus, a reduction of sliding (kinetic) friction was observed on surfaces vibrating laterally (in-plane), vertically (out-of-plane) or ellipsoidally (both in-plane and out-of-plane components). Sliding a sharp tip of an Atomic Force Microscope (AFM) against a vibrating surface revealed near-zero friction. Whereas friction is reduced, we find that a dramatic increase in wear can occur when contact resonances are excited. Thus, contact resonance can be used for ultrasonic machining.

1. Introduction

The friction between two sliding surfaces can be substantially reduced by vibrating the surfaces at ultrasonic frequencies (> 20 kHz) but the mechanism of ultrasonic lubrication, or referred to as sonolubrication, is not well understood. Even though several reports on this phenomenon have appeared in the past [1,2,3,4] it has not been extensively studied. It has not been used in industrial applications because the lubricity effect is rather weak at relatively high normal loads. With decreasing size of mechanical parts in microengineering and decreasing normal forces, ultrasound is being considered to actively control friction [5,6]. Objective of this study was to understand the mechanism of sonolubrication on the microscale.

B. Bhushan (ed.), Tribology Issues and Opportunities in MEMS, 463-469.

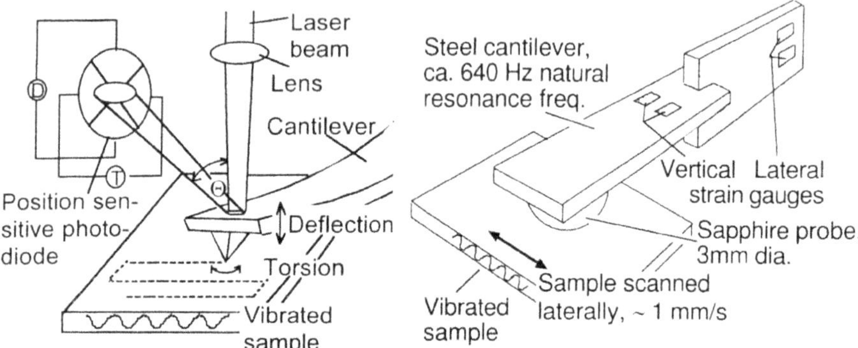

Figure 1: AFM with optical lever detection scheme, $F_N \approx 100$ nN, split-photodiode detectors for detecing vertical deflection (D) and torsion (T) of the cantilever

Figure 2: Pin-on-Flat test apparatus with strain gauge detection for lateral and vertical deflection of the cantilever beam, $F_N = 1$-500 mN

2. Experiments and Results

An atomic force microscope (AFM) was used for microscale studies and a classical pin-on-flat (POF) test apparatus was used for macroscale studies. Both AFM and POF trace a surface with a tip attached to a cantilever beam. Lateral forces acting between sample surface cause a torsion or lateral deflection of the cantilever which can be detected and converted into units of friction force. Since the normal forces F_N applied to the samples can be controlled, one is able to obtain the coefficient of kinetic friction $\mu = F_N/F_F$ with F_F being the friction force. In the AFM, a square pyramidal silicon nitride probe with a spherical shape tip of about 50 nm radius was used. In the POF, a sapphire ball of 3 mm diameter was used. Due to its sensitivity in the cantilever deflection (about 0.1 nm) by using the optical lever technique (Fig. 1), the AFM is capable to control normal forces in the nano-Newton range. When measured in ambient, capillary forces acting between probe tip and sample surface, limit the force control to about 100 nN. The normal loads applied with the POF can be as low as several mN, limited by the sensitivity of the strain gauge detection scheme (Fig. 2).

To study the effect of surface vibration, samples are insonified by being coupled to ultrasonic transducers. In-plane and out-of-plane surface oscillations at MHz-frequencies can be introduced by choosing shear or longitudinal wave transducers as indicated in Fig. 3a,b. For the lateral surface motion, we used a broadband 5 MHz center frequency shear wave transducer (V 157, Panametrics, Waltham, MA) with a piece of a silicon wafer mounted on top of the transducer as a specimen. The same way, a silicon wafer piece was mounted on a 5 MHz longitudinal wave transducer (V 1091, Panametrics) in order to obtain the vertical surface deflection. The voltage applied to the transducers was on the order of 1 V_{AC} and the surface deflections for both out-of-plane and in-plane are in the nm-regime. A third kind of surface deflection, containing both in-plane and out-of-plane components at the same time, Surface Acoustic Waves (SAW) can be generated. Splitfinger electrodes deposited on a piezoelectric substrate lead to an

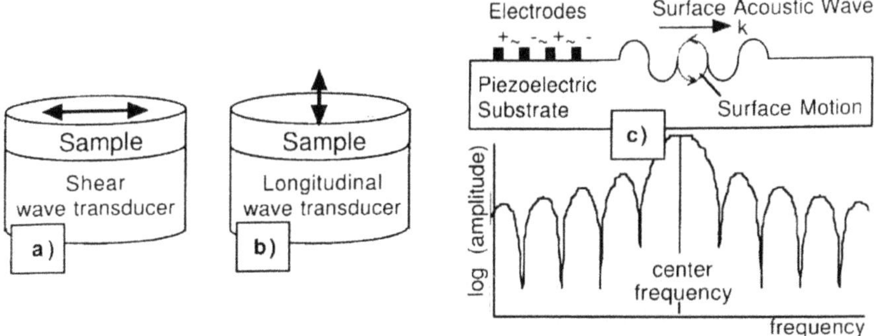

Figure 3: Sample insonification for a) in-plane, b) out-of-plane displacements and c) generation of surface acoustic waves with amplitude response (schematic) as a function of frequency

alternating mechanical stress field in the PZT material and elastic surface waves evanate from the electrode pattern and travel along the surface. The electrode spacing determines the wavelength and therewith the centre frequency of the SAW generated by these so-called Interdigital Transducers (IDT) [7]. The surface amplitude on an IDT follows a $[(\sin f)/f]^2$ function with f being the frequency of the ac voltage applied to the alternating electrodes, as indicated in Fig. 3c.

The "macro"-scale friction was measured using the POF test apparatus where the amount of lateral bending of the cantilever is measured using the strain gauge beams to

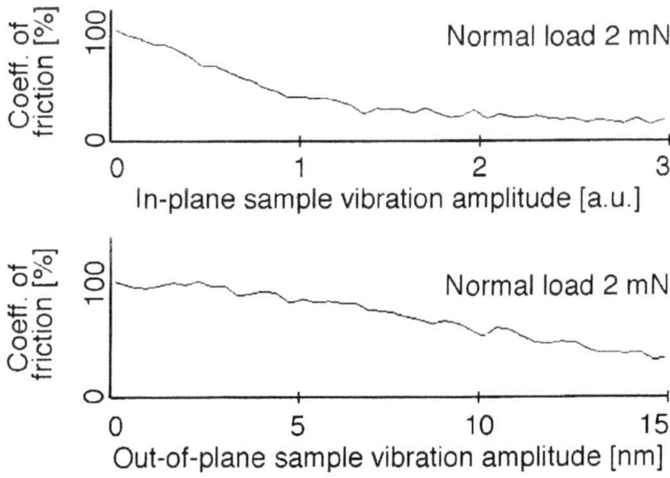

Figure 4: Reduction of friction, observed with the Pin-on-Flat test apparatus at a normal load of $F_N = 2$ mN while surface is vibrated at about 1 MHz. Upper graph with in-plane and bottom graph with out-of-plane vibration. Note that in-plane vibration amplitude is of arbitrary scale. 100 % coefficient of friction corresponds to 0.2.

obtain friction force. The coefficient of friction is calculated by dividing the friction force with an applied normal force. The coefficient of friction measured with the POF apparatus on a static i.e., non-vibrating surface was tyipcally 0.2. For simplicity, the coefficient of friction in this study is given as a fraction of the value of the static surface i.e., 100 %. Fig. 4 shows the reduction of the friction coefficient, measured with a POF when the surface is vibrated in-plane (top graph) and out-of-plane (bottom graph) at a frequency of about 1 MHz. The out-of-plane surface amplitude was measured using a path-stabilized Michelson-Interferometer [8]. In-plane vibrations are not easily measurable and were not quantified, hence, the scale is arbitrary. We assume that, in our studies, no pure in-plane motions were excited and parasitic out-of-plane components are superimposed. But we expect the in-plane component to be on the same order i.e., 10 nm max. For both experiments a substantial reduction of friction can be observed, up to a fraction of 20 % in case of in-plane and approximately 40 % for out-of-plane vibrations.

In order to understand the mechanism of sonolubrication, quasi-single asperity experiments were carried out employing an AFM. The results are shown in Fig. 5a,b. Again, we observe a reduction of friction, obtained by measurement of the amount of

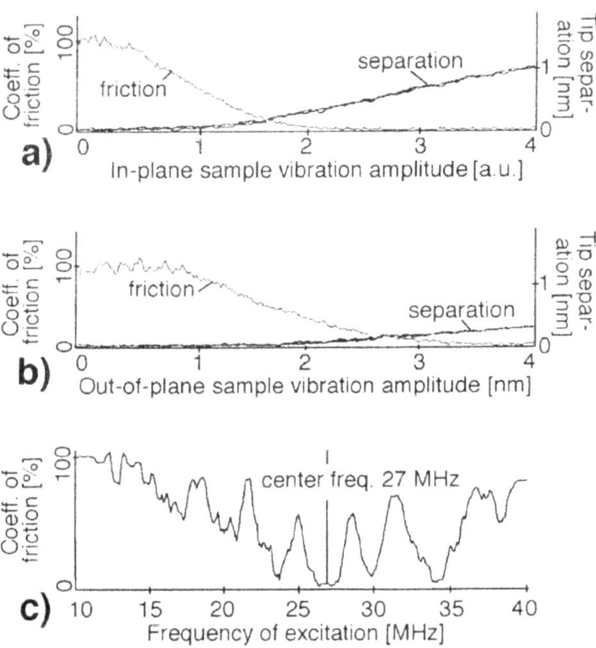

Figure 5: Reduction of friction, measured with AFM at a normal loads of $F_N = 100$ nN and average tip separation as a function of surface amplitude on polished silicon subjected to a) in-plane and b) out-of-plane vibrations at about 1 MHz and c) friction as a function of frequency applied to an IDT. Note that in-plane vibration amplitude is of arbitrary scale.

cantilever torsion, when the surface is vibrated. Note that the normal forces are now as low as 100 nN, about four orders of magnitude lower than with the POF test apparatus. Simultaneously with the friction data, the tip-sample separation, hereafter referred as to lift-off, was recorded with the split photodiode, as the surface amplitude was increased linearly. There is a non-linear increase in lift-off, with both in-plane and out-of-plane surface vibrations. In the experiments, the force exerted by the cantilever was held constant to 100 nN. To introduce a lift-off, there must be an additional repulsive force resulting from vibrating the surface. In the case of vibrating the surface out-of-plane, there are two sources of the additional force. The first one we call the mechanical diode effect. A cantilever beam system can be excited to vibrations only within a range of frequencies. In between the resonances, the cantilever cannot follow the surface motion [9]. Hence, the average cantilever height or lift-off height, is at a distance d_e from the surface, which is given by

$$d_e = \frac{A}{2} - d_p - d_{attr} \quad (1)$$

where A is the surface vibration amplitude and d_p the penetration depth of the tip into the sample and d_{attr} the deflection due to attractive forces. Recent studies have shown that d_p is a function of cantilever tip geometry, elastic properties of tip and surface material [9,10]. Furthermore, attractive forces such as adhesive and capillary forces will also decrease the lift-off by d_{attr}. Second, any two surfaces, separated by a viscous fluid (liquid or gaseous) and being in relative motion build up a so-called squeeze-film i.e., a fluid film can develop between two bodies in relative motion and the film acts like a lubricant where friction decreases linearly with increasing film thickness. The squeeze-film thickness is a function of the viscous properties of the fluid, geometry of the surfaces, normal load, velocity and direction of motion. In the present case, we assume water adsorbed at the surface to be the liquid responsible for the squeeze effect. The presence of a squeeze-film would lead to a lift-off and, as a consequence, to a reduction of friction. We can observe in Fig. 5a,b that reduction of friction occurs with no obvious lift-off. We interpret this with the presence of squeeze-film lubrication in the case of out-of-plane vibration and elastohydrodynamic lubrication (EHL) for in-plane vibrations [11]. Significant lift-off occurs at greater surface amplitudes and we believe that the tip jumps intermittently out of contact. This regime has been described by Hess and Soom as a contact resonance effect, nonlinear with the tip-sample separation, leading to a reduced effective contact area and thus, reduced friction [12]. It strongly depends on the cantilever stiffness and would not have relevance for applications other than cantilevers on vibrating surfaces.

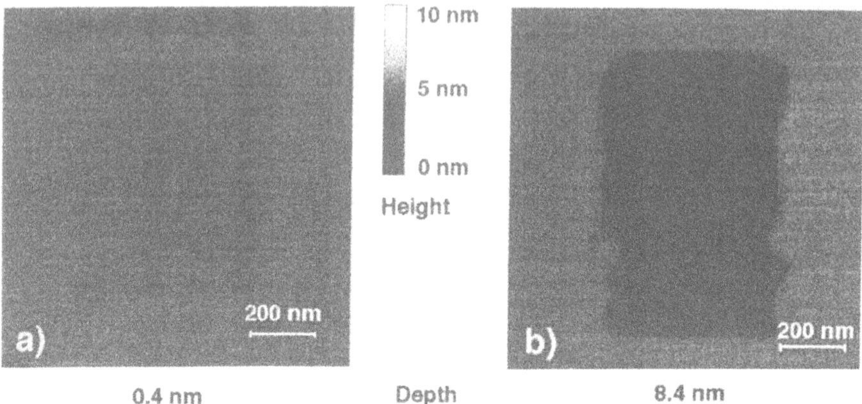

Figure 6: Wear mark on a mica surface created at a normal load of about 100 nN on an a) static surface and b) surface vibrating close to a contact resonance, here 1.34 MHz. Wear depth increases from 0.4 nm to about 8 nm when the cantilever is at resonance.

One might wonder about the freuqency-limit in surface vibrations to cause sonolubrication. Friction data on an IDT with 27 MHz centre frequency showed a reduction of friction when the surface amplitude was on the order of 1 nm (out-of-plane component). As we sweep the frequency from 10 to 40 MHz, we clearly observe the $[(\sin f)/f]^2$ pattern of the IDT amplitude from the friction data. Thus, we can say that higher harmonics of cantilever resonances, of which severals are contained within the frequency varies, do not play a major role any more. That indicates that lubrication is performed not by contact resonance effects but by the build-up of a lubricating film.
So far, we have not considered, that the cantilever can resonate while it is in contact with the surface. However, when the surface vibration frequency matches contact resonance, the cantilever can be excited to vibrations with amplitudes much greater than the surface amplitude itself. In this case, the cantilever performs chaotic oscillations and can apply temporarily high peak-loads to the sample. An example is given in Fig. 6 where a wear mark was created on a freshly cleaved mica surface. The wear depth at a normal load of 100 nN reveals a depth of 0.4 nm, corresponding to a single atomic layer or maybe a surfactant monolayer. With the same average normal load, (100 nN) but a surface vibrating out-of-plane at a contact resonance freuqency (1.34 MHz), the wear depth is increased by the twentyfold to about 8 nm. We believe, however that not only the resonating tip, but also high impact velocity leads to increased wear, an effect observed and theoretically described earlier [13] and applied in ultrasonic machining [14].

3. Summary

The AFM studies have shown that there are generally two reasons for sonolubrication. A lift-off caused by the mechanical diode effect of the cantilever beam leads to non-steady contact between tip and sample when the surface is vibrated out-of-plane. The friction is reduced due to unsteady contact. But at lower surface amplitudes or, when the surface is vibrated in-plane, a squeeze-film or an elastohydrodynamic lubrication film can support the tip and reduce kinetic friction. If, however, the tip-cantilever-sample system is driven at resonance i.e., when the surface is vibrated at a particular frequency, increased plastic deformation and wear can be introduced.

4. Acknowledgment

The authors thank D. Hurley and J.W. Wagner of Johns Hopkins University at Baltimore for supporting the interferometric measurements. Technical discussions with U. Rabe of the IZFP and B.J. Hamrock of The Ohio State University are gratefully appreciated. VS acknowledges financial support from German Federal Ministry of Education and Research, DAAD stipend HSP III.

5. References

1. Fridman, H.D. and Levesque, P.: Reduction of static friction by sonic vibration, *J. Appl. Phys* **30** (1959), 1572-1575
2. Seireg, A. and Weiter, E.J.: Frictional interface behavior under dynamic excitation, *Wear* **6** (1963), 66-77
3. Pohlmann, R. and Lehfeldt E.: Influence of ultrasonic vibration on metallic friction, *Ultrasonics* (1966), 178-185
4. Lenkewicz, W.: The sliding friction process - effect of external vibrations, *Wear* **13** (1969), 99-108
5. Frederick, J.R.: *Ultrasonic Engineering*, Wiley, New York, 1965
6. Tam, A.C. and Bhushan, B.: Reduction of friction between a tape and a smooth surface by acoustic excitation, *J. Appl. Phys.* **61** (1987), 1646-1648
7. Oliner, A.A.: *Acoustic Surface Waves*, Springer-Verlag, Berlin, 1978
8. Wagner, J.W.: Optical detection of ultrasound, in W.P. Mason (ed.): *Physical Ultrasound,* Academic Press, **19** (1990), 201
9. Rabe, U., Janser, K., and Arnold, W.: Vibrations of free and surface-coupled atomic force microscope cantilevers: theory and experiment, *Rev. Sci. Instrum.* **67** (1996), 3281-3293
10. Hirsekorn, S., Rabe, U., and Arnold, W.: Theoretical description of the transfer of vibrations from a sample to the cantilever of an atomic force microscope, *Nanotechnology* **8** (1997), 57-66
11. Hamrock, B.J.: *Fundamentals of Fluid Film Lubrication*, McGraw-Hill, New York, 1994
12. Hess, D.P. and Soom, A.: Unsteady friction in the presence of vibrations, in I.L. Singer and H.M. Pollock: (eds.) *Fundamentals of Friction: Macroscopic and Microscopic Processes*, Kluwer Academic Publishers, Dordrecht, Netherlands, 1992
13. Nayak, P.R.: Contact vibrations, *Journal of Sound and Vibration* **22** (1972) , 297-322
14. Markov, A.I.: in E.A. Neppiras (ed.): *Ultrasonic Machining of Intractable Materials*, Iliffe Books Ltd, London, 1966

CONTACT RESISTANCE MEASUREMENTS AND MODELING OF AN ELECTROSTATICALLY ACTUATED MICROSWITCH

Sumit Majumder*, N.E. McGruer*, Paul M. Zavracky*, George G. Adams**, Richard H. Morrison*, and Jacqueline Krim***
*Department of Electrical and Computer Engineering
**Department of Mechanical, Industrial and Manufacturing Engineering
***Department of Physics
Northeastern University, 360 Huntington Avenue, Boston, MA 02115

1. Abstract

Micromechanical switches have been fabricated in electroplated nickel using a four-level surface micromachining process. The simplest devices are configured with three terminals, a source, a drain, and a gate and are 30 μm wide, 1 μm thick, and 65 μm long. A voltage applied between the gate and the source closes the switch, connecting the source to the drain. Devices operate for more than 10^9 cycles in a nitrogen ambient before failure and have an initial contact resistance of less than 50 mΩ. The threshold voltages for these devices are between 30 V and 200 V. Long lifetimes are achieved with a current of 30 mA. The breakdown (stand-off) voltage between the source and the drain is greater than 100 V and the off-impedance exceeds 10^{12} Ω. The modeling of these switches includes a structural model of the beam and a contact resistance model for the electrical contact. Preliminary contact resistance measurements are presented and compared with the modeled contact resistance characteristics.

2. Introduction

Micromechanical relays are receiving increased attention recently as our community begins to realize the benefits of integration of micromechanical structures with electronics [1]-[5]. Unlike other micromechanical devices, the microrelay performs a purely electronic function. Two modes of operation can be envisioned. In one, the microrelay is a four terminal device [6]. Two terminals are used for actuation while the other two are switched. This mode is equivalent to a conventional electromagnetic relay and hereafter will be referred to as a microrelay. The second configuration is as a three terminal device. In this configuration, an electrostatic field applied between the beam (source) and the gate actuates the device. Switch closure shorts the beam tip to its counter electrode (the drain) thereby electrically connecting the source and drain. This second configuration will be referred to as a microswitch. When configured as microswitches, these devices are three terminal field effect switches.

The microrelay has obvious advantages over conventional relays in being smaller and consuming less power. However, what is most attractive is that the microrelay can be integrated with other devices on a single die. Micromachined relays can be

B. Bhushan (ed.), Tribology Issues and Opportunities in MEMS, 471-480.

fabricated in large numbers on a single die which may contain other electronic devices. The lack of high temperature steps in the fabrication process described here means that the relays can be included as post-process additions to a conventional integrated circuit. Complex switching arrays and devices designed to handle high frequency signals with low insertion loss are natural extensions of the work described here.

In addition to replacement of conventional relays, microswitches and microrelays have the potential to replace traditional solid state devices in significant market areas. For instance, a micromechanical switch has potentially a much broader operating temperature range than silicon electronic devices. Both low and high temperature applications may be considered. If long lifetimes (> 10^{13} cycles) can be established for these devices, they may be used to perform simple electronic logic functions in harsh environments. In this paper, we report our results on the development of micromechanical switches.

3. Fabrication Procedure

The process sequence that is illustrated in Fig. 1 for a three terminal device begins with the preparation of an insulating substrate, typically glass or silicon coated with 1 μm of SiO_2. A thin layer of chromium (200Å) followed by 2000Å of gold is sputter deposited. These layers are photolithographically patterned to form the source, gate and drain electrodes and the interconnection to bond pads. A sacrificial layer of copper is deposited next and patterned twice. The first patterning is used to define the contact regions which are etched isotropically into the copper. The feature size, which may be as small as one micron, expands during etching to form a hemispherical indentation in the copper sacrificial layer. The radius of the hemispherical feature is less than the thickness of the copper layer. Therefore a layer of copper remains over the entire substrate. A second patterning defines the beam base vias. These vias are etched completely through the copper layer to the Cr/Au source electrode. The wafer is patterned for the fourth and final time to define the beams. Gold is first deposited to form the contact, then the beams are electroplated in a commercial nickel plating solution. Care is taken to minimize stress by controlling the average current and duty cycle in our pulsed plating system. The additional photomasking step and insulator layer that are required to make a microrelay are not shown in the figure. The sacrificial layer is then etched away and the devices are released as described previously in [7]. SEM micrographs of devices fabricated using the process described above are shown in Fig. 2.

4. Principle of Operation

Fig. 3 shows the geometry of the basic switches and relays that have been fabricated. When a voltage is applied to the gate electrode, the beam is pulled down by the electrostatic force until the switch closes. When the gate voltage is removed, the restoring force of the beam returns it to its original position. In this section, a simple numerical model is developed.

Ignoring fringing fields, the electrostatic force per unit area acting on the beam is given by [8]

$$f_e(x) = -\frac{\varepsilon_o \cdot V^2}{2(d_o - v(x))^2}, \qquad a < x < b \tag{1}$$

Deposit first metal and define the source, gate and drain regions.

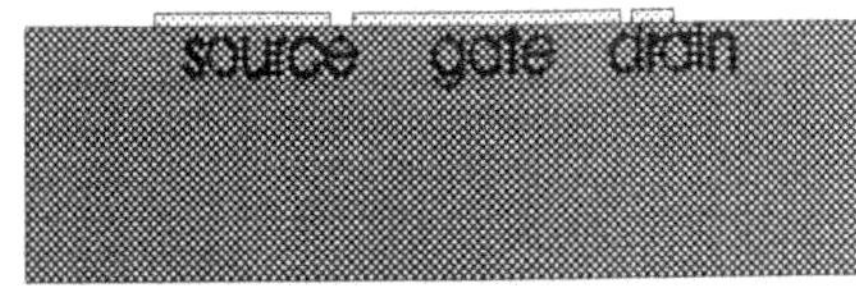

Deposit sacrificial layer and pattern tip region.

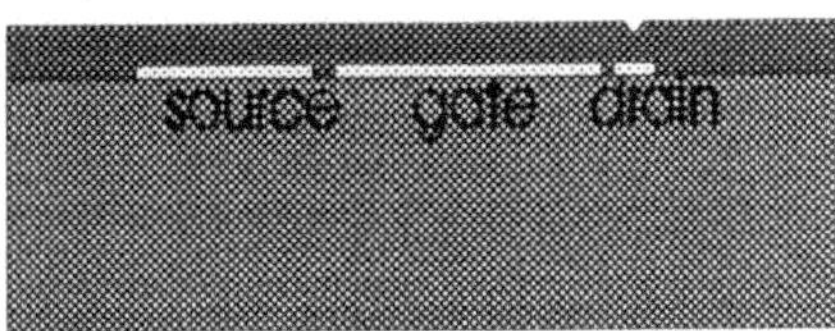

Pattern sacrificial layer and open via to source metal.

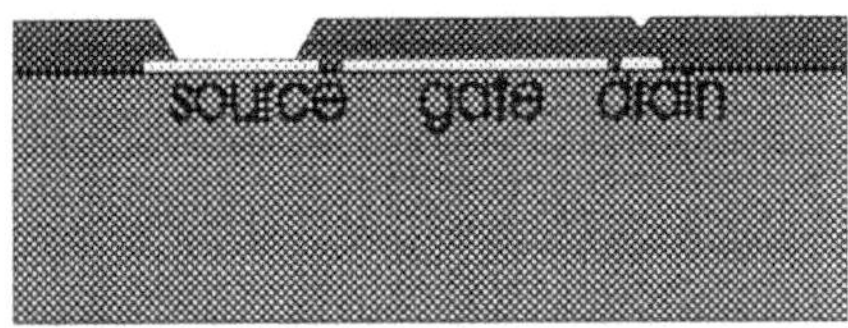

Pattern to define beam shape. Deposit tip contact material and plate beam material.

Strip resist and remove sacrificial layer.

Figure 1. Process flow for the fabrication of a micromechanical switch

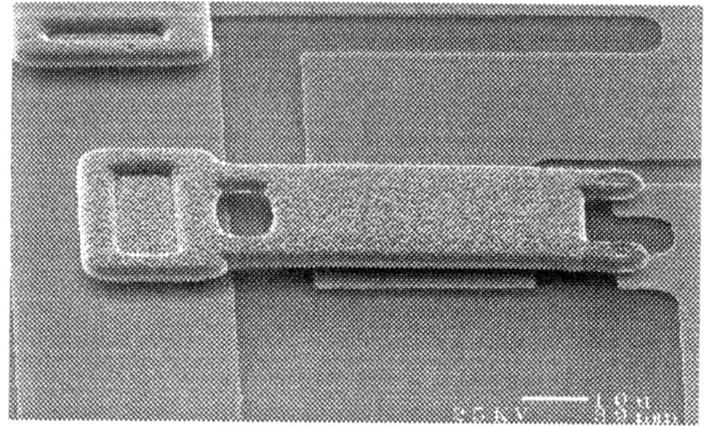

Figure 2. SEM micrograph of a microswitch. The scale bar corresponds to 10 microns.

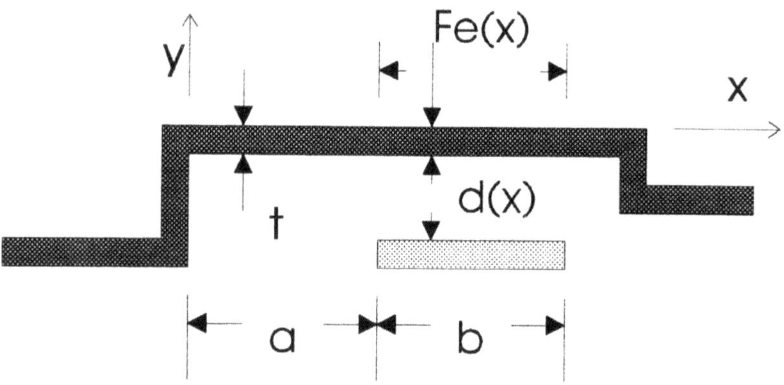

Figure 3. A cross-sectional view of a microswitch acted upon by an applied electric field, $F_e(x)$.

where V is the voltage applied to the gate, d_o is the initial spacing between the beam and the gate, ε_0 is the permitivity of free space, and $v(x)$ is the beam deflection at a position x along the beam axis. The beam deflection is obtained by solving [9]

$$\frac{d^2v(x)}{dx^2} = \frac{M(x)}{EI(x)}, \qquad v(0) = 0, \quad \frac{dv}{dx}(0) = 0, \tag{2}$$

where $M(x)$ is the internal bending moment, E is the Young's modulus, and $I(x)$ is the second moment of cross-sectional area. Eqns. (1) and (2) are solved iteratively until convergence is achieved. This process is repeated at different voltages to obtain a deflection profile for the beam $v(x)$ for typical beam parameters [10]. The numerical solution shows an instability in the position of the beam at a critical voltage.

5. Contact Resistance Model

Modeling of the contact resistance can be broadly divided into 3 successive steps - determining the contact force as a function of the applied gate voltage; determining the distribution and sizes of the areas in contact, as a function of the contact force; and determining the contact resistance as a function of the distribution and sizes of the contact areas. Based on this approach, we develop a simple contact resistance model in this section. More details can be found in [11].

The contact force can be determined as a function of the gate voltage using a simple finite difference analysis of the cantilever beam in its closed position [12]. The cantilever beam is assumed to be perfectly clamped at its fixed end, and simply supported at its (formerly) free end. The beam shape is calculated as a function of the electrostatic force, and the new electrostatic force determined as a function of the beam shape iteratively, until the solution converges.

The determination of the nature of the contact area at the interface necessitates a model of the surface profiles of the contact bump and of the drain electrode. On the basis of SEM micrographs of the contact surfaces (Fig. 4), and STM scan profiles of the deposited films, we conclude that the surface of the contact bump is much rougher than the surface of the bottom gold electrode, so that the bottom gold surface can be considered as a flat plane. The SEM and STM scans also suggest that the contacting

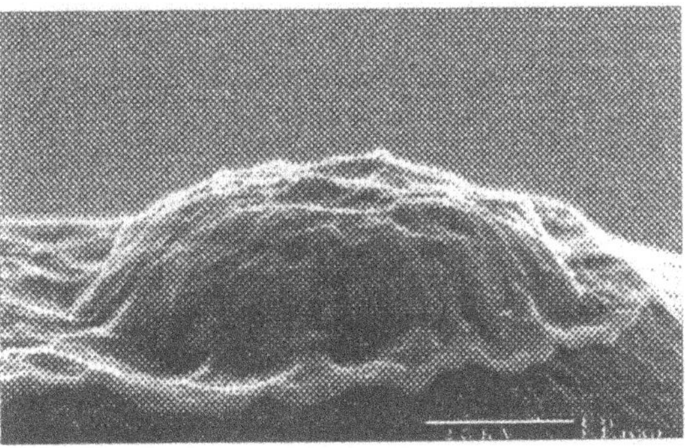

Figure 4. SEM micrograph of a contact bump. The scale bar corresponds to 1 micron.

asperities on the contact bump are likely to be few in number (less than 10), and have tip radii of 50-200 nm.

A rigorous modeling of the surface profile of the contact would require an accurate surface profile of the contact bump. However, it is instructive to study the contact resistance characteristics on the basis of some simple surface profile models. Accordingly, we consider the following 3 surface profiles, each consisting of a certain number of asperities of different heights, but with identical spherical tips of 100 nm radius, indenting a flat plane. Profile "A" consists of a single asperity of height 0.25 μm. Profile "B" consists of 3 identical asperities approximately 0.4 μm apart, of heights 0.25 μm, 0.23 μm, and 0.21 μm. Profile "C" consists of 5 asperities approximately 0.2 μm apart, and heights 0.25 μm, 0.23 μm, 0.21 μm, 0.19 μm and 0.17 μm. It has been reported in the literature that thin films, in general, have higher values of hardness than the corresponding bulk material. We use a value of 2 GPa for the hardness of gold, as reported in [13].

The number of contact spots and their sizes can be determined on the basis of the elastic-plastic model for contact of a rough surface with a flat plane, proposed by Chang, Etsion and Bogy [14]. This model is a refinement of the well-known "asperity based model" introduced by Greenwood and Williamson [15], in which a rough surface is represented by a collection of asperities with identical end radii, whose heights have a statistical distribution. The asperities are assumed to be independent of each other, in that the load acting on one asperity does not affect the deformation of another. The area of contact for each asperity in the Greenwood-Williamson model is calculated from the Hertz theory of elastic deformation. The model proposed by Chang, *et al.* calculates the elasto-plastic deformation of an asperity. The plastic deformation is determined on the basis of volume conservation of a certain control volume of the asperity. In the following paragraphs, we briefly explain the equations governing the above model, and then apply the model to the previously defined surface profiles.

Let the moduli of elasticity of the contacting surfaces be denoted by E_1 and E_2, respectively, and their Poisson's ratios by ν_1 and ν_2 respectively. The effective modulus of elasticity is defined as

$$\frac{1}{E'} = \frac{1-\nu_1^2}{E_1} + \frac{1-\nu_2^2}{E_2} \tag{3}$$

For an elastically deformed asperity, the radius of the contact spot (r), is related to the force acting on the asperity (F), by

$$r = \left(\frac{3FR_t}{4E'}\right)^{1/3} \tag{4}$$

where R_t denotes the summit radius of curvature of the asperity. Also, the contact radius r is related to the deformation of the asperity (α), by

$$r = \sqrt{R_t \alpha} \tag{5}$$

The onset of plastic yielding is assumed to occur when the maximum pressure at the contact interface exceeds $0.6\ H$, where H is hardness of the contacting material. The deformation of the asperity at this point is given by

$$\alpha_c = \left(\frac{0.3\pi H}{E'}\right)^2 R_t \tag{6}$$

In the plastic region, the average pressure acting on an asperity is assumed to be *0.6 H,* so that the contact force on the asperity is

$$F = 0.6\pi H r^2 \tag{7}$$

The contact radius is determined in terms of the deformation (α) by invoking conservation of volume of the deformed asperity [14], *i.e.*

$$r = \sqrt{R_t \alpha \left(2 - \frac{\alpha_c}{\alpha}\right)}, \qquad \alpha > \alpha_c \tag{8}$$

Hence, for a given asperity deformation, the force on each asperity as well as the radius of the corresponding contact spot can be determined. Now consider a rough surface with N asperities, each with an end radius of curvature R_t, and heights $z_1 > z_2 > z_i > z_N$. Let the separation between the reference planes be d for a given contact force F, such that $z_n > d > z_{n+1}$. Then asperities 1,...,n come into contact. The deformation of asperity i is given by

$$\alpha_i = z_i - d \tag{9}$$

For a given separation between the reference planes, the force on each asperity, and the radii of the corresponding contact spots can be obtained using the previously stated equations (see [11] for details).

The remaining step is to determine the contact resistance, given a certain number of contact spots of known radii and separation from each other. We first consider the contact resistance of a single spot of a given radius r, separating 2 semi-infinite bodies of resistivity ρ. If the radius r is smaller than the electron mean-free-path ℓ_e of the material, the resistance is given by the Sharvin model [16]

$$R = \frac{4\rho\ell_e}{3\pi r^2} \tag{10}$$

If the radius is larger than the mean-free-path length, the resistance is the Maxwell spreading resistance [17]

$$R = \frac{\rho}{2r} \tag{11}$$

If the contact force is large enough for the creation of more than one contact spot, then the Maxwell spreading resistance model is used to calculate the effective contact resistance (the error arising from this approximation is small [11]).

A lower bound can be obtained for the net contact resistance by assuming that the contact spots are independent and conduct in parallel (which is equivalent to the exact solution when the radii of the contact spots are small compared to the separation between the spots),

$$R_{lb} = \frac{\rho}{2\sum_{i=1}^{n} r_i} \tag{12}$$

An upper bound can be obtained on the net contact resistance by considering the radius r_{eff} of an equivalent circular spot of area equal to the total area of all the conducting spots combined. This upper bound is equivalent to the exact solution when all the conducting spots become large enough to merge into a single conducting spot

$$R_{ub} = \frac{\rho}{2r_{eff}} \tag{13}$$

It is now possible to determine the contact resistance as a function of the contact force. Fig. 5 shows the variation of contact resistance with the gate voltage for the 3 surface profiles "A", "B", and "C". Expectedly, all the contact resistance characteristics are identical at low gate voltages (low contact forces), when a single asperity is in contact. The single asperity characteristic, and the upper bounds on contact resistance for the multi-asperity profiles, are very similar at higher gate voltages as well. This behavior is because if all the asperities in contact deform plastically, the total contact area is independent of the number of asperities.

6. Measurements

Preliminary contact resistance measurements were carried out using switch geometries which permitted a four-point probe measurement of the contact resistance. The contact resistance was measured over five successive cycles. During each cycle, the gate voltage was increased in steps up to a maximum value, and the contact resistance recorded at each step. The gate voltage was then decreased in steps, and the contact resistance recorded, until the switch opened. The current passing through the contact was 0.1 mA. Fig. 6 shows the measured contact resistance characteristics of a microswitch during the first and the fifth cycle. Comparison with the modeled characteristics indicates that the model underestimates the contact resistance significantly. The discrepancy in threshold voltage between the modeled characteristics and measurements is probably due to a residual curvature in the cantilever. A noteworthy feature is the hysteretic nature of the contact resistance *vs.* gate voltage characteristics - the contact resistance is significantly lower and varies less sharply when the gate voltage is being reduced than when it is being increased. The hysteretic nature of the characteristics suggests the presence of adhesive forces at the interface of the contact. Another effect that may need to be taken into consideration is the fact that the bottom electrode is a thin film (0.1 μm of gold) on an insulating SiO_2 substrate (at present, the model assumes the contacts to be semi-infinite bodies). Second-order effects that have not yet been studied include the effect of insulating films or impurities at the contact interface, time-dependent effects, and the effect of resistive heating when current passes through the contact.

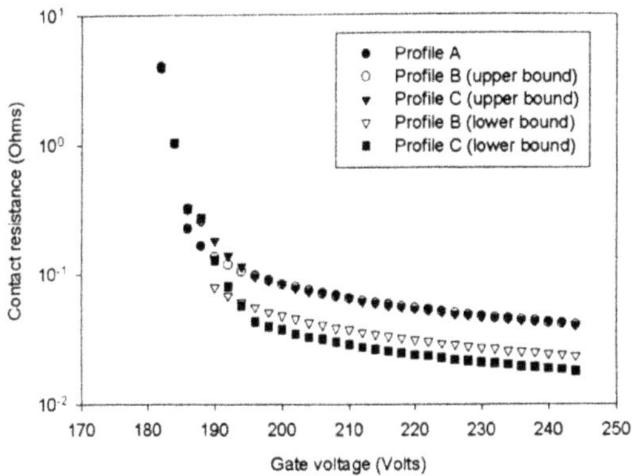

Figure 5: Modeled variation of contact resistance with gate voltage

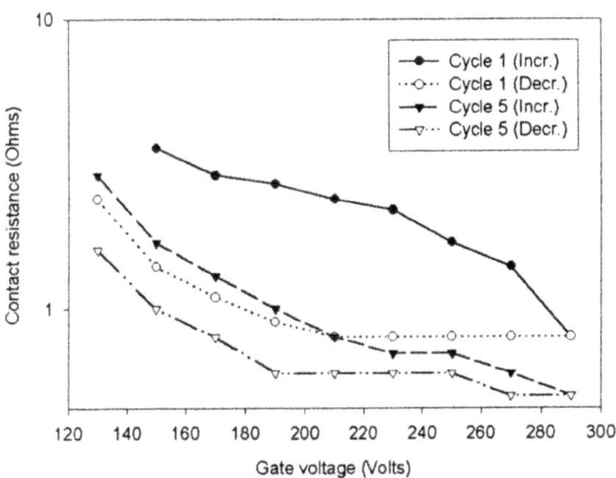

Figure 6: Measured variation of contact resistance with gate voltage.

7. Conclusions

Contact resistance modeling has been performed in order to predict contact force in terms of gate voltage, contact area from contact force, and contact resistance from contact area. The final outcome, contact resistance *vs.* gate voltage has been determined for various conditions. Contact resistance measurements demonstrate the same qualitative trends, but show significant quantitative differences with the model. These discrepancies are no doubt due to the effects of initial beam curvature, adhesive forces in the contact region, and insulating (impurity) films at the contact interface.

8. Acknowledgement

The authors would like to thank Analog Devices for their continued support of the microswitch program at Northeastern University. In particular, we are grateful to Curtis Davis and Richard Payne for their advice and support. We would also like to acknowledge the valuable contributions of Greg Jenkins and Weilin Hu in the fabrication of the devices.

9. References

1. Sakata, M. (1989) An Electrostatic Microactuator for Electro-Mechanical Relay, *Proc. IEEE MEMS Workshop '89,* Salt Lake City, UT, pp. 149-151.
2. Hosaka, H., Kuwano, H., and Yanagisawa, K. (1993) Electromagnetic Microrealys: Concept and Fundamental Characteristics, *Proc. IEEE MEMS Workshop '93*, Fort Lauderdale, FL, pp. 12-17.
3. Drake, J., Jerman, H., Lutze, B., and Stuber, M. (1995) An Electrostatically Actuated Micro-Relay, *Transducers '95,* Eurosensors IX, Stockholm, Sweden, **2**, pp. 380-383.
4. Gretillat, M., Thiebaud, P., Linder, C., and de Rooij, N. (1995) Integrated Circuit Compatible Electrostatic Polysilicon Microrelays, *J. of Micromech. Microeng.*, **5**, pp. 156-160.
5. Yao, J. and Chang, M. (1995) A Surface Micromachined Miniature Switch for Telecommunications Applications with Signal Frequencies from DC Up to 4 GHz, *Transducers '95, Eurosensors IX,* Stockholm, Sweden, pp. 384-387.
6. Petersen, K.E. (1979) Micromechanical Membrane Switches on Silicon, *IBM J. of Research and Development,* **23**, pp. 376-385.
7. Zavracky, P.M. , Adams, G.G., and Aquilino, P.D. (1995) Strain Analysis of Silicon-on-Insulator Films Produced by Zone Melting Recrystallization, *J. of Microelectromechanical Systems,* **4** , pp. 42-48.
8. Veidner, R.T. and Sells, R.L. (1965) *Elementary Classical Physics,* **2**, Allyn and Bacon, Inc., Boston, MA.
9. Fuller, C.E., Johnston, W.A. (1919) *Applied Mechanics*, **2**, John Wiley & Sons, Inc., New York.
10. Zavracky, P.M., McGruer, N.E., and Majumder, S. (1997) Micromechanical Switches, *IEEE/ASME J. of Microelectromechanical Systems*, **6**, pp. 3-9.
11. Majumder, S., McGruer, N.E., Zavracky, P.M., Adams, G.G., Morrison, R.H., and Krim, J. (1997) Measurement and Modeling of Surface Micromachined, Electrostatically Actuated Microswitches, *Transducers '97, 1997 International Conference on Solid State Sensors and Actuators,* Chicago, pp. 1145-1148.
12. Majumder, S. (1997) *Electrostatically Actuated Micromechanical Switches*, Master of Science Thesis, Depeartment of Electrical and Computer Engineering, Northeastern University, Boston, MA.
13. Oliver, W.C. Hutchings, R., and Pethica, J.B. (1986) Measurement of Hardness at Indentation Depths as Low as 20 Nanometers, *Microindentation Techniques in Materials Science and Engineering,* ASTM STP 889, Philadelphia, pp. 90-108.
14. Chang, W.R., Etsion, I., and Bogy, D.B. (1988) An Elastic PlasticModel for the Contact of Rough Surfaces", *ASME J. of Tribology,* **109**, pp. 257-263.
15. Greenwood, J.A. and Williamson, J.B.P. (1966) Contact of Nominally Flat Surfaces, *Proceedings of the Royal Society (London),* **A295,** pp. 300-319.
16. Jansen, A.G.M., van Gelder, A.P. and Wyder, P. (1980) Point Contact Spectroscopy in Metals, *J. of Physics,* **C13**, pp. 6073-6118.
17. Holm, R. (1967) *Electric Contacts*, Springer Verlag (New York).

METROLOGY FOR MEMS TRIBOLOGY

Dr. NORM V. GITIS
Center for Tribology
625-A Clyde Avenue, Mountain View, CA 94043, USA

ABSTRACT

Microtribology is a new fast-growing field which combines a number of related disciplines, like physics of solids, physical chemistry and chemical physics, mechanics and electricity, etc. It is dealing with surfaces and surface properties (as versus bulk material properties), light loads and masses (as versus heavy loads) in application to micromechanical devices, as well as magnetic recording, surface/atomic force microscopy, surface microfinish, etc.

Theoretical and experimental microtribology data for the friction phenomenon on microscale is presented. Performance of numerous micromechanical devices is limited by wear and stiction. The model of wear combines adhesive, abrasion, and fatigue processes, as well as their dependencies on surface roughness and lubrication. The model of stiction includes processes of deformation, adhesion, and viscous flow, all related to perpendicular micro-displacements in the interface. The latter is critical for the modeling of friction-induced auto-oscillations, closely related to stiction. These models may serve as the basis for designing low-wear, stiction-free devices.

To verify the models and to study the friction, adhesion, and wear processes in depth, a new bench-top micro-tribometer mod. UMT has been designed. It has precision rotary and linear motions of both upper and lower specimens, with in-situ measurements of friction force and coefficient, adhesion, wear and micro-wear, acoustic emission and electrical resistance of contact. All the force, acoustics, and electrical sensors and amplifiers are of the original design, which ensures the high accuracy and low drift. It can work in pin-on-disc, ball-on-disc, flat-on-flat, 3-ball, 4-ball, screw-in-nut, and many other modes of operation, reproducing all types of lubrication conditions, from dry and boundary friction up to hydrodynamic lubrication. This apparatus can accommodate a wide variety of sample sizes (from a 1-mm ball up to an 8-inch wafer), sample shapes (round, rectangular, triangle, etc.) and materials (metals, ceramics, plastics, paper, etc.). Its variable speed can be as low as micron per second, which is critical for the studies of stiction and stick-slip mechanisms. It has fully automated motor-control and data-acquisition.

In this work, new results have been obtained on the micro-tribometer with greases, solid lubricants and unlubricated contacts. They have confirmed the theoretical models and have established the basis for building a database of microtribology data for MEMS.

B. Bhushan (ed.), Tribology Issues and Opportunities in MEMS, 481-486.

1. INTRODUCTION

Microtribology is a new fast-growing field which combines a number of related disciplines, like physics of solids, physical chemistry and chemical physics, mechanics and electricity, etc. It deals with surfaces and surface properties (versus bulk material properties), light loads and masses (versus heavy loads) in application to magnetic recording and micromechanical devices, as well as surface/atomic force microscopy, bio-tribology, surface microfinish, etc.

A magnetic head-disc interface is of prime importance in modern micro-tribology. Even the conventional interface with 50 to 100 nm of a nominal air-bearing film between the head and the disc has numerous wear and stiction problems when periodically shuts-down and starts-up. The tendency toward lowering the head-disc separation to near-contact (under 50 nm) and in-contact (under 2 nm) recording necessitates a very smooth and uniform disc surface. Conventional recording requires extremely low wear rates of less than 3 nm/year for either head (pole tips) or disc surface (overcoat). In other frictional pairs this is usually achieved by making smooth well-protected (thick overcoats and lubricant layers) surfaces. It has not been applied, however, to the head-disc interface, because thicker overcoats and lubricants increase the magnetic head-disc separation, reducing the signal, and also because both liquid lubricant and smooth discs increase stiction by contributing a meniscus and Van-der-Waal's forces, respectively.

So, the head-disc interface should be treated in a special way. Several techniques have been proposed, for instance: lubricant bonding to the disc, reduction in slider size and mass, load reduction, hydrodynamic fluid lubrication. In this work, the advanced test equipment for both reliability and micro-tribology testing have been developed and utilized to confirm and theoretical considerations, and their practical applications for very smooth surfaces.

2. RELIABILITY TESTING

Macrogeometry of the heads was measured with optical digital phase-shifting interferometry.
Microgeometry (morphology) of the disc and slider surfaces was characterized with both an atomic force microscope (which gives high resolution, but on very limited surface area) and a surface stylus profilometer (which has less resolution, but makes longer traces).
Head-disc separation was measured using three-beam monochromatic interferometry (with limited resolution and on ideal glass discs), capacitance (higher resolution, but requires a sensor embedded onto the head), and magnetic read-back signal (high resolution, but cannot measure roll and pitch).

Despite the availability of the above and other related measurement procedures, the ultimate technique to estimate the head-disc reliability is a contact start-stop (CSS) test. In this work, advanced CSS testers have been designed and utilized. To precisely emulate the operating characteristics of actual disc drives, they employ full size discs mounted on a disc-drive spindle-motor assembly. Disc speed (up to 12,000 rpm), acceleration (0.5 to 25 s) and deceleration (0.5 to 25 s) can be varied to suit a variety of drives, including both linear and non-linear acceleration and deceleration curves, as well as back-cogging simulation at startup. Above the rotating disc is a magnetic head on its suspension, which may remain at a fixed radius or be programmed for seeking (track accessing) between any disk radii, at the frequencies corresponding to the real drive seek times. Head-suspension assembly height and length, skew angle and other geometric variables can be easily adjusted. The testers offer the options of either ramp loading or landing on the disc.

A number of important parameters is measured continuously during the test, namely: static friction (stiction), dynamic friction, take-off and landing velocities and energies, acoustic emission, magnetic signal amplitude, etc. The friction strain-gauge sensors and amplifiers have a force resolution of 10 micrograms (though the noise in the working tester allows to distinguish signals only above 10 milligrams). The original acoustic emission sensors and amplifiers have a bandwidth up to 5 MHz. The mechanical frequency of force sensors can be varied from 0.1 to 5 kHz, the electronic data acquisition rate can be varied from 1 to 20 kHz.

The computerized motor control and data acquisition provide for easy setup of different test procedures:

- contact start-stop test (on either fixed or varied disk radius),
- fly-stiction test (with a flexible sequence of head seeking, resting, and flying modes),
- rest-stiction test (with head resting on the disc for given park times),
- flying avalanche test (with measuring friction and acoustic emission speed dependencies),
- flyability test (with measuring friction, acoustic emission, and magnetic amplitude over time),
- altitude test (with sealed testers representing an altitude chamber, and a pump in the system),
- elevated-temperature test (with sealed testers having built-in heaters),
- tribo-chemical test (with heated contaminant holders inside the testers),
- temperature/humidity test (in the environmental chamber).

Though no one of the tests is sufficient to evaluate head-disc reliability, the above combination of them allows for the comprehensive and correct evaluation. As it was shown in [1], simultaneous measurements of friction and acoustic emission avalanches versus head-disc separation (that can be tested by varying either disc speed or altitude), per the procedure suggested in work [2], allows to distinguish the following three flying zones:

1- hydrodynamic friction with no surface interactions, where the slider and disc are fully separated by air;
2- hydrodynamic-with-partial-microcontacts friction with minor surface interactions, where a few local breakthroughs of the air film by asperities or debris occur;
3- mixed friction with substantial surface interactions on significant real contact areas.

Preliminary testing in all three zones has revealed that the acoustic emission avalanche did not represent a durability threshold, and its crossing did not lead to any appreciable loss of durability [1]. Then, extensive reliability testing on the above testers, using the entire combination of test procedures, has confirmed that most of the second flying zone (which is between the friction and acoustic emission avalanches) can be utilized safely, as long as there is a "safety buffer" above the friction avalanche. This "safety buffer" should be from as much as 60% for the twin-rail sliders on regular discs to as little as 5% for the carbon-overcoated tri-pad sliders on the advanced nitrogenated/hydrogenated carbon-overcoated discs.

The newly-developed CSS tester and test procedures have allowed us to obtain an experimental prove, without expensive drive tests, of the reliability of an earlier recommended head-disc interface with lower separation, higher magnetic signal and higher storage density. The laboratory and field testing of numerous disc drives, by several major drive manufacturers, have confirmed that our bench-level testing is capable to predict reliability correctly.

3. MICRO-TRIBOLOGY TESTING

The smooth disks and sliders help to reduce the head-disk separation, but exhibit problems of stiction and stiction-induced auto-oscillations.

A number of studies of the stiction phenomenon has been done [3]. Various contact processes have been investigated, including solid-to-solid and liquid-mediated adhesion, viscous flow, deformations, etc. In the absence of liquids, the viscous and meniscus components are absent, but the adhesion component due to Van-der-Waal's forces may be high on very smooth surfaces. In the presence of very thin films of liquid lubricants, which is common for thin-film magnetic disks, the viscous component is still negligible, but adhesion due to both Van-der-Waal's and meniscus forces is significant. In the presence of liquid lubricant layers, which was common for particulate magnetic disks, the viscous component might be of major concern. These processes produce mechanical and molecular components of the friction force that act in both tangential and perpendicular to the contact directions. Thus, the friction force should be considered as a two-dimensional sum of its components over all the contact segments [4].

A two-dimensional model of friction has been developed [4, 5], which combines all the above processes and relates them to slider perpendicular micro-displacements. During sliding, stopping, and starting, the slider undergoes small downward and upward displacements normal to the plane of sliding. Those micro-displacements are caused by changes in the degree of engagement of real contact spots. The friction force always varies, in both magnitude and direction, though on a large scale it appears to be constant. As the motion stops on a macro-scale, the slider motion continues on a smaller scale even during the seemingly stationary contact. A mobile lubricant re-distributes itself under the resting slider, with the lubricant moving toward the menisci from surrounding disk areas; this causes a growth of the meniscus force, pushing the slider down onto the disk. The slider descends with growing dwell time, increasing the deformation response force, which resists the micro-descent. The deformation increases the real contact area and brings new asperities into contact, which, in turn, increases adhesion. The increased adhesion leads to a further micro-descent. This self-perpetuating process stops only when the pulling and pushing forces become equal. The rate of this process and the time till equilibrium depend on the properties of the mated surfaces and lubricant and are governed mostly by surface diffusion of thin lubricant films.

To investigate the stiction and frictional oscillation processes, CSS testers have the following limitations:
- when head is mounted on the suspension, stiction is measured at the loosely-controlled angle,
- stiction cannot be measured in the direction perpendicular to surfaces,
- head-suspension assemblies exhibit slightly different characteristics, which limits test accuracy,
- the speed is typical for disc drives, but too high for a stick-slip analysis,
- lubrication regimes are typical for disc drives and not easily adjustable.

To resolve the drawbacks, a new micro-tribometer series UMT has been designed. It has precision motions of samples with in-situ measurements of friction and stiction, adhesion and wear, acoustic emission and electrical resistance (capacitance) of contact. This apparatus can accommodate a wide variety of sample sizes (from a 1-mm ball up to a 6-inch wafer), shapes and materials (metals, ceramics, plastics, paper, etc.). In particular, it can hold magnetic heads without suspension, which eliminates a lot of test uncertainties, increases the test accuracy, and allows to precisely measure stiction in both tangential and perpendicular directions. Its variable speed can be as low as microns per second, which is crucial for the studies of stiction and stick-slip mechanisms. Automated motor-control and data-acquisition allow easy changes in test procedures, including from boundary to hydrodynamic lubrication regimes.

The micro-tribometer is useful for the following projects:

- studies of the fundamentals of friction, wear, lubrication, stick-slip;
- precision measurements of friction and adhesion forces and the coefficient of friction,
- precision measurements of fatigue, abrasive, and adhesive wear,
- measurements of contact resistance, capacitance, acoustic emission and temperature,
- friction and wear testing of metals, plastics, ceramics, and coatings,
- tribology testing of solid lubricants, lubricating fluids, oils, and greases.

The micro-tribometer can utilize a variety of samples, for example: pin on disc, ball on disc, three balls, four balls, pin on V-block, ring on block, screw in nut, flat on flat, etc. It has been used successfully with oils (for motors and machine-tool slideways), greases (for ball-bearings), solid lubricants (for orthopedic devices), and unlubricated contacts (for brakes and clutches), including for the advanced MEMS.

The micro-tribometer was used in this work to measure stiction and wear in a head-disc interface. As was shown in [1], a dependence of stiction on relative lubricant thickness (the difference between the lubricant thickness and the surface roughness height) is complicated. When the lubricant thickness is lower than the roughness height, the curve is quite steep, any growth in lubricant thickness increases the degree of filling the hollows of the microrelief with lubricant (and consequently the stiction area) and leads to a significant stiction increase, while on the other hand, any small reduction in lubricant thickness leaves some surface asperities unprotected and leads to a wear problem. The stiction-lubricant graph has a maximum after the complete separation of the mated surfaces by the lubricant film, i.e., when the stiction area reaches its maximum. With the further increase in lubricant thickness within the area of liquid lubrication, stiction decreases (probably due to the reduction in the solid surface effect) and slowly reaches its low value; wear does not change.

Another interesting application of the UMT-tribometer is for scratch-resistance measurements of thin coatings. In this work, coatings of 4 nm to 40 nm were tested in both pin-on-disc and ball-on-disc modes. The repeatability of results was within 10% for both friction and acoustic noise parameters. The force resolution was 1 mg. The tests allowed to clearly distinguish the wear resistance not only of different coatings, but also of coatings of the same material (e.g., carbon) with small additions of hydrogen, nitrogen, etc. The test data was confirmed with the chemical materials analysis.

Magnetic discs have a dual lubricant with both the bonded layer that provides better resistance to lubricant removal and the liquid layer which provides replenishment capability. The optimum bonded-to-liquid percentage is shown to differ depending on the surface roughness, with low values for rough surfaces and high, up to near 100% values for smooth surfaces.

Two ways of stiction reduction have been studied with the micro-tribometer:

1 - no intentional liquid lubricant, with only solid or bonded lubricant covering the surfaces. Its usefulness is limited by the fact that the capillary force can be caused by any liquid, like condensed water or organic vapors, or bearing oils penetrated from a spindle assembly;

2 - relatively thick liquid lubricant, which completely covers the disk roughness, thus keeping both wear and stiction low, but is not too thick to be spun off during rotation.

4. CONCLUSION

The progress in micro-tribology is limited by the development of both tribometers capable of measuring micro-tribological friction and wear processes and testers correctly simulating the real machines on the bench-top level.

Two new tester types are suggested, micro-tribometer UMT for precision friction and wear studies in diversified applications, and CSS tester for simulating disc drives on the component level.

The micro-tribometer allows for accurate measurements of both static and dynamic friction, as well as adhesion, wear, scratch-resistance and other fundamental characteristics of both bulk materials and coatings.

The new testers allow scientists to gain new insights into the friction and wear mechanisms on the micro-level, including a number of molecular and mechanical processes.

REFERENCES

[1] Gitis, N. and Gerber, C.: Study of Low Flying Height Using Acoustic Emission and Friction Techniques, *Adv. Inform. Storage Systems*, Vol. 6, pp. 107-120 (1995).
[2] Kragelskii, I., Gitis, N., Schavelin, V.: Use of Acoustic Emission Method to Optimize Surface Microrelief, *Soviet Journ. Friction and Wear*, Vol. 5, pp. 1-5 (1984).
[3] Gitis, N. and Volpe, L.: Nature of Static Friction Time Dependence, *Journ. of Physics D: Applied Physics*, Vol. 25, pp. 605-612 (1992).
[4] Gitis, N., Lapidus, A., and Chizhov, B.: Investigation into the Characteristics of Mixed Friction in Slideways, *Soviet Engineering Research*, Vol. 7, pp. 67-72 (1987).
[5] Kragelskii, I., Gitis, N.: *Friction-Induced Vibrations*, (in Russian), Science Press, Moscow (1987).

MECHANICAL PROPERTY DETERMINATION USING NANOINDENTATION TECHNIQUES

CARL J. McHARGUE
Center for Materials Processing,
University of Tennessee
Knoxville, TN 37996-2350, USA

1. Introduction

Data obtained from indentation testing can provide important information about the near-surface mechanical properties and deformation behavior of solids. Properly executed and interpreted, the technique can provide information regarding hardness, elastic modulus, compressive yield strength (metals), fracture toughness (ceramics) and residual stresses. The apparent simplicity of the test procedure, however, often belies the difficulties of interpreting the data to give quantitative information on properties of interest for a specific application.

Hardness is a poorly defined term. The hardness value gives an indication of the resistance to local contact deformation. However, it is not a fundamental property of a material but is a complex combination of properties such as elastic deformation, yield strength, ductility, work-hardening, etc. It is recognized that the resistance to penetration involves both plastic and elastic parameters. The hardness number or index is very much a function of the manner in which the test is conducted and the character of the test material.

Static indentation hardness is conventionally determined by making a permanent indentation in the surface of a test sample and measuring the size of the residual indentation after removal of the load. The mean contact pressure is calculated by dividing the maximum applied load by the area of the residual indent. This procedure is often adequate to provide a quantitative or semi-quantitative measure of the mechanical properties of metals. Tabor [1951] has shown that this mean contact pressure is approximately three times the compressive yield stress for fully work-hardened metals, i.e., samples that do not experience work-hardening during the deformation caused by the test.

Most of the information regarding plastic and elastic deformation arising from indentation comes from studies of large scale indentations, i.e., with dimensions of

B. Bhushan (ed.), Tribology Issues and Opportunities in MEMS, 487-508.

about 100 μm. As interest in the properties of thin films and surface layers has increased, the use of indentation tests has been used on a ever finer scale, e. g., nanoindentation and picoindentation. The understanding and models from macroindentation have been extrapolated to interpret the measurements. Samuels [1986] noted that the principle of geometric similarity is essential for this purpose but that this principle has not been tested for nanometer sized impressions. We should not be surprised to find unexpected effects, for example of surface tension, friction, etc., at such a scale.

2. Nanoindetation Techniques

A major limitation of conventional hardness tests is the ability to accurately measure the size of the residual indent. Measurement by optical microscopy is limited to impressions having diameters or diagonals larger than about 10μm. The depth of the indent with a 10μm diagonal is about 1.4 μm in the Vickers test and about 0.33 μm in the Knoop geometry. Identifying the true edges of the indent, due to pile-up of the deformed material or elastic recovery effects, also limit the accuracy of measurement in both optical and scanning electron microscopy. These difficulties and the interest in studying the properties of regions (layers, films) on a finer scale have lead to the developing of load--depth-sensing devices that continuously measure force and displacement on a very fine scale as an indentation is being made and as the indenter is withdrawn [Pethica, 1982; Newey, Wilkens and Pollock, 1982; Loubet, Georges, Marchesini and Meille, 1984]. The load-displacement data can be used to determine certain mechanical properties even when the residual indentations are too small to be imaged conveniently or accurately.

There are at least three commercially produced instruments: (a) the Nanoindenter™ produced by Nano Instruments, Inc., Knoxville, TN, USA, based on the design of Pethica, Hutchings and Oliver [1983]; (b) an apparatus based on work at the University of Lancaster by Newey, Pollock and Wilkens [1982]; and (c) the UMIS-2000 Ultra-microindentation System developed at the Division of Applied Physics, CSIRO, Australia [Bell, Bendel, Field, Swain and Thwaite, 1991/92]. Additionally, there have been a number of instruments developed and built in various laboratories.

The Nanoindenter™ generally employs a triangular pyramid-shaped (Berkovich) diamond indenter which is forced into a specimen's surface by use of a coil and magnet assembly, Fig. 1. Other penetrater geometries (e. g., cube-corner, spherical) can be used. The mass of the loading column is balanced on springs and the indentation force resolution is 0.3 μN. The position of the indenter is determined from a three-plate capacitance system having a displacement resolution of 0.16 nm. The motion of the indenter is damped by air flow around the center plate of the capacitor, which is attached

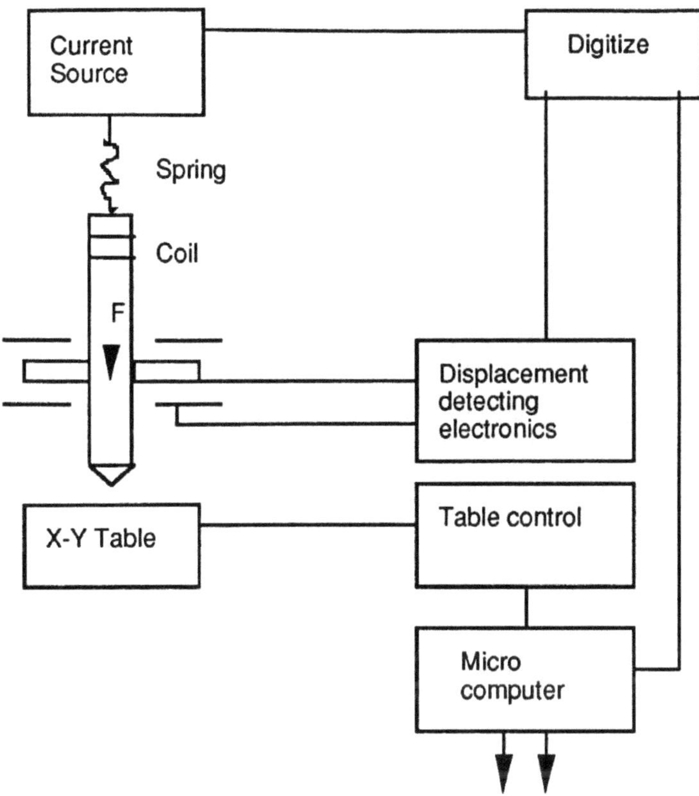

Figure 1. Schematic drawing of Nanoindenter™.

to the loading column.

Testing can be conducted at a constant loading rate or at a constant displacement rate. The test procedure involves moving the indenter towards the surface of the specimen at a constant rate, detecting surface contact as a change in velocity. The load is then increased to maintain constant velocity (in the constant displacement rate mode) or changed to maintain other deformation criteria, e.g., constant loading rate. The system continuously records the displacement and the force for some pre-set sequence. A common sequence is to load to a given depth, unload some fraction, reload to a deeper depth, hold for a period to detect thermal drift, then unload completely.

These nanoindentation devices may be able to obtain hardness and elastic modulus values from impressions as shallow as 20 to 50 nm if all sample and test conditions are ideal. Attempts to extend the measurements to more shallow depths have used sharp tips mounted on stiff cantilever beams in atomic force microscopes (AFM);

for example see the proceedings of a recent NATO Advanced Study Institute [Bhushan, 1997]. Hysitron, Inc. makes an accessory for the AFM that allows an indentation to be made and immediately imaged.

The exact shape of the indenter and residual impression are critical issues in determining values for the properties of interest.

3. Response of Various Types of Solids to Indentation

The indentation response of a material is strongly influenced by its mechanical nature. The deformation is fully reversible in an ideally elastic solid and is described by models for classical elasticity. Most solids, however, exhibit flow and/or fracture beneath the indenter. The stress distribution under a Vickers indenter, for example, can be modeled as a hydrostatic region immediately beneath the contact area that is surrounded by a region undergoing plastic deformation that is in turn surrounded by an elasticity deformed region [Bishop, Hill and Morr, 1945]. Brittle materials (ceramics, glass) may experience some plastic flow in the highly constrained volume just beneath the contact and fracture may be initiated by the tensile stresses generated upon release of the load.

In the case of polymeric materials, viscoelastic effects may accompany the loading and unloading. The deformation is time-dependent and the indentation relaxes upon unloading. In fact, the impression may entirely disappear with time. The elastic/plastic deformation may also be accompanied by densification of the test material in amorphous or glassy substances.

In order to relate the load and the penetration depth to indentation pressure (hardness) one must note that the elastic deflection of the surface extends the actual contact area, as shown in Fig. 1 for a spherical indenter of radius R. This elastic deflection of the surface must be taken into account in obtaining the actual penetration from a measurement of the total displacement of the indenter in load--depth-sensing tests.

The response of an ideally elastic solid, such as rubber, is fully reversible and totally elastic. Since the impression vanishes upon removal of the load, "hardness" must be defined in terms of the loaded condition, e.g., the load needed to produce a give penetration depth.

An ideally rigid-plastic solid undergoes no deformation until some critical value of stress (Y) is reached. Plastic flow then occurs at this constant stress. The slip-line field treatment for a flat, two-dimensional punch leads to the relation

$$H \sim 2.6\ Y \text{ to } 3\ Y \tag{1}$$

depending upon the yield criterion (Tresca or von Mises) used. In general, $H = cY$, where c depends upon the geometry of the interface and any effect of interfacial friction.

An elastic-plastic solid deforms elastically until the yield stress (Y) is reached

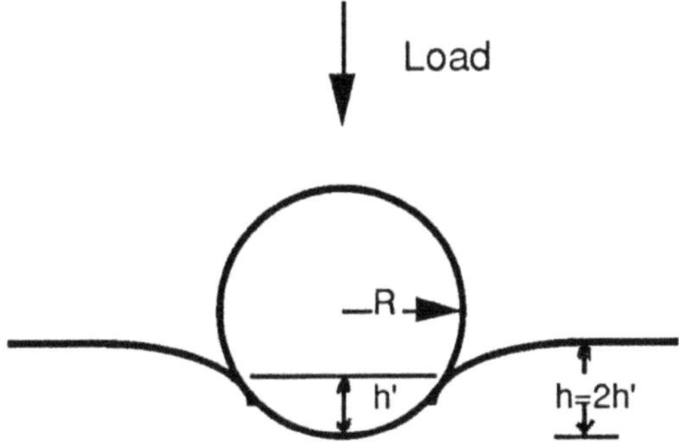

Figure 2. Schematic diagram of a spherical indenter in contact with a surface.

and then plastically at a constant yield or flow stress. In the simple case, work-hardening does not occur during the movement of the indenter into the solid. This behavior may be approached in a fully work-hardened, polycrystalline metal. For a spherical indenter, plastic deformation is initiated in a small volume for a mean contact pressure [Tabor, 1951]

$$p = H \sim 1.1\ Y \qquad (2)$$

and spreads to full plasticity at a mean contact pressure of

$$p \sim 3\ Y. \qquad (3)$$

This approximation is widely used to estimate yield stress from hardness measurements.

General plastic flow will be initiated immediately upon contact of a conical or pyramidal indenter with the surface. In this case, the hardness value depends upon the ratio of elastic modulus to yield strength (E/Y) [Marsh, 1964]. For polymers, with values of E/Y of about 10, $H < 1.5$ Y. Most metals have values of $E/Y > 100$, and $H \sim 3Y$.

Elastic-brittle solids such as glasses and ceramics may undergo brittle fracture in the elastic range during loading or upon release of the load due to the tensile stresses under the indenter at the moment of load release. Ideally, these materials would deform elastically until cracking is initiated at some critical value of the tensile component. The critical value occurs when the released elastic strain energy is sufficient to drive a crack, i.e., when the released strain energy is equal to or greater than the energy required to form the crack surfaces. Plastic deformation does occur, however, in regions of high hydrostatic constraint where crack nucleation is inhibited.

The loading-unloading response of these various types of solids during nanoindentation will be described in another section of this paper.

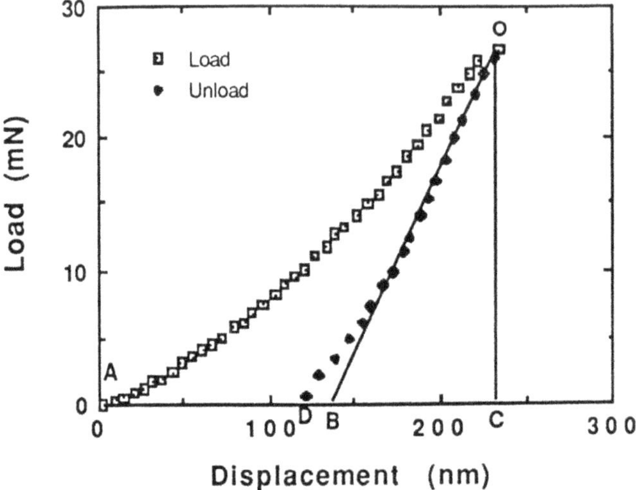

Figure 3. Load - Displacement curve for sapphire for one load/unload cycle. AC is the total displacement at maximum load; BC is elastic displacement; AB is plastic displacement; BD is due to loss of contact during lift off of indenter.

4. Load-Displacement Curves for Typical Solids

An example of the data for one load-unload cycle is shown in Fig. 3 for a material that exhibits a large elastic component with plasticity (sapphire). The total displacement under maximum load is indicated by the distance AC; the elastic component is given by BC; the plastic component is AB. The distance BD could indicate some anelastic effect or be due to loss of contact of the indenter with the surface during the final stages of unloading.

The loading portion of the curve is controlled by both elastic and plastic deformation and contains information about both E and H [Lawn and Howes, 1981]. The unloading portion is controlled only by the elastic properties. Doerner and Nix [1986] reported that the initial slope of the unloading curve is often linear and directly gives the stiffness (S = dp/dh) from which the elastic (Young's) modulus can be derived. With knowledge of the exact geometry of the diamond indenter and the system compliance, the data can be processed to give both hardness (H) and Young's modulus (E) as functions of depth from the surface. Difficulties arise from this need to know accurately the shape of the indenter tip on a nanometer scale.

Since the loading portion contains information about both elastic and plastic properties, there have been recent attempts to obtain hardness. elastic modulus and indenter shape from this portion of the data. Loubet, Georges and Meille [1986] described Vickers indentations with a relationship of the form

$$P = c\delta^n \tag{4}$$

where P is the load, δ is the displacement, and c and n are materials constants. Hainesworth, Chandler and Page [1996] followed a similar path and found that for six materials and the Berkovich geometry, the relationship was well-described as

$$P = K_m \delta^2 \tag{5}$$

where

$$K_m = E(\phi\sqrt{\frac{H}{E}} + \psi\sqrt{\frac{E}{H}})^{-2} \tag{6}.$$

These treatments requires the value of either H or E to be known in order to determine the other one.

Figure 4 shows the load-displacement response for a sample of annealed, polycrystalline copper. The behavior is typical of a material that experiences a large amount of plastic deformation relative to elastic deformation during the test. The unloading portion is linear for most of the range, allowing the stiffness to be determined accurately. If the machine stiffness has been previously determined and if the shape of the indenter tip is known, the elastic modulus can be calculated. Since there is little elastic recovery upon unloading, an approximate hardness is given by the maximum load divided by the projected contact area at maximum penetration.

Figure 5 illustrates a totally elastic response. The specimen was a high-quality single crystal of α-Al_2O_3 (sapphire) indented in a direction parallel to the c-axis. The loading and unloading curves are superposed for an indentation depth of 80 nm [Page, Oliver, McHargue]. High resolution scanning electron microscopy (SEM) and transmission electron microscopy (TEM) showed that samples exhibiting such loading-unloading behavior contained no residual indentation in the surface and that no damage, e.g., dislocations, could be detected at the point of contact.

It is often necessary, and usually desirable, to conduct auxiliary examinations to determine the nature of any unusual loading or unloading effects. For example, increasing the load on the specimen shown in Figure 5 produced a displacement discontinuity in the loading curve, Figure 6. Such a discontinuity (AB) could be caused

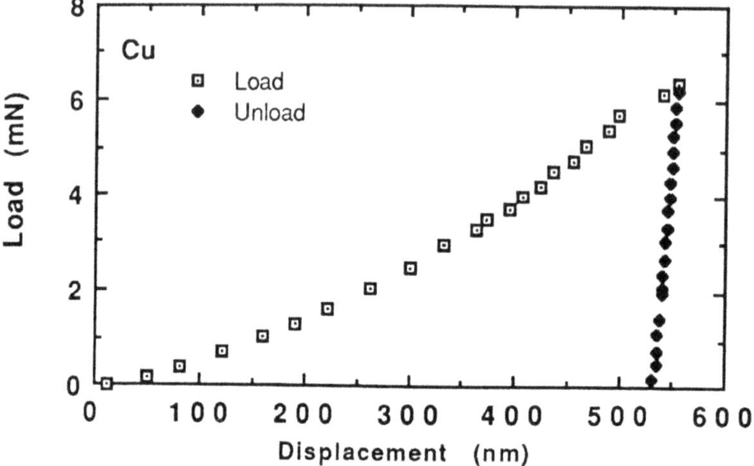

Figure 4. Load-displacement curve for polycrystalline copper.

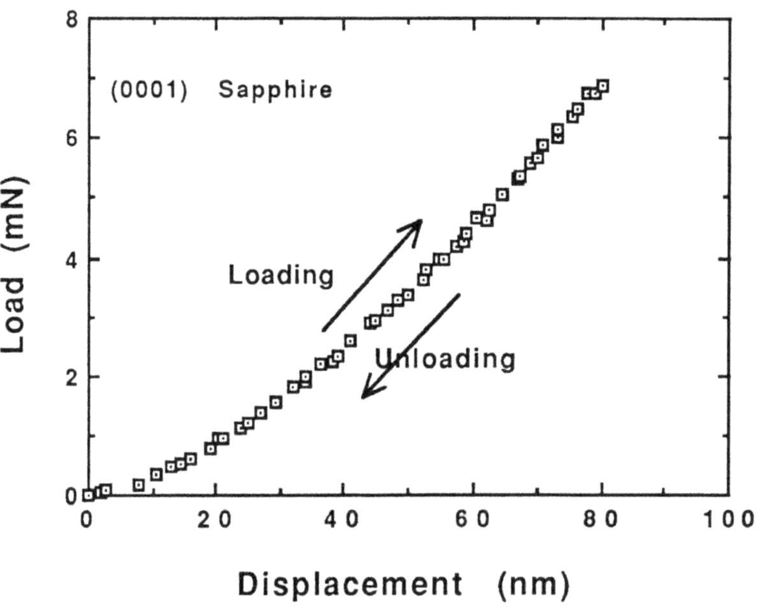

Figure 5. Load-unload curves for sapphire showing only elastic indentation to a depth of 80 nm.

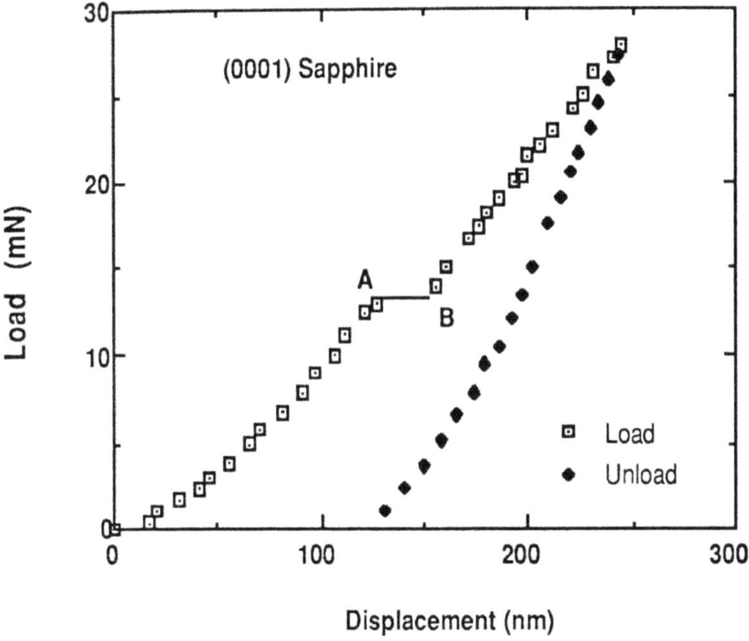

Figure 6. Displacement discontinuity (AB) in loading curve for sapphire.

by a yield point effect (i.e., the sudden onset of plastic flow), twinning, cracking or a phase transformation. Transmission electron microscopy confirmed in this case that it was due to a yield point effect. Samples taken from specimens loaded to point A contained no residual indentations and no dislocations at the site of contact, whereas, those loaded to point B contained a permanent impression with profuse dislocations.

The unloading curves for single crystal silicon specimens often exhibit an unusual behavior, illustrated by Fig. 7. There is a sharp upward inflection at point A, appearing as if there is a sudden push of the indenter from the surface. Examination by TEM revealed only a limited, localized deformation structure after an indentation to a depth of 350 nm (loaded). There was considerable structural rearrangement and amorphization. The latter appears to be the result of a pressure-induced phase transformation (and densification) caused by the high hydrostatic pressures beneath the indenter (10 - 20 GPa) [Pharr, Oliver and Clarke, 1990; Page, Oliver and McHargue, 1992].

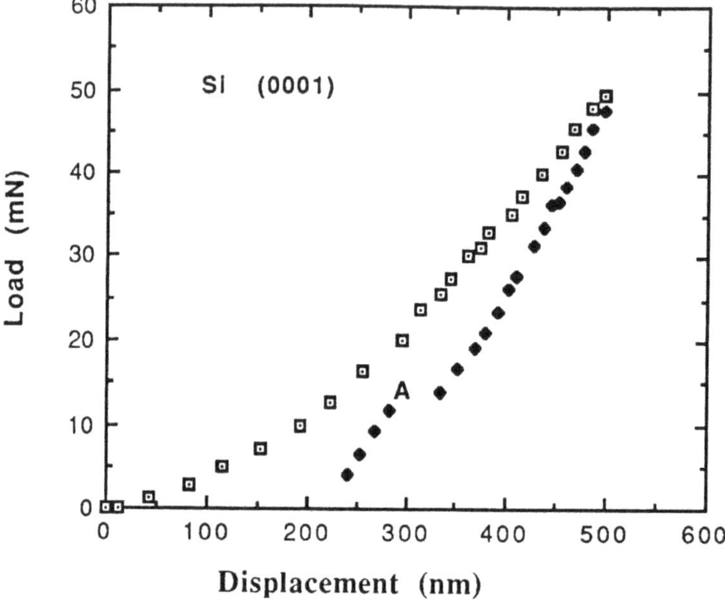

Figure 7. Load-displacement curves for one load-unload cycle for Si.

5. Analysis of Load-Displacement Curves

The adaptation of models used to describe the elastic contact of an indenter (punch) with a surface [Beussinesq, 1985; Hertz, 1882; Lowe, 1929; Sneddon, 1965] to the analysis of load-displacement curves has been described in detail [Loubet, Georges, Marchesini and Meille, 1984; Doerner and Nix, 1986; Oliver and Pharr, 1992].

If the initial slope of the unloading curve is linear, this slope directly gives the elastic (Young's) modulus if the geometry of the indenter tip is accurately known:

$$S = \frac{dP}{dh} = \frac{2}{\sqrt{\pi}} E_r \sqrt{A} = \text{Stiffness} \tag{7}$$

where

$$\frac{1}{E_r} = \frac{(1-\upsilon^2)}{E} + \frac{(1-\upsilon_i^2)}{E_i} \tag{8}$$

with E and υ being the Young's modulus and Poisson's ratio for the specimen and E_i and υ_i are the same for the indenter, P is the load, h is the indenter displacement, and A is the projected area of elastic contact.

In this case, the extrapolation of the initial slope of the unloading curve to zero load gives the plastic depth of the indentation (h_p). Hardness is then given as:

$$H = \frac{P_{max}}{A} \qquad (9)$$

where P_{max} is the maximum load and A is the projected area of the impression. For a perfect Vickers or Berkovich indenter, $A = 24.5\ h^2$.

Knowledge of the exact shape of the pyramidal diamond indenter is critical to the determination of values for E and H. At the nanometer scale of these measurements, the Vickers and Berkovich diamonds usually have some blunting or rounding at the tip as well as other irregularities. This shape is generally experimentally determined.

Dorner and Nix [1986] adopted the procedure used by Pethica, Hutchings and Oliver [1983] of examining by TEM carbon replicas of impressions made to various depths in METGLAS® 2826 and annealed α-brass. A second order polynomial was used to fit the area vs. plastic depth values determined by their straight line unloading extrapolation. The measured and ideal shapes converged at depths of ~1000 nm. At lower depths, however, the hardness calculated from the idea shape gave overestimates of the hardness value by as much as 240%.

Oliver and Pharr [1992] analyzed the unloading curves for six materials (fused silica, soda-lime glass, and single crystals of Al, W, quartz, and sapphire) and concluded that a simple power law relationship of the form

$$\mathbf{P = B\ (h - h_f)^m} \qquad [10]$$

described the data for all six materials. In this expression, h is the total displacement of the indenter and h_f is the residual depth. Values of B, m, and h_f were determined from a least squares fitting procedure. An area function, A(h), was used to describe the actual area as a function of distance from the tip for a given diamond indenter. The calibration is made from a series of indentations at different loads into Al and fused quartz. The area function was expressed as:

$$A(h) = 24.5h^2 + c_1 h + c_2 h^{1/2} + c_3 h^{1/4} \ldots c_8 h^{1/128} \qquad (11)$$

where c_1 ... c_8 are constants.

Several groups are exploring the use of a "plastic work' concept to characterize the response of surfaces to contact damage. Tabor [1948] proposed the use of the plastic work of indentation to analyze the "dynamic hardness" for the case of a ball-bearing dropped onto a sample and rebounding. By dividing the work of indentation by the plastic volume of indentation, he obtained a dynamic hardness which was dimensionally equivalent to stress.

The load-unload curve from a depth-sensing test can also be analyzed in terms of work done. The total area at maximum load (AOCA in Fig. 3) represents the work of indentation, W_T, which is comprised of elastic work, W_E (BOCB) and plastic work, W_P (AOBA) [Shorshov et al., 1981].

The difficulty of using this concept is the determination of the plastic volume, i.e., the volume in which the work is done. Pollock, Maugis and Barquins [1986] assumed that the residual indentation retains the indenter geometry and defined a plasticity index. This assumption may not be valid if there is nonunifom elastic recovery; however, in many cases the elastic recovery only occurs in the direction normal to the surface and the projected area of the indentation remains unchanged [Stilwell and Tabor, 1961]. Sakai [1993] used such an approach to analyze the energy of surface deformation of brittle materials. Twigg, McGurk, Hainesworth and Page [1996] reported that this method gave good results for PVD coatings of TiN deposited on M2 tool steel.

Loubet and co-workers [1986] reported a strong correlation between the ratios of W_p/W_T and h_p/h_T (plastic depth of indentation to total depth) for alumina, MgO, 5100 hardened steel and aluminum (99.9% pure). Further study of the work terms may provide additional insights into the deformation processes and may prove to be a useful tool for characterizing the response of materials to contact deformation.

6. Indentation Size Effect (ISE)

According to the principle of geometric similarity, the hardness index or number should not vary with the size of the indent. However, many materials exhibit an indentation size effect (ISE) wherein the hardness appears to increase as the indentation size (or load) decreases [Onitsch, 1974]]. The effect is particularly pronounced in ceramics.

Many explanations have been proposed to explain the ISE, including surface artifacts, "quantized" slip or deformation band formation, anelastic recovery of small indents, and the low probability of finding mobile dislocations or other deformation-initiating flaws in a small volume.

Sargent [1986] has discussed the many factors that may affect the measurements at the scale of micro- or nanoindentation and concludes that there are too many distinct effects for physically based models to be useful at our current state of understanding.

7. Coatings, Films, and Surface Layers

One of the driving forces for the development of nanoindentation techniques is the desire to measure the properties of thin surface-modified layers and thin films and coatings. Examples of surface-modified regions include ion-implanted and laser-treated surfaces.

A major problem in accurately determining the properties of the coating is the elimination of substrate effects on the measurements. The deformation zone under an indenter has been modeled as a hydrostatic core surrounded by a plastic zone which in turn is surrounded by an elastically deformed region. The plastic/elastic boundary expands as a hemisphere as the indenter is forced into the surface and extends to distances several times the depth of indentation. The "rule-of-thumb" cited in many text books requires the film thickness to be ten times the indentation depth.

Sargent [1986] proposed a simple rule of mixtures to analyze the measured (effective) hardness, H_{eff}, in terms of the hardness of the film, H_f, hardness of the substrate, H_s, and the fraction of the deformed volume contained in the film, V_f, and in the substrate, V_s. Thus,

$$H_{eff} = \frac{H_s V_s + H_f V_f}{V_s + V_f} \tag{12}$$

Burnett and Page [984] used a modified approach with the assumption that the volume of material is contained in a spherical cap of diameter equal to the diagonal of a Vickers indentation and incorporated the indentation size effect. They also added a "weighting factor" which is an empirical parameter that allows the data to be fitted to the model.

The hardness and elastic modulus of film and substrate have been included in the analyses in order to differentiate between soft films on hard substrates and hard films on soft substrates. Lebouvier et al. [1984] developed a plain strain model that indicates a factor of ten is required for cases where the strength of the film is much greater than that of the substrate, but that a factor of four is sufficient for softer films on hard substrates. A closed-form elasticity solution for the displacement of the surface beneath a region of uniformly distributed pressure has been proposed to describe the effective modulus for a high modulus film on a low modulus substrate [Stone, Wu, Alexopoulos and DeFoontaine, 1989]. This model predicts observable effects for ratios of depth to thickness of 0.1 to 10.

Burnett and Rickerby [1987] modified the above treatment of Burnett and Page to include an empirical factor (χ) that describes the deviation of the plastic zone from the idealized hemispherical geometry. For a soft film on a hard substrate the expression for hardness takes the form:

$$H = \frac{H_f(\chi^3)V_f + H_sV_s}{V_{total}} \tag{13}$$

and for a hard film on a softer substrate:

$$H = \frac{H_fV_f + H_s(\chi^3)V_s}{V_{total}} \tag{14}$$

The factor χ contains hardness and modulus, such that

$$\chi = \left(\frac{E_fH_s}{H_fE_s}\right)^n \tag{15}$$

Experimentally, values of n range from 0.33 to 0.5.

Batacharya and Nix used a finite element analysis to calculate the variation of hardness with indenter depth for 1 μm films on semi-infinite substrates. For soft films on harder substrates, they found that the measured hardness, H, varied as

$$H = H_s + \left(H_f - H_s\right)\exp\left[-\frac{\left(\sigma_f / \sigma_s\right)}{\left(E_f / E_s\right)}\left(\frac{h_t}{t}\right)^2\right] \tag{16}$$

where H = hardness, σ = yield stress, E = elastic modulus, h_t = indenter penetration, and t = film thickness, and the subscripts denote film (f) and substrate (s). The use of such an expression requires information on several properties of the film that are not usually known. One can use various fitting procedures if there are data for a range of conditions for a given film/substrate combination.

Examples of apparent hardness and elastic modulus are shown as functions of penetrater depths in Figs. 8 - 10. Figure 8 shows the measured hardness values for a 155 nm-thick amorphous layer on the surface of sapphire prepared by implantation at 77 K [Oliver and McHargue, 1988]. The hardness is expressed as the ratio of the as-implanted surface to the virgin crystal. The effect of the substrate on the measured value is

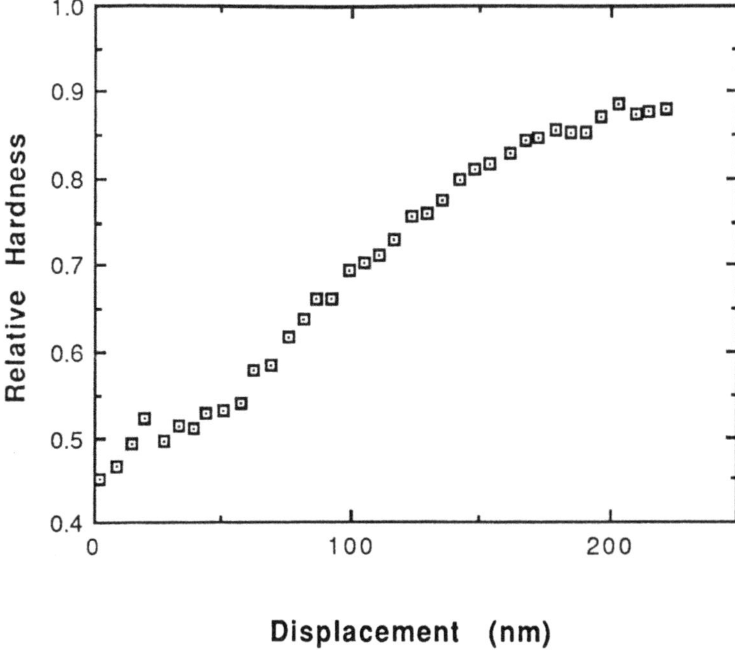

Figure 8. Relative hardness vs. indenter displacement for an amorphous surface layer produced by ion implantation of sapphire at 77 K. The amorphous layer is 155 nm thick.

apparent at indenter depths as low as 20 nm, i.e., about 12% of the thickness of the soft layer. An effect of the soft surface layer on the measured hardness value is still apparent to indenter depths as large as 150% of its thickness.

A second example of softer films on harder substrates that also illustrates the effects of film adhesion on the measurements is given in Figure 9 [Joslin, O'Hern, McHargue and Clausing, 1989]. One hundred nanometer thick films of diamond-like carbon were deposited onto sapphire substrates under conditions that produced adherent films (number 10) and non-adherent films (2, 3, 5). The substrate hardness was 25 GPa (corrected for the indentation size effect) and its modulus was 400 GPa. The hardness of each film is about 6 GPa at 28 nm indenter displacement, i.e., at 20% of the film thickness. The measured values for the non-adherent ones show little variation with depth of indentation, but there is a strong variation in the measured hardness value for the adherent film. The measured value does not reach the substrate values at penetrations that extend to the DLC/sapphire interface. The modulus measurements for both adherent and non-adherent films are affected by the substrate over the range of depths used (28 to 102 nm), with a stronger effect for the adherent films. The elastic modulus is more sensitive to the normal stresses and both the adherent and non-adherent films

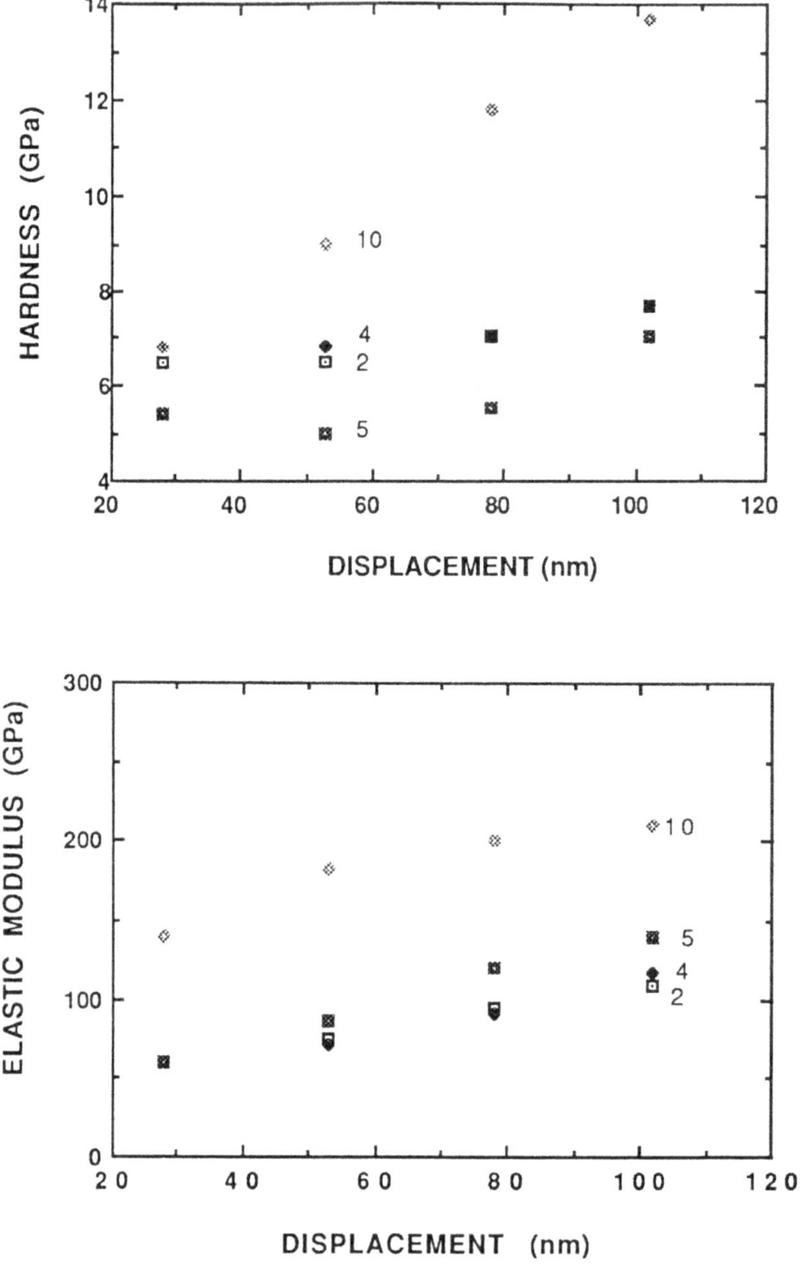

Figure 9. Hardness and elastic modulus of DLC films deposited on sapphire.

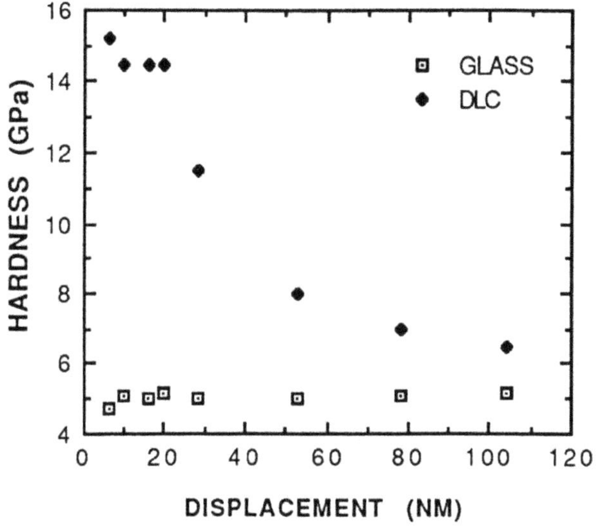

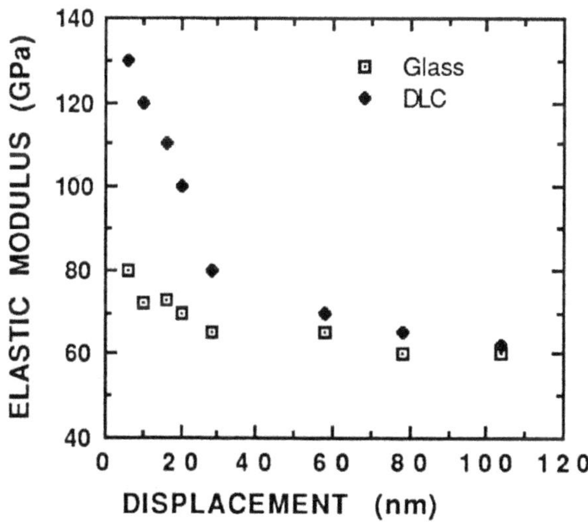

Figure 10. Hardness and elastic modulus of DLC films on glass substrates.

show the influence of the substrate over the entire range of indentation depths. The shear component of the stress is not transmitted for non-adherent films and the hardness values were independent of the depth of indenter penetration.

An example of a harder film (DLC) on a softer substrate (glass) is given in Figure 10 which shows the apparent hardness and elastic modulus values as functions of indenter penetration in a 100 nm thick adherent DLC film on a glass substrate [Joslin, et al.]. The measured hardness value remains approximately constant at about 15 GPa for indenter depths of 6 to 28 nm, then falls to a value only slightly higher than that of the substrate at 78 to 104 nm. The value of the glass substrate is approximately constant at 5 GPa. The measured elastic modulus, however, begins to decrease toward the substrate value for penetrations as low as 10 nm.

The relative elastic and plastic properties of coatings and substrates are particularly important for hard (and often brittle) coating on much softer substrates. The response to indentation testing will be influenced by the manner in which the applied stress is partitioned between the coating and substrate. If the coating can deform elastically, it may merely flex as the substrate elastically and plastically deforms and the membrane stresses in the coating may support a portion of the load. The coating recovers its original shape once the load is removed but the substrate may be permanently deformed. The coating may then debond from the substrate [Knight, Page and Hutchings, 1989]. Such behavior has been modeled as a plate on an elastic foundation by Ramsey, Chandler and Page [1991].

In some instances, the high stress beneath the indenter will both flex the coating into the substrate and initiate plastic flow in the coating The plastic zone may eventually cross into the substrate where the plastic zone may grow faster than in the coating. This leads to increasing flexure and then to cracking in the coating.

In cases where the coating is brittle, the initial flexing may cause extensive cracking in the coating. In this case, most of the plastic flow occurs in the substrate with the fractured pieces of the coating acting as inert, hard transmitters of the load.

8. Summary

Nanoindentation techniques that allow information to be deduced from the penetration depth vs. load curves during both loading and unloading extends the range of information that indentation testing can provide. Interpretation of the data often requires auxiliary examinations, particularly microscopy. Results to date have provided new insights into the response of solid surfaces to contact deformation. The measurement systems have high resolution and measurement of depth and load can be made with very high accuracy. There is the temptation to use such accurate and often reproducible values to calculate hardness and modulus values to many decimal places. One must keep in mind the assumptions and models involved in the interpretation as well as the problems introduced by the surfaces themselves. Tabor has admonished us to "not pursue precision which is not there" [1986].

9. References

Battacharya, A.K. and Nix, W.D. (1988), *Int. J. Solids Structures* **24**, 1287.

Bell, T.J., Bendel, A.,Field, J.S., Swain, M.V. and Thwaite, E.G. (1991/92) *Metrologia* **28**, 463.

Beussinesq, J. (1885) *Applications des Potentiels a l'estude de equilibre dt du mouvement des solides elastiques,* Gauthier-Villers, Paris.

Bhushan, B., (ed) (1997) *Micro/Nanotribology and Its Applications,* Kluwer Academic Publ., Dordrecht.

Bishop, R.F., Hill, R. and Mott, N.F. (1945) The Theory of Indentation Hardness Tests, *Proc. Phs. Soc. (London)* **57**, 147.

Burnett, P.J. and Page, T.F. (1984) Surface Softening in Silicon by Ion Implantation, *J. Mater. Sci.* **19**, 845.

Burnett, P.J. and Rickerby, D.S. (1987) The Mechanical Properties of Wear Resistant Coatings, I, *Thin Solid Films* **148**, 41.

Burnett, P.J. and Rickerby, D.S. (1987) The Mechanical Properties of Wear Resistant Coatings, II, *Thin Solid Films* **148**, 51.

Doerner, M.F. and Nix, W. (1986)Method for Intrepreting the Data from Depth Sensing Indentation Instruments, *J. Mater. Res.* **1**, 601.

Hanieswoth, S.V., Chandler, S.W. and Page, T.F. (1996) Analysis of Nanoindentation Load- Displacement Curves, in *Plastic Deformation of Ceramics III*, J. Routbort, R.C. Bradt and C. Brooks, Plenum Press, NY.

Hertz, H. (1882), *J. reine und angewandte Mathematik* **92**, 156.

Joslin, D.J., O'Hern, M.E., McHargue, C.J. and Clausing, R.E. (1989) Hardness of Amorphous Had Carbon Films Determined by the Ultra-Low Load Microindentation Technique, in R.L. Caswell (ed) *Infrared Systems and Components, III , SPIE Conf. Proc.* **1015**, pp 218-225.

Knight, J.C., Page, T.F. and Hutchings, I.M. (1989) Surface Deformation Behavior of TiC and TiN Coated Steels I. Indentation Response, *Surface Engineering* **5**, 213.

Lawn, B.R. and Howes, T.R. (1981) Elastic Recovery at Hardness Indentations, *J. Mater. Sci.* **16**, 2745.

Lebouvier, D., Gilormin, P.G. and Ferber, E. (1984) Modelling of Indentations, *J. Physics D* **18**, 199.

Loubet, J.L., Georges, J.M., Marchesini, D. and Meille, G. (1984) Vickers Indentation Curves of MgO, *J. Tribology* **106**, 43.

Loubet, J.L., Georges, J.M. and Meille (1986) Vickers Indentation Curves of Elasto-Plastic Materials, in P. Blau and B.R. Lawn (eds), *Microindentation Techniques in Materials Science and Engineering*, American Society for Testing and Materials, Philadelphia, PA., 72-79.

Lowe, A.E.H. (1929) *Phil Trans. A* **228**, 377.

Marsh, D.M. (1964) Plastic Flow in Glass, *Proc. Roy. Soc. (London)* **A279**, 420.

Newey, D., Wilkens, M.A. and Pollock, H.M. (1982) An Ultra-Low Load Penetration Hardness Tester, *J. Phys. E* **15**, 119.

Newey, D., Pollock, H.M. and Wilkens, M.A. (1982) The Ultra-microhardness of Ion Iron and Steel at Submicron Depths and Its Correlation with Wear Resistance, in in V. Ashworth, W. Grant and R. Proctor (eds), *Ion Implantation into Metals*, Pergamon Press, Oxford, pp 157-166.

Oliver, W.C. and McHargue, C.J. (1988) Characterization of the Hardness and Elastic Modulus of Thin Films Using a Mechanical Properties Microprobe, *Thin Solid Films*, **161**, 117.

Oliver, W.C. and Pharr, G.M. (1992) An Improved Technique for Determining Hardness and Elastic Modulus using Load and Depth Sensing Indentation Experiments, *J. Mater. Res.* **7**, 1564.

Onitsch, E.M. (1947) *Mikroskopie* **2**, 345.

Page, T.F., Oliver, W.C. and McHarue, C.J. (1992) The Unusual Deformation Behavior of Ceramic Crystals Subjected to Very Low Load Indentation, *J. Mater. Res.* **7**, 450.

Pethica, J.B. (1982) Microhardness Tests with Penetration Depths Less than the Ion-Implanted Layer Thickness, in V. Ashworth, W. Grant and R. Proctor (eds), *Ion Implantation into Metals*, Pergamon Press, Oxford, pp 147-156.

Pethica, J.B., Hutchings, R. and Woliver, W.C. (1983) Hardness Measurements at Penetration Depths as Small as 20 nm, *Phil Mag. A* **48**, 593.

Pharr, G. and Oliver, W.C., and Clarke, D.R. (1990) The Mechanical Behavior of Silicon during Small Scale Indentation, *J. Electronic Mater.* **19**, 881.

Pollock, H.M., Maugis, D. and Barquins, M. (1986) Characterization of Submicrometer Surface Layers by Indentation, in P. Blau and B.R. Lawn (eds), *Microindentation Techniques in Materials Science and Engineering*, American Society for Testing and Materials, Philadelphia, PA. pp 47-71.

Ramsay, M, Chandler, H.W. and Page, T.F. (1991) Modelling of the Contact Response of Coated Systems, *Surfacesand Coatings Tech.* **49**, 504.

Sakai, M. (1993) Energy Principle of Indentation-induced Inelastic Surface Deformation, *Acta Met. et Mater.* **41**, 1751.

Samuels, L.E. (1986) Microindentations in Metals, in P. Blau and B.R. Lawn (eds), *Microindentation Techniques in Materials Science and Engineering*, American Society for Testing and Materials, Philadelphia, PA. pp 5-25.

Sargent, P.M. (1986) Use of the Indentation Size Effect on Microhardness for Materials Characterization, in P. Blau and B.R. Lawn (eds), *Microindentation Techniques in Materials Science and Engineering*, American Society for Testing and Materials, Philadelphia, PA. , 1160-174.

Shorshorov, M.Kh., Bulychev, S.I. and Alcklin, V.P. (1981) Work of Plastic and Elastic Deformation During Indentation, *Soviet Phsics Doklady* **26** , 769.

Snedddon, I.N. (1965) Relation between Load and Penetration in the Axisymmetric Boussineq Problem for a Punch with Arbitary Profile, *Int. J. Eng. Sci.* **3**, 47.

Stilwell, N.S. and Tabor, D. (1961) Elastic Recovery of Conical Indentations, *Proc. Roy. Soc. (London)* **78**, 169.

Stone, D.S., Wu, T.W., Alexopoulos, P.S. and DeFontaine, W.R. (1989) in J.C. Bravman W.D. Nix, D.M. Barnett and D.A. Smith (eds), Materials Research Society, Pittsburgh, PA, pp 105-110.

Tabor, D. (1948) A Simple Theory of Static and Dynamic Hardness, *Proc. Roy. Soc. (London)* **A192**, 247.

Tabor, D. (1951) *The Hardness of Metals*, Clarendon Press, Oxford.

Tabor, D. (1986) Indentation Hardness and Its Measurement, in P. Blau and B.R. Lawn (eds), *Microindentation Techniques in Materials Science and Engineering*, American Society for Testing and Materials, Philadelphia, PA. pp 129-159.

Twigg, P.C., McGurk, M.R., Haniesworth, S.V. and Page, T.F (1996) Apparent Indentation Plasticity in Ceramic Coated Systems, in *Plastic Deformation of Ceramics III*, J. Routbort, R.C. Bradt and C. Brooks, Plenum Press, NY.

STRESS IN THIN FILMS FOR MEMS ACTUATORS

PAUL H. HOLLOWAY, JONATHAN GORRELL AND KENNETH SHANNON, III
Department of Materials Science and Engineering
University of Florida
Gainesville, FL 32611-6400

Abstract

The use of commercial aluminum alloys (T201, 2090 and 5052) plus an aluminum intermetallic phase (Al_3Ni) has been studied for thermally actuated MEMS. It was shown that strengthening mechanisms commonly employed in bulk Al alloys (solid solution strengthening, precipitation hardening and grain stabilization) can be used to increase the residual stress in bimetallic couple (Si and metal) by 50% to 400%. In addition, stress relaxation rates measured between 50°C and 350°C were dramatically lower for the commercial alloys up to 150°C. For Al_3Ti, no stress relaxation was detected even at 350°C. At least two mechanisms lead to stress relaxation, and the mechanism operating over times of <2 hours was suggested to be dynamic strain aging due to formation of precipitates. The second mechanism with a time constant of about 30 hours was speculated to be logarithmic creep. Oxidation was shown to change the stress in 2090 alloys at temperatures above 150°C.

1. Introduction

Microelectromechanical structures (MEMS) have been the subject of a number of conferences and reviews [1-6]. Uses envisioned for MEMS include high speed digital switches, adaptive optics, optical displays, miniature robots, etc., and all require some mechanism of mechanical actuation. Actuation has been achieved based on a number of principles, including electrostatic, electromagnetic, piezoelectric, magnetostrictive, and thermal actuation including thermopneumatic, shape memory, and bimetallic structures [5]. The emphasis in the present work has been on bimetallic structures actuated thermally, which has often been used in actual products [5] including a micro-valve [7].

The principle of actuation using a bimetallic element is shown in Fig. 1 for a cantilevered strip. In this case, Al with a high coefficient of thermal exansion (CTE = 25 ppm/°C) is

B. Bhushan (ed.), Tribology Issues and Opportunities in MEMS, 509-518.

shown as a thin film (thickness of ≈ 1 μm) deposited onto Si with a low CTE (2.3 ppm/°C). Stress in the bimetallic couple causes the cantilevered beam to deflect either down (compressive in Al versus Si) or up (tensile in Al). The origin of the stress in the couple is either defined to be intrinsic or extrinsic, depending upon whether it results from the deposition and processing steps to deposit the Al or thin the Si, or whether it results from temperature effects such as heating and cooling the couple, respectively. Intrinsic stress from deposition of thin films has been discussed by Thorton and Hoffman [8,9] and Dorner and Nix [10]. These stresses may be varied by changing the conditions of deposition of the thin films [8-10]. However, stress in bimetallic couples are routinely dominated by extrinsic stresses, especially for thermal actuation over a temperature range which causes plastic deformation [10,11]. In this actuation method and assuming the stress of the bimetallic couple is zero at ambient room temperature (RT), heating the couple will cause the Al to expand at a rate greater than Si and will cause the beam to deflect downward, as shown in Fig. 1b. The curvature of the cantilevered beam is related to the geometry of the couple and moduli of the materials as described by Timoshenko [12].

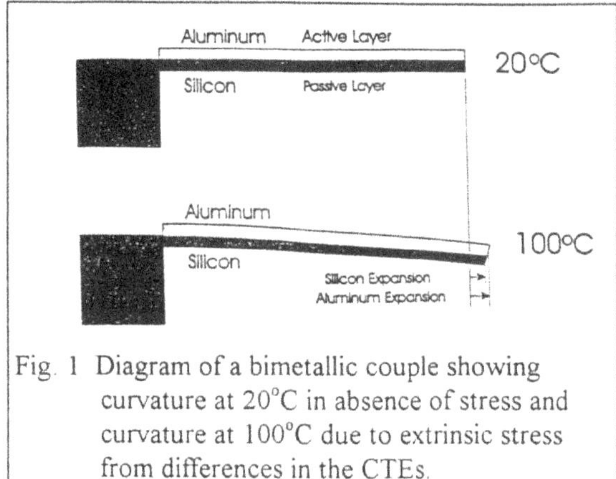

Fig. 1 Diagram of a bimetallic couple showing curvature at 20°C in absence of stress and curvature at 100°C due to extrinsic stress from differences in the CTEs.

The expression for curvature from Timoshenko assumes perfectly elastic behavior from both materials in the bimetallic couple. As shown by a number of investigators [10, 13-17], real materials often exhibit plastic deformation during thermal cycling, especially if the yield stress is low and the difference in CTE is high, as for Al alloys. Plastic deformation limits the level of stress that can be achieved in thermal actuation and therefore limits the force and displacement that can be achieved in either micro-relays or valves. As a result, it would be desirable to increase the yield strength of the thin film metallization to increase the achievable force and deflection. In addition, Draper and Hill [14] have shown that the biaxial stress in bimetallic couples will relax at temperatures near RT. It would also be desirable to reduce stress relaxation in order to maintain a constant force or deflection. As a result, the current study was conducted to determine if mechanism used to strengthen bulk alloys might be used to increase the strength and reduce the relaxation in thin films of Al alloys.

2. Experimental Procedures

Thin films ≈ 1 μm thick were sputter deposited onto (100) oriented 4" diameter Si wafers with a 100 nm thermally grown SiO_2 layer on the surface. The wafers were cleaned using the RCA procedure prior to sputter deposition. The samples were loaded into a vacuum system and pumped to less 3 x 10^{-7} Torr, followed by deposition from 8" DC planar magnetron sources at 5KW in 8 mTorr of Ar with the wafers at 100°C. Films of pure Al, Al-Si-Cu, or commercial alloys designated as T201, 5052 or 2090 were deposited. The composition and properties of these alloys are shown in Table 1. Sputter targets for the commercial alloys were cut from sheet stock. In the case of Al_3Ti intermetallic thin films, alternating layers of pure Al and Ti were deposited using electron beam evaporation from hearths containing 99.999% pure Al or Ti. Electron beam evaporation was performed in a diffusion pumped, liquid nitrogen trapped vacuum system operating at a base pressure of 5 x 10^{-7} Torr. During deposition, the pressure typically rose to 5 x 10^{-6} Torr and the substrate was heated to 150°C.

For a general discussion of Al alloys, and those used in this study, see reference [21]. Alloys of Al-Si-Cu have the nominal composition shown in Table 1. The Al-Si-Cu alloy is the standard metallization for VLSI interconnects, where the Si is added to avoid solubilization of the Si wafer which may result in "bird's beaks", and Cu is added to avoid electromigration [20]. The concentration of 1.5%Si is soluble in Al at temperatures of ≈ 500°C, but form either a metastable solid solution or precipitates of pure Si in the grain boundaries at RT. The Cu forms non-coherent θ phase Al_2Cu precitates mainly in the grain boundary. Thus neither Si or Cu contribute significantly to strengthening of the Al-Si-Cu alloy. T201 alloys are strengthened by a dispersed θ Al_2Cu precipitate stabilized to higher temperatures with the addition of Ag, Mn and Mg. Additional second phase may form from the Mg and Mn alloying elements. Ag also contributes to solid solution strengthening and Ti pins the grain boudaries to maintain a smaller grain size and consequent higher yield strength. 5052 alloys are strengthened by formation of

Table 1: Composition and Properties of Al Alloys

Alloy	Composition (%)	Melting Point (°C)	Young's Modulus (GPa)	CTE (ppm/°C)
Pure Al	99.99	650	69	25
Al-Si-Cu	1.5Si-2.5Cu	620	70	23
T 201	4.6Cu-0.57Ag-0.36Mn-0.2Mg-0.27Ti	620	70	22.7
5052	2.5Mg-0.25Cr	610	70	24.8
2090	2.57Cu-2.1Li-0.12Zr	610	70	23
Al_3Ti	25Ti	1340	100*	14

*-estimated

metastable Al_2Mg and stable Al_3Mg_2 precipitate, and grain boundary pinning by the Cr. 2090 Al alloys are some of the newest alloys, and strengthened by Al_2Li-Cu precipitates with Zr added to stabilize grain boundaries and maintain a smaller grain size. The Al_3Ni is an ordered intermetallic phase with high yield and ultimate tensile strengths.

Stress and stress relaxation in the bimetallic couples were determined using a FleXus model 2320 from Tencor, Inc. This instrument uses an optical lever to measure the curvature of the sample before and after deposition of the metal thin film, and before and after cooling to as low as -60°C or heating to as high as 500°C. Stoney's formula [18] was used to calculate the average biaxial stress, σ, in the thin metallization:

$$\sigma = (E\,h^2) / (\{1 - \nu\}\ 6\,R\,t)$$

where R is the curvature of the sample, E is Young's modulus, h is the substrate thickness, ν is Poisson's ratio, and t is the film thickness. The FleXus was also used to measure the isothermal stress relaxation by heating to 350°C at a rate of 1°C/min and holding at 350°C for 30 min, cooling to RT at 1°C/min, then heating again at 1°C/min to the test temperature and measuring the sample curvature at least every 15 min for times up to 48 hours.

The crystalline structure and microstructure was determined using X-ray diffraction on a Philips APD 3720 controlled by a PC using Philips PW 1877 Automated Powder Diffraction software. A Cu anode was operated at 40 KV and 2 mA with a Ni filter. A JEOL 200 CX analytical transmission electron microscope was used to examine plan view and cross sections of the metallization before and after heating. The surface morphology and topography was determined using a JEOL 6400 scanning electron microscope or a Digital Instruments Nanoscope III atomic force microscope (AFM). The AFM was used in the tapping mode and the Digital Instruments computer software used to quantitate the RMS surface roughness. Auger electron spectroscopy from a Physical Electronics Model 660 scanning Auger microprobe was used to determine the surface composition over broad areas and for spots of dimensions $\approx$ 1-10 μm. In cases where the composition below the surface was desired, 3 KeV Ar^+ ions were used for sputter depth profiling [19].

3. Results and Discussion

3.1 STRESS VERSUS TEMPERATURE

The stress versus temperature between RT and 250°C is shown in Fig. 2 for pure Al, Al-Si-Cu, T201 and 2090 alloys. Pure Al exhibits a residual tensile stress of $\approx$ 150 MPa at RT, which becomes compressive above $\approx$ 90°C and maintains the compressive yield strength of $\approx$ 25 MPa between $\approx$ 120°C to 250°C. Upon cooling, the stress becomes tensile until reaching the tensile yield strength of $\approx$ 50 MPa at $\approx$ 220°C. The tensile stress increases with decreasing temperature because of increasing yield strength and work hardening at lower temparatures, until the RT stress of ≈ 150 MPa is again reached. The behavior of Al-Si-Cu is similar except that the RT stress is ≈ 220 MPa, which causes the cross over from tensile to compressive stress to occur at ≈ 150°C. On the other hand, the behavior of the T201 and 2090 alloys are quite different. Not only are their RT stresses higher at 340 and 500 MPa, respectively, but they do not exhibit plastic limits and deformation either in compression or in tension (over this temperature range). In addition, the T201 alloys crosses over from tensile to compressive stresses at ≈ 180°C, while the 2090 alloys remains tensile even at 250°C.

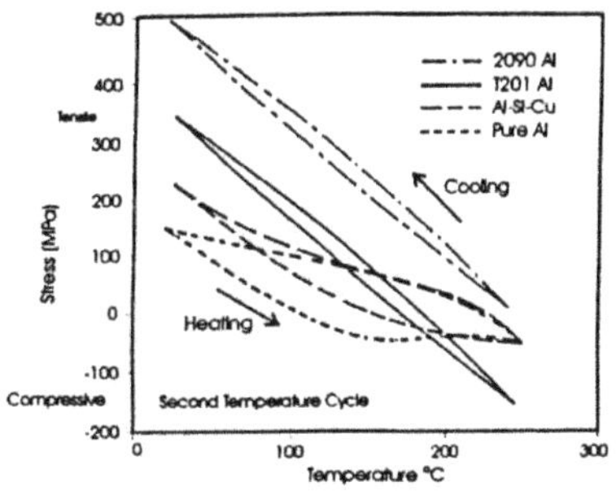

Fig. 2 Stress versus temperature between RT and 250°C for pure Al, Al-Si-Cu, T201 and 2090 alloys.

The behavior of 5052 alloy between RT and 350°C is shown in Fig. 3, and the RT tensile stress is similar to T201. Above 200°C, it exhibits compressive plastic flow with recovery upon cooling from 350°C back to RT. If T201 and 2090 alloy films were heated above ≈ 250°C and ≈ 300°C, respectively, they also exhibited high temperature plastic limits in compression.

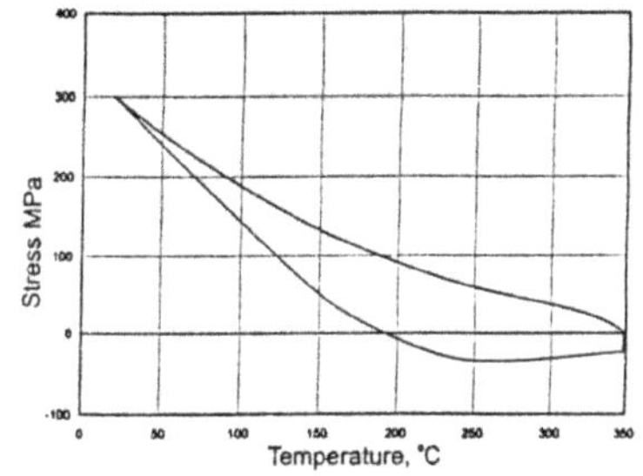

Fig. 3 Stress versus temperature up to 350°C for 5052 alloys showing higher RT stress and plastic deformation above ≈ 200°C.

The stress versus temperature up to 500°C is shown in Fig. 4, for both the first and second cycle of heating of an Al_3Ti thin film. Upon the first cycle of heating, the film consolidated and the layered structure from electron beam deposition interdiffused to form the Al_3Ti intermetallic, therefore the stress started low (≈ 200 MPa) and became compressive until it increased to zero at ≈ 500°C. Upon cooling to RT, the stress became increasingly tensile, reaching 800 MPa. Upon reheating to 500°C, the sample showed plastic behavior and did not become compressive, even at this high temperature.

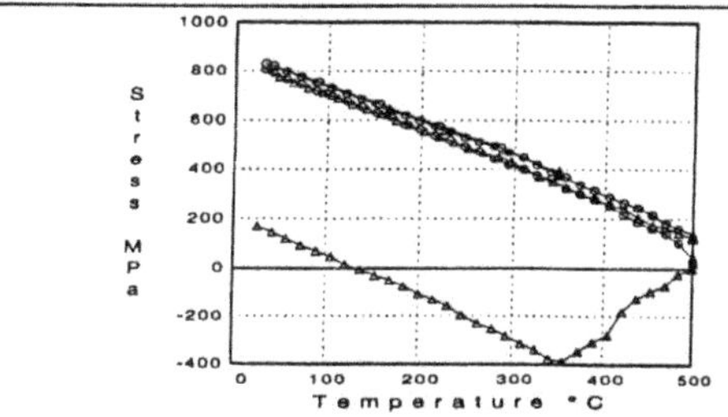

Fig. 4 Stress versus temperature up to 500°C for Al_3Ti intermetallic thin films. The behavior for both the first and second heating cycle is shown; see text for description.

In summary, the stabilized precipitates, solid solution

strengthening, and stabilization of grain size used in T201, 5052 and 2090 alloys are effective for increasing the RT residual stress and thereby allowing greater operating temperature ranges for thermally actuated MEMS. The residual stress in the Al_3Ti intermetallic film is highest of all the materials studied, and no plastic deformation was observed up to a temperature of 500°C.

3.2 STRESS RELAXATION

Upon heating the samples to an elevated temperature (e.g. 350°C) and cooling to RT, then measuring stress versus time at RT up to 150°C, stress relaxation was observed. The residual stress versus time at 50°C is shown in Fig. 5 for pure Al, Al-Si-Cu and T201. In times of ≈ 1000 min, the stress had decreased to ≈ 80% of the initial value for pure Al and Al-Si-Cu. However, T201 only relaxed a few percent of the initial residual stress. Note from Fig. 2 that the initial residual stress for T201 was twice that of pure Al and about 60% greater than for Al-Si-Cu. Therefore not only did a lower fraction of the stress relax for T201, this small relaxation was observed at a much higher value of stress.

The stress relaxation curve for 5052 alloy at 125°C is shown in Fig. 6, and the data can be fitted by two exponential curves at short times (<200 min) and long times (T>200 min). Note that the initial residual stress at 125°C was 106 MPa, and this relaxed to a value of ≈ 85 MPa after 3000 min. The fitted exponential equation indicates that the stress should relax to a value of 80 MPa after infinite time. The requirement of two exponential equations to fit the stress relaxation data suggest that two mechanisms are controlling this process, one at $t < 200$ min and another at $t > 200$ min, as discussed below. The fitting data to describe stress relaxation for T201 and 5052 are shown in Table 2. An entry of zero indicates that this exponential was not observed under the conditions specified. Specifically, no short time constant was observed for T201 at 125 °C, no residual stress or short time constant for 5052 at 150 °C, and no relaxation was observed for Al_3Ti at 350 °C. Stress relaxation data for pure Al, Al-

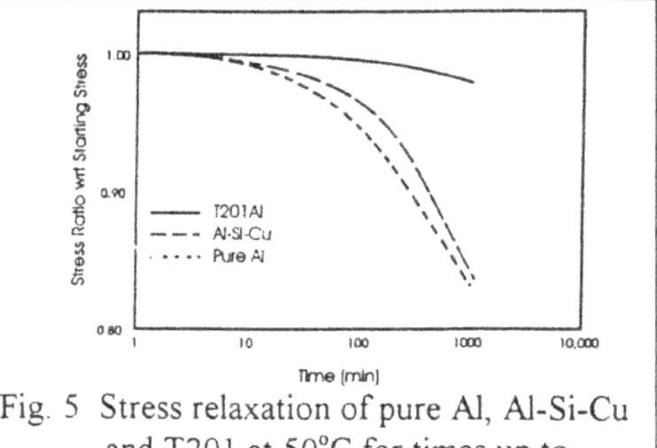

Fig. 5 Stress relaxation of pure Al, Al-Si-Cu and T201 at 50°C for times up to 1000 minutes.

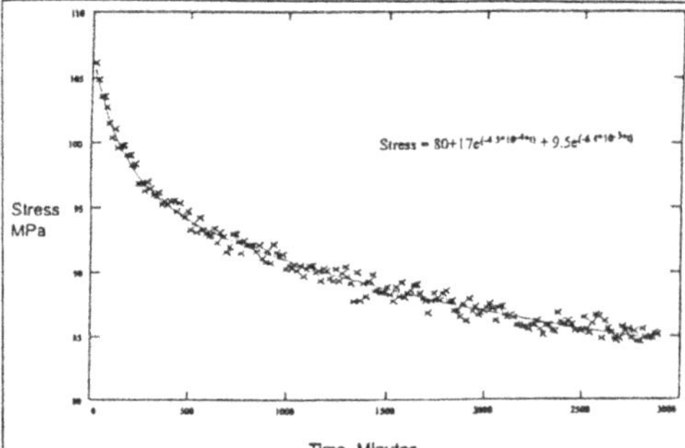

Fig. 6 Stress relaxation curve for 5052 alloy at 125°C. The X are data points and the solid line is a curve fit expression consisting of two exponential curves as labeled in the figure.

Table 2 Residual stress after infinite relaxation time, σ_∞, change in stress at short ($\Delta\sigma_<$) and long ($\Delta\sigma_>$) times, and the time constants (τ) for the two exponential equations.

Alloy/Temp. (°C)	σ_∞(MPa)	$\Delta\sigma_<$ (MPa)	$\tau_<$ (min)	$\Delta\sigma_>$ (MPa)	$\tau_>$ (min)
T201/50	330	9.2	196	49	1820
/75	240	13.8	220	26	1940
/100	160	608	220	29	1400
/125	33	0	0	81	960
/150	40	-3	200	0	0
5052/50	240	0	0	30	2800
/75	150	12	200	33	1500
/100	110	19	170	29	1380
/125	80	10	160	17	2200
/150	0	0	0	58	50000
Al_3Ti/350	380	0	0	0	0

Al-Si-Cu and 2090 were incomplete, but pure Al and Al-Si-Cu data were similar to that reported by Draper and Hill [14]. The data for 2090 was similar to that from T201 alloy [22]. Note that for T201 at 150°C no relaxation was observed. Instead the stress became slightly more tensile over a period of a few hours. This was the result of oxidation of the alloy under these conditions. In fact, oxidation caused dramatic fluctuations in stress in other materials. While the details of oxidation induced stress will not be discussed here, this effect is very important in thermally actuated MEMS devices [22].

In order to clarify the mechanism of stress relaxation, the amount and change of precipitates were carefully measured using XRD of 5052 alloys. Precipitates of metastable Al_2Mg were identified [24], but the signal/noise ratio was poor and no significant change could be detected versus relaxation. A second order diffraction peak from the single crystal Si substrate was initially confused with the possible formation of equilibrium Al_3Mg_2 precipitates. While no increased second phases were detected, the lattice constant of the alloys was observed to decrease for samples thermally cycled to 350°C and returned to RT, versus one held at 50°C or 125°C for two hours. After the initial decrease in lattice parameter over the first two hours, it remained constant for up to 32 hours. Consistent with this observation, the surface roughness of the alloys increased from an RMS value of about 4 nm at zero relaxation to 6 nm for samples relaxed for 2 hours at either 50°C or 125°C [22]. Again, the surface roughness was constant for relaxation times > 2 hours. Since no new or increased precipitation could be determined, these changes in lattice parameter and surface roughness could either be associated with an undetected lattice precipitate or with Andrade low temperature creep.

4. Discussion

As is obvious from Figs. 2-4, there is a large difference in the RT residual stress between the alloys tested. An attempt was made to develop a universal figure of merit (FOM) for comparison of one alloy to another, but all of them had too many qualifying parameters. In considering FOMs, the following properties were considered: (i) tensile and yield strength, both at RT and as a function of temperature, (ii) RT residual stress, (iiii) homologous temperature for the maximum operating temperature (homologous temperature is defined to be the temperature of interest divided by the melting point, both in °K), (iv) CTE, (v) Young's Modulus. Depending upon whether the primary requirement is force or displacement of the MEMS structure, a first order FOM would be the RT residual stress (see Figs. 2-4) multiplied by either Young's Modulus or by the CTE, respectively. These data are shown in Table 2. Based on these FOMs, the Al_3Ti film is superior to all others for generation of force because of its' high modulus and residual stress. It is also good for displacement, however its' low CTE reduces its' superiority. In fact the displacement FOM for T201 is equal to that of Al_3Ti. Note from Table 1 however, that the melting point of T201 is considerably below that of Al_3Ti, and based upon the fact that atomic processes become very fast at homologous temperatures above 0.5, a much more limited temperature range is expected for T201 than for Al_3Ti. In fact, a homologous temperature of 0.5 corresponds to a real temperature of 175°C and 535°C for T201 and Al_3Ti, respectively. For thermal actuation above these temperatures, less force and displacement and rapid stress relaxation would be expected. In addition, oxidation of the alloys could be more rapid at these temperatures which could dramatically modify the residual stress, especially for 2090 [22].

Table 3 RT residual stress and FOM for force or displacement for the alloys tested.

Alloy	RT Residual Stress (MPa)	Force FOM[1]	Displacement FOM[2]
Al	150	1.1	3.8
Al-Si-Cu	220	1.6	5.3
T201	500	3.5	11.3
5052	300	2.1	7.4
2090	350	2.5	8.1
Al_3Ti	800	8.0	11.2

1-RT residual stress times Young's Modulus/10^4
2-RT residual stress times CTE/10^3

The mechanisms responsible for stress relaxation in these thin film bimetallic actuators is unclear. As demonstrated in Fig. 6 and in Table 2, there appears to be one mechanism operative over the first 200 minutes, followed by a second mechanism with a time constant for relaxation of about 2000 minutes. There is a lower limit of stress below which relaxation will not occur. This lower limit is dependent upon temperature, but it is unclear as to whether this results from the fact that the stress is also temperature dependent. The known mechanisms of relaxation are well recognized to be temperature and stress dependent [11,23]. A major problem in trying to determine the mechanisms in these thin films is the inability to vary stress and temperature independently in order to verify or disprove a particular mechanism of relaxation.

The most widely characterized methods of stress relaxation are dynamic strain aging, anelastic creep, Herring-Nabarro-Coble creep, logarithmic or low temperature creep, and high temperature or Andrade creep [11,23]. In dynamic strain aging, changes in the microstructure can result in modification of the stress state. An example would be formation or solubilization of precipitates. A second example would be modification of Cottrell atmospheres restricting dislocation glide [23]. In this case, the relaxation would clearly depend upon temperature, but stress would play a secondary role in the process. In the case of anelastic creep, the most common example would be motion of interstitial impurities to regions of low stress with a time constant of hundreds to thousands of minutes. This would be dependent primarily upon the stress, with the requirement of a homologous temperature high enough for interstitial diffusion. In the case of Herring-Nabarro creep, the stress creates a net flux of vacancies from regions under high tensile to those with low tensile or even compressive stress, with a concomitant flux of atoms in the reverse directions. This typically requires both high stress and high temperature for vacancy lattice diffusion. Coble creep extends this same concept to lower temperatures by recognizing that grain boundaries are commonly regions for faster diffusion at lower temperatures. Finally, Andrade creep is the classical model in which mobile dislocations glide on slip systems until depleted, then continued relaxation requires dislocation climb and even dislocation multiplication in order to observe continued stress relaxation. The initial stage of Andrade creep (dislocation glide) can be observed at low temperature and higher stress, but the latter stages of this mechanism require both higher temperatures and stress.

The data reported above for rapid changes in lattice constant and surface roughness for times of less than 2 hours suggest that the origin of the first rapid mechanism is glide of mobile dislocations to reduce the strain energy. Initially we thought that dynamic strain aging from precipitation of the equilibrium Al_3Mg_2 was occurring, since the density of this precipitate was 2.22 g/cm^{3}versus the average density of 5052 of 2.68 g/cm^3. Thus precipitation should lower the tensile stress in the bimetallic structure. However, if precipitation occurred, it was not detectable by our XRD data. As a result, we are left to speculate that this initial rapid relaxation which changes the lattice parameter and surface roughness is due to mobile dislocation glide, which is rapid, relatively independent of stress, and has a low activation energy. The second longer time constant process is believed to result from Andrade creep where limited dislocation climb allows continued slip and consequent stress relaxation. However the data to support these mechanism are lacking, and definitive conclusions cannot be reached. Further work is required.

5. Conclusions

The use of commercial aluminum alloys and an intermetallic Al phase for metallization of MEMS has been studied in thermally actuated structures. The performance of T201, 2090 and 5052 Al alloys and Al_3Ti was compared to that of pure Al and Al-Si-Cu sputtered deposited films. The commercial alloys resulted in a factor of 50% to 200% increase in stress, which translates into larger force or larger displacement in MEMS devices. In addition, the stress relaxation is much smaller for the commercial alloys at temperatures up to 150°C. In the case of Al_3Ti, the stress was four time larger than for Al-Si-Cu films, and no stress relaxation was observed even at a temperature of 350°C. There was evidence that oxidation changed the stress state of 2090 films at 150°C, but no significant oxidation effects were observed for Al_3Ti, even at 350°C. The mechanism of stress relaxation was discussed and dynamic stain aging due to formation of precipitates in the lattice was concluded to occur within the first two hours for T201 and 5052 alloys. A stress relaxation mechanism was observed to occur with a time constant of $\approx$ 2000 min. and speculated to be Andrade creep, but the atomistic mechanism for this relaxation could not be identified with certainty.

6. Acknowledgements

This work was supported by DARPA contract DABT63-95-C-0038 through ICSensors. Contributions from Hal Jerman (formerly at ICSensors), Vladimir Vaganov and Nicolai Belov are gratefully acknowledged.

7. References

1. Hogan, H. New Scientist **150**, 28 (1996).
2. Frank, R., IC&S, April 1997, p. 41.
3. Angell, J.B.,Terry, S.C and Barth, P.W., Scientific American **248**, 44 (1984).
4. Beardsey, T., Scienctific American **275**, 32 (1996).
5. Madou, M, *Fundamentals ofMicrofabrtication* (CRC Press, Baca Raton LLC, NY, 1997).
6. Maseeh, F., Solid State Technology, October 1995, p. 50.
7. Jerman, H, J. Micromach. Microeng. **4**, 210 (1994).
8. Thorton, J.A. and Hoffman, D.W., Thin Solid Films **171**, 5 (1989).
9. Hoffman, R.W., Thin Solid Films **34**, 185 (1976).
10. Dorner, M.F. and Nix, W.D., Crit. Rev. Solid State Materials Sciences **14**, 225 (1988).
11. Ohring, M, *The Materials Science of Thin Films* (Academic Press, NY, 1992).
12. Timoshenko, S. J. Opt. Soc. Am. **11**, 23 (1925).
13. Gardner, D.S, and Flinn, P.A., J. Appl. Phys. **67**, 1831 (1990).
14. Draper, B.L. and Hill, T.A., J. Vac. Sci. Technol. **B9**, 1956 (1991).
15. Korhonen, M.A., Black, R.D. and Li, C.-Y., J. Appl. Phys. **69**, 1748 (1991).
16. Baldwin, F., Holloway, P.H., Bordelon, M. and Watkins, T.R., in *Stress-Induced Phenomena in Metallization*, Ed. by P.S. Ho, J. Bravman, C.-Y. Li and J. Sanchez, (AIP, NY, 1996), AIP Conf. Proc. 373, p. 81)
17. Baldwin, F., Holloway, P.H., Bordelon, M. and Watkins, T.R., *Mat. Res. Soc. Symp. Proc.* **309**, 261 (1994).
18. Stoney, G.G., Proc. Royal Soc. London **82**, 172 (1909).
19. Hudson, J.B., *Surface Science: An Introduction* (Butterworth-Heinemann, Boston, 1992).
20. Mayer, J.W. and Lau, S.S., *Electronic Materials Science: For Integrated Circuits in Si and GaAs*, (Macmilland, NY, 1990).
21. Metals Handbook, Volume 2: *Properties and Selection: Nonferrous Alloys and Special Purpose Materials*, Tenth Edition (American Society for Metals, Metals Park, OH, 1992).
22. Gorrell, J.F., "Thin Film Metallization for Micro-Bimetallic Actuators", PhD Dissertation, University of Florida, December, 1997.
23. Cahn, R.W. and Haasen, P., *Physical Metallurgy*, 3rd Edition (North-Holland, NY, 1983).

ASM Handbook, Volume 3: *Alloy Phase Diagrams*, Tenth Edition (American Society for Metals, Metals Park, OH, 1992.

RELIABILITY AND FATIGUE TESTING OF MEMS

CHRISTOPHER MUHLSTEIN AND STUART BROWN
Failure Analysis Associates, Inc.
Three Speen Street
Framingham, MA 01701

Abstract

Microelectromechanical structures (MEMS) utilize brittle materials such as polycrystalline silicon (polysilicon) under potentially severe mechanical and environmental loading conditions. These structures may be subjected to high frequency, cyclic loading conditions, accumulating large numbers of cycles in relatively short periods of time. Failure Analysis Associates has developed a technique for characterization of fatigue crack initiation and growth in MEMS materials. Preliminary results show that fatigue cracks can initiate and grow in polysilicon MEMS devices, and that water vapor is important in sub-critical crack advance. This research provides a basis for characterizing long-term durability and stability of micron-scale structures.

1. Introduction

State-of-the-art inertial guidance systems, airbag deployment sensors, and active control surfaces for aircraft integrate a variety of often brittle materials such as polycrystalline silicon (polysilicon) and nitride films under potentially severe mechanical and environmental loading conditions. These structures may be subjected to cyclic loading conditions with of kilo and megahertz frequencies, accumulating large numbers of cycles in relatively short periods of time. The application of thin films of materials such as polysilicon in safety-critical devices and structures represents a significant challenge for conventional bulk materials characterization and design approaches.

Macro-scale fatigue crack initiation and growth characterization techniques (Dowling 1996; Saxena and Muhlstein 1996) have developed largely due to the needs of the aerospace industry. The fracture and fatigue behavior of a variety of bulk ordered intermetallics and ceramics have been evaluated for potential aerospace and biomedical applications.(Evans 1980; Ewart and Suresh 1986; Dauskardt, Yu et al. 1987; Rao, Kim et al. 1995) These and related studies have established the importance of subcritical crack growth in brittle materials and design methodologies to mitigate the risks associated with placing such materials in service. Thin film polycrystalline silicon is one of the most common materials used in microelectromechanical systems (MEMS) to date. Consequently, the mechanical properties of this brittle material are critical to reliability and performance of MEMS devices.

B. Bhushan (ed.), Tribology Issues and Opportunities in MEMS, 519-528.

Basic mechanical properties of thin films such as Young's modulus (E), Poisson's ratio (υ), and tensile strength are not completely characterized, and the applicability of standard test techniques and specimen geometries remain undetermined. Current studies have remained limited to "microtensile" testing (Sharpe, Vaidyanathan et al. 1996; Sharpe, Yuan et al. 1996; Sharpe, Yuan et al. 1997) and a variety of novel test structures.(Johansson and Schweitz 1988; Hong, Weihs et al. 1989) A similar lack of understanding is found in failure modes traditionally associated with long-term durability such as static and cyclic fatigue as well as time-dependent properties such as creep. With the exception a few investigators (Bhaduri and Wang 1983; Chen and Leipold 1986; Connally and Brown 1992; Connally and Brown 1992; Brown, Arsdell et al. June 16-19, 1997), subcritical crack growth in many MEMS materials remains unexplored and is a crucial issue which must be addressed. The anecdotal evidence from investigators who have resonated microdevices without noticing changes in performance must not be considered proof of long term stability.

This investigation sought to develop a specimen geometry and test structure for characterization of fatigue crack initiation in thin films and to use this geometry to explore fatigue crack initiation in polycrystalline silicon. Furthermore, the geometry developed during the course of this investigation was intended to be a standard technique which is both robust and amenable to a variety of materials and processing platforms.

2. Materials and Experimental Procedures

Fatigue crack initiation studies are typically conducted per the recommendations of ASTM E 466. This standard addresses appropriate specimen design and techniques for bulk metallic materials. In fatigue crack initiation testing, specimens are usually subjected to uniaxial, cyclic loads until the accumulation of a preselected plastic strain (i.e. 1%) or until complete separation of the specimen occurs. The mean and amplitude of the loading may be varied to probe various aspects of the material behavior and to reflect actual service conditions. The time to failure, or "life" of the specimen is then represented as a function of the range of applied cyclic stress or strain. Provided the specimen has been properly designed, this data may be used to estimate component life and to minimize the risk of fatigue failures.

Fatigue specimens must be designed to fail in the gauge section under well-defined loading conditions. Furthermore, material anisotropy, residual stresses, and processing limitations must also be considered. Finally, the actual service conditions of the engineering components must be considered to ensure the fatigue life data ma be used for remaining life predictions and engineering design. The extreme difference in scale between MEMS and conventional structural components is a significant barrier to material property characterization. Testing geometries must be MEMS-scale due to size effects related to the sensitivity of brittle materials to flaws. In addition, the geometry must minimize the effects of residual stress, be conveniently loaded, and carefully monitored. Consequently, the materials and processing constraints common to MEMS play an important role in the development of the specimens used in this study.

2.1. MATERIAL

Thin film polysilicon has been shown to display limited ductility (Sharpe, Vaidyanathan et al. 1996; Sharpe, Yuan et al. 1996; William N. Sharpe, Yuan et al. 1996; Sharpe, Yuan et al. 1997) and fracture toughness (Rybicki 1988) at room temperature. Due to the brittle nature of polysilicon, it is crucial that specimens reflect the loading conditions and scale expected in service, particularly because etching or release processes may cause variations in surface roughness that in turn affect strength and fatigue resistance.

The polycrystalline silicon tested in this study was manufactured at MCNC during runs 11 and 13 of the multi-user MEMS process (MUMPs). The nominally 2 μm thick polysilicon was deposited by low pressure chemical vapor deposition (LPCVD) in a three-layer polysilicon surface micromachining process. The specimen was fabricated from the POLY1 layer of the MUMPS process. Microstructural characterization studies of MUMPs polysilicon suggest that MUMPs polysilicon consists of approximately 0.5 μm diameter columnar grains which extend through the thickness of the film.(Koester 1997)

2.2. SPECIMEN GEOMETRY AND PREPARATION

The specimen geometry used for this investigation is the in-plane, resonant cantilever structure pictured in Figure 1. The specimen is fixed at the base of a short beam that is in turn attached to a large plate that serves as the resonant mass. Opposite sides of the plate include comb drive structures. One side is for electrostatic actuation; the other side provides capacitive sensing of motion. Applying an alternating voltage to the actuation comb drive at the appropriate frequency induces a resonant response in the plane of the figure.

A stress concentration is introduced in the beam through the mask set, as shown in Figure 1. The radius of the stress concentration and the remaining beam ligament were selected to ensure that the specimen could be broken immediately at resonance. The longer-term fatigue response can then be measured by exciting the specimen at some fraction of the short time breaking amplitude. It should be emphasized that the stress concentration presented here is used to investigate predominantly crack initiation. This geometry has been widely distributed and used on a wide variety of platforms including multiple polycrystalline silicon, single crystal silicon, and aluminum fabrication lines. Another similar device using a deliberately introduced crack is used to evaluate crack propagation.(Arsdell 1997) For many MEMS, the most relevant issue is initiation and not growth because once a crack has initiated it rapidly propagates to failure.

The 30 μm wide, 250 μm long cantilever resonant structure was nominally 2 μm thick. The specimen resonates in the plane of the figure, generating fully reversed (maximum load/minimum load = R = -1) loading conditions at the notch. The root radius and remaining ligament of the beam were selected to ensure that failures occurred within the gauge section, and that a representative volume of material was tested at the maximum stress amplitude. The cantilever beam geometry provided well-characterized boundary conditions not possible with multiply-supported structures. The on-chip device also ensured proper specimen alignment, well characterized loading conditions, and that data was representative of MEMS-scale structures.

The sacrificial oxide from the micromachining process was removed from the fatigue specimens using standard hydrofluoric acid release procedures. Devices were placed in 49% HF for 2.5 minutes followed by rinsing for several minutes in deionized water. The structures were then rinsed in alcohol and dried for 10 minutes in air at 110°C. Specimens were then die and wire bonded in ceramic dual inline packages (DIPs) to facilitate testing.

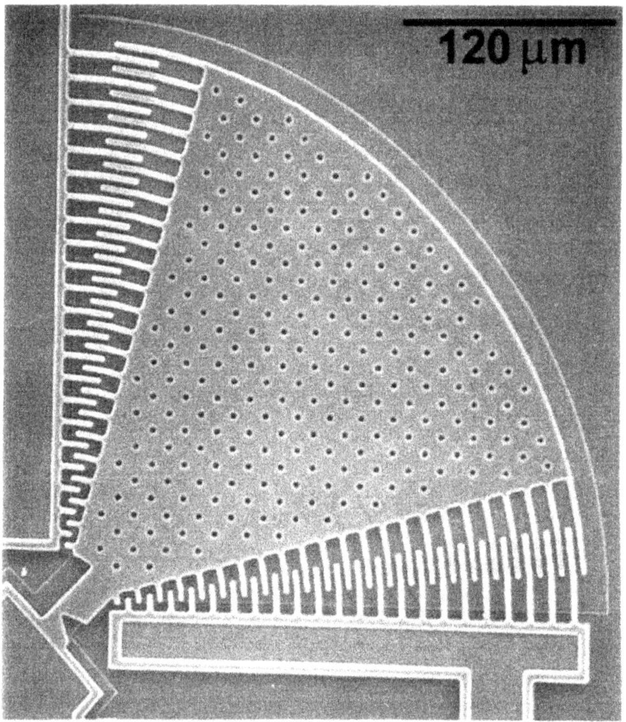

Figure 1. Fatigue crack initiation specimen.

2.3. TESTING METHODOLOGY

The device is driven at resonance using the following control scheme. The first mode resonant response of the specimen is determined by sweeping a range of frequencies around the expected response and monitoring the amplitude of response. The peak amplitude is selected by fitting a second order polynomial to the peak and extracting the maximum. The specimen is excited at the peak frequency at a defined excitation voltage for a period of time. The frequency response is then again evaluated by sweeping around the excitation frequency. Over time this permits measuring any change in resonant frequency and consequently any change in the mechanical response of the specimen.

Tests were conducted under controlled environmental conditions (27±0.1°C, 75 ± 1% relative humidity) and laboratory air until complete separation occurred at the notch. Specimens were allowed to resonate for 1 to 5 minutes followed by recharacterization of the resonant frequency. The notched beams were loaded with a sinusoidal wave form from 70 to 80 V_{rms} ± 1% at one half the resonant frequency (approximately 40 kHz). These conditions generated fully reversed (minimum load/maximum load = R = -1), constant amplitude loads at the notch. As stated earlier, changes in resonant frequency were monitored by sweeping the drive frequency ± 50 Hz and recording the amplitude response. The peak response from a given sweep was fit with a second order polynomial, and the maximum response was used as the drive frequency for the next test interval. Changes in resonant frequency may be correlated with accumulation of damage in the material or with changes in the dynamic response of the device due to accumulation of oxides, debris, and moisture. Furthermore, changes in the environment such as temperature and relative humidity may also effect the material and the dynamics of the device. The resonant frequency of the devices used in this study will decrease approximately 1 Hz per nanometer of crack growth.

3. Results and Discussion

The notched cantilever resonant specimens used in this study provided insight to the strength and fatigue crack initiation behavior of polysilicon thin films. Although the material exhibited high strength, the initiation of fatigue cracks is instrumental in the durability and stability of MEMS devices.

Figures 2 and 3 summarize the results of this fatigue crack initiation study. Seven constant load amplitude, fully reversed tests were conducted at 27 ± 0.1°C and 75 ± 1% relative humidity. Two devices stopped resonating, but did not fail at the notch. Six tests were conducted in laboratory air at ambient conditions on MUMPs-11 polycrystalline silicon. Load levels are normalized by the excitation level which resulted in immediate failure of the specimen.

These initial results show that fatigue is a critical issue for ensuring device reliability and stability. The typical decrease in device resonant frequency associated with accumulation of fatigue damage is shown in Figure 4.

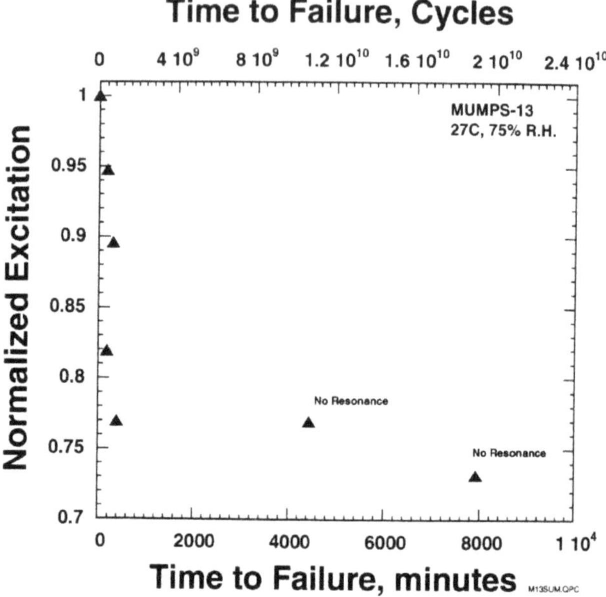

Figure 2. MUMPs-13 polycrystalline silicon fatigue initiation life at 27°C, 75% R.H.

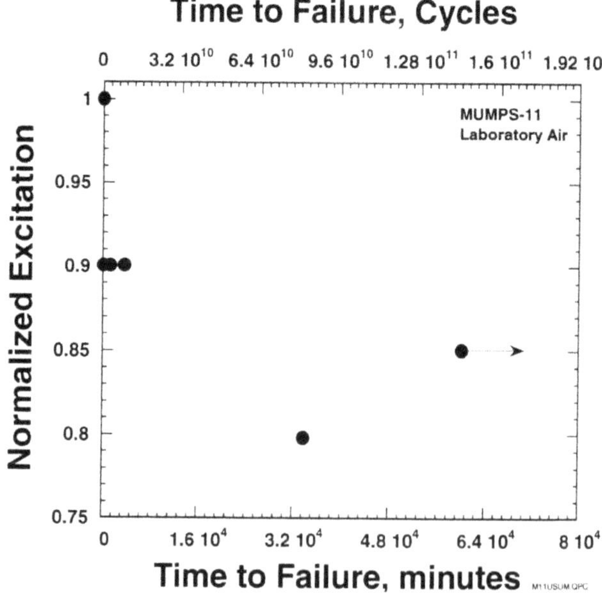

Figure 3. MUMPs-11 polycrystalline silicon fatigue initiation life in laboratory air.

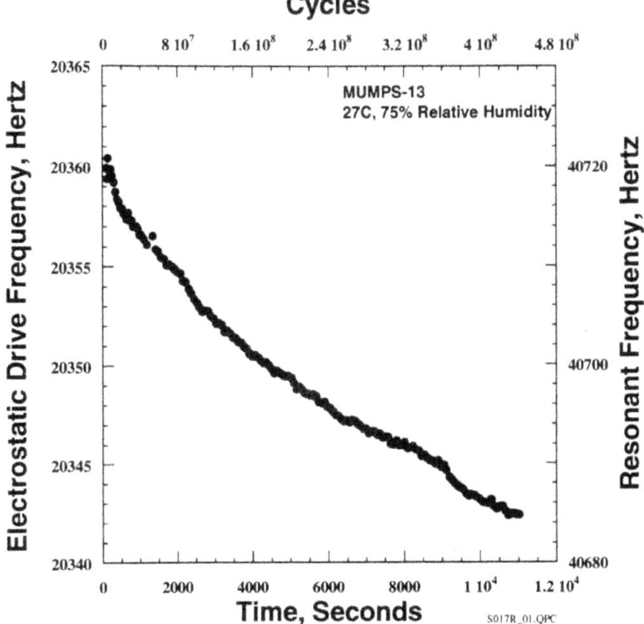

Figure 4. Change in resonant frequency of a MUMPs-11 polycrystalline silicon fatigue crack initiation specimen.

Testing of devices at progressively lower levels of excitation eventually leads to no short-term change in resonant frequency with time. This may represent a lower bound for initiation of fatigue damage in the material of interest. An upper bound strength and lower bound stable device response may be used to rapidly characterize the bounds for which fatigue damage is an important consideration. Changes in resonant frequency are intimately linked to the accumulation of damage in the polycrystalline silicon. Comparison of lives with uncontrolled laboratory air tests suggests that polycrystalline silicon device lives are shorter in moist air (Figure 5).

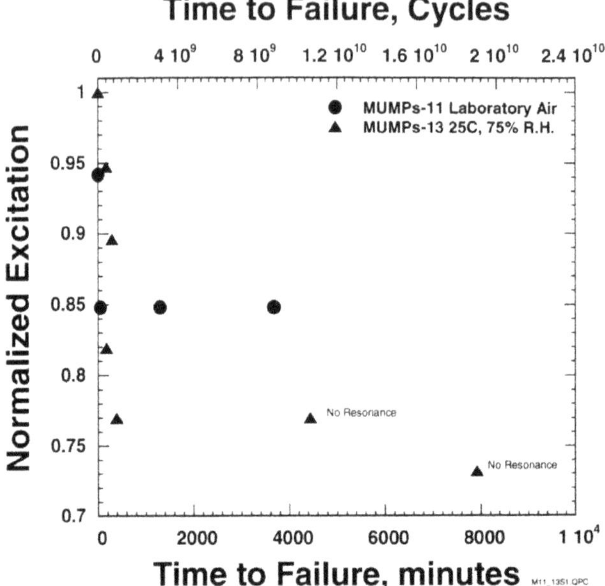

Figure 5. Effect of relative humidity on polycrystalline silicon fatigue crack initiation.

The dynamic response of the resonant fatigue structure is altered by changes in material and damping characteristics of the device. Consequently, it is important to monitor the frequency response of the device, not just the resonant frequency. Steady-state oxide layers are quickly formed on polycrystalline silicon exposed to laboratory air, and the effects of debris and moisture may be eliminated by providing clean, controlled testing environments.

4. Conclusions

This preliminary characterization of fatigue crack initiation in polycrystalline silicon provides information crucial to design of reliable, stable MEMS devices. Evidence of fatigue crack initiation under cyclic loading conditions in moist air. This study demonstrates the viability of MEMS-scale structures for fracture and fatigue characterization. This particular specimen geometry may be used for characterization of modulus, strength, fatigue crack initiation and growth, and fracture toughness. The geometry is applicable to conductive thin films amenable to standard IC photolithographic processing. This specimen and other micro and nano-scale test structures represent a unique platform for materials characterization and fundamental studies in both MEMS and traditional structural materials.

5. Acknowledgements

This research was funded by DARPA DABT69-95-C-0143 The contributions of Will Van Arsdell, Nosh Medora, Drew Diamond, Paulo Correia, and Eckart Jansen to the development of the test technique are appreciated.

6. References

Arsdell, W. W. V. (1997). Subcritical Crack Growth in Polysilicon MEMS. Department of Mechanical Engineering. Cambridge, Massachusetts Institute of Technology: 179.

Bhaduri, S. B. and F. F. Y. Wang (1983). Slow crack growth studies in silicon. Fracture Mechanics of Ceramics, Plenum Press: 327-336.

Brown, S. B., W. V. Arsdell, et al. (June 16-19, 1997). Materials Reliability in MEMS Devices. Transducers '97, Chicago, Illinois, IEEE.

Chen, C. P. and M. H. Leipold (1986). "Crack Growth in Single Crystal Silicon." NASA Tech Brief **10**(3): Item No. 106.

Connally, J. A. and S. B. Brown (1992). "Micromechanical Fatigue Testing." Experimental Mechanics(June): 81-90.

Connally, J. A. and S. B. Brown (1992). "Slow Crack Growth in Single-Crystal Silicon." Science **256**: 1537-1539.

Dauskardt, R. H., W. Yu, et al. (1987). "Fatigue Crack Propagation in Transformation-Toughened Zirconia Ceramic." Journal of the American Ceramic Society **70**: C248-C252.

Dowling, N. E. (1996). Estimating Fatigue Life. ASM Handbook. S. R. Lampman. Materials Park, OH, ASM International. **19 Fatigue and Fracture:** 250-262.

Evans, A. G. (1980). "Fatigue in Ceramics." International Journal of Fracture **16**: 485-498.

Ewart, L. and S. Suresh (1986). Journal of Materials Science Letters **5**: 774.

Hong, S., T. P. Weihs, et al. (1989). Measuring the strength and stiffness of thin film materials by mechanically deflecting cantilever microbeams. Materials Research Society, Materials Research Society.

Johansson, S. and J. Schweitz (1988). "Fracture testing of silicon microelements in situ in a scanning electron microscope." Journal of Applied Physics **10**(63): 4799-4809.

Koester, D. (1997). MCNC Polysilicon TEM Characterization.

Rao, K. T. V., Y.-W. Kim, et al. (1995). "Fatigue-Crack Growth and Fracture Resistance of a Two-Phase (g + a2) TiAl Alloy in Duplex and Lamellar Microstructures." Materials Science and Engineering A **192/193**: 474-482.

Rybicki, G. C. (1988). Indentation Plasticity and Fracture in Silicon, NASA.

Saxena, A. and C. L. Muhlstein (1996). Fatigue Crack Growth Testing. ASM Handbook. S. R. Lampman. Materials Park, OH, ASM International. **19 Fatigue and Fracture:** 168-184.

Sharpe, W. N., K. R. Vaidyanathan, et al. (1996). A New Technique for Measuring Poisson's Ratio of MEMS Materials. Materials Research Society Symposium Proceedings, Boston, MA.

Sharpe, W. N., B. Yuan, et al. (1996). New test structures and techniques for measurement of mechanical properties of MEMS materials. SPIE's 1996 Symposium on Micromachining and Microfabrication, 14-15 October 1996.

Sharpe, W. N., B. Yuan, et al. (1997). Measurements of Young's Modulus, Poisson's Ratio, and Tensile Strength of Polysilicon. MEMS 1997, Tenth IEEE International Workshop on Microelectromechanical Systems, Nagoya, Japan.

William N. Sharpe, J., B. Yuan, et al. (1996). New test structures and techniques for measurement of mechanical properties of MEMS materials. SPIE's 1996 Symposium on Micromachining and Microfabrication.

SURFACE MODIFICATION AND MECHANICAL PROPERTIES OF BULK SILICON

MATTHIAS SCHERGE AND JUERGEN A. SCHAEFER
Institut für Physik, TU Ilmenau
PF 100565, D-98684 Ilmenau

Abstract. Bulk silicon samples prepared with different surface chemistry (hydrophobic and hydrophilic) as well as different morphologies (polished and corrugated) were subjected to friction and adhesion tests using a novel micro-triboscopy tester. This tester allows the evaluation of planar as well as curved samples in the mN to nN force range. The samples do not have to be transparent and can vary in thickness and size. Hydrophobic surfaces show lower adhesion but clear stick/slips in friction experiments. In contrast, hydrophilic surfaces adhere stronger but show lower friction. Surface modification by geometric means revealed that the application of a corrugated surface is the most effective way to reduce friction and adhesion.

1. Introduction

The tribology of Micro-Electro-Mechanical Systems (MEMS) is characterized by the physical contact of millimeter or micrometer sized samples. When the roughness of the samples is small (1 nm and smaller) the asperity contact of both planes can be neglected. We assume that both substrates are separated by a water layer and that all frictional and adhesional properties are governed by boundary lubrication effects. The thickness of the water layer defines the regime of sliding, that means solid-like sliding for thin layers (3 to 5 mono-layers) and liquid-like sliding for thicker layers [1]. The common way of evaluating topography, friction and lubrication properties of micro structures is Atomic Force Microscopy (AFM) [2, 3, 4, 5]. However, the tests are always limited to the interaction of tip and surface (asperity contact) and the force range is confined to pN and nN [6]. Force distance curves show that the very sharp tip penetrates the water layer after the jump to contact [7]. To overcome this problem, i.e., to measure only

B. Bhushan (ed.), Tribology Issues and Opportunities in MEMS, 529-537.

the top most layer, the cantilever can be modified to increase the contact area [8, 9]. In [10] the sliding of nanocrystals on a single-crystal surface was described. With a crystal area of 20.000 nm^2 a lateral force of about 23 nN was observed. Interpolating those results the interaction of millimeter sized samples will lead to forces in the μN range. Moreover, the experimental force calibration of the cantilever, i.e., the exact determination of the spring constant is very often impossible or at least a complicated process [11, 12]. Thus, AFM is the ideal tool to determine roughness and topography or to perform measurements in the single asperity regime [13]. For microparts, however, larger sample areas have to be tested and AFM can be applied only in a limited way.

Another way to measure friction and adhesion of larger samples in the mN and μN range is to use a Surface Force Apparatus (SFA) [14, 15]. SFA's are applied to understand the shear dynamics of confined liquid films on the molecular level. Disadvantageous is that these testers require thin transparent mica samples to be glued on cylindric sample holders. Opaque, i.e., micro-machined silicon parts cannot be evaluated.

Our experiments with MEMS structures (micro grippers) showed that friction and/or adhesion have the magnitude of μN [16]. Tests on surface modified p and n silicon were carried out. Surface modification was achieved by altering surface geometry and chemistry. The effects were monitored using a novel micro-triboscopy tester. A roughness and/or adhesion induced deflection of a leaf spring is detected using a high resolution laser interferometer leading to a force resolution of 5 nN in both tangential and normal direction. The samples can have a dimension of several millimeters to micrometers in lateral and vertical direction. The design and the size of the leaf spring allow a simple and reliable determination of the spring constant. Since the tester can be operated in air as well as under high vacuum conditions, the environment can be controlled over a wide range. Results presented here were obtained in experiments under ambient conditions.

2. Experimental

The tribological behavior of bulk silicon samples depends strongly on the amount of dissolved water. Even when manufactured under controlled conditions the surface of the MEMS device is covered with at least one monolayer of water. Since volume and mass of the devices are very small, molecular attraction forces have the main influence on friction and adhesion. By means of wet-chemical surface modification the thickness of the water layer was controlled.

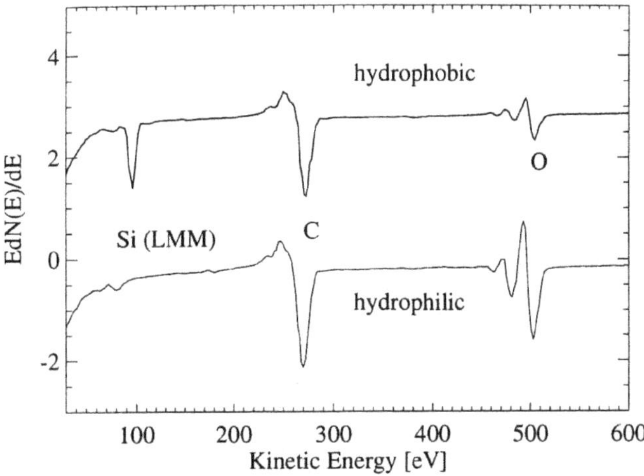

Figure 1. Auger spectra of hydrophobic and hydrophilic silicon.

2.1. SAMPLE PREPARATION

Polished (100) silicon samples (p and n type) with an area of 3 mm^2 were used in all experiments. Before the mechanical tests the samples were rinsed in acetone and alcohol followed by a 2 min. dip in hydrofluoric acid (HF) or hydrogen peroxide (H_2O_2) to generate either a hydrophobic or a hydrophilic surface. HF etching leads to hydrogen termination of the dangling bonds of silicon. As a result less water is adsorbed. The treatment with H_2O_2 generates a thin oxide layer receptive for water. These steps were accompanied by Auger Electron Spectroscopy (AES) to monitor the chemistry of the surface. AES can be used as a measure of the oxide coverage, thus being an indirect indication for the degree of hydrophobicity. From the spectra it is deduced that the oxygen intensity of the hydrophilic sample is about 3 times larger than of the hydrophobic sample, indicating the oxide layer. The small silicon peak in the spectrum of hydrophilic silicon is characteristic for the replacement of every second silicon atom by oxygen. The Auger line shape can be used to confirm that silicon was oxidized by the H_2O_2 treatment. The line shapes for core-valence-valence transition represent the weighted self-convolution of the valence band density of states. This can be shown in the LMM transition at about 90 eV. The shift to lower energies (from 92 eV for silicon to 78 eV for oxidized silicon) and the line shape are typical for oxidized silicon.

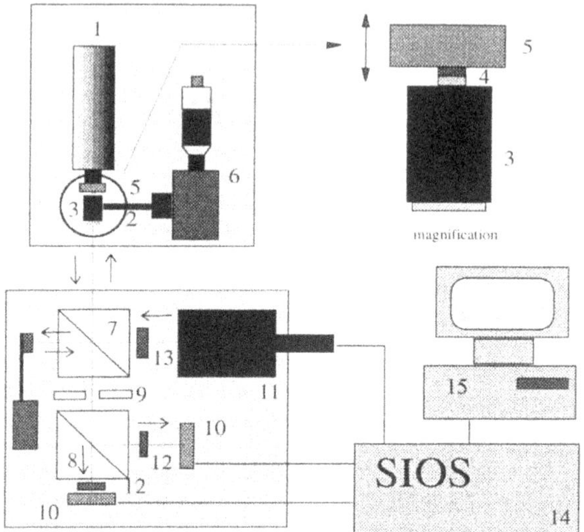

Figure 2. Schematic of the micro-triboscopy tester. Top left: mechanical part. Bottom left: optical part. Top right: magnification of the sample holders. Bottom right: control unit and PC. The numbers have the following meaning: 1) piezoelectric stack, 2) leaf spring, 3) sample holder with reflector, 4) samples, 5) two-axis sample holder, 6) micrometer, 7), 8) beam splitters, 9) aperture, 10) photo diodes, 11) laser input, 12) analyzer, 13) polarizer 14) SIOS controlling unit with laser, 15) PC

2.2. THE ADHESION TESTER

Figure 2 is a top view of the tester in adhesion mode. (A detailed description of the tester used to measure friction can be found in [16, 17].) The mechanical part consists of a piezoelectric stack (1) for dynamic application with a maximum expansion of 20 microns powered by a low voltage (-10 ...150 V) amplifier. The amplifier is operated by an input voltage ramp. The piezoelectric stack has an M3 thread to mount a two-axis sample holder (5) for the first sample. This special holder allows leveling, thus parallel engagement. The samples (4) are glued with their back sides on the sample holders (3) and (5). The cantilever is realized as a 35 micron thick molybdenum leaf spring (2). To obtain the spring constant k the leaf spring was excited to perform damped oscillations monitored by the interferometer. Taking the period time T and the amplitude of two following periods (u_n and u_{n+1}) k can be calculated by:

$$k = m(\omega^2 + \delta^2) = 5\,\frac{N}{m} \tag{1}$$

with m the mass of the cantilever, ω the frequency and δ the damping constant.

$$\omega = \frac{2\pi}{T}, \quad \delta = \frac{1}{T} ln \frac{u_n}{u_{n+1}} \tag{2}$$

Using $F_a = kx$ the adhesion force can be determined by measuring the corresponding change in distance x. A micrometer (6) is used to approach both samples. This procedure is monitored with a microscope. The final engagement is accomplished expanding the piezoelectric stack. As soon as the motion of the piezo is reversed, an adhesion induced deflection of the leaf spring takes place. As long as the stiction is larger than the spring force both samples stay together and the laser interferometer measures the traveling distance of the piezoelectric stack. At a certain point when the spring force has become large enough the sample attached to the spring jumps back to its initial position.

In friction mode the samples are moved from its vertical position (related to the direction of the laser beam) to a horizontal one. Now the two-axis sample holder holds the lower sample. The opposite sample is attached to the leaf spring via micro cardan joint to obtain parallel engagement. All frictional parameters are measured during the expansion of the piezo. When the lower sample holder starts to move, a friction induced deflection of the main spring sets in. As long as the frictional force is larger than the spring force both planes stick together and the laser interferometer measures the traveling distance of the piezoelectric stack. The tangential force is proportional to the traveling distance of the lower sample holder $F_t \sim x$. As friction force the length between zero and the first slip times spring constant is taken $F_t = kx_{1.slip}$.

For the purpose of vibrational isolation (acoustic and sub-acoustic frequencies) the whole setup is placed on a concrete block with a weight of 35000 N. Further isolation is provided by placing the frame of the instrument on rubber pads. All these precautions lead to ultra low noise of the interferometer ($\approx \pm 1$ nm). The optical and mechanical parts are positioned on a steel plate which is covered by a hood. To avoid any contribution by electrostatic charges, both samples, the piezoelectric stacks and the frame of the tester were grounded.

3. Results

3.1. ADHESION TESTING

Figure 3 shows the adhesion force curve as a function of the distance between both samples. Depending on the molecular conditions of the surface (adsorbates, water layer) the sticking of the samples is more or less intense. It was observed that the HF treated silicon shows lower stiction than the

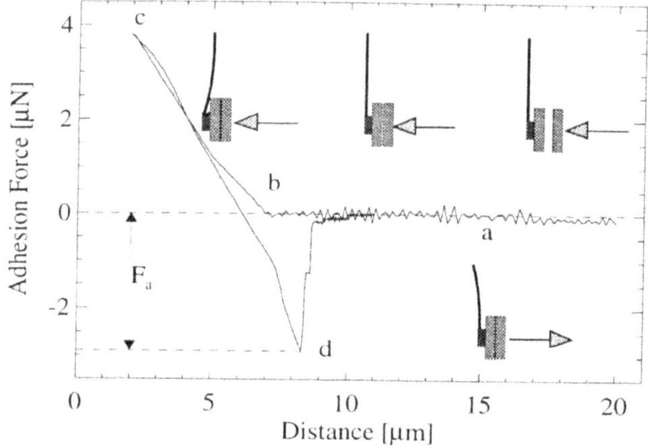

Figure 3. Adhesion force versus distance. At the beginning the samples are not in contact **a**. At **b** the samples get in contact. At **c** the motion of the stack is reversed. From now on both samples are pulled apart. At a certain point **d** the backdriving force becomes larger than the force that holds both samples together and the sample attached to the leaf spring jumps back. This force is called adhesion force.

H_2O_2 treated samples (about one order of magnitude). The obtained force-distance curves are comparable to curves measured by AFM. In contrast to AFM no sudden attractive force (jump to contact [18]) was observed for the point where both planes are in close proximity. AFM experiments applying tips of different materials (W, Al_2O_3, Si_3N_4, SiC) on SiO_2 showed attraction forces ranging between 0.13 to 13.9 μN [19]. The nominal tip radii were about 5 nm. These values seem to be large compared with our results (2.9 μN). The large forces experienced by the tip can be caused by the very intimate contact of the surface atoms of sample and tip after penetration of the water film. In comparison, the planar interface shows a larger spacing due to the separation by water and the short range Van der Waals forces are not in effect [20].

To measure the influence of topography and water layer thickness adhesion as a function of time was measured for polished and corrugated silicon ($R_{a,polished} = 1$ nm, $R_{a,corr} = 7$ nm). The corrugated surface was prepared by sputter-epitaxy of additional Si generating to a sinusoidally shaped topography, see inset of Figure 4 [8]. Both samples were HF etched right before the first test and then exposed to the lab environment (relative humidity $\approx 60\%$) for the following tests. The result is shown in Figure 4. Most obviously is that the adhesion between corrugated and polished surface is about two orders of magnitude lower than the adhesion force between both polished surfaces. Another important fact is that the surface gathers wa-

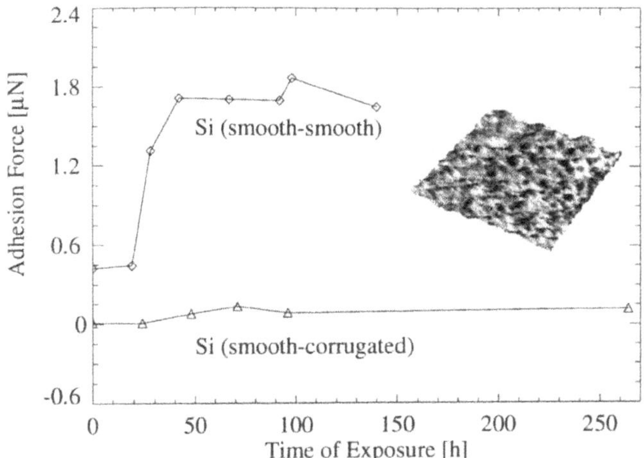

Figure 4. Adhesion as a function of the time of exposure to ambient conditions. Inset: Topography of corrugated silicon. Sample area 5 by 5 microns. The peak to valley ratio is about 50 nm.

ter and the adhesion increases gradually until a saturation is reached. We assume that the water coverage goes along with the oxidation of the surface.

3.2. FRICTION TESTING

For the friction experiments were performed also with hydrophobic and hydrophilic silicon. The hydrophobic samples showed higher friction force and distinct stick/slips. After the transition from static to dynamic friction a stick/slip frequency can be designated. Further experiments showed that the frequency increases with increasing sliding velocity until a point is reached where the periodic fluctuation becomes erratic. As the sliding velocity increases the amplitude of the stick/slips decreases [16]. No sharp stick/slips were observed for hydrophilic silicon. It is assumed that caused by a thicker water layer boundary lubrication effects set in, decreasing the values of static and dynamic friction force. According to models dealing with solid liquid transitions [1, 21, 22] stick/slip can be described by periodic melting and freezing processes. Although these findings stem from experiments using hydrocarbons water probably also forms quasi crystalline thin layers. For hydrophilic silicon we assume that this effect is covered by the contribution of additional water. The thicker water film acts as buffer damping the introduced shearing force.

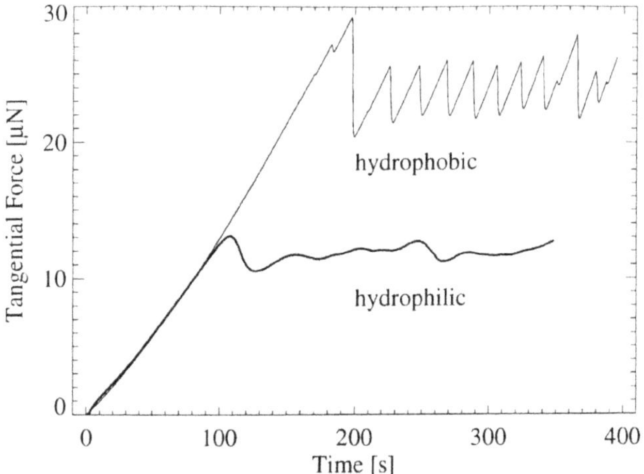

Figure 5. Tangential force as a function of time for hydrophobic and hydrophilic surface conditions.

4. Conclusions

Chemical surface modification is a means to alter the frictional and adhesional properties of silicon samples. The different properties are caused by the amount of dissolved water. HF treatment suppresses the formation of thick water layers whereas H_2O_2 makes the surface receptive for water. The friction experiments show that only thin water layers lead to clear stick/slip cycles. When the water film gets thicker the upper sample seems to float on its lower counterpart. These phenomena should be discussed further in the light of the theories describing solid liquid transitions. It was shown that adhesion is smaller for HF treated samples than for H_2O_2 treated samples. This behavior is probably caused by wetting phenomena. Subtle influences like crystal orientation or the type of doping play only a subordinate role. In comparison with AFM experiments the measured forces in our experiments are lower. This can be caused by the difference in spacing between the tip and the surface on the one hand (subnanometer range) and the spacing between the two planar substrates on the other hand.

5. Acknowledgments

This work has been supported by the Deutsche Forschungsgemeinschaft and by the Volkswagen Stiftung.

References

1. Gee, M. L., McGuiggan, P. M., Israelachvili, J. N. and Homola, A. M., Liquid to solidlike transitions of molecularly thin films under shear, *J. Chem. Phys*, **93**(1990)3, 1895-1906
2. Bhushan, B., Handbook or Micro/Nano Tribology, CRC Press, 1995
3. Fujihara, M. and Morita, Y., Atomic force microscopy and friction force microscopy of chemically modified surfaces, *J. Vac. Sc. Technol.*, **B12**(1994)3, 1609-1613
4. Ascoli, C., Dinelli, F., Frediani, C., Petracchi, D., Salerno, M., Labardi, M., Allegrini, M. and Fuso, F., Normal and lateral forces in scanning force microscopy, *J. Vac. Sc. Technol.*, **B12**(1994)3, 1642-1645
5. Grafström, S., Ackermann, J., Hagen, T., Neumann, R. and Probst, O., Analysis of lateral force effects on the topography in scanning force microscopy, *J. Vac. Sc. Technol.*, **B12**(1994)3, 1559-1564
6. Ruan, J. and Bushan, B., Atomic-scale and microscale friction studies of graphite and diamond using friction force microscopy *J. Appl. Phys.*, **76**(1994)1, 5022-5035
7. Ruan, J. and Bushan, B., Atomic-Scale Friction Measurements Using Friction Force Microscopy: Part I — General Principles and New Measurement Techniques, *J. of Tribology*, **116**(1994), 378-388
8. Vögeli, B. Känel, H. von, The Influence of Surface Topography on Adhesion and its Application to Micro Parts Handling, ECASIA 97
9. Pedersen, G. P., Høj, J. W. and Engell, J., Measuring forces between α-Al_2O_3 surfaces using the atomic force microscope, Fourth Euro Ceramics, **2**(1995), 31-38
10. Sheehan, P. E., Lieber, C. M., Nanotribology and Nanofabrication of $MoSO_3$ Structures by Atomic Force Microscopy, *Science*, **272**(1996), 1158-1161
11. Linnemann, R., Gotszalk, T., Rangelow, I. W., Dumania, P., Oesterschulze, E., Atomic force microscopy and lateral force microscopy using piezoresistive cantilevers, *J. Vac. Sc. Technol.*, **B14**(1996)2, 856-860
12. Lu, C. J., Jiang, Z., Bogy, D. B., Miyamoto, T., Development of a New Tip Assembly for Lateral Force Microscopy and Its Application to Thin Film Magnetic Media, *J. of Trib.*, **117**(1995), 244-249
13. Meyer, E., Lüthi, R., Howald, L., Güntherodt, Friction Force Microscopy, Forces in Scanning Probe Methods, NATO-ASI E, **286**1995, 285
14. Van Alsten, J., Granick, S., *Tribol. Trans.*, **33**(1990), 436-446
15. Israelachvili, J. N., *Chemtracts Anal. Phys. Chem*,**1**(1989), 1-12
16. Scherge, M. and Schaefer, J. A., Microtribological Investigations of Stick/Slip Phenomena using a novel Oscillation Friction and Adhesion Tester, to appear in: *Tribology Letters*
17. Scherge, M., Büchner, H., Jäger, G. and Schaefer, J. A., Interferometric detection of adhesion induced nano-deflections, to appear in: *Journal of Optics*
18. Blackman, G. S., Mate, C. M., Philpott, M. R., Interaction Force of a Sharp Tungsten Tip with Molecular Films on Silicon Surfaces, *Phys. Rev. Lett.*, **65**(1990)18, 2270-2273
19. Miyamoto, T., Kaneko, R., Ando, Y, Interaction Force Between Thin Film Disk Media and Elastic Solids Investigated by Atomic Force Microscope, *ASME J. Tribol.*, **112**(1990), 567-572
20. Weldon, M., K., Marsico, V. E., Chabal, Y. J., Hamann, D. R., Christman, S. B., Chaban, E. E., Infrared spectroscopy as a probe of fundamental processes in microelectronics: silicon wafer cleaning and bonding, *Surface Science*, **368**(1996), 163-178
21. Carlson, J. M., Batista, A. A., A constitutive relation for the friction between lubricated surfaces, unpublished results
22. Persson, B. N. J., Theory of friction: On the origin of the stick-slip motion of lubricated surfaces, *Chem. Phys. Lett.*, **254**(1996)114-121

MICROFRICTION AND MICROWEAR EXPERIMENTS ON METAL CONTAINING AMORPHOUS HYDROCARBON HARD COATINGS USING AN ATOMIC FORCE MICROSCOPE

Mechanisms, Models and Micro- versus Macrotests

K. I. SCHIFFMANN
Fraunhofer Institut für Schicht- und Oberflächentechnik
Bienroder Weg 54E, D-38108 Braunschweig

Abstract

Metal containing amorphous hydrocarbon films (Me-C:H) consist of nanometer sized metallic particles embedded in a highly cross-linked hydrocarbon matrix. The coatings have excellent tribological properties and an adjustable electrical conductivity, therefore being of high interest for applications in MEMS devices. In this study microtribological properties of tungsten-C:H and gold-C:H films with metal contents ranging from 0 to 50 at% have been investigated by means of atomic force microscopy (AFM) methods.
Friction force microscopy experiments have been performed and the load dependence of friction was analysed by means of Hertz theory of elastic contact. This analysis results in effective friction coefficients $\mu^*=F_{frict}/(R_{tip}F_{load})^{2/3} \sim S \cdot E^{*-2/3}$ (E^* = red. Young's modulus) which are used to determine shear stress S as a function of metal content and type of metal in the film. Comparison with results from macroscopic pin-on-disk tests shows a strong correlation with start values of friction, before wear or material transfer begin to influence frictional behaviour.
Microwear tests have been performed, using a diamond tip on a stainless steel cantilever. Real-time observation of the wear process shows periodic material break-off after a critical number of wear cycles, indicating material fatigue as an important wear mechanism. A strong influence of the columnar growth structure and the percolation of metallic nanoparticles inside the film on fatigue and wear resistance was found. Load dependence and time dependence of wear have been studied and described by simple models considering low cycle fatigue, mechanical properties of the film and change of contact area respectively. A reasonable fit of experimental data is obtained and fit parameter are in the order of magnitude predicted by the model. Microscopic (AFM) and macroscopic (pin-on-disk) wear tests partly show comparable results. Differences are attributed to different effective contact pressures due to abrasive particles in the interface during the pin-on-disk test.

B. Bhushan (ed.), Tribology Issues and Opportunities in MEMS, 539-558.

1 Introduction

A problem in micro electro mechanical systems (MEMS) still is the reliability and lifetime of moving parts due to non optimised pairs of surfaces moving against each other, resulting in high friction, stiction and wear. A possibility to reduce or solve such tribological problems is to use very thin coatings with optimised friction and wear properties. Candidates for such coatings are the metal containing amorphous hydrocarbon films (Me-C:H). Like diamond-like carbon (DLC) Me-C:H consists of a highly three-dimensionally cross-linked hydrocarbon matrix, in which metallic clusters of nanometer size are embedded [1, 2, 3]. From macroscopic applications it is known that Me-C:H films have a high hardness of about 5 to 20 GPa, high ratios of hardness to Young's modulus (≈ 0.1) [4], low friction coefficients of 0.1 to 0.2 [5], and abrasive and adhesive wear rates 5 to 100 times lower than e.g. uncoated steel surfaces [6, 7]. These good tribological properties are attributed to the strong covalent bonds of diamond like carbon and the formation of sliding films of hydrocarbon and/or small graphite clusters on the counterbody [8, 9, 10, 11] which reduce friction and wear. Additionally, due to their metal content, Me-C:H films show an electrical conductivity adjustable over several orders of magnitude [12, 13] which makes them of high interest for e.g. microscopic electrical sliding contacts. However, Me-C:H coatings may also be very useful for other tribological applications e.g. microscopic sliders, motors, pumps, gears, etc..
The microscopic structure of the films has already been investigated in great detail [14, 15, 16], indicating (i) a columnar growth structure of the films with column diameters of 200 - 300 nm, (ii) an increasing particle size with increasing metal content (0 ... 50 at%) in the range of e.g. 10 to 100 Å for gold containing films and 10 to 30 Å for tungsten containing films, (iii) the formation of pure metallic particles in the case of noble metals (e.g. gold: Au), while in the case of carbide forming metals always carbidic particles (e.g. tungsten: WC_{1-x}) are formed.
On the other hand, little is known about the microscopic mechanisms of friction and wear of Me-C:H coatings and about the influence of the microstructure of films on their tribological properties. The atomic force microscope (AFM) may be an instrument to answer these questions, because it combines a number of advantages for a fundamental study of friction and wear processes:

- a single asperity contact with a well defined contact area,
- high lateral and vertical resolution, to study the influence of the microstructure of samples and to perform tribological tests on ultrathin films down to some tens of nanometers or on very small areas of microactuators or microsystems,
- real-time observation of dynamic tribological processes,
- separation of topographic and *chemical* effects in friction force microscopy,
- in-situ three-dimensional determination of small wear volumes using the AFM-tip.

Several authors have made use of this new approach for tribological investigations [17, 18, 19, 20, 21, 22, 23, 24], especially for the investigation of thin hard coating materials. In the following AFM friction and wear results on Me-C:H films will be compared to theoretical models and to macroscopic tribological tests.

2 Experimental

2.1 PREPARATION OF COATINGS

The preparation of coatings has been performed in a combined physical and chemical vapour deposition process in a RF diode reactor. The process has been described in detail by several authors [25, 1, 26], and typical deposition parameters can be found at [14]. Coatings of about 1 micron thickness were prepared on silicon substrates containing tungsten or gold as metallic component (W-C:H, Au-C:H), with metal concentrations ranging from 0 to 50 at%.

2.2 FRICTION FORCE EXPERIMENTS

FFM experiments were performed in ambient air using a commercial stand-alone AFM/FFM-head[1], which detects lateral forces between the silicon tip and the sample by laser beam deflection due to cantilever torsion. Technical details of the FFM-head and the calibration of normal and lateral forces can be found elsewhere [27, 28, 29]. Tip radii of curvature were determined by scanning electron microscopy (SEM) before and after each sample sequence to exclude that wear alters the contact area of the tip during experiments. The determination of adhesion forces between tip and sample (pull-off force in force vs. distance curves) and the calibration of normal and lateral forces was performed before and after each friction measurement. A fixed scan speed of about 1 μm/sec was used, while the load was varied in the range of F_L = 3 ... 35 nN, including adhesion forces. Typical scan ranges were 200 to 400 nm.
Since Me-C:H samples are not atomically flat and since sample topography influences the lateral force signal [30, 31] the *friction loop:*

$$F_F = 1/2\ (F_{forward} - F_{backward}) \tag{1}$$

was computed by pixelwise subtraction of forward and backward scanned images taking into account a manually determined shift vector which considers the friction hysteresis due to change of scan direction. This procedure results in a minimisation of topographic cross-talk in FFM images.

2.3 MICROWEAR EXPERIMENTS

The AFM[2] microwear experiments were performed using a pyramidal diamond tip of 1 μm radius of curvature mounted on a stainless steel micro cantilever. The force constant of the cantilever was adjusted via the length of the free part of the lever to about 300 - 350 N/m. Possible modifications of the tip geometry or material transfer to the tip during wear experiments were controlled by SEM and energy dispersive X-ray spectroscopy. After each series of experiments a cleaning of the tip in an ultrasonic bath of de-ionised water for 30 sec was applied to removed possibly transferred material.

---

[1] Atomic Scale Tribometer, Centre Suisse Electronique et de Microtechnique, Neuchatel, Switzerland

[2] Park Scientific Instruments, Sunnyvale, CA94089, USA

For wear testing, the tip oscillates linear with constant load, thus creating a line-like groove on the sample surface. The length of the trace and the travel speed were held constant at L = 5 μm and V = 10 μm/sec respectively. Parameters varied were: the number of wear cycles N = 10 ... 1000, the load F = 100 ... 4000 μN, the metal content (0 ... 50 at%) and the metal type (W, Au) inside the coating.
The wear grooves were imaged with the same tip at a minimum load of 50 μN. The wear volumes determined from these images were normalised by the length of the trace, so wear corresponds to the cross-sectional area of the groove. For the study of dynamic tribological processes a procedure proposed by Loubet et. al. [32] was used. It consists of taking an AFM image during wear experiments, which gives the depth profile of the wear trace along sliding direction as a function of time, i.e. Z(X,t) is recorded in a cycle-by-cycle manner. In the following this kind of wear map is called *on-line wear image.*

2.4 MACROSCOPIC FRICTION AND WEAR TESTS

Macroscopic tribological tests have been performed in ambient environment, using an Al_2O_3-ball of 10 mm in diameter sliding over the sample surface at a fixed load of F_L = 5 N. For wear measurements a *linear* oscillating movement (speed = 50 mm/sec) was used in analogy to the microwear experiments. In contrast, for friction force measurements a *rotating* pin-on-disk test (speed = 200 mm/sec) was applied to avoid change between static and dynamic friction due to reversal of scan direction. The friction coefficient was monitored during the whole time from beginning of the test until failure of the coating which takes from 3 to more than 60 minutes depending on the wear resistance of the film.

3 Friction : Results and Discussion

3.1 FRICTION FORCES

Fig. 1 shows a Au-C:H sample containing 15 at% gold. In the topographic image (1a) nearly spherical gold particles are visible, enlarged due to the convolution by the finite tip radius of curvature. Fig. 1b represents the friction force image after computation of friction loop (eq. 1). The dark, nearly circular areas inside the picture can be associated with the gold particles in the topographic image whereas the brighter surrounding areas are attributed to the amorphous hydrocarbon matrix. A partially coverage of particles with monolayers of hydrocarbon matrix lead to continuous transitions from the friction value of the matrix to that of the metal. The mean friction value of the two-phase nanocomposite material depends on the area fraction of both phases in the topmost layer.
In the case of W-C:H due to the smaller particles and a larger tip radius of curvature individual WC particles could not be resolved in topography or friction force images.

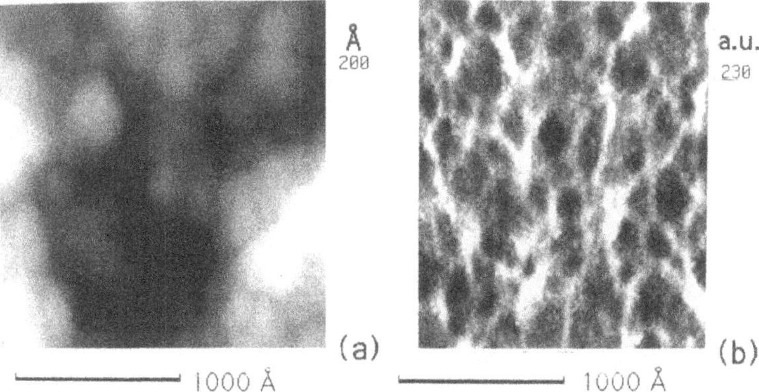

Figure 1. FFM image of a Au-C:H coating with 15 at% Au at a load of 12 nN. (a) topography (b) lateral force image after subtraction of forward and backward scanned images. Mean friction force F_L = 1.5±0.5 nN. There is a slight shift between topographic and friction image, since images are not taken simultaneous.

Fig. 2 show examples of the load dependence of friction forces for W-C:H and Au-C:H samples, averaging over an area of 0.16 or 0.04 μm^2 respectively. In the case of W-C:H a non-linear behaviour of friction versus load is clearly visible, while in the case of Au-C:H linear as well as non-linear models may be applicable.

For Au-C:H samples of more than 10 - 20 at% gold wear can be observed at loads larger than 10±2 nN by a change of slope in the friction vs. load curve (see fig. 2) and in smearing effects within the AFM images. In such cases only loads below 10 nN were used for curve fitting. For Au-C:H with lower metal concentrations and for W-C:H films no wear was observed within the load-range investigated. This is consistent with results of microwear investigations below, which show that wear of Au-C:H increases strongly above 20 vol% gold, while wear of W-C:H stays low over the whole range of metal concentrations.

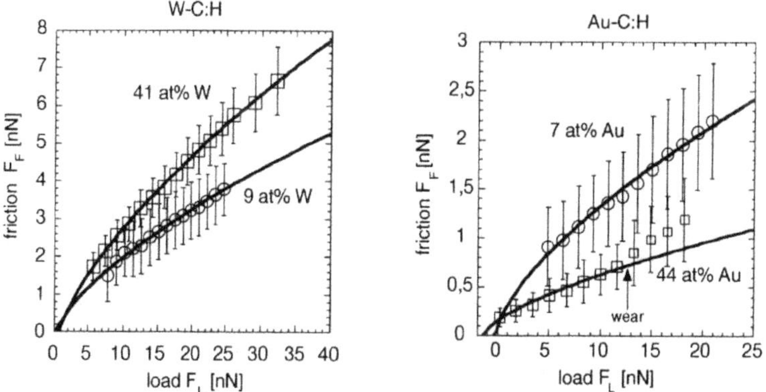

Figure 2. Dependence of friction forces on the total load (cantilever- and adhesion-forces) determined by AFM for W-C:H and Au-C:H samples with different metal contents. The continuous line correspond to the fit with eq. 4.

For the analysis of data it is assumed that friction is proportional to the real contact area A [33, 34]:

$$F_F = S \bullet A \tag{2}$$

where F_F = friction force and S = shear stress of the material, which is the friction force per unit area. For a single asperity contact with a spherical tip apex the most simple approach for the contact area is given by Hertz model of elastic contact of a sphere and a flat [35]:

$$A_{Hertz} = \pi \left(\frac{3R_t F_L}{4E^*} \right)^{2/3} \tag{3}$$

with R_t = tip radius of curvature, F_L = total load, $E^* = [(1-\nu_1^2)/E_1+(1-\nu_2^2)/E_2]^{-1}$ is the reduced Young's modulus and E_i, ν_i = Young's moduli and poisson ratio of tip and sample. Combining eq. 2 and 3 results in

$$F_F = \mu^* (R_t \cdot F_L)^{2/3} \tag{4}$$

with

$$\mu^* = \pi (3/4)^{2/3} \cdot S \cdot E^{*-2/3} \tag{5}$$

The *effective friction coefficient* μ^* only depends on the material properties of the tip and the sample. A more general model considers a pressure dependence of the shear stress S which is found in a number of experiments [36, 37, 38]:

$$S = S_0 + \alpha P \tag{6}$$

where $P = F_L/A$ is the pressure and α = constant. Up to now there is no physical model which predicts whether a pressure dependence applies or not. As Israelachvili [39, 40] pointed out, the pressure dependence may appear if the surface is damaged during friction experiments while in a really wear-free experiment a constant shear stress is found. Inserting eq. 6 and 3 into eq. 2 the following extended expression for the load dependence of friction results:

$$F_F = \mu^* (R_t \cdot F_L)^{2/3} + \alpha F_L \tag{7}$$

with $\mu^* = S_0 \pi (3/4E^*)^{2/3}$.

Applying both models to the experimental friction vs. load curves of Me-C:H coatings, in some cases eq. 7 gives a slightly better fit to the data, while in other cases eq. 4 fits best. To get comparable results, in the following the basic expression eq. 4 is used for the quantitative evaluation of all samples.

Fig. 2 displays the fit of eq. 4 to some experimental data using tip radii of R_t = 20 nm for Au-C:H samples and R_t = 30 nm for W-C:H samples, as determined by SEM. A small offset force F_L^0 is allowed in the fit to compensate for uncertainties in determination of adhesion forces [41, 42]. The results demonstrate that Hertz model gives a reasonable description of the experimental situation even if other models, as discussed above, cannot be strictly excluded.

3.2 FRICTION AND METAL CONTENT

As a result of fitting eq. 4 to the experimental $F_F(F_L)$-curves fig. 3a displays the effective friction coefficients μ^* as a function of metal content of the films. A strong dependence of the friction coefficient on the type and concentration of metal in the film is found. In both materials, Au-C:H and W-C:H, the metal-free layers show almost the same effective friction coefficient of $\mu^* \approx 0.034$ GPa$^{1/3}$. This confirms that tip radius determination by SEM and the consideration of R_t in eq. 4 is reasonable. In the case of Au-C:H the friction coefficient then decreases with increasing incorporation of gold down to $\mu^* = 0.0035$ GPa$^{1/3}$ for pure Au-films. In contrast to this, for W-C:H the friction coefficient increases with increasing metal content reaching $\mu^* = 0.062$ GPa$^{1/3}$ for 50 at% W which corresponds to 100 vol% WC, i.e. a pure tungsten-carbide film. Hence successively increasing incorporation of nanometer size metallic particles in a hydrocarbon matrix gradually changes the mean friction coefficient of the material from that of the pure matrix to that of the pure metal.

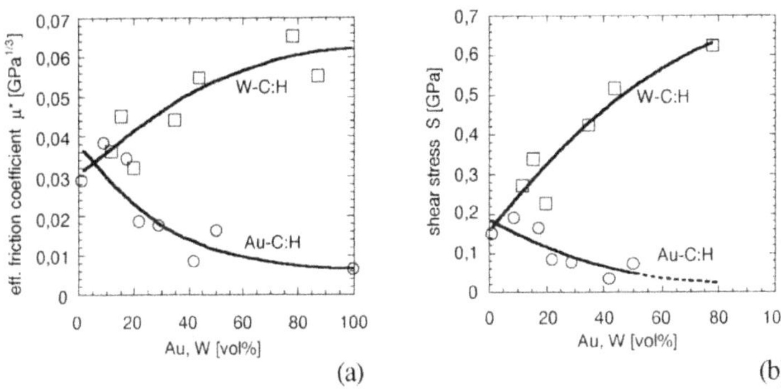

Figure 3. Effective friction coefficient μ^* defined by eq. 4 and shear stresses S computed using eq. 5 for Au-C:H and W-C:H coatings as a function of volume fraction of Au or WC respectively, determined by FFM.

Fig. 3b shows the shear stress S of Me-C:H computed from effective friction coefficients μ^* employing eq. 5 and Young's moduli of Me-C:H samples determined by bulk nanoindentation measurements using a Philips nanometer indentor (indentation depth ≈ 0.1 - 0.5 μm). Typically Young's moduli E of Au-C:H decrease with increasing metal content from 58 to 43 GPa, while for W-C:H they increase from 76 to 145 GPa [16]. The shear stress S of a-C:H films (zero metal content for Au-C:H and W-C:H) is found to be about S = 200 MPa. With increasing metal content a continuous change of S is observed up to S = 650 MPa for W-C:H, while it decreases down to S = 30 MPa for Au-C:H.

This latter value is quite typical for soft metals. E.g. for Pb and Sn values of S = 25 - 40 MPa have been published [37]. Schwarz [43] has measured the effective friction coefficients and shear stresses for different carbon modifications using an amorphous carbon coated tip. For amorphous hydrogen-free carbon films measured in dry argon atmosphere and in ambient air he found values of $\mu^* = 0.16$ GPa$^{1/3}$ and 0.45 GPa$^{1/3}$ respectively, and

S = 750 MPa (in argon). As far as we know, for Me-C:H coatings no reference measurements of μ^* and S exist.

There are a number of uncertainties in the present determination of shear stresses of Me-C:H coatings:

1. The assumption that shear stress is independent of contact pressure seems to be valid for single asperity contacts in AFM [39, 44, 45, 46, 34], but as already mentioned above, it may become invalid when wear of the surfaces occurs. This could be a reason for deviations from eq. 4 in the case of Au-C:H samples with high Au contents.
2. There may be deviations from the exact spherical shape of the tip and a certain error in tip radius determination (approx. ± 10%) which propagates into the error of the shear stress determination.
3. For the calculation of S, the bulk elastic properties of tip and sample have been used. Probably these values are modified in the vicinity of the surfaces due to (a) oxide formation on the silicon tip and (b) thin contamination films on the samples.
4. Also the surface topography of the films, which is not atomically flat, and thin films of water at the interface may influence the shear stress determination by AFM.

Notwithstanding these systematic sources of errors the *relative* variation of S for films with different metal contents should be of much higher accuracy since they are measured with the same tips, under identical conditions and using the same parameters for evaluation.

3.3 COMPARISON OF MICRO- AND MACROFRICTION

For comparison with macrofriction the conventionally defined friction coefficient $\mu = F_F/F_L$ was computed from FFM friction data (see fig. 4a). It has to be kept in mind that μ in this case depends on the load and the tip radius of curvature. The results agree well with reference data from Burger [47] and Koinkar [23]. For a pure Au-film, using a silicon tip and loads between 1 and 10nN, Burger found $\mu = 0.01$ while Koinkar using a Si_3N_4 tip of 50 nm radius of curvature and a load of 300 nN to investigate amorphous

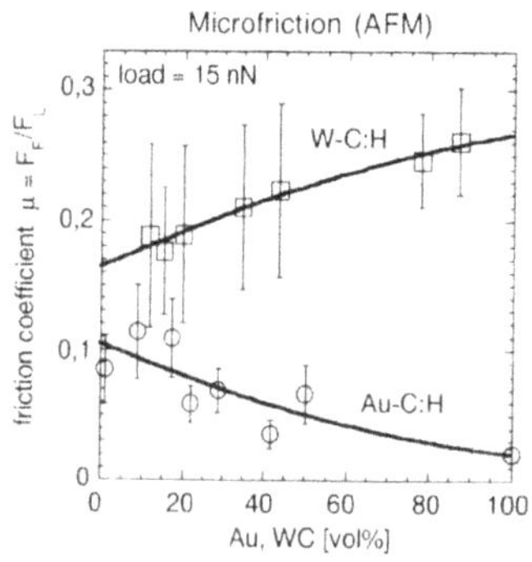

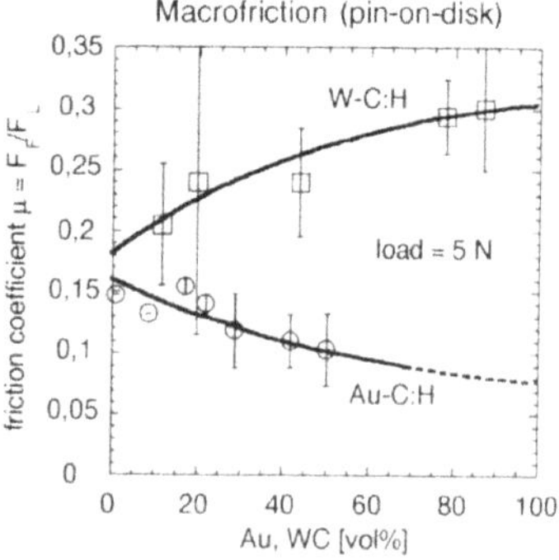

Figure 4. Conventionally defined friction coefficient μ for Au-C:H and W-C:H coatings determined (a) by AFM (load = 15 nN, silicon tip, R_t = 20 resp. 30 nm) and (b) by pin-on-disk test (load = 5 N, Al_2O_3-ball, 10mm diameter).

(metal-free) hydrocarbon films, found $\mu = 0.06$-0.07. For WC no FFM reference data has been found.

In contrast to FFM experiments, in macroscopic pin-on-disk tests friction is always accompanied by wear of the sample surface and the counterbody which influences the friction measurement. At the beginning of friction experiments for instance, in most cases material transfer from the hydrocarbon sample to the Al_2O_3-ball is observed which has a lubricating effect and therefore reduces the friction coefficient. On the other hand, due to the progress of wear during friction test, abraded particles are accumulated in the interface leading to higher local contact pressures and increasing wear, which in turn may enlarge the friction coefficient.

To get best comparison with FFM measurements, only the start values of macrofriction were taken into consideration since these values represent the minimum influence of material transfer or wear on the friction coefficient. These very first more or less wearless friction coefficients are displayed in fig. 4b. As in the microscopic case with increasing metal content, an increasing friction coefficient is observed for W-C:H, while for Au-C:H the corresponding inverse behaviour is found. W-C:H has already been investigated by several authors. All of them have found the same trend of increasing friction with increasing incorporation of tungsten-carbide, whereas part of the absolute values differs significantly. For metal concentrations between 0 and 50 at% tungsten Wang [48] found $\mu = 0.05 - 0.12$. Grischke [6] gave values of $\mu = 0.18 - 0.55$ while van Duyn [49] found $\mu = 0.12 - 0.93$. These differences may be due to different experimental conditions in friction measurement, but differences in preparation conditions of the films may also be responsible for part of these variations. For Au-C:H no reference values from literature are known. The high degree of correspondence between absolute values of micro- and macrofriction coefficients is somewhat arbitrary since microfriction changes with load and tip radius while macrofriction may also depend on a number of experimental parameters. But the correspondence of *relative* variation with metal content can be regarded as significant.

4 Wear : Results and Discussion

4.1 WEAR MECHANISM AND MATERIAL STRUCTURE

Fig. 5 shows an example of wear traces created on a W-C:H surface. Only little mounded material can be seen. The material is either swapped out of the imaged area by the tip or is transferred to the tip itself. Additionally, material densification inside the groove may appear as a material loss. Using the technique of on-line wear imaging as explained in the experimental section, the influence of the nanostructure of Me-C:H coatings and the wear mechanism itself were studied. An example of an on-line wear image of a W-C:H film (12 at% W) can be seen in fig. 6. The continuous removal of material by the tip alternates with a periodical breaking-off of material after a certain *critical number of wear cycles* N_c. This can be interpreted in terms of material fatigue: The periodic compression of the surface by the tip results in an accumulation of plastic deforma-

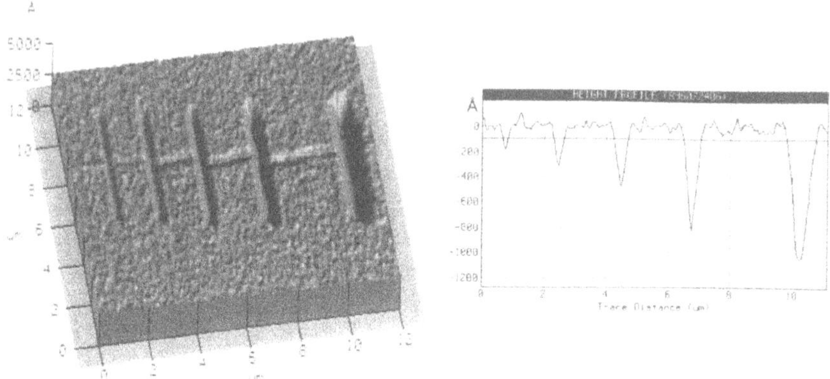

Figure 5. 3D image (12x12μm²) and cross-section of wear traces on a W-C:H film (7.3 at%W). Number of wear cycles N = 50, load = 1.3, 1.9, 2.5, 3.2, 3.8 mN.

tion and/or densification of the material at the surface. Reaching the critical shear stress of the coating microcracks are formed, finally leading to a sudden removal of material. This static *critical shear stress* τ may not be identical with the dynamic *interfacial shear stress* S determined in the FFM experiments, even though a certain relation between both quantities probably will exist.

The sudden loss of material due to material fatigue is strongly influenced by the micro- and nano-structure of the films:

I) It is well known that Me-C:H coatings have a typical columnar growth structure resulting in columns of 200 - 300 nm in diameter and of length corresponding to the respective film thickness. The ends of these growth columns can be observed as small hillocks at the sample surface and their evolution during wear can be followed in on-line wear images. Fig. 7 displays an enlarged section of an on-line wear image of a W-C:H film showing in more detail the evolution over time of individual columns during material fatigue. It can be observed that preferentially single growth columns break off while neighbouring columns persist for a certain time until they also fail. Probably this also occurs in fig. 6, only that neighbouring columns break at almost the same time due to a

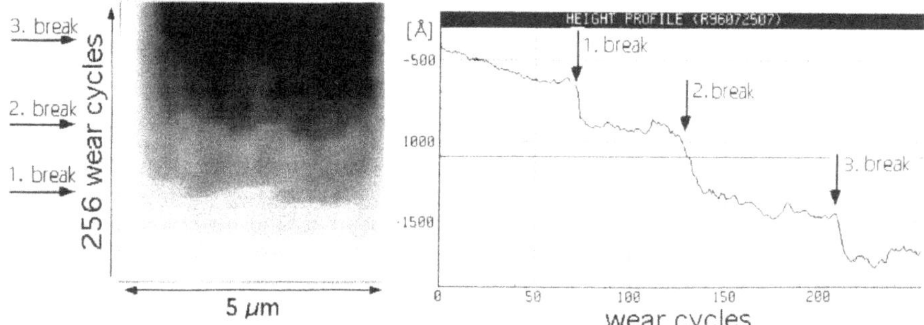

Figure 6. On-line wear image Z(X,t) of a W-C:H film (12 at%W). Black-to-white grey-scale = 1800 A height difference. Load = 1905 μN, number of wear cycles N = 256, trace length = 5 μm. The height profile is a cross-section along the marked line in the image.

shorter fatigue period. Thus it can be concluded that the growth column interfaces are the most weak parts of the film structure where material breaks first under cyclic wear loading.

II) Investigating the critical number of wear cycles N_c for W-C:H at a load of 1905 μN for different metal contents one finds that material fatigue only occurs below the percolation threshold of the metallic particles, which for tungsten-C:H is at about 15 at% W [48] (see fig. 8). Below this threshold, N_c decreases with increasing metal concentration, indicating that more or larger particles of tungsten carbide incorporated in the film worsen its wear behaviour. This is probably due to the reduction of volume fraction of highly cross-linked diamond like carbon matrix. Above the percolation threshold on the other hand, the critical number of wear cycles is infinite, i.e. no material fatigue is observed at all and only the slower continuous wear occurs. Obviously the coalescence of carbidic particles inside the film causes an additional three-dimensional cross-linking which reinforces the structure of the coating and stabilises the inherent weakness at the growth column interfaces. At higher loads of $F \geq 3200$ μN material fatigue can also be observed above the percolation threshold of 15 at% W, i.e. higher pressures are able to break the polycrystalline particle network.

On Au-C:H films material break-off due to fatigue is observed likewise, showing a similar decrease of the critical number of wear cycles N_c with increasing metal content. Since the percolation threshold of Au-C:H is at about 50 at% [2, 13] no reinforcement of the film structure due to particle coalescence is found in the low metal regime. N_c approaches zero at about 14 at%, i.e. one single pass of the tip suffices to damage the film surface, and severe wear is found at higher metal concentrations. No samples have been investigated with metal contents above the percolation threshold, but in the case of gold it is believed that particle coalescence wouldn't reinforce the film structure due to the low hardness and high plasticity of gold.

Beside columnar growth, particle percolation is the second important microstructural factor influencing the wear of Me-C:H.

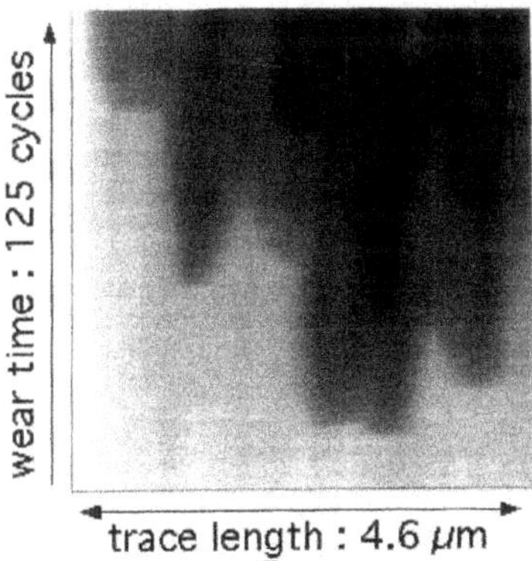

Figure 7. On-line wear image Z(X,t) of a W-C:H film (9.4 at% W). Load 1.9 mN, number of wear cycles N = 125, trace length = 4.6 μm, black-to-white greyscale = 900Å.

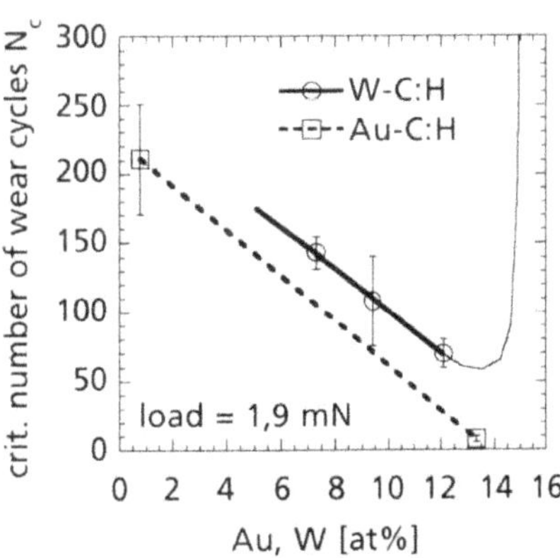

Figure 8. Critical number of wear cycles N_c as a function of metal content for W-C:H and Au-C:H films in the low metal content range at a load of 1.9 mN.

4.2 THE LOAD DEPENDENCE OF WEAR

The load dependence of the critical number of wear cycles N_c was determined from online wear images as e.g. fig. 6. As result fig. 9 displays N_c as a function of load for a W-C:H film (12 at% W) and an almost metal-free Au-C:H film. As expected, an inverse dependence of N_c on the applied load is found, i.e. the higher the load the faster the critical shear limit of the material is surpassed and the formation of microcracks is expedited.

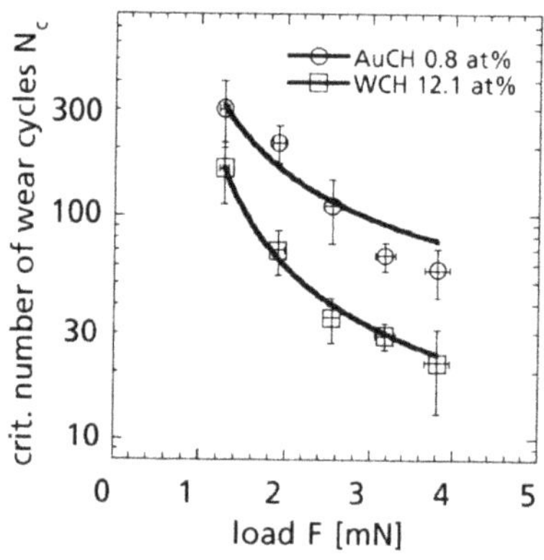

Figure 9. Load dependence of the critical number of wear cycles for a W-C:H films with 12.1 at% W and an almost metal-free Au-C:H film with only 0.8 at% Au. Full lines correspond to a fit using eq. 8, fit-parameter see table 1.

Zum Gahr [50] has analysed wear processes from a microscopic point of view and classified four major wear mechanisms: cutting, ploughing, material fatigue and crack formation. Based on the Manson-Coffin relation for low cycle fatigue [51] he gives the following equation for the load dependence of N_c for a pyramidal scratch diamond:

$$N_c = \frac{1}{2}\left(\frac{\varphi_{\lim} \cdot e^{\beta}}{\varphi_s}\right)^m = \frac{1}{2}\left[\frac{C_1}{2\ln\left(C_2 F^{1/3}\right)}\right]^m \tag{8}$$

with: $$C_1 = \varphi_{\lim}\, e^{\beta}\ , \quad C_2 = \frac{E^{2/3}\sqrt{1+10\mu^2}\,\tan\vartheta}{6.16 H_{def} R^{2/3}}$$

and N_c = critical number of wear cycles, F = load, $\varphi_{\lim}$ = critical deformation where the shear limit τ is surpassed, $\varphi_s = 2\ln(C_2 F^{1/3})$ = maximum deformation at the surface, $\beta \geq 1$, m = 1.4 ... 2, E = Young's modulus, H_{def} = hardness of the deformed surface $\approx$ 2H, μ = friction coefficient, ϑ = half opening angle of the tip, R = tip radius of curvature. Using exponent m = 2, equation (8) yields a reasonable match to the experimental data with fit parameters summarised in table 1. Since the critical deformation $\varphi_{\lim}$ is not

TABLE 1. Fit parameter for the load dependence of the critical number of wear cycles N_c applying eq. 8.

sample	C_1(exp)	C_2(exp) [$N^{-1/3}$]	C_2(theory) [$N^{-1/3}$]
W-C:H (12 at%)	8.19	11.6	5.6
Au-C:H (0.8 at%)	18.06	13.3	19.8

known for Me-C:H films, experimental values of C_1 cannot be compared with theory. However, C_2 can be calculated from known mechanical data of the samples and the tip geometry, yielding values which fit the right order of magnitude (see table 1). Considering the rather simple approach of the model and the uncertainty in E, H, μ etc. this is a relatively good approximation.

For the integral wear rate, experimentally, always a more then linear increase with load is found (see fig. 10). Miyake et al. [19, 52] and Bushan et al. [21] found similar load behaviour for silicon, a-C:H, a-Si-C:H and a-Si-F-C:H coatings. Since the basic wear models of ploughing and cutting [50] predict only linear load dependence this again indicates that material fatigue which enhances wear should be present. The model of Zum Gahr combines the linear increase with load due to ploughing and cutting with the material fatigue mechanism by dividing through the load dependent critical number of wear cycles (eq. 8) weighted by an exponent n ranging from 1.2 to 8. In a simplified version it may be written as:

$$W \propto \left(\frac{\varphi_S}{\varphi_{\lim}}\right)^n \cdot F = C_3\left[\ln\left(C_2 F^{1/3}\right)\right]^n \cdot F \qquad (9)$$

W = integral wear per unit trace length, C_3 = constant. Since nothing is known about the exponent n and the constant of proportionality C_3, fits were performed using a me-

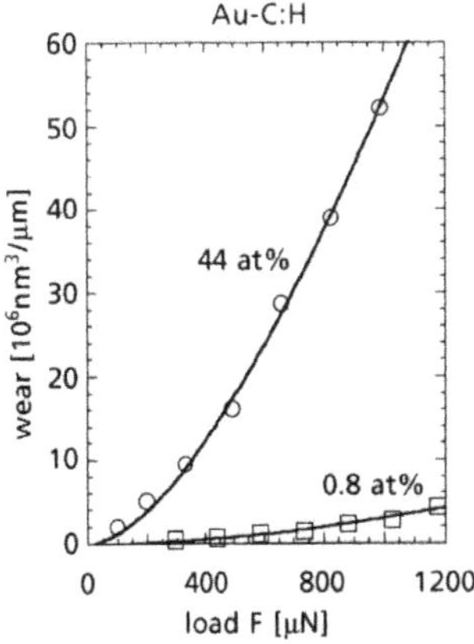

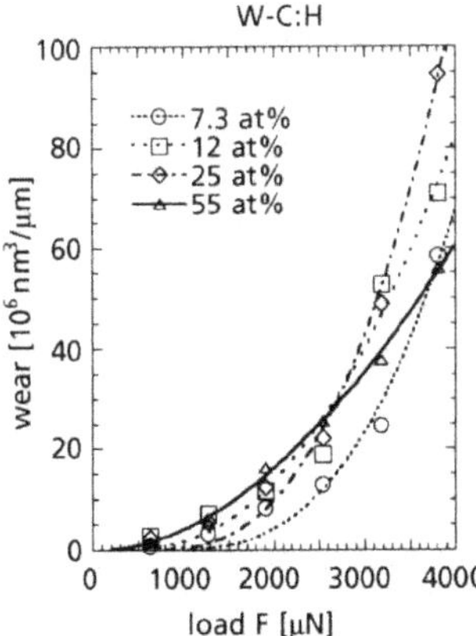

Figure 10. Integral wear as a function of load for Au-C:H with 44 and 0.8 at% Au and W-C:H with metal contents of 7.3, 12, 25 and 55 at% W. Full lines correspond to a fit with equation (9).

dium value of n = 5 and C_2 and C_3 as free parameters yielding a satisfactory match to the experimental data. Compared to the theoretical values the constant C_2 hits the right order of magnitude, but due to the arbitrary assumption made on the value of n, a quantitative comparison of theory and experiment is not recommended. Looking in detail at the fit curves, deviations from the model may be found in the low load range due to the fact that during the 50 wear cycles used in this experiment not enough fatigue periods occur for fatigue dominated wear. Furthermore, our simplified version of the model of Zum Gahr predicts zero wear in the case of finite load but absence of material fatigue ($N_c \sim (\varphi_{lim}/\varphi_s)^m \rightarrow \infty$ results in $W \rightarrow 0$), which is not adequate. The full model of Zum Gahr has an additive term which should be better, but doesn't fit with the data at all.

4.3 THE TIME DEPENDENCE OF WEAR

Measuring *integral* wear rate over longer wear periods of up to 1000 cycles, the small steps in depth profile due to material break-off are not visible any more. Instead, a decrease of wear rate with increasing number of wear cycles is found (see fig. 11). Similar reductions in wear rate were also found by other authors e.g. Miyake et al. [53, 19] for a-C:H, a-Si-C:H, c-BN and h-BN, or by Jiang et al. [22] for a-C:H. In those papers not the wear volume per trace length but the depth of the wear trace is considered, so results are not directly comparable with the ones presented here.

The decrease in wear rate can be attributed to an increase in contact area of the tip which gradually penetrates the surface during the wear process. Assuming that wear is proportional to the contact pressure:

$$W \sim F/A \tag{10}$$

(W = wear, F = load, A = contact area) time dependence may be described approximately by the following iterative model: During the first cycle the contact area A_1 can be ap-

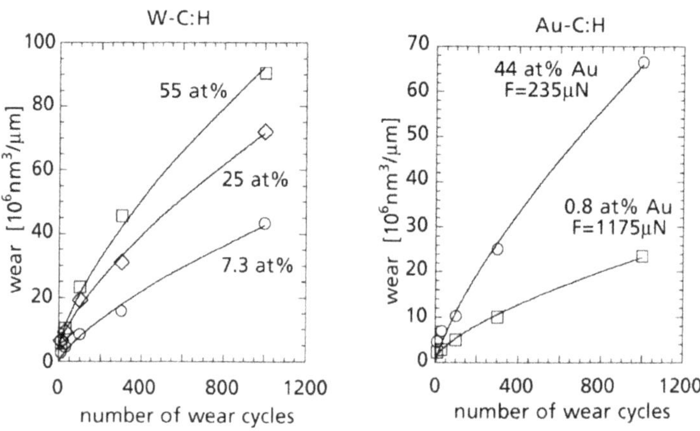

Figure 11. Integral wear as a function of time [in units of wear cycles] for W-C:H (load = 1905 μN) and Au-C:H samples at different loads (235μN, 1175 μN) due to their different wear rate. Solid lines are fits of eq. 14.

proximated by the Hertzian model of elastic contact of a sphere and a flat surface (eq. 3):

$$A_1 = \pi \left(\frac{3R_t F}{4E^*} \right)^{2/3} \tag{11}$$

(R_t = tip radius of curvature, E* reduced Young modulus of tip and sample). For each further wear cycle the semi-axis of contact across sliding direction increases roughly proportional to the width of the wear groove, while semi-axis paralell to the sliding direction will not change very much. Since wear was defined as cross-sectional area of the groove whose shape is nearly independent of the number of wear cycles (see fig. 5) the increase of contact area can be described by the increase of transversal semi-axis which is proportional to the square root of the accumulated wear until the n-th cycle:

$$A_n = A_1 + c_2 \sqrt{\sum_{i=1}^{n-1} W_i} \tag{12}$$

The combination of (10), (11) and (12) results in an recursion formula for the wear of the n-th cycle

$$W_n = \frac{c_1 F}{\pi \left(\frac{3R_t F}{4E^*} \right)^{2/3} + c_2 \sqrt{\sum_{i=1}^{n-1} W_i}} \tag{13}$$

(W_n = differential wear during the n-th cycle, n>1, $W_1 = c_1 F / A_1$, c_1, c_2 = constants). The *integral* wear up to the N-th cycle is then given by:

$$W(N) = \sum_{n=1}^{N} \frac{c_1 F}{\pi \left(\frac{3R_t F}{4E^*} \right)^{2/3} + c_2 \sqrt{\sum_{i=1}^{n-1} W_i}} \tag{14}$$

While the *wear parameter* c_1 indicates the wear per cycle and unit contact pressure, the *area parameter* c_2 defines how fast contact area grows and therefore wear rate decreases. Due to the recursive relation both parameters are coupled in a complex way. High wear rates can be caused by high values of c_1 or low values of c_2. Therefore the ratio $K = c_1/c_2$ may be proposed as a *wear coefficient* indicating the relative strength of wear chiefly independent of the number of wear cycles. The solid lines in Fig 11 represent the fit with the recursion formula and table 2 gives the corresponding fit parameters for different samples. As mentioned above, the parameters c_1, c_2 are not very significant on their own, while the wear coefficient K represents the relative strength of wear of different samples in a proper way.

TABLE 2. Fit parameter c_1, c_2 and wear coefficient K for the recursion formula (14) describing the time dependence of wear for different samples.

material	F [μN]	c_1 [Å^3/μm/GPa]	c_2 [μm]	K=c_1/c_2 [Å^3/GPa]
Au-C:H 0.8 at%	1175	94.0	75 x 10^3	1,64
Au-C:H 44 at%	235	16.4	1.9 x 10^3	3,64
W-C:H 7.3 at%	1905	14.4	3.6 x 10^3	4.0
W-C:H 25 at%	1905	412.	62 x 10^3	6.6
W-C:H 55 at%	1905	414.	42 x 10^3	9.8

4.4 WEAR AND METAL CONTENT

In fig. 12 the integral wear of W-C:H and Au-C:H is displayed as a function of metal content in the film. From what is already known about wear mechanisms and material structure it can now easily be understood:

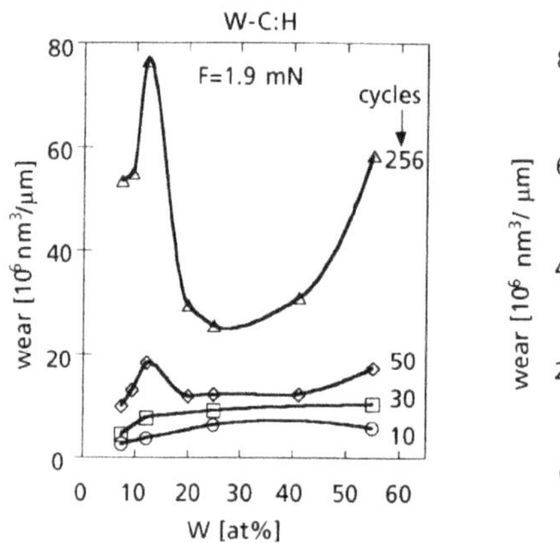

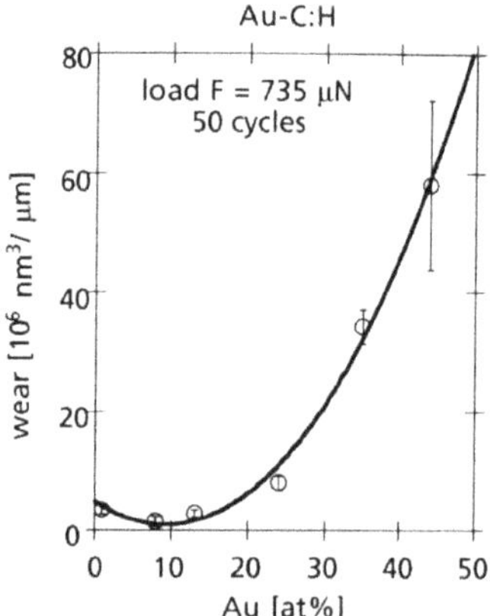

Figure 12. Integral wear of W-C:H and Au-C:H films as a function of metal content in the film. Wear of W-C:H has been determined at a load of 1905 μN for N = 10, 30, 50 and 256 wear cycles. Wear of Au-C:H has been determined at a load of 735 μN for a fixed number of 50 wear cycles.

For W-C:H staying below the critical number of wear cycles (N = 10 - 30 cycles) wear is found more or less constant, independent of the metal content of the film. I.e. if no material fatigue occurs only a weak influence of the metallic phase on the wear behaviour is found. However, surpassing the critical number of wear cycles (N = 50 - 256 cycles) three different regimes can be distinguished: (I) below the percolation threshold (< 15 at% W) due to the onset of material fatigue high wear is observed. Since N_c decreases with increasing metal content (see fig. 8) wear is at maximum just below percolation threshold (12 at% W). (II) Surpassing the percolation threshold the cross-linking of carbidic particles prevents material fatigue and wear drops by a factor of three to four. (III) approaching 50 at% W, i.e. 100% tungsten carbide wear raises again due to the omission of the lubricating hydrocarbon matrix which results in an increase of the friction coefficient of the film (see fig. 3,4).

For Au-C:H the minimum of wear is found at metal contents of less than 10 at% gold. Since the percolation threshold for Au-C:H is much higher than for W-C:H at about 50 at%, no reinforcement of the film structure due to coalescence of particles is found and wear increases nearly exponentially by a factor of 40 with increased incorporation of the soft gold. The slight increase of wear for almost metal-free Au-C:H films can be attributed to a change of the chemical structure of the hydrocarbon matrix. IR-spectroscopy has shown [1, 3] that with decreasing metal content covalent C-C-bonds are progressively replaced by -C:H_x-groups, i.e. the degree of three-dimensional cross-linking of the matrix is reduced, thus reducing the wear resistance of the film.

To compare wear of W-C:H and Au-C:H quantitatively, the results of W-C:H are scaled to the same load of 735 μN used in the Au-C:H measurements applying eq. (10) and the corresponding fitparameter (see fig. 13).

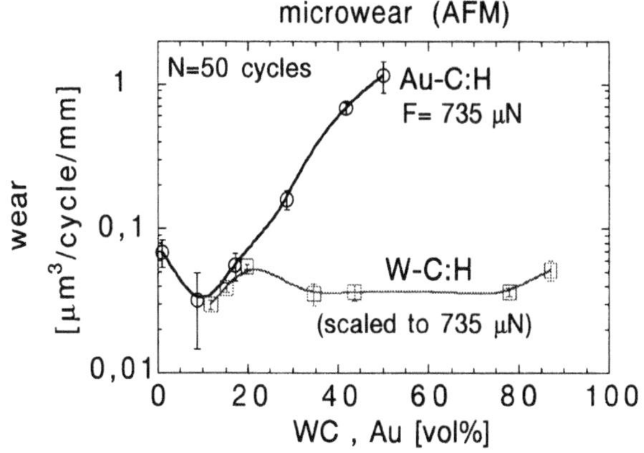

Figure 13. Wear of Au-C:H and W-C:H for N = 50 wear cycles scaled to the same load. Atomic fractions of metal are converted into volume fractions of metal respective metal carbide. Wear is normalised to the number of wear cycles.

An interesting correspondence is found for metal concentrations below 20 vol% where wear of Au-C:H and W-C:H is nearly equal. Since gold is a very soft and strongly wearing material, while tungsten carbide is very hard and wear resistant, it can be concluded that wear in the metal regime below 20 vol% is determined exclusively by the amorphous hydrocarbon matrix. Hence decrease of wear resistance due to increasing incorporation of metallic particles is independent of the sort of the particles. Only when particles percolate to a three dimensional network the properties of the metallic component becomes important, leading to a reduction of wear as in the case of the hard and rigid tungsten carbide particles.

4.5 COMPARISON OF MICRO- AND MACROWEAR

For comparison of micro- (AFM-) and macro- (pin-on-disk-) tests results are scaled to the same units [volume per cycle and per unit trace length] and metal contents are given in vol% metal or metal carbide respectively (see fig. 14). There are two remarkable points with the results: 1. the macrowear is 10 to 100 times larger than the microwear, even though the mean contact pressure in the microtest is higher than in the macrotest (see table 3). 2. for W-C:H, the shape of the curves differs in the middle range of concentrations.

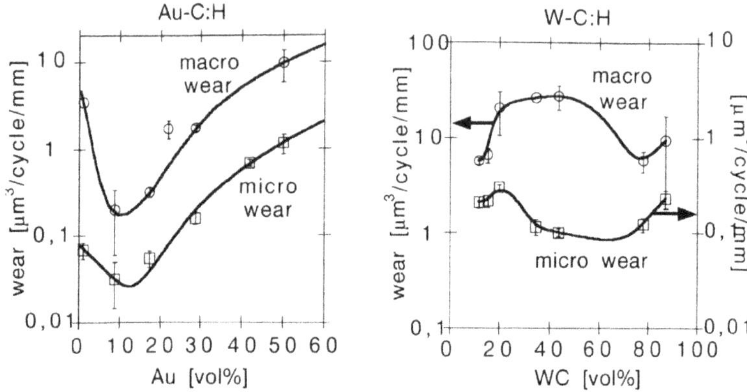

Figure 14. Comparison of micro- and macrowear tests as a function of metal content for Au-C:H and W-C:H. Please notice different scales for micro- and macro-wear in the case of W-C:H.

Both effects are supposed to be caused by the fact that in the macroscopic test abraded particles stay in the interface between ball and sample, clearly detectable by AFM inspection of the worn surface, while they are transported out of the wear groove in the microscopic case. These abrasive particles lead to pointlike, very high contact pressures increasing the wear rate in the macro wear test. Moreover, these higher pointlike contact pressures probably surpass the fatigue limit even for metal concentrations above the percolation threshold, thus resulting in the observed high wear rates in macrotest for W-C:H in the middle concentration range.

TABLE 3. Contact areas and contact pressures in micro- and macrowear experiments, computed by Hertz model of elastic contact. It is also applicable in the macroscopic case since surface roughness of sample and Al_2O_3-ball are only 3 nm and 6 nm respectively.

	microwear			**macrowear**		
material	load [µN]	area [µm²]	pressure [GPa]	load [N]	area [µm²]	pressure [GPa]
Au-C:H	735	0.14	4.7	5	17000	0.29
W-C:H	1905	0.17	10.9	5	10700	0.47

5 Conclusion and Outlook

Methods for the investigation of friction and wear at low loads and on small areas have been presented, and it has been proven that reliable results can be obtained even on not atomically flat surfaces.

The influence of the tip radius of curvature on quantitative friction force microscopy results has been separated from the material dependent friction properties (shear stress, Young's modulus) of tip and sample and collected in an effective friction coefficient μ^*. Hence, μ^* can be used to characterise the friction behaviour of materials on the microscopic scale in analogy to the conventional friction coefficient for the macroscale. More experiments have to be performed to prove the general applicability of the proposed relation and to compare determined shear stresses with independently determined material parameters.

It has been shown that on-line wear imaging is a very useful technique to investigate material fatigue phenomena and the influence of the microstructure of samples. Attempts have been made to describe load- and time-dependence of microscopic wear by theoretical models. Even though results are very promising the models should be refined and improved by more experimental data. E.g. the recursion formula for the time dependence of wear may be extended to consider the discontinuous material break-off and the critical number of wear cycles. Possibly, relations between fit-parameters and material properties can be established. For the load dependence of microwear the inconsistency of zero wear in the absence of fatigue has to be solved and more material data is needed to compare theory with experiment.

For comparison of micro- and macrotribological tests the main differences to be considered are: (i) single asperity contact versus multi asperity contact, (ii) wear-free versus wearing friction measurement and (iii) absence versus presence of abraded particles in the interface.

Acknowledgements

I would like to thank R. Thyen and H. Hübsch for the preparation of Me-C:H coatings, K. Taube for nanoindentation measurements and M. Reck for the performance of macroscopic friction and wear tests. This work was supported by the Volkswagenstiftung, Hannover.

6 References

[1] M. Grischke, Fortschrittberichte VDI, Reihe 5, No 179 (1989)
[2] H. Köberle, Thesis at the University of Hamburg, Dept. of Phys. (1989)
[3] M. Fryda, Fortschrittberichte VDI, Reihe 5, No 303 (1993)
[4] K. Taube; Fortschrittberichte VDI, Reihe18, No 102 (1991)
[5] C.P. Klages, R. Memming; Material Science Forum Vols. 52&53 (1989) 609, TransTech publication, Switzerland
[6] M. Grischke, K. Bewilogua, H. Dimigen; Mat. Manufact. Processes 8(4&5) (1993) 407
[7] H. Dimigen, C.P. Klages; Surf. Coat. Techn. 49 (1991) 543
[8] S. Miyake; Surf. Coat. Techn. 54/55 (1992) 563
[9] H. Ronkainen, J. Koskinen, J. Likonen, S. Varjus, J. Vihersalo; Diam. Rel. Mat. 3 (1994) 1329
[10] C. Donnet, M. Belin, J.C. Augé, J.M. Martin, A. Grill, V. Patel; Surf. Coat. Techn. 68/69 (1994) 626
[11] Y. Liu, A. Erdemir, E.I. Meletis; Surf. Coat. Techn. 82 (1996) 48
[12] C.P. Klages, H. Köberle, M. Bauer, R. Memming; Proc. 1. Int. Symp. „Diamond and Diamond-Like Films" , Edt. J.P. Dismukes, Elec-Chem. Soc. Proc. vol. 89-12 (1989) 225
[13] E. Boettger; Diploma Thesis at the University of Hamburg, Dept. Phys.-Chem. (1990)
[14] K.I. Schiffmann, M. Fryda, G. Goerigk; Microchim. Acta 125 (1997) 107
[15] K.I. Schiffmann, M. Fryda, G. Goerigk, R. Lauer, P. Hinze; Ultramicroscopy 66 (1996) 183
[16] K. I. Schiffmann, Thesis at the University of Hamburg, Dept. of Phys. (1997)
[17] C.M. Mate; Surf. Coat. Techn. 62 (1993) 373
[18] I. Fujiwara, T. Kamaei, K. Tanaka; J. Appl. Phys. 78(6) (1995) 4189
[19] S. Miyake, S. Watanabe, M. Murakawa, R. Kaneko, T. Miyamoto; Thin Solid Films 212 (1992) 262
[20] S. Miyake, R. Kaneko; Thin Solid Films 212 (1992) 256

[21] B. Bhushan, V.N. Koinkar, J.A. Ruan; Proc. Instn. Mech. Engrs. 208 (1994) 17
[22] Z. Jiang, C.J. Lu, D.B. Bogy, C.S. Bhatia, T. Miyamoto; Thin Solid Films 258 (1995) 75
[23] V.N. Koinkar, B. Bhushan; J. Appl. Phys. 79(10) (1996) 8071
[24] R. Kaneko, T. Miyamoto, Y. Andoh, E. Hamada; Thin Solid Films 273 (1996) 105
[25] H. Dimigen, H. Hübsch; Philips Techn. Rev. 41 (1983) 186
[26] J.T. Harnack; Thesis at the University of Hamburg, Dept. of Phys. (1993)
[27] M. Binggeli, R. Christoph, H.E. Hintermann, O. Marti; Surf. Coat. Technol. 62 (1993) 523
[28] M. Hipp, H. Bielefeldt, J. Colchero, O. Marti, J. Mlynek; Ultramicroscopy 42-44 (1992) 1498
[29] J. Colchero; „Reibungskraftmikroskopie", Thesis at the University of Konstanz, Germany, Dept. of Phys. (1993)
[30] S. Grafström, M. Neitzert, T. Hagen, J. Ackermann, R. Neumann, O. Probst, M. Wörtge; Nanotechnology 4 (1993) 143
[31] S. Fujisawa, Y. Sugawara, S. Morita; Microbeam Analysis 2 (1993) 311
[32] J.L Loubet, M. Belin, R. Durand, H. Pascal; Thin Solid Films 253 (1994) 194
[33] U.D. Schwarz, H. Bluhm, H. Hölscher, W. Allers, R. Wiesendanger; in „Physics of Sliding Friction" (Edt. B.N.J. Persson) Nato ASI Series, Kluwer, Dordrecht (1996)
[34] A. Berman, J. Israelachvili; in „Micro/Nanotribology and its Applications" Edt.: B. Bhushan, Nato ASI Series E, Vol. 330, Kluwer Academic Publishers (1997) 317
[35] K.L. Johnson; „Contact Mechanics", Cambridge University Press (1985)
[36] I.L. Singer; in „Fundamentals of Friction: Macroscopic and Microscopic Processes" Edt. Singer , Pollock; Nato ASI Series E, Appl. Sci. Vol. 220 (1992) 237
[37] T.H.C. Childs; in „Fundamentals of Friction: Macroscopic and Microscopic Processes" Edt. Singer , Pollock; Nato ASI Series E, Appl. Sci. Vol. 220 (1992) 209
[38] B.J. Briscoe; in „Fundamentals of Friction: Macroscopic and Microscopic Processes" Edt. Singer , Pollock; Nato ASI Series E, Appl. Sci. Vol. 220 (1992) 167
[39] J.N. Israelachvili; in „Fundamentals of Friction: Macroscopic and Microscopic Processes" Edt. Singer , Pollock; Nato ASI Series E, Appl. Sci. Vol. 220 (1992) 351
[40] A. Berman, J.N. Israelachvili; in „Micro/Nanotribology and Its Applications" Edt. Bhushan; Nato ASI Series E, Appl. Sci. Vol. 330 (1997) 317
[41] H.K. Christenson; J. Colloid Interface Sci. 121 (1988) 170
[42] U.D. Schwarz, P. Köster, R. Wiesendanger: „Quantitative analysis of lateral force microscopy experiments" ; Rev. Sci. Instr., in press (1996/97)
[43] U.D. Schwarz, personal communication (1997), to be published
[44] M.A. Lantz, S.J. O'Shea, M.E. Welland, K.L. Johnson; Phys. Rev. B 55(16) (1997) 55
[45] R.W. Carpick, D.F. Ogletree, M. Salmeron; Appl. Phys. Lett (1997) in press
[46] U.D. Schwarz, O. Zwörner, P. Köster, R. Wiesendanger; Phys. Rev. B (1997) in press
[47] J. Burger, M. Binggeli, R. Christoph, H.E. Hintermann, O. Marti; personal communication (1994)
[48] M. Wang; research center Jülich, report no. Jül-2595, ISSN 0366-0885 (1991)
[49] W. van Duyn, B. van Lochem; Thin Solid Films 181 (1989) 497
[50] K.H. Zum Gahr; „Microstructure and Wear of Materials", Tribology Series No 10, Elsevier, New York (1987)
[51] S.S. Manson, Discussion to: J.F. Tavernelli, Coffin, Jr; Trans. ASME, J Basic Eng. 84 (1962) 533
[52] S. Miyake, T. Miyamoto, R. Kaneko, T. Miyazaki; IEICE Trans. Electron. E78-C(2) (1995) 180
[53] S. Miyake, R. Kaneko, T. Miyamoto; Diam. Films. Technol. 1(4) (1992) 205

Formation of Carbon Films on Ceramic Carbides by High Temperature Chlorination

M. McNALLAN, *Y. GOGOTSI, I. JEON
CME Dept., M/C 246
University of Illinois at Chicago
842 W. Taylor St.
Chicago, IL 60607

**Department of Mechanical Engineering, M/C 251*

Abstract

Carbon films can be formed on SiC and other metal carbides by selective chlorination at high temperature and atmospheric pressure because of the volatility of compounds formed between chlorine and silicon or other metals, and because of the low thermodynamic stability of the corresponding carbon-chlorine compounds. An apparatus has been constructed in which carbon films are formed on metal carbides by reaction with flowing argon-chlorine-hydrogen gas mixtures in a fused silica furnace tube. The structure of the carbon films can be manipulated by controlling the temperature and composition of the reactive gas mixtures. Electron microscopy and Raman spectroscopy have been used to characterize the carbon films formed on β-SiC in Ar-Cl_2-H_2 gas mixtures at temperatures between 600 and 1000°C.

1. Introduction

Hard ceramic compounds with carbon, such as SiC and TiC, have been used for their wear resistance in a number of tribological applications. [1,2] These materials are also important in microelectronic technology and have potential applications in Microelectromechanical Systems (MEMS). Although these ceramics perform well in many tribological applications, even better performance can be expected from ceramics coated by diamond or diamond like carbon films.[3]

Diamond and diamond like carbon coatings are generally produced by chemical vapor deposition from hydrocarbon gases at low pressures in plasma environments.[4] Although the high hardness of the films and their low friction coefficients give them superior tribological properties, they may fail due to loss of adhesion with the underlying substrate. In addition, films produced by these techniques may be unsuitable for many

B. Bhushan (ed.), Tribology Issues and Opportunities in MEMS, 559-565.

MEMS applications because these deposition processes are essentially line of sight and may not penetrate into deep pores or undercuts during microfabrication processing.

An alternative method for formation of diamond or diamond like carbon films may be considered when the substrate is a carbide ceramic. Many metal chlorides have high vapor pressures at elevated temperatures, while carbon chlorine compounds are generally not thermodynamically stable under these conditions. When SiC is corroded at elevated temperature in chlorine containing gases with low oxygen potentials, the silicon component is preferentially attacked and a carbon rich layer builds up on the surface. [5,6] By controlling the gas composition during the etching, it may be possible to control the structure of the carbon film to optimize its tribological properties. Because the chlorination reaction is performed at atmospheric pressure without plasma activation, the film can coat shapes with pores and undercuts produced by other etching reactions during microfabrication. This paper addresses the thermodynamics and kinetics of carbon formation on several metal carbide surfaces, which may be useful in MEMS devices.

2. Thermodynamic Considerations

The high thermodynamic stability of volatile metal chlorides and the low thermodynamic stability of carbon-chlorine compounds at high temperatures favor selective chlorination of metal species from carbides. For example at 1200 K, the Gibbs Free Energies of formation of $SiCl_4$, $TiCl_4$, and CCl_4 are -500, -480, and +67.4 kJ/mol respectively.[7] From this information, chemical reactions of the form:

$$TiC(s) + 2\ Cl_2(g) = TiCl_4(g) + C(s) \tag{1}$$

are thermodynamically favorable, and would generally be expected to produce carbon with volatilization of the metal chloride.

Reaction [1] represents a somewhat oversimplified view of the chlorination process because factors such as the formation of compounds containing metals with lower valence states may affect the stoichiometry and reaction sequence. The presence of hydrogen in the system is a further complicating factor. The presence of hydrogen favors the formation of diamond in carbon films formed in CVD, and also ameliorates the effects of oxygen impurities, which are often present in industrial grades of Cl_2. Complex thermodynamic equilibria in multicomponent systems are most easily handled by using free energy minimization software such as the HSC program marketed by Outokumpo Oy [8]. In these programs, the gas and solid phase compositions at equilibrium are calculated for a given temperature and molar input of reactants by minimization of the overall Gibbs Free Energy. Figure 1 is an example of the equilibrium compositions calculated for an input of one mole TiC, one mole Cl_2, one mole H_2, and 0.1 mole O_2 at a total pressure of one atm., as a function of temperature between 300 and 1300°C. Although a variety of reaction products are formed, carbon is predicted to be the predominant solid reaction product at all temperatures below 1000°C, with a maximum yield at approximately 800°C. The oxygen impurities lead to formation of TiO_2, but this need not interfere with the carbon formation.

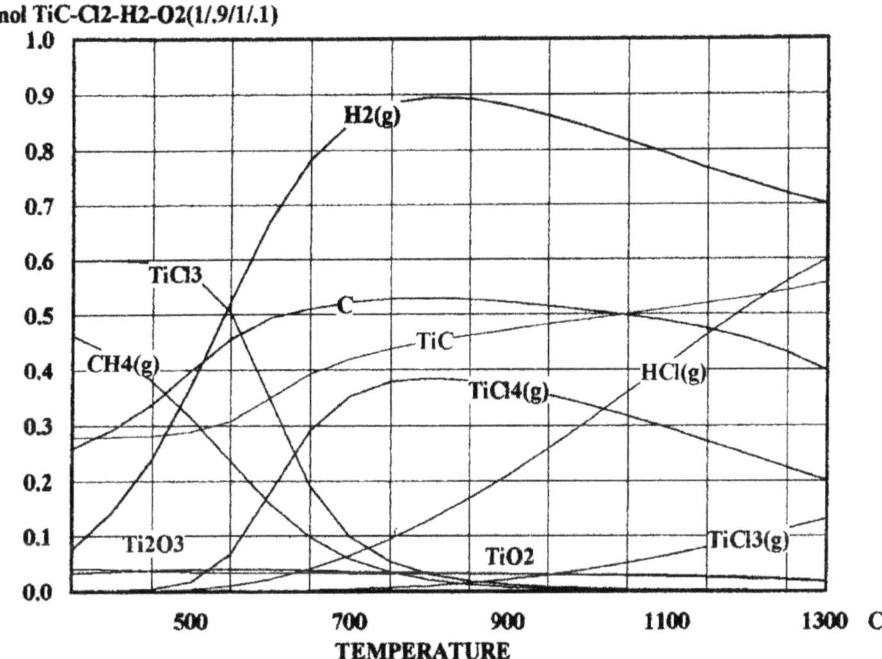

Fig. 1. Thermodynamic equilibrium compositions for TiC-Cl_2-H_2-O_2 System in molar ratio 1:1:1:0.1.

3. Experimental Details

The thermodynamic calculations predict the formation of carbon, but do not provide information on the rates of the reactions or on the structure of the carbon films produced. In order to assess these factors, a series of experiments have been performed on metal carbide powders in which they have been exposed to flowing gas mixtures of chlorine and hydrogen in a fused silica tube at temperatures between 400 and 1000°C. The apparatus and procedures used in these experiments have been described elsewhere.[9] Samples of metal carbide powders were suspended in platinum crucibles in flowing gas mixtures of reagent grade argon, hydrogen, and chlorine for periods of 4 to 72 hours. Thermogravimetric measurements performed during some of the experiments indicated decreases in the specimen mass during the exposures, consistent with reactions of the form of equation (1). In order to assess the structure of the carbon films produced in the reactions, the exposed carbide powders have been analyzed using Raman spectroscopy and transmission electron microscopy to determine the nature and phases of the carbon films.

4. Results and Discussion

Figure 2 shows the Raman spectra produced from TiC powder before and after exposure to Ar-3.5%Cl_2 gas mixture at 800°C for two hours. The small peaks at 1344 and 1568 cm^{-1} are indications of slight graphite contamination in the unexposed TiC. After exposure, however, the peaks are much stronger, indicating that a large amount of graphite was formed during the reaction. The small peaks at 429 and 598 cm^{-1} demonstrate that some rutile (TiO_2) formed in the powder from reactions with the oxygen contamination in the chlorine.[10] These results demonstrate formation of carbon by high temperature chlorination of TiC at atmospheric pressure.

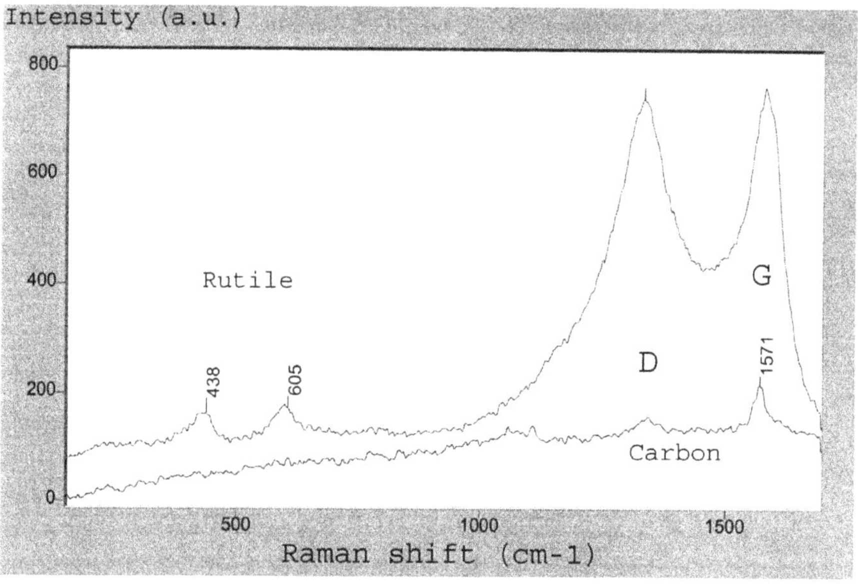

Fig. 2. Raman Spectra for TiC powder as received and after exposure to Ar-3.5% Cl_2 at 800°C for 2 hours.

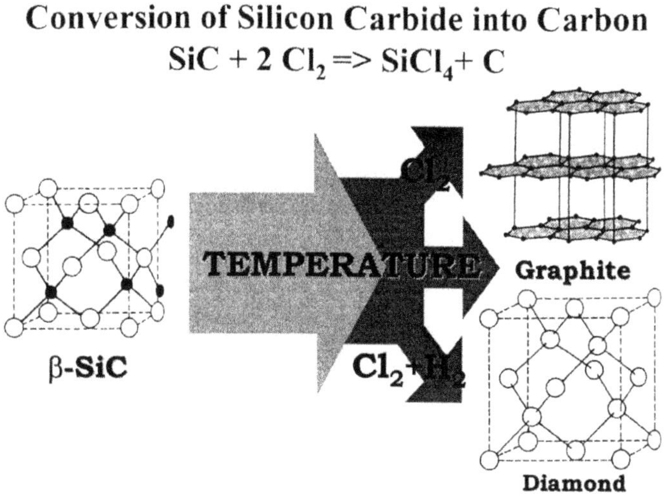

Fig. 3. Schematic diagram of formation of diamond structure on β-SiC.

The TiC experiments did not produce diamond or diamond like carbon. Two possible means to improve the likelihood of formation of these valuable coatings, are changes in the structure of the substrate and in the composition of the gas mixture. Nucleation of diamond on a surface is encouraged if the crystal structure of the substrate is similar to the tetrahedral cubic structure of diamond. β silicon carbide has a Zincblende structure, which is very similar to the diamond cubic structure except that Si atoms occupy half of the lattice sites, as shown in Figure 3. Therefore, diamond films would be more easily nucleated on β SiC than on non-cubic ceramics. In addition, titanium forms a particularly stable oxide, and may be very sensitive to the presence of oxygen impurities in the chlorine. Titanium oxide would be expected to form at the gas-solid interface, with carbon forming between the titanium oxide and the carbide.[11] The addition of hydrogen to the gas mixture may help by suppressing oxidation and may also contribute to the formation of diamond structure by a similar mechanism as in CVD processes.

Figure 4 shows the Raman spectra obtained on a sample of β-SiC that had been exposed to Ar-2.6%Cl_2-1.3%H_2 at 950°C for 24 hours. Rather than having the two peaks characteristic of graphite, the spectra show a broad disordered peak centered at 1350 cm^{-1}. This Raman pattern indicates that the carbon has a non-crystalline carbon structure characteristic of diamond like carbon films. This has been confirmed by transmission electron microscopy. Similar Raman spectra have been produced by carbon films obtained on β-SiC powders exposed to other gas mixtures containing both chlorine and hydrogen at temperatures of 600°C and above and in some cases evidence of formation of crystalline diamond has been observed.[9]

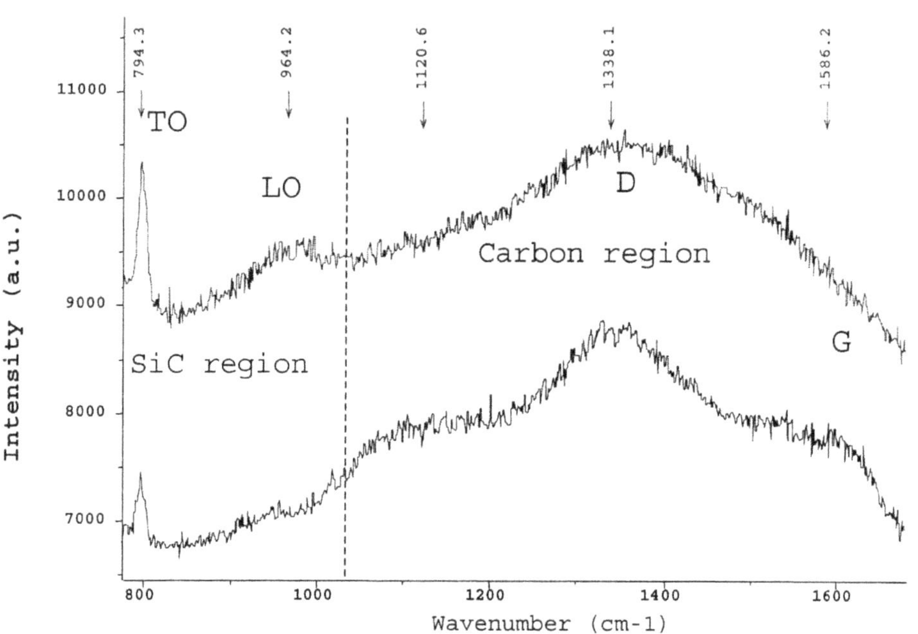

Fig. 4. Typical Raman spectra for β-SiC powder exposed to Ar-2.6%Cl_2-1.3%H_2 at 950°C for 24 hours.

5. Summary and Conclusions

Carbon films can be produced on metal carbide ceramics by exposure to chlorine containing gases at elevated temperatures. When the gas mixture also contains hydrogen, the films have structures characteristic of diamond like carbon with mixed sp^2/sp^3 bonding. These films can be expected to provide tribological benefits similar to those obtained from diamond like carbon films deposited from hydrocarbon gases. Chlorination may be a suitable technique for production of carbon films in MEMS devices because it does not require use of low-pressure plasmas and has the potential for penetrating deep into pores and overhangs produced during microfabricatiion.

ACKNOWLEDGEMENT

Research supported by the National Science Foundation under Grant # CMS-9617452.

References

1. D. Cranmer, (1985) Friction and Wear Properties of Monolithic Silicon Based Ceramics, Journal of Materials Science, **20**, 2029-2037.
2. G.R. Fenske, (1989) Nitride and Carbide Coatings for High Speed Steel Cutting Tools, STLE Tribology Transactions, **32**, 339-345.
3. A. Erdemir, C. Bindal, G.R. Fenske, C. Zuiker, R. Csenesits, A.R. Krauss, D.M. Gruen, (1996) Tribological Characterization of Smooth Diamond Films Grown in Ar-C_{60} and Ar-CH_4 Plasmas, Diamond Films and Technology, **6**, 31-47.
4. S.J. Bull (1995) Tribology of Carbon Coatings: DLC, Diamond and Beyond, Diamond and Related Materials, **4**, 827-836.
5. J.E. Marra, E.R. Kreidler, N.S. Jacobson, D.S. Fox, (1988) Reactions of Silicon-Based Ceramics in Mixed Oxidation Chlorination Environments, Journal of the American Ceramic Society, **71**, 10167-1073.
6. D.S. Park, M.J. McNallan, C. Park, W. W. Liang, (1990), Active Corrosion of Sintered α-Silicon Carbide in Oxygen-Chlorine-Gases at Elevated Temperature, Journal of the American Ceramic Society, **73**, 1323-1329.
7. JANAF Thermochemical Tables, 3rd Edition, (1995)
8. HSC Chemistry for Windows, Version 2.0, (1994) Outokumpu Research, P.O. 60 Finland.
9. Y.G. Gogotsi, I.D. Jeon, M.J. McNallan, (1997) Carbon Coatings on Silicon Carbide by Reaction with Chlorine Containing Gases, Journal of Materials Chemistry, 7, 1841-1848.
10. J.C. Parker, R.W. Siegel, (1990) Raman Microprobe Study of Nanophase TiO_2 and Oxidation-Induced Spectral Changes, Journal of Materials Research, **5**, 1246-1252.
11. S. Shimada, T. Ishii, (1990) Oxidation Kinetics of Zirconium Carbide at Relatively Low Temperatures, Journal of the American Ceramic Society, **73**, 2804-2808.

DEPOSITION AND CHARACTERIZATION OF DIAMOND-LIKE MATERIALS

T.C. OVAERT, R. MESSIER, L.J. PILIONE, R. COLLINS, J. LEE, W. OTAÑO-RIVERA, AND J. A. ZAPIEN
The Pennsylvania State University
Materials Research Laboratory, University Park, PA 16802

1. Introduction

Providing a low-friction, wear-resistant coating on silicon is a potential scheme for increasing the performance and life of micro electro mechanical systems (MEMS). In theory, there is no set of guidelines for selection of specific coatings (or coating properties) for these systems, however, larger scale systems have benefited in many industrial applications through the incorporation of very hard deposited surface layer(s), minimizing contact deformation and/or adhesion and its associated sliding system damage and energy dissipation. This talk will discuss work in progress at The Materials Research Laboratory, Penn State University, on the deposition and characterization of three hard coatings of potential interest for MEMS: nano-crystalline diamond, SiC, and the cubic phase of boron nitride (c-BN).

While the deposition of diamond coatings is well established, new methods and techniques have been developed to produce nano-crystalline coatings with excellent adhesion and mechanical properties. The novelty of these films is their enhanced smoothness and toughness, as compared to previous diamond films whose deposition methodology produced a highly-faceted rough surface that often required polishing to reduce its abrasive tendencies. In addition, real-time spectroscopic ellipsometry (in the visible region) has been utilized to monitor diamond film sp^2 C and void volume fractions, the temperature of the top 500 Å of the substrate under growth conditions, bulk layer diamond film thickness during the growth process, and film growth rate.

In addition, recent investigations have focused on the deposition of SiC films using a DC magnetron sputtering technique. With the availability of new doped SiC targets, very high deposition rates can be obtained. These films exhibit Si-C, Si-N, and C-N bonding as determined from infrared measurements, resulting from variations in the argon/nitrogen ratio in the plasma. Also, optical and electronic properties were measured and exhibit unique variations with argon/nitrogen concentration.

B. Bhushan (ed.), Tribology Issues and Opportunities in MEMS, 567-578.

Another candidate hard coating material for MEMS applications is c-BN. Films containing high concentrations of c-BN have been deposited on Si wafers using unbalanced magnetron sputtering with pulsed DC substrate biasing, and characterized using FTIR. Our results confirm that there exists a threshold value for the momentum imparted to the growing film per arriving boron atom necessary for stabilization of the cubic phase. Variations in the BN phase also exist near the substrate interface, consisting of both an amorphous and hexagonal phase.

2. Nano-Crystalline Diamond Films

Enhanced CVD techniques employing hydrocarbon gases (e.g., CH_4) highly diluted in H_2 are the most common methods for producing diamond films on substrates, which must be maintained in the 800 °C temperature range [1,2]. More recently, it has been shown that diamond films grown on diamond-powder-seeded Si wafers and Si wafers having a deposited nano-crystalline diamond thin-film, from CO rich mixtures of CO and H_2, exhibit high growth rates (up to 2.5 μm/hr) in a narrow temperature range of 350 °C to 500 °C, using a relatively low power (300 - 500 W) microwave plasma-enhanced CVD reactor (MPECVD) [3]. In this process, the diamond phase appears to develop more directly from the C- and O-containing precursors (for example, H_xCO radicals or reaction products of surface CO with excited H_2 [4]), rather than from hydrocarbon radicals [5]. In addition, incorporation of the precursor into the diamond surface appears to be thermally activated with an activation energy of 8 kcal/mol [6,7], which accounts for decreasing growth rates with decreasing temperature.

The use of real-time spectroscopic ellipsometry (RTSE) for process monitoring of the diamond film growth process is well established [8] and an extremely useful tool. A typical process-monitoring arrangement combining MPECVD with RTSE can be seen in Figure 1.

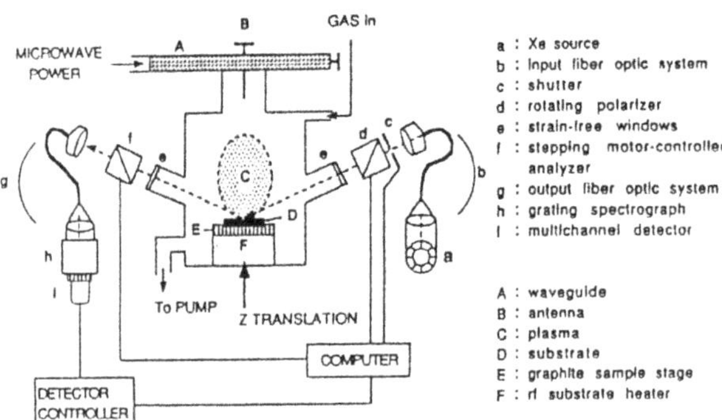

Figure 1. Schematic of MPECVD reactor with RTSE

With a multichannel instrument, single pairs of successive Ψ,Δ spectra were collected in the 1.5 to 4.5 eV range with a 2.6 s collection time. After least squares regression analysis determination of the real and imaginary parts of the dielectric constant, effective medium models have been incorporated that allow one to deduce film growth parameters. Substrate temperature values are true temperatures of the near surface of the Si substrate obtained from RTSE-based calibrations [9].

In Figure 2, one can see the effect of temperature on film growth rate for various mixtures of $[CO]/[H_2]$, as well as a comparison with a $[CH_4]/[H_2]$ mixture. The near-linear curves for the $[CH_4]/[H_2]$ and lowest $[CO]/[H_2]$ mixtures correlate with the established activation energy of 8 kcal/mol. The growth rate is expressed in terms of the mass thickness as a function of time, calculated using the volume fraction of diamond in the bulk layer as well as the surface layer thickness. The growth rate behavior at high $[CO]/[H_2]$ ratios suggests a new type of growth mechanism for diamond.

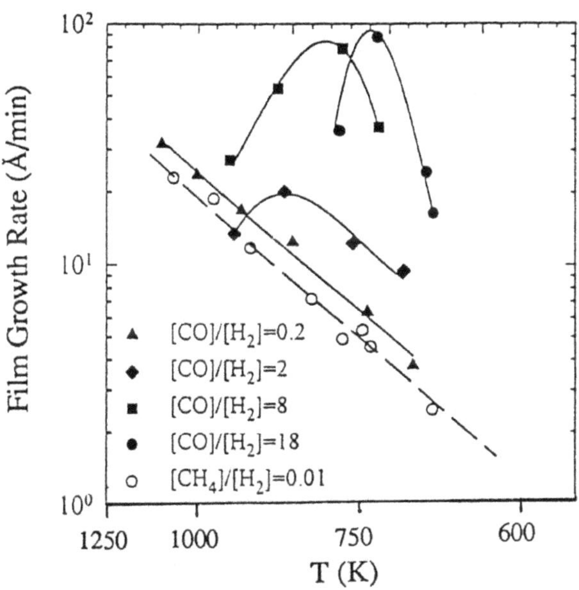

Figure 2. Growth rate of MPECVD diamond on diamond film substrates

RTSE also allows one to observe the structural evolution of the films, in terms of the change in bulk layer thickness, surface layer thickness, void fraction, and sp^2 C fraction as a function of time, as can be seen in Figure 3. The surface roughness of these films has been measured by AFM and exhibits maximum peak-to-valley values in the 100 to 200 Å range (grain sizes 100 to 500 Å) consistent with data in Figure 3.

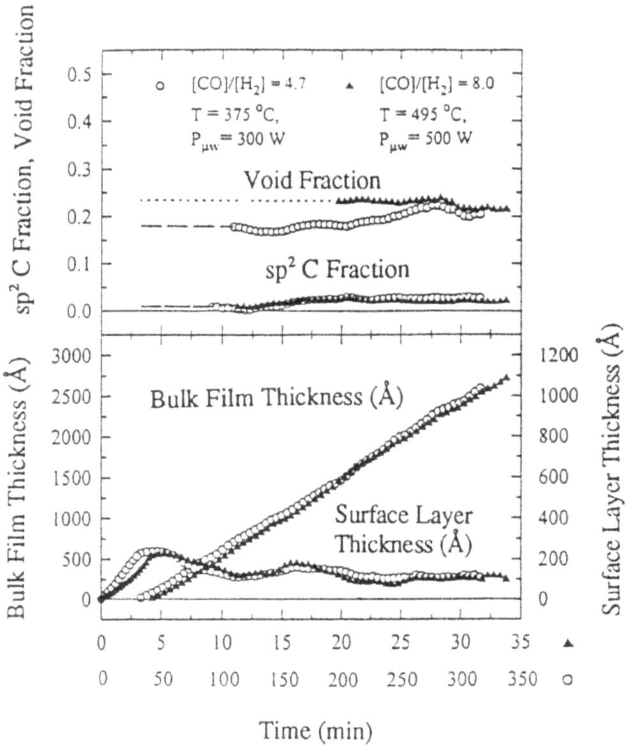

Figure 3. Structural evolution of diamond films by RTSE

The quality of the nano-crystalline diamond films is high and compares to the best nano-crystalline films grown by conventional processes at higher temperatures. Figure 4 shows a micro-Raman spectrum of a diamond film prepared at 500 °C, 350 Å/min growth rate, with a $[CO]/[H_2]$ ratio equal to 8. The peaks at approximately 1138 and 1336 cm^{-1} are evident, and compare with those exhibited by films obtained CH_4/H_2 mixtures at 800 °C.

The success of diamond films in MEMS applications will largely be determined by their to meet performance requirements for particular configurations. The advantages of diamond, namely enhanced thermal conductivity, in addition to high hardness, low friction/low wear, and environmental stability, make nano-crystalline diamond a potential candidate. Furthermore, the utilization of hydrocarbon-free gas mixtures combined with RTSE process monitoring techniques will likely improve diamond film quality and production consistency, while at the same time maximizing film growth rates and minimizing substrate temperatures. These will be important considerations when examining the economics of production process design and selection.

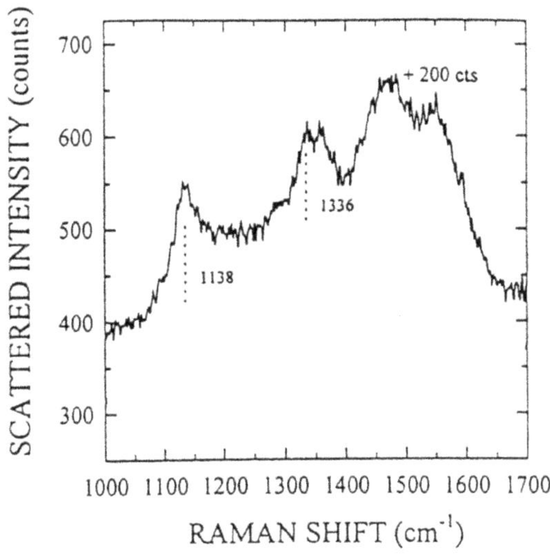

Figure 4. Micro-Raman spectrum of diamond film, [CO]/[H_2]= 8

3. Films of SiC prepared by balanced DC magnetron sputtering

SiC thin films have been deposited using balanced DC magnetron sputtering at very high deposition rates (500 Å/min) using new doped SiC sputtering targets (Carborundum Hexoloy® SG-90 and SA). Films were sputtered in argon or argon and nitrogen at 2.5 and 10 mTorr. Target powers or 150, 100, and 50 W were provided by the DC power supply, and a target-substrate distance of 50 mm was maintained. Borosilicate glass was used as a substrate. Film thickness measurements were made with an Alpha-Step 200 profilometer, and ranged from 150 to 400 nm. Film roughness (R_a) was 20 Å or less, the resolution of the instrument. Optical constants were obtained between 40 and 100 nm wavelengths by determining the transmission (Cary 2300 spectrophotometer) and reflection (Beckman DK-2A integrating sphere spectrophotometer). A computer subroutine was used to determine the index of refraction, extinction coefficient, and absorption coefficient at selected wavelengths. In addition, a four-point probe system was used to measure the surface current and voltage characteristics of the deposited films. Electrical surface resistivities were determined from the current, voltage, separation and size of the probes, and the film thickness.

Resistivity values for the SA and SG-90 targets can be seen in Figure 5. It is apparent that large resistivity changes can occur as a function of the N_2 concentration in the plasma, for both the SA and SG-90 target materials. The effect of pressure on the resistivity values is less pronounced. In general, the film growth rates were highest (on the order of 400 Å/min) for the 2.5 mTorr total pressure (the sum of the Ar and N_2 partial pressures), dropping to approximately 275 Å/min at 10 mTorr for the SA target. SG-90 target growth rates were slightly higher. In addition, increasing the N_2 concentration from 0 to 10% increased the deposition rate by roughly 25%; the growth rate remained relatively constant between 10% and 40% N_2. These overall trends were observed for both target materials. This information may be useful when designing devices that must function in a semi-conducting or insulating mode, depending on the application.

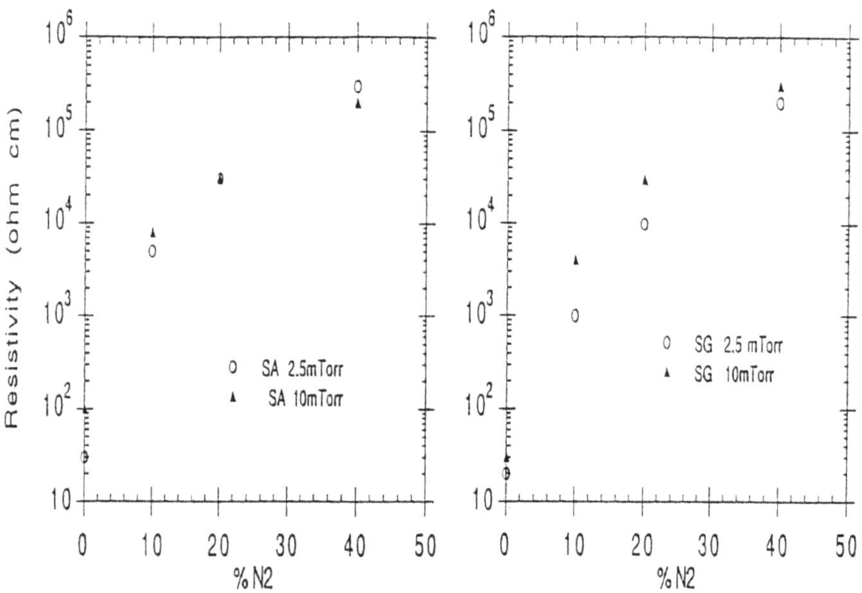

Figure 5. Resistivity of SA and SG-90 films as a function of $\%N_2$ and pressure

Optical properties are also affected by nitrogen concentration. Film index of refraction and absorption coefficient as a function of $\%N_2$ and wavelength for the SA and SG-90 targets at 10 mTorr can be seen in Figures 6 and 7. Similar trends are observed at 2.5 mTorr.

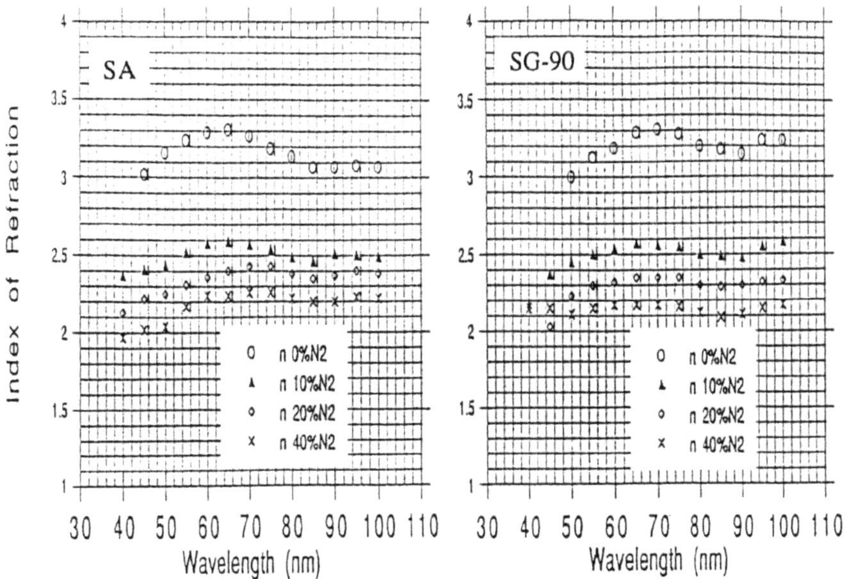

Figure 6. Index of refraction of SA and SG-90 films vs. $\%N_2$ at 10 mTorr

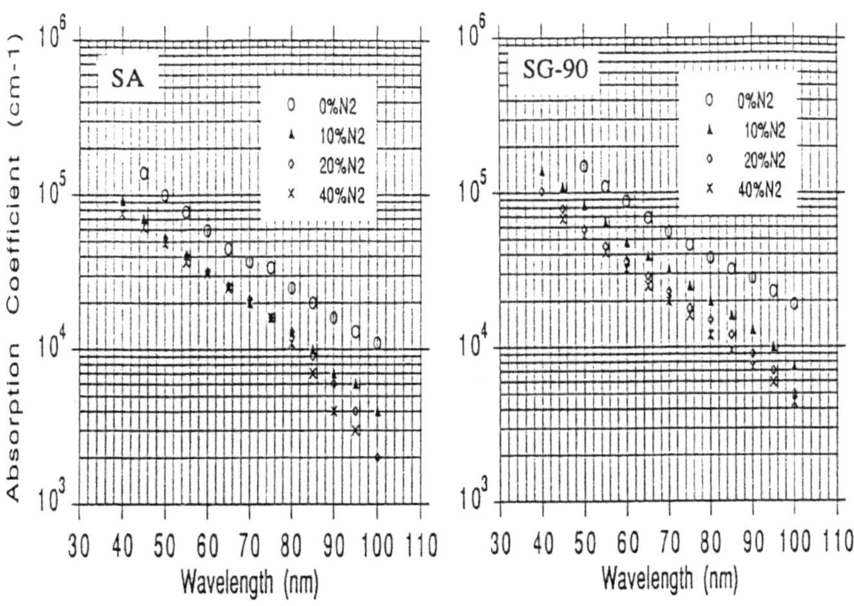

Figure 7. Absorption coefficient of SA and SG-90 films vs. $\%N_2$ at 10 mTorr

It is apparent from these results that the optical properties of the films are affected by the nitrogen concentration. In general, the index of refraction decreases with increasing $\%N_2$, for both target materials over the range of wavelengths tested. A similar trend is observed for the absorption coefficient, however, the effect of $\%N_2$ is less pronounced above 10% N_2. This aspect may be important when engineering devices that must meet certain optical or combined optical/electrical design requirements (e.g. transparent conductors).

3.1. FTIR SPECTRA OF THE Si-C-N DEPOSITED FILMS

Figure 8 shows the Fourier transform infrared (FTIR) spectra for a film grown on crystalline Si using the SG-90 target at 10 mTorr Ar (0% N_2). The film was 3600 Å thick, with a resistivity value of 30 Ω-cm. The strong absorption at 750 cm^{-1} indicates Si-C bonding. No other absorptions are evident in the 400 cm^{-1} to 4000 cm^{-1} range. A similar spectra was observed for the SA target at 0% N_2.

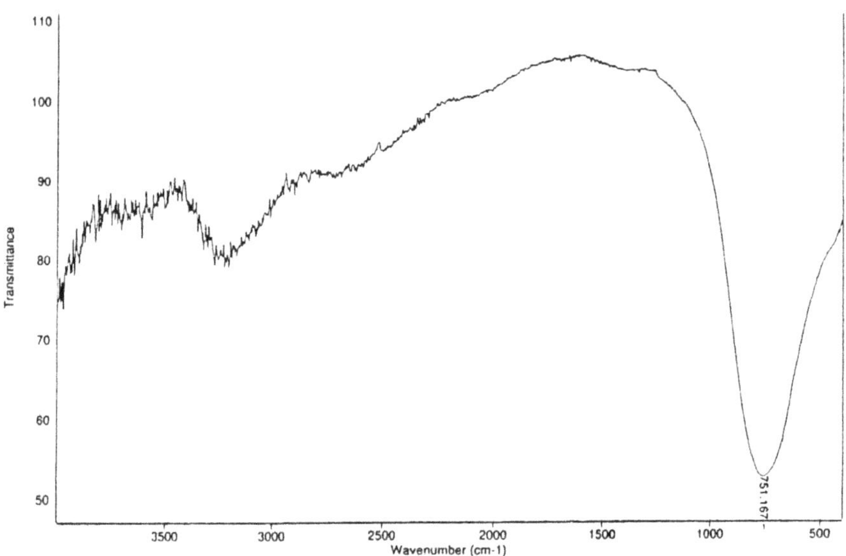

Figure 8. FTIR spectra for a film grown at 10 mTorr, SG-90 target, 0% N_2

Figures 9 and 10 show the FTIR spectra of SG-90 and SA target films, respectively, deposited at 2.5 mTorr, at 40% N_2. At this concentration level, the absorptions at approximately 490, 590, 900-1000 cm^{-1} are due to silicon nitride.

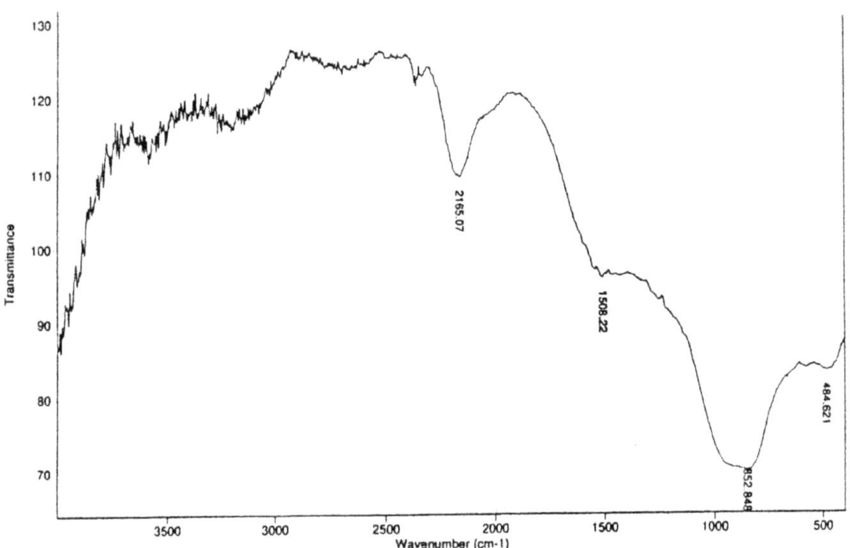

Figure 9. FTIR spectra for a film grown at 2.5 mTorr, SG-90 target, 40% N_2

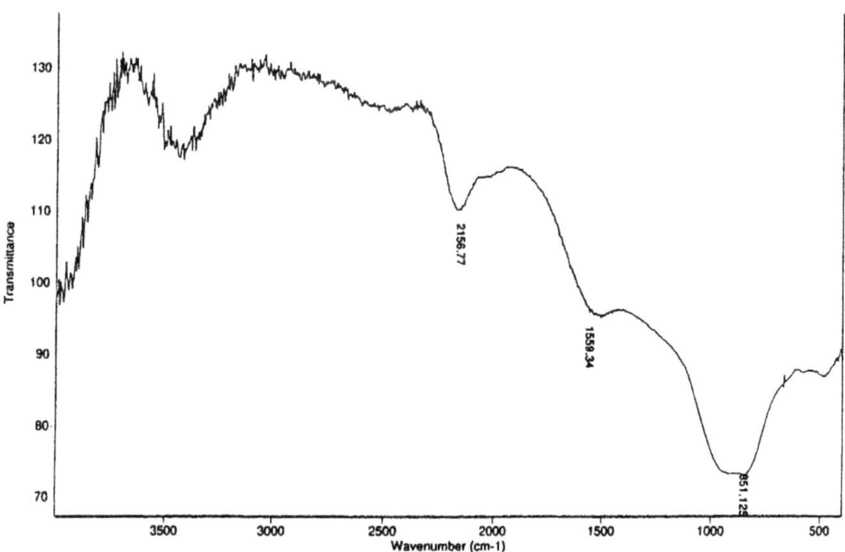

Figure 10. FTIR spectra for a film grown at 2.5 mTorr, SA target, 40% N_2

The absorption at approximately 1550 cm^{-1} is the stretch mode corresponding to conjugated aromatic C=C sp^2 carbon. The absorption at approximately 2150 cm^{-1} is likely due to a carbon-nitrogen triple bond. What is missing from these spectra is the 750 cm^{-1} line attributed to SiC. It turns out that at the relatively low deposition temperatures used here (≈ 100 °C), Si_3N_4 is very stable. Thus, at 40% N_2, is appears that the films do not contain any SiC. Future FTIR investigations of specimens prepared at 10% and 20 % N_2 are planned, and the results should elaborate on this phenomenon further.

Based on these investigations, it is apparent that new sputtering targets exist that enable high-deposition-rate balanced DC magnetron sputtering of films in the Si-C-N bond system. By varying the N_2 concentration, one can obtain compositions that range from SiC to Si_3N_4, or intermediate compositions thereof, enabling the design of engineered films for particular mechanical, optical, and/or electrical properties within MEMS applications.

4. Cubic Boron Nitride (c-BN)

Unlike diamond with its covalent bond structure, the cubic phase of boron nitride (c-BN) contains both ionic and covalent bonding. It is extremely hard, capable of operating at higher temperatures than diamond, has high oxidation resistance, and is highly chemically inert. For these reasons, there are considerable research efforts being undertaken at the present time aimed at understanding the mechanics of formation of metastable c-BN. The common themes in the deposition of c-BN are bombardment stabilization, temperature (i.e., threshold values), and stoichiometry (1:1 B/N ratio). Thus far, the major obstacles have been excessive film stress and low adhesion.

Applications of c-BN coatings include cutting tools, wear resistant surfaces, and n- and p-type doped semiconductors. It is likely that certain MEMS applications may benefit from its extreme properties as a hard coating material, particularly when operating in relatively extreme environments.

There exists a wide variety of deposition methods; and deposition of metastable c-BN through ion bombardment stabilization using unbalanced magnetron sputtering with DC pulsed substrate biasing has been the method of choice at Penn State. This method offers the potential for large-scale economical production environments. The reason for using the pulsed DC power supply to bias the substrate is that in the DC pulsed mode, an asymmetrical (sawtooth) signal is produced. The forward bias provides the accelerating ion bombardment voltage while the reverse bias portion compensates for the buildup of charge on the (insulating) film. The pulse frequency can be optimized to maximize ion current, for BN this is 218 kHz. At this frequency, the ion transfer time is small compared to the period of oscillation, and thus the ions respond to an instantaneous substrate potential [10]. The maximum negative bias voltage, V_{max}, is used to control the deposition process.

Thin films of BN were deposited at different substrate bias voltages, and subsequently analyzed for phase content using FTIR. The spectra of a film with the hexagonal phase is characterized by a B-N stretch peak at 1390 cm^{-1}, a small peak at 780 cm^{-1}, associated with the B-N-B bending mode [11]. Films with a predominant cubic phase exhibit a peak in the 1055 to 1100 cm^{-1} region. The phase composition is estimated by calculating the intensity ratios at the 1060 cm^{-1} and 1400 cm^{-1} absorption peaks. Films that show a shift toward 1100 cm^{-1} are generally adherent to the substrate (crystalline Si), however, they are at a high state of stress. If the film spalls, the peak tends to shift toward the 1060 cm^{-1} value as the stress is relieved. Figure 11 shows the effect of negative bias voltage on film composition. It is apparent that a threshold value for cubic phase formation exists at approximately 72 V. Films deposited at bias voltages above 109 are roughly 70% c-BN or better.

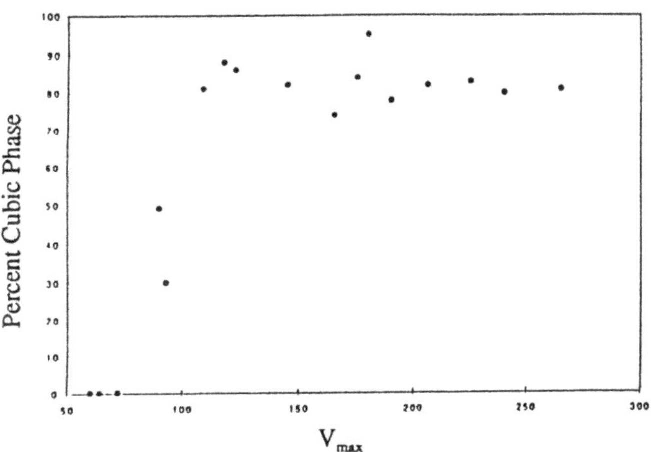

Figure 11. Percent cubic phase vs. V_{max}

Figure 12 shows the film growth rate as a function of substrate negative bias voltage. The increase in deposition rate with increase in pressure follows from the increase in the sputtering rate of boron due to an increase in plasma density and resulting sputtering current. The threshold values of V_{max} shift toward higher values as the pressure increases, thus, a c-BN deposition "window" will depend on the negative substrate bias voltage, V_{max}.

Films have also been deposited on AlN and diamond, and analyzed using FTIR and XRD. Due to the small crystallite size, absorption peaks associated with c-BN are undetectable. In general, films deposited on diamond coated Si had greatly improved adhesion and bond ordering, probably due to enhanced microstructural growth at the BN-diamond interface. This is in sharp contrast to the well-known observation that BN films grown on Si generally consist of a cubic layer superimposed on hexagonal and amorphous layers, each approximately 5 nm thick [12]. It has been proposed that these hexagonal and amorphous layers build a stress that is necessary to begin nucleation of the cubic phase [13], however, they and the associated are detrimental to film adhesion, which has been a major obstacle preventing use of these hard coatings. Future work will continue to investigate the effects of interlayers on c-BN nucleation, stress, and adhesion.

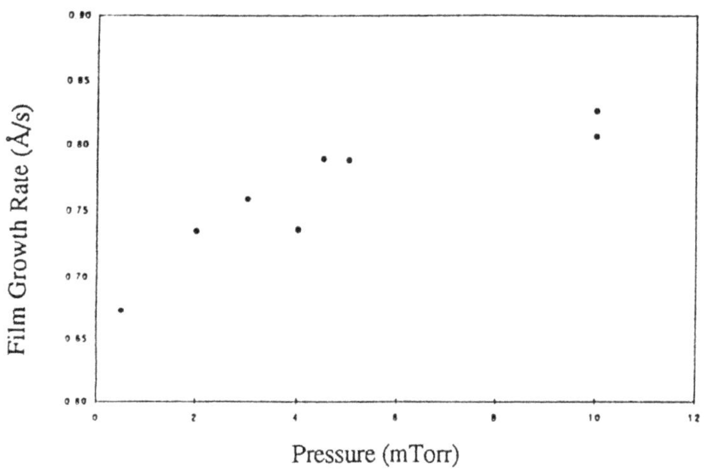

Figure 12. Film growth rate vs. pressure

5. References

1. Angus, J.C., and Hayman, C.C. (1988) *Science*, **241**, p. 913.
2. Yarbrough, W.A., and Messier, R. (1990) *Science*, **247**, p. 605.
3. Lee, J., Collins, R.W., Hong, B., Messier, R., and Strausser, Y.E. (1997) *J. Vac. Sci. Tech.*, **A** 15(4), p. 1929.
4. Johnson, C. E., Weimer, W.A., Cerio, F. M. (1992) *J. Mater. Res.*, **7**, p. 1427.
5. Kondoh, E., Ohta, T., Mitomo, T., and Ohtsuka, K. (1991) *Appl. Phys. Lett.*, **59**, p. 488.
6. Tsuda, M., nakajima, M, and Oikawa, S. (1986) *J. Amer. Cer. Soc.*, **108**, p. 5780.
7. Harris, S. J., and Martin, L.R. (1990) *J. Mater. Res.*, **5**, p. 2313.
8. Collins, R. W. (1990) *Rev. Sci. Instrum.*, **61**, p. 2029.
9. Lee, J., Hong. B., Messier, R., and Collins, R.W. (1996) *J. Appl. Phys.*, **80** (11), p. 6489.
10. Lieberman, M.A., and Lictenberg, A.J. (1994) Principles of Plasma Dischages and Materials Processing, J. Wiley & Sons.
11. Jäger, S., Bewilogua, K., and Klages, C.-P. (1994) *Thin Solid Films*, **245**, p. 50.
12. Kester, D.J., Ailey, K.S., and Davis, R.F. (1994) *Diamond and Related Materials*, **3**, p. 332.
13. Köhler, K., Horne, D.E., and Coburn, J.W. (1985) *J. Appl. Phys.*, **58** (9), p. 3350.

WEAR AND NANOMECHANICAL STUDIES OF SILICON OXIDE AND SILICON NITRIDE THIN FILMS FOR MEMS APPLICATIONS

Z. RYMUZA, Z. KUSZNIEREWICZ, M. MISIAK
Warsaw University of Technology
Institute of Micromechanics and Photonics
ul.Chodkiewicza 8, 02-525 Warszawa, Poland
E-mail kup_ryz@mp.pw.edu.pl
K. SCHMIDT-SZAŁOWSKI, Z. RŻANEK-BOROCH, J. SENTEK
Warsaw University of Technology
Faculty of Chemistry
ul. Noakowskiego 3, 00-664 Warszawa, Poland

Abstract Thin films produced by PE-CVD techniques are often used in microelectronic and MEMS applications. New kind of discharges, glow discharges so called Atmospheric Pressure Glow (APG) can be otained when a dielectric barrier is used. The process of discharges stabilized by a dielectric barrier having non-uniform character (similar to the „silent" discharges) was used to produce thin films of silicon oxide and silicon nitride applicable in MEMS technology. The substrate temperature (silicon or metal) was usually 150-250°C. Such film deposited using polycondensation of tetraethoxysilane (TEOS) and hexamethylodisilazane (HMDS) for silicon oxide and silicon nitride respectively were tested using a special microtribometer for wear studies and Hysitron Inc.Triboscope ™ which has the capability to carry out indentations at very low loads and to make load-displacement measurements with subnanometer indentation depth sensitivity. The wear and nanomechanical behaviors of the films were found to be very sensitive to the thickness (films up to 300 nm thick have been investigated) and the material composition.

1.Introduction

The rapid progress in MEMS technology gives promise for new designs of integrated sensors, actuators and other devices [1]. Both surface and bulk micromachining are employed to make MEMS devices [1-5]. At typical IC-LPCVD conditions, polysilicon and other films used to produce MEMS elements are found to be in residual compressive state. This is one of the important disatvantages of such micromachining techniques [6, 7].

To avoid the disatvantages of the aforementined techniques of fabrication processes, the frequently used materials such as silicon oxide and silicon nitride can be deposited by very simple technique at atmospheric pressure and at low temperature by so called

B. Bhushan (ed.), Tribology Issues and Opportunities in MEMS, 579-589.

Atmospheric Pressure Glow (APG) discharges [8,9]. Sliding structures and rotating elements on such structural or structural/isolation materials are applicable [6,10] but further studies are needed to be done on friction, wear and nanomechanical behavior of the materials used in MEMS applications [11-12].

The aim of the paper is a presentation of the above mentioned APG techniques used for depositing thin films of silicone oxide and silicone nitride on silicon (or metal substrate) and to discuss their wear and nanomechanical behaviours.

2.Deposition technique

The APG discharges technique was used for the thin surface layer deposition when the discharges were stabilized by a dielectric barrier. Such technique is different from the process of a simple APG discharges since the used discharges are of a non-uniform character and are similar to the „silent" discharges. The discharge was stabilized by means of a dielectric barrier- a 0.1-1 mm thick glass or polymer (e.g. poly(ethylene therephtalate) - PETP). At very thin barrier the glow discharge at atmospheric pressure can be obtained at relatively low voltage of about 2-3 kV. The schematic of the reactor used for the deposition of thin films is presented in Fig. 1.

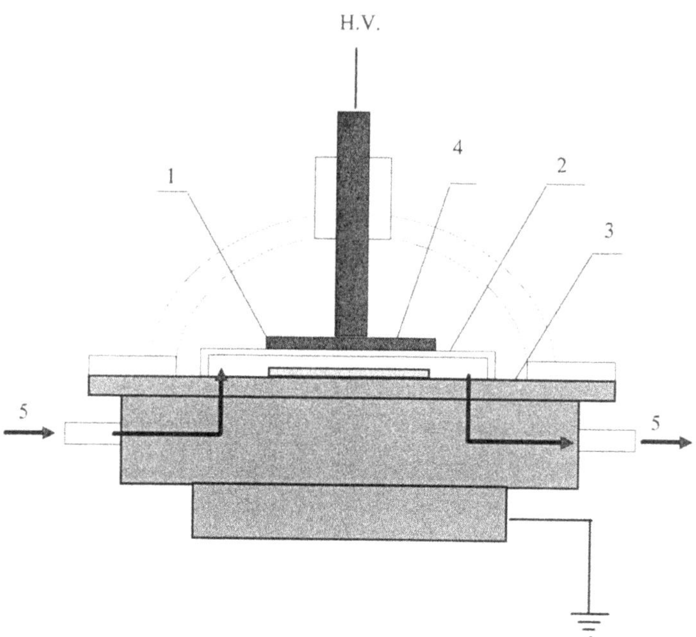

Fig.1. Reactor for thin deposition from gas reagents under plasma conditions in electrical discharges stabilized with a dielectric barrier under atmospheric pressure; 1 - high-voltage electrode, 2 - dielectric barrier, 3 - grounded electrode, 4 - electrode with substrate, 5- gas inlet and outlet holes.

The whole system used for depositing thin films is shown in Fig. 2.

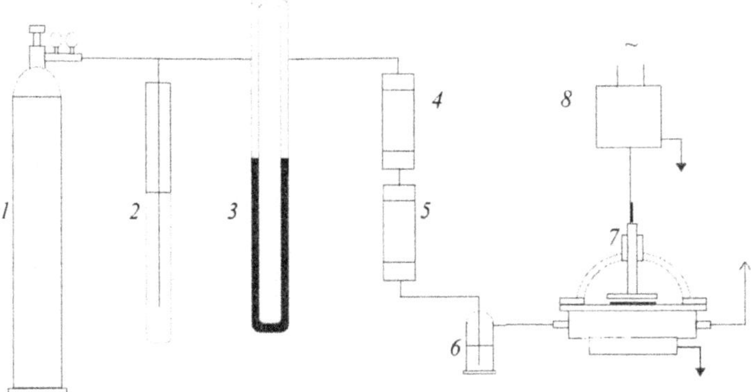

Fig.2. Set-up used for deposition of thin films of silicon oxide or silicon nitride on silicon or steel substrates; 1 - gas cylinder, 2- manostat, 3- flow-meter, 4- oxygen absorber, 5- desiccant, 6- vessel with TEOS (for depositing silicon oxide) or HMDS (for depositing silicon nitride), 7- reactor (see Fig.1), 8- high-voltage transformer.

The mixture of gases was made by flow of a transport gas, argon or helium, through the vessel containing tetraethoxysilane (TEOS) or hexamethylodisilazane (HMDS) which chemical structures are presented below:

TEOS

$$\mathrm{Si}(\mathrm{OC_2H_5})_4: \quad \begin{array}{ccc} H_5C_2O & & OC_2H_5 \\ & \diagdown\ \diagup & \\ & \mathrm{Si} & \\ & \diagup\ \diagdown & \\ H_5C_2O & & OC_2H_5 \end{array}$$

HMDS

$$\begin{array}{ccccccc} & & CH_3 & & CH_3 & & \\ & & | & & | & & \\ CH_3 & - & Si & - N - & Si & - & CH_3 \\ & & | & | & | & & \\ & & CH_3 & H & CH_3 & & \end{array}$$

In several cases to the above mentioned mixture of gases also oxygen or ammonia were additionaly introduced. To obtain uniform film it was necessary to control flow intensity of the gases through the gap between the electrodes (Fig. 1). The gas was therefore introduced to the reactor through the holes uniformly distributed on the base plate. The temperature in the reactor was controlled by a heater up to 200°C.The reactor was supplied from the transformer by the current of frequency 50 Hz or from the power supply built of an inverter and transformer with the frequency 1300 Hz. In the case of the frequency 1300 Hz it is possible to obtain easier the glow discharge with the high distributed structure.

The rate of film deposition can be varied by the selection of the composition of the gas mixture and the power supply to the reactor. The applied voltage is important , however at to high voltage the non-uniformity of the film structure is high. The deposition rate is very effected by the gap between electrodes. The narrow gap is advantageous but to high intensity of the glow discharges is not profitable. The optimum gap is about 2 mm. The deposition rate depends also on the substrate temperature and time of deposition and is usually about 1-3 nm/min.

The conditions of the film deposition process are very important. The film composition can be varied by the composition of the gas mixture and the temperature of the substrate. For instance at low temperature when the gas mixture does not contain oxygen and depositing film is rich of the polycondensation products of TEOS with high quantitity of organic compounds, the films are soft and even quasi-liquid and during the storage transform their properties. At higer temperature and higher energy of glow discharge and in the presence of reactive components , such as oxygen, the film are more uniform , contain less carbon and hydrogen but more silicon oxide [13,14]. Additional annealing can reduce the contents of organic compounds [14].

3. Experimental

The films produced by the above described technique were tested using a special microtribometer (for wear investigation) [15] and the Hysitron Inc.(U.S.A.) Triboscope™ which has the capability to carry out indentations at very low loads and to make load-displacement measurements with subnanometer indentation depth sensitivity [16,17]. The tested films of silicon oxide and silicon nitride were produced at various conditions and they were in the range of mechanical properties from „soft" to „hard.".

The load range employed in nanomechanical studies of the films was 1 μN to 10 mN and the load resolution 100 nN. The displacement resolution was 0.1 nm. The *in situ* image of indent was provided by AFM at resolution of 1nm in lateral direction and a vertical resolution was 0.2 nm. A three-sided Berkovitch indenter with tip radius of about 100 nm has been used for the measurements. The load-displacement data were acquired and then hardness, elasticity modulus and stiffness were obtained.

The schematic of the microtribometer used to make wear tracks is shown in Fig.3. The counterface is made of diamond with the flat surface of the diameter of 0.028 mm. The wear tracks are made as circles. The diameter of the circles can be varied from 0 to 4 mm.The shaft 18, to which the diamond holder 1is eccentrically fixed, rotates and the diamond produces wear grooves in the tested film deposited on the substrate (sample 12) (Fig.3). The shaft 18 is rotated through the torque measurement unit 3,7,23 by the DC electric motor 32 with the belt transmission 33, 30. The load is applied by the springs 16, 23 or by external dead-weight (not shown). The wear scars produced on the circle of diameter 2.4 mm were tested by profilometric method using Tencor profilometer.

The wear experiments were carried out at the rotational speed (of the shaft 18) 6 rev/min, i.e. the sliding speed was 0.7536 mm/s. The load applied was 2 N. The test duration was 1 hour.

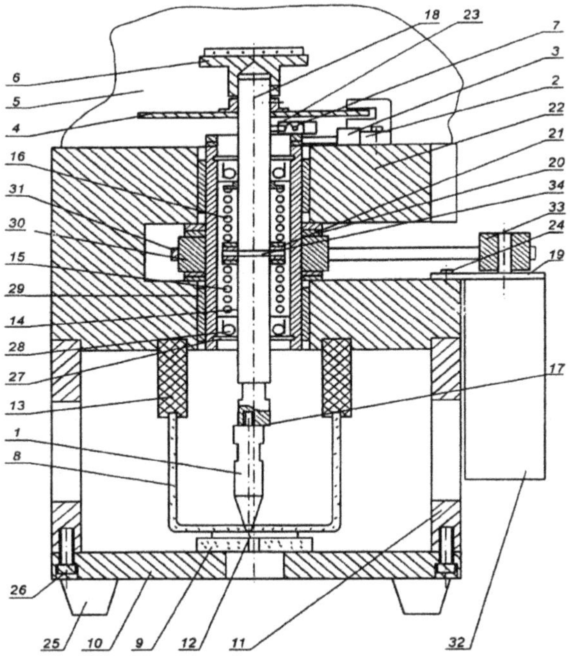

Fig.3. Microtribometer used in wear tests (description of numbers see in text).

The films were deposited on p^+-type single-crystal silicon (100) wafer. The film thickness was measured using ellipsometry. The composition of the films were investigated by FTIR and XPS techniques. To enter into the material inside the film the surface was etched *in situ* by bombarding with argon ions. The composition was determined on several layers under the surface. The energy of argon ions was 4 kV and the time of etching was 5, 10, 20 and 35 minutes.

4. Experimental results and discussion

The films were amorphous and various as relates to their mechanical and wear properties. They were very soft demonstating relatively high wear rate and they were also hard with relatively low wear rate, however it was found that the wear resistance of the silicon oxide and silicon nitride films is not simply correlated with hardness.

The FTIR spectra for relatively soft and relatively hard silicon oxide films are shown in Fig. 4 a and b respectively. The difference between soft and hard films is not distinct. The XPS analysis showed however that in the case of soft films they contain more organic substances and the hard films contain more silicon and oxygen i.e. silicon oxide. The films produced in the presence of oxygen had less contents of organic substances that the films obtained without oxygen.

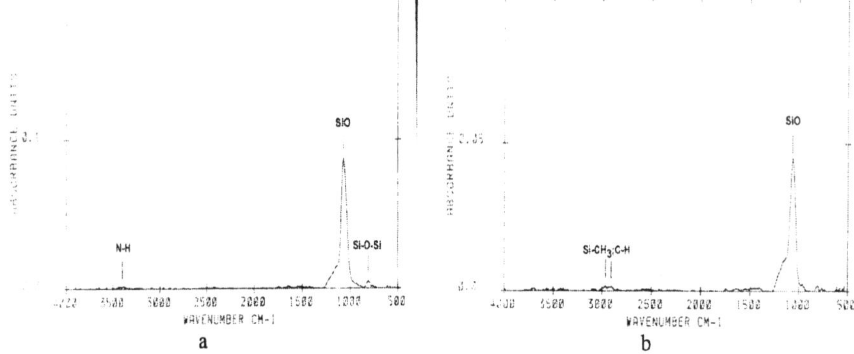

Fig.4. FTIR spectra of relatively hard (a) and relativeley soft (b) films of silicon oxide.
a - film produced at 295°C frequency 1300Hz, volatage 2.05kV, quartz dielectric, gap 1.5mm, time 60 minutes, mixture TEOS+Ar+10%NH_3, thickness 84nm,b - film produced at 212°C frequency 1300Hz, volatage 2.12kV, quartz dielectric, gap 1.5mm, time 60 minutes, mixture TEOS+Ar+O_2(550ppm), thickness 100nm.

The representative load displacement plots of the indentation made on relatively soft and relatively hard silicon oxide films are shown in Fig.5 a and b respectively.The elastic behaviour of the film can be clearly observed in Fig. 5 a. The hysteresis is characteristic for the hard film.

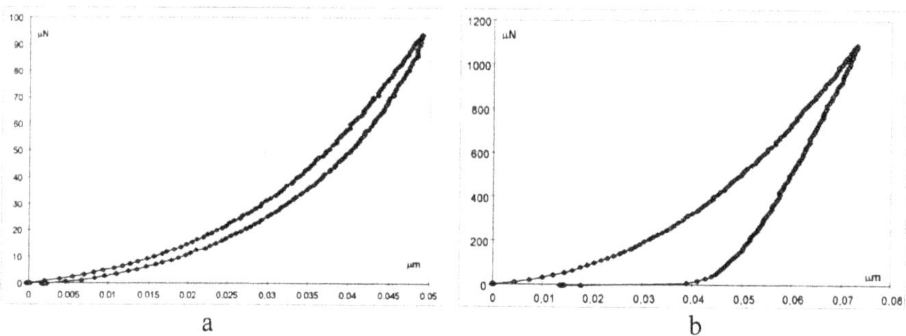

Fig.5. Load versus displacement plots of the indentations for soft (a) and hard (b) films. a - silicon nitride film produced at 300°C, frequency 1300Hz, voltage 2.23kV, qyartz dielectric, gap 1.5, time 60 minutes, mixture HMDS+Ar+O_2 (550ppm), thickness 176 nm; b - silicon oxide film (see captions to Fig.4a), thickness 84 nm. Load+unload period 15s.

The hardness of the soft and hard films, which the load displacement plots were shown in Fig. 5, are presented in Fig.6. The elasticity modulus of the same films can be compared in Fig. 6 a and b.

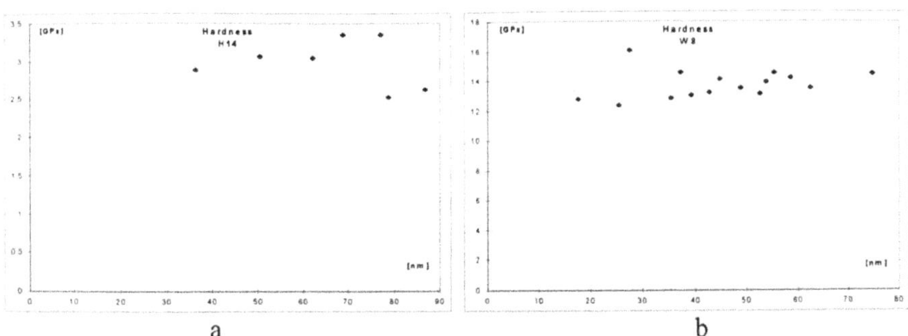

Fig.6. Hardness versus indentation depth for same films as decribed in captions to Fig. 5 a and b respectively.

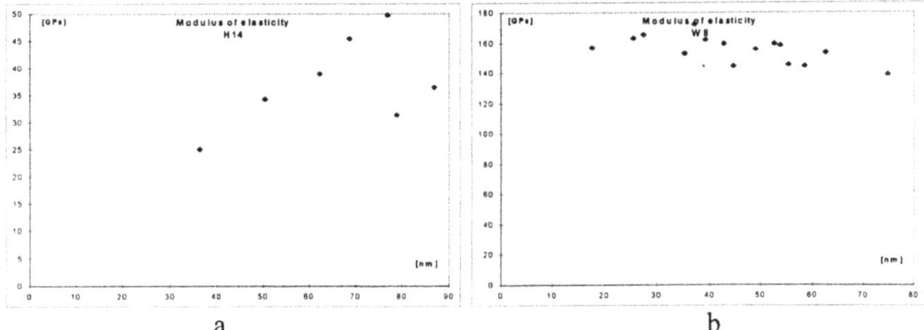

Fig.7. Elasticity modulus versus indentation depth for same films described in captions to Fig.5 a and b respectively.

Hardness and elasticity modulus of soft films are higher at higher depth of indentation. The realtively soft surface layers are not protection during the glow discharge and the deeper layers are exposed for the glow discharge longer time.The hardness and elasticity modulus for hard films do not change dramatically as a function of depth of indentation.

The wear rates of the tested silicon oxide and silicon nitride films are not straightly related to hardness. Fig. 8 shows the profilogram of silicon oxide film (the same as in caption to Fig. 5 b) after the wear test. The wear track is formed mainly by ploughing, resulting in considerable pile-up at the side of the track. The wear rate is relatively high. The profilogram of the bare silicon wafer after the wear test is shown in Fig. 9.

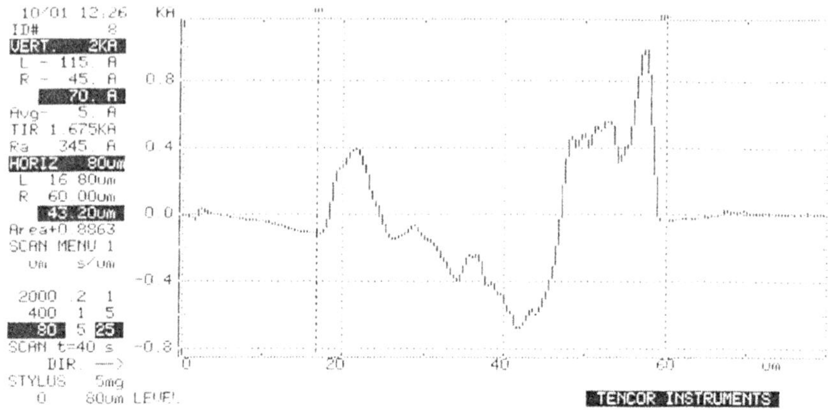

Fig.8. Profilogram after wear test for silicon oxide (same sample as decribed in caption to Fig. 5 b).

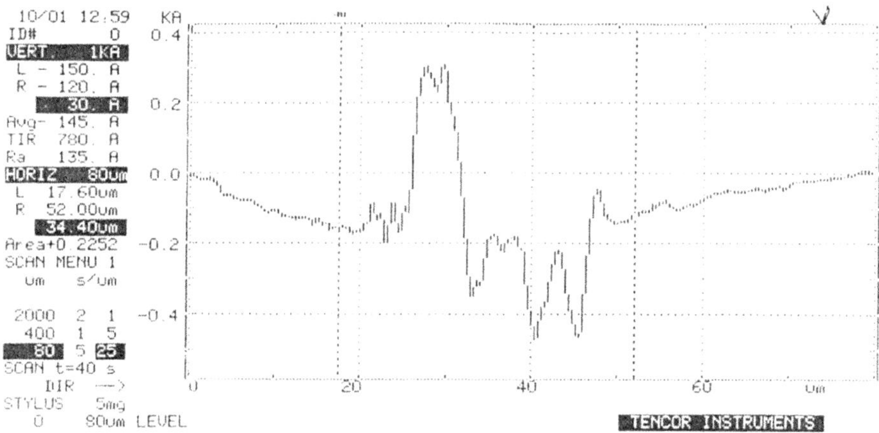

Fig. 9. Profilogram of bare silicon wafer after wear test.

The relatively wear resistant films were these made of silicon nitride (Fig. 10). The film which wear scar is shown in Fig.10 was produced in the presence of nitrogen and ammonia at temperature 300°C.

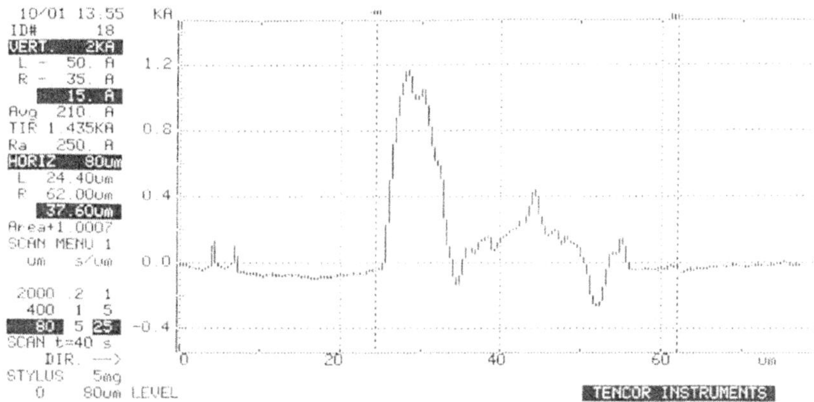

Fig. 10. Profilogram of silicon nitride film after wear test. Film was produced at 300 °C , frequency 1300 Hz, voltage 2.5 kV, quartz dielectric, gap 1.5 mm, time 18 min, mixture HMDS +nitrogen + 30% ammonia; film thickness 5 nm.

The comparison of the mechanical properties and wear behaviour of the tested films shows that generally higher hardness is accompanied by higher wear resistance but also lubricating properties of the film are important. The organic compounds of the film can play a lubricant role. This aspect of the tested films is very important in MEMS applications. The triboengineering of the films is possible i.e. the design of the films for well-defined requirements.

Wear process is accompanied by plastic deformation of the material. Surface damage during sliding is caused by high stresses which exceed the flow strength of the film material. The flow and transfer of the material during rubbing was characteristic for all tested samples.

Silicon nitride films demonstrate relatively good wear resistance as compared with the silicon oxide films, however the mechanical properties and wear behaviour depends strongly on the deposition process. Argon and oxygen presence during the deposition of silicon oxide films is more effective than when only argon or oxygen is applied. Silicon nitride films are more wear resistant when they are deposited in the presence of ammonia as compared with the films deposited in the presence of nitrogen.

5. Conclusions

The films of silicon oxide and silicon nitride deposited on silicon by glow discharge at atmospheric pressure demonstrate good mechanical prpoerties and are enough wear resistant. When the hardness increases the wear rate generally decreases but some lubricating properties of organic compounds of the films can play important role during rubbing. Silicon nitride films exhibit better wear resistance as compared with silicon oxide films however it depends also on the deposition process of the films.

The glow discharge (stabilized by a dielectric barrier) deposition technique applied at atmospheric pressure and at low temperature can be an alternative for other expensive techniques to produce wear resistant films on silicon or other substrates . Such films are also interesting as lubricating films since the organic compounds can control the friction and wear processes i.e. triboengineering of the films is possible to fulfil the requirements for e.g. reliable and durable sliding contacts embodied in MEMS.

Acknowledgments

The authors are grateful for Dipl.-Ing. Jerzy Wyrobek for inviting one of us (M.Misiak) to spend some months in Hysitron Inc. laboratories and investigate of nanomechanical properties of our samples. We are also thankful for Thomas Wyrobek , Dr.Ashok Kulkarni and the staff of the lab. The help of Dr. Jacek Jagielski from the Institute of Technology of Electronic Materials (Warsaw, Poland) in profilometric studies of the worn samples is greatly appreciated.

References

[1] Muller, R.S. (1997) Microdynamical systems, in B.Bhushan (ed.), *Micro/Nanotribology and Its Applications,* Kluwer Academic Publishers, Dordrecht, pp. 579-599.
[2] Mehregany, M and Tai, C. (1991) Surface micromachined mechanisms and micromotors, *Journal of Micromechanics and Microengineering* **1**, 73-85.
[3] Mehregany, M., Senturia, S.D., Lang, J.H.and Nagarkar,P (1992)., Micromotor fabrication, *IEEE Transactions on Electron Devices 39,* 2060-2069.
[4] Benitez, A., Esteve, J. and Bausells, J.(1995), Bulk silicon microelectromechanical devices fabricated from commercial bonded and etched-back silicon-on-insulator substrates, *Sensors and Actuators* **A 50**, 99-103.
[5] Miyajima, H. and Mehregany, M. (1995), High-aspect-ratio photolithography for MEMS applications, *Journal of Microelectromechanical Systems* **4**, 220-228.
[6] Frangoul, A.G. and Sundaram, K.B., (1995), Design and fabrication process for electrostatic side-drive motors, *Journal of Micromechanics and Microengineering* **5**, 11-17.
[7] Bühler, J., Steiner F.P. and Baltes, H, (1997), Silicon dioxide sacrificial layer etching in surface micromachining, *Journal of Micromechanics and Microengineering* **7**, R1-R13.
[8] Okazaki, S., Kogoma, M., Uehara, M. and Kimura, Y., (1993), Appearance of stable glow discharge in air , argon, oxygen and nitrogen at atmospheric pressure using a 50 Hz source, *Journal of Physics D: Applied Physics* **26**, 889-892.
[9] Schmidt-Szałowski, K., Fabianowski, W., Rżanek-Boroch, Z. and Gutkowski, M.,(1996), Thin surface layers of SiO2 obtained from tetraethoxysilane (TEOS) in electric discharges stabilized by a dielectric barrier, in *Proceedings of International Symposium on High Pressure Low Temperature Plasma Chemistry „HAKONE 5", Contributed Papers,* Milovy, Czech Republik, September 2-4, 1996, pp. 190-194.
[10] Bhushan, B. and Venkatesan, S.(1993), Mechanical and tribological properties of silicon for micromechanical applications, *Advances in Information Storage Systems* **5**, 211-239.
[11] Komvopoulos, K. (1996), Surface engineering and microtribology for microelectromechanical system, *Wear* **200**, 305-327.
[12] Li, Y. and Danyluk, S.(1996), Dynamic measurements of damage generation in single crystal silicon due to sliding contact with a spherical diamond, *Wear* **200**, 238-243.

[13] Schmidt-Szałowski, K., Fabianowski, W., Rżanek-Boroch, Z. and Gutkowski, M.(1997), Surface films SiO2 produced by plasma glow discharge using tetraethoxysilane , *Nowe Materiały, Prace Naukowe Politechniki Warszawskiej, Seria Inżynieria Materiałowa Z.6*, 91-98, in Polish.
[14] Schmidt-Szałowski, K., Fabianowski, W., Rżanek-Boroch, Z., Sentek, J. and Gutkowski, M.,(1998) Depositing thin films using silicon compounds by plasme glow discharge under atmospheric pressure,*Prace Naukowe Politechniki Warszawskiej, Seria Inżynieria Materiałowa,*, in Polish, to be published.
[15] Kusznierewicz, Z., Misiak, M. and Rymuza, Z.(1997), Construction of tribotester for investigation of new materials, biomaterials, implanted materials or/and thin coatings - mechanical part, *Nowe Materiały, Prace Naukowe Politechniki Warszawskiej, Seria Inżynieria Materiałowa* **Z 6**, 111-123, in Polish.
[16] Bhushan, B. (1997), Friction, scratching/wear, indentation and lubrication on micro- to nanoscales, in B.Bhushan (ed.), *Micro/Nanotribology and Its Applications*, Kluwer Academic Publishers, Dordrecht, pp. 169-191.
[17] Bhushan, B., Kulkarni, A.V., Bonin, W. and Wyrobek, J.T.(1996), Nano/picoindentation measurements using capacitive transducer system in Atomic Force Microscopy, *Philosophical Magazine* **A 74**, 1117-1128.

a-SiC Thin Films Deposited Using Pulsed Laser Ablation of Graphite and Magnetron Sputtering of Silicon onto Steel Substrates at Room Temperature

Josekutty J. Nainaparampil
AFRL/MLBT, Materials Directorate, Bldg. 654, 2941 P St. Ste. 1, Wright Patterson AFB, OH 45433-7750

1. ABSTRACT

Typically SiC films are deposited on substrates at elevated temperatures either by sputtering SiC targets or using other conventional deposition methods. To maintain substrate properties (temper, dimensional tolerance etc.) a low process temperature is required. In this work, silicon carbide is formed from simultaneous sputtering of silicon and laser ablation of graphite onto substrates at room temperature. The advantage in this method lies in the individual selection of the species accelerated to optimum energy which permits the formation of selected phases in the film. Films are formed on M50 steel substrates at biasing varied from -100 to -300 Volts resulting in different deposition rates for carbon and silicon and changes in bonding strength. The shift in the binding energies of carbon and silicon obtained from XPS analysis corresponds to the silicon carbide bonds. Chemical bonding data is correlated to microstructure and mechanical properties. Special emphasis is given to film stoichiometry control and Si/C composition ratio to achieve optimum tribological properties.

2. INTRODUCTION

Silicon carbide is a material that is well characterized by its hardness, high thermal stability, chemical inertness and stiffness which is comparable to beryllium. Because of its mechanical strength and chemical inertness silicon carbide is a good protective coating. Several deposition techniques are used to produce silicon carbide in its crystalline and amorphous form. Most of the crystalline films are fabricated using chemical vapor deposition with substrates held at very high temperature. Recently it is shown that CVD process lets formation of crystalline SiC at comparatively low substrate temperatures. (Ulrich S. et.al) (1997), (He J. L et. Al.)(1996). a-SiC formation on different substrates at low temperature from plasma chemical decomposition of methyltrichrolosiline is reported by Berezhinski L.I and et.al.(1989). Amorphous form of silicon carbide annealed at 800 °C can act as a lubricant for comparatively long life (Sugimoto I. et.al) (1989). Room temperature deposition of silicon carbide using laser ablation of silicon carbide target has been accomplished. (Capano M. A.) et. Al. (1994) The purpose of this work is to investigate the formation of amorphous silicon carbide on 440C steel substrates at room temperature using a novel hybrid technique. A hybrid deposition technique combining magnetron sputtering of silicon and pulsed laser ablation of graphite was used for the deposition. Since both species are formed separately, individual selection of the species accelerated to optimum energy permits the formation of selected phases in the film. Chemical binding energy of these films were measured using X ray photoelectron spectroscopy and micro hardness was determined using Nanoindentation.

3. EXPERIMENTAL

The deposition system used is schematically represented in Figure 1. A pulsed laser beam was focused onto a graphite target to make partially ionized carbon plume. Along with carbon ablation silicon was magnetron sputtered onto the substrate. Targets were oriented such that the carbon plume and magnetron sputtered silicon intersected on the substrate surface. The distance between the magnetron target and the

B. Bhushan (ed.), Tribology Issues and Opportunities in MEMS, 591-596.

substrate was around 15 cm and the PLD target - substrate distance was around 5 cm. Ultrasonically cleaned M50 steel substrates were used in this study. The base chamber pressure was kept around 10^{-7} Torr, and 200 mJ energy pulses from a Lamda Physic LPX 110I eximer laser source were focused on the target to get a power density of 10^9 W/cm[8]. Carbon plumes formed under these conditions expanded in a direction normal to the target towards the substrate with kinetic energies of 1.5 keV in the leading edge.[9] The magnetron was in

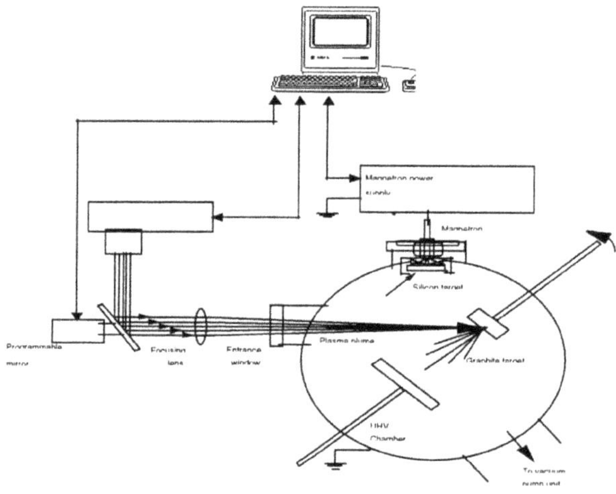

Figure 1. A Schematic diagram of the hybrid system used for the deposition. The magnetron supply and laser source are controlled by a computer.

the balanced configuration to produce Si atoms with kinetic energies of several electron volts. The reactive sputtering pressure maintained inside the system was around 4-5 $\times$ 10^{-3} Torr. Room temperature deposition was made on the substrate kept at biasing potentials varied from -100 to -350 Volts.

The independent operation of Si and C sources provides ability to control the energy and composition of each species arriving at the substrates. Since the Si atoms coming from the magnetron is ionized the kinetic energy can be imparted by the substrate biasing. Since most of the C atoms ablated from PLD targets are neutral the target biasing cannot make much change in the kinetic energy of these species.

At low substrate temperature the deposition rate is determined by self sputtering at the surface of the substrate. The energetics of a hybrid deposition technique like this is significantly different from conventional techniques. Ionized silicon atoms can be accelerated by negatively biased substrate implying that the sputtering effect of these atoms can be substantial on neutral atoms. It has also been shown that the ionized species have higher sticking coefficient than neutral species. Considering the 1.5 keV laser ablated C atoms, they should have the higher sputtering ability than the silicon atoms. However these highly energetic atoms are only present in the plumes leading edge, while the rest of the species within the plume may have considerably lower energies in the range of 10 to 100 eV. Furthermore, in a low pulse rate deposition, the plume presence is relatively short, 10^{-6} s, compared to the interval, 10^{-1} - 10^{-2} s, between the pulses. This leaves more exposure time for C atoms to be sputtered from the surface.

4. SAMPLE CHARACTERIZATION

Surface chemistry was studied with a Surface Science Instruments (SSI) M-probe XPS instrument operated at a base pressure of 3×10^{-7} Pa. Using an Al anode, a 400 X 1000 μm line spot, and a 25 eV pass energy, the full width at half maximum (FWHM) of the Au $4f_{7/2}$ peak was 0.71 eV. Binding energy positions were calibrated against the Au $4f_{7/2}$ peak and energy separations were calibrated using the Cu 3s and Cu $2p_{3/2}$ peaks at 122.39 and 932.47 eV, respectively. Crystal structures were determined using data acquired by a Rigaku D/max -1B diffractometer equipped with a thin film attachment and a monochrometer. High resolution scanning electron microscopy (SEM) was performed on the films using a Leica 360 field emission SEM equipped with an energy dispersive x-ray analyzer.

Room temperature friction and wear data from coated specimens were collected using a ball-on-flat tribometer run with the specimens in the horizontal position. The ball was held in a lever arm, and the friction force was measured using strain gauge circuitry. The normal load was computed from the dead weight hung in the cantilever assembly. The percentage relative humidity was measured by a high performance sensor inserted in the chamber. A 6.35 mm diameter 440C steel ball was used as the counter face. For each friction track, the rotational speed was adjusted to get a constant sliding speed of 17.4 cm/s. The majority of the friction measurements were made at a normal load of 1 N. For steel-on-steel, a normal load of 1 N on a 6.35 mm diameter ball corresponds to a Hertzian contact pressure of 675 MPa. The relative humidity in the current study was maintained at 15 % - 20 %, and measurements were made at room temperature.

Microhardness and elastic moduli of the film were determined using a Nanoindenter® II microprobe. A Berkovich indenture was used to make indents of 50 nano meter depths. Procedures developed by Oliver and Pharr are applied for both instrument calibration and data analysis at three depths which allows the minimization of indentation size effect at small penetration depths.[12]

5. RESULTS AND DISCUSSION

Results form the XPS analysis showed that most of the films formed with biasing -100 V- 200 V have Si 2p peak energy and C 1s peak energy similar to that of crystalline SiC. Si rich samples showed more or less Si

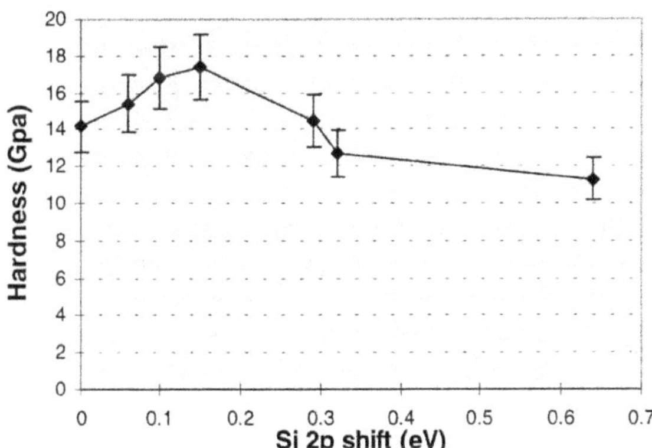

Figure 2. Variation of hardness of a-SiC thin films with Si 2p shift, Si 2p shift is the difference of Si 2p peak of the deposited films and the average of the values available in the literature.

type bonding energy for silicon atoms. The higher negative biasing of the substrate was less helpful in the formation of SiC like bonding. XRD and Raman analysis did not show any significant crystalline features of SiC showing that the films are either nanocrystalline or amorphous in nature.

Figure 2 shows the variation of hardness of these films as a function of shift in energy for Si 2p peak from that of SiC standard. The hardness is around 15 Gpa for samples with a peak shift values around 0.1 eV means these samples has a SiC like bonding. The comparatively low value of hardness is probably due to the amorphous nature of the films. When samples become less SiC like the hardness decreases. Samples with higher shifts for chemical binding energy, Si 2p peak was divided into two peaks, of which one of

them was more shifted towards value for a Si-Si bonding signifying the presence of more silicon type material in those samples.
Figure 3 describes Lower critical force measured on a scratch tester working under variable load form zero to 100 Newtons. Test sample was traveling at a rate of 3.3 mm per minute. A conical diamond tip was used as the probe to make the scratch. It is evident from the figure that the toughness of the films depend on the chemical binding energy of these species. Films formed with species bonded like SiC were having high critical load up to 40 N. But films formed with Si like structures were very brittle and began to break at very low load like 10 N - 12 N.

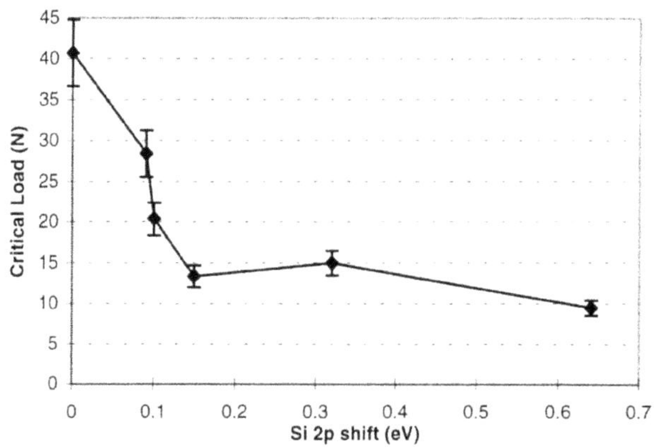

Figure 3. Variation of critical load with Si chemical bonding energy. The critical load drastically drops when the Si bonding energy departs from that of the SiC

Another interesting property measured was the steady state friction and wear of these films with 440 C steel balls. The condition for this testing were given in the characterization section. The result is given in Figure 4. showing the variation of friction with Si fraction in films. The COF seems to be lowered to 0.3 to 0.4 range at Si fraction slightly above the stoichiometric value for SiC. At silicon fraction around 0.7 the friction goes back to 0.6 and stays there for Si rich films. For these films having larger fraction of Si the bonding energy also has a larger shift form that of SiC. The nature of these films are similar to silicon films forming abrasive wear debris and worn out very easily. The evidence for the Si type behavior of these films are obtained in the SEM analysis given in the next section.

5.1 SEM characterization

Scanning Electron Microcopy of as deposited films did not show any significant features of the topography. Even at very high magnification the films were featureless and very smooth. The significant changes are observed for tshe wear scar and scratches in the scratching tests. Figure 5(a) and 5(b) shows the micrographs of wear scars of films formed from SiC like bonded species and Si like bonded species respectively. The films were deposited with -150 V and -200 V biasing with 100 watts magnetron power and 200 mJ of laser pulse energy. In both films the Si/C ratio is around 0.58. But the Si bonding energy for the sample shown in Figure 5(a) is very closer to that of crystalline SiC than the sample shown in Figure 5(b). The wear scar on the SiC like bonded film is very smooth without any cracks or formation of any abrasive debris. But in Figure 5(b) the wear scar is very rough forming particulate wear debris. In Figure

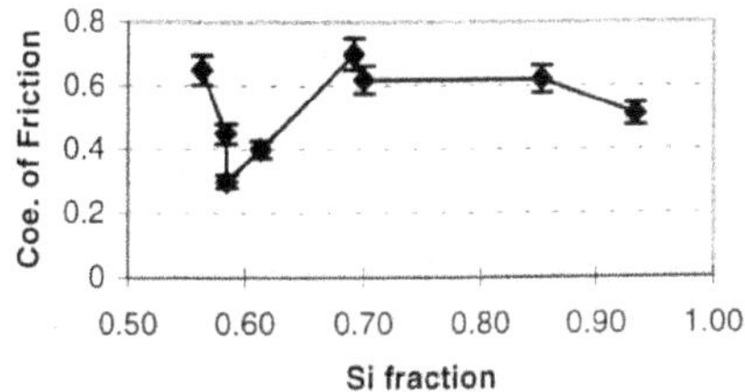

Figure 4. Room temperature friction (steady state) as a function of Si fraction as measured by a pin on disk tribometer. The measurements were made in ambient (RH 15 % - 20 %)

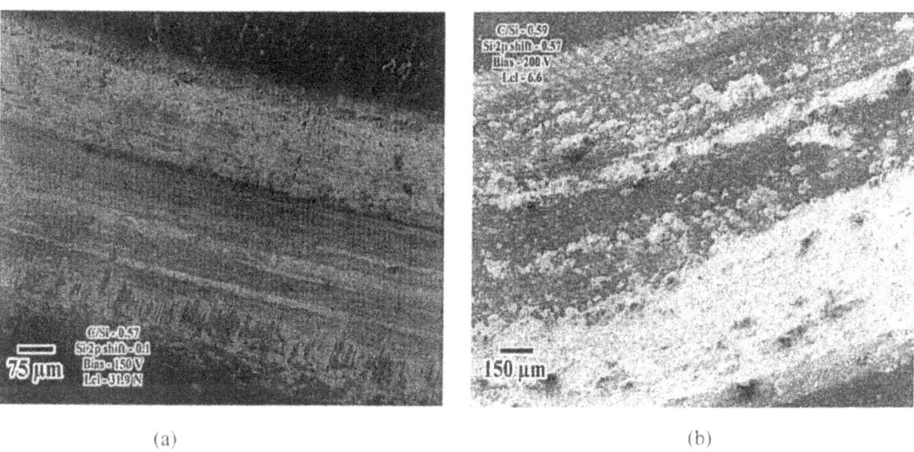

Figure 5. SEM micrographs showing wear scars. (a) Film formed form SiC like bonded species, wear scar is smooth and with no abrasive wear debris. (b) Film formed from Si like bonded species, wear scar is rough with abrasive wear debris.

5(b) the ion probing of worn out white region detected the presence of O_2 and Si showing that SiO_2 has formed in tribochemical reactions. The increase in the wear life of SiC like bonded structure in Figure 5(a) may be due to the better hardness and more stable bonding between the Si and C atoms.

Figure 6 gives the micrograph of track formed as a result of scratch testing on the same samples that are used in friction measurements. A variable force scratch was made on both films. The force was varied from 0 -100 N in both measurements. Figure 6 (a) shows the high toughness films with SiC like bonding and the Figure 6 (b) gives scratch made on film formed with Si like bonded structures. The lower critical force measured on the sample of Figure 6 (a) is 31.9 N and that of the 6 (b) is only 6.6 N. The sample with high toughness shows no cracking behavior while the film with low toughness shows crack propagation towards the both sides of the track. The structure seen in the Figure 6 (a) is not a crack but it is metal flake rolled over the side of the track.

SUMMARY

Amorphous silicon carbide films were formed on room temperature steel substrates from independent sources for carbon and silicon. The effect of substrate biasing in the formation of mechanical properties of PLD deposited a-SiC films is demonstrated. It is shown that comparatively hard and tough silicon carbide films can be formed on room temperature substrates by PLD deposition. It is also noted that biasing and arrival rate of species can change chemical binding energy of these films.

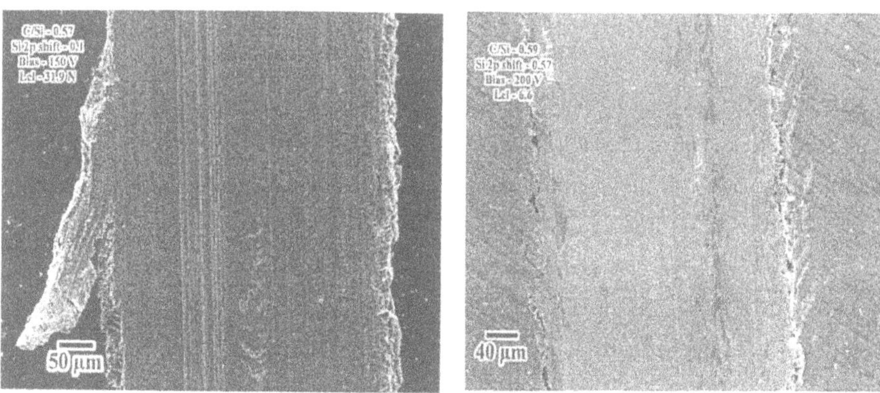

Figure 6. Micrograph showing the magnified scratch. (a) Film formed with SiC like bonded structures. Cracks are not formed even at the high end of the track showing high toughness. (b) Film formed with Si like bonded structures. Cracks are well developed showing very low toughness.

RFERENCES

1. L. I. Berezhinski, S. I. Vlaskina, M. P. Lisitsa, G. I. Lyashenko and V. E. Rodionov, Films of Amorphous Silicon Carbide on Foreign Substrates, Sov. Tech. Phys. Lett 15(11) (1989) 850-851
2. M. A. Capano, S. D. Walck, and P. T. Murray, Pulsed Laser Deposition of Silicon Carbide at Room Temperature, Appl. Phys. Lett. 64 (25) (1994) 3413-34-15
3. J. L. He, M. H. Hon, L. C. Chang, Properties of Amorphous Silicon Carbide film deposited by PECVD on Glass, Materials Chemistry and Physics, 45 (1996) 43- 49
4. I. Sugimoto and S. Miyake, Lubricating Performance Enhancement of Amorphous Silicon Carbide Film by Annealing Effects and Microbeam Analysis of the Tribological Interface, J. Appl. Phys. 66 (2) (1989) 596 - 604
5. S. Ulrich, T. Theel, J. Schwan, V. Batori, M. Scheib, H. Ehrhardt, Low Temperature Formation of β Silicon Carbide, Diamond and Related Materials 6 (1997) 645- 648

LUBRICATION OF POLYSILICON MICROMECHANISMS WITH ALKYLSILOXANE SELF-ASSEMBLED MONOLAYERS: COEFFICIENT OF STATIC FRICTION MEASUREMENTS

U. SRINIVASAN[a], R.T. HOWE[b], and R. MABOUDIAN[a]
Berkeley Sensor and Actuator Center
[a]Department of Chemical Engineering
[b]Department of Electrical Engineering and Computer Sciences
University of California, Berkeley, CA 94720, USA

Abstract

We have investigated the lubricating effects of alkylsiloxane and fluoroalkylsiloxane self-assembled monolayers (SAMs) by measuring static friction in a poly-crystalline silicon (polysilicon) surface-micromachined device. This test structure, based on the design of Lim *et al.* [1], was fabricated in a four-mask process and coated with SAMs following the release etch. The coefficient of static friction, μ_s, was found to decrease from 2.3 ± 0.8 for the SiO_2 coating to 0.13 ± 0.01 and 0.10 ± 0.007 for the octadecyltrichlorosilane (OTS) and 1H,1H,2H,2H-perfluorodecyltrichlorosilane (FDTS) SAM films, respectively. A similar trend has been observed for the apparent works of adhesion of these coatings. Low standard deviations in the μ_s data for SAM treatments indicate uniform coating coverage. The μ_s values were shown to be independent of apparent contact area for the three surface coatings.

1. Introduction

Friction and wear are critical issues for microactuators with joints and bearings [1-4]. Studies of electrostatic micromotors indicate that bearing friction consumes a significant portion of the motive torque [1]. In particular, high static friction, or stiction, can cause device lock-up, seriously limiting reliability [3,4]. For this reason, there is a strong motivation to quantify the static coefficient of friction at the microscale and control this parameter accurately.

In micro-electro-mechanical systems (MEMS), contacting surfaces are lightly loaded (≈μN), placing friction in a regime where the contribution from adhesion outweighs those from asperity deformation and ploughing [5,6]. Between oxide-coated surfaces, this adhesion arises from water capillary attractions which often exceed the actuation force. Coating with self-assembled monolayers of hydrocarbon and fluorocarbon

B. Bhushan (ed.), Tribology Issues and Opportunities in MEMS, 597-606.

molecules can eliminate these attractions and decrease the adhesional energy significantly. SAMs derived from octadecyltrichlorosilane and 1H,1H,2H,2H-perfluorodecyltrichlorosilane give adhesion reductions of over three orders of magnitude [7-9].

Tribological studies on SAM-coated surfaces have been conducted at the nanoscale using atomic force microscopy (AFM) on smooth surfaces such as silicon (100) and mica [10]. The difficulty in applying results from these types of frictional studies to surface-micromachined devices is that the real area of contact between the relatively rough polysilicon surfaces in MEMS is not easily known. In addition, variability in the topography of LPCVD films makes standardization of surface phenomena of adhesion and friction difficult. At the macroscale, measurements of friction coefficients on SAM-lubricated and oxide-coated Si(100) have been taken using conventional tribometers [10,11]. In these experiments, however, the contact pressure is typically on the order of mN, and adhesion from meniscus formation is no longer critical.

Several experiments have been done to determine a range for dynamic coefficient of friction, μ_d, for contacting surfaces of oxide-coated polysilicon in micromotors. Bart *et al.* reported a μ_d range of 0.26-0.4, and similar ranges have been reported for polysilicon rubbing against silicon [3,13,14]. Recently, dynamic friction has been investigated for Teflon-coated micromotors, giving $\mu_d = 0.07$ [15,16]. Only one study, however, has been done to quantify static friction using a surface-micromachined test structure. Lim *et al.* found the coefficient of static friction for oxide-coated polysilicon surfaces to be extremely high with a large standard deviation: $\mu_s = 4.9 \pm 1.2$ [1]. Using a test device based on the design of Lim *et al.*, we have investigated static friction between polysilicon contacts coated with OTS and FDTS SAMs, as well as oxide for reference. While the coefficient values presented here are unique to the topography of the particular polysilicon used, the results give an indication of the degree of reduction afforded by SAMs.

2. Experimental

Figure 1a shows the layout of the static friction testing microstructure taken from Lim *et al.* [1]. The shuttle is maintained 2 μm above the ground plane by a folded beam suspension. The underlying electrode is used to exert an electrostatic normal force on the shuttle. Bumpers on the underside of the shuttle land on an electrically separate, grounded polysilicon surface, thereby determining the frictional contact area. The electrostatic comb drive attached to the shuttle is used to determine the spring constant of the folded beam suspension.

The friction test structures were fabricated in a 4-mask process. The polysilicon film deposition conditions are as follows: 590°C/ground plane, 585°C/structural layer, 500 mtorr, and 3.2×10^{-3} mole fraction PH_3 in SiH_4. The 1 μm deep bumper molds are defined in the sacrificial phosphosilicate glass layer using a timed BHF wet etch. Figure

1b shows a cross section of the deposited films in the bumper region of the shuttle before the sacrificial PSG etch. To assess frictional dependence on contact area, microstructures with varying numbers of bumpers and total bumper surface area were fabricated. A Topometrix atomic force microscope was used to image the contacting surfaces and study their textures.

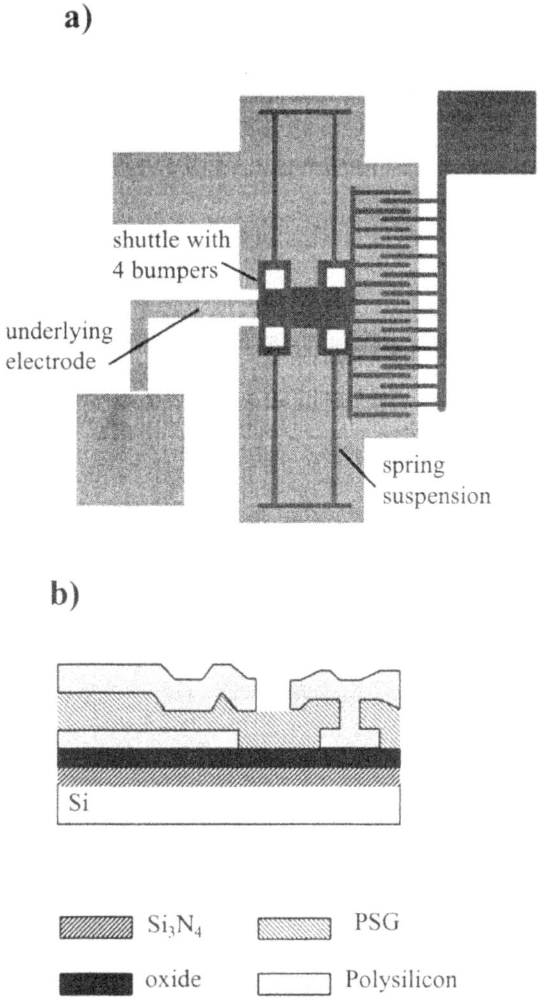

Figure 1. a) Layout of friction testing microstructure and **b)** cross section of deposited and patterned films in the bumper region of the friction testing microstructure prior to sacrificial layer etch.

The structures were released in concentrated HF and then SAM-coated. The procedure for SAM coating microstructures is shown in Table 1 [7]. The iso-octane, hexadecane, and chloroform used in the SAM coating are anhydrous (Aldrich Chemical Co.), and the remaining chemicals are standard solvent grade. The OTS SAM solution is 0.5 mM of octadecyltrichlorosilane (Aldrich Chemical, 95% purity) in a 4:1 vol. mixture of hexadecane and chloroform. This solution was prepared and used under a fume hood. The fluorinated SAM solution, 1 mM FDTS (PCR Chemicals, 98% purity) in iso-octane, was mixed and used in a N_2-filled drybox to prevent bulk polymerization caused by its highly water-sensitive headgroup. The structures were immersed in the OTS and FDTS SAM solutions for 20 and 10 minutes, respectively. Solvent changes were accomplished by dilution rinses for the water soluble steps, and by quickly moving the wafer pieces to the next solvent for the organic solvent changes. Teflon dishes cleaned in 5:1 H_2SO_4/H_2O_2 and deionized water are used for the monolayer self-assembly step for both types of SAMs. Structures with oxide-coated surfaces, used as a benchmark for the static friction studies, were released using supercritical CO_2 drying following the H_2O_2 reoxidation. Contact angle measurements were taken with a Ramé-Hart goniometer using water and hexadecane as test liquids.

TABLE 1. SAM coating procedures for polysilicon microstructures. The OTS SAM deposition is done on a fume hood while in the FDTS treatment, the iso-octane washes and SAM coating steps are performed in an N_2-filled dry box.

Oxide Coating	OTS SAM	FDTS SAM
Sacrificial Etch HF (10 min.)	same as oxide coating	same as oxide coating
Surface Reoxidation H_2O (5 min.) H_2O_2 (15 min.) H_2O (5 min.)	same as oxide coating	same as oxide coating
Drying MeOH (15 min.) Supercritical CO_2 dry	*Water Removal* IPA (15 min.) Iso-octane (10 min.)	same as OTS SAM
	SAM deposition 1 mM OTS in 3:1 v/v hexadecane: chloroform (20 min.)	1 mM FDTS in iso-octane (10 min.)
	Excess Precursor Wash Iso-octane (10 min.) IPA (5 min.) H_2O (5 min.)	same as OTS SAM

The static friction testing procedure taken from Lim [1] consists of four steps. Initially, all voltages are set to zero. (1) The shuttle is laterally displaced using a tungsten probe tip or the electrostatic comb drive, and the spring suspension supplies a restoring force. (2) The shuttle's bumpers are electrostatically brought into contact with the grounded landing pad by applying a voltage to the underlying electrode (50 and 20V for SAM-

and oxide-coated contacts, respectively). (3) The lateral force is removed, whereupon the friction between the contacting surfaces maintains the displacement. (4) The electrostatic clamping force is decreased slowly by ramping down the voltage until the lateral suspension force can overcome the static frictional force, and the shuttle starts to move back to its equilibrium position. The voltage at which slippage occurs, V_s, is used to determine the coefficient of static friction, μ_s, through a force balance:

$$F_n = \mu_s F_t \tag{1}$$

Here, F_n and F_t are the net normal and tangential forces exerted on the shuttle at the onset of slippage. F_n consists of the electrostatic clamping force minus the spring suspension's restoring force in the z-direction, and F_t is the restoring force in the x-direction.

$$F_n = F_{elect} - F_{kz} \tag{2}$$

$$F_t = F_{kx} = k_x \Delta x \tag{3}$$

F_{kz} is also equal to the electrostatic force caused by the application of the voltage, V_p, that is needed to pull the shuttle down and overcome the spring's z-direction restoring force. Then, treating the shuttle and the underlying electrode as a parallel plate capacitor, the net normal force can be written as:

$$F_n = \frac{\alpha_z \varepsilon_0 w l (V_s^2 - V_p^2)}{2z^2} \tag{4}$$

where α_z is the correction factor to account for fringing effects, l and w are the length and width of the underlying electrode, z is the gap distance, ε_0 is the permittivity of air and V_s and V_p are the potentials at which the shuttle begins to slip and where it can overcome the spring's z-restoring force, respectively.

The spring constant k_x is calculated empirically by resonating the electrostatic comb drive and using Rayleigh's method [1,17]:

$$k_x = 4\pi^2 f_r^2 (M_0 + 0.25M_t + 0.343M_b) \tag{5}$$

where f_r is the fundamental resonant frequency of the suspended structure, and M_0, M_t, and M_b are the masses of the shuttle, the outer connecting truss and the beams, respectively. SEM is used to determine the physical dimensions of the device precisely and density of silicon is taken as 2.3 g/cm^3. With M_0 = 0.03 μg, M_t = 0.005 μg and M_b

= 0.012 μg, w = 10 μm, l = 120 μm, z = 0.95 μm and α_z = 1.1 [1], eq. 1 can be written as eq. 6. For each structure, f_r, V_S and V_P are measured and μ_s is calculated.

$$\mu_s = \frac{0.2 f^2 \Delta x}{(V_s^2 - V_p^2)} \tag{6}$$

Testing is done at a probe station fitted with an optical microscope with 1000x maximum magnification. Signals are fed to the test dice through series resistors (≥ 1 MΩ) for short-circuit protection. Conventional electronic equipment is used, including DC voltage sources with 50V maximum voltage output and function generators with 1 MHz sinewave capability and 10 V zero-to-peak amplitude. To resonate the comb drives, a DC bias of 15V is applied to the moveable comb and a AC sinewave signal with a zero-to-peak amplitude of 5V is fed to the stationary comb. The amplitude of shuttle vibration at resonance is ≈10 μm, so visual determination of resonant frequency is possible using an optical microscope.

3. Results

Figure 2a is an SEM micrograph of a static friction test structure. The beams of the folded spring suspension are 200 μm long, and the bumper dimensions are 10x10 μm^2. A close-up of the shuttle is shown in Figure 2b. An AFM image of the ground plane polysilicon (rms = 3 nm) is shown in Figure 3. Contact angle data given in Table 2 confirms that well-packed SAMs were formed [7,8].

Figure 2.a) SEM of a static friction testing microstructure, consisting of a spring-suspended shuttle attached to an electrostatic comb drive.

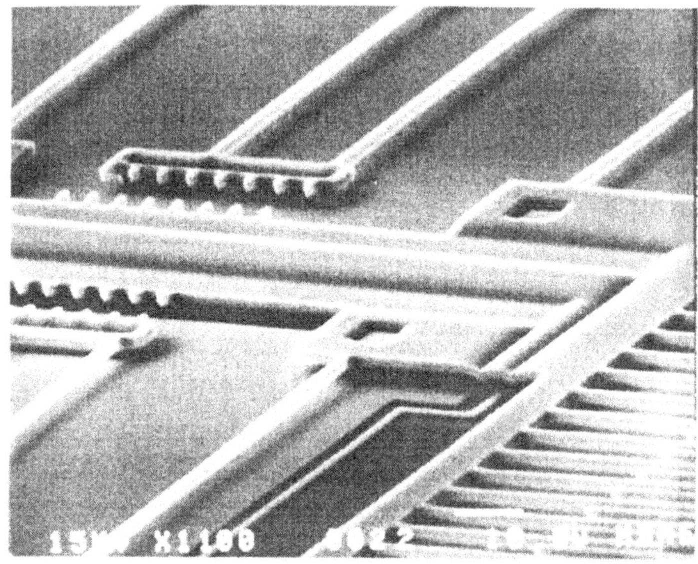

Figure 2.b) SEM close-up of the shuttle.

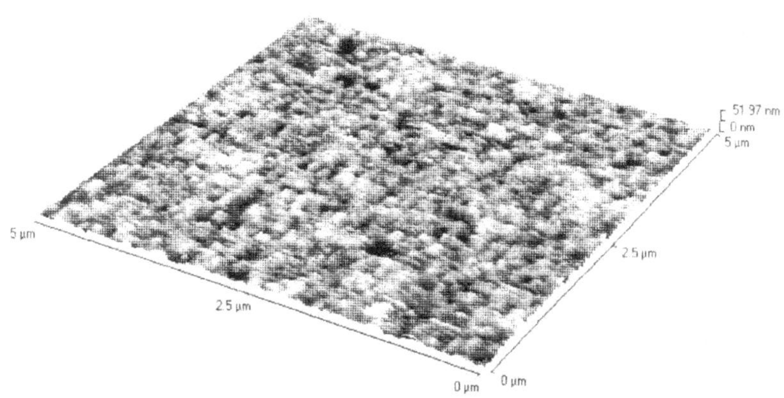

Figure 3. 5×5 µm² AFM image of polysilicon surface from the ground plane. The maximum height is 52 nm and rms roughness is 3 nm.

TABLE 2. Contact angle data for SAM-coated polysilicon.

SAM Coating	Water (°)	Hexadecane (°)
OTS	112	38
FDTS	115	68

Table 3 gives the data taken from friction experiments performed at room temperature in ambient conditions (RH = 50%) on SiO_2 and SAM-coated structures from one wafer. V_S and V_P are averaged over multiple trials, f_r is observed visually through a microscope and Δx is measured by comparing vernier scales on the shuttle and ground plane. The results shown in Table 4 indicate that in this testing situation, the static coefficient of friction is independent of apparent contact area. Stick-slip motion is observed upon gradual decrease of clamping voltage to V_S for both SAM- and oxide-coated structures.

TABLE 3. Averaged values from static friction experiments performed at room temperature in ambient conditions (RH = 50%). The total bumper area is 300 μm² and the shuttle displacement is 12 μm.

	f (Hz)	V_s (V)	V_p (V)	μ_s
oxide	8520	9.6	4.2	2.0 ± 0.8
OTS	9700	40.4	4.7	0.14 ± 0.02
FDTS	7167	37.7	3.5	0.095 ± 0.01

TABLE 4. Static coefficient of friction data taken from test structures with total bumper apparent areas of 144 and 300μm².

	A = 144 μm²	A = 300 μm²
oxide	2.3 ± 0.8	2.0 ± 0.8
OTS	0.13 ± 0.01	0.14 ± 0.02
FDTS	0.10 +/- 0.007	0.09 +/- 0.01

4. Discussion

The experiments performed on SAM-coated test structures give μ_s values of 0.13 ± 0.01 and 0.10 ± 0.007 for OTS and FDTS films, respectively. These values, similar to μ_d results reported by Cléchet obtained using a tribometer (OTS SAM-coated Si(100), 0.12 [11]), are approximately 20 times lower than that from oxide-coated structures, 2.3 ± 0.8. This reduction is to be expected, considering that the static friction between two surfaces depends on the level of adhesion present, and the apparent adhesive energy of FDTS SAM coated polysilicon is more than four orders of magnitude lower than that between oxide-coated surfaces. This difference can be attributed to the lack of water capillary forces between hydrophobic SAM surfaces. The variation in the μ_s values between the two SAM coatings may be caused by the lower polarizability of perfluorinated surfaces versus hydrogenated ones, given that the adhesional energy between SAM-coated surfaces is ideally due to van der Waals attractions [7]. The results indicate that in this testing situation, static friction does not depend on total bumper area, as expected [1,18,19].

Using similar test structures, Lim, *et al.* reported higher μ_s values of 4.9 ± 1.2 for oxide-coated polysilicon contacts. The difference in μ_s may be caused by the drying technique used, i.e., methanol rinse followed by air drying by Lim, *et al.* versus supercritical drying with ultra pure CO_2 in this paper. The high standard deviations in μ_s(oxide) data given here as well as by Lim, *et al.* can be attributed to a number of factors. First, the oxide surface has a high surface energy which favors the adsorption of polar contaminants from the ambient atmosphere, resulting in a highly variable surface layer. Second, water menisci form around the asperities and therefore adhesion depends strongly on the surface topography. Also, capillary condensation is highly sensitive to slight changes in humidity. Low standard deviations in the μ_s data for the SAM films indicate uniform coating coverage and a lack of capillary attractions.

The slick-slip behavior observed may be caused by the roughness of the polysilicon surfaces sliding over one another [19,20].

5. Conclusions and Future Work

This work shows that alkylsiloxane SAMs provide significant decrease in static friction, resulting in a factor of 20 reduction with respect to the oxide coating. SAMs are therefore recommended for lubrication purposes in MEMS. These coatings may also be applied in the vapor phase following release using another process. Future work includes the measurement of dynamic friction [20], and the wear characteristics of these coatings.

6. Acknowledgments

This research has been supported by Sandia National Laboratories. The authors acknowledge additional support from the National Science Foundation in the form of a graduate fellowship (US) and a young investigator award (RM). Additional support from The Arnold and Mable Beckman Foundation in form of a young investigator award (RM) is gratefully acknowledged. The authors would like to thank Dr. Michael Houston for his help in fabricating the test structures as well as the Berkeley Microfabrication Staff for their assistance.

7. References

1. Lim, M.G., Chang, J.C., Schultz, D.P., Howe, R.T., and White, R.M., *Proceed. IEEE MEMS Workshop,* Napa Valley, CA, U.S.A., 1990, pp. 82-88; Lim, M.: Masters Thesis, University of California at Berkeley, Berkeley CA, 1990.
2. Deng, K., Collins, R.J., Mehregany, N., and Sukkenik, C.N., *Proceed. IEE MEMS Workshop,* Amsterdam, the Netherlands, 1995, pp. 368-375; Deng, K., Ph.D.: Thesis, Case Western Reserve University, Cleveland OH, 1994.
3. Zarrad, H. *et al.*: *Sensors and Actuators A,* **46/47**, 598-600 (1995).

4. Garcia, E.J. and Sniegowski, J.J.: *Journal of Sensors and Actuators,* **A48**, 203 (1995).
5. Bhushan, B.: *Tribology and Mechanics of Magnetic Storage Devices* (Springer, New York, 1990).
6. Komvopoulos, K.: Personal Communication.
7. Srinivasan, U., Houston, M.R., Howe, R.T. and Maboudian, R.: *Proceedings of the 1997 International Conference on Solid State Sensors and Actuators*, Chicago IL, USA 1399-1402 (1997).
8. Ulman, A.: *Introduction to Ultra-Thin Organic Films: From Langmuir Blodgett to Self-Assembly*, Academic Press, San Diego, CA, 1991.
9. Houston, M.R., Maboudian, R. and Howe, R.T.: *Proc. IEEE Solid-State Sensor Actuator Workshop,* Hilton Head, SC, USA (1996).
10. Bhushan, B., Israelachvili, J. and Landman, U.: *Nature*, **374**, 607-616 (1995).
11. Cléchet, P., Martelet, C., Belin, M., Zarrad, H., Jaffrezic-Renault, N. and Fayeulle, S.: *Sensors and Actuators A,* **44,** 77 (1994).
12. Deng, K., Ko, W.H. and Michal, G.M.: *Proceedings of the 1991 International Conference on Solid State Sensors and Actuators*, San Francisco, CA, USA 213-216 (1991).
13. Tai, Y-C. and Muller, R.S.: *Sensors and Actuators A*, **21-23**, 180-183 (1990).
14. Gabriel, K.J., Behi, F., Mahadevan, R. and Mehregany, M.: *Sensors and Actuators A*, **21-23**, 184-188 (1990).
15. Smith, B.K., Sniegowski, J.J. and LaVigne, G.: *Proceedings of the 1997 International Conference on Solid State Sensors and Actuators*, Chicago IL, USA 245-248 (1997).
16. Mastrangelo, C.H. and Hsu, C.H.: *Proc. IEEE Solid-State Sensor Actuator Workshop,* Hilton Head, SC, USA, 208 (1992).
17. Tang, W.: Ph.D. Thesis, University of California at Berkeley, Berkeley CA, 1990.
18. Bowden, F.P. and Tabor, D.: *Friction and Lubrication,* Methuen & Co., 1967.
19. Rabinowicz, E.: *Friction and Wear of Materials*, John Wiley & Sons, 1965.
20. Srinivasan, U., Howe, R.T. and Maboudian, R.: work in progress.

TRIBOLOGICAL PROPERTIES OF MODIFIED MEMS SURFACES

V. V. Tsukruk [1], *T. Nguyen* [1], *M. Lemieux* [1], *J. Hazel* [1],
W. H. Weber [2], *V. V. Shevchenko* [3], *N. Klimenko* [3], *E. Sheludko* [4]

[1] College of Engineering & Applied Sciences, Western Michigan University, Kalamazoo, MI, USA
[2] Physics Department, Ford Motor Co., Dearborn, MI, USA
[3] Institute of Macromolecular Chemistry, National Academy of Science, Kiev, Ukraine
[4] Institute of Bioorganic and Petrochemistry, National Academy of Science, Kiev, Ukraine

Abstract

This paper discusses molecular layer films, composed of rigid/soft polymers tethered to the functionalized MEMS surface as an alternative method to lubricate these systems. Surfaces with NH_2, and SO_3H terminal groups were obtained by direct chemisorption of silane based compounds on silicon wafers and polysilicon microvibromotors followed by adsorption of a polymer layer. The properties of the 0.4 - 1.0 nm thick composite films were probed by ellipsometry, contact angle measurements, and scanning probe microscopy (SPM). Surface topography, adhesive forces, and friction coefficients were examined. SPM imaging revealed fine grainy surface morphology of silicon nitride MEMS with an average microroughness of 2 - 5 nm. We observed significant increase of contact angle, decrease of adhesive forces, sharp reduction of friction coefficient along with a virtually unchanged surface texture after surface modification of silicon nitride MEMS.

Introduction

Various soft and solid materials are used for modification of surface tribological properties of microelectromechanical systems (MEMS). [1] Ultrathin molecular films (Langmuir-Blodgett (LB) films and self-assembled monolayers (SAMs)) are prospective boundary lubricants for such devices. [2, 3] However, weak adhesion of LB films and velocity dependent friction behavior of SAMs restrict their usefulness. Novel polymeric materials suitable for surface modification on the molecular scale should be sought to control tribological properties of MEMS surfaces.

The *composite molecular films* can be formed by a combination of self-assembling chemisorption and physical adsorption of organic molecules/polymers with different substrate. [4] With such an approach, the first stage consists of surface modification with appropriate chemical functionality (e. g., NH_2 and OH terminal groups) via chemical self-

B. Bhushan (ed.), Tribology Issues and Opportunities in MEMS, 607-614.

assembly. The second stage involves selective adsorption of various polymer "building blocks" and their tethering to the functionalized surface through specific interactions (e. g., Coulombic, covalent, or hydrogen binding). In the present communication, we report the results of the study of the tribological properties of such molecular layers composed of SAM with functional terminal groups coated by molecular polymer layers.

Experimental

The substrates were silicon wafers of the {100} orientation and polysilicon surfaces of microvibromotors.[1,5] The preliminary process before surface modification of MEMS was the removal of the sacrificial layer. The vibromotor substrates were secured on a Teflon holder to prevent element damage during the sacrificial release or surface modification.

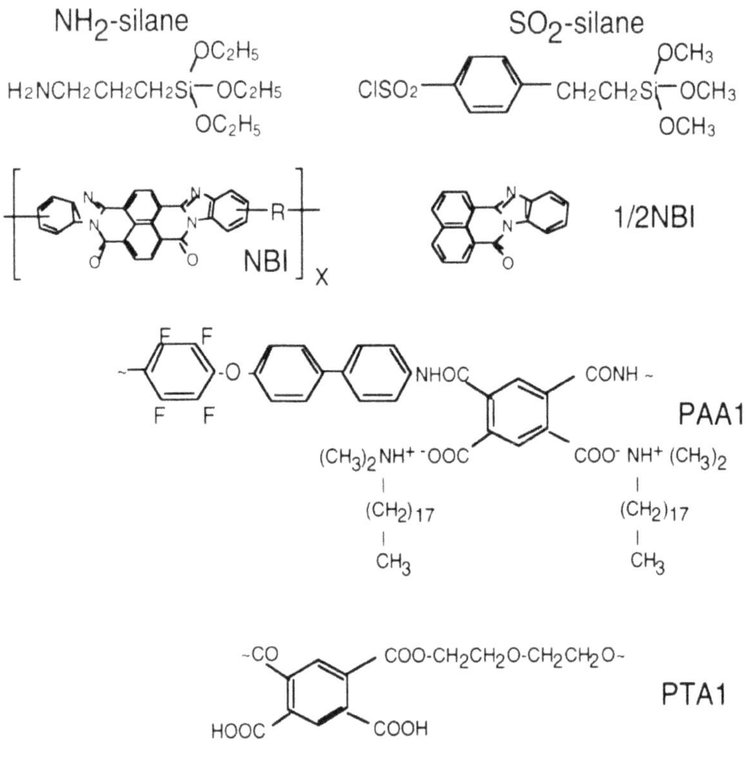

Figure 1. Chemical formulas of some compounds used

The NH_2 and SO_3H terminated SAMs (see Figure 1 for chemical formulas) were prepared according to the known self-assembly procedure.[3] We prepared SAM solutions by dissolving silanes in ethanol : water solution. The cleaned and hydrophilized substrates were then transferred from the ethanol bath into the solution for 10-15 minutes with continuous gentle stirring. Finally, the substrates were submerged in ethanol and ultrasonificated to remove any remaining molecule aggregates. The modified substrates were then rinsed several times with ethanol and dried using dry nitrogen.

In the second stage, the substrates with functionalized SAMs were submerged in the polymer solution in 1-methyl-2-pyrrolidinone for 2 hours. After the modification treatment, they were placed in an ultrasonic bath to reduce possible aggregate formation on the surface. The substrates were then washed with Milli-Q water and dried by dry nitrogen.

Several different sets of rigid polymers and polyions were chosen initially as candidates for possible composite layer films for surface modification of microelectromechanical vibromotors: 1) monomeric units and macromolecules of ladder polymers, polyheteroarylenes, and their complexes with stearic acid (NBI and NBI-St); 2) polyester acid - poly [2,4,7-trioxaheptylpyromellitic acid] and its potassium salt (PTA#); 3) oligo- and polyamide acids and their salts - N,N-bis[1,4-phenylene-1,4-tetrafluorophenylenedioxa-1,4-phenyleneimino-methylene-4-hydroxyphenyl) pyromellitic-amidoacid and poly[bis(N,N-dime-thyloctadecylammonium)-1-tetrafluoroethoxy-2,4-phenylene-bis(p-aminoben-zamide)-benzenephenonedicarboxamide dicarboxylate] (PAA#) (see chemical formulas for some of the representative polymers in Figure 1). One NBI compound and one PTA polymer (Figure 1, 1/2NBI and PTA1) were chosen from this group for complete testing on MEMS surfaces due to their high stability, uniformity, and complete coverage.

The NBI compounds possess stable complexes with fatty acids and form LB films at solid surfaces. [6] Synthesis of PAA polymers as well as fabrication of Langmuir monolayers were described in previous works. [7] PTA polymers were obtained by the chemical reaction of pyromelitic anhydride and diethyleneglycols of different molecular weights in a dry nitrogen atmosphere at a temperature gradually rising to 150°C. The mixture was kept for about an hour at this temperature, cooled, and dissolved in N,N-dimethylformamide. PTA polymer precipitated and dried at 70°C in a vacuum. Content of COOH groups has been groups has been probed by titration (found 26.7%, calculated 27.8%) and characterized by IR-spectroscopy [1730 cm-1 (C=O)]. Potassium salt of this polymer was obtained by treatment of its alcohol solution with potassium acetate alcohol-water solution (1:1) followed by drying at 50°C in vacuum; IR-spectrum (KBr): 1730, 1721, 1706, 1590 cm-1 (C=O, COO).

The contact angle of the substrate was measured by the sessile drop method. [3] The sessile drop method is known as a simple method accurate within ± 2°. Modified surfaces were probed on a custom-designed optical microscopic system. Milli-Q water droplets were placed over the studied surface and their shape was observed by a microscope equipped with a CCD camera.

Scanning probe microscopy (SPM) [8] was used to characterize modified surfaces and probe their adhesive properties on the Nanoscope IIIa - Dimension 3000 microscope. Surfaces were probed at several randomly selected locations with scan sizes varying from 100 μm x 100 μm to 200 nm x 200 nm. Microroughness was calculated for a 1 μm x 1 μm scan size. To evaluate adhesion interactions, we measured pull-off forces from force-distance curves at several randomly selected locations. A cross-section of surface topography and a variation of torsion deflections (a friction loop) were detected simultaneously. [9] Loading curves (friction forces versus normal load) were measured for the range of normal loads from 1 to 300 nN . A fundamental resonant frequency was used for cantilever spring constant evaluation according to the procedure described earlier. [10]

Results and discussions

Surface morphology of the composite films.

Composite films deposited on silicon wafers possess very smooth surface morphology with homogeneous topography and friction forces. Microroughness of these films is very close to microroughness values of underlying SAMs and lies in the range of 0.2 - 0.5 nm (Table 1).

Table 1. Roughness and contact angles for samples.

Surface	Micro roughness, nm	Contact Angles, °
SiO_2	0.15	<5
NH_2 SAM	0.43	38
NH_2/NBI-St	0.50	48
NH_2/PAA1	0.18	56
NH_2/PAA2	0.27	58
NH_2/PAA3	0.32	37
SO_3 SAM	0.41	34
SO_3/PTA2	0.16	42
SO_3/PTA1	0.50	45

Ellipsometry data provides total thickness of organic layers deposited on silicon wafers (Table 2). The thickness of the silicon oxide layer for silicon wafers used here was 2.1 ± 0.05 nm. Thickness of adsorbed organic overlay determined from this data is about 0.6 nm and about 0.4 nm for NBI and PTA, respectively. These values are close to ones determined from SPM "scratch" test and are slightly lower than the extended length of the molecules. Comparison of film thicknesses with geometrical dimensions of NBI complexes and independent data for Langmuir monolayers allows drawing conclusions about the spatial arrangement of NBI molecules adsorbed on NH_2 terminated SAMs. Indeed, thickness of NBI monolayer complexes with tilted stearic acid molecules should be in the range of 1.7 - 2.0 nm as was demonstrated for Langmuir films from these compounds.[4] On the other hand, estimation of transverse dimensions of NBI molecules gives 0.4 nm for flat-on orientation and 0.8 - 0.9 nm for edge-on orientation of highly anizometric molecules. Therefore, predominantly flat-on positioning of NBI molecules can be concluded for composite films studied here.

Table 2. Thicknesses of surface layers.

Surface	Thickness, nm
Silicon oxide	2. 1
NH_2 SAM	0.4
NH_2/1/2NBI-St	0.95
SO_3 SAM	0.3
NH_2/PAA1	0.34
SO_3/PTA1	0.67

Macromolecules of PTA1 form a very thin layer with a thickness close to the cross-section of macromolecular chains (0.4 nm) that indicates complete spreading of flexible macromolecular chains over SAM surface. Spreading of polymer chains with the thickness close to a single macromolecular chain cross-section is usually observed for flexible polymers bearing ionizable groups capable of strong electrostatic interaction with the supporting substrate.[11]

Contact angle characterizes the wettability of NH_2 or SO_3H terminated SAM surfaces and polymer layer on the surface. Table 1 shows the result of the contact angle measurement. Bare silicon oxide is highly hydrophilic with a contact angle close to zero. The NH_2 and SO_3H terminated layers are partially hydrophobic with their contact angles in the range of 30^0 to 40^0 depending on the environment. If the surface has the additional polymer layer, its contact angle is usually larger than for NH_2 or SO_3H terminated surfaces (Table 1).

Tribological properties of composite layers.

Loading curve F_f (F_n) for silicon oxide surface of silicon wafers is shown in Figure 2 along with linear regression analysis according to the generalized Amontons law:

$$F_f = f_o + \mu F_n$$

where μ is a friction coefficient defined as $\mu = \partial F_F / \partial F_n$. [12]

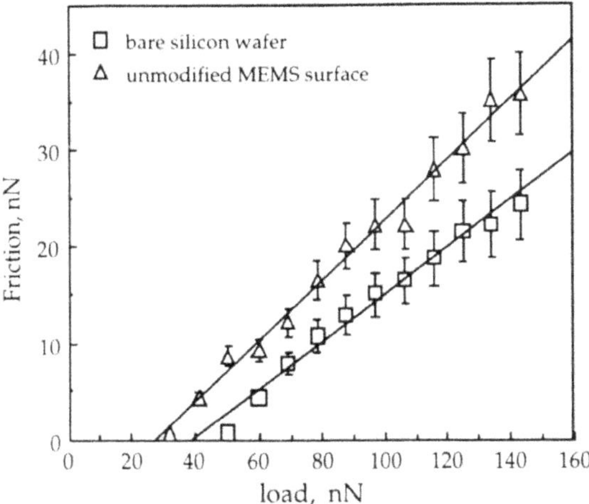

Figure 2. Loading data for two surfaces.

Friction coefficients determined from the slopes of linear fits are presented in Table 3 for different surfaces. Friction coefficient of a clean silicon oxide surface varies (μ = 0.2 - 0.35) depending on the environment. Friction coefficient remains virtually unchanged after surface modification with NH_2 and SO_3H terminated SAMs. The friction coefficient and adhesion interactions of surfaces coated with composite layer are found to be much lower than for a bare silicon wafer (0.06 - 0.1, Table 3).

Table 3. Composite layers at silicon wafers.

Surface	Adhesion, nN	μ
Silicon oxide	112	0.3
NH_2- SAM	292	0.23
NH_2/1/2NBI-St	105	0.07
NH_2/PAA1	124	0.06
NH_2/PAA2	82	0.1
SO_3H SAM	261	0.2
SO_3H/PAA1	92	0.07

MEMS surfaces

Figure 3 shows topographical and friction images of MEMS surfaces at different scales.

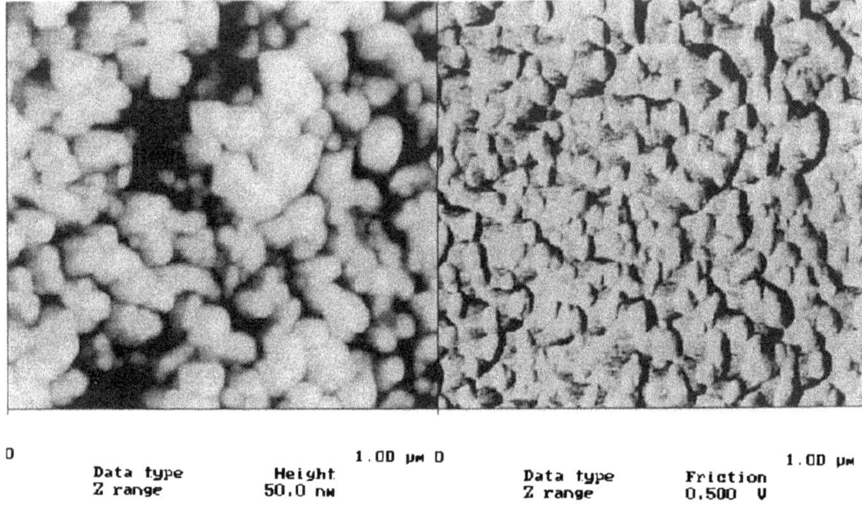

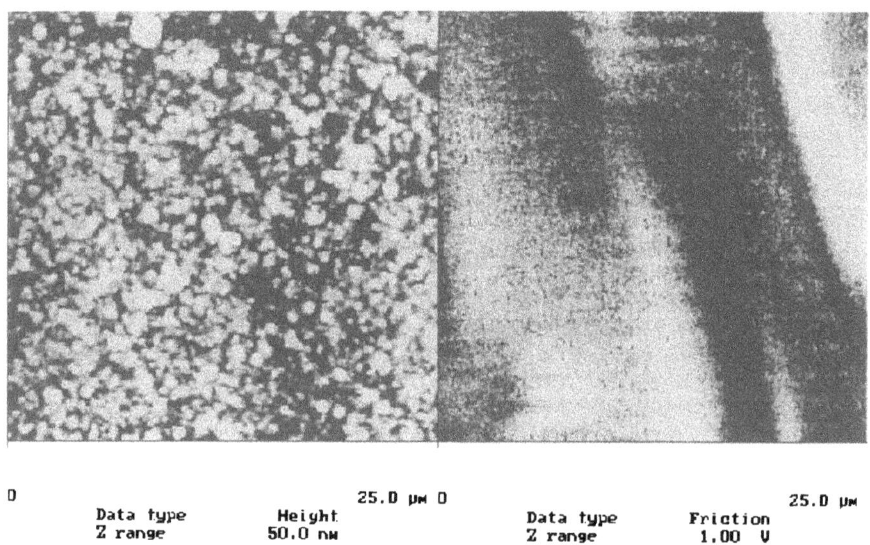

Figure 3. Topographical (left) and friction (right) images of MEMS surfaces at different scales.

Different locations at MEMS surfaces (slider, base, resonators, etc.) have a very uniform type of surface texture. At high magnification, fine grainy surface morphology is observed with uniform sized grains less than 100 nm across and 5 - 10 nm in height. Significant clustering is visible at lower magnification for overall very flat surface areas. Lateral sizes of these grain clusters are in the range of 0.5 - 1 μm. The microroughness of MEMS surfaces in 1 μm x 1 μm areas is determined to be 3.5 ± 1.5 nm. This value is much higher than the ones on the silicon wafer surface (Table 1). Surface texture remains virtually unchanged after surface modification. Microroughness for modified MEMS surfaces is in the range 2 - 6 nm which is essentially the same as for unmodified surfaces. Therefore, adding of the molecular polymer layer does not heal nanoscale surface roughnessity. This important observation indicates that the modification procedure proposed does not increase total contact area of the modified surface.

Friction distribution is very even at a microscale which indicates very uniform chemical and physical composition of the surface (Figure 3, note "geometrical contribution" across the grain boundaries). However, at a large scale, we observe a micron scale uneven patterning of friction distribution for MEMS surface which can probably be related to local stresses produced during release process (Figure 3).

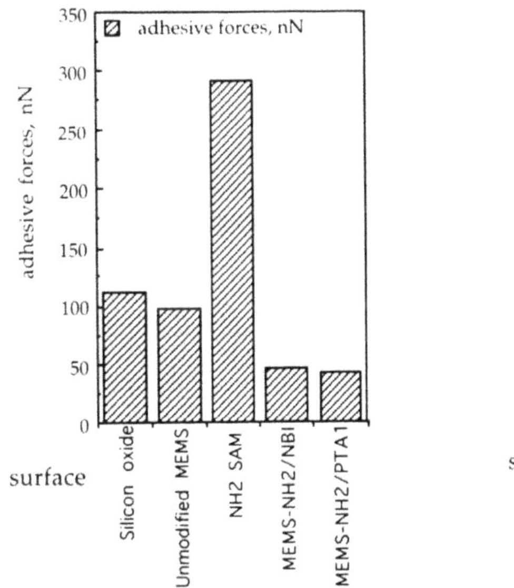

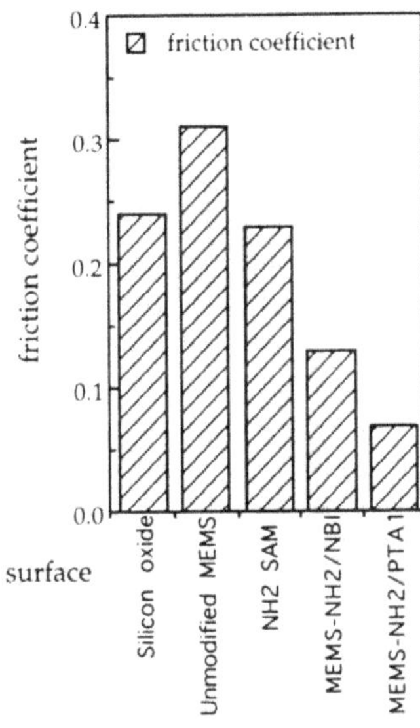

Figure 4. Adhesion interactions and friction coefficients for various substrates.

The adhesive forces and friction coefficients measured for the bare silicon wafers, released and unmodified MEMS, NH_2 terminated SAM on silicon wafer, and MEMS coated with two different composite layers are presented in Figure 4. Adhesion forces of the silicon wafers and unmodified MEMS surfaces are close to each other (about 100 nN). Adhesion interaction is much higher for NH_2 modified silicon wafers approaching 300 nN (Figure 4a). A sharp increase of the adhesion interaction for this surface is observed despite higher level of hydrophobicity of modified surfaces, which significantly reduces capillary forces (Table 1). Strong adhesion is usually observed between silanol groups of silicon substrates and silylamine surface groups of silicon nitride tips and is related to strong Coulombic interactions.[13, 14]

"Capping" NH_2 modified MEMS with rigid (NBI) and soft (PTA1) polymer layer results in a great reduction of surface adhesion to a level well below the initial one for unmodified surfaces (Figure 4a). Apparently, a number of factors contribute to the observed changes such as altering of chemical composition, increasing hydrophobicity, and increasing stiffness in the case of NBI layer.

Friction coefficients for MEMS surfaces were determined by a linear regression analysis of the loading curves as discussed above (see Figure 2 for an example of loading data for unmodified MEMS surface). As we discussed before, the friction coefficient for bare silicon wafers and MEMS varies in the range of 0.2 - 0.35 (Figure 4b). Modification of MEMS surfaces with NH_2 SAMs does not change the friction coefficient significantly. However, capping of this surface with polymers results in a sharp reduction of the friction coefficient to a 0.05 - 0.1 range (Figure 4b). These composite films are also exceptionally stable under shear stresses introduced by the SPM tip scanning with very high forces. This is due to strong chemical tethering of the polymer layer to the underlying functionalized surface. Low friction coefficients along with much reduced adhesion and good wear stability make the composite layers discussed promising for the monitoring of MEMS surface properties without changing their surface texture.

Acknowledgments

Funding for this research from the Surface Engineering & Tribology Program, National Science Foundation, CMS-94-09431 Grant and CMS-96-10408 and Ford Motor Co. is gratefully acknowledged. International collaboration was supported by NATO Linkage Grant OUTR.LG 951524 and NATO CNS-970320. The authors thank R. Hipwell and R. Muller for providing microvibromotor chips, I. I. Ponomarev for providing NBI monomers and polymers, J. Wu for technical assistance and sample preparation, and Dr. V. Bliznyuk for useful discussions.

References

1. Muller, R. S., in: *Micro/Nanotribology and Its Applications*, B. Bhushan, Ed., Kluwer Press, 1997, p. 579.
2. Komvopoulos, K., *Wear, 200*, 305, 1996.
3. Ulman, A., *An Introduction to Ultrathin Organic Films*, Academic Press, Inc., 1991.
4. Bliznyuk, V. N., Everson, M., Tsukruk, V. V., *J. Tribology*, 1997, accepted
5. Kassing, R., Rangelow, I. W., in: *Micro/Nanotribology and Its Applications*, B. Bhushan, Ed., Kluwer Press, 1997, p. 601.
6. Tsukruk, V. V., Bliznyuk, V. N., Reneker, D. H., *Thin Solid Films, 244*, 745, 1994.
7. Sheludko, E., Mishchenko, N., Golod, L., Kovtun, G., Tsukruk, V. V., Petrash, S., *Ukr. Chem. J., 58*, 906, 1992.
8. Sarid, D. *Scanning Force Microscopy*, Oxford University Press, NY, 1991.
9. Tsukruk, V. V., Ratner, B., Eds. "*Scanning Probe Microscopy in Polymers*", ACS Symposium Series, 1997, in press
10. Hazel, J., Tsukruk, V. V., *J. Tribology*, submitted
11. Tsukruk, V. V., Bliznyuk, V. N., Visser, D. W., Campbell, A. L., Bunning, T., Adams, W. W., *Macromolecules*, accepted
12. *Micro/Nanotribology and Its Applications*, B. Bhushan, Ed., Kluwer Press, 1997.
13. Noy, I. A., Vezenov, D. V., Lieber, C. M. *Annu. Rev. Mater. Sci. 27*, 381, 1997.
14. Tsukruk, V. V., Bliznyuk, V., Wu, J., Visser, D. *Polymer Prepr., 37* (2), 575, 1996.

BREAKOUT SESSION REPORT: Tribology Research in MEMS

Dr. Marc Madou, CMR Scholar
Center for Material Research, The Ohio State University
Columbus, OH 43210-1178

The breakout session on Tribology Research in MEMS was organized to address the following points:

1) What are the most challenging problems-issues?

2) What are the best approaches-expertise needs?

3) Recommendations-directions for research

We started by asking the following question: Is stiction in MEMS still an issue or did industry solve it already and is keeping it a secret ? The answer which became clear is that stiction remains a very big issue even for industry (good input was given here from Analog Devices represented at the meeting).

1. STICTION

1) What are the most challenging problems-issues?
At this point in time, being able to avoid stiction in a surface micromachined device might determine if a company will make it or not. The yield in production is very much determined by stiction and knowing all of the different options on how to prevent it remains very important.

2) What are the best approaches-expertise needs?
Some approaches to the stiction study were suggested. Destructive evaluation, designing stiction out of the device, more fundamental understanding by testing more parameters (at this stage the fundamental understanding is mainly about isolated cases), correlate topography and chemistry contributions.

3) Recommendations-directions for research
We ranked product areas in MEMS that are challenging and would benefit greatly from MEMS and tribologists working together. We also added as a criterium that it would be an application area with a great market potential.

B. Bhushan (ed.), Tribology Issues and Opportunities in MEMS, 615-617.

2. RANKING of APPLICATION AREAS (1 to 3, 1 is top)

1) Accelerometers or comb drives in general (1) (Aim in this structure is to avoid contact). See also stiction above.
2) Micro relays (1) (Intentional contact). This was seen as a very important area where tribologists have contributed before and where micromachining sofar has failed. It also involves stiction. There is also a very large market perceived for such a micro relay justifying a large investment in the tribology of it.
3) Optical switch (3)
4) Valves (2) (wear issues - contact, erosion, corrosion?)
5) Bearing supported devices (problem - low torque - requires gearing to get to mesoscopic region) (3)
6) Micromirrors (solved by TI?, when a torsion is involved stiction is easier to avoid) (2)
7) Micro fluidics (3)
8) Turbines (3) (thought provoking but for the future)
9) Inchworms (3) (want reliable friction)
10) Gear Train (3) (new tribomaterials)
11) Hinges (3)
12) Data Storage Devices (1) (AFM)
This was also rated very high as the recent breakthroughs with AFM data storage research warrant an in-depth investigation if tribologists can contribute as much here as they have contributed in the hard disc arena.
13) Grippers (3) (want reliable friction - gripping) (don't want adhesion)

In conclusion the three top ranked application areas are:

1. Comb drives (e.g. accelerometers, stiction)
2. Relay
3. Data storage.

3. SCALING

It was pointed out over and over how important it is to recognize scaling of forces in all the various applications where tribology science will be applied. Traditional tribologists often do not know how stiction, friction, heat transfer, etc. scale in the microdomain. It seems almost mandatory that besides learning the MEMS jargon, a study of scaling in tribology phenomena would be undertaken most useful for micromachinists and tribologists alike.

4. MEMS Devices to help Tribology Research

It was pointed out that making MEMS devices to test tribology would be another good arena to foster collaborations between these two fields. Moreover one could envision making micromachines to help in tribology issues, e.g. make MEMS devices to help tribology applications (e.g. a micro dispenser of lubricant).

5. General Recommendations

1) Research in stiction should be directed at the fundamental issues. More parameters that influence stiction need to be investigated.

2) There is a communication issue here in that the jargon of tribologists and MEMS people makes understanding the REAL issues more difficult. Tribologists have worked more on the macroscale and a good understanding of scaling of tribology phenomena in the micro-domain would be very good for both communities.

3) The top research areas should be comb drives (avoid contact stiction), relays (contact and stiction) and data storage.

4) MEMS can be used to study tribology and MEMS can help in tribology applications.

BREAKOUT SESSION REPORT: MECHANISMS OF NANO-SCALE TRIBOLOGY AND INSTRUMENTATION FOR MEMS DEVICES

K. KOMVOPOULOS
Department of Mechanical Engineering
University of California
Berkeley, CA 94720, USA

1. Introduction

Microelectromechanical systems (MEMS) is a rapidly growing interdisciplinary field dealing with the design and manufacturing of miniaturized micromachines using mainly semiconductor technology. Recent demands have led to significant research efforts aiming to the integration of electrical, mechanical, optical, fluidic, thermal, and biological microdevices into versatile microsystems capable of performing complex sensing, control, and computing functions. It is predicted that by the end of the century MEMS device revenues will exceed ten billion dollars. Early indications suggest that, while significant opportunities exist for MEMS, a large number of challenging issues are presently preventing the evolution of MEMS from the laboratory to the application world.

To further advance the state-of-the-art in MEMS and, thus, expand their practical utility, basic understanding of the different phenomena inhibiting fabrication and long-term use of microdevices is imperative. Surface interactions occurring during micromachine processing and operation are of particular concern. In submicron-scale tribology, the physical and chemical properties of surfaces are primary factors rather than the bulk properties. In view of the relatively low stiffness and large surface areas of micromachines, excessive surface forces may result in premature failure and reduced operation lives. To overcome such problems, it is essential to obtain insight into the chemical and mechanical properties of MEMS interfaces and to elucidate the prevailing trobophenomena. This can be accomplished through the development of new experimental and analytical approaches that will enable the identification and study of micro-scale surface interactions. Measurements of very small forces and relative displacements and difficulties in controlling the shape and size of interfacial contact regions can only be overcome through innovations in experimental methodologies and instrumentation.

It is clear, therefore, that interfacing the MEMS and tribology communities would be especially beneficial since it will provide the technological basis for alleviating many of the present obstacles that limit the application range and reliability of several MEMS devices. The purpose of this article is to present an assessment of the important research issues and expertise needs in nano-scale tribology and instrumentation for MEMS and provide a vision of plausible research directions.

B. Bhushan (ed.), Tribology Issues and Opportunities in MEMS, 619-625.

2. Research Issues

Previous research and development has been primarily focused on the microfabrication aspects of MEMS. However, there is a number of challenging research issues (Table 1) that must be addressed before miniaturized devices can be fully commercialized. Among challenging problems is the identification of the nature of submicron-scale processes encountered at MEMS interfaces during fabrication and operation, the standardization of methods for objective evaluation of the nano/micro-mechanical properties of structural materials, and the fabrication of microstructures suitable for performing friction, wear, and fatigue testing at scales and conditions similar to those encountered at MEMS interfaces.

TABLE 1. Critical research issues in MEMS

- Transfer of basic tribology knowledge to microsystem applications
- Identification of actual scale of problems involved
- Methods yielding quantitative insight into micromachine-related problems
- Standard methods for testing:
    - Failure analysis (adhesion (stiction), wear, fracture, and fatigue)
    - Material properties
    - Reliability
    - Process monitoring
- Scale-dependence of mechanical properties
- Dependence of measurements on instrument and environment
- Measurement tools in industrial and research environments
- Basic understanding of micro/nano-mechanical properties:
    - Thin solid films (e.g., Si, Al, Ni, and W)
    - Thin liquid films (self-assembled monolayers)
    - Self-lubricating surfaces
    - Smart materials
- Special micromachines for friction, wear, fracture, and fatigue testing
- Difficulties in fabricating consistent microstructures for testing
- Early indication of feasibility of process:
    - In situ sensing
    - Feedback

With regard to the study of the various tribological interactions at the micromachine level, the transfer of basic tribology knowledge acquired for similar-scale technologies, such as information storage devices, is an attractive possibility. Understanding of microscopic tribophenomena has been greatly advanced in recent years largely due to rapid developments in high-resolution surface probe microscopy (SPM). It is not apparent, however, how such basic information can be used to solve tribological problems occurring in MEMS devices. The transfer of microtribology knowledge to the MEMS field has began to occur only recently. It is well established that the kinetic energies, forces, and torques available for micromachine operation are often much smaller than adhesive interfacial forces. Research directed toward the identification of the causes of such high adhesive forces and the development of new methods for obtaining quantitative information is therefore expected to lead to the discovery of techniques enabling the enhancement of micromachine reliability.

The design of robust MEMS devices able to achieve long operation lives requires the establishment of standard testing methodologies. At the present time, there are no standard codes for failure analysis, material property characterization, reliability evaluation, and process monitoring in the MEMS field. As a consequence, the majority of material, life, and reliability data obtained at the microdevice level are biased by the instrument, environment, and experimental scale used to obtain the measurements. Of particular importance is the material behavior at different scales. There is very little known about the relationship between elastic-plastic material behavior and fatigue resistance at the nanometer and micrometer scales. Nevertheless, measurements of the elastic modulus, hardness, coefficient of friction, and wear rate are often made with very sharp microprobes, i.e., at scales approximately an order of magnitude less than the actual scales. In addition, differences between measurement tools used in industrial and research environments may inhibit direct comparisons between testing results. It is clear therefore that significant effort should be devoted to the development of the basic guidelines and standards for testing in the MEMS field. This should provide the vehicle to overcome the aforementioned difficulties and to minimize uncertainties in the measurements.

In view of the relatively small thickness of most MEMS devices, reliable mechanical property characterization of structural materials must be obtained at the micron and submicron scales. Specifically, the elastic-plastic, fracture, fatigue, and tribological properties of thin films (e.g., 2-10 μm thick) consisting of single-crystal and polycrystalline silicon, aluminum, nickel, and tungsten are of particular interest. There is very little known about the effect of processing parameters (e.g., doping level (or implantation dose), annealing temperature, and deposition time), microstructure, and specimen size on the micro/nano-mechanical behavior of MEMS structural materials. In addition, the rheological properties of thin liquid films (e.g., evaporated and self-assembled monolayers) used to minimize micromachine stiction and wear must be studied at environments (e.g., operation temperature and humidity) and contact conditions (e.g., relative speed, oscillation frequency, surface roughness, and mean contact pressure) that are typical of MEMS. Basic research on the durability of thin liquid films compatible with microfabrication processes (e.g., OTS and HMDS ultra-thin layers applied before the etch release-drying process) is expected to yield effective means for increasing the wear and fatigue resistance of micromachines. The accumulation of basic knowledge in the above research areas should provide the basis for developing self-lubricating and smart material surfaces that can be tailored to meet the specific application requirements.

In addition to the simple specimens used to measure the micro/nano-mechanical properties of thin films, there is also a need for testing under conditions resembling those of MEMS devices. This suggests that micromachines specifically fabricated for tribological and fracture/fatigue testing must be designed and fabricated in a consistent manner. In situ sensing and real-time feedback during testing should provide important insight into the evolution of surface and subsurface wear and fatigue damage. Very little research has been performed in this critical area.

Understanding of micromachine operation requires continuous recording of various data. Table 2 lists the various types of in situ measurements that are possible to obtain during micromachine testing.

TABLE 2. Types of in situ measurements

- Surface topography
- Force, torque, deflection
- Resonance frequency, ampl itude
- Real contact area
- Electrical contact resistance
- Geometrical parameters
- Thermal conductivity
- Chemical interactions
- Magnetic resonance
- Potential function
- Time-of-flight (SIMS)

In addition to the surface topography imaging, other commonly recorded signals include the normal and lateral (friction) forces, torque, deflection, resonance frequency, amplitude, and electrical contact resistance. In-process monitoring of the variation of these signals with operation time can yield important insight into the evolution of friction, wear, and fatigue damage during micromachine operation. The change of thermal, electrical, and magnetic properties with time can be observed with relatively more complex methods which may require further research and development.

3. Approaches and Expertise Needs

In view of the above challenging research issues, it is necessary to consider some of the available special approaches and expertise needs (Table 3). It is apparent that the design of reliable and robust MEMS devices can greatly benefit from the implementation of the knowledge derived from the traditional and contemporary subfields of tribology. A relatively large data base exists for bulk mechanical properties of MEMS structural materials. This valuable information should be considered in bulk micromachining processes (e.g., LIGA and HexSil (for polysilicon)) where the microfabricated devices are relatively thick. In surface micromachining, however, the devices are a few micrometers thick and, thus, mechanical property characterization should be performed at the micro/nano-scales. Contemporary tribology (i.e., micro/nano-tribology) is a fairly new subfield of tribology in which rapid developments have been observed over the past decade, largely due to advances in high-resolution SPM and pressing demands for high-density recording in the information storage industry. Although micro/nano-tribology is still at its infancy, it may provide invaluable insight into tribological problems observed with MEMS devices, mainly due to the similarity of the scales involved and the ability to accurately measure and probe surface properties. Understanding of surface phenomena, such as stiction and wear due to adhesion and contact fatigue mechanisms, that affect the operation life requires the use of in situ and lab-on-chip diagnostics.

Due to the high oscillation frequencies of most microdevices, it is essential to employ appropriate instrumentation in order to obtain real-time information about the surface condition. Examples of such measurements include surface topography maping, electrical and thermal conductivity, potential function differences, and magnetic resonance.

The analysis of the recorded data is expected to reveal important information about the change of surface properties due to contact interaction. Recent advances in optical, thermal, magnetic, and chemical surface microscopy should be beneficial to the development of techniques for obtaining such information. The chemical state (e.g., hydrophilicity) and composition of interacting micromachine surfaces can be analyzed in light of contact angle measurements, atomic force microscopy (AFM), x-ray photoelectron spectroscopy (XPS), secondary ion-mass spectroscopy (SIMS), and Raman spectroscopy, which are standard methods used in thin-film characterization.

TABLE 3. Approaches and needs for developing reliable MEMS d evices

- Utilization of knowledge from traditional and contemporary tribology
- In situ diagnostics and lab-on-chip diagnostics for life to failure detection
- Electrical and thermal conductivity measurements for failure analysis
- Potential function
- Magnetic resonance
- Thermal/magnetic/chemical surface microscopy
- Contact angle measurement
- AFM, XPS, SIMS, Raman, ellipsometry
- Critical dimensioning
- Scaling
- Simple testing methods
- FEM analysis of MEMS devices excited by friction forces
- Patterned and shaped surfaces
- Surface property modification and tailoring
- Self-replenishing thin lubricant layers

Despite information obtained based on the above mentioned approaches, there are several important needs that must be considered in the context of new-generation microdevice requirements. In particular, dimensional control and scaling are of paramount importance to the design of accelerometers, micromirrors, gear trains, and micromotors. It is well known that the material strength increases as the scale decreases. For example, the strength of brittle materials is scale-dependent due to the lower probability of flaws and higher-quality surfaces. Thus, polysilicon may exhibit higher fracture strength at small scales. Miniaturized air bearings can support higher loads because the surface area-to-weight ratio of a rotor increases with decreasing scale. There is also a need to identify simple testing methods devoid of environmental and human factors. Information about the variation of the friction force with operation time for a given material/environment system can be used to determine the dynamics of the actual microdevice by performing finite element simulations. The common approach has been to consider the excitation of the device by an invariant lateral (friction) force and to neglect changes in the interfacial contact conditions resulting from surface rubbing. Characteristic examples of such changes include the removal of native oxide layers, the development of triboelectricity (mechanical probing is often used to remove the electric charge built-up), and the smoothening of rotor edges and gear teeth during operation. Such changes can be interpreted in terms of the evolution of electrical contact resistance, potential function, and surface topography statistics, respectively.

To achieve long lives to failure, patterning and shaping of MEMS interfaces is critical. Surface texturing (roughening) and the fabrication of "microbumps" for reducing the real area of contact and, in turn, the magnitude of adhesive forces and "stoppers" to prevent crushing of accelerating parts on side-wall surfaces have been used in various recent MEMS applications. Optimization of these surface patterns is essential for each design. In addition, the modification of surface properties through the deposition of ultra-thin solid films compatible with standard microfabrication technology (e.g., wear-resistant, hydrophobic, thin films of amorphous or diamond-like carbon) and low surface energy liquid monolayers (e.g., OTS and HMDS) and the means of replenishing such liquid lubricant layers during operation are important research areas.

4. Research Directions

In view of the above analysis of the various challenging issues, available approaches, and research needs for MEMS, the following main research directions are envisioned:

(1) Fundamental understanding of material properties at different scales is essential for elucidating the dominant tribological and fracture/fatigue mechanisms at micromachine interfaces. Very little is known about the nature and evolution of these mechanisms in terms of processing conditions, resulting microstructures, and testing environments. Consequently, the operation lives of micromachines have been mainly determined based on empirical approaches.

(2) New measurement techniques suitable for MEMS testing and evaluation at the micron and submicron scales must be identified. It is imperative that testing is performed under conditions and environments resembling those of the MEMS application under consideration.

(3) Standards for mechanical property testing at the MEMS scale must be introduced in order to facilitate the transfer of results from the research laboratory to the application world. This can only be accomplished through close interactions and collaborations between researchers and engineers in the tribology and MEMS communities.

(4) Advances in micromachine design can be achieved through the development of the necessary data base of micro/nano-tribological properties of structural materials used in MEMS and the establishment of the infrastructure needed for tribologically-enhanced MEMS design.

(5) Government support for the implementation of basic tribology knowledge into the MEMS field appears to be fundamentally important to the reinforcement of the effort aiming to reveal potential applications of MEMS devices.

5. Summary

There is a growing consensus that MEMS will eventually lead to a new design technology with impacting effects on society. The interdisciplinary nature of MEMS has brought together experts from mechanical, electrical, chemical, and biomedical fields and is presently generating many opportunities and challenges requiring innovations and exploitation of effective means leading to more micromachine applications. Some of the commercial areas where long-term potential of MEMS has been recognized are in the automotive, information storage, digital mirror display, optical systems, and biomedical industries. As MEMS growth accelerates, more demands for greater reliability and increased useful life will be encountered. However, tribological obstacles are currently inhibiting the evolution of MEMS devices from the laboratory to the application world. In this article, some important challenges and expertise needs were identified and plausible approaches and future research directions were proposed. From the information presented herein, it may be concluded that there is a strong need for contributions in research and development from both industry and universities. This is essential for defining the boundaries to fabrication technologies, materials, instrumentation, and system functionality. This should lead to increased productivity and significant broadening of the application range of MEMS devices.

BREAKOUT SESSION REPORT: LUBRICANTS, OVERCOATS AND SURFACE MODIFICATION TECHNIQUES

D. A. RIGNEY
Materials Science and Engineering
The Ohio State University
2041 College Road
Columbus, OH 43210-1179

1. Introduction

This group consisted of scientists and engineers with diverse training and from varied working environments. There was a good mix of scientists and engineers, tribologists and MEMS specialists, and researchers from universities, industry and government laboratories. As a result, the discussion ranged widely. Discussion in the afternoon session returned to some of the topics highlighted in the morning, e.g., stiction and characterization, but with the addition of new opinions provided by a number of newcomers to the session.

2. Issues

A major topic--one which came up early and to which the group returned often--was stiction. While some tribologists object to the use of the term itself, it is well established in the MEMS community, and the discussion group did not waste time debating word use.

It was pointed out that the nature of the stiction problem depends on whether the motion involves rolling or sliding and whether the application requires occasional or frequent motion (e.g., in sensors, accelerometers, actuators), large numbers of cycles, and high frequencies (10^5 to 10^6 rpm).

The following questions were raised: "Do we have any design guidelines for controlling stiction?" And, "Have sufficient data accumulated to help us to select systems with reduced stiction?" There was general agreement that a fundamental and perhaps ultimate guideline is surface energy itself. What we already know about surface energy tells us the types of coatings and lubricants to try, but our accumulated knowledge also provides some information on limits to reducing stiction by this approach alone. It was suggested that a complementary approach was to design appropriate surface roughness, but the route to an optimum roughness seems to have progressed mainly by trial and error.

The discussion did not focus separately on liquid and solid lubricants, because there was general recognition that the materials used might not exhibit bulk properties for the size scales involved. An important question raised, but not answered, was, "Is there an optimum thickness for a lubricant used in a MEMS application?" A goal for overcoats

B. Bhushan (ed.), Tribology Issues and Opportunities in MEMS, 627-628.

on sliders is for continuous films without pinholes, but this is difficult to achieve when the average thickness is less than ~50 Å.

There was some concern that the processes used for application of a given coating might coat parts which should *not* be coated if they are to move freely when in use. Suggestions for avoiding this problem included using selective deposition techniques, taking advantage of throwing power and rotating the part if possible.

A reference to self-assembled monolayers (SAM's) led to comments about overemphasis of SAM's in discussions of MEMS. There was some preference expressed for a more generic term which would include multi-layers, e.g., "self-assembled systems." Too often one is shown "cartoons" of perfect SAM's with 100% coverage, a situation not realized in practice. Real layers may have much less than 100% coverage, with about 30% thought to be common. Island structures and layers with trapped or buried defects (e.g., hydroxyls) were also mentioned.

Another major topic was environmental conditions during processing, storage and use. A principal concern was exposure to elevated temperature. Devices must survive conditions encountered during packaging. A header package approach helps, but this technology is not easily automated and is costly. Current technology for packaging requires 2 to 12 minutes at ~400°C in air, ~330°C in nitrogen and ~175°C for plastic packaging processes. Storage temperatures may be ~150°C. Although use temperatures are commonly in the range 0-100°C, devices need to survive ~200°C for 10 days. For applications in space or in polar climates, low temperatures create special problems which were not addressed during this discussion.

Surface charge effects were mentioned as potential problems, during processing as well as during operation, for non-conducting materials. This is an electrical issue which might require a surface passivation step to control unwanted effects from charged species.

Characterization of MEMS lubricants and coatings was considered important at all stages, i.e., during production of the device, at completion of manufacturing, and during use. Improved techniques are needed for measuring film thickness, for chemical analysis of organics with better than 10 μm resolution (chemical force microscopy may accomplish this soon), chemical composition of other components with better than 1 μm resolution, orientation of molecules (e.g., those comprising a SAM), and, of course, micromechanical and tribological properties.

There was general agreement that one should design for steady-state operation. As part of this, self-repair mechanisms can be included in surface-film design. These would rely on surface mobility, out-gassing, formation of appropriate reaction products, etc. "Bad actors" such as Au, Fe and Cu were mentioned, but it was felt that these could be controlled by using diffusion barriers.

It was mentioned several times that the MEMS field need not be dependent on silicon-based devices, despite the advanced state of silicon technology.

Finally, it was recognized that computational limits are being extended, so that computer simulations are improving. A high priority is for better simulations of combinations of organics with inorganics. It was felt that the MEMS field could benefit from the expertise available in the petroleum industry, where there has been considerable progress on related topics.

In summary, the problem of "stiction" remains a high priority subject, with surface energy values providing obvious constraints. MEMS components and devices must survive processing conditions encountered, e.g., during packaging, *and/or* more benign processing/packaging techniques should be developed. Convenient characterization, e.g., of film thickness and chemical composition with adequate resolution, is desirable.

BREAKOUT SESSION REPORT: MECHANICAL PROPERTY MEASUREMENTS

CARL J. McHARGUE
Center for Materials Processing
University of Tennessee
Knoxville, Tennessee 37996-2350 USA

1. Introduction

Approximately 20 of the workshop attendees participated in the two breakout sessions devoted to discussions of the needs for mechanical property data and issues concerning measurements of the properties in small volumes. Since the devices under discussion are required to perform some sort of mechanical function, successful design, manufacture and performance require information on the mechanical properties of the material(s) of choice. Just as important is a knowledge of changes in properties caused by the manufacturing procedures and the operating environments.

Most efforts to date have concerned devices made of silicon using the manufacturing procedures developed for integrated circuits production. As the applications of MEMS devices increases, it is anticipated that all classes of materials (metals, ceramics, polymers, composites) will become important. There are certain unique features of the mechanical properties of each class that require the testing procedures to vary from one to another.

The properties considered to be important for the design of MEMS devices include: strength (yield, ultimate, fatigue), moduli, toughness, friction, adhesion, damping, Poisson's ratio, and residual stresses. From a manufacturing view, mechanical properties such as strength, hardness, scratch resistance, and chemo-mechanical behavior are important. Especially important is the effect of the manufacturing procedure on properties. Properties such as R-curve (crack propagation, toughness), friction, wear, fatigue, adhesion, and the change in properties with corrosion/oxidation/erosion during service will determine the lifetime of a component. Since many components will be subjected to high frequency vibrations, damping and other dynamic properties may be important for many applications.

B. Bhushan (ed.), Tribology Issues and Opportunities in MEMS, 629-632.

2. Issues and Research Needs

2.1. HOW DO PROPERTIES AND PERFORMANCE SCALE WITH SIZE?

It has been known for many years that the measured indentation hardness value often increases as the load or depth of indentation is decreased. This Indentation Size Effect (ISE) is still not understood although it often appears in tests of different classes of materials (e. g., metals and ceramics). Among the many explanations suggested are: surface artifacts such as oxide or adsorbed films; nonunifom elastic recovery of small indents; quantizied slip or deformation band formation; and the low probability of finding mobile dislocations or other deformation-initiating flaws in a small volume of material.

This effect becomes more important as the sampled volume approaches the nano- or picometer scale. Whether the response of microcomponents to applied stresses exhibit similar behavior is an important issue to be resolved. Mechanical tests on fabricated MEMS components may provide insights into this question. The question of whether properties should scale with surface area or with volume becomes important as the components become very small.

2.2. SURFACE PROPERTIES VS. BULK PROPERTIES

Any surface structure or composition that differs from the bulk may dominate the measurements at the nanometer scale. The surface may be significantly different from the underlying material due to segregated species (Gibbsian segregation), oxide or adsorbed films, surface reconstruction, etc. The surfaces of MEMS components are often given special chemical treatments in order to control the tendency to adhere to other components or to control the interaction with lubricants. It has been known for thirty years that the presence of moisture strongly affects the measured hardness value of many ceramics. It was reported at this workshop that the test atmosphere (humidity) could change the wear rate of silicon nitride by a factor of 100. Other surface treatments encountered intentionally or unintentionally during manufacturing or service may also have strong effects on surface mechanical properties. In order to be of value, measurements must be made for surface conditions that are similar to those encountered in service, recognizing that both the surface condition and properties may change with time.

Likewise, as the scale of the component decreases, the relative importance of the surface structure and properties increases and may control the performance. Understanding of these effects is essential.

2.3. PROPERTIES OF THIN FILMS-SUBSTRATE SYSTEMS

Many of the efforts to develop nano- or picoindentation devices have been directed to the determination of the properties of very thin films or coatings. The ability to deconvolute measurements into properties of the surface film or layer and those of the substrate is required. Measurements on the in situ film are necessary since the microstructure, composition, and residual stress state of films often differs from that in the bulk. Since components in service will consist of surfaces and substrates, the ability to use the data to predict the behavior of the system as a whole is an important issue. There has been progress in modeling the response of coatings on various substrates in recent years. Most systems studied have consisted of hard coatings, often load bearing, on softer substrates. Such experimental and modeling efforts are required for the material combinations and conditions encountered in MEMS applications.

2.4. TESTING (EQUIPMENT AND TECHNIQUES)

Many of the desired mechanical properties can be determined by the new generation of load - depth-sensing indentation apparatuses that are now commercially available. In principle, such properties as hardness, elastic moluli, creep, fracture toughness, stress-induced phase transformations, etc. can be determined with high apparent precision. Adaptation of the equipment to obtain additional property information (scratch testing, micro-bend bar strengths, etc.) appears to be relatively straightforward and in progress. However, a fundamental understanding of the meaning of the measured values and the intrinsic properties requires additional studies, particularly with attention to the factors discussed herein.

Further studies to determine exactly what combination of mechanical properties is measured by "hardness" is required for the various classes of materials. Since the loading portion of the load-displacement data obtained from load - depth-sensing tests contains information about both the elastic and plastic response of the surface to contact deformation, additional efforts to analyze this portion of the data may give additional valuable information. The concept of using work of deformation (both elastic work and plastic work) should be explored as a useful tool for characterizing surface deformation and mechanical behavior.

In situ probes which may be developed using MEMS techniques are desirable to obtain information regarding the changes of properties during service. Papers presented in this workshop indicate that progress is being made to develop appropriate probes for friction and fatigue measurements.

Standardized testing procedures are highly desired. Test atmosphere is an important variable. The rate of loading is another factor that influences the results of load - depth-sensing tests, particularly in materials having a high degree of strain-rate sensitivity. This parameter should be very important for tests conducted using indenters on cantilever beams in the AFM. The "jump-to-contact" behavior observed as the indenter approaches the test surface may result in uncontrolled, high rates of initial loading.

At the pico- scale, transfer of material between the indenter and the test specimen surface may become a limiting factor in the ability to obtain accurate data. Theoretically modeling has suggested that significant material transfer and distortion of both indenter and surface may occur before actual contact is made. The attractive force between contacts is typically in the range of 5 to 10 μN. Clearly, these interactions must be taken into account for measurements with applied forces of this order.

The role of subsurface damage in determining results of friction and wear measurements needs better definition for various classes of materials. Techniques for determining the kind and amount of subsurface damage are required to characterize surfaces after manufacturing procedures and under service conditions.

2.5. RESIDUAL STRESSES

Determination of the role of residual stresses on the response of surfaces to applied loads requires additional efforts. Methods to measure the residual stresses and their recovery or relaxation are required for MEMS scale components.

PANEL DISCUSSION REPORT:
TRIBOLOGY ISSUES AND OPPORTUNITIES IN MEMS

C.R. FRIEDRICH
Michigan Technological University
Houghton, MI 49931

1. Introduction

A panel discussion and open forum was held among approximately 100 members of the tribology and MEMS communities to discuss issues and opportunities for research collaboration among these groups. The discussions by the panel members and the workshop participants are hereby summarized The panel members were:

Craig Friedrich, Moderator - Michigan Technological University
Steve Danyluk, Georgia Institute of Technology
Long-Sheng Fan, IBM Almaden Research Center
Michael Gardos, Hughes Aircraft
Frank Michel, IMM-Mainz

2. Panel Presentations

An informal poll was taken of the open forum audience and the makeup of the conference, as indicated by the poll, was 75%-80% tribologists and 20%-25% MEMS-based researchers. Each panelist then made a brief presentation of their perspectives on issues and opportunities in MEMS tribology.

2.1 Steve Danyluk - Tribolgy solutions to MEMS technology

Many MEMS devices have moving parts. These moving parts come into contact and the surfaces sustain damage and wear. This degradation and wear causes shortened life of the components, and failure. The key to preventing this degradation in macrosystems is to keep the surfaces separated by lubricating fluid or air films. This is not always possible in MEMS devices, especially in the

B. Bhushan (ed.), Tribology Issues and Opportunities in MEMS, 633-640.

startup and shutdown of sliding and rotating parts. In addition, the use of lubricants cannot be implemented in MEMS devices because of capillarity in fluids. New solutions to these problems will need to be devised by the joint research of the MEMS and tribology communities. The reliability and usefulness of MEMS devices will depend on the success of this research effort.

This workshop is the first attempt to expose tribologists and MEMS technologists to the problems and solutions in their particular areas. For example, there is a good deal known in the tribology community about gas and fluid film lubrication of a wide variety of surfaces, and the physical properties of these fluids on the improvement of reliability of moving parts that come into contact. Also, there is a good deal known about the wear mechanisms of metals, coatings, semiconductors and ceramics, and how to reduce the surface and subsurface damage in these materials. This knowledge is key to the sliding contact, wear and reliability of sliders and rotating hubs and gears.

2.2 Long-Sheng Fan - MEMS opportunities/tribology issues for data storage

Today's computer data storage device is the non-volatile magnetic hard disk drive (HDD), in which magnetic transducers are mechanically positioned on spinning magnetic disks to write and read data with an areal density of a few Gb/in^2 and a data rate up to a few 10 MB/s. In the short term, MEMS technology offers a viable option to conventional magnetic data storage to go beyond 10 Gb/in^2 by increasing the tracking bandwidth and positioning accuracy through insertion of MEMS devices in the servo mechanical loop and thus enabling a drastic change of the track-to-bit aspect ratio. These MEMS devices include various position and inertial sensors and actuators. They are batch fabricated with precise feature definition. They can be appropriately small for specific applications, cheaper, better in dynamic performance, and have more functionally. Applications include shock sensors, accelerometers, etc. for write inhibition and vibration control, piggy-back fine actuators in a two-stage servo control scheme, and a skew-angle adjustment actuator (from ID to OD) in a rotary VCM system.

Because friction and stiction is a largely unknown territory for MEMS devices, applications involving contacting motion are typically avoided. The major devices in development are flexure-type devices, which are designed to minimize the problems of external friction/stiction. However, tribology issues still remain. First, we need to better understand the internal friction issue of the material to control the dynamic response of the actuator. Second, stiction might occur in the fabrication process, although some solutions such as the super-critical CO_2 release process can be used. Third, the HDD's operating environment is very dynamic and occasional in-use contact is unavoidable. To minimize the maximum stress and possible irreversible changes which can occur during such dynamic contact, a reasonably large effective contact area is required; therefore, one will need to consider minimizing the stiction for these contacts and further understanding in dynamic contact stiction is needed.

2.3 Craig Friedrich - Issues and opportunities for micromechanical systems

Micromechanical systems capable of producing outputs (kinetic, elastic, thermal, fluidic, etc.) suitable for interfacing with the macro-world have largely not been developed. Perhaps the closest approach has been the development of microfluidics for control. There is a vast difference between microelectronic applications for sensing and subsequent output signal amplification and mechanical outputs. Sensors and electronics perform analog and logic operations but mechanical systems transform energy and perform a greater array of functions. Mechanical amplification relies on stored mechanical energy which can require significant volume and mass. There are electronic primitives, the most fundamental is the electron or photon but here is meant logic gates (transistors), but few mechanical primitives and the smallest would be components capable of sustaining or transmitting some level of useful energy. Additionally, hierarchical microelectronic systems are easily synthesized based on the electrical primitives, but mechanical systems quickly become exceedingly complex and are also influenced by secondary effects such as stress, wear, friction, fatigue, and corrosion, for example. Unresolved issues and opportunities exist in defining mechanical primitives which will allow uncomplicated hierarchical mechanical microsystems to be realized. Such primitives will have no sliding contact unless it can be effectively lubricated and they will also be as reliable as their macro-scale counterparts.

Micromechanical systems can also be distinguished into two very general categories. One category is those mechanical devices, systems, and phenomena which fundamentally benefit from miniaturization. The other category is for those which may suffer or see no fundamental benefit except for the miniaturization itself. An example of the first category is micro-thermal devices where the surface area to volume ratio gets large as the characteristic dimensions get small. Localized, rather than bulk, heat transfer can make many chemical processes more efficient. Using massively parallel thermal microdevices, high efficiency macroscale energy transfer is possible. An example of the second category is micromotors and gear trains. These systems have relatively fixed losses due to friction, for example, and making them massively parallel to transfer macroscale energy will not reduce these losses and fundamental benefits will not be realized. Surely, motors and power transmission microcomponents can be used to provide motion at the microscale, but may not be the best choice because of the complexity eluded to earlier. The issues and opportunities here are centered around which fundamental phenomena can be truly helped by miniaturization. Systems and instrumentation needs to be developed to better measure the in-situ performance and failures of the applications and the phenomena which drive these applications.

2.4 Michael Gardos - Wanted: realistic tribotest methods and bearing materials for MEMS moving mechanical assembly applications

Tribologists and MEMS technologists need to take a hard look at atomic force microscopy/friction force microscopy (AFM/FFM) and molecular tribometry (MT) as realistic test machines for characterizing and screening likely MEMS bearing, gear, rotor and sliding rail materials. The performance of load-carrying MEMS sliding-rolling interfaces is controlled by adhesion, friction and wear in a load range where Amonton's law is followed. In nanotribometry, represented by AFM/FFM, the friction force is not linearly proportional to the normal load. The surface behavior of those MEMS microelements which move (but just bend or twist) or even if they touch a counterface (e.g., in sensors, membrane pumps, or microtweezers) is vastly different from those that operate under surprisingly high Hertzian stresses (e.g., in micromotors and actuators). Therefore, at interfaces where the proper degree of adhesion ("stiction" or repulsion) of more-less statically contacting surfaces is the critical value for MEMS operation, AFM/FFM appears to be useful. Even MT might be made acceptable, if the specimen materials were changed from the ubiquitous mica to more realistic MEMS-friendly candidates. Approximating MEMS-moving-mechanical-assembly bearing surface behavior at a bench-top level needs more suitable test machines. What and where are they?

An equally troublesome problem is the use of silicon, nickel, polymers and LIGA-built (cast metal or metal-plated) MEMS tribohardware without adequate lubrication methods. With the primitive (actually, practically nonexistent) lubricant application techniques presently used for dynamic MEMS, the useful life of sliding-rotating micromachines fabricated from these materials (especially from Si, an extremely poor bearing material), is short. The root cause of this problem is excessive wear exacerbated by high coefficients of friction measured during actual micromachine operation and bench-top tribometry. Poor results are obtained even under the relatively mild environmental stresses of room-ambient conditions, much less under extreme conditions. It is seldom mentioned that today's micromotors and the associated gear assemblies are not true power generating and transfer devices: they drive only themselves. They have no real load-carrying capacity. What/where are the MEMS bearing materials and lubrication techniques of tomorrow?

2.5 Frank Michel - Requirements for tribology research for non- silicon micro actuators

MEMS no longer consist only of silicon. A widespread mixture of lithography and precision engineering techniques (LIGA, EDM, milling etc.) allows the use of a large variety of materials, e.g. metals, ceramics, glasses or polymers. The micro actuators assembled from these components range from silicon MEMS to precision engineered motors and can deliver considerable performance suitable for moving more than themselves.

There are, similar to silicon MEMS, stiction problems to be solved including for instance relay contacts, motion stopping structures, valves, or when releasing gripped parts during micro assembly. Currently, the friction in bearings is the main requirement for triboloby research in non-silicon micro actuators.

On the one hand, traditional bearing materials and fluid lubricants are still in use. Due to scaling effects, friction losses in micro actuators increase considerably and it becomes extremely difficult to keep the lubricant inside the bearing system. Ageing of fluid lubricants limits the life-time, particularly of micro actuators working in an intermittent mode. Consequently, micro structured coatings like epilams should be explored in order to guide the lubricants. Alternatively, fluid lubricants should be replaced by solid lubricant layers or coatings.

On the other hand, new bearing materials, processed for example by LIGA, open new possibilities in bearing design and functionality. Efforts should focus on research of bulk material, coatings, and surface treatments to improve the hardness and to reduce wear and friction. External influences (extreme temperature, humidity) have been paid little attention by MEMS-developers and should be another topic of joint research of MEMS- and tribology scientists.

Many industrial applications of non-silicon micro actuators are just starting. Hence, it is the right time to bring MEMS and tribology scientists together. The workshop has initiated further co-operative work to improve micro actuators. On the other hand MEMS developers can provide tribology research with suitable micro test equipment.

3. Audience/panel discussion summary

The overwhelming consensus was that the conference was beneficial to the tribology and MEMS communities as each was generally unaware of many of the problems, issues, and opportunities afforded by the other. Particular interest was paid to the following specific issues and these, along with the discussion given below, should serve as general recommendations and guidelines for needed future research efforts.

- manufacturability of MEMS which can provide real work:

current mechanical micromachines can only operate themselves and can not produce useful mechanical output, MEMS actuators and bearings for microgyroscopes, for example, are far from being reality.

- effects of assembly and processing on micromechanism reliability and wear:

material properties, and therefore performance and reliability are dependent on processing variables and must be adequately understood.

- second only to reliability is the issue of sticking surfaces during processing and prolonged dormancy, even in currently marketed products:

sticking surfaces is the major processing-related issue and there needs to be a significant effort to understand and solve this problem.

- reliable lubrication of micro-bearing surfaces:

sliding contact of surfaces goes against every tenet of reliable and low friction/wear mechanical design and this approach must be eliminated by proper lubricating films or surfaces.

- MEMS has the potential to have a significant impact on data storage with micro-actuators which can withstand large shock >1000 g's:

high reliability and repeatable MEMS devices under operating shock loads measured in 1000s of Gs are needed to fulfill the operating requirements for small high density data storage systems.

- satellite and weapons systems must demonstrate absolute reliability:

although the market numbers are relatively low, these devices must have proven reliability under the most extreme environmental conditions and prolonged dormancy.

- machines for in-situ and non-destructive microtribology testing are needed and MEMS may provide many needed solutions:

MEMS test and measurement systems may provide the tribology community with methods for quantifying and standardizing many of the surface and performance issues in MEMS applications, destructive tests are not desirable and current methods, such as AFM, are too big to be used in-situ and under operating conditions.

In more general terms, there was discussion and consensus that MEMS is constantly evolving and therefore requires continued collaboration at least among the communities represented. Based on this, follow-on conferences are needed. Most of all, each other's language must be better understood. While some of the MEMS community indicated they had some familiarity with tribology, the opposite was generally less prevalent. Strong statements were made by the NSF that truly innovative ideas and approaches are required. The current approach to solving microsystem problems are far too costly and efforts should be focused on the good ideas which have already been developed. The term "gears-for-years" was used as an example of the lack of innovative approaches. It was felt that many of the solutions required for MEMS, such as tailoring surface chemistry and ways to reduce friction and wear have already been developed and documented among the tribology literature and need to be applied.

Reliability of MEMS under adverse environmental conditions, prolonged or intermittent use, and guaranteed start-up after long dormancy are issues which repeatedly surfaced. Sticking of surfaces was sighted as the greatest problem and one where tribology will have significant impact. It was felt that devices must be designed for the specific task at hand, rather than building, or building upon, standardized subsystems or components. In concert with this was the feeling that fabrication processes and materials must be selected based on the application rather than on what has been demonstrated or what is considered as standard.

The feeling among many of the MEMS participants was the full scope of microsystems development was not brought forth in this conference, mainly due to the relatively small representation by the MEMS community. It was expressed that there are systems currently in many phases of development, some in industrial use, and some about to be introduced to the market place. For this reason, it was felt the perception of what MEMS are and what they can or should do was not adequately presented. Questions were raised about the apparent reliance on silicon. The response was, apart from the fact that MEMS was born out of proximity to the semiconductor industry, that in the early developmental days of MEMS it was felt the mechanical and electrical functions would have to be inseparable and so silicon was used to fulfill the electronics need. As MEMS has progressed, this may be less of a driver and for in-situ biomedical applications it was stated that integrated electronics is not desirable. Also, the process of lithography was recognized for its mass production capabilities but there were questions whether large numbers of MEMS are being required to fulfill actual markets and applications. Another comment was made that in general MEMS with linear actuation are making money, such as accelerometers, micromirror devices, and microvalves, and tribology may not be a major issue for these applications. Rotary applications still have considerable technical hurdles to be overcome for prolonged and reliable operation, such as bearing surfaces, lack of sticking, etc.

One of the major themes of the discussion by the tribology community was that processing affects the surface chemistry, material properties, performance, and reliability and must be better understood. There was a feeling among the tribologists that the MEMS community has largely ignored this concern leading to unreliable rotary devices or devices which exhibited unacceptable levels of wear under useful loads at rolling contact interfaces, such as meshing gears, or sliding contact surfaces at bearings surfaces.

4. Conclusions

The open forum allowed many opposing viewpoints to be presented and debated. The overwhelming feeling was that this is necessary and healthy for diverse, yet related, research communities to progress and take advantage of developed expertise. Future conferences were proposed and verbally supported. Efforts need to be undertaken to increase awareness and foster collaboration among these groups. Funding agencies can greatly help by emphasizing collaborative and innovative research in materials, processes, and system development and characterization.

List of Participants

Dr. Phillip Abel
NASA-Lewis Research Center
21000 Brookpark Rd., M.S. 23-2
Cleveland, OH 44135

Prof. George G. Adams
Dept. of Mechanical Engr., 334 SN
Northeastern University
Boston, MA 02115

Mr. Ankur B. Agarwal
428 Cross Street
Ann Arbor, MI 48104

Dr. Orlando Auciello
Argonne National Laboratory
Materials Science Division
9700 S. Cass Ave., Bldg. 212
Argonne, IL 60439-4838

Prof. Charles L. Bauer
Dept. of MSE
Carnegie Mellon University
Pittsburgh, PA 15213

Mr. Michael H. Beggans
New Jersey Institute of Technology (NJIT)
Dept. of Physics
Unviersity Heights
Newark, NJ 07102

Mr. Bertrand B. Bellaton
C.S.E.M. Instruments
Jaquet-Droz 1
CH 2007 Neuchatel
Switzerland

Prof. B. Bhushan
Dept. of Mechanical Engineering
Ohio State University
206 W. 18th Avenue
Columbus, OH 43210

Dr. Stuart B. Brown
Failure Analysis Associates
Three Speen Street
Framingham, MA 01701

Mr. Sameera K. Chilamakuri
Dept. of Mechanical Engineering
Ohio State University
206 W. 18th Avenue
Columbus, OH 43210

Dr. Sergei A. Chizhik
Metal-Polymer Research Inst.
NBAS of Belarus
32A, Kirov Street
Gomel 246652
Belarus

Dr. Unchung Cho
National Institute of Standards & Technology
Building 223, Room A256
Gaithersburg, MD 20899

Dr. Sean Corcoran
Hysitron, Incorporated
2010 East Hennepin Avenue
Minneapolis, MN 55413

Mr. Chetan A. Dandavate
Dept. of Mechanical Engineering
Ohio State University
206 W. 18th Avenue
Columbus, OH 43210

Prof. Steven Danyluk
School of Mechanical Engineering
Georgia Institute of Technology
813 Ferst Drive, MARC Suite 311
Atlanta, GA 30332-0405

Major Hugh C. De Long
AFOSR/NL
110 Duncan Ave., Ste. B115
Bolling AFB, D.C.
20332-8050

Mr. Derik C. DeVecchio
Dept. of Mechanical Engineering
Ohio State University
206 W. 18th Avenue
Columbus, OH 43210

Dr. Michael T. Dugger
Sandia National Laboratories
P.O. Box 5800, MS 0340
Albuquerque, NM 87185-0340

Dr. John Ehmke
Raytheon TI Systems
P.O. Box 655474 MS 245
Dallas, TX 75265

Prof. Norman S. Eiss
Mechanical Engineering
Virginia Tech
Blacksburg, VA 24061-0238

Mr. Ali Erdemir
Argonne National Laboratory
9700 S. Cass Avenue, ET/212
Argonne, IL 60439

Dr. L.S. Fan
IBM Almaden Research Center
650 Harry Road
San Jose, CA 95120-6099

Prof. Gary K. Fedder
Dept. of Electrical & Computer Engr.
Carnegie Mellon University
Pittsburgh, PA 15213-3890

Prof. Traugott E. Fischer
Dept. of Materials Science & Engr.
Stevens Institute of Technology
Hoboken, NJ 07030

Mr. F.J. Franklin
Dept. of Mechanical Engineerning
University of Sheffield
Mappin Street, Sheffield S1 3JD
UK

Prof. Craig R. Friedrich
MEEM Dept.
Michigan Technological University
1400 Townsend Drive
Houghton, MI 49931

Dr. Satoru Fujisawa
Mechanical Engineering Laboratory
Namiki 1-2, Tsukuba City
Ibaraki 305, Japan

Dr. Michael N. Gardos
Hughes Aircraft Company
P.O. Box 902; E1/F150
El Segundo, CA 90245

Dr. Richard S. Gates
DOC/NIST
I270 & Quice Orchard Rd.,
Bldg. 223, Rm. A256
Gaithersburg, MD 20899

Prof. Andrew J. Gellman
Dept. of Chemical Engineering
Carnegie Mellon University
5000 Forbes Ave.
Pittsburgh, PA 15213

Dr. Norm V. Gitis
Center for Tribology, Inc.
625-A Clyde Avenue
Mountain View, CA 94043

Prof. Yury Gogotsi
University of Illinois at Chicago
842 W. Taylor Street, M/C 251
Chicago, IL 60607

Dr. Valeri V. Gorbunov
Metal-Polymer Research Institute NBAS of Belarus
32A, Kirov Street
Gomel 246652
Belarus

Prof. Steve Granick
University of Illinois
105 South Goodwin Avenue
Urbana, IL 61801

Prof. K. Gupta
Head, ITMMEC
Indian Institute of Technology
Hauz Khas, New Delhi 110016
India

Prof. Bernard J. Hamrock
Dept. of Mechanical Engineering
Ohio State University
206 W. 18th Avenue
Columbus, OH 43210

Prof. Judith A. Harrison
U.S. Naval Academy
572 Holloway Road
Annapolis, MD 21402

Mr. John L. Hazel
Western Michigan University
1057 Khorman Hall
Kalamazoo, MI 49006

Dr. Lou Hermans
IMEC
Kapeldreef 75
B-3001 Leuven
Belgium

Prof. Peter J. Hesketh
Univ. of Illinois at Chicago
EECS Dept. (M/C 154)
Chicago, IL 60607

Dr. Sabine Hild
Experimental Physics
University of Ulm
D89069 Ulm, Germany

Prof. Paul H. Holloway
Dept. of Materials Science & Engr.
University of Florida
Gainesville, FL 32611-6400

Dr. Tim R.P. Hsia
DiCon Fiberoptics, Inc.
1331 8th Street
Berkeley, CA 94710

Dr. Hsing-Sen S. Hsiao
Dept. of Mechanical Engineering
The Ohio State University
206 W. 18th Avenue
Columbus, OH 43210-1107

Dr. Meng-Hsiung Kiang
DiCon Fiberoptics, Inc.
1331 Eight Street
Berkeley, CA 94710

Dr. Stephen Hsu
NIST
RM A265, Building 223
Gaithersburg, MD 20899

Prof. Jack W. Judy
UCLA
56-125 B. Engineering IV
Los Angeles, CA 90095-1594

Prof. Leon M. Keer
Northwestern University
Dept. of Civil Engineering
2145 Sheridan Road
Evanston, IL 60208

Prof. Tom. W. Kenny
Dept.. of Mechanical Engineering
Stanford University
Stanford, CA 94305-4021

Dr. Barbara J. Kinzig
Surfaces Research
8330 Melrose Drive
Lenexa, KS 66214-1630

Dr. Vilas Koinkar
Rodel, Inc.
451 Bellevue Road
Newark, DE 19713

Prof. Kyriakos Komvopoulos
Dept. of Mechanical Engineering
University of California
Berkeley, CA 94720

Prof. Martin E. Kordesch
Ohio University
Athens, OH 45701-2979

Dr. A.R. Krauss
Argonne National Laboratory
Materials Science & Chemistry Divisions
9700 S. Cass Ave., Bldg. 200
Argonne, IL 60439

Prof. Jacqueline Krim
Physics Dept.
110 Dana Bldg.
Northeastern University
Boston, MA 02115

Dr. Jorn Larsen-Basse
National Science Foundation
4201 Wilson Blvd., Rm. 545
Arlington, VA 22230

Mr. Russell A. Lawton
Jet Propulsion Laboratory
M.S. 303-220
4800 Oak Grove Drive
Pasadena, CA 91109-8099

Mr. Cin-Young Lee
University of Michigan at Ann Arbor
2350 Hayward
2250 G.G. Brown, Rm. #2205
Ann Arbor, MI 48109

Dr. Ho Shang Lee
DiCon Fiberoptics, Inc.
1331 8th Street
Berkeley, CA 94710

Prof. Si C. Lee
Dept. of Mechanical Engineering
Ohio State University
206 W. 18th Avenue
Columbus, Ohio 43210

Mr. Melburne LeMieux
Western Michigan University
427 Ackley Hall
Kalamazoo, MI 49008

Dr. Xiaodong Li
Dept.. of Mechanical Engineering
Ohio State University
206 W. 18th Ave.
Columbus, OH 43210

Prof. Liwei Lin
University of Michigan
2350 Hayward Street
Ann Arbor, MI 48105-2125

Dr. Jing Luo
Analog Devices
MS511
831 Woburn Street
Wilmington, MA 01887

Mr. Xiaoding Ma
Dept.. of MSE
Carnegie Mellon University
Pittsburgh, PA 15213

Prof. Roya Maboudian
Dept. of Chemical Engineering
University of California at Berkeley
201 Gilman Hall
Berkeley, CA 94720-1462

Prof. Marc J. Madou
Dept. of Mat. Sci. & Engr.
Ohio State University
2041 College Road
Columbus, OH 43210

M.J. Mamin
IBM Almaden Research Center
K13/E3, 650 Harry Road
San Jose, CA 95120-6099

Mr. James Marchetti
IntelliSense Corporation
16 Upton Drive
Wilmington, MA 01887

Dr. Jack R. Martin
Analog Devices, Inc.
Mail Stop 4210
804 Woburn Street
Wilmington, MA 01887

Dr. Fariborz Maseeh
IntelliSense Corporation
16 Upton Drive
Wilmington, MA 01887

Prof. Carlos H. Mastrangelo
University of Michigan
2405 EECS Bldg.
1301 Beal Avenue
Ann Arbor, MI 48109-2122

Prof. Carl J. McHargue
University of Tennessee
100 Estabrook Hall
Knoxville, TN 37996-2350

Prof. Michael J. McNallan
CME Dept., M/C 246
University of Illinois at Chicago
842 W. Taylor Street
Chicago, IL 60607

Mr. Mehran Mehregany
Dept. of Electrical Engr. &
Applied Physics
Case Western Reserve University
Cleveland, OH 44106

Dr. Frank Michel
Institut für Mikrotechnik Mainz
GmbH
Carl-Zeiss-Str. 18-20, D-55129 Mainz
Germany

Mr. S.A.M Mirmiran
Mechanical & Aeronautical
Engineering Dept.
Western Reserve University
Kalamazoo, MI 49008

Dr. Eric L. Montei
Read Rite Corp.
345 Los Coches Street
Milpitas, CA 95035

Dr. John Moreland
Nanoprobe Imaging for Mag. Tech.
Proj., Electromagnetic Tech. Div.
NIST, Mail Stop 814.05
325 Broadway
Boulder, CO 80303

Dr. Ted A. Morgan
The Dow Chemical Company
1776 Building
Midland, MI 48674

Mr. Alex E. Moser
Dept. of Chemistry
University of Nebraska-Lincoln
HaH Box E-4
Lincoln, NE 68588-0304

Dr. Josekutty J. Nainaparampil
Wright Patterson Air Force Base
WL/MLBT
2941 P Street, Suite 1
Dayton, OH 45433

Prof. Nobuo Ohmae
Faculty of Engineering
Osaka University
2-1 Yamadaoka, Suita,
Osaka 565, Japan

Prof. Timothy C. Ovaert
Penn State University
328 Reber Bldg.
University Park, PA 16802

Prof. Leslie M. Phinney
Dept. of Mech. Ind. Engr.
Univ. of Illinois at Urbana-Champaign
140 Mechanical Engr., Bldg. MC-244
1206 West Green Street
Urbana, IL 61801

Dr. Braham Prakash
Indian Insitute of Technology Delhi
ITMMEC IIT Delhi Hauz Khas
New Delhi, India 110 016

Dr. Gouri Radhakrihnan
Mail Stop MS-271
The Aerospace Corporation
P.O. Box 92957
Los Angeles, CA 90009-2057

Mr. Nick N. Randall
CSEM Instruments
Jaquet-Droz 1
Ch-2007 Neuchatel
Switzerland

Prof. David A. Rigney
Dept. of Materials Sciences & Engr.
Ohio State University
2041 College Road
Columbus, OH 43210

Dr. Zygmunt Rymuza
Warsaw University of Technology
Chodkiewicza 8/623
02-525 Warszawa
Poland

Mr. Hermann A. Rump
Max-Planck-Institut für
Mikrostrukturphysik
Dresdner Platz 5, 3.10
D-72 760 Reutlingen, Germany

Dr. Vikas Sachan
Rodel, Inc.
451 Bellevue Road
Newark, DE 19713

Prof. Muhammed T. Saif
Mechanical & Industrial Engineering
University of Illinois at Urbana-Champaign
224 MEB, MC-244, 1206 West Green St.
Urbana, IL 61801

Dr. Erich Santner
Bundesanetalt fuer Material
forschung und -pruefung
BAM, Division "Tribology & Wear
Protection"
D-12200 Berlin
Germany

Dr. Volker Scherer
Fraunhofer Institute for
Nondestructive Testing
University Bldg. 37
D-66123 Saarbruecken
Germany

Dr. Matt Scherge
Technische Universität Ilmenau
Inst. of Physik, PF 100565 TU Ilmenau
Weimarer St. 32
D-98684 Ilmenau
Germany

Mr. Kirsten I. Schiffmann
Fraunhofer Institute für Schicht- und Oberflächentechnik
Bienroder Weg 54E
D-38108 Braunschweig
Germany

Dr. Steve Shaffer
Battelle Memorial Institute
505 King Avenue
Columbus, OH 43201

Mr. S.K. Sharma
U.S. Air Force, Wright-Laboratory
WL/MLBT, 2941 P St. Ste. 1
2941 P Street, Suite 1
Dayton, OH 45433

Mr. Edward N. Sickafus
Ford Research Laboratory
MD 3028, RM 1513
P.O. Box 2053
Dearborn, MI 48121-2053

Dr. Jeffry J. Sniegowski
Sandia National Laboratories
P.O. Box 5800, MS 1080
Albuquerque, NM 87185-1080

Prof. S. Mark Spearing
MIT
MIT Rm 33-315
77 Massachusetts Avenue
Cambridge, MA 02139

Dr. Shane C. Street
The University of Alabama
Dept. of Chemistry
Box 870336
Tuscaloosa, AL 35487-0336

Mr. Brian D. Strom
Quantum
500 McCarthy Blvd.
Milpitas, CA 95035

Mr. Robert E. Sulouff
Analog Devices
Micromachined Products Division
21 Osborne Street
Cambridge, MA 02139

Mr. Sriram Sundararajan
Dept. of Mechanical Engr.
Ohio State University
206 W. 18th Avenue
Columbus, OH 43210

Mr. Christian Thurigen
Faulhaber Motoren
Daimlerstraße 23
D-71101 Schonaich
Germany

Prof. John A. Tichy
Dept. of Mech. Engr.
Aeronautical Engr. & Mechanics
Rensselaer Polytechnic Institue
Troy, NY 12180-3590

Prof. Norman C. Tien
School of Electrical Engineering
Cornell University
402 Phillips Hall
Ithaca, NY 14853-5401

Prof. Kristian Tonder
Norwegian University of Science and Technology
N.7034 Trondheim
Norway

Mr. John H. Tripp
SKF Engineering & Research Centre B.V.
Postbus 2350
3430 DT Nieuwegein
The Netherlands

Prof. Vladimir V. Tsukruk
College of Engineering & Applied Sciences
Western Michigan University
Kalamazoo, MI 49008

Prof. William N. Unertl
University of Maine
LASST/SERC
Orono, ME 04469

Dr. Kathryn J. Wahl
Naval Research Laboratory
Code 6176
Washington, DC 20375-5342

Prof. Mark E. Walter
Applied Mechanics Section
Ohio State University
209 Boyd Lab.,
155 W. Woodruff Ave.
Columbus, OH 43210-1181

Dr. Robert O. Warrington
Michigan Technological University
College of Engineering
RL. Smith Bldg., 1400 Townsend Drive
Houghton, MI 49931-1295

Dr. Donald R. Wiff
Materials Directorate
Air Force Research Lab.
WL/MLBT
Wright-Patterson AFB, OH 45433-7702

Mr. Kirt R. Williams
Research Staff Scientist
Lucas NovaSensor
1055 Mission Ct.
Fremont, CA 94539-8203

Prof. Bjong W. Yeigh
Oklahoma State University
1917 N. Skyline Street
Stillwater, OK 74075

Dr. J.S. Zabinski
Air Force Research Laboratory
Materials Directorate
WL/MLBT Bldg. 654
Wright Patterson AFB, OH 45433-7702

Prof. Xiao C. Zeng
University of Nebraska-Lincoln
536 Hamilton Hall
Lincoln, NE 68588-0304

Dr. Zheming Zhao
Dept. of Mechanical Engineering
The Ohio State University
206 W. 18th Avenue
Columbus, OH 43210-1107

EDITOR's VITA

Dr. Bharat Bhushan, a pioneer of tribology and mechanics of magnetic storage devices, is an internationally recognized expert in the general fields of conventional tribology and micro/nanotribology and one of the most prolific authors. He is presently an Ohio Eminent Scholar and The Howard D. Winbigler Professor in the Department of Mechanical Engineering and the Director of Computer Microtribology and Contamination Laboratory at the Ohio State University, Columbus, Ohio. He received a MS in mechanical engineering from the Massachusetts Institute of Technology, a MS in mechanics and a Ph.D. in mechanical engineering from the University of Colorado at Boulder, an MBA from Rensselaer Polytechnic Institute at Troy, NY, Doctor Technicae from the University of Trondheim at Trondheim, Norway, and a Habilitate Doctor of Technical Sciences from the Warsaw University of Technology at Warsaw, Poland. He is a registered professional engineer (mechanical). He has authored 4 technical books, 15 handbook chapters, more than 300 technical papers in referred journals, and more than 60 technical reports, edited more than 15 books, and holds 6 US patents. He is founding editor-in-chief of World Scientific Advances in Information Storage Systems Series, CRC Press Mechanics and Materials Science Series and Journal of Information Storage and Processing Systems. He has given more than 175 invited presentations on five continents including several keynote addresses at major international conferences. He is an accomplished organizer. He organized the first symposium on Tribology and Mechanics of Magnetic Storage Systems in 1984 and the first international symposium on Advances in Information Storage Systems in 1990 and continues to organize and chair the AISS symposia annually. He founded an ASME Information Storage and Processing Systems Division and is the founding chair. His biography has been listed in over two dozen Who's Who books including Who's Who in the World. He has received more than a dozen awards for his contributions to science and technology from professional societies, industry, and U.S. government agencies. He is a foreign member of Byelorussian Acedemy of Engineering and Technology, fellow of ASME and the New York Academy of Sciences, a senior member of IEEE, and the member of STLE, NSPE, Sigma Xi and Tau Beta Pi. Dr. Bhushan has worked for the Department of Mechanical Engineering at the Massachusetts Institute of Technology, Cambridge, MA; Automotive Specialists, Denver, CO; the R & D Division of Mechanical Technology Inc., Latham, NY; the Technology Services Division of SKF Industries Inc., King of Prussia, PA; the General Products Division Laboratory of IBM Corporation, Tucson, AZ; the Almaden Research Center of IBM Corporation, San Jose, CA, and the Department of Mechanical Engineering at the University of California, Berkeley, CA.

He is married and has two children. His hobbies are music, photography, hiking, and traveling.

INDEX

GPSR Compliance
The European Union's (EU) General Product Safety Regulation (GPSR) is a set of rules that requires consumer products to be safe and our obligations to ensure this.

If you have any concerns about our products, you can contact us on

ProductSafety@springernature.com

In case Publisher is established outside the EU, the EU authorized representative is:

Springer Nature Customer Service Center GmbH
Europaplatz 3
69115 Heidelberg, Germany

www.ingramcontent.com/pod-product-compliance
Ingram Content Group UK Ltd.
Pitfield, Milton Keynes, MK11 3LW, UK
UKHW021933200726
13853UKWH00010B/1388

* 9 7 8 9 4 0 1 1 5 0 5 1 4 *